Lecture Notes in Computer Science 16535

The series Lecture Notes in Computer Science (LNCS), including its subseries Lecture Notes in Artificial Intelligence (LNAI) and Lecture Notes in Bioinformatics (LNBI), has established itself as a medium for the publication of new developments in computer science and information technology research, teaching, and education.

LNCS enjoys close cooperation with the computer science R & D community, the series counts many renowned academics among its volume editors and paper authors, and collaborates with prestigious societies. Its mission is to serve this international community by providing an invaluable service, mainly focused on the publication of conference and workshop proceedings and postproceedings. LNCS commenced publication in 1973.

Hyungsoo Jung · Tianzheng Wang ·
Masashi Toyoda · Hyuk-Yoon Kwon ·
Jae-woong Lee
Editors

Database Systems for Advanced Applications

31st International Conference, DASFAA 2026
Jeju, South Korea, April 27–30, 2026
Proceedings, Part I

Springer

Editors
Hyungsoo Jung
Seoul National University
Seoul, Korea (Republic of)

Masashi Toyoda
The University of Tokyo
Tokyo, Tokyo, Japan

Jae-woong Lee
Kangwon National University
Chuncheon-si, Korea (Republic of)

Tianzheng Wang
SFU Computing Science, ASB10070
Simon Fraser University
Burnaby, BC, Canada

Hyuk-Yoon Kwon
SeoulTech
Seoul, Korea (Republic of)

ISSN 0302-9743 ISSN 1611-3349 (electronic)
Lecture Notes in Computer Science
ISBN 978-981-92-0362-8 ISBN 978-981-92-0363-5 (eBook)
https://doi.org/10.1007/978-981-92-0363-5

This Springer imprint is published by the registered company Springer Nature Singapore Pte Ltd.
The registered company address is: 152 Beach Road, #21-01/04 Gateway East, Singapore 189721, Singapore

Preface

It is our great pleasure to present the proceedings of the 31st International Conference on Database Systems for Advanced Applications (DASFAA 2026), held on April 27–30, 2026, in South Korea. DASFAA is a leading annual international conference in the database field, showcasing state-of-the-art research and development in database systems and advanced applications. It serves as a premier forum for technical presentations and discussions among researchers, developers, and users from both academia and industry.

This year, DASFAA received 834 research paper submissions. Following the long-standing DASFAA tradition, we conducted a rigorous double-blind review process. To support this process, we assembled a large and dedicated Program Committee consisting of 31 Senior Program Committee (SPC) members and 245 Program Committee (PC) members, including emergency reviewers. Each valid submission was reviewed by mostly three PC members and meta-reviewed by one SPC member. Based on the recommendations of the SPC members, we, the PC co-chairs, carefully examined each submission together with its reviews before making the final decisions. As a result, 228 full papers were accepted, yielding an acceptance rate of 27.3%. The review process was supported by the Microsoft CMT system.

During the three main conference days, the accepted full papers were presented across 44 research sessions. The dominant keywords among the accepted papers included language models, graph, recommendation, knowledge, efficient, data, framework, network, prediction, detection, neural, and federated, reflecting several of the major research trends currently shaping the database community. In addition to the research track, the conference program also included 16 industry papers, 11 demo papers, and 5 tutorials.

To highlight important emerging directions in the field, DASFAA 2026 featured four invited keynote presentations by Victor Leis (Technical University of Munich), Wenjie Zhang (University of New South Wales), Meihui Zhang (Beijing Institute of Technology), and Wook-Shin Han (Pohang University of Science and Technology).

Five tutorials were selected by the Tutorial Co-chairs:

1. Continual Recommender Systems, by Seunghan Lee (Korea University), Seunghyun Baek (Korea University), Dojun Hwang (Korea University), Hyunsik Yoo (University of Illinois Urbana-Champaign), and SeongKu Kang (Korea University);
2. Efficient Compression and Queries of Large Graphs, by Fan Zhang (Guangzhou University), Qingshuai Feng (Great Bay University), and Kai Wang (Shanghai Jiao Tong University);
3. Advances in Real-Time Processing of Longitudinal Data: From Statistical and Deep Learning Methods to Applications, by Ying-Ren Chien (National Taipei University of Technology), Pavel Loskot (Zhejiang University-University of Illinois Urbana-Champaign Institute), and Yu Gao (Midea Group);
4. Trustworthy Foundation Models with a Data-Centric Approach, by Wenjie Fang (University of Science and Technology of China), Dan Li (Sun Yat-sen University), and Jian Lou (Sun Yat-sen University);

5. Reliable Visual Analytics with Dimensionality Reduction: Quality Evaluation and Interpretation of Projections, by Jeon Hyeon (Seoul National University), Takanori Fujiwara (University of Arizona), and Rafael M. Martins (Linnaeus University).

Two workshops were selected by the Workshop Co-chairs to be held in conjunction with DASFAA 2026: the 10th International Workshop on Graph Data Management and Analysis (GDMA 2026) and the 12th International Workshop on Big Data Management and Service (BDMS 2026). The papers accepted to these workshops are included in a separate proceedings volume, also published by Springer in the Lecture Notes in Computer Science (LNCS) series.

We are deeply grateful to the General Chairs, Sang-Won Lee (Seoul National University, South Korea), Bin Cui (Peking University, China), and Zi (Helen) Huang (University of Queensland, Australia) for their outstanding leadership and support. We also sincerely thank all SPC members, PC members, and external emergency reviewers for their tremendous dedication, time, and expertise. Finally, we would like to express our appreciation to all members of the Organizing Committee and the many volunteers for their invaluable contributions to the success of DASFAA 2026.

April 2026

Hyungsoo Jung
Tianzheng Wang
Masashi Toyoda

Organization

Steering Committee Members

Chair

Lei Chen	Hong Kong University of Science and Technology (Guangzhou), China

Vice Chair

Stéphane Bressan	National University of Singapore, Singapore

Treasurer

Yasushi Sakurai	Osaka University, Japan

Secretary

Kyuseok Shim	Seoul National University, South Korea

Members

Aamir Cheema	Monash University, Australia
Arnab Bhattacharya	IIT Kanpur, India
Atsuyuki Morishima	University of Tsukuba, Japan
De-Nian Yang	Academia Sinica, Taiwan
Eun-Jun Hwang	Korea University, South Korea
Guoliang Li	Tsinghua University, China
Sang-Won Lee	Seoul National University, South Korea
Sourav Bhowmick	Nanyang Technological University, Singapore
Yang-Sae Moon	Kangwon National University, South Korea
Zhanhuai Li	Northwestern Polytechnical University, China
Zhiyong Peng	Wuhan University, China
Zi Huang	University of Queensland, Australia

Organizing Committee

General Co-chairs

Bin Cui	Peking University, China
Zi (Helen) Huang	University of Queensland, Australia
Sang-Won Lee	Seoul National University, South Korea

Program Chairs

Hyungsoo Jung	Seoul National University, South Korea
Tianzheng Wang	Simon Fraser University, Canada
Masashi Toyoda	University of Tokyo, Japan

Industry PC Chairs

Lei Chen	Hong Kong University of Science and Technology, China
Min-Soo Kim	KAIST, South Korea

Demo Chairs

Byungchul Tak	Kyungpook National University, South Korea
Siqiang Luo	Nanyang Technological University, Singapore

Local Chairs

Jongwuk Lee	Sungkyunkwan University, South Korea
Buru Chang	Korea University, South Korea

Panel Chair

Seung-won Hwang	Seoul National University, South Korea

Workshop Chairs

Joyce Jiyoung Whang	KAIST, South Korea
Wenjie Zhang	University of New South Wales, Australia

Tutorial Chairs

Shenghua Liu	Chinese Academy of Science, China
Yasushi Sakurai	Osaka University, Japan
Kijung Shin	KAIST, South Korea

Publicity Chairs

Jaeyoung Do	Seoul National University, South Korea
Yi Cai	South China University of Technology, China
Atsuyuki Morishima	Tsukuba University, Japan

Web/Slack Chairs

Suan Lee	Semyung University, South Korea
Sun-Young Ihm	Pai Chai University, South Korea

Publication Chairs

Jae-woong Lee	Kangwon National University, South Korea
Hyuk-Yoon Kwon	Seoul National University of Science and Technology, South Korea

Registration Chair

Hyunsouk Cho	Ajou University, South Korea

Treasurer

Jonghyeok Park	Korea University, South Korea

Sponsorship Chairs

Kyongha Lee	Korea Institute of Science and Technology Information, South Korea
Kyoungsook Kim	AIST, Japan
Xin Wang	Tianjin University, China

SC Liaison

Yang-Sae Moon	Kangwon National University, South Korea

Advisory Board

Kyuseok Shim	Seoul National University, South Korea
Yang-Sae Moon	Kangwon National University, South Korea
Young-Koo Lee	Kyunghee University, South Korea
Ha-Joo Song	Pukyong National University, South Korea

Program Committee

Research Track

Chairs

Hyungsoo Jung	Seoul National University, South Korea
Tianzheng Wang	Simon Fraser University, Canada
Masashi Toyoda	University of Tokyo, Japan

Senior Program Committee

Alkis Simitsis	Athena Research Center, Greece
Byung Suk Lee	University of Vermont, USA
Cheng Long	Nanyang Technological University, Singapore
Da Yan	Indiana University Bloomington, USA
Dong Wen	University of New South Wales, Australia
Hideyuki Kawashima	Keio University, Japan
Jianliang Xu	Hong Kong Baptist University, China
Jieming Shi	Hong Kong Polytechnic University, China
Jun Miyazaki	Institute of Science Tokyo, Japan
Ke Wang	Simon Fraser University, Canada
Kyoung-Sook Kim	AIST, Japan
Kyuseok Shim	Seoul National University, South Korea
Leong Hou U	University of Macau, China
Nikos Bikakis	Hellenic Mediterranean University and ATHENA Research Center, Greece
Sang-Wook Kim	Hanyang University, South Korea
Satoshi Oyama	Nagoya City University, Japan
Sebastian Link	University of Auckland, New Zealand
Sen Wang	University of Queensland, Australia
Sibo Wang	Chinese University of Hong Kong, China
Steven E. Whang	KAIST, South Korea

Verena Kantere	University of Ottawa, Canada
Wei-Shinn Ku	Auburn University, USA
Wen Hua	Hong Kong Polytechnic University, China
Wenjie Zhang	University of New South Wales, Australia
Xiangyu Ke	Zhejiang University, China
Xin Cao	University of New South Wales, Australia
Xin Wang	Tianjin University, China
Yanfeng Zhang	Northeastern University, China
Yixiang Fang	Chinese University of Hong Kong, Shenzhen, China
Yon Dohn Chung	Korea University, South Korea
Zeke Wang	Zhejiang University, China

Program Committee

Alexander Zhou	Hong Kong Polytechnic University, China
Anne Laurent	University of Montpellier, France
Aziz Nasridinov	Chungbuk National University, South Korea
Beibei Li	Chongqing University, China
Bin Wang	Northeastern University, China
Bo Xu	Dalian University of Technology, China
Buru Chang	Korea University, South Korea
Byoungwook Kim	Gangneung-Wonju National University, South Korea
Byungchul Tak	Kyungpook National University, South Korea
Cai Xu	Xidian University, China
Changdong Wang	Sun Yat-sen University, China
Chao Zhang	Renmin University of China, China
Chen Shen	Megagon Labs, USA
Chengcheng Mai	Nanjing Normal University, China
Chengliang Chai	Beijing Institute of Technology, China
Chenhao Ma	Chinese University of Hong Kong, Shenzhen, China
Chenxu Wang	Xi'an Jiaotong University, China
Chenyang Wang	Aalto University, Finland
Cheqing Jin	East China Normal University, China
Chihyun Park	Kangwon National University, South Korea
Chongjun Wang	Nanjing University, China
Christos Doulkeridis	University of Piraeus, Greece
Dan He	University of Queensland, Australia
Dan Li	Sun Yat-sen University, China
Debarati B. Chakraborty	University of Hull, UK

Kwanghyun Park	Yonsei University, South Korea
Kyong-Ha Lee	Korea Institute of Science and Technology Information, South Korea
Kyoungsoo Bok	Wonkwang University, South Korea
Kyungyong Lee	Hanyang University, South Korea
Ladjel Bellatreche	ISAE-ENSMA, France
Lanjun Wang	Tianjin University, China
Lei Li	Hong Kong University of Science and Technology (Guangzhou), China
Li Jiajia	Shenyang Aerospace University, China
Libin Zheng	Sun Yat-sen University, China
Lijun Chang	University of Sydney, Australia
Linglin Yang	Peking University, China
Lingwei Chen	Rochester Institute of Technology, USA
Lingyang Chu	McMaster University, Canada
Lizhen Cui	Shandong University, China
Lu Chen	Swinburne University of Technology, Australia
Marwan Hassani	TU Eindhoven, Netherlands
Meng Liu	National University of Defense Technology, China
Mengxuan Zhang	Australian National University, Australia
Mijin An	SAP, South Korea
Ming Zhong	Wuhan University, China
Minghao Zhao	East China Normal University, China
Minghe Yu	Northeastern University, China
Mingyue Cheng	University of Science and Technology of China, China
Min-hwan Oh	Seoul National University, South Korea
Mo Li	Liaoning University, China
Muhammad Haris	Liaoning University, China
Nannan Wu	Tianjin University, China
Ning Liu	Shandong University, China
Ning Wang	Beijing Jiaotong University, China
Ning Yang	Sichuan University, China
Ningning Cui	Chang'an University, China
Norio Katayama	National Institute of Informatics, Japan
O-Joun Lee	Catholic University of Korea, South Korea
Ozan Kahramanogullari	Free University of Bozen-Bolzano, Italy
Peng Cai	East China Normal University, China
Peng Cheng	Tongji University, China
Peng Liu	Guangxi Normal University, China
Pengpeng Zhao	Soochow University, China

Tangpeng Dan	Wuhan University of Technology, China
Tao Qiu	Shenyang Aerospace University, China
Tao Zhao	National University of Defense Technology, China
Tianzi Zang	Nanjing University of Aeronautics and Astronautics, China
Tieke He	Nanjing University, China
Tiezheng Nie	Northeastern University, China
Tomoki Yoshihisa	Shiga University, Japan
Wei Hu	Nanjing University, China
Wei Li	Harbin Engineering University, China
Wei Liu	Sun Yat-sen University, China
Wei Song	Wuhan University, China
Wei Emma Zhang	University of Adelaide, Australia
Wentao Li	University of Leicester, UK
Wolf-Tilo Balke	TU Braunschweig, Germany
Woohwan Jung	Hanyang University, South Korea
Woong-Kee Loh	Gachon University, South Korea
Xiang Ao	Institute of Computing Technology, CAS, China
Xiang Lian	Kent State University, USA
Xiang Zhao	National University of Defense Technology, China
Xiangguo Sun	Southeast University, China
Xiangmin Zhou	RMIT University, Australia
Xiangtao Li	Jilin University, China
Xiangyu Song	Chang'an University, China
Xiao Pan	Shijiazhuang Tiedao University, China
Xiaofeng Gao	Shanghai Jiao Tong University, China
Xiaoling Wang	East China Normal University, China
Xiaoou Ding	Harbin Institute of Technology, China
Xiaoyang Wang	University of New South Wales, Australia
Xin Huang	Hong Kong Baptist University, China
Xinqiang Xie	Neusoft, China
Xuanhe Zhou	Shanghai Jiao Tong University, China
Yajun Yang	Tianjin University, China
Yan Zhang	Peking University, China
Yang Chen	Fudan University, China
Yang-Sae Moon	Kangwon National University, South Korea
Yanhao Wang	East China Normal University, China
Yanmei Hu	Chengdu University of Technology, China
Yanmin Zhu	Shanghai Jiao Tong University, China

Yeon-Chang Lee	Ulsan National Institute of Science and Technology, South Korea
Yeonsu Park	Kangwon National University, South Korea
Yingxia Shao	Beijing University of Posts and Telecommunications, China
Yishu Wang	Northeastern University, China
Yohan Jo	Seoul National University, South Korea
Yong Zhang	Tsinghua University, China
Yongchao Liu	Ant Group, China
Yonghong Yu	Nanjing University of Posts and Telecommunications, China
Yongpan Sheng	Southwest University, China
Young-Koo Lee	Kyung Hee University, South Korea
Young-Kyoon Suh	Kyungpook National University, South Korea
Yu Hua	Huazhong University of Science and Technology, China
Yu Liu	Huazhong University of Science and Technology, China
Yu Sun	Nankai University, China
Yu Yang	Education University of Hong Kong, China
Yuanbo Xu	Jilin University, China
Yuankai Fan	Institute of Artificial Intelligence (TeleAI), China Telecom, China
Yuchen Li	Singapore Management University, Singapore
Yue Kou	Northeastern University, China
Yunjun Gao	Zhejiang University, China
Yunpeng Chai	Renmin University of China, China
Yunyong Ko	Chung-Ang University, South Korea
Yurong Cheng	Beijing Institute of Technology, China
Yuxiang Zeng	Hong Kong University of Science and Technology, China
Zhao Zhang	East China Normal University, China
Zhaojing Luo	Beijing Institute of Technology, China
Zhaonian Zou	Harbin Institute of Technology, China
Zhengyi Yang	University of New South Wales, Australia
Zhihui Wang	Fudan University, China
Zhiwei Zhang	Beijing Institute of Technology, China
Zhixu Li	Renmin University of China, China
Zhongnan Zhang	Xiamen University, China
Zi Chen	Wuhan University of Technology, China
Zimu Zhou	City University of Hong Kong, China
Zirui Zhuang	Beijing University of Posts and Telecommunications, China

Industry Track

Chairs

Lei Chen	Hong Kong University of Science and Technology, China
Min-Soo Kim	KAIST, South Korea

Demo Track

Chairs

Byungchul Tak	Kyungpook National University, South Korea
Siqiang Luo	Nanyang Technological University, Singapore

Program Committee

Chenhao Ma	Chinese University of Hong Kong, Shenzhen
Chun-Hee Lee	Kyungpook National University, South Korea
Dong-Hyuk Im	Kwangwoon University, South Korea
Ergute Bao	Mohamed bin Zayed University of Artificial Intelligence, UAE
Ki Yong Lee	Sookmyung Women's University, South Korea
Kisung Lee	Louisiana State University, USA
Ling Li	Shanxi University, China
Meng Li	Nanjing University, China
Minsu Cho	Kwangwoon University, South Korea
Peng Fang	Huazhong University of Science and Technology, China
Sahil Suneja	IBM Research, USA
Weiping Yu	Nanyang Technological University, China
Xiaolin Han	Northwestern Polytechnical University, China
Yon Dohn Chung	Korea University, South Korea

Contents

Search and Recommendation

Search and Recommendation

GeoPhrase Tree: An Efficient Index for Frequent Phrase Query over Spatio-Temporal Ranges

Chenning Wu[1], Chang Lu[2], Yinan Jing[1], Zhenying He[1(✉)], Zhihui Yang[3,4], and X. Sean Wang[1]

[1] Fudan University, Shanghai, China
wucn23@m.fudan.edu.cn, {jingyn,zhenying,xywangCS}@fudan.edu.cn
[2] Stevens Institute of Technology, Hoboken, NJ, USA
clu13@stevens.edu
[3] School of Software Technology, Institute of Fundamental and Transdisciplinary Research, Zhejiang University, Hangzhou, China
zhyangcs@zju.edu.cn
[4] Hangzhou High Tech Zone (Binjiang) Institute of Blockchain and Data Security, Hangzhou, China

Abstract. The rapid growth of social media has led to massive volumes of posts tagged with locations and timestamps, offering valuable insights into local events and trends. Identifying frequently mentioned phrases within specific spatial regions and time periods is key to uncovering such localized topics and user interests. In this paper, we study the problem of querying frequent phrases within user-defined spatio-temporal ranges over large text collections. The main challenge lies in efficiently matching and aggregating phrases across vast and various documents. Existing approaches either incur high computational overhead or yield incomplete results. We propose a novel hybrid index, `GeoPhrase Tree`, that accelerates phrase matching by pre-constructing Frequency Suffix Trees for spatio-temporal textual objects, indexing them with R-Tree, and applying bound-based pruning strategies during online traversal. To further optimize performance in large-scale and memory-constrained scenarios, we introduce a partial storage strategy. Experiments on real-world datasets demonstrate that our method achieves superior efficiency, scalability, and semantic expressiveness compared with state-of-the-art baselines.

Keywords: Spatio-temporal indexing · Frequent phrase query · Hybrid index structure

1 Introduction

The increasing popularity of mobile devices and social media flourishes the generation of spatio-temporal posts. Users publish reviews containing text descriptions, geographical information, and timestamps of events that occur around

H. Jung et al. (Eds.): DASFAA 2026, LNCS 16535, pp. 3–19, 2026.
https://doi.org/10.1007/978-981-92-0363-5_1

them and catch their attention. Such posts offer first-hand information on local events of varying types and scales [9], which provides a valuable source for data analytical problems such as topic trends detection [21,22] and trajectory similarity search [19].

To understand local hot events and people's opinions, detecting frequent terms from posts within a specified spatio-temporal region is an existing practical way [3,4,8,17,26]. However, individual terms are too broad to provide accurate and comprehensive information. We therefore propose a more general and expressive query, **Frequent Spatio-Temporal Phrase (FSP) Query**, which retrieves frequent multi-word expressions within a given spatio-temporal range, offering richer and more interpretable insights than conventional **Frequent Spatio-Temporal Term (FST) Query**. We use a concrete example from real-world social media data to illustrate the advantages of our proposed query below.

Example Scenario. During the Australian Open 2025[1], Alice, a visitor to Melbourne, seeks trending events and attractions near Melbourne Park (Fig. 1) during her stay. An example FSP query (in region R, during Jan 10–17, 2025) successfully returns meaningful phrases like "Rod Laver" and "tennis" as top-2 results, whereas a FST query yields fragmented terms (e.g., "tennis" and "Rod"). Unlike existing platforms like Yelp that summarize predefined items (e.g., restaurants or entertainments), FSP queries reveal diverse public interests within specific spatio-temporal ranges.

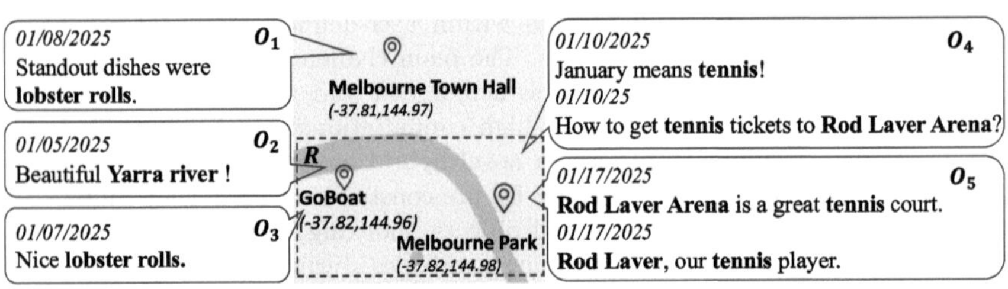

Fig. 1. An example of frequent phrase query over a spatio-temporal range.

Our proposed FSP query supports a wide range of applications. Beyond personal travel planning, it can also assist public services. For instance, monitoring surges in social media discussions during epidemic outbreaks can help health authorities assess situations and coordinate timely responses. Moreover, frequent phrase extraction serves as a fundamental operator in quality phrase mining [25,32], emphasizing the importance of query efficiency.

A straightforward approach for FSP queries is to apply FST queries on pre-segmented text [17]. However, traditional pre-segmentation tools struggle with

[1] The Australian Open is a tennis tournament held annually at Melbourne Park in Melbourne, Australia.

irregular social media language [15,33]. While LLMs [1,11] handle such irregularities, their high inference costs make them impractical for large-scale, frequent queries [6]. To ensure complete phrase detection, especially for emerging entities and buzzwords, we formulate FSP as a *sub-string matching problem* across documents, utilizing **Frequency Suffix Trees** for efficient frequent phrase detection.

Other existing alternatives, such as combining FST results to form phrases or applying Sequential Pattern Mining (SPM) [14,25], suffer from poor scalability due to excessive candidate generation. This inefficiency arises from the continuity and diversity of natural phrases in long text. To address this challenge, we incorporate *upper and lower bound* constraints during the traversal of pre-constructed Frequency Suffix Trees to enable early pruning and therefore significantly improve query performance.

To put it more precisely, this paper introduces a novel combined index structure, `GeoPhrase Tree`, designed to accelerate phrase matching in user-specified spatio-temporal document collections. The approach integrates pre-constructed Frequency Suffix Trees with corresponding R-Tree nodes, leveraging spatio-temporal post information for efficient indexing. This design greatly accelerates querying while allowing flexibility for users to specify the spatio-temporal ranges. Query speed is optimized through a pruning strategy that utilizes upper and lower bounds on phrase occurrences to terminate the traversal early, which significantly enhances performance. Furthermore, we explore various storage strategies for `GeoPhrase Tree` and propose a partial storage model to reduce runtime memory usage, ensuring efficient query processing even for large-scale text collections when memory is constrained.

The contributions of the paper can be summarized as follows:

- We formulate a new research question, Frequent Spatio-Temporal Phrase (FSP) Query, which detects frequent phrases in posts within user-specified spatio-temporal regions. This question is a more informative generalization of existing FST queries and has a broader range of application scenarios.
- We propose `GeoPhrase Tree`, a new hybrid index structure for efficient spatio-temporal documents organization, and implement algorithms for two common query modes, λ-threshold FSP Query and Top-k FSP Query.
- We further extend it to `S-GeoPhrase Tree` and `I-GeoPhrase Tree` to optimize query speed and storage overhead on large-scale datasets.
- Extensive experiments on three real-world datasets show that our approach achieves an order-of-magnitude improvement in query efficiency compared to state-of-the-art methods.

2 Related Work

Sequential Pattern Mining (SPM). Sequential Pattern Mining (SPM) [2] has been studied for decades, focusing on discovering frequently occurring ordered event sequences from sequential data. Classical SPM methods [2,23] identify frequent sequential patterns based solely on their frequency values. Later extensions like high-utility sequential pattern mining (HUSPM) incorporated item weights and quantities [13,30]. Unlike SPM, which extracts arbitrary sequences, our work focuses on *consecutive terms* (phrases) appearing in documents.

Phrase Mining. Phrase extraction has been widely studied in both the NLP [10, 15,33] and data mining communities [20,25,29,31,32]. Approaches in NLP community range from early rule-based systems [10,33] (suffering in domain adaptation [20]) to recent LLMs [1] (computationally expensive). Data mining approaches [20,25,32] typically utilize statistical frequent pattern mining. However, these methods focus on static corpora and overlook query efficiency over user-specified regions. In contrast, our framework enables efficient user-specified spatio-temporal phrase queries based on a flexible hybrid index.

Frequent Spatio-temporal Term Queries. Recent research has focused on identifying the prevalent topics and interests of local users through frequent spatio-temporal term analysis [8,17,26]. focuses on finding frequent *terms* within spatial or temporal regions. Early studies address frequent spatio-temporal term queries in main memory [26], while later research improves scalability and efficiency by updating the underlying data structures [8,17]. Although these methods effectively capture local trends, single-term results often lack contextual clarity [5], which we discuss further in the Introduction and Experiments. We extend this line of research to frequent *phrases*, enabling richer and more discriminative representations within spatio-temporal regions.

3 Problem Formulation

Spatio-temporal Documents. FSP queries are conducted on a collection of *spatio-temporal documents* $\mathscr{D} = \{o_1, \ldots, o_n\}$, where each document $o = \langle S, \rho(x, y), t\rangle$ consists of a set of term lists S, a geographic location $\rho(x, y)$ (latitudelongitude), and a timestamp t.

Phrase Frequency. Within $\mathscr{D}$, a phrase is any contiguous sequence of words occurring in a term list. The *frequency* of a phrase p, denoted $f(p, \mathscr{D})$, is the total number of times p appears across all term lists in $\mathscr{D}$. This frequency captures how salient a phrase is within a chosen spatio-temporal scope.

FSP Queries. Given a spatial region R represented as an axis-aligned rectangle on a two-dimensional map, $R = \{(x, y) \mid x_1 \leq x \leq x_2, y_1 \leq y \leq y_2\}$; a temporal interval I that specifies the time window of interest, $I = \{t \mid t_s \leq t \leq t_e\}$. The subset of documents within this spatio-temporal range forms the search space of an FSP query. The goal is to retrieve phrases that frequently occur in the document subsets. Two query modes are commonly used:

Definition 1 (λ-Threshold and Top-k FSP Queries).

- ***λ-Threshold FSP*** *($q(R, I, \lambda)$): Given a frequency threshold λ, return all phrases p with $f(p, \mathscr{D}) \geq \lambda$ within spatial region R and temporal region I on documents collection $\mathscr{D}$.*
- ***Top-k FSP*** *$q(R, I, k)$: Given an integer k, return the k phrases with the highest frequencies in $\mathscr{D}$ within spatial region R and temporal region I on documents collection $\mathscr{D}$.*

For instance, within the spatial range R in Fig. 1 and time interval $I =< 01/10/2025, 01/17/2025 >$, both a top-2 and 3-threshold query return phrases "Rod Laver" and "tennis".

Top-k and λ-threshold queries serve complementary purposes. The top-k query [34] retrieves a fixed number of results, offering concise and interpretable outputs but may miss longer and less frequent yet informative phrases. In contrast, the λ-threshold query [27] returns all phrases exceeding a frequency threshold, ensuring comprehensive coverage for analytical tasks like data quality assessment. However, it requires prior knowledge of data distribution to choose an appropriate λ. A detailed case study is provided in Sect. 5.3.

Following the convention in prior work [3], we illustrate our approach in a two-dimensional space for simplicity in the following sections, which can be naturally extended to higher dimensions by adding the timestamp to R-Tree.

4 Methodology

To efficiently process FSP queries, we propose a hybrid index `GeoPhrase Tree`, which integrates two well-established structures, **R-Tree** [16] and **Suffix Tree** [28], in a novel way. We first introduce the **Frequency Suffix Tree (FST)** to enable fast phrase matching without pre-segmentation (Sect. 4.1). We then integrate FST with the R-Tree to form `GeoPhrase Tree`, which performs online traversal over pre-built FSTs to efficiently retrieve frequent phrases within a user-specified spatio-temporal region (Sect. 4.2). Furthermore, we enhance `GeoPhrase Tree` along two key dimensions: the `S-GeoPhrase Tree` accelerates query execution through bound-based pruning (Sect. 4.3), and the `I-GeoPhrase Tree` optimizes storage efficiency by partially aggregating internal nodes (Sect. 4.4).

4.1 Frequent Phrases Detection

Detecting frequent phrases requires counting all contiguous term sequences that occur frequently in the target documents. A straightforward approach is to rely on pre-trained segmentation tools or large language models to parse text, but they either struggle with phrases deviating from strict language rules [25], which is common in social media, or incur prohibitively high deployment and inference cost [24], making them impractical for large-scale spatio-temporal phrase queries.

Instead of relying on pre-segmentation, we adopt an index-based pattern-matching approach. Specifically, we employ a prefix-matching mechanism implemented through the **Suffix Tree** [28], a compact trie that indexes all suffixes of a string, where each path from the root to a leaf represents a distinct suffix of a string. Building on this structure, we develop a new variant **Frequency Suffix Tree (FST)**, which augments each node with the occurrence frequency of the phrase represented by its root-to-node path, enabling efficient discovery of frequent phrases without depending on pre-trained segmentation tools or language models. Figure 2b shows a Frequency Suffix Tree constructed on the spatio-temporal document object o_4 in Fig. 1.

To execute FSP queries efficiently, we propose breadth-first traversal on the Frequency Suffix Tree, enabling early pruning under the *Downward Closure Property*. Formally, for any node n and its parent $parent(n)$, $freq(parent(n)) \geq freq(n)$ holds. This property arises naturally from the suffix tree structure where each parent node represents a phrase that is a prefix of its children's phrases. Thus, when a parent's frequency falls below the current threshold, its entire subtree can be pruned, allowing efficient level-wise exploration of frequent phrases. The detailed algorithm is presented in Algorithm 1 and explained in the next section.

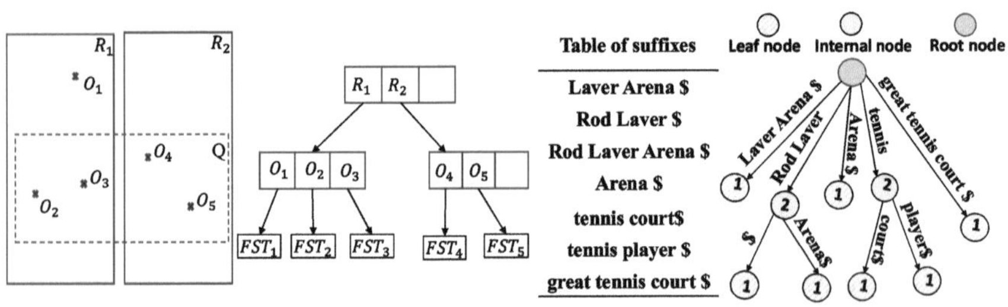

(a) R-Tree of spatio-temporal documents. (b) Frequency Suffix Tree of o_4.

Fig. 2. An example of `GeoPhrase Tree`

FSTs are constructed using a linear-time online algorithm (e.g., Ukkonen's method [28]), updating node frequencies as each document string is processed. The resulting tree compactly stores all repeated phrases on edges and their count on nodes. The total time complexity for construction is $O(n)$, where n is the combined length for all strings in S.

4.2 Hybrid Index Framework: The `GeoPhrase Tree`

As the query region R_Q expands, constructing Frequency Suffix Trees (FSTs) on demand becomes inefficient. We therefore pre-build FSTs offline and organize them with an R-Tree, forming a hybrid index, `GeoPhrase Tree`, as shown in Fig. 2. Each leaf node in the R-Tree corresponds to a spatio-temporal document and stores its pre-constructed FST, while internal nodes maintain aggregated metadata for pruning during spatial filtering.

Algorithm 1: FSP Query over `GeoPhrase Tree`

Input: Query region R_Q, top-k, threshold λ, root r of `GeoPhrase Tree`
Output: Set of frequent phrases $\mathcal{P}$

1 $\mathcal{N} \leftarrow$ RTreeFilter(R_Q, r);
2 Initialize queue Q with roots of FSTs in $\mathcal{N}$;
3 **while** Q *not empty* **do**
4   pop node n from Q;
5   **if** $n.freq \geq minK$ *(or* $n.freq \geq \lambda$ *for threshold query)* **then**
6   add ($n.phrase, n.freq$) to result list $\mathcal{P}$;
7   push children of n into Q;
8   update $minK$ as the smallest frequency in current top-k list
9   **end**
10 **end**
11 **return** $\mathcal{P}$

Algorithm 1 illustrates the unified framework for executing both top-k and λ-threshold FSP queries over `GeoPhrase Tree`. It proceeds in two phrases: (1) Region filtering (line 1). The R-Tree retrieves all leaf nodes intersecting the query region. (2) Phrase aggregation (line 4–8). It then performs a Breadth-First Search(BFS) over the corresponding FSTs, aggregating phrase frequencies level-by-level. For top-k queries, the algorithm maintains a dynamic threshold $minK$, which is the frequency of the current k^{th} ranked phrase. Nodes with frequencies below $minK$ are pruned. For λ-threshold queries, the similar pruning process is applied with a fixed threshold λ. This unified BFS-based approach efficiently combines region filtering, textual pattern aggregation, and frequency-based pruning, achieving high performance across both query types.

4.3 Speed Optimization: S-Geophrase Tree

In real-world corpora, most phrases occur infrequently while only a few are highly repetitive, causing the Frequency Suffix Tree, which only supports depth-width pruning, to traverse many low-frequency branches before reaching meaningful results. To further improve query efficiency, we extend `GeoPhrase Tree` into an enhanced `S-GeoPhrase Tree`, which extends the Frequency Suffix Tree (FST) into a **Sorted Frequency Suffix Tree (SFST)** that prioritizes high-frequency nodes and supports width-wise pruning across spatial-temporal nodes. In the SFST, nodes are linked in descending order of phrase frequency, enabling sequential access to the most frequent nodes and early termination during query processing. Figure 3 illustrates the SFSTs constructed from documents posted at Melbourne Park.

We formulate top-k and λ-threshold FSP queries over multiple SFSTs as retrieving phrases with the highest aggregate frequencies. Leveraging the sorted structure, we adapt a variant of the *No Random Access (NRA)* algorithm [12], which sequentially scan nodes and applies early pruning with *upper and lower bounds*. For each phrase p observed in a subset of SFSTs $\{T_1, \ldots, T_l\}$ with frequencies $f_1, \ldots, f_l$, the frequency bounds are:

$$\text{LB}(p) = \sum_{i=1}^{l} f_i, \qquad \text{UB}(p) = \sum_{i=1}^{l} f_i + \sum_{T_j \notin \{T_1,\ldots,T_l\}} freq(\text{current node in } T_j). \tag{1}$$

Intuitively, the **Lower Bound (LB)** represents the sum of all currently observed frequencies. The **Upper Bound (UB)** estimates the maximum possible frequency of a phrase by adding the highest potential contributions from unseen trees to the LB, that is, the current node frequency in each SFST visited in descending order of frequency.

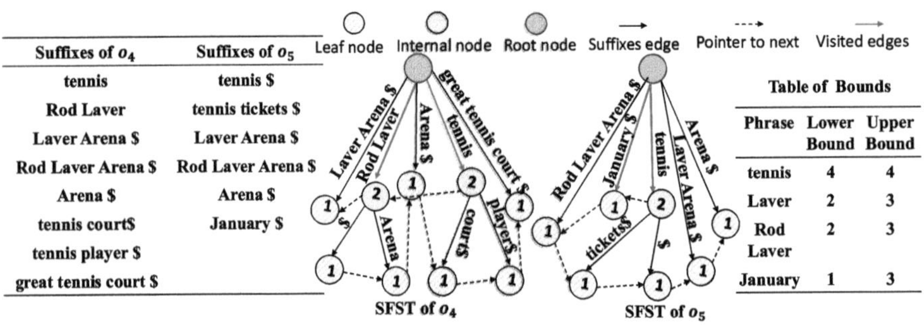

Fig. 3. An example of Sorted Frequency Suffix Tree query process.

Algorithm 2: Multi-SFST Top-k Phrase Query

Input: Number of output phrases k, candidate nodes set $\mathcal{N}$ in `GeoPhrase Tree`
Output: top-k frequent phrases list L

$L \leftarrow \emptyset$, $t \leftarrow 1$, $minK \leftarrow 0$;
repeat
  foreach *SFST* $T \in \mathcal{N}$ **do**
    $n \leftarrow$ GetNode(t,T) ;
    foreach *phrase p on path(root→ n)* **do**
      $\text{LB}(p) \mathrel{+}= \text{freq}_T(p)$
      $\text{UB}(p) = \text{LB}(p) + \sum_{T' \notin \text{seen}(p)} \text{freq}_{T'}(\text{current\_feq})$;
    end
  end
  $L \leftarrow$ top-k by LB; $minK \leftarrow \min_{p \in L} \text{LB}(p)$; $t \leftarrow t+1$;
until $\max_p \text{UB}(p) \leq minK$;
return L

Algorithm 2 sequentially scans SFSTs stored on candidate nodes in `S-GeoPhrase Tree` within given region. At iteration t, it visits the tth node of each SFST, updates for each encountered phrase p its lower bound and upper bound, and maintains the top-k list by lower bound. The query terminates when the k-th largest lower bound exceeds all remaining upper bounds. Figure 3 demonstrates the query process when $t = 2$.

For λ-threshold queries, the termination occurs when the smallest UB $< \lambda$ and we revisit all SFSTs for phrases with $UB > \lambda$ and $LB < \lambda$ to find its exact count. This extra verification remains efficient, since each lookup for phrase p in SFST takes linear time in phrase length $O(m)$, and only a small number of boundary candidates require verification.

4.4 Storage Optimization: I-Geophrase Tree

`S-GeoPhrase Tree` performs well on small query ranges but degrades as the number of SFSTs increases, since more trees introduce higher upper bounds, delaying effective pruning and increasing bound maintenance costs. Empirically, it performs best with *fewer but broader* SFSTs, motivating the design of `I-GeoPhrase Tree`, which aggregates child SFSTs into internal R-tree nodes rather than storing them only at the leaves, thereby improving pruning efficiency.

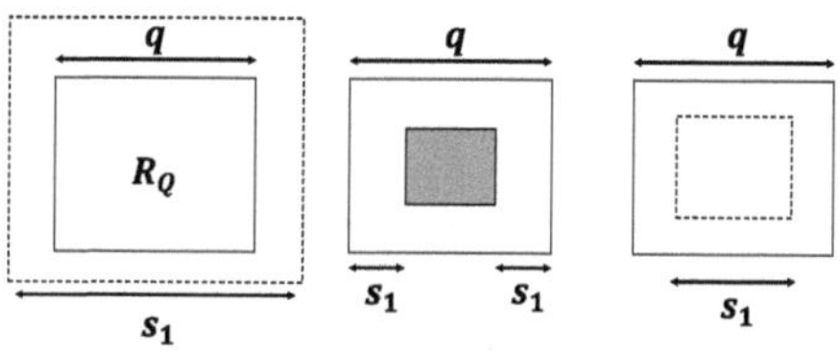

Fig. 4. Analysis of query region and node sizes.

In `I-GeoPhrase Tree`, each internal node maintains an aggregated SFST representing the union of its children. The aggregation is fundamentally an edge-based merging and frequency accumulation process. The edge integration handles three label-matching cases: (1) **Exact matches** trigger recursive frequency aggregation; (2) **Partial prefix matches** split existing edges to insert intermediate nodes; and (3) **No matches** directly attach the edge as a new branch of parent SFST.

Query Workflow. During the FSP query traversal, `I-GeoPhrase Tree` utilizes a top-down pruning strategy.

- **Containment Check:** If an internal node's MBR (i.e., the smallest rectangle that completely encloses the data points within it) is fully contained within the query range R_Q, its aggregated SFST is treated as a direct candidate source for FSP Query.
- **On-demand Merge:** If the query requires the tth entry where $t \leq \rho$, it returns the stored node directly. If $t > \rho$ (exceeding partial storage), the algorithm recursively descends to child nodes to retrieve finer-grained entries based on top-k query on children SFSTs.

Balancing Space and Efficiency. A small ρ reduces storage overhead but increases merge latency. The optimal ρ should align with Algorithm 2's termination point to minimize runtime merges. Adopting the framework from [3], we estimate ρ by first calculating the expected candidate nodes $(m_1, \ldots, m_h)$ involved in FSP queries at each level of `I-GeoPhrase Tree`, and then deriving the expected rank of the termination phrase via Zipf's Law [18].

Given an R-tree of height h and fan-out f storing N objects, the number of nodes at level i is :$N_i = \frac{N}{f^{i-1}}$, and the height of the tree is $h = 1 + \left\lceil \log_f \frac{N}{f} \right\rceil$. In a d-dimensional space, each node at level i covers an average region with edge size s_i, which is recursively estimated as

$$s_i = \left(f^{1/d} - 1\right) \cdot \frac{1}{\left(N_{i-1}\right)^{1/d}} + s_{i-1} \tag{2}$$

The expected number of nodes at level i intersecting a query region of edge length q is $intersect(N_i, s_i, q) = N_i \cdot (s_i + q)^d$. Nodes fully contained in the query region are determined by the relative size of q and s_i (shown in Fig. 4):

- Case 1 ($s_i > q$): no node is contained, go to next level;
- Case 2 ($q \geq 2s_i$): the node inside out rectangle is the node intersects the inner rectangle, $intersect(N_i, s_i, q - 2s_i)$;
- Case 3 ($s_i \leq q < 2s_i$): only one node can be contained in the query region.

Suppose each spatio-temporal object contains n phrases and there are N objects in total, the corpus size is Nn. Under Zipf's law [18], given the pre-computed Zipf parameter c, the frequency of the p-th ranked phrase is $freq(p, Nx) = \frac{cNx}{p}$. Within a query region of size q^d, the local frequency of the pth phrase becomes $\frac{c_q q^d Nx}{p}$. The top-k FSP Query terminates when the k-th lower bound exceeds the highest possible upper bound of all other items. To simplify the computation, we conduct a conservative estimation by assuming that each term appears in only one list, then the upper bound of unseen items is approximated by $\theta(\rho, Nn, M) = \sum_{i=1}^{h} m_i \frac{c_i Nn}{N_i \rho}$, and the optimal ρ satisfies

$$\min_{\rho} \theta(\rho, Nn, M) \leq \frac{c_q q^d Nn}{k}. \tag{3}$$

Example. Consider $N = 10^6$ spatio-temporal objects with average terms of 1,000. R-tree fan-out $f = 100$. Zipf parameter $c = 0.1$. Query region size $R_Q = 0.1^3$, and $k = 20$. The R-tree height is $h = 3$, with $N_1 = 10^4$ leaf nodes, $N_2 = 10^2$ inner nodes, and one root. Average node edge sizes are $s_3 = 1.0$, $s_2 = 0.1$, and $s_1 = 0.01$. Nodes intersecting the query region are $m_3 = 0$, $m_2 = 1$, and $m_1 = intersect(N_1, s_1, q - 2s_1) = 81$. Given $freq(20, Nn) = \frac{0.1 \times 10^6}{20} = 5{,}000$, the minimum ρ satisfying $\theta(\rho) \leq 5{,}000$ is 75.

4.5 Discussion on Index Updates

In this paper, we focus on queries over existing documents, but `GeoPhrase Tree` can be adapted to streaming spatio-temporal data as well. When new documents arrive, updates proceed step by step: (1) R-tree update, (2) SFST construction, and (3) integration. The most critical and time-consuming step arises during R-tree node splits, which necessitate redistributing and remerging inner-level SFSTs, posing significant computational challenges. To reduce update latency, a two-phase scheme can be adopted: a *synchronous phase* that immediately builds and attaches new SFSTs upon receiving a new document, and an *asynchronous phase* that reconstructs R-Tree and internal SFSTs in batch to minimize the impact on overall system latency.

5 Experimental Study

We evaluate `GeoPhrase Tree` on three real-world datasets to assess query efficiency, index storage, and semantic richness.

5.1 Experimental Settings

Hardware. All experiments are conducted on Ubuntu 18.04 (64-bit) with an Intel Xeon E5-2620 v2 @ 2.10 GHz × 24 CPU, 64 GB RAM, and 2 TB disk. All methods were implemented in C++ (7.5.0).

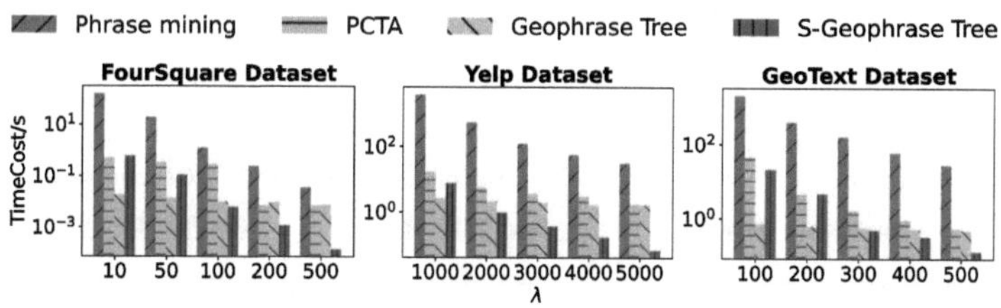

Fig. 5. Performance comparison of algorithms under different thresholds λ.

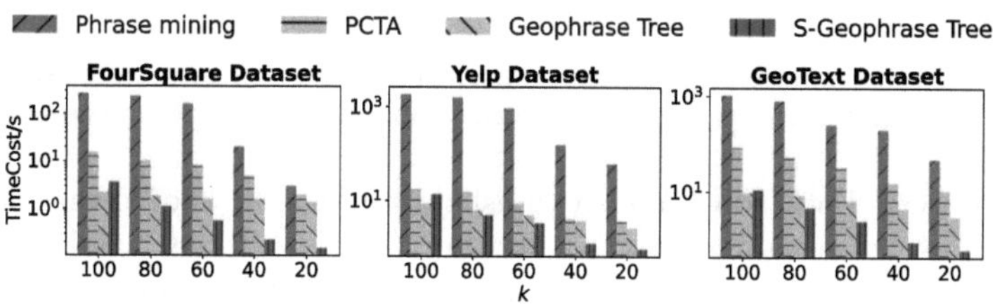

Fig. 6. Performance comparison of algorithms under different top-k.

Datasets. We use three public datasets: **FourSquare**[2] (New York restaurants check-ins and tags); **Yelp**[3] (over 7 million geo-tagged reviews with timestamps), which serves as a benchmark for scalability; and **GeoText**[4] (over 377k microblogs), containing informal language to test `GeoPhrase Tree`'s ability to extract frequent phrases from nonstandard text.

Baselines. We compare our Top-k and λ-threshold FSP queries based on `GeoPhrase Tree` against: (1) **Phrase Mining** [25], a data-driven phrase mining approach based on the frequent pattern mining principle. (2) **PCTA** [17], a Top-k FST Query method using prefix cubes. To obtain phrase results, we modified the algorithm by concatenating its query outputs. (3) **Pre-segmentation** [7]. In our case study, we compare the query results obtained by performing FST Queries after pre-tokenizing with Natural Language Toolkit(NLTK) [7] to the results produced by our proposed method.

Implementation. The fan-out factor f of R-Tree is 0.8 and the node size is 4KB which corresponds to a maximum of 32 entries. Frequency Suffix Trees are built via Ukkonen's algorithm. Text data is preprocessed by converting to lowercase, removing stop words, and encoding tokens numerically to minimize storage overhead.

5.2 Query Efficiency

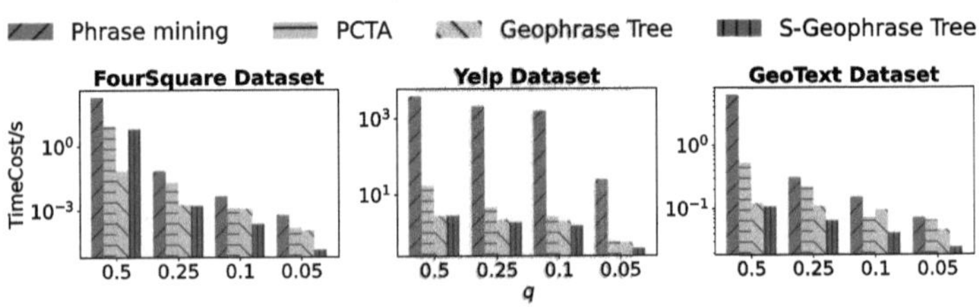

Fig. 7. Performance comparison of algorithms under different range sizes q.

Across datasets, both top-k and λ-threshold queries on `GeoPhrase Tree` and `S-GeoPhrase Tree` significantly outperform state-of-the-art baselines in efficiency. Furthermore, under varying parameters of k, λ, and q (i.e., query region edge size), we observe that `GeoPhrase Tree` and `S-GeoPhrase Tree` exhibit complementary strengths. Generally, `GeoPhrase Tree` excels in large-range queries with big result sets due to its lightweight aggregation strategy,

[2] https://sites.google.com/site/yangdingqi/home/foursquare-dataset.
[3] https://www.yelp.com/dataset.
[4] http://www.ark.cs.cmu.edu/GeoText.

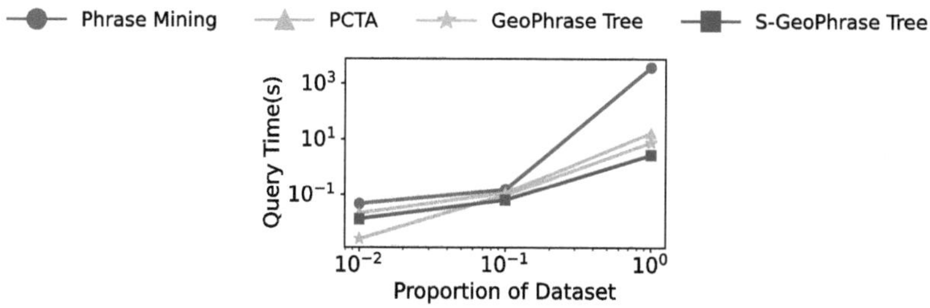

Fig. 8. Performance comparison of algorithms across datasets of different sizes.

while `S-GeoPhrase Tree` performs better in smaller query ranges or under stricter frequency thresholds, benefiting from its aggressive pruning strategy.

As shown in Fig. 5, when λ is small, the lightweight `GeoPhrase Tree` performs better, since it avoids the overhead of bound computation. As λ increases, `S-GeoPhrase Tree` rapidly surpasses all baselines, because its bound pruning strategy eliminates low-frequency branches early, reducing runtime by up to one to two orders of magnitude. Similar trends are observed for top-k queries (shown in Fig. 6). `GeoPhrase Tree` is more efficient for large k values, while `S-GeoPhrase Tree` excels when k is small, where tighter bounds yield earlier termination.

Figure 7 compares the algorithms' performance with various query range sizes. When the query range is wide, `GeoPhrase Tree`'s lightweight aggregation performs best; as the range narrows, `S-GeoPhrase Tree` gains dominance due to fewer participating trees and tighter bounds.

We evaluate the scalability of our proposed algorithms by running top-k FSP queries with fixed k and R_Q while varying the **Yelp** dataset size. As shown in Fig. 8, both `GeoPhrase Tree` and `S-GeoPhrase Tree` scale nearly linearly with data volume, demonstrating strong suitability for large-scale applications. `S-GeoPhrase Tree` achieves increasing performance gains as the data grow, reflecting the effectiveness of its bound-based pruning.

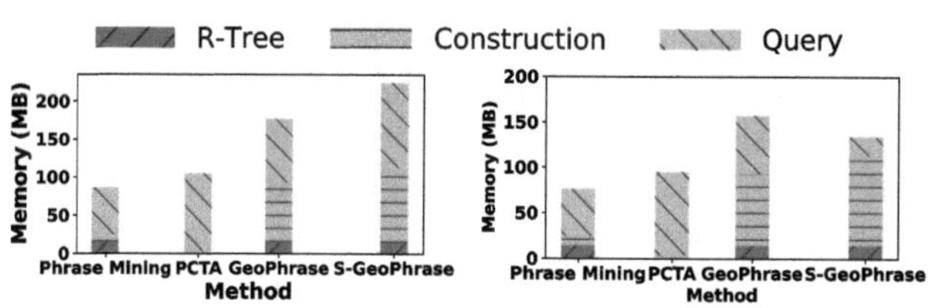

(a) $\lambda = 1000$ (b) $\lambda = 3000$

Fig. 9. Run-time memory consumption of different methods.

5.3 Case Study

Table 1. Comparison of query results on Yelp dataset (Case Study)

λ-threshold Query ($\lambda = 20$)			Top-k Query ($k = 50$)		
FSP	FST	NLTK+FST	FSP	FST	NLTK+FST
service	service	great job	service	service	great job
le schwab tire center	**schwab**	**yuki song**	**le schwab**	salad	**yuki song**
aa window gutter cleaning	table	**different place**	**gutter cleaning**	**schwab**	**different place**
round table	**gutter**		**customer service**	**customer**	**round table**
yuki song	**window**		highly recommend	pizza	snowball
...	...	...	...	...	...

We conduct a case study on the **Yelp** dataset to demonstrate FSP's semantic advantage. Table 1 compares the results of the proposed FSP query against FST query and FST query after pre-segmentation using NLTK with query parameters $\lambda = 20, k = 50$ in a static region. FSP queries outperform baselines by identifying distinctive multi-word expressions (e.g., "Le Schwab Tire Center"), offering richer and more accurate semantics. Notably, λ-threshold FSP queries favor specific and long phrases, while top-k FSP queries highlight general and high-frequency expressions. In contrast, pre-segmentation tools struggle with λ-thresholds, as they tend to generate long and sparse phrases that often fail frequency constraints.

5.4 Index Efficiency

We evaluate the indexing efficiency of `GeoPhrase Tree` in terms of construction time, storage overhead, and query performance improvement.

As reported in Table 2, `GeoPhrase Tree` incurs a 4.1–4.3× storage expansion. This is a typical trade-off for text indices, which is well compensated for by the orders-of-magnitude efficiency gains (detailed in Sect. 5.2). Since construction is offline, it has negligible runtime impact. In terms of memory consumption (Fig. 9), `GeoPhrase Tree` consumes more space for trees kept in memory to boost query speed, which is well justified by the substantial gains in query performance. Notably, `S-GeoPhrase Tree` memory usage peaks at a small λ due to ineffective pruning, whereas `GeoPhrase Tree` maintains stable consumption.

Table 2. Index efficiency for various datasets

Dataset	Size	Expansion	Build Time	Reduction/query	Efficiency Improvement
FourSquare	100M	4.1	15.5 s	35 s	10^3
Yelp	10G	4.3	1729 s	812 s	10^3
GeoText	1000M	4.1	62.8 s	320 s	10^3

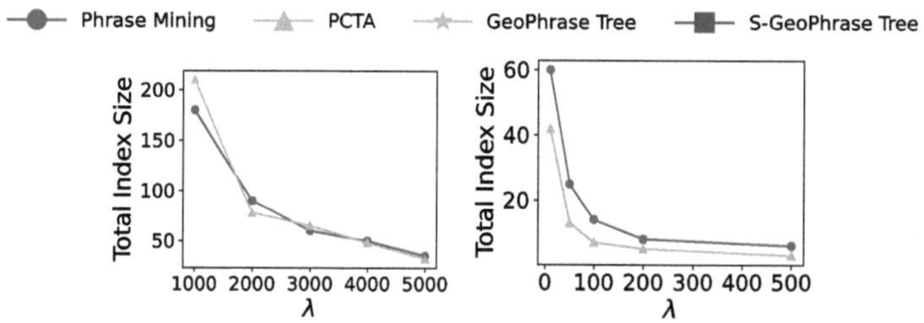

(a) **Yelp** dataset (b) **FourSquare** dataset

Fig. 10. Estimated versus actual internal node size.

5.5 Partial Storage Model Validation

We compare the estimated partial index size (i.e., the calculated values of ρ for the parent nodes of the leaf-level SFSTs, see Sect. 4.4) with the actual runtime sizes and find them closely aligned, confirming the accuracy of our partial storage model. As shown in Figs. 10a and 10b, the model is more accurate on large datasets like **Yelp**, while smaller datasets such as **FourSquare** exhibit greater fluctuation due to non-uniform spatio-temporal objects distributions and deviation from Zipf's law. This limitation is acceptable in practice, as partial storage is primarily needed for large datasets where full in-memory indexing is infeasible and the model performs reliably.

6 Conclusions

In this paper, we propose `GeoPhrase Tree`, a hybrid indexing framework that integrates Frequency Suffix Trees with R-tree nodes for efficient processing of Frequent Spatio-Temporal Phrase (FSP) queries. We develop two enhanced variants, `S-GeoPhrase Tree` for query-speed optimization and `I-GeoPhrase Tree` for storage efficiency to address distinct scalability challenges in large datasets. By combining Frequency Suffix Trees, frequency-bound pruning, and a partial storage strategy, `GeoPhrase Tree` supports both λ-threshold and top-k queries with reduced memory usage and faster convergence. Extensive experiments on real-world datasets demonstrate that `GeoPhrase Tree` achieves substantial gains in both efficiency and result quality, providing a scalable foundation for phrase-centric spatio-temporal analytics.

Acknowledgements. The authors would like to thank the anonymous reviewers for their insightful comments and suggestions. This work was supported by NSFC (Grant No. 62272106).

References

1. Achiam, J., et al.: Gpt-4 technical report. arXiv preprint arXiv:2303.08774 (2023)
2. Agrawal, R., Srikant, R.: Mining sequential patterns. In: Proceedings of the Eleventh International Conference on Data Engineering, pp. 3–14. IEEE (1995)
3. Ahmed, P., et al.: Efficient computation of top-k frequent terms over spatio-temporal ranges. In: Proceedings of the 2017 ACM International Conference on Management of Data, pp. 1227–1241 (2017)
4. Ahmed, P., et al.: Reverse spatial top-k keyword queries. VLDB J. **32**(3), 501–524 (2023)
5. Ahonen, H., et al.: Applying data mining techniques for descriptive phrase extraction in digital document collections. In: Proceedings IEEE International Forum on Research and Technology Advances in Digital Libraries-ADL'98-, pp. 2–11. IEEE (1998)
6. Aryan, A., et al.: The costly dilemma: generalization, evaluation and cost-optimal deployment of large language models. arXiv preprint arXiv:2308.08061 (2023)
7. Bird, S., Klein, E., Loper, E.: Natural language toolkit (NLTK). In: Proceedings of the ACL 2009 Workshop on Software Demonstrations, pp. 139–142 (2009)
8. Chen, L., et al.: Location-aware top-k term publish/subscribe. In: IEEE 34th International Conference on Data Engineering (ICDE), pp. 749–760. IEEE (2018)
9. Chen, L., et al.: Top-k term publish/subscribe for geo-textual data streams. VLDB J. **29**(5), 1101–1128 (2020)
10. Deane, P.: A nonparametric method for extraction of candidate phrasal terms. In: Proceedings of the 43rd Annual Meeting of the Association for Computational Linguistics (ACL'05), pp. 605–613 (2005)
11. Devlin, J., et al.: Bert: pre-training of deep bidirectional transformers for language understanding. In: Proceedings of the 2019 Conference of the North American Chapter of the Association for Computational Linguistics: Human Language Technologies, vol. 1 (long and short papers), pp. 4171–4186 (2019)
12. Fagin, R., Lotem, A., Naor, M.: Optimal aggregation algorithms for middleware. In: Proceedings of the Twentieth ACM SIGMOD-SIGACT-SIGART Symposium on Principles of Database Systems, pp. 102–113 (2001)
13. Gan, W., et al.: ProUM: projection-based utility mining on sequence data. Inf. Sci. **513**, 222–240 (2020)
14. Geng, M., et al.: RNP-miner: repetitive nonoverlapping sequential pattern mining. IEEE Trans. Knowl. Data Eng. **36**(9), 4874–4889 (2023)
15. Gu, X., et al.: UCPhrase: unsupervised context-aware quality phrase tagging. In: Proceedings of the 27th ACM SIGKDD Conference on Knowledge Discovery & Data Mining, KDD '21. Virtual Event, Singapore, pp. 478–486 (2021)
16. Guttman, A.: R-trees: a dynamic index structure for spatial searching. In: Proceedings of the 1984 ACM SIGMOD International Conference on Management of Data, pp. 47–57 (1984)
17. He, Z., et al.: Efficiently answering top-k frequent term queries in temporal-categorical range. Inf. Sci. **574**, 238–258 (2021)
18. Li, W.: Zipf's Law everywhere. Glottometrics **5**(2002), 14–21 (2002)
19. Lin, Y., et al.: UniTE: a survey and unified pipeline for pre-training spatiotemporal trajectory embeddings. IEEE Trans. Knowl. Data Eng. **37**(3), 1475–1494 (2024)
20. Liu, J., et al.: Mining quality phrases from massive text corpora. In: Proceedings of the 2015 ACM SIGMOD International Conference on Management of Data, pp. 1729–1744 (2015)

21. Lu, W., et al.: Detecting research topic trends by author-defined keyword frequency. Inf. Process. Manag. **58**(4), 102594 (2021)
22. Mondal, M., et al.: A survey of multimodal event detection based on data fusion. VLDB J. **34**(1), 9 (2025)
23. Pei, J., et al.: Mining sequential patterns by pattern-growth: the prefixspan approach. IEEE Trans. Knowl. Data Eng. **16**(11), 1424–1440 (2004)
24. Samsi, S., et al.: From words to watts: benchmarking the energy costs of large language model inference. In: IEEE High Performance Extreme Computing Conference (HPEC), pp. 1–9. IEEE (2023)
25. Shang, J., et al.: Automated phrase mining from massive text corpora. IEEE Trans. Knowl. Data Eng. **30**(10), 1825–1837 (2018)
26. Skovsgaard, A., Sidlauskas, D., Jensen, C.S.: Scalable top-k spatio-temporal term querying. In: International Conference on Data Engineering (2014)
27. Tomaszuk, D., et al.: Threshold queries in theory and in the wild. Proc. VLDB Endow. **34**(4), 49 (2022)
28. Ukkonen, E.: On-line construction of suffix trees. Algorithmica **14**(3), 249–260 (1995)
29. Yongliang, W., et al.: ParsingPhrase: parsing-based automated quality phrase mining. Inf. Sci. **633**, 531–548 (2023)
30. Yin, J., Zheng, Z., Cao, L.: USpan: an efficient algorithm for mining high utility sequential patterns. In: Proceedings of the 18th ACM SIGKDD International Conference on Knowledge Discovery and Data Mining, pp. 660–668 (2012)
31. Zhang, X., Li, Q., Li, C., Chen, H.: Automated context-aware phrase mining from text corpora. In: Jensen, C.S., et al. (eds.) DASFAA 2021. LNCS, vol. 12682, pp. 20–36. Springer, Cham (2021). https://doi.org/10.1007/978-3-030-73197-7_2
32. Zhang, X., et al.: Conphrase: enhancing context-aware phrase mining from text corpora. IEEE Trans. Knowl. Data Eng. **35**(7), 6767–6783 (2022)
33. Zhang, Z., Gao, J., Ciravegna, F.: Semre-rank: improving automatic term extraction by incorporating semantic relatedness with personalised pagerank. ACM Trans. Knowl. Disc. Data (TKDD) **12**(5), 1–41 (2018)
34. Zhao, J., et al.: On efficiently diversified top-k geo-social keyword query processing in road networks. Inf. Sci. **512**, 813–829 (2020)

Cluster-Guided Disentangled Representation for Cold-Start Cross-Domain Recommendation

Huping Yu[1,2], Yuhan Wang[1,2], Qing Xie[1,2(✉)], Mengzi Tang, Lin Li[1,21,2], and Yongjian Liu[1,2]

[1] Wuhan University of Technology, Wuhan 430070, Hubei, China
{yuhuping,wyh0520,felixxq,cathylilin,liuyj}@whut.edu.cn
[2] Engineering Research Center of Intelligent Service Technology for Digital Publishing, Ministry of Education, Wuhan, China

Abstract. Cross-domain recommendation (CDR) aims to alleviate the user cold-start problem of recommender systems by transferring information from auxiliary domains. Existing CDR methods can be divided into two main paradigms: the mapping-based paradigm and the joint learning paradigm. Although these works have achieved promising results, the former often fails to distinguish domain-specific features, while the latter suffers from suboptimal performance under data imbalance when one domain has limited data. Meanwhile, research into enhancing CDR through clustering techniques is also advancing, but they typically adopt fixed cluster centers, which restrict the adaptability of user representations and neglect the correlations between items and user clusters. In this paper, we propose a Cluster-Guided Cross-Domain Recommendation (CGCDR) framework to address these limitations. We introduce learnable cluster centers in each domain to dynamically capture evolving user intents and jointly optimize clustering and recommendation. Then, through soft assignment and weighted aggregation, we decompose each user's interaction sequence into disentangled cluster-level intent before knowledge transfer. Finally, we design a cluster-assisted contrastive loss to enhance representation robustness in the target domain. Extensive experiments on benchmark datasets demonstrate that CGCDR outperforms state-of-the-art CDR baselines in cold-start scenarios. Our source code is available at https://github.com/yuhuping/CGCDR.

Keywords: Cross-Domain Recommendation · Cluster Learning · Disentangled Representation

1 Introduction

Recommendation systems based on collaborative filtering (CF) face the challenging cold-start problem [25,27], as they cannot provide accurate recommendations

H. Jung et al. (Eds.): DASFAA 2026, LNCS 16535, pp. 20–36, 2026.
https://doi.org/10.1007/978-981-92-0363-5_2

for users with insufficient interaction data. As a promising solution, cross-domain recommendation (CDR) [19,20] aims to alleviate this problem by transferring information from auxiliary domains. Over the years, CDR methods have evolved into two primary paradigms: the mapping-based paradigm and the joint learning paradigm for user representations, as shown in Fig. 1. Mapping-based paradigm (Fig. 1 (a)), represented by early works such as EMCDR [15], focuses on transferring user preferences by learning a direct mapping function from source domain embeddings to target domain ones. This paradigm is praised for its simplicity and effectiveness in alleviating data sparsity. However, by relying solely on a global mapping, it overlooks domain-specific characteristics [12] that reflect users' unique preferences and behavioral patterns within each domain, resulting in homogeneous user representations and negative transfer.

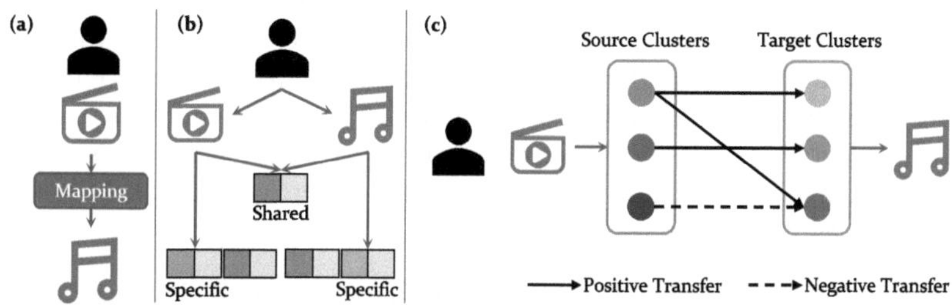

Fig. 1. An illustration of (a) mapping-based paradigm; (b) joint learning paradigm; (c) our CGCDR.

To address the above issue, the joint learning paradigm (Fig. 1(b)) has gained increasing attention for its capability to distinguish domain-shared and domain-specific features, thereby capturing user preferences more comprehensively. For instance, CDRIB [2] optimizes embeddings across domains to disentangle shared information such as *Category* in the Movie–Music scenario, while preserving domain-specific signals like *Film Actors* and *Audio Features*. Despite these advantages, the joint learning paradigm tends to perform less effectively in domains with smaller data scales (as confirmed in Sect. 5.2), as it struggles to balance learning when significant data imbalance exists across domains.

Meanwhile, clustering algorithms [4,24] have also demonstrated promising results in recommendation systems by effectively grouping similar users into the same cluster. CDIE-C [22] extracts cluster-level correlations across domains using co-clustering and filters out noise. UCMF [16] utilizes user clustering to find user-group information and to guide user feature learning. MD-CSKT [26] improves the target domain recommendation from multiple auxiliary domains and ensures consistency of cluster-level knowledge through the design of regularization constraints. These methods have advanced CDR by incorporating group-level insights to improve diversity and accuracy. Nevertheless, existing CDR approaches that incorporate clustering suffer from two key limitations: (1)

cluster centers are not explicitly modeled as learnable user intent prototypes for joint optimization with recommendation; and (2) some learned intent prototypes inevitably capture domain-specific features, potentially leading to negative transfer.

In this paper, we introduce CGCDR, a Cluster-Guided Cross-Domain Recommendation framework that addresses the aforementioned limitations through a structured, cluster-guided design. Different from previous works, we initialize cluster centers as learnable neurons within each domain and design a **user alignment with cluster center** module to form well-separated intent prototypes. Building upon this foundation, we propose a **cluster-guided for disentanglement** module to avoid negative transfer by distinguishing cluster centers, as shown in Fig. 1(c). Specifically, we introduce a soft clustering-based sequence decomposition mechanism that encodes each user's interaction history as a weighted aggregation of cluster centers, each corresponding to a distinct interest. This weighted representation enables effective disentanglement between domain-shared and domain-specific features. Furthermore, we introduce a **cluster-assisted contrastive loss** that pulls user representations closer to positive items while pushing them away from the target domain cluster center nearest to each negative sample, thereby enhancing mapping robustness. Our contributions can be summarized as follows:

- To the best of our knowledge, this is the first work to achieve effective feature disentanglement in cross-domain recommendation by leveraging learnable cluster centers as structural guidance.
- We propose a novel cluster-guided framework that transfers domain-shared cluster centers across domains while filtering domain-specific ones to mitigate negative transfer.
- Extensive experiments on the real-world datasets demonstrate that CGCDR outperforms state-of-the-art models in most scenarios. We also conduct comprehensive ablation studies to validate the effectiveness of each component.

2 Related Work

2.1 Cross-domain Recommendation

CDR methods transfer knowledge from source domains to target domains with sparse or cold-start users. Existing CDR tasks can be categorized into two types: Intra-domain CDRs [7,11,18] aim to compensate for sparse interactions in the target domain by leveraging a user's interaction interests in the source domain. CMF [18] proposed collective matrix factorization to enable cross-domain recommendation via shared latent factors. CoNet [7] and DDTCDR [11] enhance the information transfer process by cross-connection networks and a deep dual-transfer framework, respectively. On the other hand, inter-domain CDRs aim to learn the representation of the target domain without interaction records by using the interaction information of the user in the source domain. Early mapping-based models (e.g., EMCDR [15]) learn linear or neural transformations between user/item embeddings of different domains. SSCDR [9] employs sharpness-aware minimization to seek flat minima, enhancing generalization and adversarial robustness. CDRIB [2] leverages the variational information bottleneck to constrain representation learning for domain-shared information extraction. However, they either fail to distinguish domain-specific features or cannot effectively transfer relationships between domains.

2.2 Disentangled Representation Learning

After demonstrating its potential in computer vision, Disentangled Representation Learning has begun to be applied in decomposing user representations, effectively distinguishing underlying interpretive factors within user intent. For instance, CDRIB(a VAE-based model) utilizes a variational graph encoder to capture the user-item interaction graph and leverage the information bottleneck [21] principle to learn domain-shared representations. DGCDR [23] proposed a GNN-enhanced encoder-decoder framework with anchor-based supervision for supervised disentanglement to improve cross-domain recommendation. Such representations help alleviate entanglement and improve transferability. CGCDR adopts a cluster-guided decomposition that disentangles user behaviors into cluster-level features, enabling fine-grained alignment across domains.

2.3 Cluster Learning for Recommendation

Clustering techniques have been widely applied to extract group-level prototypes from items or users. ELCRec [13] unified behavior representation learning into an end-to-end learnable clustering framework, for effective and efficient Recommendation. In the cross-domain scenario, UCMF [16] enhances cross-domain recommendation by integrating user-clustering with GNNs and multidimensional information fusion via an attention mechanism to learn more robust user representations. Inspired by these, CGCDR learns cluster centers in the source domain and uses them to guide knowledge transfer to avoid negative transfer.

3 Problem Setting

In this work, we use $\mathcal{S} = \{\mathcal{U}^s, \mathcal{V}^s, \mathcal{E}^s\}$ and $\mathcal{T} = \{\mathcal{U}^t, \mathcal{V}^t, \mathcal{E}^t\}$ to denote the source domain and target domain, respectively. Specifically, $\mathcal{U}$, $\mathcal{V}$, and $\mathcal{E}$ represent the user set, item set, and interaction set in each domain. The binary interaction matrix is represented as $A \in \{0,1\}^{|\mathcal{U}|\times|\mathcal{V}|}$, where $A_{ui} = 1$ indicates an observed interaction between user u and item i; otherwise, $A_{ui} = 0$. For model training, we construct triplets $(u, i^+, i^-) \in \mathcal{E}^s$ where i^+ is a positive item that u has interacted with ($A_{ui^+} = 1$), and i^- is a negative item that u has not interacted with($A_{ui^-} = 0$). Moreover, we denote the overlapping user subset as $\mathcal{U}^o = \mathcal{U}^s \cap \mathcal{U}^t$, which contains users who have interactions in both domains.

To model diverse user intents, we introduce learnable cluster centers (intent prototypes) in each domain. We use $\mathcal{C}^s = \{\mathbf{c}_1^s, \ldots, \mathbf{c}_{K_s}^s\}$ to denote a set of K_s cluster centers in the source domain, where each $\mathbf{c}_k^s \in \mathcal{R}^d$ represents a distinct intent prototype. Similarly, $\mathcal{C}^t = \{\mathbf{c}_1^t, \ldots, \mathbf{c}_{K_t}^t\}$ denotes K_t cluster centers in the target domain.

Definition (Cold-Start Cross-domain Recommendation). Given the observed interaction data $\{\mathcal{E}^s, \mathcal{E}^t\}$ from both domains, the cold-start cross-domain recommendation task aims to generate personalized item recommendations in the target domain for cold-start users. Formally, given a cold-start user u and their source domain interaction history $I_u^s = \{i_1, i_2, \ldots, i_L\} \subseteq \mathcal{V}^s$, our goal is to predict a ranked list of Top-K items from the target domain $\mathcal{V}^t$ that the user is most likely to interact with.

4 Methodology

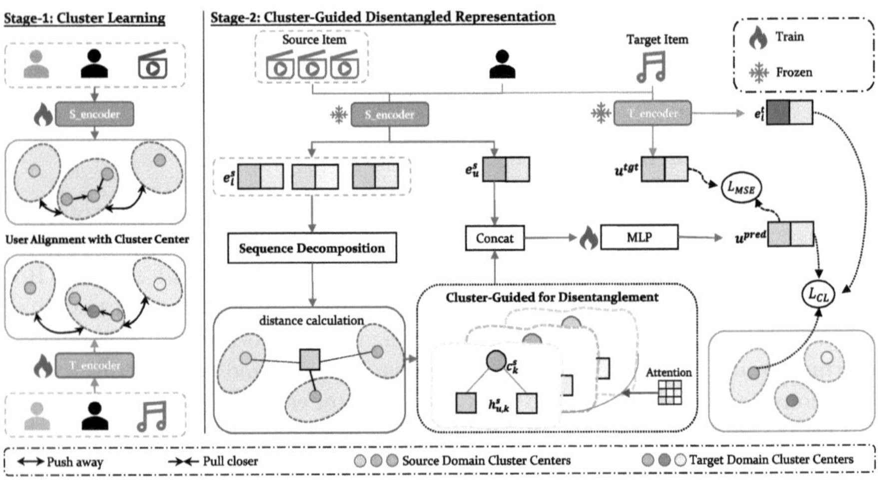

Fig. 2. The two-stage training architecture of CGCDR.

In this section, we introduce our model CGCDR, a Cluster-Guided Cross-Domain Recommendation framework, as shown in Fig. 2. CGCDR can be divided into two stages: (1) Cluster Learning in a Single Domain, and (2) Cluster-Guided Disentangled Representation.

4.1 Cluster Learning in a Single Domain(Stage-1)

At this stage, we aim to learn the cluster centers for individual domains to fully capture the intent prototype of user groups within that domain.

Matrix Factorization Encoding. We denote the MF encoders for the source domain and target domain as $\mathcal{F}_s, \mathcal{F}_t$, respectively. Each encoder consists of two embedding layers that transform user and item IDs into dense vector representations. Formally, given a user $u^s \in \mathcal{U}^s$ and an item $i^s \in \mathcal{V}^s$ in the source domain, we obtain their embeddings through:

$$E_u^s, E_i^s = \mathcal{F}_s(U^s, V^s), \quad E_u^t, E_i^t = \mathcal{F}_t(U^t, V^t), \tag{1}$$

where $E_u^s \in \mathcal{R}^{|\mathcal{U}^s| \times d}$ and $E_i^s \in \mathcal{R}^{|\mathcal{V}^s| \times d}$ represent the user and item embedding matrices in the source domain, with d denoting the embedding dimension. Similarly, $E_u^t \in \mathcal{R}^{|\mathcal{U}^t| \times d}$ and $E_i^t \in \mathcal{R}^{|\mathcal{V}^t| \times d}$ are the corresponding embeddings for the target domain.

For a specific user u in the source domain, we extract the user embedding $\mathbf{e}_u^s \in E_u^s$ and item embedding $\mathbf{e}_i^s \in E_i^s$. The predicted interaction score between user u and item i is computed via inner product: $\hat{y}_{u,i} = \mathbf{e}_u^s \cdot \mathbf{e}_i^s$. Then, we employ Bayesian Personalized Ranking (BPR) loss [17] to optimize embedding quality, assuming that users prefer interacted items over non-interacted ones:

$$\mathcal{L}_{bpr}^s = - \sum_{(u,i^+,i^-) \in \mathcal{E}^s} \ln \sigma(\hat{y}_{u,i^+} - \hat{y}_{u,i^-}), \tag{2}$$

where $\mathcal{E}^s$ denotes the user-item interaction set in the source domain, with i^+ representing a positive (interacted) item and i^- a negative (non-interacted) item for user u, and $\sigma(\cdot)$ is the sigmoid function. The same loss formulation applies to the target domain as $\mathcal{L}_{bpr}^t$.

User Alignment with Cluster Center. Taking the source domain as an example. To capture diverse user intents in the source domain, we introduce K_s learnable cluster centers $\mathcal{C}^s = \{\mathbf{c}_1^s, \mathbf{c}_2^s, \ldots, \mathbf{c}_{K_s}^s\}$ in the source domain, where each $\mathbf{c}_k^s \in \mathcal{R}^d$ represents a distinct intent prototype. These prototypes are jointly optimized with user and item embeddings to form well-separated, interpretable intent clusters.

To ensure the learned cluster centers are meaningful and discriminative, we impose two complementary regularization terms. First, the compactness loss encourages users to be close to cluster centers, promoting tight user-cluster space:

$$\mathcal{L}_{compact}^s = \frac{1}{K_s |U^s|} \sum_{u \in U^s} \sum_{k=1}^{K_s} \|\bar{e}_u^s - \bar{c}_k^s\|_2^2, \tag{3}$$

where $\bar{\mathbf{e}}_u^s = \mathbf{e}_u^s / \|\mathbf{e}_u^s\|_2$ and $\bar{\mathbf{c}}_k^s = \mathbf{c}_k^s / \|\mathbf{c}_k^s\|_2$ denote the L2-normalized user embedding and cluster center, respectively. Second, the separation loss enforces inter-cluster distinctiveness by maximizing distances between different cluster centers:

$$\mathcal{L}_{separate}^s = -\frac{2}{K_s(K_s - 1)} \sum_{k=1}^{K_s} \sum_{l=k+1}^{K_s} \|\bar{c}_k^s - \bar{c}_l^s\|_2^2, \tag{4}$$

where we maximize the distance between each cluster center and all other centers (excluding themselves) during updates. Together, these two terms prevent cluster collapse and ensure that each cluster center captures unique information reflecting user intents.

Final Loss of Stage-1. In summary, the overall training objective for Stage-1 combines the recommendation loss with cluster regularization:

$$\mathcal{L}^s_{stage1} = \mathcal{L}^s_{bpr} + \alpha \cdot (\mathcal{L}^s_{compact} + \mathcal{L}^s_{separate}). \tag{5}$$

where α is a hyperparameter controlling the strength of cluster regularization. An analogous loss $\mathcal{L}^t_{stage1}$ is used for the target domain.

4.2 Cluster-Guided Disentangled Representation(Stage-2)

After learning well-separated intent prototypes in Stage-1, we decompose user representations by leveraging specific items from user interactions through the cluster center. This stage addresses the core challenge of cold-start recommendation: how to accurately transfer source domain user features to the target domain while avoiding negative transfer.

Soft Cluster-Based Sequence Decomposition. Given a source-domain interaction sequence $I^s_u = \{i_1, i_2, \ldots, i_L\}$ for user u, where L denotes the sequence length (with sequence padded or truncated if necessary). We first obtain the item embeddings $\{\mathbf{e}^s_{i_1}, \mathbf{e}^s_{i_2}, \ldots, \mathbf{e}^s_{i_L}\}$ from the learned E^s_i, with each $\mathbf{e}^s_{i_l} \in \mathcal{R}^d$.

To associate each item with intent prototypes, we compute the similarity score between item i_l and each source-domain cluster center $\mathbf{c}^s_k$ ($k = 1, \ldots, K_s$). This score is then converted into a normalized cluster-assignment weight via a temperature-scaled softmax, as defined by the following expressions:

$$s_{l,k} = -\left\|\mathbf{e}^s_{i_l} - \mathbf{c}^s_k\right\|^2, \quad w_{l,k} = \frac{\exp\left(s_{l,k}/\tau\right)}{\sum_{k'=1}^{K_s} \exp\left(s_{l,k'}/\tau\right)}, \tag{6}$$

where $\tau > 0$ is a temperature hyperparameter controlling the distribution sharpness. The resulting weight $w_{l,k} \in [0,1]$ represents the degree to which item i_l belongs to cluster k, with $\sum_{k=1}^{K_s} w_{l,k} = 1$. Then, we aggregate the item embeddings to form cluster-level intent summaries. For each cluster k, the user's cluster-level intent representation is computed as:

$$h^s_{u,k} = \sum_{l=1}^{L} w^T_{l,k} \cdot e^s_{i_l}, \tag{7}$$

where $\mathbf{h}^s_{u,k} \in \mathcal{R}^d$ captures the user's intent related to the k-th cluster center. This decomposition yields K_s cluster-level components $\{\mathbf{h}^s_{u,1}, \mathbf{h}^s_{u,2}, \ldots, \mathbf{h}^s_{u,K_s}\}$ that represent the user's various intents, enabling fine-grained disentanglement of user preferences across different interest dimensions.

Cluster-Guided for Disentanglement. With the decomposed cluster-level intent summaries, the current focus is to leverage the explicit cluster structure to guide the mapping process, ensuring that each intent prototype is transferred independently while preserving the meaning of its cluster center.

For each cluster $k \in \{1, 2, \ldots, K_s\}$, we construct a comprehensive input representation by concatenating three complementary components: the cluster-level intent summary $\mathbf{h}^s_{u,k}$, the corresponding cluster center $\mathbf{c}^s_k$, and the base user embedding $\mathbf{e}^s_u$:

$$\mathbf{x}_{u,k} = [\mathbf{h}^s_{u,k} \oplus \mathbf{c}^s_k \oplus \mathbf{e}^s_u] \in \mathcal{R}^{3d}, \tag{8}$$

where $\oplus$ denotes concatenation. This triple concatenation serves distinct purposes: $\mathbf{h}^s_{u,k}$ encodes the user's personalized interaction degree within cluster k, $\mathbf{c}^s_k$ provides the raw cluster center of that intent prototype, and $\mathbf{e}^s_u$ captures coarse-grained features of users in the source domain. By combining these three types of information, we employ a two-layer neural generator $\mathcal{G}(\cdot)$ with GELU activation to transform this enriched representation into the target domain:

$$\mathbf{g}_{u,k} = \mathcal{G}(\mathbf{x}_{u,k}) \in \mathcal{R}^d, \tag{9}$$

This generator $\mathcal{G}(\cdot)$ learns the complex mapping function that translates source-domain intent prototype to their target-domain prototype. By independently processing each of the K_s clusters through $\mathcal{G}$, we obtain a set of disentangled representation in the target domain $\{\mathbf{g}_{u,1}, \mathbf{g}_{u,2}, \cdots, \mathbf{g}_{u,K_s}\}$, where each $\mathbf{g}_{u,k}$ represents a specific intent prototype transferred to the target domain.

To aggregate these multiple intent candidates into a unified user representation in the target domain, we employ an attention mechanism that dynamically weights each cluster's contribution. We use a learnable scoring function $f(\cdot) : \mathcal{R}^d \to \mathcal{R}$ to compute attention weights and obtain user representations through the following:

$$\alpha_{u,k} = \frac{\exp(f(\mathbf{g}_{u,k}))}{\sum_{k'=1}^{K_s} \exp(f(\mathbf{g}_{u,k'}))}, \quad \mathbf{u}^{pred} = \sum_{k=1}^{K_s} \alpha_{u,k} \cdot \mathbf{g}_{u,k} \in \mathcal{R}^d, \tag{10}$$

where $\alpha_{u,k}$ reflects the relevance of cluster center k for user u and u^{pred} is the final disentangled representation of the user predicted by the model. This cluster-guided approach enables knowledge transfer by explicitly modeling how each source-domain cluster center corresponds to target-domain preferences.

Cluster-Assisted Contrastive Loss($\mathcal{L}_{CL}$). To train the generator $\mathcal{G}(\cdot)$ on overlapping users $\mathcal{U}^o = \mathcal{U}^s \cap \mathcal{U}^t$, we employ two complementary objectives. First, following previous work [9,15,30], we enforce that the predicted representation $\mathbf{u}^{pred}$ aligns with the ground-truth user embedding $\mathbf{u}^{tgt}$ (obtained from E^t_u) via mean squared error (MSE):

$$\mathcal{L}_{\text{MSE}} = \frac{1}{|\mathcal{U}_o|} \sum_{u \in \mathcal{U}_o} \left\| \mathbf{u}^{pred} - \mathbf{u}^{tgt} \right\|_2^2. \tag{11}$$

While $\mathcal{L}_{MSE}$ provides direct supervision, it does not explicitly encourage discriminative item-level distinctions. To address this, we introduce a cluster-assisted contrastive loss that leverages target-domain cluster centers as hard negatives. For each user u in a mini-batch $\mathcal{B}$, we sample a positive item i^+ from the user's target-domain interactions and a negative item i^- that the user has not interacted with, obtaining their embeddings $\mathbf{e}_{i^+}^t$ and $\mathbf{e}_{i^-}^t$ from E_i^t. Rather than directly using the negative item embedding $\mathbf{e}_{i^-}^t$, we construct a cluster-level hard negative by identifying the target-domain cluster center closest to i^-:

$$\mathbf{c}^- = \arg \min_{\mathbf{c}_j^t \in \mathcal{C}^t} \left\| \mathbf{e}_{i^-}^t - \mathbf{c}_j^t \right\|_2^2, \tag{12}$$

where $\mathcal{C}^t = \{\mathbf{c}_1^t, \ldots, \mathbf{c}_{K_t}^t\}$ denotes the target-domain cluster centers learned in **Stage-1**. Using the cluster center $\mathbf{c}^-$ instead of the negative item $\mathbf{e}_{i^-}^t$ yields better results because they capture the characteristics of the entire group rather than individual random samples, thereby forcing the model to learn more discriminative knowledge transfer. The contrastive loss is formulated with temperature τ and cosine similarity $\text{sim}(\cdot, \cdot)$ as following:

$$\mathcal{L}_{\text{CL}} = -\frac{1}{|\mathcal{B}|} \sum_{u \in \mathcal{B}} \log \frac{\exp\left(\text{sim}(\mathbf{u}^{\text{pred}}, \mathbf{i}^+)/\tau\right)}{\exp\left(\text{sim}(\mathbf{u}^{\text{pred}}, \mathbf{i}^+)/\tau\right) + \exp\left(\text{sim}(\mathbf{u}^{\text{pred}}, \mathbf{c}^-)/\tau\right)}, \tag{13}$$

which pulls the predicted representation closer to positive items while pushing it away from the negative cluster center, thereby enhancing more precise and robust representations.

Final Loss of Stage-2. The overall training objective for Stage-2 combines the reconstruction loss and the cluster-assisted contrastive loss:

$$\mathcal{L}_{stage2} = \mathcal{L}_{MSE} + \beta \cdot \mathcal{L}_{CL}. \tag{14}$$

where β is a hyperparameter controlling the relative importance of contrastive learning. During inference, for cold-start users who have no interactions in the target domain, we directly use the predicted representation $\mathbf{u}^{pred}$ obtained from Eq. (10) to compute the Top-K recommendations for items in the target domain.

5 Experiments

In this section, we conduct extensive experiments to validate the performance of CGCDR across different scenarios. The experiments are designed to address the following questions:

RQ1 Is CGCDR more effective than existing models?
RQ2 What are the respective contributions and roles of each module in CGCDR?
RQ3 Does CGCDR alleviate the negative transfer problem?
RQ4 How do the hyper-parameter settings of CGCDR impact the performance?

5.1 Experimental Settings

Datasets. We conduct experiments on four cross-domain scenarios from the large-scale public Amazon dataset [5], namely Game-Video, Music-Movie, Cloth-Sport, and Phone-Electronic (Elec for short). Following previous work, we convert ratings into implicit feedback, meaning that interactions are counted regardless of whether the rating value is high or low. We filter out users and items with fewer than 5 interactions from each dataset to ensure data quality, following established practices [3,29]. To better simulate realistic cold-start conditions, we further define overlapping users as those with at least 10 interactions in the source domain and at least 1 interaction in the target domain. Among these, 20% are randomly selected as cold-start users, meaning their target domain interaction records are removed during the validation and testing phases. Detailed information about the dataset is provided in the Table 1.

Table 1. The statistics of datasets.(*Overlap denotes the number of overlapping users across domains, and *Avg indicates the average length of source domain interaction sequences for overlapping users.)

Year	Domain	$\|U\|$	$\|V\|$	Interaction	*Overlap	Valid	Test	*Avg.len	Sparsity
2018	Game	208,180	78,772	1,828,971	8,169	5,573	5,573	13.71	99.9888%
	Video	55,223	17,408	497,577		5,464	5,465	14.18	99.9482%
2018	Music	112,396	73,714	1,443,755	28,223	24,319	24,320	14.43	99.9826%
	Movie	297,530	60,176	3,410,019		26,547	26,548	14.19	99.9810%
2014	Sport	35,599	18,358	296,337	3,908	1,319	1,320	13.33	99.9547%
	Cloth	39,388	23,034	278,677		1,482	1,483	13.86	99.9693%
2014	Phone	27,879	10,429	194,439	13,682	6,735	6,735	14.36	99.9331%
	Elec	192,403	63,001	1,689,188		8,711	8,712	12.80	99.9861%

Baselines. We compare CGCDR with the following representative cross-domain and single-domain baselines. For *cross-domain baselines*, we compare our CGCDR with SOTA baselines, including: (1) **CMF** [18] for cross-domain recommendation can be regarded as a variant of MF, where data from two domains are trained simultaneously with different weight adjustments. (2) **EMCDR** [15] employs Embedding Mapping for Cross-Domain Recommendation by learning linear or nonlinear mapping functions to transfer knowledge between domains. (3) **SSCDR** [9] trains a cross-domain mapping function to encode such distance information by exploiting both overlapping users as labeled data and all the items as unlabeled data. (4) **CDRIB** [2] considers both domain interactions jointly for de-biasing with an additional contrastive information regularizer. (5) **CUT** [10] leverages target domain user similarity to filter user collaborative information from the source domain. *Single-domain baselines* refer to training solely on the target domain dataset without utilizing the source domain: (6) **BPR-MF** [17] combines MF with Bayesian Personalized Ranking optimization to model user-item interactions through pairwise ranking loss. (7) **CML** [6] learns a joint metric space where users are closer to their preferred items, effectively capturing collaborative relationships.

Evaluation. We adopt the leave-one-out evaluation strategy to assess the performance of all methods [1,2,7,9]. Specifically, for each cold-start user $u \in \mathcal{U}^s$ with a ground-truth interaction (u, i) in the target domain $\mathcal{T}$, we randomly sample 999 items from the target item set $\mathcal{V}^t$ that the user has not interacted with as negative instances. We employ two widely-adopted ranking metrics to evaluate the top-10 recommendation quality: **Hit Ratio@10 (HR@10)**, which measures whether the ground-truth item appears in the top-10 ranked list, and **Normalized Discounted Cumulative Gain@10 (NDCG@10)** [8], which evaluates ranking quality by assigning higher weights to top-ranked positions.

Table 2. Recommendation performance (expressed in %) of CDR methods. The best performance is in **bold** and the runner-up is underlined.

Dataset	Metric@10	Single-Domain Methods		Cross-Domain Methods					Ours
		BPRMF	CML	CMF	EMCDR	SSCDR	CDRIB	CUT	CGCDR
Game	HR	2.70±0.19	6.22±0.30	7.84±0.47	9.66±0.31	6.70±0.30	13.01±0.48	10.86±0.35	**14.69±0.32**
	NDCG	1.30±0.08	2.93±0.19	4.19±0.27	4.52±0.17	3.49±0.10	6.56±0.21	5.31±0.10	**7.57±0.14**
Video	HR	3.13±0.10	6.52±0.59	9.18±0.11	7.36±0.17	5.01±0.33	9.40±0.36	9.17±0.14	**13.35±0.27**
	NDCG	1.64±0.04	2.89±0.28	4.57±0.07	3.47±0.10	2.40±0.15	4.66±0.24	4.60±0.09	**6.94±0.08**
Music	HR	1.89±0.03	7.66±0.35	12.35±0.36	16.45±0.24	11.78±0.17	15.67±0.32	17.19±0.15	**18.18±0.40**
	NDCG	0.89±0.02	3.58±0.16	6.34±0.19	8.68±0.22	5.64±0.11	8.58±0.22	8.77±0.15	**9.80±0.31**
Movie	HR	6.05±0.21	10.72±0.82	17.07±0.47	16.39±0.42	15.16±0.40	18.01±0.51	14.15±0.39	**21.20±0.57**
	NDCG	3.01±0.10	4.79±0.42	8.85±0.30	7.75±0.26	7.62±0.28	9.33±0.36	7.07±0.19	**11.53±0.38**
Sport	HR	1.64±0.08	8.06±0.45	8.85±0.22	9.24±0.27	1.30±0.09	**12.76±0.33**	9.39±0.15	12.23±0.59
	NDCG	0.92±0.04	3.75±0.26	5.39±0.26	4.61±0.07	0.59±0.07	**6.62±0.19**	4.78±0.15	6.22±0.36
Cloth	HR	0.80±0.13	4.90±0.39	2.02±0.13	5.91±0.34	0.76±0.05	12.46±0.20	7.43±0.26	**12.58±0.55**
	NDCG	0.38±0.05	2.50±0.34	1.12±0.06	2.86±0.18	0.36±0.02	6.91±0.12	3.83±0.11	**6.96±0.43**
Phone	HR	2.53±0.05	10.41±0.82	16.27±0.46	16.12±0.39	0.94±0.09	12.46±0.20	17.44±0.23	**19.25±0.32**
	NDCG	1.24±0.04	4.94±0.37	8.59±0.34	8.09±0.22	0.45±0.05	6.42±0.15	8.85±0.08	**10.20±0.14**
Elec	HR	4.47±0.22	11.41±0.29	19.19±0.46	16.20±0.28	2.37±0.09	22.21±0.12	16.23±0.27	**23.58±0.20**
	NDCG	2.15±0.16	4.91±0.14	10.96±0.29	7.66±0.20	1.18±0.06	12.66±0.09	8.56±0.11	**13.47±0.22**

Implementation Details. In our experiments, apart from CDRIB and CUT, which rely on official implementations, all other baselines are reproduced based on RecBole-CDR [28] and seamlessly integrated into our implementation framework. This setup ensures consistency across all stages, including data processing, random seed initialization, and negative sample generation during evaluation. For fairness, we standardize the main parameters for each model implementation, such as the embedding dimension to 64, the L2 regularizer weight is $1e^{-5}$, and the batch size to 2048. For our model CGCDR, we set the number of cluster centers K from $\{128, 256, 348\}$ based on the data size in each domain. We tune the parameter α in $\{1e^{-4}, 1e^{-3}, 5e^{-3}, 1e^{-2}, 1e^{-1}\}$ and the β in $\{1e^{-5}, 1e^{-4}, 1e^{-3}, 5e^{-3}, 1e^{-2}\}$, the learning rate in $\{1e^{-3}, 5e^{-3}, 1e^{-2}\}$. For all these methods, we conduct five runs with different random seeds and report the mean and standard deviation of the best HR@10 results.

5.2 Performance Analysis(RQ1)

Table 2 presents the mean and standard deviation of metrics across all methods in four scenarios, revealing the following key findings:

- Cross-domain methods outperform single-domain methods, particularly in domains with sparse data, confirming the existence of domain-shared features across domains and demonstrating that CDR methods effectively alleviate the cold-start issue.
- Joint learning paradigm, exemplified by CDRIB, outperforms the mapping-based paradigm (i.e., EMCDR and SSCDR) due to its ability to distinguish domain-specific features. Yet, it exhibits suboptimal performance in the domain with a smaller data scale under data imbalance (e.g., Video, Music, Phone). While CUT consistently improves performance by filtering out irrelevant source domain collaboration information, it still has bottlenecks.
- Our proposed CGCDR demonstrates superior performance compared to CDRIB and CUT. Compared to CDRIB's implicit disentanglement via variational inference, CGCDR employs explicit learnable cluster centers to guide fine-grained interest decomposition. Unlike CUT's coarse filtering that may discard useful signals, our cluster-guided approach preserves valuable collaborative signals while filtering out irrelevant information. These results demonstrate that CGCDR achieves more robust knowledge transfer under data imbalance.

5.3 Ablation Study(RQ2)

To validate the effectiveness of each core component in CGCDR, we conduct ablation experiments by removing key modules and evaluating their impact on recommendation performance. We design four variants: (1) **w/o CG** removes the cluster-guided sequence decomposition $\mathbf{h}^s_{u,k}$, which denotes the user's personalized interaction degree within cluster k; (2) **w/o CC** removes the raw cluster center of that intent prototype $\mathbf{c}^s_k$; (3) **w/o U_src** removes the coarse-grained features of users $\mathbf{e}^s_u$ from the source domain; (4) **w/o CL** excludes the cluster-assisted contrastive learning loss, retaining only the MSE loss for knowledge transfer.

Table 3 presents the ablation results across four scenarios. The results clearly demonstrate that all components contribute significantly to the overall performance. When removing cluster-guided decomposition (w/o CG), the model's performance significantly decreases, highlighting the importance of cluster-guided capture in identifying users' fine-grained features and distinguishing

Table 3. Ablation studies of CGCDR under different scenarios, with the best result shown in **bold**.

Scenarios	Metrics@10	Model Variants				
		w/o CG	w/o CC	w/o U_src	w/o CL	CGCDR
Game	HR	11.70±0.46	13.64±0.39	11.32±0.44	14.18±0.34	**14.69±0.32**
	NDCG	6.12±0.28	7.09±0.22	5.95±0.28	7.27±0.16	**7.57±0.14**
Video	HR	11.06±0.20	13.01±0.14	11.86±0.20	13.24±0.26	**13.35±0.27**
	NDCG	5.74±0.15	6.80±0.09	5.98±0.02	6.87±0.06	**6.94±0.08**
Music	HR	16.81±0.36	18.11±0.27	18.00±0.40	**18.58±0.36**	18.18±0.40
	NDCG	8.97±0.31	9.75±0.2	9.72±0.31	**10.06±0.30**	9.80±0.31
Movie	HR	20.68±0.50	21.07±0.56	20.82±0.58	20.97±0.59	**21.20±0.57**
	NDCG	11.23±0.35	11.46±0.38	11.32±0.41	11.38±0.39	**11.53±0.38**
Sport	HR	7.45±0.55	10.14±0.64	9.59±0.66	12.14±0.43	**12.23±0.59**
	NDCG	3.31±0.24	4.80±0.31	4.46±0.37	6.08±0.29	**6.22±0.36**
Cloth	HR	8.00±0.15	10.84±0.33	11.06±0.41	12.52±0.54	**12.58±0.55**
	NDCG	3.97±0.13	5.97±0.29	6.08±0.32	6.59±0.43	**6.96±0.43**
Phone	HR	16.18±0.28	18.56±0.42	16.59±0.44	18.83±0.42	**19.25±0.32**
	NDCG	8.64±0.12	9.74±0.17	8.87±0.11	9.97±0.17	**10.20±0.14**
Elec	HR	21.83±0.28	23.32±0.24	23.06±0.22	23.42±0.22	**23.58±0.20**
	NDCG	12.51±0.29	13.30±0.27	13.17±0.18	13.38±0.24	**13.47±0.22**

domain-specific features. Meanwhile, the results of w/o CC and w/o U_src also demonstrate the role of raw cluster centers and users' coarse-grained features, respectively. Notably, the performance of w/o CL indicates that replacing negative samples with clusters for contrastive learning yields superior results in most scenarios.

5.4 Representation Visualization(RQ3)

To investigate whether CGCDR effectively transfers knowledge without negative transfer, we visualize the learned user representations and cluster centers on the Game→Video task.

User Representation Visualization. We randomly select 2,000 overlapping users and project their predicted representations alongside ground-truth target domain embeddings using t-SNE [14]. As shown in the left part of Fig. 3, blue points denote ground-truth user embeddings in the target domain, while orange and red points represent the predicted embeddings from CGCDR and its w/o CG variant (without cluster-guided disentanglement), respectively. The CGCDR representations exhibit a distribution that is more closely aligned with the ground-truth embeddings, whereas the w/o CG variant shows a more dispersed distribution. Quantitatively, the average distance between predicted and

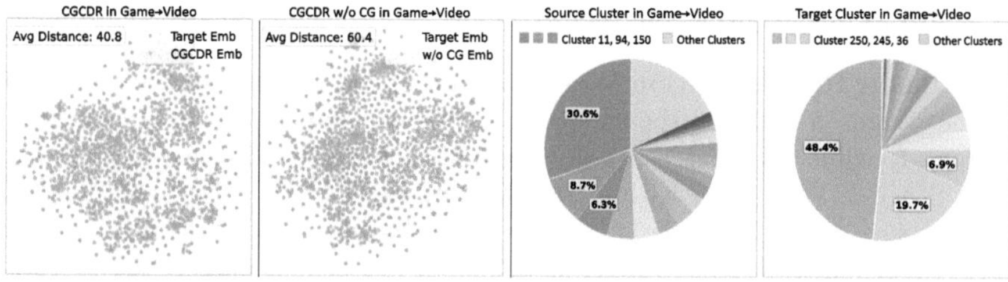

Fig. 3. The visualization of user representations and cluster centers on the Game→Video task. Left: t-SNE projections of user representations. Right: The frequency of cluster centers in the source domain and the distribution of cluster centers in the target domain.

ground-truth embeddings is 40.8 for CGCDR and 60.4 for w/o CG, confirming that CGCDR achieves more accurate and consistent knowledge transfer through cluster-guided disentanglement.

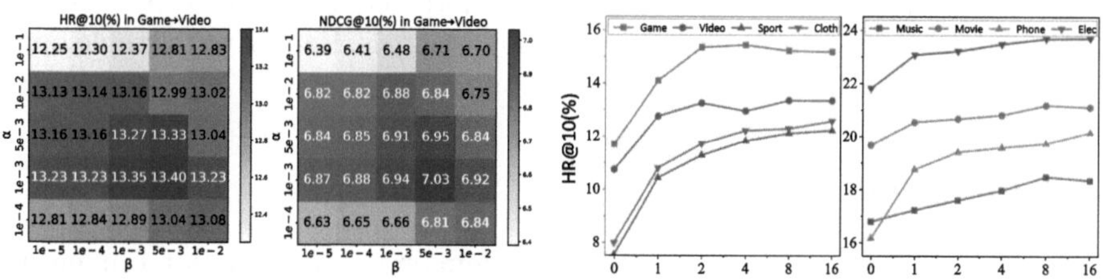

(a) (b)

Fig. 4. Parameter analysis results. (a): Effect of cluster regularization weight α and contrastive loss weight β (b): Effect of user interaction sequence length L.

Cluster Center Visualization. To further validate the role of cluster centers in mitigating negative transfer, we visualize the frequency of source domain cluster centers appearing during cluster-guided disentanglement. As shown in the right part of Fig. 3, the frequency of source domain centers appearing during disentanglement varies significantly. Centers with high frequency are likely perceived by the model as containing effective information, potentially representing domain-shared features, while those with low frequency may be regarded as domain-specific features. For comparison, we also visualize the distribution of cluster centers in the target domain, which is consistent with the popularity bias existing in the dataset, confirming the rationality of our clustering operation.

5.5 Hyper-parameter Analysis(RQ4)

The Cluster Regularization Weight α and the Contrastive Loss Weight β. The weight α controls the strength of cluster compactness and separation in Stage-1, while the contrastive loss weight β balances discriminative learning in Stage-2. We vary α within the range of $\in \{1e^{-4}, 1e^{-3}, 5e^{-3}, 1e^{-2}, 1e^{-1}\}$ and $\beta \in \{1e^{-5}, 1e^{-4}, 1e^{-3}, 5e^{-3}, 1e^{-2}\}$ on the Game→Video task as shown in Fig. 4(a). When α is too small (such as $1e^{-4}$), the cluster centers fail to form well-separated intent prototypes. Similarly, extremely low β values provide insufficient contrastive supervision, weakening the model's ability to distinguish positive items from negative cluster centers. We recommend setting α and β within the range of $\{1e^{-3}, 5e^{-3}\}$ to achieve optimal performance.

Effect of User Interaction Sequence Length L. The sequence length L determines how many recent user interactions are used for sequence decomposition. We evaluate $L \in \{0, 1, 2, 4, 8, 16\}$ across all domain pairs. Figure 4(b) shows that performance consistently improves as L increases from 0 to 8, indicating that longer sequences provide richer information for capturing user fine-grained features. However, performance tends to saturate or slightly decline when $L > 8$, as most overlapping users have fewer than L interactions(Table 1). As a result, longer sequences may introduce sparse or noisy signals, leading to unstable optimization.

6 Conclusion

We propose CGCDR, a cluster-guided cross-domain recommendation framework for cold-start users. By introducing learnable cluster centers as intent prototypes, CGCDR enables fine-grained interest decomposition and robust feature disentanglement. Extensive experiments demonstrate its consistent superiority over state-of-the-art methods. Future work will explore non-overlapping user settings and multi-domain recommendations.

Acknowledgments. This research is partially supported by National Natural Science Foundation of China (Grant No. 62572365) and National Key Research and Development Program of China (2024YFF0907002).

References

1. Cao, J., Li, S., Yu, B., Guo, X., Liu, T., Wang, B.: Towards universal cross-domain recommendation. In: WSDM 2023, pp. 78–86 (2023)
2. Cao, J., Sheng, J., Cong, X., Liu, T., Wang, B.: Cross-domain recommendation to cold-start users via variational information bottleneck. In: ICDE 2022, pp. 2209–2223. IEEE (2022)
3. Han, T., Wang, P., Niu, S., Li, C.: Modality matches modality: pretraining modality-disentangled item representations for recommendation. In: WWW '22, pp. 2058–2066. ACM (2022)

4. Han, Y., et al.: Guesr: a global unsupervised data-enhancement with bucket-cluster sampling for sequential recommendation. In: International conference on database Systems for Advanced Applications, pp. 286–296. Springer, Heidelberg (2023). https://doi.org/10.1007/978-3-031-30672-3_19
5. Hou, Y., Li, J., He, Z., Yan, A., Chen, X., McAuley, J.: Bridging language and items for retrieval and recommendation. arXiv preprint arXiv:2403.03952 (2024)
6. Hsieh, C.K., Yang, L., Cui, Y., Lin, T.Y., Belongie, S., Estrin, D.: Collaborative metric learning. In: WWW 2017, pp. 193–201 (2017)
7. Hu, G., Zhang, Y., Yang, Q.: Conet: collaborative cross networks for cross-domain recommendation. In: CIKM 2018, pp. 667–676 (2018)
8. Järvelin, K., Kekäläinen, J.: Cumulated gain-based evaluation of IR techniques. ACM Trans. Inf. Syst. **20**(4), 422–446 (2002)
9. Kang, S., Hwang, J., Lee, D., Yu, H.: Semi-supervised learning for cross-domain recommendation to cold-start users. In: CIKM 2019, pp. 1563–1572 (2019)
10. Li, H., et al.: Aiming at the target: filter collaborative information for cross-domain recommendation. In: SIGIR 2024, pp. 2081–2090 (2024)
11. Li, P., Tuzhilin, A.: DDTCDR: deep dual transfer cross domain recommendation. In: WSDM '20, pp. 331–339 (2020)
12. Liu, X., Li, B., Chen, Y., Li, X., Xu, S., Yin, H.: Disentangled representations for cross-domain recommendation via heterogeneous graph contrastive learning. In: International Conference on Database Systems for Advanced Applications, pp. 35–50. Springer, Heidelberg (2024). https://doi.org/10.1007/978-981-97-5555-4_3
13. Liu, Y., Zhu, S., Xia, J., Ma, Y., Ma, J., Liu, X., Yu, S., Zhang, K., Zhong, W.: End-to-end learnable clustering for intent learning in recommendation. NeurIPS **2024**(37), 5913–5949 (2024)
14. Maaten, L.V.D., Hinton, G.: Visualizing data using t-sne. J. Mach. Learn. Res. **9**(Nov), 2579–2605 (2008)
15. Man, T., Shen, H., Jin, X., Cheng, X.: Cross-domain recommendation: an embedding and mapping approach. In: IJCAI, vol. 17, pp. 2464–2470 (2017)
16. Nie, J., Zhao, Z., Huang, L., Nie, W., Wei, Z.: Cross-domain recommendation via user-clustering and multidimensional information fusion. IEEE Trans. Multimedia **25**, 868–880 (2021)
17. Rendle, S., Freudenthaler, C., Gantner, Z., Schmidt-Thieme, L.: BPR: bayesian personalized ranking from implicit feedback. arXiv preprint arXiv:1205.2618 (2012)
18. Singh, A.P., Gordon, G.J.: Relational learning via collective matrix factorization. In: Proceedings of SIGKDD, pp. 650–658 (2008)
19. Sun, L., Liu, J., Zhang, S., Nie, W., Liu, A., Su, Y.: Beyond users: denoising behavior-based contrastive learning for disentangled cross-domain recommendation. In: International Conference on Database Systems for Advanced Applications, pp. 163–178. Springer, Heidelberg (2024). https://doi.org/10.1007/978-981-97-5779-4_11
20. Tang, J., Feng, X., Chen, X.: Fairness-aware cross-domain recommendation. In: International Conference on Database Systems for Advanced Applications, pp. 293–302. Springer, Heidelberg (2024). https://doi.org/10.1007/978-981-97-5555-4_19
21. Tishby, N., Pereira, F.C., Bialek, W.: The information bottleneck method. arXiv preprint physics/0004057 (2000)
22. Wang, Y., Feng, C., Guo, C., Chu, Y., Hwang, J.N.: Solving the sparsity problem in recommendations via cross-domain item embedding based on co-clustering. In: WSDM 2019, pp. 717–725 (2019)

23. Wang, Y., Xie, Q., Bao, Z., Tang, M., Li, L., Liu, Y.: Enhancing transferability and consistency in cross-domain recommendations via supervised disentanglement. In: RecSys 2025, pp. 104–113 (2025)
24. Xiaojun, L.: An improved clustering-based collaborative filtering recommendation algorithm. Clust. Comput. **20**(2), 1281–1288 (2017)
25. Zawia, J.M., et al.: Comprehensive review of meta-learning methods for cold-start issue in recommendation systems. IEEE Access **13**, 24622–24641 (2025)
26. Zhang, H., Kong, X., Zhang, Y.: Cross-domain recommendation with multi-auxiliary domains via consistent and selective cluster-level knowledge transfer. Expert Syst. Appl. **223**, 119861 (2023)
27. Zhang, L., et al.: A unified framework for adaptive representation enhancement and inversed learning in cross-domain recommendation. In: International Conference on Database Systems for Advanced Applications, pp. 115–130. Springer, Heidelberg (2024). https://doi.org/10.1007/978-981-97-5555-4_8
28. Zhao, W.X., et al.: Recbole: towards a unified, comprehensive and efficient framework for recommendation algorithms. In: CIKM '21, pp. 4653–4664 (2021)
29. Zhou, X., Lin, D., Liu, Y., Miao, C.: Layer-refined graph convolutional networks for recommendation. In: ICDE 2023, pp. 1247–1259. IEEE (2023)
30. Zhu, Y., et al.: Personalized transfer of user preferences for cross-domain recommendation. In: WSDM '22, pp. 1507–1515 (2022)

Causal Cross-Domain Sequential Recommendation with Preference Evolution

Jiaxuan Ma[1], Yue Kou[1(✉)], Dong Li[2], Derong Shen[1], Xiangmin Zhou[3], and Tiezheng Nie[1]

[1] Northeastern University, Shenyang 110004, Liaoning, China
2301930@stu.neu.edu.cn, {kouyue,shenderong,nietiezheng}@cse.neu.edu.cn
[2] Liaoning University, Shenyang 110036, Liaoning, China
dongli@lnu.edu.cn
[3] RMIT University, Melbourne, Melbourne, Australia
xiangmin.zhou@rmit.edu.au

Abstract. Cross-Domain Sequential Recommendation (CDSR) has attracted growing interest for alleviating data sparsity through multi-domain knowledge integration. While progress has been made, semantic, temporal, and behavioral heterogeneity across domains poses key challenges for effective knowledge transfer. Existing CDSR approaches exhibit limitations in modeling cross-domain attribute relationships, capturing both domain-specific and cross-domain interest dynamics, and addressing activity frequency bias that distorts true preferences. Therefore, we propose a novel model, ExcelRec (**E**volution-aware mi**X**ture of experts for **C**ausal cross-domain s**E**quentia**L** **REC**ommendation). Specifically, we first propose an LLM-augmented co-occurrence miner, which enriches item representations with structured semantic augmentation by leveraging LLMs. Then we propose an evolution-aware cross-domain preference model to address temporal heterogeneity in cross-domain interest transitions, using a Temporal-Domain dual-conditioned Mixture of Experts (TD-MoE) to capture both fast and slow preference dynamics. Finally, we propose a causality-enhanced preference learning strategy to eliminate activity-induced bias by disentangling spurious activity patterns from genuine preferences. Experimental results demonstrate the high effectiveness of our proposed ExcelRec model.

Keywords: Cross-Domain Sequential Recommendation · Mixture of Experts · Semantic Augmentation · Causal Preference Disentanglement

1 Introduction

Sequential Recommendation (SR), which aims to predict the next item in a user's interaction sequence, has garnered significant attention [7]. However, in

This work was supported by the National Natural Science Foundation of China (62472204).

H. Jung et al. (Eds.): DASFAA 2026, LNCS 16535, pp. 37–53, 2026.
https://doi.org/10.1007/978-981-92-0363-5_3

real-world applications, data sparsity remains a critical bottleneck: most users exhibit limited interactions within each domain, leading traditional SR models to suffer from overfitting and poor generalization.

To address this challenge, Cross-Domain Sequential Recommendation (CDSR) has emerged as a promising approach by leveraging auxiliary knowledge from various domains to mitigate data scarcity. Unlike traditional single-domain recommendation, which treats domains in isolation, CDSR dynamically integrates sequential behavioral patterns across domains, enabling more precise and contextually relevant suggestions. For instance, as illustrated in Fig. 1, a user's sequential interactions in Cloth (e.g., shirts, sneakers) and Electronic (e.g., smartwatches) may appear disjointed when analyzed separately. However, a cross-domain perspective reveals an underlying preference for "sporty lifestyle" and "high-performance tech gear", which a single-domain model would fail to capture.

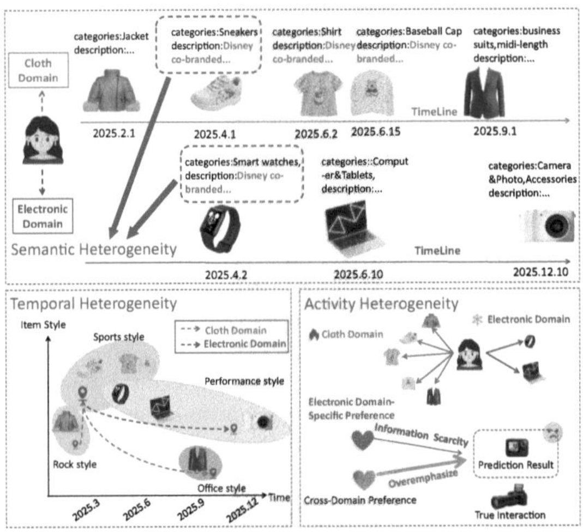

Fig. 1. Cross-domain heterogeneity in semantics, temporal dynamics, and user activity.

Given a target user u and his/her interaction histories in multiple domains, our goal is to simultaneously predict the next items that u is likely to interact with in each respective domain. As shown in Fig. 1, CDSR faces cross-domain heterogeneity in semantics, temporal dynamics, and user activity, making it crucial to bridge these gaps for effective knowledge transfer. Correspondingly, three key issues need to be addressed. First, to bridge the semantic gap in cross-domain recommender systems, we need models capable of identifying latent attribute correlations across domains. Taking the fitness scenario as an example, when Disney-co-branded sports shoes (Cloth) and Disney smartwatches (Electronic) frequently co-occur in user behaviors, the model should recognize their shared implicit attributes (e.g., "Disney-themed professional fitness gear")—the critical foundation for accurate cross-domain recommendations. Second, we need to design a cross-domain preference model to accurately capture the evolution of user interests within and across domains. The core challenge lies in modeling interest evolution from cross-domain sequences, where evolution rates in each domain is different. For instance, in Cloth domain, user preferences may exhibit rapid iteration (e.g., transitioning from rock-style → sports-style → office-style within a year), where such visible style shifts often occur within short time. Conversely, in Electronic domain, although user sequentially purchases different categories (e.g., smart watches → high-performance computer → professional camera), their underlying core demand (e.g., "performance-oriented") typically remains stable. Therefore, an ideal cross-domain preference model should dynamically adapt to these heterogeneous temporal scales (e.g., capturing quarterly style shifts in Cloth and year-spanning performance needs in Electronic) through unified temporal encoding. Finally, we need to effectively address user activity-induced bias in cross-domain sequential recommendation. Consider a user who frequently interacts with clothing (e.g., weekly sportswear purchases) but rarely engages with electronic items (e.g., twice-yearly smartwatch browsing). If the model overweights activity frequency, it may overemphasize cloth preferences while overlooking genuine electronic interests, despite potential cross-domain connections (e.g., "sports" style preferences). A robust solution must disentangle cross-domain interest transfer from both domain-specific and cross-domain activity patterns to prevent frequency-biased recommendations.

Existing CDSR methods can be categorized into two paradigms: relying solely on collaborative information and leveraging large language models (LLMs). The first category employs traditional techniques such as gating mechanisms [12], contrastive learning [11] and graph neural networks (GNNs) [2] to align cross-domain user-item interactions. With the development of LLMs, recent CDSR methods have begun to harness their extensive knowledge and reasoning capabilities, either by directly applying them for recommendation tasks [17] or using LLMs to enhance performance [10]. However, these methods are limited by their inability to model latent cross-domain attribute correlations, capture domain-specific and cross-domain interest dynamics, and overcome activity frequency bias that masks true cross-domain preferences. To address the limitations of existing cross-domain sequential recommendation, we propose a novel model, ExcelRec (**E**volution-aware mi**X**ture of experts for **C**ausal cross-domain s**E**quentia**L** **REC**ommendation). Specifically, we leverage LLMs to explicitly uncover latent attribute correlations, thereby establishing meaningful cross-domain semantic connections beyond superficial co-occurrence patterns. Then we design an evolution-aware cross-domain preference model to dynamically fuse domain-specific interaction patterns with temporal-contextual signals while preserving cross-domain representation consistency under multi-scale temporal dynamics. Finally, a causal preference enhancement strategy is proposed, thereby eliminating activity-induced bias. We summarize our contributions as follows.

- We propose an LLM-augmented co-occurrence miner that leverages LLMs to uncover implicit item-item relationships from historical interactions, enabling structured semantic augmentation of item representations.
- We propose an evolution-aware cross-domain preference model to address the critical challenge of temporal heterogeneity in interest transitions across domains. The model adopts a Temporal-Domain dual-conditioned Mixture of Experts (TD-MoE) mechanism that effectively captures both rapidly evolving and slowly changing preference patterns through unified representations.
- Building upon the unified evolution-aware representations, we propose a causality-enhanced preference learning strategy to eliminate activity-induced bias by disentangling spurious activity patterns from genuine preferences, ensuring robust cross-domain recommendation.
- We conduct extensive experimental evaluation on four real datasets. The results validate the high effectiveness of our approach.

2 Related Work

Cross-Domain Sequential Recommendation. Cross-domain sequential recommendation aims to alleviate data sparsity [4] and enhance recommendation performance across domains. Early approaches like Π-Net [12] employ gating mechanisms for cross-domain information sharing. Subsequent work, such as Tri-CDR [11], introduces triple-contrastive learning to weight domain-specific features for knowledge transfer. Recent advances leverage Graph Neural Networks (GNNs) [2] to model item-item and user-item relationships. However, these methods primarily focus on collaborative signals while overlooking auxiliary features. With the rise of LLMs, studies like [17] directly utilize LLMs as recommenders. Others [10] design prompts to generate auxiliary information or extract features via LLMs, further improving recommendation performance.

LLMs for Sequential Recommendation. LLMs' strong semantic understanding has promoted their use in recommender systems [24]. Current LLM-based sequential recommendation (SR) approaches fall into two categories: 1) LLMs as SR, where LLMs directly generate recommendations (e.g., TALLRec [1]), and 2) LLMs as SR Enhancers, leveraging LLMs to augment conventional SR models. The latter includes methods like [19] that employ LLMs as encoders for user/item representation learning. To unify collaborative and semantic signals, [9] proposes a dual-view modeling framework for deep modality integration.

Debiasing for Recommendation. Debiasing methods aim to mitigate biases inherent in user behavior data. Common approaches include: 1) Inverse propensity scoring [16] for exposure bias correction through reweighting; 2) Causal modeling techniques, where backdoor adjustment (e.g., PDA [23]) handles observed confounders in both traditional and session-based recommenders, while front-door adjustment (e.g., PACIFIC [3]) addresses unobserved confounders like product quality. Recent cross-domain extensions include constrained IPS variants [8] and the doubly robust models (e.g., AMID [21]).

3 Our Proposed Model: ExcelRec

We propose a novel model (called ExcelRec) for the CDSR task. For each user $u \in U$, we observe interactions in a source domain $S^X = \{(d_i^X, t_i^X)\}_{i=1}^p$ and a target domain $S^Y = \{(d_j^Y, t_j^Y)\}_{j=1}^q$, where t denotes interaction timestamps. These are merged into a mixed sequence $S^M = \{(d_k^M, t_k^M)\}_{k=1}^{p+q}$, where each d_k^M belongs to $\{S^X \cup S^Y\}$. The goal is to predict the next items d_{p+1}^X and d_{q+1}^Y.

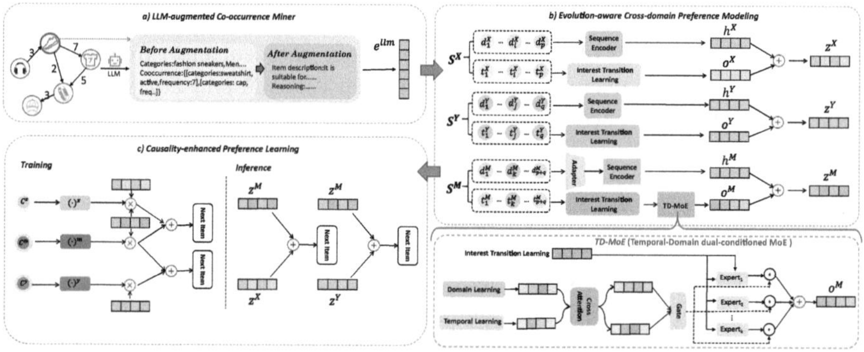

Fig. 2. Framework of ExcelRec.

As illustrated in Fig. 2, ExcelRec integrates three key components to address cross-domain recommendation challenges: a) LLM-augmented Co-occurrence Miner leverages LLMs to analyze cross-domain item co-occurrence patterns and generate enriched item descriptions. b) Evolution-aware Cross-domain Preference Modeling explicitly captures domain-specific and cross-domain temporal dynamics through a Temporal-Domain Dual-Conditioned Mixture of Experts (TD-MoE). This component adaptively recalibrates interest transitions by jointly fusing domain semantics and absolute temporal contexts. c) Causality-enhanced Preference Learning mitigates confounding biases from users' activities via backdoor adjustment.

3.1 LLM-Augmented Co-occurrence Miner

Existing item auxiliary information (e.g., categorical labels) often lacks detailed descriptions, which limits recommendation performance. For example, in Amazon's Cloth domain, a product like a 'basketball cap' might only be tagged with generic categories such as 'Clothing → Women/Men → Accessories', failing to provide sufficient contextual information. To address this limitation, we leverage LLMs to enrich item descriptions by integrating behavioral co-occurrence patterns across domains. Previous work [15,17] has demonstrated LLMs' capability to generate item summaries from raw data. We extend this paradigm by incorporating co-occurrence patterns to model functional item relationships. Our

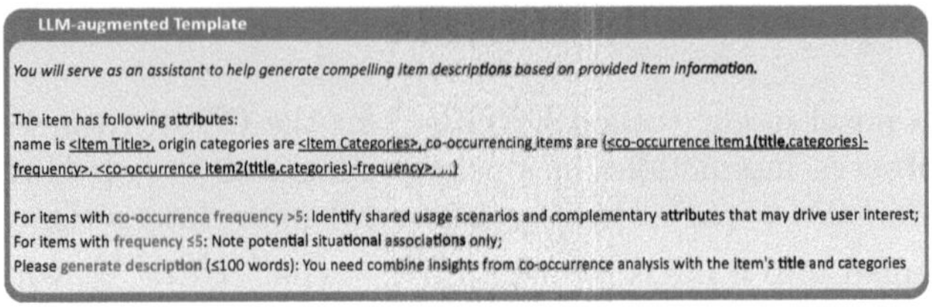

Fig. 3. Illustration of an item co-occurrence prompt.

method operates in two phases: 1) Cross-domain co-occurrence frequency calculation, where the co-occurrence count between items A and B is incremented upon their joint appearance in any user sequence; 2) LLM-augmented description generation, employing carefully engineered prompts that prioritize each item's top-five co-occurring partners. To ensure description quality, we implement differentiated strategies: for high-frequency pairs, we focus on identifying shared attributes that drive user engagement, while for low-frequency pairs, we emphasize contextual scenario associations.

Our designed prompts enable the generated descriptions to better capture items' distinctive characteristics. We then employ LLMs to produce frozen item embeddings $E^{llm} \in \mathbb{R}^{L \times d_{|llm|}}$, where L represents the total number of items and $d_{|llm|}$ denotes the embedding dimension. The prompt is shown in Fig. 3. Notably, to prevent data leakage issues, our LLM-augmentation module exclusively utilizes training data, and during the testing phase, the generated frozen item embeddings can be directly employed.

3.2 Evolution-Aware Cross-Domain Preference Modeling

Building upon the LLM-augmented co-occurrence relationships, we model user interest evolution across both cross-domain and domain-specific sequences. User interest evolution refers to the dynamic transformation of preferences over time, shown by gradual shifts in how users interact with different types of items. Existing studies in sequential recommendation [13] reveal that inter-event time intervals characterize interest transition patterns: short intervals suggest gradual drift while long intervals indicate abrupt shifts. However, such temporal dynamics remain under-explored in CDSR [5,20]. Our solution adopts the TD-MoE module that captures both cross-domain and domain-specific interest evolution. Specifically, it learns multi-scale transition patterns while preserving domain-heterogeneous temporal semantics through dual-path modeling.

Domain-Specific Interest Evolution. To model interest evolution within individual domains, we jointly analyze behavioral sequences and temporal interaction patterns. Building upon the LLM-generated item embeddings E^{llm}, we first perform Principal Component Analysis (PCA) to derive compact local embeddings E^{local} that preserve dominant semantic features.

For domain-specific pattern extraction, we employ a SASRec-based encoder [7] with self-attention mechanisms. Taking domain X as an example, we construct an input matrix $A^X \in \mathbb{R}^{p\times d}$ (where d denotes the embedding dimension) containing the item semantic embeddings. The model then computes: $Q = A^X W^Q$, $K = A^X W^K$ and $V = A^X W^V$, where W^Q, W^K, $W^V \in \mathbb{R}^{d\times d}$ are learnable projection matrices. The resulting domain-specific preference representation H^X captures the evolutionary patterns through:

$$H^X = Attention(Q, K, V) = Softmax\left(\frac{QK^T}{\sqrt{d}}\right)V \tag{1}$$

Given the timestamp sequence $\{t_1, t_2, ...t_p\}$ from cross-domain interactions, we first compute relative time intervals: $\Delta^X = \{0, t_2 - t_1, ..., t_p - t_{p-1}\}$. Following [22], we then process these intervals through a hierarchical temporal encoding to generate intervals embeddings:

$$O^X = Pos^X \cdot (\mu \log(\Delta^X + 1)) \tag{2}$$

where μ controls the interval position granularity and $Pos^X \in \mathbb{R}^{p\times d}$ is the temporal transformation matrix that maps normalized intervals to the embedding space. The final interest evolution representation Z^X for domain X is obtained by concatenating: $Z^X = [H^X || O^X]$. An analogous procedure yields the domain Y evolution representation Z^Y.

TD-MoE for Cross-Domain Interest Evolution. To model cross-domain interest evolution, we analyze patterns in the mixed interaction sequence S^M. Following [10], we first establish semantic item relationships through a unified LLM representation module. Since the original LLM embeddings E^{llm} may have incompatible dimensions for recommendation tasks, we employ a dimension alignment adapter. For each item embedding $e_i^{llm} \in E^{llm}$, the transformation is performed as:

$$\mathbf{e}_i^{global} = W_1(W_2 \cdot \mathbf{e}_i^{llm} + \mathbf{b}_2) + \mathbf{b}_1 \tag{3}$$

where $W_1 \in \mathbb{R}^{d\times d_{llm}/2}$, $W_2 \in \mathbb{R}^{d_{llm}/2\times d_{llm}}$, $b_1 \in \mathbb{R}^d$, $b_2 \in \mathbb{R}^{d_{llm}/2}$. Here they are learnable parameters for dimensionality adaptation. During training, we freeze E^{llm} to preserve the original information from the LLM.

Our framework processes behavioral sequences and temporal intervals using global item embeddings $E^{global} = \{\mathbf{e}_i^{global}\}_{i=1}^{p+q}$, where Eq. 1 yields unified representations H^M and Eq. 2 generates temporal embeddings $\hat{O}^M$. Unlike single-domain scenarios, cross-domain sequences require special consideration of two key characteristics: domain-dependent semantic interpretation of identical time intervals and heterogeneous temporal pattern distributions across domains. To address these challenges, we propose TD-MoE, a dual-conditioned mechanism, that integrates absolute timestamps and domain indicators as gating conditions. TD-MoE unifies cross-domain temporal patterns into a shared latent space.

We decompose temporal signals into multi-granularity components and construct unified chronological representations through: year embedding $E^{Year} \in \mathbb{R}^{N_y \times d}$, month embedding $E^{Mon} \in \mathbb{R}^{N_m \times d}$ and day embedding $E^{Day} \in \mathbb{R}^{N_d \times d}$. Here, N_y, N_m, and N_d denote the temporal spans in years, months, and days respectively, which are set to 30, 12, and 31 in our implementation. For timestamp t, the absolute temporal feature is calculated as follows:

$$t^{abs} = W_3 \cdot (\text{Relu}(W_4 \cdot [e^{Year}; e^{Mon}; e^{Day}] + b_3)) + b_4 \tag{4}$$

where $W_3 \in \mathbb{R}^{d \times d}$, $W_4 \in \mathbb{R}^{d \times 3d}$ and $b_3, b_4 \in \mathbb{R}^d$ are learnable parameters. To guide cross-domain interest transfer, we use binary domain identifiers $\delta \in$ 0,1 (0=source, 1=target) as semantic anchors. Each identifier is mapped to its corresponding embedding $e^{dm} = D[\delta]$ through a learnable embedding matrix $D \in \mathbb{R}^{2 \times d}$, which captures domain-specific behavior patterns.

The TD-MoE router dynamically computes expert selection weights through cross-attention fusion of domain characteristics e^{dm} and temporal patterns t^{abs}. To establish a universal temporal reference frame while preserving domain-specific semantics, we designate domain signals as query and absolute temporal signals as key and value, applying an attention mechanism $f_{atten}(\cdot)$ to capture their joint influence. The gating weights are computed as ($W^{gate} \in \mathbb{R}^k$):

$$W^{gate} = \text{Softmax}(W_5 \cdot f_{atten}(e^{dm}, t^{abs}) + b_5) \tag{5}$$

where $W_5 \in \mathbb{R}^{2d \times k}$ and $b_5 \in \mathbb{R}^k$ are learnable parameters, and k denotes the number of experts. The TD-MoE computes the cross-domain interval embedding of each item through expert-weighted aggregation:

$$o_i^M = \sum_{j=1}^{k} W_j^{gate} \cdot (W_j^{expert} \odot \hat{o}_i^M) \tag{6}$$

where $W_j^{expert} \in \mathbb{R}^{1 \times 2d}$ is the trainable weight vector for expert j, $\odot$ is element-wise product, and W_j^{gate} represents the routing weight of the j-th expert. Then we can obtain final interest evolution representation $Z^M = [H^M || O^M]$ ($O^M = \{o_i^M\}_{i=1}^{p+q}$). Through this mechanism, we effectively model the evolution of user interests across domains.

3.3 Causality-Enhanced Preference Learning

Although the above model effectively addresses the CDSR task, the learned user preferences still suffer from activity-level bias. Domains with higher interaction frequencies disproportionately influence preference modeling, creating skewed estimates. This bias stems from an observed inverse relationship between activity levels - users' intensive interactions in one domain typically correspond to sparse activity in the other, compromising recommendation validity when relying on single-domain observations. Inspired by [14,23], we propose a causality-enhanced preference learning strategy. As illustrated in Fig. 4 (taking domain X as an example), it includes the following nodes and edges:

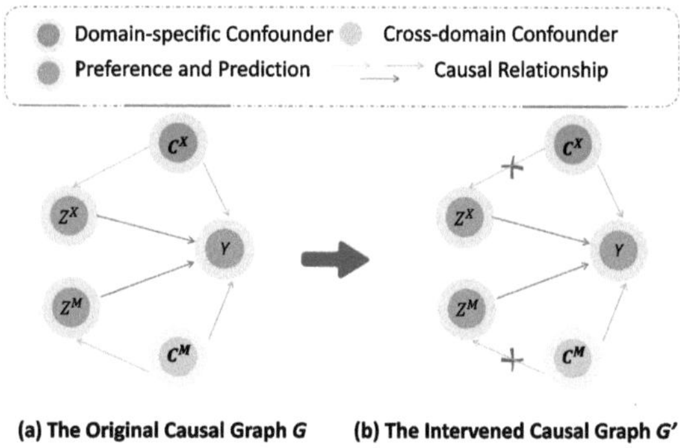

Fig. 4. Causality-enhanced preference learning.

- Node Z^X and Node Z^M: represents domain-specific and cross-domain user dynamic preference.
- Node C^X and Node C^M: represents user domain-specific and cross-domain activity.
- Node Y: represents the interaction between the user and the target item.
- Edge $Z^X \rightarrow Y$, $Z^M \rightarrow Y$ represents how domain-specific and cross-domain user preferences influence user's interaction with the target item.
- Edge $C^X \rightarrow Z^X$, $C^M \rightarrow Z^M$ represents how user activity influences the modeling of user preferences.
- Edge $C^X \rightarrow Y$, $C^M \rightarrow Y$ represents how user activity influences the final interaction.

The causal structure of CDSR involves two key confounders: domain-specific activity C^X and cross-domain activity C^M,which introduce spurious correlations through two bias pathways: 1) the performance bias path $Z^X \leftarrow C^X \rightarrow Y$, which inflates accuracy for highly active users, and 2) the head-user bias path $Z^M \leftarrow C^M \rightarrow Y$, which skews recommendations toward frequent interactors. These confounding links $C^X \rightarrow Z^X$ and $C^M \rightarrow Z^M$ obscure genuine user interests by coupling representations with activity frequency rather than true preference signals.

To mitigate confounding effects, we apply backdoor adjustment [14] to estimate causal effects $P(Y|do(Z^X), do(Z^M))$, where the do-operator severs the confounding paths (Fig. 3(b)). Let G denote the original causal graph and G' denote the intervened graph. The adjusted causal effect is computed as:

$$
\begin{aligned}
&P(Y \mid \mathrm{do}(Z^X), \mathrm{do}(Z^M)) \\
&\overset{(a)}{=} \textstyle\sum_{C^X} \sum_{C^M} P_{G'}(Y \mid Z^X, Z^M, C^X, C^M) \\
&\quad \times P_{G'}(C^X \mid Z^X) P_{G'}(C^M \mid Z^M) \\
&\overset{(b)}{=} \textstyle\sum_{C^X} \sum_{C^M} P_{G'}(Y \mid Z^X, Z^M, C^X, C^M) \\
&\quad \times P_{G'}(C^X) P_{G'}(C^M) \\
&\overset{(c)}{=} \textstyle\sum_{C^X} \sum_{C^M} P(Y \mid Z^X, Z^M, C^X, C^M) P(C^X) P(C^M) \\
&\overset{(d)}{=} f(z^X, z^M) E(Z^{\gamma_m}) E(Z^{\gamma_x})
\end{aligned}
\tag{7}
$$

- (a) is because of Bayes' theorem;
- (b) blocks backdoor paths via $do(Z^X)$ and $do(Z^M)$. Thus, $P_{G'}(C^X|Z^X)$ and $P_{G'}(C^M|Z^M)$ change into $P_{G'}(C^X)$ and $P_{G'}(C^M)$.
- (c) The equality $P(Y|Z^X, Z^M, C^X, C^M)P(C^X)P(C^M) = P_{G'}(Y|Z^X, Z^M, C^X, C^M)$ holds because the causal mechanism $\{Z^X, Z^M, C^X, C^M\} \rightarrow Y$ is invariant after intervening on $C^X \rightarrow Z^X$ and $C^M \rightarrow Z^M$. The prior distributions $P(C^X) = P_{G'}(C^X)$ and $P(C^M) = P_{G'}(C^M)$ also remain unchanged.
- (d) $E(Z^{\gamma_m})$ and $E(Z^{\gamma_x})$ denote the expectation of Z^{γ_m} and Z^{γ_x}. Since $P(Y|do(Z^X), do(Z^M))$ is used to rank items for a user, the existence of $E(C^{\gamma_m})$ and $E(C^{\gamma_x})$ don't change the ranking.

This adjustment marginalizes over the confounders C^X and C^M, blocking their spurious influence on the preference representations Z^X and Z^M.

3.4 Training and Inference

We estimate the causal effect $P(Y|do(Z^X), do(Z^M))$ using $\hat{R}^X = f(z^X, z^M) E(Z^{\gamma_m})E(Z^{\gamma_x})$, where $f(\cdot)$ is implemented via linear fusion. The hyperparameters γ_m and γ_x control activity smoothing during training. To handle dynamic user activity levels, we partition data into temporal windows and measure activity through interaction counts per window. The binary cross-entropy loss for domain X is defined as:

$$
\mathcal{L}_X = -\sum_{i \in \mathcal{U}} \left[r_i^X \log(\hat{r}_i^X) + (1 - r_i^X) \log(1 - \hat{r}_i^X) \right] \tag{8}
$$

where r_i^X denotes ground truth labels. The loss $\mathcal{L}_Y$ for domain Y is computed analogously, yielding the final training objective: $\mathcal{L}_{total} = \mathcal{L}_X + \mathcal{L}_Y$. During inference, we first leverage LLMs to generate item embeddings for all items, then feed domain-specific interaction sequences into ExcelRec to produce the final recommendations.

4 Experiment

4.1 Experimental Settings

Dataset. We build CDSR scenarios from the public Amazon dataset, creating two domain pairs: Movie-Book and Cloth-Electronic. Following prior work [11, 18], we filter out users/items with fewer than 10 interactions and keep sequences with at least 3 items per domain. Using a leave-one-out evaluation, the last item is for testing and the penultimate for validation. All reviews are considered positive feedback. Dataset statistics are in Table 1.

Table 1. The statistics of the datasets.

Dataset	# Users	# Items	Sparsity	Avg.len
Movie	22,143	37,020	96.92%	31.41
Book		73,722	97.01%	
Cloth	21,253	20,568	96.89%	14.87
Electronic		27,209	96.15%	

Table 2. Performance comparison on Movie and Book datasets.

Method	Movie				Book			
	N@5	H@5	N@10	H@10	N@5	H@5	N@10	H@10
Bert4Rec	0.4078	0.4734	0.4084	0.4754	0.2468	0.3185	0.2581	0.3519
GRU4Rec	0.4254	0.5306	0.4370	0.5452	0.2497	0.3200	0.2751	0.3661
SASRec	0.4285	0.5224	0.4316	0.5296	0.2215	0.3173	0.2626	0.3739
LLMESR	0.4692	0.5471	0.4709	0.5600	0.2879	0.3469	0.3275	0.4857
RLM	0.4767	0.5678	0.4791	0.5748	0.3083	0.4148	0.3338	0.4917
C2DSR	0.4437	0.5396	0.4505	0.5502	0.2408	0.3389	0.2838	0.3704
TriCDR	0.4502	0.5301	0.4545	0.5537	0.2377	0.3402	0.2677	0.3929
AMID	0.4670	0.5437	0.4697	0.5624	0.2585	0.3408	0.2846	0.4119
URLLM	0.4559	0.5322	0.4707	0.5620	0.2979	0.3941	0.3417	0.4576
LLM4CDSR	0.4907	0.5560	0.4996	0.5976	0.3655	0.4651	0.3753	0.5436
Ours	**0.5116**	**0.5870**	**0.5240**	**0.6217**	**0.3853**	**0.4949**	**0.3987**	**0.5828**
Impr(%)	4.26	3.38	4.88	4.03	5.42	6.41	6.24	7.21

Table 3. Performance comparison on Cloth and Electronic datasets.

Method	Cloth				Electronic			
	N@5	H@5	N@10	H@10	N@5	H@5	N@10	H@10
Bert4Rec	0.2090	0.2818	0.2335	0.3569	0.2513	0.3465	0.2882	0.4604
GRU4Rec	0.2086	0.2895	0.2420	0.3722	0.2486	0.3496	0.2885	0.4733
SASRec	0.2020	0.2841	0.2376	0.3745	0.2270	0.3149	0.2629	0.4265
LLMESR	0.2379	0.3002	0.2623	0.3978	0.2776	0.4122	0.3079	0.5026
RLM	0.2440	0.3390	0.2706	0.4199	0.2906	0.4013	0.3338	0.5349
C2DSR	0.2202	0.3051	0.2478	0.3796	0.3365	0.3782	0.3828	0.4882
TriCDR	0.2277	0.2977	0.2441	0.3801	0.3557	0.3929	0.3901	0.4699
AMID	0.2240	0.3021	0.2502	0.3889	0.3711	0.3793	0.3998	0.4706
URLLM	0.2412	0.3102	0.2664	0.4007	0.3482	0.4225	0.4067	0.4906
LLM4CDSR	0.2402	0.3363	0.2674	0.4339	0.3891	0.4773	0.4161	0.5611
Ours	**0.2592**	**0.3479**	**0.2881**	**0.4476**	**0.4194**	**0.4901**	**0.4403**	**0.5802**
Impr(%)	6.23	2.63	6.47	3.16	7.79	2.68	5.82	3.40

Baselines. To validate the effectiveness of our model, we compare it with state-of-the-art baselines from two categories. Single-domain sequential recommendation (SR) models include: GRU4Rec [6] (using GRU), Bert4Rec [18](using bidirectional self-attention), SASRec [7](using self-attention), RLM [15](aligning semantic and collaborative spaces), and LLMESR [9](an LLM-enhanced dual-view framework). Cross-domain sequential recommendation (CDSR) models include: C2DSR [2](using GNNs for item relationships), TriCDR [11] (using fine-grained contrastive learning), AMID [21](using a doubly robust estimator to mitigate bias), URLLM [17] (leveraging LLMs for attribute enrichment and recommendation), and LLM4CDSR [10](a tri-thread LLM framework for CDSR).

Evaluation Metrics. We adopt two widely used metrics: Normalized Discounted Cumulative Gain (N@K) and Hit Rate (H@K), where $K \in \{5, 10\}$. To ensure robust evaluation, we conduct triplicate tests with random seeds $\{42, 43, 44\}$ and report the average results.

Implementation Details. All experiments are implemented in Python 3.12.0 and PyTorch 2.5.1 on Intel(R) Xeon(R) Platinum 8358P CPUs and RTX 4090 GPUs, utilizing CUDA 12.4. We employ DeepSeek-V3's API for item information summarization and LLaMA-3-8B for generating LLM embeddings. The model architecture uses 2 self-attention layers with a hidden dimension of 64. We train with a batch size of 128, learning rate of 0.005, and the Adam optimizer, applying early stopping. Key hyperparameters include: expert count in TD-MoE selected from {2,4,6,8,12}, γ_x, γ_y and γ_m searched in [0.02, 0.22] with step 0.04, $\mu = 100$, and temporal activity window set to 90 days. Our code are available at: https://github.com/ExcelRec/ExcelRec.

4.2 Performance Comparison

Table 2 and 3 shows the comparative performance across methods in both 'Movie-Book' and 'Cloth-Electronic' CDSR scenarios, where bold and underlined values indicate the best and second-best results respectively.

Regarding single-domain SR baselines, their underperformance stems primarily from two factors: 1) data sparsity issues, and 2) insufficient semantic modeling. In contrast, LLM-enhanced SR models demonstrate superior performance, due to their sophisticated alignment mechanisms that effectively bridge semantic and collaborative information. For CDSR baselines, the performance of various models generally surpasses that of single-domain SR approaches, validating that introducing cross-domain data can alleviate data sparsity issues. In particular, traditional collaborative filtering-based CDSR methods underperform LLM-based SR models, highlighting the critical importance of semantic information. Specifically, LLM4CDSR leverages features extracted from LLMs to infer users' cross-domain preferences and establish inter-domain connections, achieving competitive performance. Our proposed model further advances CDSR by explicitly modeling interest evolution while incorporating LLM-augmented auxiliary information. Through causality-enhanced preference learning, it effectively addresses the limitations of existing methods (e.g., URLLM's lack of causal analysis and temporal learning) and achieves state-of-the-art performance across both single-domain SR and CDSR benchmarks.

4.3 Causal Analysis

To examine the underlying causal relationships, we conduct stratified analysis by partitioning users from both Movie and Book domains into eight distinct groups based on their activity levels (ordered from least to most active). As shown in Fig. 5, we then systematically evaluate and compare the performance of various LLM-based recommendation approaches across these user segments. The analysis reveals two critical insights:

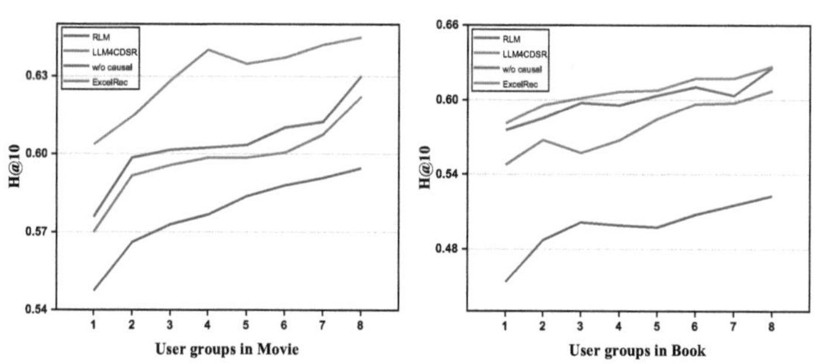

Fig. 5. Causal analysis.

1) All methods exhibit performance gains with increasing user activity, with ExcelRec consistently dominating across all groups; 2) ExcelRec achieves substantial improvements for inactive users, demonstrating its unique capability in addressing cross-domain user activity heterogeneity through causal modeling. We performed a sensitivity analysis by simulating an unobserved confounder (sampled from a normal distribution) with increasing influence strength (τ from 0.02 to 0.22). The results(as shown in Table 4) demonstrate robustness, as the H@10 metric changed by a maximum of —Δ— = +0.4% (Movies) and —Δ— = +0.9% (Books), which are negligible deviations.

Table 4. Simulated confounding stress tests for sensitivity analysis.

Stress(τ)	ExcelRec(0.00)	0.02	0.06	0.10	0.14	0.18	0.22	$\|\Delta\|$
Movie (H@10)	0.6217	0.6219	0.6222	0.6218	0.6221	0.6217	0.6214	0.4%
Book (H@10)	0.5828	0.5826	0.5819	0.5822	0.5822	0.5823	0.5827	0.9%

4.4 Ablation Analysis

To validate the design of ExcelRec, we conduct ablation studies by removing key components: 1) **w/o LLM**: Ablate the LLM-augmented Co-occurrence Miner. 2) **w/o transition**: Remove domain-specific and cross-domain interest transition patterns. 3) **w/o TD-MoE**: Exclude TD-MoE. 4) **w/o causal**: Disable causality-enhanced preference learning. 5) **w/o da**: Ignore domain-specific activity bias. 6) **w/o ca**: Remove cross-domain activity bias. Results in Table 5 reveal that all components contribute significantly. Key findings are: 1) Modeling interest transitions is most critical, highlighting the importance of capturing dynamic preferences and temporal evolution modeling in CDSR. 2) Addressing both activity biases is necessary, as they play complementary roles. 3) The LLM enhances item semantics, while TD-MoE effectively unifies slow/fast-evolving preferences through disentangled representations.

Table 5. Results for ablations.

Variant	Movie				Book			
	N@5	H@5	N@10	H@10	N@5	H@5	N@10	H@10
w/o LLM	0.4998	0.5802	0.5102	0.6113	0.3794	0.4880	0.3902	0.5782
w/o transition	0.4809	0.5571	0.4932	0.5887	0.3427	0.4561	0.3629	0.5445
w/o TD-MoE	0.4977	0.5792	0.5020	0.5933	0.3698	0.4796	0.3810	0.5721
w/o causal	0.4882	0.5641	0.4969	0.5927	0.3666	0.4800	0.3784	0.5603
w/o da	0.4924	0.5733	0.5029	0.6016	0.3702	0.4907	0.3881	0.5717
w/o ca	0.4903	0.5728	0.4997	0.5981	0.3752	0.4883	0.3824	0.5699
ExcelRec	**0.5116**	**0.5870**	**0.5240**	**0.6217**	**0.3853**	**0.4949**	**0.3987**	**0.5828**

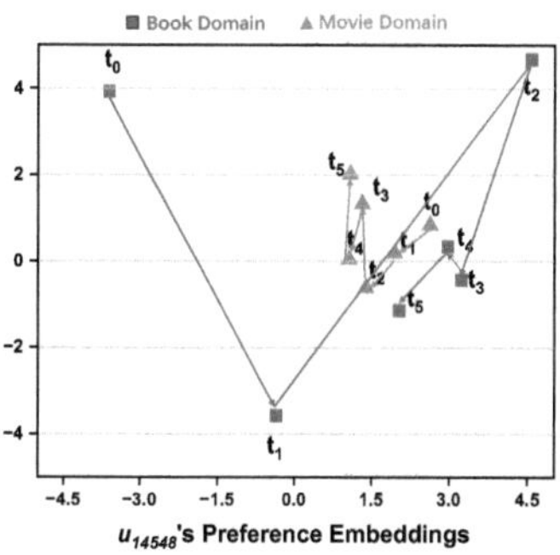

Movie Domain

step	Interest type
t_0	Science Fiction
t_1	Science Fiction
t_2	Comedy
t_3	Science Fiction
t_4	Fantasy
t_5	Science Fiction

Book Domain

step	Interest type
t_0	Animals
t_1	Arts
t_2	Literature
t_3	Mystery
t_4	Mystery
t_5	Science Fiction

Fig. 6. Case study on cross-domain interest evolution.

4.5 Case Study

Figure 6 illustrates the user u_{14548}'s cross-domain (Movie-Book) interest evolution through a two-dimensional projection of his/her last six interaction timesteps. The projection reveals distinct transition patterns: the user's Movie interests remain stable (e.g., in *Science Fiction*), while Book interests shift rapidly across diverse categories. Semantically similar interests from both domains are projected into proximate latent areas, validating our method's effectiveness, particularly TD-MoE, in capturing and unifying these cross-domain dynamics.

4.6 Hyperparameter Analysis

Our comprehensive analysis of key hyperparameters (Fig. 7) reveals that both the MoE expert count k and debiasing coefficients $\{\gamma_x, \gamma_m, \gamma_y\}$ exhibit distinct optimal operating regimes. The MoE expert count k has a concave relationship with N@10, peaking at $k = 4$ before overfitting. The debiasing coefficients $\gamma_m, \gamma_x, \gamma_y$ exhibit similar optimal configurations across both domains: $\gamma_m = 0.1$, $\gamma_x = 0.14$, and $\gamma_y = 0.1$.

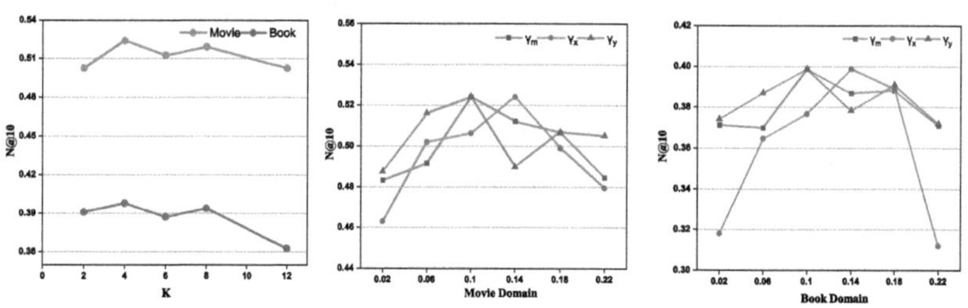

Fig. 7. The influence of hyperparameters.

5 Conclusion

In this paper, we propose ExcelRec, an evolution-aware mixture of experts for causal cross-domain sequential recommendation that addresses three key challenges in CDSR: semantic heterogeneity, temporal dynamics, and activity bias. First, we propose an LLM-augmented co-occurrence miner to establish deep semantic connections across domains through structured attribute analysis. Second, we design a TD-MoE mechanism to capture multiscale interest evolution while maintaining cross-domain representation consistency. Finally, we propose a causality-enhanced learning strategy to disentangle genuine preferences from spurious activity patterns. Future work will focus on temporal alignment techniques to address cross-domain time drift and enhance dynamic interest tracking.

References

1. Bao, K., Zhang, J., Zhang, Y., Wang, W., Feng, F., He, X.: Tallrec: an effective and efficient tuning framework to align large language model with recommendation. In: Proceedings of the 17th ACM Conference on Recommender Systems, pp. 1007–1014 (2023)
2. Cao, J., Cong, X., Sheng, J., Liu, T., Wang, B.: Contrastive cross-domain sequential recommendation. In: Proceedings of the 31st ACM International Conference on Information & Knowledge Management, pp. 138–147 (2022)
3. Chen, J., et al.: PACIFIC: enhancing sequential recommendation via preference-aware causal intervention and counterfactual data augmentation. In: Proceedings of the 33rd ACM International Conference on Information and Knowledge Management, pp. 249–258 (2024)
4. Chen, S., Xu, Z., Pan, W., Yang, Q., Ming, Z.: A survey on cross-domain sequential recommendation. In: Proceedings of the Thirty-Third International Joint Conference on Artificial Intelligence (2024)
5. Guo, L., Zhang, J., Tang, L., Chen, T., Zhu, L., Yin, H.: Time interval-enhanced graph neural network for shared-account cross-domain sequential recommendation. IEEE Trans. Neural Netw. Learn. Syst. **35**(3), 4002–4016 (2022)
6. Hidasi, B.: Session-based Recommendations with Recurrent Neural Networks. arXiv preprint arXiv:1511.06939 (2015)
7. Kang, W.C., McAuley, J.: Self-attentive sequential recommendation. In: 2018 IEEE International Conference on Data Mining, pp.197–206 (2018)
8. Li, S., et al.: Debiasing learning based cross-domain recommendation. In: Proceedings of the 27th ACM SIGKDD Conference on Knowledge Discovery & Data Mining, pp.3190–3199 (2021)
9. Liu, Q., et al.: LLM-ESR: large language models enhancement for long-tailed sequential recommendation. In: Advances in Neural Information Processing Systems, pp. 26701–26727 (2024)
10. Liu, Q., et al.: Bridge the domains: large language models enhanced cross-domain sequential recommendation. In: Proceedings of the 48th International ACM SIGIR Conference on Research and Development in Information Retrieval, pp.1582–1592 (2025)
11. Ma, H., et al.: Triple sequence learning for cross-domain recommendation. ACM Trans. Inf. Syst. **42**, 1–29 (2024)

12. Ma, M., Ren, P., Lin, Y., Chen, Z., Ma, J., Rijke, M.D.: π-net: a parallel information-sharing network for shared-account cross-domain sequential recommendations. In: Proceedings of the 42nd International ACM SIGIR Conference on Research and Development in Information Retrieval, pp. 685–694 (2019)
13. Park, S., Yoon, M., Choi, M., Lee, J.: Temporal linear item-item model for sequential recommendation. In: Proceedings of the Eighteenth ACM International Conference on Web Search and Data Mining, pp. 354–362 (2025)
14. Pearl, J.: Causality. Cambridge university press, Cambridge (2009)
15. Ren, X., et al.: Representation learning with large language models for recommendation. In: Proceedings of the ACM Web Conference 2024, pp. 3464–3475 (2024)
16. Schnabel, T., Swaminathan, A., Singh, A., Chandak, N., Joachims, T.: Recommendations as treatments: debiasing learning and evaluation. In: International Conference on Machine Learning, pp. 1670–1679 (2016)
17. Shen, T., et al.: Exploring user retrieval integration towards large language models for cross-domain sequential recommendation. arXiv preprint arXiv:2406.03085 (2024)
18. Sun, F., et al.: BERT4Rec: sequential recommendation with bidirectional encoder representations from transformer. In: Proceedings of the 28th ACM International Conference on Information and Knowledge Management, pp. 1441–1450 (2019)
19. Xi, Y., et al.: Towards open-world recommendation with knowledge augmentation from large language models. In: Proceedings of the 18th ACM Conference on Recommender Systems, pp. 12–22 (2024)
20. Xiao, S., Chen, R., Han, Q., Lai, R., Song, H., Li, L.: Proxy-aware cross-domain sequential recommendation. In: International Joint Conference on Neural Networks (IJCNN), pp.1–8 (2023)
21. Xu, W., et al.: Rethinking cross-domain sequential recommendation under open-world assumptions. In: Proceedings of the ACM Web Conference 2024, pp. 3173–3184 (2024)
22. Zhang, S., Chen, L., Shen, D., Wang, C., Xiong, H.: Hierarchical time-aware mixture of experts for multi-modal sequential recommendation. In: Proceedings of the ACM on Web Conference, pp. 3672–3682 (2025)
23. Zhang, Y., et al.: Causal intervention for leveraging popularity bias in recommendation. In: Proceedings of the 44th International ACM SIGIR Conference on Research and Development in Information Retrieval, pp. 11–20 (2021)
24. Zhao, W.X., et al.: A survey of large language models. arXiv preprint arXiv:2303.18223 (2023)

Syntax-Aware Dependency Parsing for Dual-Origin Noisy Correspondence in Text-Based Person Search

Xinpan Yuan[1,4], Wenguang Gan[1], Yuqi Chen[1], Shaomin Xie[1], Chengyuan Zhang[2(✉)], and Liujie Hua[3(✉)]

[1] Hunan University of Technology, Zhuzhou, China
[2] Hunan University, Changsha, China
cyzhangcse@hnu.edu.cn
[3] Hunan Police Academy, Changsha, China
liujiehuahnp@163.com
[4] Hunan Provincial Engineering Technology Research Center for Industrial Data Intelligence, Hunan University of Technology, Zhuzhou, China

Abstract. Text-based person search (TBPS) aims to retrieve a target individual from an image gallery based on a natural language description. In practice, this cross-modal retrieval task is often affected by noisy correspondence (NC) arising from annotation inaccuracies and low-quality visual observations. Such noise introduces two key challenges: (1) textual ambiguity within the description, including unclear attribute-Adjective associations; and (2) semantic confusion across modalities, where visually similar individuals are described with overlapping language, leading to feature entanglement and retrieval ambiguity. To tackle these challenges, we propose a Dual-level Adaptive Robust Network (DARN) that enhances both syntactic and geometric robustness during training. At the textual level, a syntax-aware text (SAT) purification module leverages dependency parsing and relevance-adaptive token selection to clarify modifier scopes and suppress noisy or low-relevance words. At the alignment level, a geometry-robust aligner (GRA) applies similarity variance regularization and identity-balanced contrastive learning to prevent feature collapse among visually similar identities. Furthermore, a noise-coupled loss dynamically adjusts the contribution of each sample according to a noise coefficient derived from syntactic uncertainty and semantic inconsistency, enabling adaptive optimization under varying noise intensities. Extensive experiments on CUHK-PEDES, ICFG-PEDES, and RSTPReid under different noise injection levels (0%, 20%, and 50%) demonstrate that DARN consistently improves retrieval accuracy and alignment stability over CLIP-based baselines. Our method consistently delivers strong performance across all three datasets, achieving 76.45% R@1 and 52.00% mINP on CUHK-PEDES, and maintaining high robustness even in the presence of synthetic correspondence noise.

Keywords: Text-based Person Search · Syntactic Knowledge · Semantic Alignment · Consistent Constrain

H. Jung et al. (Eds.): DASFAA 2026, LNCS 16535, pp. 54–70, 2026.
https://doi.org/10.1007/978-981-92-0363-5_4

1 Introduction

Text-based person search (TBPS) is a cross-modal retrieval task that retrieves target pedestrians from large-scale image galleries using natural language descriptions [3,12]. It offers interpretable, flexible queries and has been widely applied in intelligent surveillance and public safety scenarios [26], outperforming traditional image-based person re-identification (Re-ID) [4,8] in terms of practicality and efficiency [6,20].

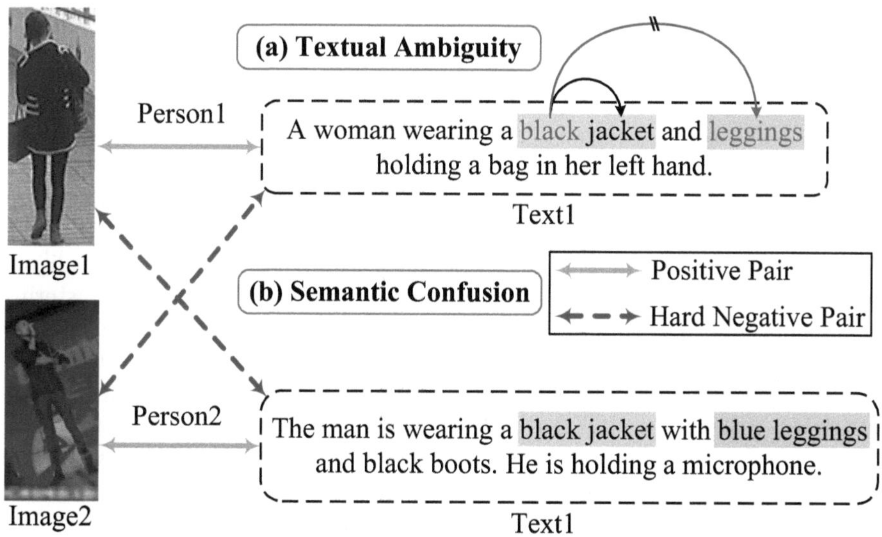

Fig. 1. Two forms of correspondence noise in text-based person search. (a) textual ambiguity (top): the modifier "black" is mis-bound in the textual structure and fails to attach to "leggings", causing incorrect grounding of the attribute–adjective pair. (b) semantic confusion (bottom): Person 1 and Person 2 exhibit highly similar appearances and overlapping textual descriptions, which entangle cross-modal features and lead to retrieval errors.

Existing TBPS models have achieved promising progress, yet still suffer from the challenge of noisy correspondence [15,16], where misalignments between visual and textual modalities degrade performance during both training and inference. This problem mainly involves two types of noise. The first is textual ambiguity, which originates from ambiguous or redundant expressions in human-written descriptions [9]. As shown in Text1 of Fig. 1, unclear references such as "black jacket and leggings" distract attention and hinder accurate alignment. The second is cross-modal semantic noise, caused by high visual similarity among identities in surveillance settings [23]. As shown in person 1 with person 2 of Fig. 1, pedestrians often wear similar clothing (e.g., "black jacket and leggings"), resulting in overlapping textual descriptions and entangled features [7]. When negative pairs exceed a similarity threshold, contrastive learning models [24] tend to misclassify them, especially under limited annotation granularity.

Recent methods have attempted to tackle the dual-source noise problem at different extent with limitation to their alignment magnitude. For example, global alignment models like CLIP [17] offer strong generalization but are sensitive to local linguistic noise. Region-word alignment models such as SSAN [5] enhance fine-grained matching but may amplify syntactic ambiguity. Text-focus methods like DEAAN [21] employ dependency parsing and token filtering to reduce noise, whilst ignoring visual similarity. Conversely, visual-focus methods strengthen cross-modal constraints without considering the linguistic adaptability. Other works attempt to integrate knowledge graphs [18] or uncertainty modeling [14], yet still treat modalities independently and thus cannot jointly model the dual-noise interaction.

To address the dual-source noisy correspondence problem in text-based person search, we propose a **Dual Adaptive Robust Network (DARN)** that systematically resolves both textual ambiguity and cross-modal semantic confusion through linguistic structure modeling and geometric alignment. The proposed framework consists of three key components. First, a syntax-aware text purifier leverages enhanced dependency parsing and relevance-adaptive token selection to eliminate ambiguous expressions. Second, a geometric robust aligner introduces cross-modal variance constraints and noise-aware identity clustering to distinguish visually similar negatives. Finally, a noise-coupled loss dynamically adjusts optimization objectives according to estimated correspondence noise, improving robustness under real-world noisy scenarios. Extensive experiments on CUHK-PEDES [12], ICFG-PEDES [5], and RSTPReid [26] datasets demonstrate that DARN achieves competitive performance among recent methods in both R@1 and mINP metrics, validating its effectiveness and generalization capability. Our contributions are threefold:

- We propose DARN, a novel dual adaptive framework that systematically models the interplay between intra-modal textual noise and cross-modal semantic noise, two critical challenges in TBPS.
- The Syntax-Aware Text Purifier (SAT) alleviates attribute-noun ambiguity through dependency parsing and relevance-based filtering, while the Geometrically Robust Aligner (GRA) disentangles visually entangled features via cross-modal variance regularization and noise-aware identity clustering.
- We introduce a Noise-Coupled Loss (NCL), which adaptively reweights sample contributions based on estimated noise intensity, thereby improving the robustness and stability of optimization under diverse noisy conditions.
- We validate DARN on CUHK-PEDES, ICFG-PEDES, and RSTPReid, where it achieves 76.45% Rank-1 accuracy and 52.00% mINP on CUHK-PEDES, strong baselines.

2 Related Work

Text-Based Person Search (TBPS). TBPS retrieves pedestrian images using free-form textual queries [3,12] and has evolved along two main routes: *global* image–text alignment and *local/region–word* alignment. CLIP-style contrastive learning provides strong global alignment and generalization [17], and recent CLIP-adapted TBPS baselines (e.g., IRRA [11] and TBPS-CLIP [1]) report consistent gains on CUHK-PEDES, ICFG-PEDES, and RSTPReid. Local alignment methods (e.g., SSAN [5], CFine [22]) enhance fine-grained grounding, while prototype/center-based objectives further stabilize matching [24,25]. Recent advances also explore local-enhanced representations [24]. Beyond single-dataset benchmarks, the field is moving toward ultra-fine textual granularity and robustness (e.g., UFineBench [27]), motivating methods that explicitly resolve attribute–noun ambiguity and handle visually similar identities.

Noisy Correspondence Learning. Noisy correspondence in TBPS arises from *intra-modal textual noise* (ambiguous or redundant phrases) and *cross-modal semantic noise* (hard negatives among visually similar identities). Syntax-aware approaches introduce structural priors to disambiguate attribute–noun bindings and suppress low-utility tokens, such as TEF [10]. On the geometric side, robustness is improved by hard-negative mining and disentanglement [2], uncertainty modeling [14], causality-inspired invariance [13], and curriculum strategies [4]. However, most prior studies treat the two noise sources separately. In contrast, our DARN framework couples *syntax-aware text purification* with *geometry-robust alignment* and introduces a *noise-coupled loss* to adapt supervision to estimated noise intensity, explicitly modeling the interaction between textual ambiguity and cross-modal confusion.

3 Method

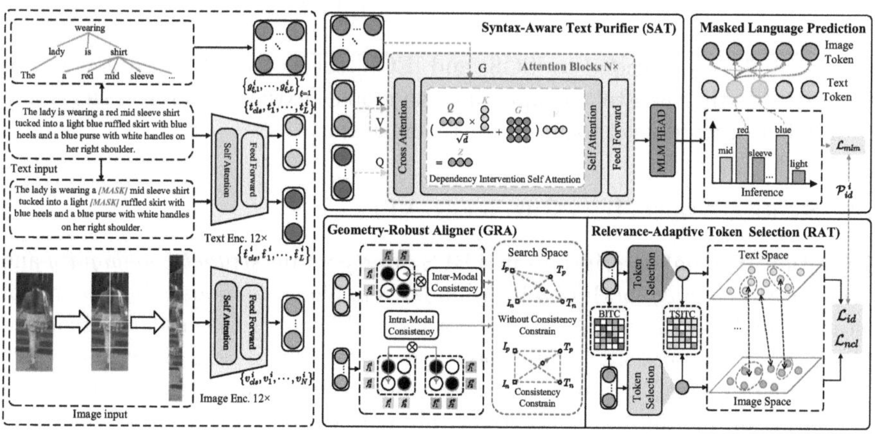

Fig. 2. Overview of the proposed DARN framework for robust text-based person search under noisy correspondence. The framework comprises three components: SAT constructs dependency-guided attention and recovers noisy tokens to enhance syntactic clarity; GRA imposes dual-space variance regularization to enforce visual-textual consistency; RAT filters irrelevant tokens based on importance and alignment confidence. The model is jointly trained with identification and noise-coupled losses for robust optimization.

To address the dual-source noisy correspondence problem in text-based person search namely, textual ambiguity and cross-modal semantic interference—we propose a unified Dual Adaptive Robust Network (DARN). DARN is composed of three major modules: a syntax-aware text purifier (SAT) to remove text-side noise, a geometry-robust aligner (GRA) to reduce visual-textual confusion caused by visually similar identities, and a noise-coupled loss (NCL) to dynamically balance learning objectives under varying noise levels. The overall framework is shown in Fig. 2.

3.1 Feature Extraction Encoder

Our framework adopts a dual-stream encoder to extract discriminative features from both visual and textual modalities, following the CLIP architecture. Given an input image $I_i \in \nu$, the visual encoder (ViT-B/16) of CLIP, based on a Transformer architecture, partitions the image into N fixed-size patches, each projected into a d-dimensional vector. These vectors are processed through multiple Transformer layers to obtain patch-level features together with a global class token:

$$f_i^v = IE(I_i) = \{v_{cls}^i, v_1^i, \ldots, v_N^i\}^\top \in \mathbb{R}^{(N+1)\times d}, \tag{1}$$

where v_{cls}^i denotes the class token representing the global image representation. For the textual branch, the input description $T_i \in \tau$ is first tokenized by the CLIP Byte-Pair Encoding (BPE) tokenizer with a vocabulary size of 49,152. The token sequence, appended with [SOS] and [EOS] tokens, is then fed into the CLIP Transformer-based text encoder:

$$f_i^t = TE(T_i) = \{t_{cls}^i, t_1^i, \ldots, t_L^i, t_e^i\}^\top \in \mathbb{R}^{(L+2)\times d}, \tag{2}$$

where t_s^i and t_e^i correspond to [SOS] and [EOS] tokens, and t_s^i (also denoted as t_{cls}^i) serves as the class token capturing the global textual semantics. Here, L is the number of word-level tokens after tokenization.

3.2 Syntax-Aware Text Purifier (SAT)

Motivation. Textual descriptions in TBPS often contain *syntactic ambiguity* and *semantic redundancy*, both of which hinder reliable cross-modal alignment. For instance, in "a man wearing black clothes and pants", the modifier scope of "black" is ambiguous. To handle these issues, we propose the **Syntax-Aware Text Purifier (SAT)**, which leverages explicit dependency structure to clarify modifier–noun relations and adaptively filter redundant or noisy tokens.

Dependency-guided attention bias. Given a tokenized sentence $T = \{w_1, \ldots, w_L\}$ consisting of L tokens, a syntactic parser produces a *dependency tree* describing grammatical relations among words. From this tree we derive:

a binary *adjacency matrix* $A \in \{0,1\}^{L\times L}$ where $A_{ij} = 1$ if w_i and w_j are directly connected; a *dependency-distance matrix* $D \in \mathbb{R}^{L\times L}$ whose element D_{ij} is the shortest-path length between w_i and w_j.

To convert these discrete relations into continuous priors, we define a Gaussian mask:

$$G_{ij} = \exp\left(-\frac{D_{ij}^2}{2\sigma^2}\right), \tag{3}$$

where σ controls the decay speed of dependency influence; a smaller σ emphasizes close dependencies. $G_{ij} \in (0, 1]$ thus encodes the syntactic proximity between tokens w_i and w_j the closer they are in the dependency tree, the larger G_{ij} (Fig. 3).

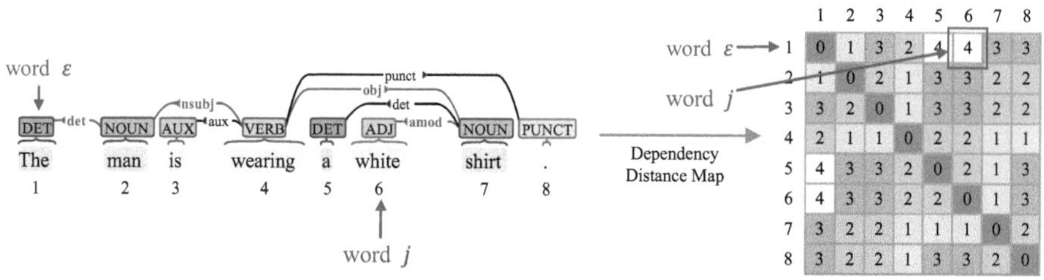

Fig. 3. Syntax-Driven Dependency Mask. The dependency matrix $\mathcal{D}$ (right) is derived from the parsed tree (left), where each entry encodes the syntactic distance between a query word ξ and a target word j. In this example, the dependency distance between "The" ($\xi = 1$) and "white" ($j = 6$) is 4.

The mask G is injected into the Transformer attention mechanism to form a syntax-aware attention:

$$\text{Attn}_{SAT}(Q, K, V) = \text{Softmax}\left(\frac{QK^\top}{\sqrt{d}} + \gamma G\right) V, \tag{4}$$

where $Q, K, V \in \mathbb{R}^{L \times d}$ are the query, key and value matrices produced by the text encoder, d is the feature dimension, and γ is a learnable scalar controlling the contribution of syntactic bias. This formulation enhances attention weights between syntactically related tokens (e.g., "red" and "bag") while reducing spurious associations.

Relevance-Adaptive Token Selection (RATS). Although dependency-aware attention clarifies modifier scopes, redundant or low-impact tokens still exist. RATS quantifies the contribution of each token by measuring its gradient-based saliency with respect to the alignment loss:

$$\alpha(w_i) = \frac{1}{d}\sum_{k=1}^{d} \left| \frac{\partial \mathcal{L}_{align}}{\partial t_{ik}} \right|, \tag{5}$$

where t_{ik} is the k-th element of the contextual embedding vector $t_i \in \mathbb{R}^d$ of token w_i, and $\mathcal{L}_{align}$ denotes the image–text alignment loss (e.g., contrastive objective). $\alpha(w_i) \geq 0$ measures the sensitivity of the alignment objective to token w_i; higher values imply greater semantic relevance.

Tokens are retained only if their importance lies within an adaptive confidence band:

$$T_{\text{keep}} = \{\, w_i \mid \kappa_l(1+\eta_t) \le \alpha(w_i) \le \kappa_h \,\}, \tag{6}$$

where κ_l and κ_h are lower and upper thresholds (empirically 0.2 and 0.6), and $\eta_t \in [0,1]$ is the estimated text-side noise level. The factor $(1+\eta_t)$ raises the threshold under heavy noise, ensuring that only confident tokens are preserved. Because η_t is computed globally across the batch, this rule avoids circular dependency.

Purified Textual Embedding. Let T_{keep} be the retained token set. Their final purified embedding $\tilde{f}^t \in \mathbb{R}^d$ is obtained by a weighted aggregation:

$$\tilde{f}^t = \sum_{i \in T_{\text{keep}}} \frac{\alpha(w_i)\, G_i\, t_i}{\sum_{j \in T_{\text{keep}}} \alpha(w_j) G_j}, \tag{7}$$

where $G_i \in \mathbb{R}^{1 \times L}$ denotes the i-th row of the Gaussian mask (representing dependencies of w_i to all other tokens). The numerator fuses syntactic proximity G_i and semantic importance $\alpha(w_i)$ for each token, while the denominator normalizes their total contribution. This produces a dependency-aware textual representation that preserves attribute–adjective coherence and discards visually irrelevant expressions, yielding a structurally purified text feature for cross-modal alignment.

3.3 Geometry-Robust Aligner (GRA)

Even with syntactically purified text features, the alignment between images and texts remains fragile when visually similar identities exhibit subtle attribute differences. For example, two pedestrians wearing black jackets may differ only by small local cues, which can cause feature entanglement in the embedding space. The **Geometry-Robust Aligner (GRA)** aims to alleviate such cross-modal semantic noise by (1) enforcing geometric dispersion among hard negatives and (2) balancing the contribution of identities with different sample frequencies.

Cross-modal variance regularization. Let $S(f_i^v, f_j^t)$ denote the cosine similarity between an image feature f_i^v and a text feature f_j^t, both normalized to unit length. For each anchor image f_i^v, we identify a set of K hard negatives $\mathcal{N}_i$—non-matching text samples with the highest similarity scores. We then regularize the similarity variance among these negatives to maintain geometric diversity:

$$\mathcal{L}_{var} = \sum_i \left(\mathrm{Var}_{j \in \mathcal{N}_i}[S(f_i^v, f_j^t)] - \delta\right)^2, \tag{8}$$

where Var computes the sample variance, and δ is a target variance that controls the expected spread (empirically $\delta = 0.5$). This formulation encourages the similarity distribution within $\mathcal{N}_i$ to maintain moderate variance—preventing collapse (all negatives too close) or excessive scattering (no structural separation).

Explanation of terms: $f_i^v \in \mathbb{R}^d$: feature of the i-th image, extracted by the visual encoder; $f_j^t \in \mathbb{R}^d$: feature of the j-th text, obtained after SAT purification;

$\mathcal{N}_i$: the hard-negative text set for anchor i; $S(\cdot,\cdot)$: cosine similarity function; δ: expected variance margin to preserve embedding dispersion.

By constraining the similarity variance around δ, GRA maintains uniform geometric separation among hard negatives, which enhances discriminability in crowded embedding spaces.

Identity-balanced contrastive learning (IBW). While variance regularization improves geometry, data imbalance across identities can bias the alignment optimization. Common identities with abundant samples dominate the gradient, whereas rare identities contribute marginally. To mitigate this issue, we reweight the image–text contrastive loss by inverse identity frequency.

The image-to-text contrastive objective is defined as:

$$\mathcal{L}_{it} = -\sum_i w_{\text{id}_i} \log \frac{\exp\left(S(f_i^v, f_i^t)/\tau\right)}{\sum_j \exp\left(S(f_i^v, f_j^t)/\tau\right)}, \tag{9}$$

where τ is the temperature parameter controlling the softness of distribution, and w_{id_i} is the weight assigned to identity id_i. Specifically,

$$w_{\text{id}_i} = \frac{C(\text{id}_i)^{-1}}{\sum_k C(\text{id}_k)^{-1}}, \tag{10}$$

where $C(\text{id}_i)$ counts the number of samples belonging to identity id_i. Hence, identities with fewer samples receive larger weights, balancing their gradient contributions.

A symmetric text-to-image loss $\mathcal{L}_{ti}$ is defined analogously by swapping the roles of image and text features. The overall contrastive alignment loss becomes:

$$\mathcal{L}_{ctr} = \tfrac{1}{2}(\mathcal{L}_{it} + \mathcal{L}_{ti}). \tag{11}$$

Joint objective of GRA. The GRA module integrates geometric variance regularization and identity-balanced contrastive learning:

$$\mathcal{L}_{GRA} = \lambda_1 \mathcal{L}_{var} + \lambda_2 \mathcal{L}_{ctr}, \tag{12}$$

where λ_1 and λ_2 are hyperparameters controlling the relative contribution of the two terms (set to 0.5 and 1.0 by default).

Interpretation: $\mathcal{L}_{var}$ ensures intra-batch geometric dispersion among negatives; $\mathcal{L}_{ctr}$ guarantees global cross-modal discrimination; the combination balances local variance stabilization and global semantic alignment.

Through this design, GRA enhances robustness against semantically overlapping samples and complements SAT by regularizing the embedding geometry in the feature space.

3.4 Noise-Coupled Loss (NCL)

Although SAT and GRA enhance syntactic clarity and geometric robustness respectively, both operate under fixed weighting schemes. In practice, the level of noise in TBPS data varies dynamically across samples and batches. Descriptions may contain uncertain grammatical constructs, incomplete attribute mentions, or visually mismatched clauses. To achieve adaptive robustness, we design the **Noise-Coupled Loss (NCL)**, which adjusts supervision strength according to the estimated noise intensity of each text–image pair.

Noise coefficient estimation. We define a noise coefficient $\eta_t \in [0, 1]$ for each textual instance, reflecting the confidence of its syntactic and semantic reliability. η_t is computed as a convex combination of two complementary signals:

$$\eta_t = \alpha_u \, H(A_t) + (1 - \alpha_u) \, Z_t, \tag{13}$$

where $H(A_t)$ measures *syntactic uncertainty*, and Z_t quantifies *semantic inconsistency*. The hyperparameter $\alpha_u \in [0, 1]$ balances their contributions (empirically α_u=0.6).

(1) Syntactic uncertainty. Given the dependency-guided attention matrix $A_t \in \mathbb{R}^{L \times L}$ from the SAT module, we measure uncertainty as its normalized entropy:

$$H(A_t) = -\frac{1}{L} \sum_{i=1}^{L} \sum_{j=1}^{L} \tilde{A}_{ij} \log(\tilde{A}_{ij} + \varepsilon), \tag{14}$$

where $\tilde{A}_{ij} = A_{ij} / \sum_j A_{ij}$ ensures row normalization, and ε is a small constant for numerical stability. Higher $H(A_t)$ indicates more diffused dependencies, implying ambiguous syntactic structures.

(2) Semantic inconsistency. For semantic reliability, we evaluate the normalized deviation of similarities between the current text feature f_t and all image features within a mini-batch:

$$Z_t = \frac{1}{B} \sum_{b=1}^{B} \frac{S(f_t, f_b^v) - \mu_S}{\sigma_S}, \tag{15}$$

where B is the batch size, $S(f_t, f_b^v)$ is the cosine similarity between text t and image b, μ_S and σ_S are the mean and standard deviation of all S values in the batch. Larger deviations correspond to semantically unstable or ambiguous alignments.

*Interpretation of Eq. (*13*)–(*15*):* η_t increases with syntactic entropy or semantic inconsistency, meaning noisier samples receive higher η_t and thus weaker direct supervision in subsequent optimization.

Noise-coupled optimization. Based on the estimated η_t, we adaptively blend the base contrastive loss $\mathcal{L}_{ctr}$ (Eq. 11) with an auxiliary alignment regularizer $\mathcal{L}_{aux}$:

$$\mathcal{L}_{NCL} = (1 - \eta_t) \, \mathcal{L}_{ctr} + \eta_t \, \mathcal{L}_{aux}. \tag{16}$$

Here, $\mathcal{L}_{ctr}$ supervises standard image–text discrimination, while $\mathcal{L}_{aux}$ provides structure-consistent regularization. Specifically, $\mathcal{L}_{aux}$ is defined as a dependency-aware token alignment loss:

$$\mathcal{L}_{aux} = \frac{1}{|T_{\text{keep}}|} \sum_{i \in T_{\text{keep}}} \Big(1 - S(t_i, f^v_{match})\Big), \tag{17}$$

where $S(t_i, f^v_{match})$ measures the cosine similarity between each preserved token feature t_i and its matched visual feature region f^v_{match} obtained via cross-attention mapping. This auxiliary loss ensures that purified textual tokens remain geometrically aligned with corresponding visual cues even when global supervision is down-weighted.

3.5 Overall Training Framework

Overview. The Dual-level Adaptive Robust Network (DARN) integrates the three proposed modules: SAT, GRA, and NCL, into a unified end-to-end training pipeline. SAT performs syntactic purification of textual features, GRA refines the geometric alignment between visual and textual embeddings, and NCL dynamically calibrates the learning process according to estimated noise intensity. All components are differentiable and jointly optimized within a CLIP-style contrastive framework.

Table 1. Performance comparison of representative methods on CUHK-PEDES under varying noise levels.

Noise	Methods	Ref	Type	Split	R-1	R-5	R-10	mAP	mINP
0%	IRRA [11]	CVPR23	G	Best	73.38	89.93	93.71	66.13	50.24
	CLIP- C [1]	AAAI24	L	Best	73.54	88.19	92.35	65.38	–
	RDE [15]	CVPR24	G	Best	75.94	<u>90.14</u>	<u>94.12</u>	<u>67.56</u>	<u>51.44</u>
	WoRA [19]	WWW25	L	Best	<u>76.38</u>	89.72	93.49	67.22	–
	DARN (Ours)	–	G	Best	**76.45**	**90.45**	**94.32**	**68.10**	**52.00**
20%	IRRA [11]	CVPR23	G	Best	69.74	87.09	92.20	62.28	45.84
				Last	69.44	87.09	92.04	62.16	45.80
	CLIP- C [1]	AAAI24	L	Best	66.41	85.15	90.89	59.36	43.02
				Last	66.10	86.01	91.02	59.77	43.57
	DECL [16]	ACM'MM22	L	Best	70.29	87.04	91.93	62.84	46.54
				Last	70.08	87.20	92.14	62.86	46.63
	RDE [15]	CVPR24	G	Best	74.46	89.42	93.63	66.13	49.66
				Last	74.53	89.23	93.55	66.13	49.63
	DARN (Ours)	–	G	Best	<u>75.61</u>	**89.90**	**93.80**	**67.52**	**50.94**
				Last	**75.83**	<u>89.73</u>	<u>93.72</u>	<u>67.40</u>	<u>50.85</u>
50%	IRRA [11]	CVPR23	G	Best	62.41	82.23	88.40	55.52	38.48
				Last	43.72	64.31	72.58	36.76	21.11
	CLIP- C [1]	AAAI24	L	Best	64.02	83.66	89.38	57.33	40.90
				Last	63.97	83.74	89.54	57.35	40.88
	DECL [16]	ACM'MM22	L	Best	65.22	83.72	89.28	57.50	40.95
				Last	65.09	83.58	89.26	57.89	41.35
	RDE [15]	CVPR24	G	Best	71.33	87.41	91.81	63.50	47.36
				Last	71.25	87.39	91.76	63.59	47.50
	DARN (Ours)	–	G	Best	**73.75**	**88.25**	**92.35**	<u>65.42</u>	**49.55**
				Last	**73.75**	<u>88.10</u>	<u>92.27</u>	**65.63**	<u>49.30</u>

Note: L = Local-based method, G = Global-based method. "–" indicates not reported. **Bold** indicates the top-ranked data. and <u>underline</u> represents the second-ranked data.

Optimization Objective. The overall objective combines all modules:

$$\mathcal{L}_{total} = \mathcal{L}_{SAT} + \mathcal{L}_{GRA} + \mathcal{L}_{NCL}. \tag{18}$$

Here, $\mathcal{L}_{SAT}$ captures syntactic purification consistency, $\mathcal{L}_{GRA}$ ensures geometric dispersion and identity balance, and $\mathcal{L}_{NCL}$ adaptively adjusts supervision strength. We optimize $\mathcal{L}_{total}$ via AdamW with learning rate 1×10^{-5}, weight decay 1×10^{-4}, and warm-up cosine scheduling.

4 Experiments

4.1 Experimental Setup

Datasets. We evaluate DARN on three widely used text-based person search benchmarks: CUHK-PEDES, ICFG-PEDES, and RSTPReid. CUHK-PEDES contains 40,206 images of 13,003 identities, each annotated with an average of 2.57 textual descriptions. ICFG-PEDES provides 54,522 image–text pairs with concise one-sentence descriptions, emphasizing fine-grained visual–semantic alignment. RSTPReid is a smaller dataset of 4,101 images and 3,768 descriptions, characterized by many visually similar pedestrians, making it ideal for testing robustness under high inter-class ambiguity.

Table 2. Performance comparison of representative methods on ICFG-PEDES and RSTPReid under varying noise levels.

Noise	Methods	ICFG-PEDES					RSTPReid				
		R-1	R-5	R-10	mAP	mINP	R-1	R-5	R-10	mAP	mINP
0%	IRRA	63.46	80.25	85.82	38.06	7.93	60.20	81.30	88.20	47.17	25.28
	CLIP-C	65.05	80.34	85.47	39.83	–	61.95	83.55	88.75	48.26	–
	RDE	67.68	82.47	87.36	40.06	7.87	65.35	83.95	89.90	50.88	28.08
	IRRA	63.46	80.25	85.82	38.06	7.93	60.20	81.30	88.20	47.17	25.28
	DARN	**68.70**	**83.10**	**87.92**	**40.91**	**8.12**	**66.62**	**84.50**	**90.32**	**51.80**	**28.65**
20%	IRRA	60.76	78.26	84.01	35.87	6.80	58.75	81.90	88.25	46.38	24.78
	CLIP-C	55.25	74.76	81.32	31.09	4.94	54.45	77.80	86.70	42.58	21.38
	DECL	61.95	78.36	83.88	36.08	6.25	61.05	80.45	86.65	47.74	25.86
	RDE	66.54	81.70	86.71	39.08	7.55	64.45	83.50	90.00	49.78	27.43
	DARN	**67.85**	**82.42**	**87.15**	**40.20**	**8.01**	**65.92**	**84.25**	**90.35**	**50.82**	**28.36**
50%	IRRA	52.53	69.62	79.41	29.05	4.43	56.65	78.40	85.65	42.41	21.05
	CLIP-C	51.00	71.69	79.31	28.76	4.33	53.45	76.80	85.50	41.93	21.17
	DECL	57.49	75.10	81.23	32.63	5.26	55.00	80.45	86.65	43.81	23.31
	RDE	63.76	79.53	84.91	37.38	6.80	62.85	83.20	89.15	47.67	23.97
	DARN	**65.30**	**80.65**	**85.77**	**38.52**	**7.05**	**64.71**	**83.92**	**89.82**	**48.73**	**24.85**

Noise Injection Protocol. To assess noise robustness, we follow a controlled noise-injection protocol. We randomly replace a percentage (0%, 20%, 50%) of image–text pairs with semantically similar but identity-mismatched captions retrieved via CLIP similarity. This setting simulates realistic annotation errors and textual ambiguities in large-scale TBPS datasets.

Implementation Details. All models are implemented in PyTorch and trained for 30k iterations on two RTX 3090 GPUs with a batch size of 32 and learning rate 1×10^{-5}. The visual encoder adopts CLIP ViT-B/16 with input size 384×128, while the text encoder uses CLIP's BPE tokenizer (vocab size 49,152) and Transformer layers to produce 512-dimensional embeddings. DARN is integrated into training as a robustness-enhancing framework with zero inference overhead.

Evaluation Metrics. We adopt standard retrieval metrics: Rank-k (R@1, R@5, R@10), mean Average Precision (mAP), and mean Inverse Negative Penalty (mINP). Higher values indicate better performance and robustness.

4.2 Main Results

Table 1 presents the performance comparison on the CUHK-PEDES dataset under three controlled noise levels (0%, 20%, and 50%). Under the clean condition (0%), DARN achieves the best R@1 = 76.45%, mAP = 68.10%, and mINP = 52.00, outperforming strong baselines such as RDE [15] and WoRA [19]. Even compared to WoRA, which explicitly models fine-grained attribute relations, DARN still achieves higher R@1 (+0.07%) and mINP (+0.88%), indicating that syntax-aware purification and geometry-robust alignment provide complementary robustness beyond relational modeling.

As the synthetic noise ratio increases to 20% and 50%, the performance of all baselines declines, while DARN remains consistently strong. At 20% noise, DARN yields 75.61% R@1 and 67.52 mAP, with only a 0.84% drop from the clean case, significantly smaller than RDE (−1.48%) and DECL (−2.0%). At the most challenging 50% noise level, DARN still retains 73.75% R@1 and 65.42% mAP, outperforming RDE by +2.42% and +1.92%, respectively. These results confirm that dependency-guided token purification and adaptive alignment weighting enable DARN to remain stable under heavy annotation noise.

Table 2 further shows results on ICFG-PEDES and RSTPReid under identical settings. On ICFG-PEDES, DARN achieves 68.70% R@1 under the clean condition and 65.30% under 50% noise, surpassing RDE by +0.94% and +1.54%. On the more challenging RSTPReid dataset, DARN outperforms RDE by +1.27% (0% noise) and +1.86% (50% noise), reaching 64.71% R@1 and 48.73 mAP. These consistent gains across datasets highlight DARN's effectiveness in maintaining semantic alignment precision and noise robustness simultaneously.

4.3 Ablation Study

We conduct a detailed ablation study on the CUHK-PEDES dataset to assess the contribution of each module in the DARN framework. As summarized in

Table 3, removing any of the key components leads to a consistent performance decline across all metrics, demonstrating that each part plays a crucial role in the model's overall robustness and discriminative capability.

Effect of Syntax-Aware Text Purification (SAT) and Relevance-Adaptive Token Selection (RATS). Removing the syntax-aware parsing module (w/o SAT) results in a 1.27% drop in R@1, from 76.45% to 75.18%, and a slight decrease in mINP from 52.00% to 51.33%, indicating that dependency parsing significantly enhances text clarity and reduces ambiguity in attribute–noun bindings. When the RATS module is removed (w/o RATS), R@1 decreases by 0.9% to 75.55% and mINP drops by 0.7% to 51.30%, showing the importance of relevance-adaptive token filtering in suppressing redundant and low-relevance phrases. When both SAT and RATS are removed (w/o SAT & RATS), the performance drops further, with R@1 decreasing to 74.13% and mINP to 50.94%, demonstrating that these two modules are complementary and work together to enhance the textual representation and align it with visual features.

Effect of Geometry-Robust Alignment (GRA). The GRA module consists of two subcomponents: similarity-variance regularization (VarReg) and identity-balanced weighting (IBW). Removing the VarReg component results in a slight decrease in R@1 (from 75.45% to 75.01%) and a 0.3 drop in mINP, showing that variance regularization plays a role in stabilizing the alignment between the visual and textual features. Excluding the IBW component causes a smaller reduction in R@1 by 0.2% and mINP by 0.2%, confirming that the identity-balanced weighting helps to mitigate class imbalance, particularly in noisy settings. When both VarReg and IBW are removed (w/o GRA), the model experiences a significant performance drop, with R@1 falling to 74.58% and mINP to 51.05%, confirming the necessity of the geometry-robust alignment in maintaining feature separability across visually similar identities.

Table 3. Ablation study of DARN components on CUHK-PEDES under the same evaluation protocol.

No.	Variant on CUHK-PEDES	R@1	R@5	R@10	mAP	mINP
1	Baseline (CLIP fine-tune)	72.13	88.92	93.10	65.45	49.78
2	**DARN (Full)**	**76.45**	**90.45**	**94.32**	**68.10**	**52.00**
3	w/o SAT (syntax-aware parsing)	75.18	90.02	93.90	66.88	51.25
4	w/o RATS (token selection)	75.55	90.15	94.08	67.02	51.30
5	w/o SAT & RATS	74.13	89.85	93.78	66.28	50.94
6	w/o VarReg (similarity variance)	75.01	90.05	94.01	66.90	51.12
7	w/o IBW (identity-balanced)	75.29	90.10	94.02	67.15	51.38
8	w/o GRA (VarReg & IBW)	74.58	89.72	93.85	66.56	51.05
9	w/o NCL (noise-coupled loss)	74.85	89.85	93.98	66.70	51.16
10	NCL (static weights)	75.10	90.00	94.05	67.01	51.27

Effect of Noise-Coupled Loss (NCL). Replacing the dynamic NCL with static weighting (NCL static) results in a 0.35% decrease in R@1 (from 75.45% to 75.10%) and a small reduction in mINP by 0.2%. Fully removing NCL (w/o NCL) causes a more substantial degradation, with R@1 dropping by 0.55% (from 75.45% to 74.90%) and mINP decreasing by 0.15% to 51.10%. This shows that dynamically adjusting the loss contribution based on the estimated noise intensity improves optimization stability and makes the model more robust to noisy correspondence during training.

Overall, the ablation results confirm that each module in DARN contributes to its overall performance. Syntax-aware text purification (SAT) and relevance-adaptive token selection (RATS) are essential for refining the textual representation and eliminating redundancy, while geometry-robust alignment (GRA) ensures stable feature mapping across modalities. Finally, the noise-coupled loss (NCL) plays a critical role in maintaining robustness under noisy conditions by dynamically adjusting supervision intensity.

4.4 Visual Analysis

To further illustrate the effectiveness of DARN, we visualize several retrieval examples on the CUHK-PEDES and ICFG-PEDES datasets, as shown in Fig. 4. Each row presents a text query along with the Rank-5 retrieved pedestrian images, where green boxes indicate correct matches and red boxes denote incorrect retrievals. Compared with the CLIP-based baseline, DARN retrieves more semantically aligned images, even when the query contains ambiguous or noisy descriptions.

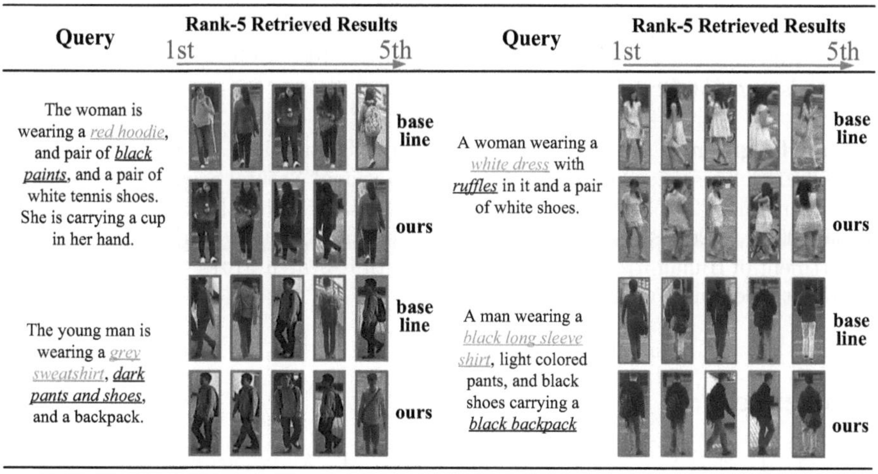

Fig. 4. Comparison of Rank-5 Retrieval Results between Baseline and DARN. DARN retrieves more accurate matches under noisy queries by suppressing redundant tokens and resisting semantic confusion. Green/red boxes indicate correct/incorrect retrievals. (Color figure online)

In particular, when the textual query includes redundant or overlapping modifiers (e.g., "a young man wearing a light gray jacket and white sneakers walking near a bench"), the syntax-aware text purifier (SAT) accurately localizes the dependency relation between "gray jacket" and "white sneakers", filtering irrelevant tokens such as "walking" or "near a bench". This selective purification produces cleaner sentence embeddings and helps the model focus on visually grounded attributes.

Moreover, DARN effectively suppresses cross-modal confusion between visually similar individuals. The geometry-robust alignment (GRA) regularizes feature variance and balances identity-level contributions, allowing the model to distinguish subtle differences such as bag color or sleeve length under high noise. As a result, the retrieved Rank-5 images exhibit greater consistency and less semantic drift across all test sets.

Finally, the visualization confirms that DARN's improvements originate not only from quantitative gains but also from qualitative robustness. It generates semantically coherent, visually interpretable alignments that remain stable even under 50% synthetic noise, validating the framework's practical reliability for real-world text-based person search.

5 Conclusion

In this paper, we proposed DARN, a dual-level adaptive robust network for text-based person search that jointly enhances syntactic clarity and geometric stability. By integrating syntax-aware purification, geometry-robust alignment, and noise-coupled optimization, DARN effectively mitigates noisy correspondence between visual and textual modalities. Extensive experiments on CUHK-PEDES, ICFG-PEDES, and RSTPReid demonstrate its superior robustness across varying noise levels, while qualitative results confirm improved semantic consistency and retrieval reliability.

Acknowledgment. National Natural Science Foundation of China 62472161, 2023JJ30169; This research was funded by the National Natural Science Foundation of Hunan Province (2025JJ70028, 2025JJ81178, 2024JJ9550, 2026JJ81383, 2026JJ81941, 2026JJ80098, 2026JJ81868, 2026JJ81154); Scientific Research Project of Education Department of Hunan Province (24A0401); Health Research Project of Hunan Provincial Health Commission (20254688).

References

1. Cao, M., Bai, Y., Zeng, Z., Ye, M., Zhang, M.: An empirical study of clip for text-based person search. In: Proceedings of the AAAI Conference on Artificial Intelligence, vol. 38, no. 1, pp. 465–473 (2024)
2. Chen, Y., Zhang, G., Lu, Y., Wang, Z., Zheng, Y.: TIPCB: a simple but effective part-based convolutional baseline for text-based person search. Neurocomputing **494**, 171–181 (2022)

3. Chen, D., et al.: Improving deep visual representation for person re-identification by global and local image-language association. In: Proceedings of the European Conference on Computer Vision (ECCV), pp. 54–70 (2018)
4. Cui, Z., Zhou, J., Wang, X., Zhu, M., Peng, Y.: Learning continual compatible representation for re-indexing free lifelong person re-identification. In: Proceedings of the IEEE/CVF Conference on Computer Vision and Pattern Recognition, pp. 16614–16623 (2024)
5. Ding, Z., Ding, C., Shao, Z., Tao, D.: Semantically self-aligned network for text-to-image part-aware person re-identification. arXiv preprint arXiv:2107.12666 (2021)
6. Ergasti, A., Fontanini, T., Ferrari, C., Bertozzi, M., Prati, A.: MARS: paying more attention to visual attributes for text-based person search. arXiv preprint arXiv:2407.04287 (2024)
7. Gao, C., et al.: Contextual non-local alignment over full-scale representation for text-based person search. arXiv preprint arXiv:2101.03036 (2021)
8. Guo, W., et al.: Lidar-based person re-identification. In: Proceedings of the IEEE/CVF Conference on Computer Vision and Pattern Recognition, pp. 17437–17447 (2024)
9. Han, X., He, S., Zhang, L., Xiang, T.: Text-based person search with limited data. arXiv preprint arXiv:2110.10807 (2021)
10. Ji, Z., Hu, J., Liu, D., Wu, L.Y., Zhao, Y.: Asymmetric cross-scale alignment for text-based person search. IEEE Trans. Multimedia **25**, 7699–7709 (2022)
11. Jiang, D., Ye, M.: Cross-modal implicit relation reasoning and aligning for text-to-image person retrieval. In: Proceedings of the IEEE/CVF Conference on Computer Vision and Pattern Recognition, pp. 2787–2797 (2023)
12. Li, S., Xiao, T., Li, H., Zhou, B., Yue, D., Wang, X.: Person search with natural language description. In: Proceedings of the IEEE Conference on Computer Vision and Pattern Recognition, pp. 1970–1979 (2017)
13. Liu, Y., Qin, G., Chen, H., Cheng, Z., Yang, X.: Causality-inspired invariant representation learning for text-based person retrieval. In: Proceedings of the AAAI Conference on Artificial Intelligence, vol. 38, no. 12, pp. 14052–14060 (2024)
14. Ning, E., Wang, C., Zhang, H., Ning, X., Tiwari, P.: Occluded person re-identification with deep learning: a survey and perspectives. Expert Syst. Appl. **239**, 122419 (2024)
15. Qin, Y., Chen, Y., Peng, D., Peng, X., Zhou, J.T., Hu, P.: Noisy-correspondence learning for text-to-image person re-identification. In: Proceedings of the IEEE/CVF Conference on Computer Vision and Pattern Recognition, pp. 27197–27206 (2024)
16. Qin, Y., Peng, D., Peng, X., Wang, X., Hu, P.: Deep evidential learning with noisy correspondence for cross-modal retrieval. In: Proceedings of the 30th ACM International Conference on Multimedia, pp. 4948–4956 (2022)
17. Radford, A., et al.: Learning transferable visual models from natural language supervision. In: International Conference on Machine Learning, pp. 8748–8763. PmLR (2021)
18. Shu, X., et al.: See finer, see more: Implicit modality alignment for text-based person retrieval. In: European Conference on Computer Vision, pp. 624–641. Springer, Cham (2022). https://doi.org/10.1007/978-3-031-25072-9_42
19. Sun, J., Fei, H., Ding, G., Zheng, Z.: From data deluge to data curation: a filtering-wora paradigm for efficient text-based person search. In: Proceedings of the ACM on Web Conference 2025, pp. 2341–2351 (2025)

20. Wang, Z., Fang, Z., Wang, J., Yang, Y.: *ViTAA*: visual-textual attributes alignment in person search by natural language. In: Vedaldi, A., Bischof, H., Brox, T., Frahm, J.-M. (eds.) ECCV 2020. LNCS, vol. 12357, pp. 402–420. Springer, Cham (2020). https://doi.org/10.1007/978-3-030-58610-2_24
21. Xia, W., Gan, W., Yuan, X.: Dependency-aware entity-attribute relationship learning for text-based person search. Big Data Cogn. Comput. **9**(7), 182 (2025)
22. Yan, S., Dong, N., Zhang, L., Tang, J.: Clip-driven fine-grained text-image person re-identification. IEEE Trans. Image Process. **32**, 6032–6046 (2023)
23. Zhang, C., Wu, L., Wang, Y.: Crossing generative adversarial networks for cross-view person re-identification. Neurocomputing **340**, 259–269 (2019)
24. Zhang, G., Chen, Y., Zheng, Y., Martin, G., Wang, R.: Local-enhanced representation for text-based person search. Pattern Recogn. **161**, 111247 (2025)
25. Zheng, Y., et al.: CPCL: cross-modal prototypical contrastive learning for weakly supervised text-based person re-identification. arXiv preprint arXiv:2401.10011 (2024)
26. Zhu, A., et al.: DSSL: deep surroundings-person separation learning for text-based person retrieval. In: Proceedings of the 29th ACM International Conference on Multimedia, pp. 209–217 (2021)
27. Zuo, J., et al.: Ufinebench: towards text-based person retrieval with ultra-fine granularity. In: Proceedings of the IEEE/CVF Conference on Computer Vision and Pattern Recognition, pp. 22010–22019 (2024)

Efficient Personalized Reranking with Semi-autoregressive Generation and Online Knowledge Distillation

Kai Cheng[1], Hao Wang[1(✉)], Wei Guo[2], Weiwen Liu[3], Yong Liu[2], Yawen Li[4], and Enhong Chen[1(✉)]

[1] University of Science and Technology of China, Hefei, China
ck2020@mail.ustc.edu.cn, {wanghao3,cheneh}@ustc.edu.cn
[2] Huawei, Shenzhen, China
{guowei67,liu.yong6}@huawei.com
[3] Shanghai Jiao Tong University, Shanghai, China
wwliu@sjtu.edu.cn
[4] Beijing University of Posts and Telecommunications, Beijing, China

Abstract. Generative models offer a promising paradigm for the final-stage reranking in multi-stage recommender systems, with the ability to capture inter-item dependencies within reranked lists. However, their practical deployment still faces two key challenges: (1) an inherent conflict between achieving high generation quality and ensuring low-latency inference, making it difficult to balance the two, and (2) insufficient interaction between user and item features in existing methods. To address these challenges, we propose a novel **P**ersonalized **S**emi-**A**utoregressive with online knowledge **D**istillation (PSAD) framework for reranking. In this framework, the teacher model adopts a semi-autoregressive generator to balance generation quality and efficiency, while its ranking knowledge is distilled online into a lightweight scoring network during joint training, enabling real-time and efficient inference. Furthermore, we propose a User Profile Network (UPN) that injects user intent and models interest dynamics, enabling deeper interactions between users and items. Extensive experiments conducted on three large-scale public datasets demonstrate that PSAD significantly outperforms state-of-the-art baselines in both ranking performance and inference efficiency.

Keywords: Recommender System · Reranking · Generative Model

1 Introduction

Large-scale online platforms widely adopt Multi-Stage Recommender Systems (MRS) to enhance user experience and optimize decision-making [5,31]. Such systems typically consist of three stages: retrieval, ranking, and reranking [34], where the reranking stage refines a small set of candidate items for final presentation [10]. Unlike the previous stages, reranking must account for both the

H. Jung et al. (Eds.): DASFAA 2026, LNCS 16535, pp. 71–87, 2026.
https://doi.org/10.1007/978-981-92-0363-5_5

individual relevance of items and their mutual dependencies to produce an optimal overall list [18]. Effectively modeling interactions between items and understanding diverse user interests remain key challenges in this field [6].

Several research efforts have been dedicated to reranking. Early studies adopted discriminative approaches, which rank items based on their individual scores, primarily model relationships within the candidate set [13,35]. Consequently, they often neglect the crucial contextual dependencies within the final reranked list, leading to suboptimal performance [3,26].

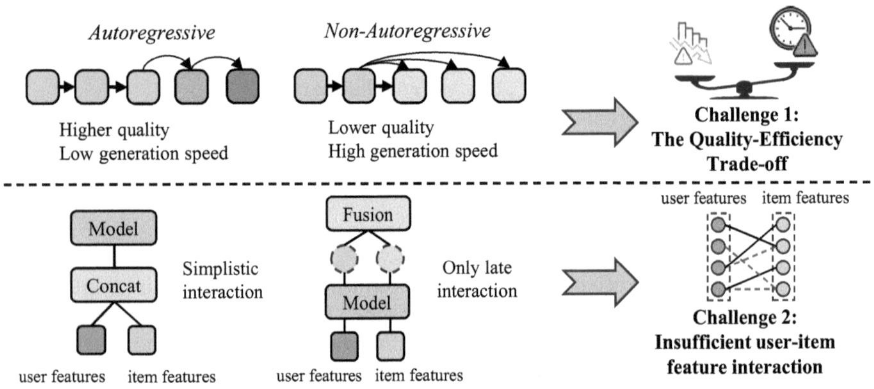

Fig. 1. Two challenges faced by generative reranking methods.

In recent years, generative models have gained significant attention for their ability to dynamically generate item permutations while explicitly modeling dependencies within the reranked list [2,11], ultimately optimizing the final display list. This approach includes both autoregressive and non-autoregressive strategies [29]: the former generates items sequentially based on previously generated ones [3,8], while the latter produces the entire sequence without strict order, focusing on global item relationships [14,21].

Despite notable progress, generative reranking methods still encounter several key challenges, are illustrated in Fig. 1: (1) **The Quality-Efficiency Trade-off**: Autoregressive models achieve high generation quality through fine-grained sequential modeling but suffer from slow inference and error accumulation due to their step-by-step generation process [21]. In contrast, non-autoregressive models offer efficient parallel generation but rely on a strong conditional independence assumption, resulting in incoherent outputs [16]. To mitigate these quality issues, researchers often employ iterative refinement [26] or two-stage paradigm [21], which in turn compromise their original efficiency and real-time capability. (2) **Insufficient User–Item Feature Interaction**: Although existing methods have explored personalized modeling [23,27], the interaction between user and item features remains insufficient. Some approaches simply concatenate user and item features [20], failing to capture the semantic variations of the same item under different user perspectives; others perform interaction only after extracting high-level representations [29], overlooking early-stage latent connections and thus limiting the modeling of complex user interest patterns.

To address these issues, we propose a novel **P**ersonalized **S**emi-**A**utoregressive with online knowledge **D**istillation (**PSAD**) framework for reranking. First, to tackle Challenge (1), we introduce a unified online knowledge distillation framework. We employ a semi-autoregressive generation paradigm whose block-wise mechanism reduces error accumulation while significantly lowering training overhead. To further reduce inference latency, we perform online knowledge distillation, jointly training a lightweight scoring network and the generator from scratch with shared encoder parameters. Unlike offline methods that require a pretrained teacher, our approach allows the scoring network to learn the generator's ranking knowledge on-the-fly without incurring significant distillation overhead, thereby approximating high performance with low latency at inference. Second, to tackle Challenge (2), we deeply inject user preferences into the model via our User Profile Network (UPN). It comprises two components: personalized gates that dynamically adapt the semantic representations of items based on user profiles, and a personalized position encoding mechanism to capture unique user interest decay patterns. In summary, our contributions are as follows:

- We developed PSAD, which is the first to effectively address latency and quality issues in generative ranking from a novel perspective of fine-grained user-item feature interactions in low-latency generative reranking.
- We design an innovative online distillation architecture: a semi-autoregressive teacher achieves high quality and training efficiency via block-wise generation, while distilling its knowledge on-the-fly to a lightweight student for significantly reduced inference latency.
- We design UPN, consisting of personalized gates and a personalized position encoding mechanism, to achieve deep fusion of user and item features.
- Extensive experiments demonstrate that PSAD significantly reduces inference latency while maintaining high-quality ranking performance.

2 Related Work

2.1 Reranking Method

Reranking methods refine an initial list of candidates to optimize overall ranking quality. Early discriminative approaches, such as the RNN-based DLCM [1], the Transformer-based PRM [20] and SetRank [19], and the GNN-based IRGPR [17], primarily focused on modeling contextual dependencies within the candidate set. In recent years, generative reranking has emerged as a new paradigm, framing the task as a process of "generating the optimal permutation." Representative methods like Seq2Slate [3], CRUM [28], and GRN [7] employ autoregressive or reinforcement learning strategies to directly generate the optimized item sequence; PIER [23] leverages hashing algorithms to filter candidates, while others like NAR4Rec [21] introduce non-autoregressive generation to reduce inference latency. However, this emerging paradigm also faces new challenges. Existing methods generally struggle to balance generation quality with inference efficiency: while autoregressive models can produce high-quality results, their inference is slow; conversely, non-autoregressive models are highly efficient but often lack coherence due to overly strong independence assumptions, making it difficult to achieve ideal performance in practical deployments.

2.2 Personalized Reranking Method

Mining user intentions from behavioral data has become a key direction for personalized reranking. Existing methods, such as PEAR [13] and MIR [29], model historical lists to capture user preferences but often overlook the full utilization of user features. Some models, like GRN [7] and PRM [20], simply concatenate user and item features, while others, such as MIR [29] and RAISE [15], perform user–item interactions only in later hidden layers. This delayed interaction limits the model's ability to capture complex user–item relationships in a timely manner, thereby constraining the effectiveness of personalized modeling.

2.3 Semi-autoregressive Generation

To improve the efficiency of autoregressive models, early NLP research introduced semi-autoregressive modeling, which generates multiple token spans per step instead of a single token [25]. Later advancements, such as parallel decoding [24] and speculative decoding [30], further refined this idea. Inspired by these developments, we apply the semi-autoregressive generation mechanism to the reranking stage of recommender systems, enabling parallel generation of multiple candidate items to enhance overall performance.

3 Preliminary

Definition 1 (Reranking Problem). The reranking stage is typically defined as follows: given a user u, the candidate sequence C_u obtained from the previous recommender stage, the user's recent interaction history H_u, and the user's profile features $\mathcal{P}_u$ (e.g., age, gender, etc.). Each item in the C_u or H_u set is characterized by sparse features (such as category ID, brand ID, etc.) and dense features (such as price, score, etc.). The reranking problem involves learning a mapping function $\mathcal{F}$, such that $\mathcal{F}(C_u, H_u, \mathcal{P}_u) \to R_u$, where R_u is the reranked list presented to the user and $R_u \subseteq C_u$. The length of H_u is denoted as N, the C_u as M, and R_u as T, which typically satisfies $T \leq M$.

Definition 2 (Generative Reranking). Generative reranking models aim to find the optimal sequence R_u by directly modeling its conditional probability. The goal is to generate the list that maximizes the likelihood given the user's context: $R_u^* = \arg\max_{R_u \in \mathcal{P}(C_u)} p(R_u|C_u, H_u, \mathcal{P}_u)$, where $\mathcal{P}(C_u)$ represents the space of all valid permutations of items from the candidate set C_u. Different modeling assumptions on $p(R_u \mid \cdot)$ lead to various generation paradigms, such as autoregressive or non-autoregressive approaches.

4 Methodology

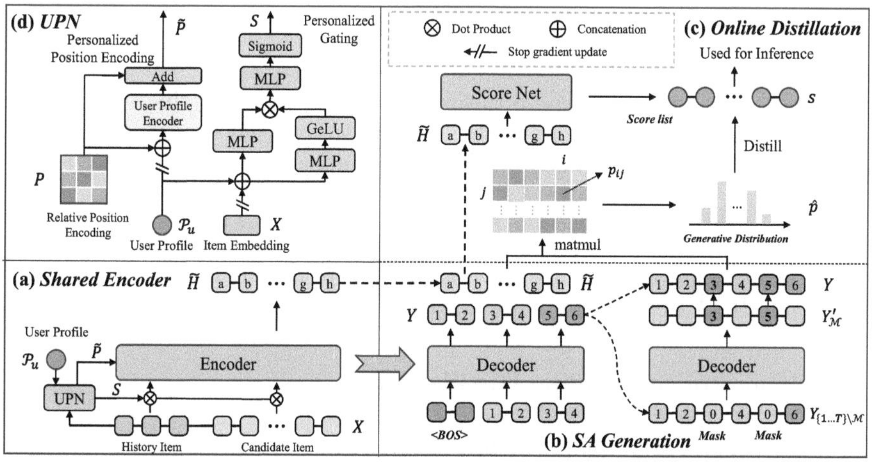

Fig. 2. The overall framework of PSAD and its sub-modules.

Existing generative reranking models often struggle to balance generation quality and inference efficiency while lacking sufficient depth in personalized modeling. To address these challenges, we propose PSAD, whose overall architecture is illustrated in Fig. 2. This section provides a detailed description of its core components: First, a shared encoder (Fig. 2(a)) processes all input sequences, as described in Sect. 4.1. Next, we address the trade-off between quality and efficiency through semi-autoregressive generation (Fig. 2(b)) and online distillation (Fig. 2(c)), detailed in Sect. 4.2. Then, the User Profile Network (UPN) (Fig. 2(d)) is introduced to enable deep personalization, as discussed in Sect. 4.3. Finally, Sect. 4.4 presents the inference process of PSAD.

4.1 Encoder

Embedding Layer. According to *Definition 1*, for each item v in the user's historical sequence H_u or the candidate set C_u, the i-th sparse feature of the item, is mapped into representation $e_{v,i}$ through the corresponding embedding table. Subsequently, we concatenate the embeddings of all sparse features with the item's dense features to form the final item embedding: $x_v = [e_{v,1} \oplus \cdots \oplus e_{v,k} \oplus$

$x_d^v] \in \mathbb{R}^{d_x}$, where k denotes the number of sparse features and x_d^v represents the dense feature vector. The operator $\oplus$ indicates vector concatenation. By stacking all item embeddings x_v, we obtain the input item matrix $X \in \mathbb{R}^{(N+M)\times d_x}$. Similarly, the embedding vector of user u is represented as $\mathcal{P}_u \in \mathbb{R}^{d_u}$.

Shared Encoder. To better capture the relationships among items, we adopt a self-attention structure, and the attention score matrix is calculated by:

$$A^l = \frac{\text{Softmax}((X^l W_Q^l)(X^l W_K^l))}{\sqrt{d_x}} + P^l, \tag{1}$$

where l denotes the l-th layer in the model, W_Q represents the learnable parameters $\in \mathbb{R}^{d_x \times d_x}$, and similarly for the others. The matrix P^l is a learnable parameter, where $P_{i,j}^l$ represents the positional encoding value for the i-th and j-th positions. The attention score is then multiplied by the attention value and passed through a Feed-Forward layer to obtain the output of the encoder:

$$H^l = A^l(X^l W_V^l), \quad \tilde{H}^l = \text{FFN}(H^l). \tag{2}$$

The encoder stacks multiple identical layers, where the input X^l of each layer is the output $\tilde{H}^{l-1}$ from the previous layer.

4.2 Generator and Online Distillation

Semi-autoregressive Generation. After the encoding stage, we employ a generator to rerank the candidate set. To overcome the inherent limitations of autoregressive and non-autoregressive models, we design a semi-autoregressive generation scheme that reduces the number of generation steps to mitigate error accumulation and improve generation efficiency, while retaining autoregressive dependencies to capture sequential relationships. Conceptually, it is analogous to a pointer network [3], where query vectors are generated to select items from the candidate set. In practice, the generator constructs blocks containing K items in parallel at each iteration step until the complete sequence is formed.

The specific process is as follows: let the target result length be T, and the block size be K, the position indices for the b-th block are defined as $\mathcal{I}_b = \{(b-1)K+1, \ldots, \min(bK, T)\}$. For each block, we generate K tokens simultaneously, conditioned on the previously generated context:

$$Y_{\mathcal{I}_b}^l = \text{CrossAttn}(H^l, Y_{<\mathcal{I}_b}^l), \tag{3}$$

where CrossAttn denotes the model cross attention, and $Y_{<\mathcal{I}_b}$ represents the result from tokens before the current block. To reduce inconsistencies inside each parallelly generated block, we employ a contextual enhancement step that refines the full sequence Y using a mask-and-refine paradigm. We randomly mask a subset of tokens $\mathcal{M}$, where $|\mathcal{M}| = \gamma$, and then update them with predictions conditioned on the unmasked context:

$$Y'_{\mathcal{M}} = \text{CrossAttn}(H^l, Y_{\{1..T\}\setminus\mathcal{M}}), \quad Y \leftarrow Y_{\{1..T\}\setminus\mathcal{M}} \cup Y'_{\mathcal{M}}. \tag{4}$$

The generation probability matrix is derived from the multiplication of the generated sequence representations and the hidden vectors of the candidate set:

$$p_{ij} = \text{Sigmoid}(\tilde{H}_i^L \cdot Y_j^L), \tag{5}$$

where L represents the total number of layers. The matrix element p_{ij} represents the probability that the i-th item in the candidate set is the j-th position in the reranked list. Then we compute the generative loss:

$$\mathcal{L}_{gen} = \sum_j \omega_j (\sum_i (-\hat{y}_i \log p_{ij}) + \max\{0, 1 - \min_{i:\hat{y}_i=1} p_{ij} + \max_{i:\hat{y}_i=0} p_{ij}\}), \tag{6}$$

where $\hat{y}_i$ is ground truth where $\hat{y}_i$ is 1 if i-th item with positive label otherwise 0. Our generative loss combines a cross-entropy term for pointwise accuracy and a hinge loss for listwise ranking order, weighted by a DCG inspired term, $\omega_j = \frac{1}{\log(j+1)}$, to prioritize early positions.

Online Knowledge Distillation. To further reduce the computational burden during inference, we propose a unified Online Knowledge Distillation framework. This framework includes a lightweight Scoring Network as the student model, which is jointly trained with the previously introduced generator serving as the teacher model. The scoring network computes a score s_i for each candidate item i and is supervised using a standard cross-entropy loss:

$$s_i = \text{Sigmoid}(\text{MLP}(\tilde{H}_i^L)), \quad \mathcal{L}_{scorer} = -\sum_i \hat{y}i \log s_i. \tag{7}$$

Unlike offline distillation methods that rely on a pre-trained teacher model, we perform on-the-fly distillation to transfer the generator's ranking knowledge to the student model, aggregating the generator's probability matrix into target scores $\hat{p}_i$ for each item using an exponential decay weighting scheme:

$$\tilde{pij} = \frac{\exp(pij)}{\sum_k \exp(p_{kj})}, \quad \hat{pi} = \frac{\sum_j \tilde{pij} \cdot e^{-j}}{\sum_j e^{-j}}, \tag{8}$$

Subsequently, to more effectively measure the consistency between the teacher's and student's output distributions, we employ Kullback-Leibler Divergence as the distillation loss:

$$\mathcal{L}_{Distillation} = \text{KL}\Big(\text{Softmax}(\hat{p}/\tau) \parallel \text{Softmax}(s/\tau)\Big), \tag{9}$$

where τ is the temperature that softens the probability distributions. Finally, the total training loss consists of the losses from the collaborative training process:

$$\mathcal{L} = \mathcal{L}_{gen} + \mathcal{L}_{scorer} + \alpha \cdot \mathcal{L}_{Distillation}\ , \tag{10}$$

where α is a hyperparameter balancing the distillation loss. By optimizing this objective, enabling the student model to emulate the teacher's complex ranking preferences with high efficiency during inference, thus achieving high generation quality while maintaining low-latency inference.

4.3 User Profile Network

Personalized Gating. Previous reranking methods typically encode items as fixed embeddings. However, due to the diversity of user interests, item representations should adapt dynamically to user-specific characteristics. To this end, we design a personalized gating unit that integrates user profile information into item representations while preserving the original item features, thereby enabling personalized item embeddings. Its working principle is as follows:

$$\mathcal{G}_1 = \mathrm{MLP}_1(\oslash X, \mathcal{P}_u), \quad \mathcal{G}_2 = \mathrm{GeLU}(\mathrm{MLP}_2(\oslash X, \mathcal{P}_u)), \tag{11}$$

where MLP refers to a fully connected layer, and GeLU is a nonlinear activation function. $\oslash$ denotes the stop gradient operation, this ensures that the module optimizes only the user profile, thereby facilitating faster convergence during training. Subsequently, we compute the interaction between the activated and non-activated representations of the item-user pair to generate the gating signal:

$$S = \mathrm{Sigmoid}(\mathrm{MLP}(\mathcal{G}_1 \odot \mathcal{G}_2)), \quad \hat{X} = S \odot X, \tag{12}$$

Table 1. Dataset statistics.

Dataset	#Users	#Items	#Records
Ad	1.06×10^6	8.27×10^5	2.04×10^6
PRM Public	7.44×10^5	7.25×10^6	1.44×10^7
Avito	1.32×10^6	2.36×10^7	5.36×10^7

where $\odot$ represents the Hadamard product. The gating signal S is applied to the item embedding X, and combined with the projected user profile to produce the personalized representation $\hat{X}$, which serves as the input to Eq. (1).

Personalized Position Encoding. In Eq. (1), the relative position encoding P^l assumes a fixed interest decay pattern for all users. To enable personalization, we propose a position-adaptive scheme that allows the model to dynamically adjust the universal position bias P^l_{ij} based on the user profile. Specifically, we generate an adaptive bias for each position pair by fusing the user preference $\mathcal{P}_u$ with the generic positional relationship P^l_{ij}. The bias is added to the original encoding to obtain the personalized positional encoding $\hat{P}^l$:

$$\hat{P}^l_{ij} = P^l_{ij} + \mathrm{Tanh}(\mathrm{MLP}(\oslash \mathcal{P}_u \oplus P^l_{ij})). \tag{13}$$

This enables the model to learn user-specific interest dynamics rather than a uniform temporal decay pattern.

4.4 Model Inference

During inference, the generator computes the probability distribution according to Eq. (5) and produces candidate sequences via beam search, which are then scored by PRM to select the highest-scoring sequence. The scoring network, in contrast, directly assigns scores based on Eq. (7) and ranks items to select the top T candidates. Evidently, inference via the scoring network is more efficient, playing a crucial role in alleviating latency in the reranking process.

5 Experiments

5.1 Experimental Setup

Datasets. To ensure the broad applicability of the proposed method, we conducted extensive experiments on public datasets, including the E-commerce Reranking dataset and Ad dataset.

- **Ad**[1]: The dataset records interactions between users and advertisements, containing 9 user features (e.g., ID, age, and occupation) and 6 item features (e.g., ID, campaign, and brand). Based on the timestamps of user interactions with advertisements, we transform each user's records into ranking lists. Items interacted with within five minutes are grouped into the same list.
- **PRM Public**[2]: A dataset includes 3 user profile features, 5 categorical features, and 19 dense features. We use the last interacted list as the target, while interactions from previous lists are used to build the history lists.
- **Avito**[3]: A dataset consists of multiple user search logs and metadata. For each user, ads on the most recent search page are used as prediction targets, while the preceding pages serve as historical behaviors.

In terms of data processing, we adopted a similar approach to previous studies [22,32], removing users and items with fewer than four interaction records and retaining only users with at least one click, ensuring sufficient interaction data to accurately capture user preferences. The detailed statistical information of the processed datasets is summarized in Table 1.

Evaluation Metrics. Reranking encompasses more diverse objectives, such as utility [1], diversity [4], and fairness [9]. This study focuses on utility-driven reranking, and we adopt the widely used NDCG@K and MAP@K [12,33]. Since the number of reranked items in real-world applications is typically in the tens [20,27], we set the length of the reranked result list to 10 for the Ad and PRM Public datasets, and accordingly set $K = 5$ and 10. Since the candidate set in the Avito dataset contains only 5 items, we set $K = 5$ for this dataset [21].

[1] https://tianchi.aliyun.com/dataset/56.
[2] https://github.com/rank2rec/rerank.
[3] https://www.kaggle.com/c/avito-context-ad-clicks/data.

Baseline Methods. We compare the proposed method with 6 state-of-the-art reranking methods. We select miDNN, DLCM, and PRM as typical baselines representing different structures, and Seq2Slate, CRUM, PIER, and NAR4Rec as typical baselines for generative models. The details are listed as follows:

- **MIDNN** [35]: MIDNN extracts mutual influence between items in the input ranking list with global feature extension.
- **DLCM** [1]: DLCM first employs GRU to sequentially encode the entire ranking list, generating item representations.
- **PRM** [20]: PRM employs self-attention to model the mutual influence between any pair of items and users' preferences.
- **Seq2Slate** [3]: Seq2Slate is based on an RNN pointer network architecture that sequentially predicts the next item to generate the reranked list.
- **CRUM** [28]: CRUM achieves high utility in pairwise reranking by using a utility evaluator to guide item pair swapping.
- **PIER** [23]: PIER utilizes a hashing algorithm to select the top-k candidates from the full permutation based on user interests. Subsequently, the generator and evaluator are jointly trained to generate more optimal permutations.
- **NAR4Rec** [21]: NAR4Rec employs a non-autoregressive generative model to mitigate inference bias and enhance efficiency.

Hyper-parameter Settings. We use Adam as the optimizer to implement our model and baseline methods, with a batch size of 256. We use the last 50 items clicked by the user as their history. According to different scenarios, the maximum candidate set length for Ad is set to 20 with a learning rate of 1e-4, while for PRM Public, the maximum candidate set length is set to 30 with a learning rate of 3e-5, and for Avito, the maximum candidate set length is set to 5 with a learning rate of 3e-5. The embedding size of sparse features is set to 16. Our source code will be made publicly available after the paper is published.

Table 2. Overall performance comparison of all methods in terms of NDCG@K and MAP@K (K=5, 10). (p-value < 0.05)

Datasets	Ad				PRM Public				Avtio	
Metric	N@5	N@10	M@5	M@10	N@5	N@10	M@5	M@10	N@5	M@5
miDNN	0.6819	0.6952	0.6025	0.6063	0.2620	0.3341	0.2842	0.2954	0.6737	0.6838
DLCM	0.6857	0.6989	0.6064	0.6100	0.2664	0.3395	0.2918	0.3023	0.6912	0.7057
PRM	0.6872	0.7015	0.6080	0.6117	0.2689	0.3430	0.2937	0.3042	0.7047	0.7174
Seq2Slate	0.6916	0.7062	0.6119	0.6155	0.2709	0.3449	0.2959	0.3061	0.7245	0.7360
CRUM	0.6899	0.7037	0.6084	0.6112	0.2703	0.3445	0.2952	0.3056	0.7110	0.7266
PIER	0.6940	0.7070	0.6123	0.6158	0.2694	0.3423	0.2941	0.3038	0.7156	0.7292
NAR4Rec	<u>0.6965</u>	<u>0.7101</u>	<u>0.6151</u>	<u>0.6188</u>	0.2701	0.3455	0.2968	0.3071	<u>0.7315</u>	<u>0.7462</u>
PSAD-G	**0.7038**	**0.7184**	**0.6216**	**0.6255**	**0.2762**	**0.3528**	**0.3039**	**0.3136**	**0.7427**	**0.7561**
PSAD-S	0.6962	0.7091	0.6146	0.6185	<u>0.2717</u>	<u>0.3464</u>	<u>0.2985</u>	<u>0.3079</u>	0.7285	0.7396

5.2 Overall Performance Comparison

For all experiments, we first train the generator and score network. Refer to Sect. 4.4, for our approach, the variant that performs inference solely with the trained generator is denoted as **PSAD-G**, while the variant that jointly trains the generator and scorer and performs inference via the online distilled scoring network is denoted as **PSAD-S**.

Performance Comparison. In this part, we compare PSAD against both discriminative and generative reranking baselines, with the results presented in Table 2. We also conduct Wilcoxon signed rank tests to evaluate the statistical significance of PSAD with the base model. We have the following observations:

(1) The generative paradigm outperforms the discriminative paradigm. Even the earlier generative model Seq2Slate surpasses strong discriminative baselines such as PRM, highlighting the importance of the research problem we address. By leveraging generative methods to model the mutual influences among items within the final reranked list, we can more effectively capture contextual dependencies and achieve superior ranking performance.
(2) Methods that model global relationships among candidate items (e.g., PRM) outperform the pointwise method MIDNN and unidirectional approaches DLRM, highlighting the effectiveness of our choice to employ attention mechanisms for capturing inter-item dependencies.

Table 3. The training and inference time comparison between PSAD and baseline methods on the PRM Public dataset. The comparison is conducted under different batch sizes, and the time is calculated by averaging the results over 10 steps.

Batch Size	64		128		256	
Model	Train(↓)	Inf.(↓)	Train(↓)	Inf.(↓)	Train(↓)	Inf.(↓)
miDNN	0.0895	0.0764	0.1425	0.1256	0.2132	0.1893
DLCM	0.2799	0.1497	0.4301	0.2409	0.6827	0.3940
PRM	0.1238	0.1014	0.1852	0.1360	0.2806	0.2037
Seq2Slate	0.3142	0.2176	0.5114	0.3993	0.8384	0.5873
CRUM	0.2021	0.1589	0.3185	0.2850	0.4415	0.3048
PIER	0.2428	0.2247	0.3608	0.3394	0.5482	0.4735
NAR4Rec	0.1903	0.1356	0.2904	0.2215	0.4649	0.3572
PSAD-G	0.1834	0.1568	0.2860	0.2376	0.4931	0.4239
PSAD-S	0.1897	0.0801	0.2943	0.1290	0.5128	0.2161

(3) The generative model PSAD-G consistently outperforms other generative methods across all datasets, demonstrating strong effectiveness and generalization ability. This is attributed to the semi-autoregressive generation strategy and contextual enhancement module, which jointly improve generation

quality. Moreover, the proposed User Profile Network efficiently leverages user representations, enhancing personalized modeling.

(4) The scoring model PSAD-S outperforms all discriminative baselines and performs comparably to the strongest generative baseline NAR4Rec. This confirms that our online knowledge distillation framework effectively compresses the generative model's item distribution into a lightweight scoring network. Furthermore, the subsequent efficiency comparison validates the framework's superior efficiency in both training and inference phases.

Efficiency Comparison. Following prior work [21], we measure the average training and inference time and compare our model with baselines under varying batch sizes on the PRM Public dataset, as shown in Table 3:

(1) Generative models exhibit significantly higher training and inference time consumption compared to discriminative models, which validates our research motivation that existing generative approaches often struggle to balance performance and efficiency, leading to substantial computational overhead.
(2) PSAD demonstrates markedly higher training efficiency than the autoregressive generative model Seq2Slate, while remaining comparable to non-autoregressive models such as NAR4Rec and CRUM. This confirms that the semi-autoregressive generation strategy adopted by PSAD effectively maintains high-quality generation without introducing excessive training costs.
(3) PSAD-S achieves faster inference than all generative baselines and even surpasses some complex discriminative models such as PRM. Meanwhile, its overall training cost shows only a marginal increase compared to PSAD-G. These results validate the effectiveness of our shared-encoder online distillation framework, which introduces a lightweight scoring model with minimal additional training cost while achieving extremely low inference latency.

Table 4. Ablation analysis of PSAD-G

Datasets	Ad		PRM Public		Avito	
Metric	N@10	M@10	N@10	M@10	N@5	M@5
w/o sa	0.7095	0.6192	0.3455	0.3054	0.7145	0.7295
w/o ce	0.7133	0.6220	0.3482	0.3096	0.7231	0.7360
w/o sace	0.7042	0.6139	0.3441	0.3052	0.7103	0.7217
w/o ppe	0.7168	0.6248	0.3523	0.3128	0.7384	0.7546
w/o pg	0.7136	0.6224	0.3507	0.3110	0.7402	0.7543
PSAD-G	**0.7184**	**0.6255**	**0.3528**	**0.3136**	**0.7427**	**0.7561**

5.3 Ablation Study

We designed multiple variants of the model to investigate the effectiveness of each component in our approach and conducted a series of experiments on public datasets. ***Variant without Semi-Autoregressive (w/o sa)*** indicates that we generate all items in one step instead of using a block-wise generation strategy, ***Variant without Contextual Enhancement (w/o ce)*** means that we remove the contextual enhancement process ***Variant without sa and ce (w/o sace)*** indicates that both the semi-autoregressive generation and the contextual enhancement module are removed , ***Variant without Personalized Position Encoding (w/o ppe)*** denotes that we remove the personalized relative positional encoding used to model interest decay, ***Variant without Personalized Gate (w/o pg)*** indicates that we remove the personalized gating network.

The comparison of the above variants and the PSAD on the three public datasets is shown in Table 4: The (***w/o sa***) shows that compared to one-shot generation, block-wise generation better preserves dependencies within the generated sequence, validating the effectiveness of our proposed semi-autoregressive generation. The (***w/o ce***) demonstrates that contextual enhancement helps enhance local coherence and inter-block consistency in the generated sequence. The (***w/o sace***) indicates that the combination of semi-autoregressive generation and contextual enhancement yields better performance. The (***w/o ppe***) reveals the importance of Personalized Position Encoding in modeling personalized interest decay, which contributes to more accurate user intent positioning in the ranking list. The (***w/o pg***) confirms that Personalized Gate improves user-item alignment by dynamically adjusting item features based on user profiles.

5.4 Hyper-parameter Study

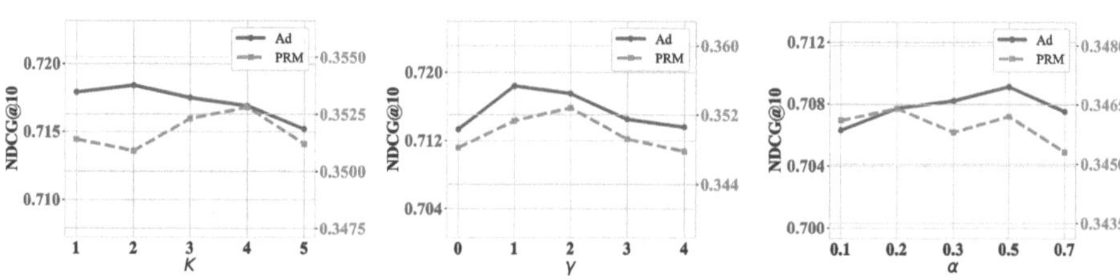

Fig. 3. Hyper-Parameter Performance

As shown in Fig. 3, we analyze the key hyperparameters of PASD on two datasets, including the parallel generation block size (K), the number of random masks (γ), and the distillation weight (α). For the block size K, the overall performance exhibits a rise-then-fall trend. A small K fails to alleviate the error accumulation from item-by-item generation, whereas an excessively large

Table 5. Comparison of different distillation methods in terms of performance (NDCG@10) and training time. The batch size is set to 256, and the time is reported as the average over 10 training steps. The time unit is seconds.

Dataset	Ad		PRM	
Teacher Model	N@10(↑)	Train(↓)	N@10(↑)	Train(↓)
AR(online)	0.7078	0.4896	0.3434	0.7372
NAR(online)	0.7046	0.3004	0.3429	0.4165
SAR(offline)	0.7083	0.6547	0.3481	0.8933
SAR(online)	0.7091	0.3752	0.3464	0.5128

K harms the internal consistency of the sequence due to overly strong independence assumptions. For the number of random masks γ, the model achieves optimal performance when $\gamma = 1$ or 2, while larger values disrupt the existing reasonable sequence structure. For the distillation weight α, PSAD-S performs best at $\alpha = 0.5$ and 0.2 on the respective datasets. As α controls the trade-off between supervision from ground-truth labels and distributional guidance from the generator, suboptimal values disrupts this balance and leads to worse results.

5.5 In-depth Analysis

Analysis of Distillation Methods. To validate the necessity of employing a semi-autoregressive generator for online distillation of the score network, we compare the effects of different teacher models, including autoregressive(AR), non-autoregressive(NAR), and semi-autoregressive(SAR), as well as offline distillation, on the final performance of the student model. According to the results in Table 5, we draw the following conclusions: (1) Using a semi-autoregressive model as the teacher achieves the best distillation results while requiring significantly less training time compared with a fully autoregressive model, confirming the rationale for adopting a semi-autoregressive generator as the teacher. (2) The traditional two-stage offline distillation takes much longer than our online approach. More importantly, online distillation enables the student to learn from a continuously improving teacher rather than static knowledge, leading to comparable or even superior performance of the student model.

Analysis of User–Item Interactions. To evaluate the personalized modeling capability of the UPN module, we define the top 20% of users based on interaction sequence length as high-activity users and the remaining users as low-activity users. We compare the performance of four interaction strategies: no interaction, concatenation [20], late fusion [29], and UPN, across two datasets. As shown in Fig. 4, results indicate that UPN consistently outperforms other interaction strategies, achieving particularly pronounced improvements for high-activity users. With sufficient personalized signals, UPN's multidimensional interaction more effectively capture user characteristics and translate them into notable improvements in ranking performance.

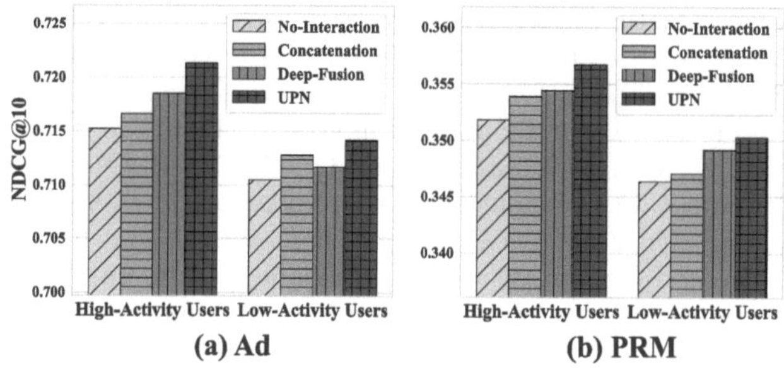

Fig. 4. Performance comparison of item–user feature interaction mechanisms across user sequences with varying activity levels.

6 Conclusion

We proposed PSAD, a reranking framework that combined a high-quality semi-autoregressive teacher model PSAD-G with a block-wise generator and user profile network and a lightweight student PSAD-S trained via online knowledge distillation. PSAD-G demonstrated superior ranking effectiveness across three large-scale datasets, while PSAD-S achieved much lower latency with comparable accuracy, closing the gap between generation quality and inference efficiency. Future work will improve the distillation to enhance transfer and generalization.

Acknowledgments. This work was supported by the National Natural Science Foundation of China (Nos. U23A20319, 62441239, 62472394, and 62441227), as well as the Anhui Province Science and Technology Inno vation Project (Nos. 202423k09020010 and 202423k09020011).

Disclosure of Interests. The authors have no competing interests to declare that are relevant to the content of this article.

References

1. Ai, Q., Bi, K., Guo, J., Croft, W.B.: Learning a deep listwise context model for ranking refinement. In: The 41st international ACM SIGIR Conference on Research & Development in Information Retrieval, pp. 135–144 (2018)
2. Ai, Q., Wang, X., Bruch, S., Golbandi, N., Bendersky, M., Najork, M.: Learning groupwise multivariate scoring functions using deep neural networks. In: Proceedings of the 2019 ACM SIGIR International Conference on Theory of Information Retrieval, pp. 85–92 (2019)

3. Bello, I., et al.: Seq2slate: re-ranking and slate optimization with RNNs. arXiv preprint arXiv:1810.02019 (2018)
4. Carraro, D., Bridge, D.: Enhancing recommendation diversity by re-ranking with large language models. ACM Trans. Recommender Syst. (2024)
5. Deng, J., et al.: OneRec: unifying retrieve and rank with generative recommender and iterative preference alignment. arXiv preprint arXiv:2502.18965 (2025)
6. Feng, Y., Gong, Y., Sun, F., Ge, J., Ou, W.: Revisit recommender system in the permutation prospective. arXiv preprint arXiv:2102.12057 (2021)
7. Feng, Y., Hu, B., Gong, Y., Sun, F., Liu, Q., Ou, W.: GRN: generative rerank network for context-wise recommendation. arXiv preprint arXiv:2104.00860 (2021)
8. Gong, X., Feng, Q., Zhang, Y., Qin, J., Ding, W., Li, B., Jiang, P., Gai, K.: Real-time short video recommendation on mobile devices. In: Proceedings of the 31st ACM international conference on information & knowledge management. pp. 3103–3112 (2022)
9. Han, F., Wang, S., Zhao, J., Wu, R., Rui, X., Wang, Z.: Fair re-ranking recommendation based on debiased multi-graph representations. In: International Conference on Advanced Data Mining and Applications, pp. 168–182. Springer, Cham (2023). https://doi.org/10.1007/978-3-031-46661-8_12
10. Hron, J., Krauth, K., Jordan, M., Kilbertus, N.: On component interactions in two-stage recommender systems. Adv. Neural. Inf. Process. Syst. **34**, 2744–2757 (2021)
11. Huzhang, G., et al.: Aliexpress learning-to-rank: maximizing online model performance without going online. IEEE Trans. Knowl. Data Eng. **35**(2), 1214–1226 (2021)
12. Järvelin, K., Kekäläinen, J.: Cumulated gain-based evaluation of IR techniques. ACM Trans. Inf. Syst. (TOIS) **20**(4), 422–446 (2002)
13. Li, Y., et al.: Pear: personalized re-ranking with contextualized transformer for recommendation. In: Companion Proceedings of the Web Conference 2022, pp. 62–66 (2022)
14. Lin, X., et al.: Discrete conditional diffusion for reranking in recommendation. In: Companion Proceedings of the ACM on Web Conference 2024, pp. 161–169 (2024)
15. Lin, Z., et al.: Attention over self-attention: intention-aware re-ranking with dynamic transformer encoders for recommendation. IEEE Trans. Knowl. Data Eng. **35**(8), 7782–7795 (2022)
16. Liu, S., et al.: Generative flow network for listwise recommendation. In: Proceedings of the 29th ACM SIGKDD Conference on Knowledge Discovery and Data Mining, pp. 1524–1534 (2023)
17. Liu, W., Liu, Q., Tang, R., Chen, J., He, X., Heng, P.A.: Personalized re-ranking with item relationships for e-commerce. In: Proceedings of the 29th ACM International Conference on Information & Knowledge Management, pp. 925–934 (2020)
18. Liu, W., et al.: Neural re-ranking in multi-stage recommender systems: a review. arXiv preprint arXiv:2202.06602 (2022)
19. Pang, L., Xu, J., Ai, Q., Lan, Y., Cheng, X., Wen, J.: Setrank: learning a permutation-invariant ranking model for information retrieval. In: Proceedings of the 43rd International ACM SIGIR Conference on Research and Development in Information Retrieval, pp. 499–508 (2020)
20. Pei, C., et al.: Personalized re-ranking for recommendation. In: Proceedings of the 13th ACM Conference on Recommender Systems, pp. 3–11 (2019)
21. Ren, Y., Yang, Q., Wu, Y., Xu, W., Wang, Y., Zhang, Z.: Non-autoregressive generative models for reranking recommendation. In: Proceedings of the 30th ACM

SIGKDD Conference on Knowledge Discovery and Data Mining, pp. 5625–5634 (2024)
22. Shen, T., et al.: P-Law: predicting quantitative scaling law with entropy guidance in large recommendation models. In: The Thirty-ninth Annual Conference on Neural Information Processing Systems
23. Shi, X., et al.: Pier: permutation-level interest-based end-to-end re-ranking framework in e-commerce. In: Proceedings of the 29th ACM SIGKDD Conference on Knowledge Discovery and Data Mining, pp. 4823–4831 (2023)
24. Stern, M., Shazeer, N., Uszkoreit, J.: Blockwise parallel decoding for deep autoregressive models. In: Advances in Neural Information Processing Systems, vol. 31 (2018)
25. Wang, C., Zhang, J., Chen, H.: Semi-autoregressive neural machine translation. arXiv preprint arXiv:1808.08583 (2018)
26. Wang, S., et al.: NLGR: utilizing neighbor lists for generative rerank in personalized recommendation systems. In: Companion Proceedings of the ACM on Web Conference 2025, pp. 530–537 (2025)
27. Weng, M., et al.: Beyond positive history: re-ranking with list-level hybrid feedback. arXiv preprint arXiv:2410.20778 (2024)
28. Xi, Y., et al.: Context-aware reranking with utility maximization for recommendation. arXiv preprint arXiv:2110.09059 (2021)
29. Xi, Y., et al.: Multi-level interaction reranking with user behavior history. In: Proceedings of the 45th International ACM SIGIR Conference on Research and Development in Information Retrieval, pp. 1336–1346 (2022)
30. Xia, H., Ge, T., Wang, P., Chen, S.Q., Wei, F., Sui, Z.: Speculative decoding: exploiting speculative execution for accelerating seq2seq generation. arXiv preprint arXiv:2203.16487 (2022)
31. Yin, M., et al.: Dataset regeneration for sequential recommendation. In: Proceedings of the 30th ACM SIGKDD Conference on Knowledge Discovery and Data Mining, pp. 3954–3965 (2024)
32. Yu, H., et al.: Thought-augmented planning for LLM-powered interactive recommender agent. arXiv preprint arXiv:2506.23485 (2025)
33. Zhang, J., et al.: Td3: tucker decomposition based dataset distillation method for sequential recommendation. In: Proceedings of the ACM on Web Conference 2025, pp. 3994–4003 (2025)
34. Zhang, L., et al.: Killing two birds with one stone: unifying retrieval and ranking with a single generative recommendation model. In: Proceedings of the 48th International ACM SIGIR Conference on Research and Development in Information Retrieval, pp. 2224–2234 (2025)
35. Zhuang, T., Ou, W., Wang, Z.: Globally optimized mutual influence aware ranking in e-commerce search. arXiv preprint arXiv:1805.08524 (2018)

GBCG: Granular Ball and Counterfactual Guided Profile Injection Attack in Recommender Systems

Yunmeng Zhao[1], Yuran He[1], Shenbao Yu[1], Ruihong Huang[1], Jiayin Li[1], Jun Shen[2], and Jiayin Lin[1(✉)]

[1] College of Computer and Cyber Security, Fujian Normal University, Fuzhou 350117, China
{yushenbao,ruihong,jy.lin}@fjnu.edu.cn
[2] University of Wollongong, Wollongong, Australia
jshen@uow.edu.au

Abstract. Recommender systems are vulnerable to shilling attacks, where adversaries inject fabricated user profiles to manipulate recommendation outcomes. Existing attacks often exhibit limited effectiveness under resource constraints, weak transferability, and poor stealthiness. This paper introduces GBCG (Granular Ball and Counterfactual Guided) attack, a generative shilling framework that achieves efficient and stealthy attacks in black-box and budget-limited settings. GBCG integrates three interdependent modules forming a unified attack pipeline. Granular-ball computing partitions the item embedding space into multi-granular clusters, providing structural priors that refine the candidate space. Counterfactual gain estimation, built upon these priors, evaluates each candidate's marginal contribution to target exposure through sampled user feedback and guides the selection of high-impact items. A WGAN-GP-based generator, conditioned on the selected items, learns real users' rating distributions to synthesize realistic fake profiles. Experiments on multiple public datasets and representative recommender systems show that GBCG consistently outperforms mainstream baselines in effectiveness, generalization, and stealthiness. Ablation and visualization analyses further confirm its validity, establishing GBCG as a robust benchmark for evaluating the security of recommender systems. Our code is available at: https://github.com/KREML0/GBCG.

Keywords: Recommender systems · Shilling attacks · Granular-ball computing · Counterfactual reasoning · Generative adversarial networks

1 Introduction

Recommender systems (RSs) play a central role as information filters in the digital ecosystem [6,12]. However, their personalization capability also makes them

H. Jung et al. (Eds.): DASFAA 2026, LNCS 16535, pp. 88–103, 2026.
https://doi.org/10.1007/978-981-92-0363-5_6

vulnerable to manipulation [20]. Recent studies show that RSs are highly susceptible to **shilling attacks** or **profile injection attacks** [8,14], where adversaries inject fabricated user profiles to distort recommendation outcomes without modifying the model architecture.

A shilling attack aims to alter the ranking or predicted ratings of target items under limited injection budgets [22]. In black-box scenarios, attackers can promote targets by fabricating plausible user histories without access to model parameters or gradients [13]. Despite growing research, existing attacks face three persistent challenges:

1. **Limited attack capability:** This is related to the difficulty in significantly improving target-item rankings under realistic constraints such as few injected users or ratings.
2. **Poor generalization:** Most attacks are tailored to a type of the single RS (e.g., graph-based [15] or contrastive-learning models [21]) and transfer poorly across systems.
3. **Low stealthiness:** Generated fake users often deviate sharply from real-user distributions, making them easily detected.

These limitations reflect a fundamental trade-off among effectiveness, generalization, and stealthness. Strengthening one aspect often sacrifices another. This motivates the research question: can the attack generation process be redesigned to ensure both structural coherence and behavioral plausibility under resource constraints?

To address this, we herein propose **GBCG** (Granular Ball and Counterfactual Guided) attack, an unsupervised framework that integrates **granular-ball computing** and **counterfactual guidance**. Granular-ball computing adaptively partitions the item embedding space into multi-scale clusters without predefined cluster counts, providing robust structural priors. Counterfactual guidance estimates item-level promotion impact by sampling real users in black-box settings. Combined with a WGAN-GPbased generator [2], GBCG synthesizes realistic, semantically coherent fake profiles that jointly enhance effectiveness and stealth.

GBCG is driven by two key insights: *(1) user preferences in embedding space exhibit hierarchical and locally stable structures, motivating adaptive granular-ball clustering; and (2) counterfactual gain estimation efficiently guides attack optimization without model access. The resulting framework is end-to-end, scalable, and model-agnostic.*

The main contributions are as follows:

1. Introducing granular-ball computing into shilling attacks to construct adaptive, noise-tolerant structural priors for low-complexity, model-agnostic attack generation.
2. Proposing a counterfactual gain mechanism that samples real users to estimate item-level promotion impact, combined with a WGAN-GP generator to synthesize realistic fake profiles.

3. Demonstrating through extensive experiments that GBCG consistently outperforms baselines in effectiveness, transferability, and stealth; ablation and visualization analyses further validate its core components.

2 Problem Formulation and Research Motivation

This section presents the fundamental terminology and problem definition of shilling attacks, followed by a discussion of the motivation behind the research.

2.1 Terminology

Attack Objective. A shilling attack aims to systematically manipulate the outputs of a recommender system by injecting fabricated user profiles. Depending on intent, attacks are typically categorized into **push attacks** (promoting target items) and **nuke attacks** (demoting them). This paper focuses on push attacks. Let $\mathcal{U}$ and $\mathcal{I}$ denote the sets of real users and items, respectively, and let $f(\cdot)$ denote the black-box recommender system. The attacker selects a subset of target items $\mathcal{T} \subseteq \mathcal{I}$ to be promoted by injecting fake-user interactions R_F into the system.

Attack Cost. The cost of an attack consists of two components: (1) the number of injected fake users, denoted by $|\mathcal{F}| = F$, bounded by a maximum budget $F_{\max}$; and (2) the number of feedback interactions per fake user, denoted by $\|r_f\|_0 = m$. The overall resource constraints are as shown in Eq. (1).

$$|\mathcal{F}| \leq F_{\max}, \qquad \|r_f\|_0 \leq m \tag{1}$$

Attacker's Prior Knowledge. We consider a black-box setting, in which the attacker could not access the internal parameters or gradients of the victim model, but might possess limited side information such as partial user histories or item embeddings derived from a surrogate model. Assuming access to public data to train a surrogate model (e.g., LightGCN [3]) is a standard practice in black-box attacks [11,22], representing a realistic scenario where attackers utilize public knowledge. In our experiments, LightGCN serves as the surrogate model to extract item representations and assist in estimating counterfactual gains.

Fake Profile Construction. A fake user profile typically consists of four categories of items:

1. **Target items** ($\mathcal{T}$): items to be promoted, assigned with maximal feedback (e.g., rating 5 or binary 1);
2. **Selected items** ($\mathcal{S}$): semantically or behaviorally related items chosen to enhance the attack's effectiveness;
3. **Filler items:** additional rated items that make the generated one more like the real profile, possibly overlapping with $\mathcal{S}$;

4. **Unrated items:** items left unscored to maintain realistic sparsity.

Each fake user $f \in \mathcal{F}$ is represented by a sparse interaction vector $r_f \in \{0,1\}^{|\mathcal{I}|}$. The binary entry $\mathbf{r}_f(i)$ equals 1 if the item i belongs to the target set $\mathcal{T}$.

2.2 Problem Definition

Let the real user-item interaction matrix be $R \in \{0,1\}^{|\mathcal{U}|\times|\mathcal{I}|}$, where $R_{u,i} = 1$ indicates that user u has given positive feedback on item i. The attacker injects fake-user interactions $R_F \in \{0,1\}^{|\mathcal{F}|\times|\mathcal{I}|}$, forming a new interaction matrix, as shown in Eq. (2).

$$R' = R \oplus R_F \tag{2}$$

where $\oplus$ denotes concatenation of fake-user profiles with real ones.

The recommender system retrains or re-evaluates on R' and generates updated predictions $\hat{R}' = f(R')$, while the pre-attack predictions are $\hat{R} = f(R)$. The attacker aims to maximize the aggregate score improvement of target items for all real users, as given in Eq. (3).

$$\max_{R_F} \mathcal{J}(R_F) = \sum_{u\in\mathcal{U}} \sum_{t\in\mathcal{T}} \left(f(R')_{u,t} - f(R)_{u,t}\right) \tag{3}$$

subject to the resource constraints in Eq. (1) and the profile definition in Eq. (2).

2.3 Motivation

Constructing effective and realistic fake profiles under black-box, budget-limited conditions is nontrivial. GBCG is guided by two complementary observations that enable a structured, robust, and interpretable attack design.

Observation 1. User preferences in embedding space form hierarchical, locally coherent clusters [33]. Global partitioning (e.g., K-means [10]) could be unstable on complex, noisy distributions. Granular-ball computing adaptively splits and merges spherical regions, yielding multi-scale, noise-tolerant clusters that act as reliable structural priors for candidate selection [28].

Observation 2. Even in black-box settings, the marginal effect of adding an interaction can be approximated by sampling a small set of users and comparing predictions before and after perturbation [19]. This counterfactual gain might provide an efficient, gradient-free signal to prioritize high-utility items under strict budgets.

Combining these ideas, a WGAN-GP [2] generator models cluster-level rating distributions, producing profiles that fuse local semantic coherence with high-gain selections. GBCG therefore aims to (1) exploit multi-granular structural priors, (2) use counterfactual gains for efficient selection, and (3) apply generative modeling for realistic, transferable fake profiles.

3 Methodology

This section presents GBCG, a shilling attack framework designed to operate effectively under black-box access and resource constraints.

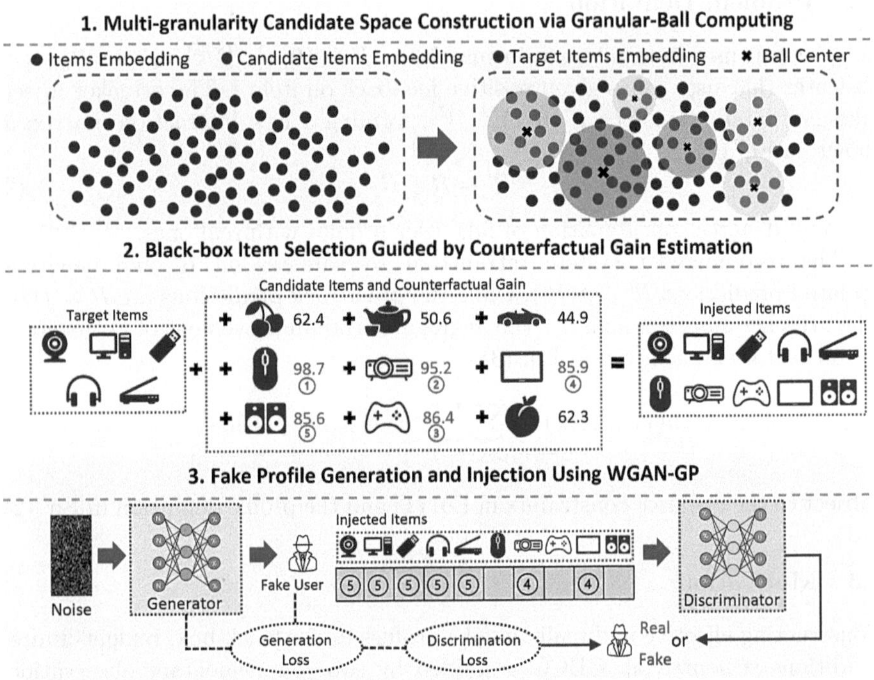

Fig. 1. Overview of the proposed GBCG framework. The pipeline integrates multi-granularity candidate-space construction via granular-ball computing, counterfactual-guided item selection under black-box constraints, and WGAN-GP based fake-profile generation and injection to realize effective and covert push attacks.

3.1 Framework Overview

As illustrated in Fig. 1, our framework comprises three main components:

1. **Multi-granularity candidate-space construction via granular-ball computing.** We apply adaptive multi-granularity clustering over the entire item embedding space to obtain semantic clusters (granular balls). Balls containing target items and their nearest neighbors are merged to form a compact, semantically coherent candidate set, thus reducing the search space and improving stability.
2. **Black-box item selection guided by counterfactual gain.** From the candidate set, a small batch of real users is used to probe the black-box

recommender. We iteratively estimate the counterfactual gain of each candidate subset, which measures the change in target scores before and after the hypothetical injection, and greedily select the items that provide the highest marginal improvement while applying a minor regularization term to preserve semantic coherence. The process continues until the resource budget is reached, forming a target-augmented item set.

3. **Fake-profile generation via WGAN-GP.** Conditioned on the selected items, a WGAN-GP model learns the empirical interaction distribution of real users sampled from related balls. The generator produces continuous score vectors that are binarized through a top-m projection while ensuring target inclusion. The resulting fake-user matrix is concatenated with the original data to evaluate or retrain the victim recommender.

The implementation details of each module are presented in the following subsections.

3.2 Multi-granularity Candidate Space Construction via Granular-Ball Computing

Item embeddings $E \in \mathbb{R}^{|I| \times d}$ are assumed to be available from a substitute model (e.g., LightGCN [3]). Granular-ball clustering is applied to E to discover unsupervised, adaptive, and multi-scale semantic structures. Unlike K-means, granular-ball clustering does not require a pre-specified number of clusters and proceeds by recursively splitting along the dimension of maximal variance to form stable spherical regions, thereby adapting to non-spherical and density-heterogeneous distributions.

Each granular ball B_j is represented by its index set $\text{idxs}(B_j)$, center $\mathbf{c}_{B_j}$, maximum radius $r_{B_j}^{\max}$, and average radius $r_{B_j}^{\text{avg}}$, as formulated in Eq. (4).

$$\mathbf{c}_{B_j} = \frac{1}{|\text{idxs}(B_j)|} \sum_{i \in \text{idxs}(B_j)} E_i, \quad r_{B_j}^{\text{avg}} = \frac{1}{|\text{idxs}(B_j)|} \sum_{i \in \text{idxs}(B_j)} \|E_i - \mathbf{c}_{B_j}\|_2 \tag{4}$$

Splitting is terminated and the ball is retained as a leaf node when $|\text{idxs}(B_j)| \leq n_{\min}$ or $r_{B_j}^{\text{avg}} \leq \tau$, where τ is set adaptively as a proportion of the global average radius. The resulting hierarchical partition is robust to noise thus it can provide a stable structural prior for the attack.

A candidate set is constructed from balls that are relevant to the target set. First, balls that contain any target item are identified as Eq. (5).

$$\mathcal{B}_T = \{B \in \mathcal{B} \mid \text{idxs}(B) \cap \mathcal{T} \neq \emptyset\} \tag{5}$$

To capture local semantic proximity, cosine similarity between ball centers is computed and the top-k nearest neighbor balls $\text{Nbr}(B)$ are selected for each ball. The related ball set and candidate item set are then defined as Eq. (6).

$$\mathcal{B}_{\text{rel}} = \bigcup_{B \in \mathcal{B}_T} (\{B\} \cup \text{Nbr}(B)), \qquad \mathcal{C} = \Big(\bigcup_{B \in \mathcal{B}_{\text{rel}}} \text{idxs}(B)\Big) \setminus \mathcal{T} \tag{6}$$

The constructed candidate set $\mathcal{C}$ thus attains local semantic coherence and global multi-granularity coverage, reducing search complexity while improving transferability and stability of the attack.

3.3 Counterfactual-Guided Item Selection Under Black-Box Constraints

Given the candidate set $\mathcal{C}$, a counterfactual gain estimator is used to approximate the marginal contribution of adding a candidate subset S to user profiles under black-box constraints. A small batch of real users $U_s \subseteq \mathcal{U}$ is sampled and the recommender's predictions on R are used as a baseline. The baseline score is defined as Eq. (7).

$$s_{\text{base}}(U_s) = \sum_{u \in U_s} \sum_{t \in \mathcal{T}} f(R)_{u,t} \tag{7}$$

After injecting S (denoted by $\mathbf{1}_S$) into these users, the counterfactual score is given in Eq. (8).

$$s_{\text{cf}}(U_s, S) = \sum_{u \in U_s} \sum_{t \in \mathcal{T}} f(R \oplus \mathbf{1}_S)_{u,t} \tag{8}$$

The counterfactual gain is defined as shown in Eq. (9).

$$\text{Gain}(S) = s_{\text{cf}}(U_s, S) - s_{\text{base}}(U_s) \tag{9}$$

To encourage semantic coherence with granular balls, a small similarity term is appended, as shown in Eq. (10).

$$\text{Gain}_{\text{GBC}}(S) = \text{Gain}(S) + \epsilon \cdot \text{Sim}_{\text{ball}}(S) \tag{10}$$

where $\text{Sim}_{\text{ball}}(S)$ denotes the average cosine similarity between items in S and the center of target-related balls, and ϵ is a small coefficient. This term is intended only to break ties and does not alter the dominant ordering induced by $\text{Gain}(S)$. A greedy selection is performed starting from $S = \mathcal{T}$ until $|S| = m$. At each iteration, a probe set $\mathcal{P} \subseteq \mathcal{C}$ is sampled and the following update is performed as Eq. (11).

$$i^* = \arg\max_{i \in \mathcal{P}} \text{Gain}_{\text{GBC}}(S \cup \{i\}), \quad S \leftarrow S \cup \{i^*\}, \quad \mathcal{C} \leftarrow \mathcal{C} \setminus \{i^*\} \tag{11}$$

This strategy balances computational efficiency and approximate optimality, yielding a structured, high-gain selection set S.

3.4 WGAN-GP-Based Profile Generation and Injection

Once S is determined, a WGAN-GP generator G and critic D are trained on the S-dimensional subspace to model the distribution of real users' ratings within the related balls. The generator $G : \mathbb{R}^{|S|} \to \mathbb{R}^{|S|}$ maps Gaussian noise $z \sim \mathcal{N}(0, I)$ to continuous scores $\tilde{x}$, which are scaled to the rating range (e.g., $[0, 5]$). The critic $D : \mathbb{R}^{|S|} \to \mathbb{R}$ outputs real-valued scores.

The standard WGAN-GP losses are employed. The critic loss is given by Eq. (12).

$$\mathcal{L}_D = \mathbb{E}_{\tilde{x}\sim p_g}[D(\tilde{x})] - \mathbb{E}_{x\sim p_r}[D(x)] + \lambda\,\mathbb{E}_{\hat{x}\sim p_{\hat{x}}}\left(\|\nabla_{\hat{x}}D(\hat{x})\|_2 - 1\right)^2 \tag{12}$$

where p_r denotes the empirical distribution of real users' S-dimensional interactions sampled from related balls, p_g denotes the generator distribution, and λ is used for the gradient penalty. The generator loss is defined as Eq. (13).

$$\mathcal{L}_G = -\mathbb{E}_{\tilde{\mathbf{x}}\sim p_g}[D(\tilde{\mathbf{x}})] \tag{13}$$

where $D(\cdot)$ is the discriminator that estimates the Wasserstein distance between the real and generated interaction distributions.

Training is performed with multiple critic updates per generator update to improve stability. After training, $|F|$ samples $\tilde{\mathbf{x}} \in \mathbb{R}^{|S|}$ are drawn from the generator G. Each continuous sample is converted into a binary interaction vector via a top-m projection while ensuring that all target items T are included, as shown in Eq. (14).

$$r_f(i) = \begin{cases} 1, & i \in \text{top-}m(\tilde{x}) \cup T \\ 0, & \text{otherwise} \end{cases} \tag{14}$$

where $m = \|r_f\|_0$ denotes the permitted number of interactions per fake user. The assembled fake-user matrix $R_F \in \{0,1\}^{|F|\times|I|}$ is concatenated with the original matrix R to form the attacked dataset R'. This attacked dataset is then used to retrain or evaluate the victim recommender, thereby realizing the promotion of the target items.

4 Experiments

In this section, we conduct extensive experiments to validate the effectiveness and stealthiness of the proposed attack. Our evaluation measures overall attack performance across diverse recommender architectures, user-level coverage and ranking shifts, the contribution of individual modules via ablation, and the detectability of injected profiles under unsupervised detectors.

4.1 Experimental Setup

Datasets. To evaluate our method, we conducted experiments on three widely used public datasets: **MovieLens-100K**[1], **Yelp**[2], and **Amazon Automotive**[3]. Following the common practice in recommender system research, we binarized the rating data by retaining only entries with rating values ≥ 4, treating them as positive feedback [4]. This setting aligns with implicit feedback scenarios, reduces noise from ambiguous ratings, and ensures comparability with prior studies. The datasets were then randomly split into training, validation, and test sets with a ratio of 8:1:1.

[1] https://grouplens.org/datasets/movielens/100k/.
[2] https://www.kaggle.com/c/yelp-recruiting/data.
[3] https://cseweb.ucsd.edu/~jmcauley/datasets.html#amazon_reviews.

Attack Baselines. To comprehensively assess our method, we compare it with eight representative attack baselines categorized by their knowledge assumptions. Black-box attacks include RandomAttack [8], BandwagonAttack [13], GSPAttack [15], PoisonRec [18], and CEFP [33]. Gray-box attacks comprise AUSH [11] and GOAT [22], which utilize GAN-based profile generation. White-box attack CLeaR [21] targets contrastive learning recommenders but it can also operate in black-box settings. This diverse set ensures a fair and comprehensive evaluation of attack performance.

Victim Recommender Systems. We evaluate GBCG against five representative victim models: GMF [7], LightGCN [3], SGL [23], SSL4Rec [30], and XSimGCL [31]. These models span matrix factorization, graph-based, and self-supervised paradigms, providing a comprehensive benchmark for evaluating attack performance.

Evaluation Metrics. To assess the practical impact of injection attacks on recommender systems, we adopt **Hit Rate (HR@K)** and **Normalized Discounted Cumulative Gain (NDCG@K)** as our primary evaluation metrics [20]. These metrics measure the extent to which the target item's ranking improves after the attack, thus reflecting attack effectiveness from multiple perspectives. Since users typically focus on the top positions of recommendation lists, we set $K = 3$ to better approximate real-world usage scenarios. Specifically, HR@K measures whether the target item appears in the user's TOP-K recommendation list. A hit is counted if the target item appears in a user's TOP-K list, and HR@K is the average hit rate over all users. In contrast, NDCG@K not only considers whether the target item is recommended but also accounts for its position in the TOP-K list, with higher ranks yielding larger gains. The combination of these two metrics provides a comprehensive evaluation of attack performance, where HR@K reflects coverage and NDCG@K emphasizes ranking quality and significance.

Parameter Settings. To highlight both the effectiveness and reproducibility, we adopt the following parameter settings and strategies in our experiments. The target items are five unpopular items. All items are sorted in ascending order based on the number of interactions. The subset consisting of the lowest quartile of the interaction counts is selected and five target items are randomly chosen from it. To ensure effective partitioning of counterfactual candidate sets and granular-ball clusters, the cosine similarity threshold between items is set to $\alpha = 0.65$. By default, each attack injects fake users equivalent to 1% of the total user population, and each fake user interacts with as many items as the average number of interactions per real user. For model training, both generator and discriminator are based on WGAN-GP, optimized with a bi-level training loop: multiple critic updates followed by one generator update in each outer iteration. We use Adam optimizers with learning rates lr_G and lr_D, both set to

Table 1. Overall Attack Performance (Hit Rate@3 $\times 10^{-4}$). The best results are highlighted in bold.

Methods	ML-100K					Yelp					Automotive				
	GMF	Light	SSL4	XSim	SGL	GMF	Light	SSL4	XSim	SGL	GMF	Light	SSL4	XSim	SGL
GBCG	**8.0**	**7.2**	**42.1**	**20.6**	**3.8**	**3.7**	**1.3**	**4.3**	**3.2**	**3.3**	**21.4**	**19.6**	**21.1**	**22.6**	**20.6**
None	0.0	0.0	8.5	0.0	0.0	1.0	0.0	4.0	2.0	1.0	3.4	2.0	9.6	2.0	6.1
Random	3.8	4.2	32.0	10.1	2.5	2.0	0.1	3.4	2.2	2.2	20.0	17.0	19.3	20.8	18.5
Bandwagon	2.5	1.3	27.8	15.1	2.1	3.4	0.7	2.9	2.4	1.9	20.3	14.6	17.6	21.8	18.1
PoisonRec	2.9	2.1	22.7	12.6	1.7	2.9	0.4	3.2	2.7	2.7	20.2	14.7	20.0	18.1	16.5
GSP	3.8	4.2	31.6	18.1	2.5	3.2	0.4	2.7	2.4	2.2	19.2	14.5	18.8	18.7	18.0
AUSH	5.1	1.3	27.8	11.8	2.5	1.9	0.7	3.3	2.0	2.3	20.2	16.2	17.9	19.2	19.2
GOAT	4.2	3.8	28.6	12.2	2.9	2.0	0.1	3.0	1.3	1.4	20.7	15.2	18.0	18.4	19.2
CEFP	4.2	0.8	20.2	9.3	2.5	2.4	0.6	2.2	2.2	1.9	19.1	15.7	19.3	18.7	18.3
CLeaR	5.5	4.2	31.1	10.9	2.5	1.9	0.0	3.3	2.6	2.2	17.9	15.3	17.2	20.8	18.5

Table 2. Overall Attack Performance (NDCG@3 $\times 10^{-4}$). The best results are highlighted in bold.

Methods	ML-100K					Yelp					Automotive				
	GMF	Light	SSL4	XSim	SGL	GMF	Light	SSL4	XSim	SGL	GMF	Light	SSL4	XSim	SGL
GBCG	**12.3**	**11.7**	**68.2**	**31.7**	**5.3**	**6.0**	**1.7**	**7.4**	**4.6**	**5.3**	**38.6**	**32.5**	**34.3**	**40.4**	**37.3**
None	0.0	0.0	13.1	0.0	0.0	1.0	0.0	7.0	4.0	2.0	6.4	2.8	20.5	3.0	9.1
Random	6.8	5.8	53.7	16.2	4.2	3.5	0.2	6.3	3.0	4.0	35.6	28.8	31.9	37.4	33.9
Bandwagon	4.2	1.7	46.6	22.9	3.2	5.7	0.9	5.5	3.6	3.0	37.1	25.8	28.3	39.9	33.9
PoisonRec	4.6	3.6	37.7	20.8	3.1	4.1	0.5	5.7	3.9	4.5	35.0	25.9	33.9	32.8	29.8
GSP	4.6	6.3	50.2	29.7	3.7	4.6	0.5	4.8	3.8	3.4	33.5	24.1	30.0	36.6	33.2
AUSH	7.8	1.5	45.7	18.3	3.7	2.5	1.0	6.2	2.9	4.0	36.7	27.1	29.6	34.7	35.2
GOAT	6.4	5.6	47.8	19.2	5.0	3.3	0.2	5.2	1.6	2.5	37.5	26.4	30.4	34.0	32.7
CEFP	6.8	1.1	32.9	14.3	3.7	4.1	1.1	4.1	3.0	2.8	34.0	26.1	31.6	34.5	33.4
CLeaR	8.8	6.5	51.4	17.5	4.1	3.4	0.0	5.8	3.6	3.5	33.7	26.2	27.9	36.9	33.5

a default of $1e^{-4}$. The generator architecture adopts an adaptive hidden width, with a minimum size of 64. As the output dimension k increases, the hidden width grows incrementally, adding four neurons for every 16 additional output units.

4.2 Overall Attack Performance

We first evaluate the overall effectiveness of GBCG under strict resource constraints across multiple recommender systems. As detailed in the parameter settings, the attack budget is highly limited, where only 1% of the total user population can be injected, and each fake user interacts with roughly the average number of items per real user. This constrained configuration highlights the challenge of performing effective attacks with minimal resources. The experimental results are summarized in Tables 1 and 2.

We observe that GBCG achieves consistently effective attacks across all three datasets. For example, in ML-100K, there are the target items that originally never appeared in the recommendation lists of models such as GMF and LightGCN (HR@3 = 0), but after our attack, they successfully entered users' top recommendations, with GBCG yielding the highest improvement in hit rate. Moreover, in the Yelp dataset, some baseline attacks even reduced the hit rate and NDCG of target items when attacking SSL4Rec and XSimGCL, whereas GBCG still improved the ranking quality. These results confirm the robustness of GBCG across heterogeneous recommender systems. The improvement largely stems from granular-ball clustering of target items and counterfactual gain, which enable originally unpopular items not only to appear in recommendation lists but also to achieve competitive ranking quality.

4.3 Attack Coverage Analysis

To further examine how the attack affects individual users, we also analyze coverage and ranking shifts. For clarity, we report both the mean and median ranking changes across all users. As shown in Fig. 2, taking the attack on the XSimGCL model with the ML-100K dataset as an example, we define ΔRank as the post-attack rank minus the pre-attack rank of the target item. A more negative ΔRank indicates a greater promotion in its ranking position. The results show that GBCG consistently promotes target items upward in most users' recommendation lists.

Other attack methods often produce higher average or median ranks, indicating that they affect only a small subset of users—yielding strong local promotion but limited global impact, and thus smaller gains in hit rate. GBCG, on the other

Table 3. Ablation study on the effectiveness of granular-ball clustering and counterfactual gain. The best results are highlighted in bold.

Victim RS	Metric	GBCG	WoGB	WoCF	K-means	DBSCAN	OPTICS
GMF	HR@3	**8.0**	3.8	4.6	2.5	3.4	4.6
	NDCG@3	**12.3**	5.8	6.8	3.2	5.3	7.6
LightGCN	HR@3	**7.2**	2.5	1.3	0.0	2.1	4.2
	NDCG@3	**11.7**	4.1	1.9	0.0	2.9	4.9
SSL4Rec	HR@3	**42.1**	25.2	37.4	27.8	34.5	26.1
	NDCG@3	**68.2**	44.0	63.6	47.4	58.0	43.7
XSimGCL	HR@3	**20.6**	16.0	16.0	20.2	16.8	15.1
	NDCG@3	**31.7**	23.9	25.2	30.5	26.1	21.9
SGL	HR@3	**3.8**	3.4	2.5	3.4	1.7	2.5
	NDCG@3	**5.3**	5.2	3.8	4.7	2.1	3.6

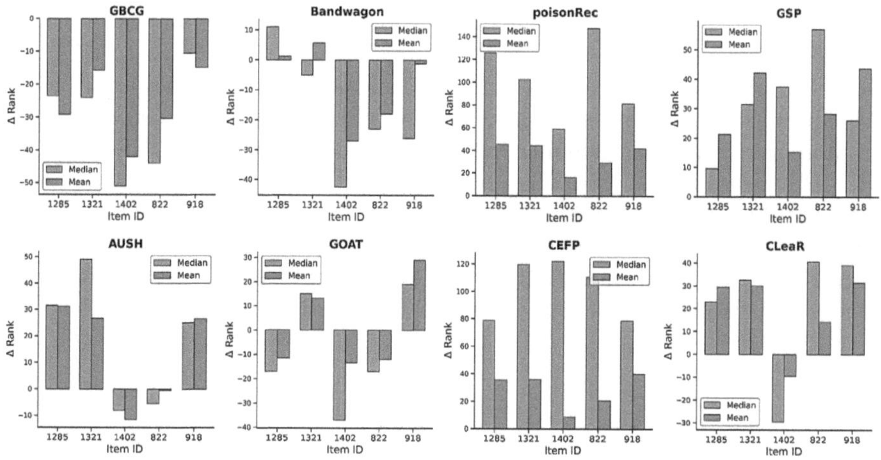

Fig. 2. Changes in target item rankings before and after the attack. Taking XSimGCL on the ML-100K dataset as an example, this Figure illustrates the distribution of ranking shifts (ΔRank) across users, where ΔRank is defined as the difference between the post-attack and pre-attack ranks of the target item. A more negative value indicates a stronger promotion effect. The green bars represent the median change in ranking, while the orange bars denote the mean change.

hand, benefits from its adaptive design: granular-ball clustering captures multi-granularity useritem semantics, while counterfactual-guided selection allocates limited perturbations to the most influential regions. This adaptivity enables broader user coverage and more stable ranking promotion, leading to superior hit rate and overall effectiveness.

4.4 Ablation Analysis

We next present the ablation study to quantify the contribution of each component in GBCG. Specifically, we remove the *granular-ball clustering* module and the *counterfactual gain estimation* module to assess their individual impact. In addition, since this work is the first to introduce granular-ball clustering into poisoning attacks, we further compare it with alternative clustering methods to validate its effectiveness. The ablation results are summarized in Table 3, where we report experiments on the ML-100K dataset for brevity. Overall, the complete GBCG framework consistently achieves the best performance across all recommender systems and evaluation metrics, confirming the effectiveness of the proposed design.

The ablation analysis further reveals the role of each component: removing granular-ball clustering (WoGB) leads to a substantial performance drop. For example, on SSL4Rec, Hit Rate@3 decreases from 42.1 to 25.2 and NDCG@3 decreases from 68.2 to 44.0, a reduction of more than 35%, indicating that granular-ball clustering is crucial for capturing local structures and semantic boundaries in user interests. Similarly, removing counterfactual gain (WoCF) also results in performance degradation, e.g., Hit Rate@3 drops to 37.4 and NDCG@3 to 63.6 on SSL4Rec, suggesting that counterfactual gain is indispensable for enhancing representation robustness and enriching contrastive signals.

Moreover, replacing granular-ball clustering with K-means, DBSCAN, or OPTICS fails to recover the full performance. Although K-means achieves competitive results in certain cases, it still lags behind GBCG overall. These results collectively demonstrate that granular-ball clustering and counterfactual gain serve as two complementary pillars of GBCG: the former ensures clustering quality and structural rationality, while the latter strengthens the effectiveness of contrastive learning.

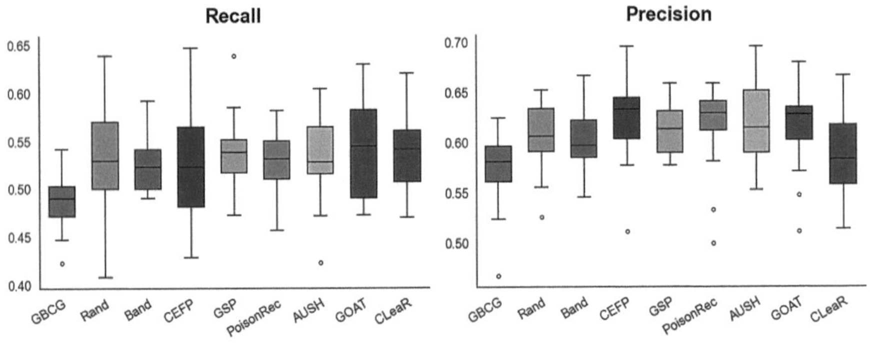

Fig. 3. Shilling detection of injected profiles on ML-100K. Lower value suggests a better stealthiness.

4.5 Stealthiness Detection

Finally, we investigate whether the attack remains stealthy while preserving effectiveness. To amplify the detection challenge, we set the injected fake user proportion to 10%. We apply a widely used unsupervised detector [32] and report **precision** and **recall**—lower values indicate better stealth. As shown in Fig. 3, GBCG exhibits strong stealthiness, attributed to three design choices: (1) granular-ball clustering enforces local semantic coherence, ensuring selected items lie naturally within real semantic neighborhoods; (2) counterfactual-guided selection produces plausible behavioral patterns rather than contrived sets; and (3) the WGAN-GP generator produces high-fidelity profiles projected to realistic discrete interactions. Together, these elements yield fake profiles statistically similar to genuine users, making them harder to isolate. This suggests that robust defenses should focus on detecting local semantic perturbations and distributional mimicry rather than relying on simple outlier removal.

5 Related Work

5.1 Granular-Ball Computing

Granular-Ball computing is an emerging theory of adaptive multi-granularity representation and computation that aims to improve the efficiency, robustness, and interpretability of artificial intelligence models [9]. The central idea of granular-ball computing is to adaptively cover the sample space with hyperspherical regions (granular balls) of varying scales, where iterative splitting and merging operations are employed to achieve representations from coarse to fine granularity [29]. Compared with conventional point-based learning methods, granular-ball computing offers higher computational efficiency, stronger noise resistance, and inherent hierarchical expressiveness. The theory has been extended to multiple domains, including classification [16], clustering [5], neural networks [25], rough sets [27], and optimization [26]. In particular, for clustering tasks, the granular-ball approach can automatically identify complex non-convex structures without requiring the number of clusters to be predefined, providing a novel perspective for structured data representation [5].

5.2 Shilling Attacks on Recommender Systems

Shilling attacks are designed to manipulate the outputs of recommender systems by injecting fabricated user profiles [14]. Early attack strategies, such as RandomAttack [8] and BandwagonAttack [13], were built on the idea of mimicking rating behaviors that could resemble those of genuine users to promote the rankings of target items. In recent years, reinforcement learning [17], generative adversarial networks (GANs) [1], and graph neural networks (GNNs) [24] have been employed to enhance attack effectiveness and stealthiness. However, these approaches often rely on extensive attack resources or access to internal model information, which limits their practicality in real-world black-box scenarios [20]. Consequently, recent studies have shifted attention toward generative attacks under resource constraints and cross-model transferability, with representative methods such as AUSH [11], GOAT [22], and CLeaR [21], providing new insights for evaluating the security of recommender systems.

6 Conclusion

This paper proposes GBCG, a generative framework designed for efficient, covert push attacks in black-box, resource-constrained recommender systems. By integrating granular-ball computing, counterfactual gain estimation, and WGAN-GP, GBCG achieves a robust trade-off between effectiveness and stealthiness. Extensive experiments across diverse datasets confirm its consistent superiority over baselines. Future work will explore granular-ball structures in defense modeling and adversarial learning to further enhance recommender system security.

Acknowledgements. This research has been carried out with the support of the National Natural Science Foundation of China, No. 62307008.

References

1. Creswell, A., White, T., Dumoulin, V., Arulkumaran, K., Sengupta, B., Bharath, A.A.: Generative adversarial networks: an overview. IEEE Signal Process. Mag. **35**(1), 53–65 (2018)
2. Gulrajani, I., Ahmed, F., Arjovsky, M., Dumoulin, V., Courville, A.C.: Improved training of wasserstein GANs. In: Advances in Neural Information Processing Systems (NeurIPS), vol. 30 (2017)
3. He, X., Deng, K., Wang, X., Li, Y., Zhang, Y., Wang, M.: LightGCN: simplifying and powering graph convolution network for recommendation. In: Proceedings of the 43rd International ACM SIGIR Conference on Research and Development in Information Retrieval, pp. 639–648 (2020)
4. He, X., Liao, L., Zhang, H., Nie, L., Hu, X., Chua, T.S.: Neural collaborative filtering. In: Proceedings of the 26th International Conference on world Wide Web, pp. 173–182 (2017)
5. Jia, Z., Zhang, Z., Pedrycz, W.: Generation of granular-balls for clustering based on the principle of justifiable granularity. IEEE Trans. Cybern. **55**(4), 1687–1700 (2025)
6. Jiang, J., et al.: Reformulating sequential recommendation: Learning dynamic user interest with content-enriched language modeling. In: International Conference on Database Systems for Advanced Applications, pp. 353–362. Springer, Cham (2024). https://doi.org/10.1007/978-981-97-5555-4_25
7. Koren, Y., Bell, R., Volinsky, C.: Matrix factorization techniques for recommender systems. Computer **42**(8), 30–37 (2009)
8. Lam, S.K., Riedl, J.: Shilling recommender systems for fun and profit. In: Proceedings of the 13th International Conference on World Wide Web, pp. 393–402. ACM (2004)
9. Li, Y., et al.: Multi-granularity open intent classification via adaptive granular-ball decision boundary. In: Proceedings of the AAAI Conference on Artificial Intelligence, vol. 39, pp. 24512–24520. AAAI (2025)
10. Likas, A., Vlassis, N., Verbeek, J.J.: The global k-means clustering algorithm. Pattern Recogn. **36**(2), 451–461 (2003)
11. Lin, C., Chen, S., Li, H., Xiao, Y., Li, L., Yang, Q.: Attacking recommender systems with augmented user profiles. In: Proceedings of the 29th ACM International Conference on Information & Knowledge Management, pp. 855–864. ACM (2020)
12. Lin, J., et al.: Deep-cross-attention recommendation model for knowledge sharing micro learning service. In: Bittencourt, I.I., Cukurova, M., Muldner, K., Luckin, R., Millán, E. (eds.) AIED 2020. LNCS (LNAI), vol. 12164, pp. 168–173. Springer, Cham (2020). https://doi.org/10.1007/978-3-030-52240-7_31
13. Mobasher, B., Burke, R., Bhaumik, R., Williams, C.: Effective attack models for shilling item-based collaborative filtering systems. In: Proceedings of the 2005 WebKDD Workshop, held in Conjuction with ACM SIGKDD, vol. 2005 (2005)
14. Nawara, D., Aly, A., Kashef, R.: Shilling attacks and fake reviews injection: principles, models, and datasets. IEEE Trans. Comput. Soc. Syst. (2024)
15. Nguyen Thanh, T., et al.: Poisoning GNN-based recommender systems with generative surrogate-based attacks. ACM Trans. Inf. Syst. **41**(3), 1–24 (2023)
16. Sajid, M., Quadir, A., Tanveer, M., Initiative, A.D.N., et al.: GB-RVFL: fusion of randomized neural network and granular ball computing. Pattern Recogn. **159**, 111142 (2025)

17. Shakya, A.K., Pillai, G., Chakrabarty, S.: Reinforcement learning algorithms: a brief survey. Expert Syst. Appl. **231**, 120495 (2023)
18. Song, J., et al.: Poisonrec: an adaptive data poisoning framework for attacking black-box recommender systems. In: 2020 IEEE 36th International Conference on Data Engineering (ICDE), pp. 157–168. IEEE (2020)
19. Tan, J., Xu, S., Ge, Y., Li, Y., Chen, X., Zhang, Y.: Counterfactual explainable recommendation. In: Proceedings of the 30th ACM International Conference on Information & Knowledge Management, pp. 1784–1793. ACM (2021)
20. Wang, Z., Gao, M., Yu, J., Ma, H., Yin, H., Sadiq, S.: Poisoning attacks against recommender systems: a survey (2024). arXiv:2401.01527
21. Wang, Z., Yu, J., Gao, M., Yin, H., Cui, B., Sadiq, S.: Unveiling vulnerabilities of contrastive recommender systems to poisoning attacks. In: Proceedings of the 30th ACM SIGKDD Conference on Knowledge Discovery and Data Mining, pp. 3311–3322. ACM (2024)
22. Wu, F., Gao, M., Yu, J., Wang, Z., Liu, K., Wang, X.: Ready for emerging threats to recommender systems? A graph convolution-based generative shilling attack. Inf. Sci. **578**, 683–701 (2021)
23. Wu, J., et al.: Self-supervised graph learning for recommendation. In: Proceedings of the 44th International ACM SIGIR Conference on Research and Development in Information Retrieval, pp. 726–735. ACM (2021)
24. Wu, S., Sun, F., Zhang, W., Xie, X., Cui, B.: Graph neural networks in recommender systems: a survey. ACM Comput. Surv. **55**(5), 1–37 (2022)
25. Xia, S., Ma, X., Liu, Z., Liu, C., Zhao, S., Wang, G.: Graph coarsening via supervised granular-ball for scalable graph neural network training. In: Proceedings of the AAAI Conference on Artificial Intelligence. vol. 39, pp. 12872–12880. AAAI (2025)
26. Xia, S., Tian, X., Yuan, S., Deng, J.D.: Efficient quantum approximate k nn algorithm via granular-ball computing (2025). arXiv:2505.23066
27. Xia, S., et al.: GBRS: a unified granular-ball learning model of Pawlak rough set and neighborhood rough set. IEEE Trans. Neural Netw. Learn. Syst. (2023)
28. Xia, S., Zheng, S., Wang, G., Gao, X., Wang, B.: Granular ball sampling for noisy label classification or imbalanced classification. IEEE Trans. Neural Netw. Learn. Syst. **34**(4), 2144–2155 (2021)
29. Xie, Q., et al.: GBG++: a fast and stable granular ball generation method for classification. IEEE Trans. Emerg. Top. Comput. Intell. **8**(2), 2022–2036 (2024)
30. Yao, T., et al.: Self-supervised learning for large-scale item recommendations. In: Proceedings of the 30th ACM International Conference on Information & Knowledge Management, pp. 4321–4330. ACM (2021)
31. Yu, J., Yin, H., Xia, X., Chen, T., Cui, L., Nguyen, Q.V.H.: Are graph augmentations necessary? simple graph contrastive learning for recommendation. In: Proceedings of the 45th International ACM SIGIR Conference on Research and Development in Information Retrieval, pp. 1294–1303. ACM (2022)
32. Zhang, Y., Tan, Y., Zhang, M., Liu, Y., Chua, T.S., Ma, S.: Catch the black sheep: unified framework for shilling attack detection based on fraudulent action propagation. In: Proceedings of the 24th International Joint Conference on Artificial Intelligence (IJCAI), pp. 2408–2414 (2015)
33. Zhao, Y., Lin, L., Huang, R., Lin, J.: Clustering-based enhancement for fake user profiles generation in recommender systems. In: International Conference on Intelligent Computing, pp. 260–272. Springer, Cham (2025)

RICE: Enhancing Transparency in Recommendation via Interaction-Based Counterfactual Explanations

Shiqin Liu[1], Minjun Zhao[2], and Jiajun Bu[1](✉)

[1] Zhejiang Key Laboratory of Accessible Perception and Intelligent Systems, Zhejiang University, Hangzhou, China
{sqliu,bjj}@zju.edu.cn

[2] College of Computer Science, Zhejiang University, Hangzhou, China
minjunzhao@zju.edu.cn

Abstract. Counterfactual explanations, which explore "what if" scenarios post-hoc, have recently attracted significant attention for clarifying the predictions of complex machine learning models. However, its application in recommender systems remains relatively unexplored, particularly in generating explanations that are both accurate and diverse. This paper introduces RICE, a novel historical interaction-based counterfactual explanation method tailored for multimodal recommendation systems. Unlike existing approaches that often fall short in providing effective and varied insights into recommendation decisions, RICE produces counterfactual explanations that not only have short distances but also offer a range of alternatives. Extensive experiments on multiple benchmark datasets and multimodal recommendation models validate the effectiveness of RICE.

Keywords: Counterfactual Explanation · User Trust and Transparency · Recommender Systems

1 Introduction

Recommender systems [4,5,19,22,35,37] have become integral components of modern online platforms, delivering personalized suggestions based on users' historical interactions. As the volume and complexity of available content increase, multimodal recommendation systems [10,13,17,26,27,33,38], which jointly model textual and visual modalities together with user interaction signals, have emerged as an effective paradigm for capturing richer user preferences. Despite their strong performance, such systems are often highly complex, making their decision-making process more opaque [30,43]. Users may question why specific items are recommended (or not), and the lack of transparency in complex models can diminish user trust [8,23,25,28]. This has led to growing interest in developing explanation methods that can clarify recommendation results and enhance interpretability. Among various explainability approaches,

H. Jung et al. (Eds.): DASFAA 2026, LNCS 16535, pp. 104–120, 2026.
https://doi.org/10.1007/978-981-92-0363-5_7

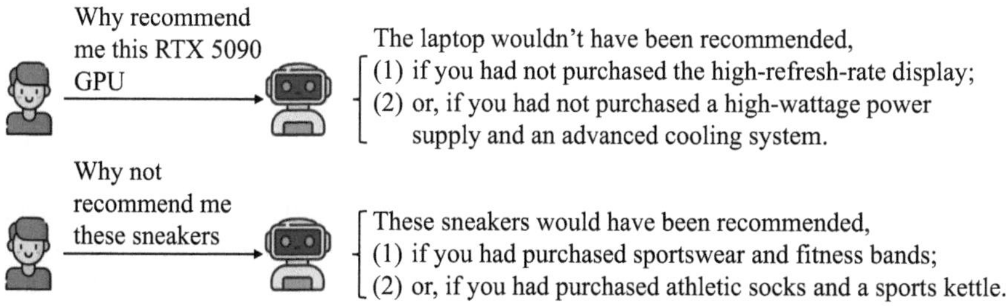

Fig. 1. Examples of counterfactual explanations produced by RICE in two recommendation cases.

post-hoc explanation methods [6,36,40] have received significant attention due to their ability to explain trained models without modifying model structure or training procedures. Counterfactual explanations [9,15,18,21,32] stand out as a compelling type of post-hoc explanation due to their intuitive "what if" framing and actionable nature. By identifying minimal modifications to a user's interaction history that would yield different recommendation results, counterfactual explanations provide clear insights into the decision process of the recommender system [1,28]. Their usefulness extends even to models traditionally considered interpretable, such as random forests and generalized linear models [16,29,34], and they have been acknowledged as satisfying interpretability requirements by some legislative bodies, such as those outlined in the GDPR [24,32].

However, existing historical interaction-based counterfactual explanation methods encounter several challenges when applied to multimodal recommendation tasks. (i) Firstly, generating counterfactual explanations with **validity** is challenging since multimodal recommenders operate over high-dimensional embeddings and extensive user-item interactions, making it computationally demanding to produce counterfactual explanations that faithfully reflect user historical interactions [1]. (ii) Secondly, previous research has shown that providing counterfactual explanations with minimal modifications [9,15,32] and diversity [18,21,24,31] is crucial for explainability. The **effectiveness** of counterfactual explanations is measured by the distance between the explanation and the original sample [15,24,28,32] as shorter modifications are generally easier for humans to comprehend, and **diversity** is also considered an important property of counterfactual explanations [24,30–32] since diverse explanations of the same prediction result help humans to have a more comprehensive understanding of the model's decision-making mechanism. However, ensuring counterfactual explanations remain proximate to the original user behavior and diverse enough to offer meaningful alternatives is challenging. Existing methods for recommender systems [1,28] have difficulty in generating short-distance counterfactual explanations, and produce only one counterfactual explanation without diversity, thereby limiting the explainability and practical usefulness of the generated explanations. (iii) Thirdly, exhaustive searches for counterfactual explanations are often impractical in real-world applications. Conventional approaches generally lack adaptive termination mechanisms, resulting in incomplete or overly large sets of explanations.

To address these challenges, we introduce **RICE**, a novel counterfactual explanation framework that aims to generate valid, minimal (i.e., effective), and diverse interaction modifications for counterfactual recommendation results. We propose the vector search-based generation algorithm and the freezing-based verification method for RICE, in which the vector search-based generation algorithm generates candidate counterfactual explanations iteratively, and the freezing-based verification method verifies the validity of generated candidates (i.e., checks whether the generated candidates are real counterfactual explanations or not) in each iteration. Specifically, the vector search-based generation algorithm begins with retrieving similar counterfactual users. Then, the algorithm uses multimodal item similarity to generate new candidate explanations based on the beam search strategy iteratively. The iterative algorithm stops until no validated counterfactual explanations exist in a new iteration. Then, we use a greedy algorithm to extract diverse and nearest counterfactual explanations from all counterfactual explanations generated during the process. Note that, since we have counterfactual explanations at the beginning of the vector search-based generation algorithm and generate better (i.e., closer) explanations in each iteration, we can terminate the algorithm at any iteration to get the best explanations so far flexibly. The freezing-based verification method freezes all item embeddings to generate approximate embeddings for candidates under verification, ensuring validity without retraining the whole recommender system. Figure 1 illustrates two typical counterfactual explanations produced by RICE. Therefore, by integrating the vector search-based generation algorithm and the freezing-based verification method, RICE is able to (i) generate valid counterfactual explanations, (ii) produce effective and diverse counterfactual explanations, (iii) support adaptive termination at any iteration.

The contributions of this paper are concluded as follows:

- We introduce RICE, the first historical interaction-based counterfactual explanation method tailored for multimodal recommender systems, producing explanations with validity, effectiveness, and diversity.
- We propose the vector search-based generation algorithm to generate counterfactual explanations iteratively and the freezing-based verification method to verify the candidate explanations. We seamlessly integrate these algorithms into RICE.
- We have conducted comprehensive experiments on multiple public datasets, demonstrating that RICE outperforms existing historical interaction-based counterfactual explanation methods in validity, effectiveness, and diversity.

2 Related Work

Explainability in recommender systems has gained increasing attention in recent years as a way to improve user satisfaction, trust, and system accountability.

Existing approaches can be categorized into intrinsic and post-hoc methods. Intrinsic methods [2,3,7,20] incorporate explainability objectives directly into the model design that tend to compromise recommendation performance. In contrast, post-hoc methods [1,28] aim to explain the pre-trained models without altering their architecture. Counterfactual explanation is a subclass of post-hoc methods, and in recommender systems, it corresponds to identifying the shortest change in a user's interaction history that would lead to an altered recommendation result. ACCENT [28] adopts influence functions to trace which user behaviors contribute most to the recommendation. LXR [1] leverages self-supervised training to learn discriminative and explainable representations.

However, these methods lack of validity, effectiveness, and diversity, and are not designed for multimodal recommendation. Thus, we propose RICE, which is a counterfactual explanation framework tailored for multimodal recommendation systems, to generate validity, minimal (i.e., effective), and diverse interaction modifications for counterfactual recommendation results.

3 Preliminaries

Counterfactual Explanation Based on Historical Interactions. In recommender systems, users may occasionally find certain recommendation results unsatisfactory or puzzling, for instance, "Why am I being recommended item y?" or "Why is my desired item not ranked in the top list?" Counterfactual explanations provide a powerful method to address such concerns by exploring "what if" scenarios: if some historical interactions were changed, would the recommendation for item y be altered accordingly? Concretely, suppose that a user u_0 is dissatisfied with the current ranking of item y (e.g., y appears in u_0's Top-K list but the user does not desire it, or y is absent from the Top-K whereas the user wishes to see it). By modifying a small subset of the user u_0's interaction records, flipping certain items, we can create a counterfactual user u^* that yields a flipped result for item y (from recommended to not recommended, or vice versa). The fewer modifications needed to achieve such a flip, the more intuitive and comprehensible the counterfactual explanation typically is.

Validity of Counterfactual Explanation in Recommender System. Since the time-consuming process of verifying historical interaction-based counterfactual explanations through retraining, existing methods [1,28] generate counterfactual explanations in an approximate way, so there may be some explanations generated that are not true counterfactual explanations, that is, the validity is not equal to 1. Therefore, the existing method only verifies the validity of counterfactual explanations through retraining in the experimental stage and calculates the validity rate. The higher the validity rate, the better the method for producing explanations.

Effectiveness of Counterfactual Explanation. The effectiveness of counterfactual explanations is measured by the distance between the explanations and the original sample [15,24,28,32] since shorter explanations tend to be more concise and user-friendly (i.e., easier to understand by humans). To quantify the degree of modifications required to achieve a counterfactual explanation, we use the Hamming distance [11,24], which measures the difference between the interaction history of the original user u_0 and the counterfactual user u^*. A smaller Hamming distance indicates that only minimal changes to the user's historical interactions are needed to achieve the desired ranking flip for item y, making the explanation more feasible and explainable. Specifically, we first define the interaction between user u_i and item y_j as $R(u_i, y_j)$. Formally,

$$R(u_i, y_j) = \begin{cases} 1, & \text{if user } u_i \text{ has interacted with item } y_j, \\ 0, & \text{otherwise.} \end{cases} \tag{1}$$

Then, given the original user u_0 and a counterfactual user u^*, the Hamming distance of the counterfactual user is:

$$\text{dist}_{\mathbf{H}}(u_0, u^*) = \sum_{i=1}^{d} \mathbf{1}\left[R(u_0, y_i) \neq R(u^*, y_i)\right], \tag{2}$$

where d is the total number of items, and $\mathbf{1}[\cdot]$ is the indicator function that returns 1 if the condition inside holds true, and 0 otherwise.

Coverage of two Counterfactual Explanations. Let A and B be two counterfactual explanations for a user u_0. Denote by $\Delta(A)$ and $\Delta(B)$ the sets of modified interactions (added or removed) for A and B, respectively. If

$$\Delta(A) \subseteq \Delta(B), \tag{3}$$

we say counterfactual B covers A. In this case, A can be considered a "smaller" modification, since its set of changes is a proper subset of those in B. Given that both A and B lead to the same prediction flip for item y, A would be regarded as more preferable.

Diversity of Counterfactual Explanations. Prior works [24,30–32] use diversity to measure the richness of counterfactual explanations generated by an approach since diverse explanations of the same prediction result help humans to understand the model's decision-making mechanism better. Thus, we define diversity in interaction-based counterfactual explanations for the recommender system as below. Given a set of counterfactual explanations $\mathcal{C} = \{C_1, C_2, \ldots, C_n\}$, if C_i covers C_j, then C_i is a superset modification of C_j and is therefore considered redundant. Therefore, we discard any explanation that covers another, and retain only non-redundant explanations. We define diversity as the number of retained explanations. A higher diversity value indicates that the method generates more mutually independent and non-redundant counterfactual explanations. For example, if three explanations $\{a, b, c\}$ are generated but a covers b, the diversity is 2, as a is redundant and discarded. Note that diversity does not mean sacrificing the effectiveness (distance) of explanations; in fact, it is usually possible to produce diverse explanations that are close to each other and do not cover each other in the experiments.

4 Methodology

4.1 Problem Formulation

Let u_0 be the target user and y_0 be the item under explanation, $\mathbf{U}$ and $\mathbf{I}$ denote all users and all items, respectively. S is the number of items presented in the user interface. We use $R(u_0, y_i)$ to represent the interaction of user u_0 and item y_i, as mentioned in Eq. 1. Denote the current recommendation result of item y_0 for user u_0 by M, where

$$M = \begin{cases} 1, & \text{if } y_0 \text{ appears in } u_0\text{'s Top-}S \text{ list}, \\ 0, & \text{otherwise}. \end{cases} \tag{4}$$

Then, we formally define the counterfactual explanation generation problem. Given u_0, y_0, S and the number of results n, our goal is to generate a set of *counterfactual explanations* $\{u_1^*, \ldots, u_n^*\}$, each of which differs from u_0's historical interactions in only a small subset, yet leads to a flipped result for y_0. Formally, for each u_i^*,

$$\text{Rec}(u_i^*, y_0) = \neg M, \tag{5}$$

where $\text{Rec}(\cdot, \cdot)$ indicates the Top-S ranking result of item y_0 for a given user. Consistent with the definitions of effectiveness (i.e., distance) and diversity (introduced in Sect. 4), we seek multiple non-covering counterfactual explanations that consider both the distance and the diversity.

4.2 Vector Search-Based Generation Algorithm

In this section, we introduce the vector search-based generation module of RICE. We use the vector search-based generation algorithm to generate counterfactual explanations iteratively. During each iteration, candidate explanations are verified using the freezing-based verification method. Based on the retrieved similar users and multimodal embeddings of items, we iteratively construct a set of virtual counterfactual users that differ from the target user only in a small subset of interaction history while yielding different recommendation results. The most similar and diverse virtual counterfactual users are used to answer the core question of how the interaction records should be modified to flip the recommendation result.

Initial Retrieval. To efficiently identify promising interaction modifications, we first employ *binary vector search*[1] to retrieve a set of users $\{u_{(1)}, u_{(2)}, \ldots, u_{(k_{start})}\}$ satisfying: (i) High interaction similarity to u_0; (ii) A *different* recommendation result for y_0. These conditions can be formalized as follows:

$$\{u_{(1)}, u_{(2)}, \ldots, u_{(k_{start})}\} = \arg\min^{k_{start}}_{\substack{u^* \in \mathbf{U} \\ \text{Rec}(u^*, y_0) = \neg M}} \text{dist}_{\text{H}}(u_0, u^*) \tag{6}$$

[1] https://github.com/facebookresearch/faiss/wiki/Binary-indexes.

These users serve as *natural counterfactual explanations*, i.e., they already exhibit an opposite recommendation result for y_0 while remaining relatively similar to u_0 in interaction space. Notably, even without further iteration, we already obtain valid explanations, although they may be relatively distant. If time or computational cost is critical, the process can stop here and return the top few users that meet these criteria.

Initialization. After retrieving the natural counterfactual explanations, we first perform initialization before conducting the iterative refinement procedure based on beam search and multimodal item embedding similarity. The goal is to progressively reduce the distance between each counterfactual user and the target user u_0 while preserving the flipped recommendation result, thereby producing more concise counterfactual explanations. We define $\mathbf{U_{new}}$ as the set of counterfactual users verified in last iteration, $\mathbf{T}$ as a temporary list of candidate (virtual) users to be tested, and $\mathbf{U}_e$ as an accumulated set of counterfactual users found across all iterations. We further define the multimodal similarity, used in both the Initialization module and the Iterative Generation module. Specifically, for an item y_j with textual embedding e_j^t and visual embedding e_j^v, the multimodal embedding e_j of the item y_j is derived by concatenating e_j^t and e_j^v as follows:

$$e_j = e_j^t \oplus e_j^v. \tag{7}$$

Next, we formally define

$$\text{sim}(y_a, y_b) = \cos(e_a, e_b) \tag{8}$$

as the *multimodal similarity* between items y_a and y_b. At this stage, we generate initial candidates based on the multimodal information of items and two observations in recommender systems: (i) *High-Similarity Non-Determinant.* If a user u_i interacts with (or misses) an item y_j that is highly similar to y_0, yet y_0 is still not recommended to (or is still recommended to) u_i, this suggests that y_j is unlikely to be a key factor determining the recommendation result of y_0. Therefore, flipping the interaction on y_j can potentially make u_i closer to u_0 while preserving the current counterfactual recommendation result. (ii) *Low-Similarity Non-Influential.* If a user u_i interacts with (or misses) an item y_j that is least similar to y_0, yet y_0 is still recommended to (or is still not recommended to) u_i, this suggests that y_j is unlikely to significantly influence the recommendation result of y_0. The reason is that removing the interaction with a dissimilar item is unlikely to push y_0 out of the Top-S list, and adding such an interaction is unlikely to bring y_0 into the Top-S list. Therefore, flipping the interaction on y_j can potentially make u_i closer to u_0 while preserving the current counterfactual recommendation result. We define the number of the most/least similar items in this stage as m.

To begin with, we assign the previously retrieved natural counterfactual explanations $\{u_{(1)}, \dots, u_{(k_{start})}\}$ to $\mathbf{U_{new}}$ and copy them to $\mathbf{U}_e$. For each $u_i \in \mathbf{U_{new}}$, let D_{u_0,u_i} be the set of items where u_0's and u_i's interactions differ:

$$D_{u_0,u_i} = \{\, y_j \in \mathbf{I} \mid R(u_0, y_j) \neq R(u_i, y_j) \}. \tag{9}$$

We then perform multimodal embedding search to find the m items s_{most} in D_{u_0,u_i} most similar to y_0. Next, to maintain the current recommendation result about item y_0 (i.e., $\mathrm{Rec}(u_i, y_0)$) while gradually reducing the distance between u_i and u_0, we first use the *High-Similarity Non-Determinant* strategy to flip items for initialization of $\mathbf{T}$. Thus, for each user u_i we generate a candidate user $u_{i,0}^0$ and put it in $\mathbf{T}$ by flipping the interactions on all items in:

$$\{y_j | y_j \in s_{most} \wedge y_j \in D_{u_0,u_i} \wedge R(u_i, y_j) = M\}. \tag{10}$$

Then, we use the *Low-Similarity Non-Influential* strategy and define s_{least} as top-m items that are least similar to y_0, from all items $\mathbf{I}$. Thus, s_{least} is the set of items unlikely to affect the result. For each $u_i \in \mathbf{U_{new}}$, we generate $u_{i,1}^0$ to $\mathbf{T}$ by flipping the interactions on all items in:

$$\{y_j | y_j \in s_{least} \wedge y_j \in D_{u_0,u_i} \wedge R(u_i, y_j) = \neg M\}. \tag{11}$$

After initialization of $\mathbf{T}$, we verify whether the users in $\mathbf{T}$ are counterfactual explanations and get $\mathbf{U}'_{\mathbf{new}} = \{u_i \in \mathbf{T} \wedge \mathrm{Rec}(u_i, y_0) = \neg M\}$. The proposed verification method will be introduced in Sect. 4.3. Then we set $\mathbf{U_{new}} = \mathbf{U_{new}} \cup \mathbf{U}'_{\mathbf{new}}$ and $\mathbf{U_e} = \mathbf{U_{new}}$.

Iterative Generation. In counterfactual explanation generation for recommender systems, the goal is to minimize modifications to the interaction records while flipping the recommendation result of the target item y_0. Therefore, we propose an iterative generation algorithm based on beam search. The algorithm generates candidate virtual users iteratively from counterfactual explanation (i.e., verified virtual users) generated in the last iteration that are closer to the target user, and retains the k_{width} counterfactual explanation users closest to the target user after verification.

Specifically, in iteration l, for each $u_i^l \in \mathbf{U_{new}}$, the set of interaction differences between u_0 and u_i^l is D_{u_0,u_i^l}. We generate t candidate virtual users $\{u_{i,2}^l, \ldots, u_{i,t+1}^l\}$ with closer distance to target user u_0 by randomly select an interaction record $R(u_i^l, y_r^l)$ from D_{u_0,u_i^l} and flip it. Since the search process may be too slow if each iteration modifies only one interaction record, it is difficult for the generated counterfactual user to effectively approach the target user u_0 within a limited number of iterations. To address this and accelerate the convergence of the counterfactual user to u_0, we modify top-s most similar items to y_r^l, which also have different interactions between u_i^l and u_0.

After generation from each counterfactual explanation in $\mathbf{U_{new}}$, we put $t|\mathbf{U_{new}}|$ candidate virtual users into $\mathbf{T}$. Then we verify whether the users in $\mathbf{T}$ are counterfactual explanations and get $\mathbf{U}'_{\mathbf{new}} = \{u_i \in \mathbf{T} \wedge \mathrm{Rec}(u_i, y_0) = \neg M\}, |\mathbf{U}'_{\mathbf{new}}| \leq t|\mathbf{U_{new}}|$. In order to avoid an overly large search space and reduce unnecessary search processes, thereby improving query efficiency, we calculate the distance between the new counterfactual explanations in $\mathbf{U}'_{\mathbf{new}}$ and the target user u_0 based on the **beam search** strategy, and retain the nearest new samples of beam width k_{width} as $\mathbf{U_{new}}$. Formally,

$$\mathbf{U_{new}} = \arg\min_{u_i \in \mathbf{U}'_{\mathbf{new}}}^{k_{width}} \mathrm{dist_H}(u_0, u_i). \tag{12}$$

Results Selection. The iterative generation keeps going until all candidate virtual users from U_{new} are not verified as counterfactual explanations. At this point, multiple counterfactual explanations have been generated and kept in $\mathbf{U_e}$ during the iteration. Then, we obtain the nearest (i.e., the most effective) and diverse counterfactual explanations in $\mathbf{U_e}$ by a **greedy algorithm**. Specifically, we sort all candidate counterfactual explanations in $\mathbf{U_e}$ by their Hamming distance to u_0, in ascending order. Then, we iteratively select explanations that do not cover any previously selected ones, where coverage is defined in Sect. 3. This ensures that each selected explanation contributes unique information. The process continues until the number of results n (given by user) diverse counterfactual explanations are selected or all candidates are exhausted.

4.3 Freezing-Based Verification Method

In counterfactual explanation tasks for recommendation, we need to verify whether modifying a user's interaction history leads to the expected change in recommender results. However, retraining the entire model for each modified user is computationally expensive. To address this, we propose the freezing-based verification method, where all model parameters remain fixed, and only the embeddings of newly introduced counterfactual users are optimized. This approach significantly reduces computation costs while maintaining the validity of counterfactual explanations. For convenience, we introduce our approach in this section based on the LightGCN model [14], which is widely used in multimodal recommendation systems [10,26,39,41,42,44].

Given a target user u_0 whose recommendation results need counterfactual explanation, we generate a candidate virtual user u^* through iterative modifications of interaction records. Then, we need to verify whether its final recommendation results align with expectations, i.e., whether it is a valid counterfactual user. In LightGCN, the embedding of the newly introduced user u^* after L layers of propagation is computed as $\mathbf{e}_{u^*} = \sum_{l=0}^{L} \alpha_l \mathbf{e}_{u^*}^{(l)}$, where the propagation rule for each layer is given by:

$$\mathbf{e}_{u^*}^{(l+1)} = \sum_{v \in \mathcal{N}_{u^*}} \frac{1}{\sqrt{|\mathcal{N}_{u^*}| \cdot |\mathcal{N}_v|}} \mathbf{e}_v^{(l)}, \tag{13}$$

where $\mathcal{N}_{u^*}$ denotes the neighbors (interacted items) of user u^*. During verification, all pre-trained user and item embeddings are frozen. Consequently, the final embedding of the virtual user can be expressed as:

$$\mathbf{e}_{u^*} = \alpha_0 \mathbf{e}_{u^*}^{(0)} + \mathbf{C}. \tag{14}$$

Accordingly, optimization is restricted to the initial embedding, eliminating the need to retrain the entire graph model and substantially reducing computational overhead. For faster convergence, $\mathbf{e}_{u^*}^{(0)}$ is initialized using the embedding of the corresponding real user. Furthermore, because each candidate virtual user is optimized independently without shared trainable parameters, multiple counterfactual candidates can be efficiently optimized in parallel through GPU-based batched computation, enabling scalable verification within each iteration.

5 Experiments

5.1 Experimental Setup

Recommender Models and Datasets. RICE is a model-agnostic framework that can be applied to common multimodal recommender models. In our experiments, we use three representative state-of-the-art multimodal recommender models: LATTICE [41], FREEDOM [44], and SOIL [26]. LATTICE [41] is the first to explicitly capture latent item structures for multimodal recommendation. FREEDOM [44] builds on LATTICE by introducing contrastive learning to enhance the stability of user representations. SOIL [26] achieves sota performance by explicitly modeling second-order interest. To evaluate the performance of the proposed RICE model, we conduct experiments on three popular datasets [12]: (a) Baby, (b) Sports and Outdoors, and (c) Clothing, Shoes, and Jewelry. These datasets provide rich multimodal information, including both textual and visual attributes of items, as well as historical user-item interactions. To ensure data quality and mitigate uncertainties in user behavior, we apply a 5-core filtering process, following previous studies [26,41,44,45]. Only users and items with at least five interactions are retained. Each item is characterized by an associated image representing its visual information and four textual attributes: title, description, category, and price.

Baseline Methods. To evaluate the effectiveness of our proposed RICE framework, we compare it against two state-of-the-art historical-interaction-based counterfactual explanation methods, i.e., ACCENT [28] and LXR [1], representing two different paradigms: influence function-based and learning-based approaches. ACCENT is an influence function-based counterfactual explanation framework. It estimates the impact of user interactions and identifies minimal modifications to change the recommendation decision. LXR is a learning-based counterfactual explanation method. It introduces self-supervised counterfactual losses to assess the relevance of user interactions. These baselines are state-of-the-art in counterfactual explanations, allowing us to evaluate RICE.

Evaluation Metrics. Following the past works [28,31], to systematically evaluate the effectiveness of our proposed counterfactual explanation method RICE, we adopt Validity, Diversity, and Distance to capture the key aspects of counterfactual explanations. Validity denotes the average percentage of generated counterfactual explanations that modify the recommendation results successfully. Diversity represents the average number of counterfactual explanations generated without coverage per query. Besides, we evaluate the distance between the nearest counterfactual explanation and the original user by measuring the average Hamming distance per query. For clearer comparison, we normalize all distances by that of RICE, setting RICE to 1 and reporting relative distances for the baselines.

Experimental Setting. All experiments in this paper are implemented in PyTorch, and run on a Dell server with Ubuntu 18.04.5 LTS, Intel Xeon Silver 4210R 2.40 GHz CPU, 128 GB RAM, and a GeForce RTX 3090 GPU.

Table 1. Overall performance of counterfactual explanation methods across three datasets and three multimodal recommenders.

Dataset	Recommender	Explanation	Validity↑	Diversity↑	Distance↓
Baby	LATTICE	LXR	0.20	1.00	3.44
		ACCENT	0.48	1.00	1.72
		RICE	**0.71**	**4.31**	**1.00**
	FREEDOM	LXR	0.31	1.00	2.70
		ACCENT	0.53	1.00	2.32
		RICE	**0.83**	**4.89**	**1.00**
	SOIL	LXR	0.37	1.00	3.35
		ACCENT	0.59	1.00	2.18
		RICE	**0.76**	**4.36**	**1.00**
Sports	LATTICE	LXR	0.14	1.00	3.87
		ACCENT	0.52	1.00	1.95
		RICE	**0.84**	**4.90**	**1.00**
	FREEDOM	LXR	0.43	1.00	3.60
		ACCENT	0.57	1.00	2.32
		RICE	**0.87**	**4.15**	**1.00**
	SOIL	LXR	0.42	1.00	5.14
		ACCENT	0.58	1.00	2.82
		RICE	**0.88**	**4.50**	**1.00**
Clothing	LATTICE	LXR	0.18	1.00	3.97
		ACCENT	0.59	1.00	2.32
		RICE	**0.83**	**4.84**	**1.00**
	FREEDOM	LXR	0.29	1.00	2.59
		ACCENT	0.51	1.00	2.54
		RICE	**0.78**	**4.73**	**1.00**
	SOIL	LXR	0.34	1.00	3.36
		ACCENT	0.63	1.00	2.42
		RICE	**0.80**	**4.11**	**1.00**

5.2 Performance Comparison

Overall Performance. To comprehensively evaluate the effectiveness of our proposed model, we conduct extensive experiments on three real-world datasets and three state-of-the-art multi-modal recommender systems. We compare our approach with two state-of-the-art historical-interaction-based counterfactual explanation methods, assessing the performance differences between these baselines and our RICE. Table 1 displays the comparison results.

It shows that our method consistently outperforms all baseline methods in validity, diversity, and distance. In these three evaluation metrics, the diversity of baseline methods, ACCENT and LXR, is 1, because these baseline methods lack diversity and can only produce a single counterfactual explanation, so the diversity of them is always 1. In terms of validity, RICE consistently outperforms all baseline methods. This improvement mainly comes from the freezing-based verification method, where all pre-trained user and item embeddings are kept fixed and only the embedding of the newly constructed virtual user is optimized. By keeping all learned embeddings unchanged during verification, the change in recommendation results is primarily driven by the modified interaction history of the virtual user. This controlled evaluation process leads to more reliable verification of whether a counterfactual explanation truly induces the intended recommendation change. For distance, RICE iteratively adjusts the counterfactual users' interaction history, gradually decreasing the number of modifications. This ensures that the final counterfactual explanations are close to the target user while still successfully flipping the recommendation result. In addition to achieving higher validity and shorter distance, RICE is naturally designed for high diversity, encouraging multiple explanation paths and ensuring that multiple counterfactual explanations are generated throughout the entire process. We use a greedy algorithm to select diverse and effective final counterfactual explanations generated in iterations, thus making the results with high diversity. These results highlight the adaptability and effectiveness of RICE across all different datasets and different recommender models.

Table 2. Evaluate the effectiveness of multimodal similarity on explanation quality across three datasets.

Dataset	Method	Validity	Diversity	Distance
Baby	w/o multimodal similarity	0.79	4.49	1.64
	RICE	0.76	4.36	1.00
Sports	w/o multimodal similarity	0.83	4.27	1.83
	RICE	0.88	4.50	1.00
Clothing	w/o multimodal similarity	0.74	4.35	1.68
	RICE	0.80	4.11	1.00

Effectiveness of Multimodal Similarity. To assess the contribution of multimodal similarity, we replace the multimodal similarity-based item search and interaction flip steps in the Initialization and Iterative Generation modules with random item selection. Table 2 reports Validity, Diversity, and Distance of the replaced method on the Baby, Sports, and Clothing datasets. Without multimodal similarity, Validity and Diversity remain comparable to the full RICE, but Distance increases substantially on all datasets. This degradation occurs because both Initialization and Iterative Generation use similarity to identify items whose flipped interactions do not change the current counterfactual recommendation result while moving virtual users closer to the target. Random flips yield candidates that fail to meet the counterfactual criterion and cause iterations to terminate prematurely before closer explanations can be found. These findings confirm that multimodal similarity is essential for RICE to produce concise explanations in multimodal recommender systems.

Table 3. Ablation study on each module in RICE across three datasets.

Dataset	Method	Validity	Diversity	Distance
Baby	w/o Initial Retrieval	0.73	4.28	1.56
	w/o Initialization	0.72	4.11	1.44
	w/o Results Selection	0.76	3.91	1.00
	RICE	0.76	4.36	1.00
Sports	w/o Initial Retrieval	0.87	4.63	1.71
	w/o Initialization	0.84	4.47	1.57
	w/o Results Selection	0.88	3.46	1.00
	RICE	0.88	4.50	1.00
Clothing	w/o Initial Retrieval	0.81	4.06	1.49
	w/o Initialization	0.78	4.10	1.53
	w/o Results Selection	0.80	3.62	1.00
	RICE	0.80	4.11	1.00

Effectiveness of Each Module. We conduct an ablation study by selectively omitting each module from RICE and evaluating its effect on Validity, Diversity, and Distance across three datasets.

To evaluate the effectiveness of the Initial Retrieval module, we replace it with randomly retrieving users with different recommendation results (i.e., w/o Initial Retrieval) and report the experimental results in Table 3. The table shows that Validity and Diversity remain relatively stable, but Distance increases. This occurs because the counterfactual users retrieved randomly are quite different from the target users, requiring more iterations, and the approximate algorithm used in iterations may miss the optimal refinement path, resulting in an increase in distance. These results indicate that finding close initial counterfactual users is essential to generate counterfactual explanations with short distance.

In addition, we evaluate the contribution of the Initialization module by removing it, and report the results in Table 3. Without Initialization, the Distance increases. This degradation arises because the iterative generation starts directly, which limits the search space explored in the refinement process.

Finally, to assess the impact of the Results Selection module, we replace this module with directly selecting the top-k similar counterfactual users (i.e., w/o Results Selection) and show the results in Table 3. As shown in the results, Validity and Distance remain unchanged after removing the Results Selection module, while Diversity exhibits a noticeable decline. This is because RICE and RICE w/o Results Selection select the same nearest counterfactual explanation, which determines both the validity of the explanation and the distance. However, directly selecting the top-k similar counterfactual users introduces redundancy among the selected explanations, some explanations cover other explanations, therefore, lower diversity. In contrast, the Results Selection module explicitly encourages diversity by employing a greedy selection strategy, which effectively filters redundant candidates and ensures a highly diverse set of RICE explanations.

6 Conclusion

In this paper, we propose RICE, an interaction-based counterfactual explanation framework for multimodal recommendation systems. By flipping minimal portions of user-item interactions, our counterfactual explanation framework helps users understand how certain items are (or are not) recommended. Comprehensive experiments demonstrate that RICE achieves high-quality explanations in terms of validity, effectiveness, and diversity.

Acknowledgments. This work is supported by the National Natural Science Foundation of China (Grant No. 62372408).

Disclosure of Interests. The authors have no competing interests to declare that are relevant to the content of this article.

References

1. Barkan, O., Bogina, V., Gurevitch, L., Asher, Y., Koenigstein, N.: A counterfactual framework for learning and evaluating explanations for recommender systems. In: Proceedings of the ACM on Web Conference 2024, pp. 3723–3733 (2024)
2. Barkan, O., Fuchs, Y., Caciularu, A., Koenigstein, N.: Explainable recommendations via attentive multi-persona collaborative filtering. In: Proceedings of the 14th ACM Conference on Recommender Systems, pp. 468–473 (2020)
3. Barkan, O., Shaked, T., Fuchs, Y., Koenigstein, N.: Modeling users' heterogeneous taste with diversified attentive user profiles. User Model. User-Adap. Inter. **34**(2), 375–405 (2024)
4. Cai, D., Qian, S., Fang, Q., Hu, J., Ding, W., Xu, C.: Heterogeneous graph contrastive learning network for personalized micro-video recommendation. IEEE Trans. Multimed. (2022)
5. Cai, X., Huang, C., Xia, L., Ren, X.: Lightgcl: simple yet effective graph contrastive learning for recommendation. arXiv preprint arXiv:2302.08191 (2023)
6. Filali Boubrahimi, S., Hamdi, S.M.: On the mining of time series data counterfactual explanations using barycenters. In: Proceedings of the 31st ACM International Conference on Information & Knowledge Management, pp. 3943–3947 (2022)

7. Gaiger, K., Barkan, O., Tsipory-Samuel, S., Koenigstein, N.: Not all memories created equal: dynamic user representations for collaborative filtering. IEEE Access **11**, 34746–34763 (2023)
8. Geng, Z., Schleich, M., Suciu, D.: Computing rule-based explanations by leveraging counterfactuals. Proc. VLDB Endow. **16**(3), 420–432 (2022)
9. Guo, H., Nguyen, T.H., Yadav, A.: Counternet: end-to-end training of prediction aware counterfactual explanations. In: Proceedings of the 29th ACM SIGKDD Conference on Knowledge Discovery and Data Mining, pp. 577–589 (2023)
10. Guo, Z., Li, J., Li, G., Wang, C., Shi, S., Ruan, B.: Lgmrec: local and global graph learning for multimodal recommendation. In: Proceedings of the AAAI Conference on Artificial Intelligence, vol. 38, pp. 8454–8462 (2024)
11. Hamming, R.W.: Error detecting and error correcting codes. Bell Syst. Tech. J. **29**(2), 147–160 (1950)
12. He, R., McAuley, J.: Ups and downs: modeling the visual evolution of fashion trends with one-class collaborative filtering. In: Proceedings of the 25th International Conference on World Wide Web, pp. 507–517 (2016)
13. He, R., McAuley, J.: Vbpr: visual bayesian personalized ranking from implicit feedback. In: Proceedings of the AAAI Conference on Artificial Intelligence, vol. 30 (2016)
14. He, X., Deng, K., Wang, X., Li, Y., Zhang, Y., Wang, M.: Lightgcn: simplifying and powering graph convolution network for recommendation. In: Proceedings of the 43rd International ACM SIGIR Conference on Research and Development in Information Retrieval, pp. 639–648 (2020)
15. Jiang, J., Leofante, F., Rago, A., Toni, F.: Formalising the robustness of counterfactual explanations for neural networks. In: Proceedings of the AAAI Conference on Artificial Intelligence, vol. 37, pp. 14901–14909 (2023)
16. Karimi, A.H., Barthe, G., Balle, B., Valera, I.: Model-agnostic counterfactual explanations for consequential decisions. In: International Conference on Artificial Intelligence and Statistics, pp. 895–905. PMLR (2020)
17. Kim, D., Park, C., Oh, J., Lee, S., Yu, H.: Convolutional matrix factorization for document context-aware recommendation. In: Proceedings of the 10th ACM Conference on Recommender Systems, pp. 233–240 (2016)
18. Ley, D., Bhatt, U., Weller, A.: Diverse, global and amortised counterfactual explanations for uncertainty estimates. In: Proceedings of the AAAI Conference on Artificial Intelligence, vol. 36, pp. 7390–7398 (2022)
19. Lin, Z., Tian, C., Hou, Y., Zhao, W.X.: Improving graph collaborative filtering with neighborhood-enriched contrastive learning. In: Proceedings of the ACM Web Conference 2022, pp. 2320–2329 (2022)
20. Melchiorre, A.B., Rekabsaz, N., Ganhör, C., Schedl, M.: Protomf: prototype-based matrix factorization for effective and explainable recommendations. In: Proceedings of the 16th ACM Conference on Recommender Systems, pp. 246–256 (2022)
21. Mothilal, R.K., Sharma, A., Tan, C.: Explaining machine learning classifiers through diverse counterfactual explanations. In: Proceedings of the 2020 Conference on Fairness, Accountability, and Transparency, pp. 607–617 (2020)
22. Rendle, S., Freudenthaler, C., Gantner, Z., Schmidt-Thieme, L.: Bpr: bayesian personalized ranking from implicit feedback. arXiv preprint arXiv:1205.2618 (2012)
23. Ribeiro, M.T., Singh, S., Guestrin, C.: Why should i trust you? Explaining the predictions of any classifier. In: Proceedings of the 22nd ACM SIGKDD International Conference on Knowledge Discovery and Data Mining, pp. 1135–1144 (2016)
24. Schleich, M., Geng, Z., Zhang, Y., Suciu, D.: Geco: quality counterfactual explanations in real time. Proc. VLDB Endow. **14**(9) (2021)

25. Scott, M., Su-In, L., et al.: A unified approach to interpreting model predictions. Adv. Neural Inf. Process. Syst. **30**, 4765–4774 (2017)
26. Su, H., Li, J., Li, F., Lu, K., Zhu, L.: Soil: contrastive second-order interest learning for multimodal recommendation. In: ACM Multimedia, 2024
27. Tao, Z., et al.: Self-supervised learning for multimedia recommendation. IEEE Trans. Multimed. **25**, 5107–5116 (2022)
28. Tran, K.H., Ghazimatin, A., Saha Roy, R.: Counterfactual explanations for neural recommenders. In: Proceedings of the 44th International ACM SIGIR Conference on Research and Development in Information Retrieval, pp. 1627–1631 (2021)
29. Ustun, B., Spangher, A., Liu, Y.: Actionable recourse in linear classification. In: Proceedings of the Conference on Fairness, Accountability, and Transparency, pp. 10–19 (2019)
30. Verma, S., Boonsanong, V., Hoang, M., Hines, K., Dickerson, J., Shah, C.: Counterfactual explanations and algorithmic recourses for machine learning: a review. ACM Comput. Surv. **56**(12), 1–42 (2024)
31. Vo, V., et al.: Feature-based learning for diverse and privacy-preserving counterfactual explanations. In: Proceedings of the 29th ACM SIGKDD Conference on Knowledge Discovery and Data Mining, pp. 2211–2222 (2023)
32. Wachter, S., Mittelstadt, B., Russell, C.: Counterfactual explanations without opening the black box: automated decisions and the gdpr. Harv. JL Tech. **31**, 841 (2017)
33. Wang, W., Feng, F., He, X., Nie, L., Chua, T.S.: Denoising implicit feedback for recommendation. In: Proceedings of the 14th ACM International Conference on Web Search and Data Mining, pp. 373–381 (2021)
34. Wexler, J., Pushkarna, M., Bolukbasi, T., Wattenberg, M., Viégas, F., Wilson, J.: The what-if tool: interactive probing of machine learning models. IEEE Trans. Vis. Comput. Graph. **26**(1), 56–65 (2019)
35. Wu, J., et al.: Self-supervised graph learning for recommendation. In: SIGIR, pp. 726–735 (2021)
36. Yan, J., Wang, H.: Self-interpretable time series prediction with counterfactual explanations. In: International Conference on Machine Learning, pp. 39110–39125. PMLR (2023)
37. Yang, Y., Wu, L., Hong, R., Zhang, K., Wang, M.: Enhanced graph learning for collaborative filtering via mutual information maximization. In: SIGIR, pp. 71–80 (2021)
38. Yi, Z., Wang, X., Ounis, I., Macdonald, C.: Multi-modal graph contrastive learning for micro-video recommendation. In: Proceedings of the 45th International ACM SIGIR Conference on Research and Development in Information Retrieval, pp. 1807–1811 (2022)
39. Yu, P., Tan, Z., Lu, G., Bao, B.K.: Multi-view graph convolutional network for multimedia recommendation. In: Proceedings of the 31st ACM International Conference on Multimedia, pp. 6576–6585 (2023)
40. Zemni, M., Chen, M., Zablocki, É., Ben-Younes, H., Pérez, P., Cord, M.: Octet: object-aware counterfactual explanations. In: Proceedings of the IEEE/CVF Conference on Computer Vision and Pattern Recognition, pp. 15062–15071 (2023)
41. Zhang, J., Zhu, Y., Liu, Q., Wu, S., Wang, S., Wang, L.: Mining latent structures for multimedia recommendation. In: Proceedings of the 29th ACM International Conference on Multimedia, pp. 3872–3880 (2021)
42. Zhang, J., Zhu, Y., Liu, Q., Zhang, M., Wu, S., Wang, L.: Latent structure mining with contrastive modality fusion for multimedia recommendation. IEEE Trans. Knowl. Data Eng. **35**(9), 9154–9167 (2022)

43. Zhang, Y., Chen, X., et al.: Explainable recommendation: a survey and new perspectives. Found. Trends® Inf. Retr. **14**(1), 1–101 (2020)
44. Zhou, X., Shen, Z.: A tale of two graphs: freezing and denoising graph structures for multimodal recommendation. In: Proceedings of the 31st ACM International Conference on Multimedia, pp. 935–943 (2023)
45. Zhou, X., et al.: Bootstrap latent representations for multi-modal recommendation. In: Proceedings of the ACM Web Conference 2023, pp. 845–854 (2023)

Inter-attribute Semantic Correlation-Guided Federated Recommender System Against Attribute Inference Attacks

Qiwen Gu, Xuhao Zhao, Yanmin Zhu(✉), Wenze Ma, Jiadi Yu, and Feilong Tang

Shanghai Jiao Tong University, Shanghai, China
{qwgu,zhaoxuhao,yzhu,mawenze991226,jiadiyu,tang_feilong}@sjtu.edu.cn

Abstract. Federated recommender systems (FedRSs) mitigate direct privacy leakage by keeping user data local and sharing only model updates with the server. Nevertheless, FedRSs remain vulnerable to Attribute Inference Attacks (AIAs), which leverage user embeddings to infer sensitive attributes. To defend against AIAs, recent studies have introduced adversarial learning into FedRSs. However, existing methods adopt single-task learning attackers that focus on single attributes in isolation, neglecting the semantic correlations among multiple attributes. This leads to suboptimal attack performance, thereby limiting the model's overall privacy protection during adversarial learning. To address this issue, we propose the Multi-Attribute Collaborative Privacy-Preserving Federated Recommender System (MACPP-FedRS). MACPP-FedRS introduces a multi-task learning attacker that leverages a shared feature extractor to exploit inter-attribute semantic correlations, thereby building stronger attackers and helping the recommendation model learn more privacy-preserving representations. In addition, imbalances in attack performance across different attributes often arise in multi-task learning, with some attributes dominating the optimization process. To promote balanced privacy protection, we design an uncertainty-weighted module that adaptively adjusts the learning emphasis across attributes. Experiments on three real-world datasets demonstrate that MACPP-FedRS reduces privacy leakage under AIAs while maintaining high-quality recommendations. Our code is available at https://github.com/Loretz1/MACPP.

Keywords: Recommender System · Federated learning · Multi-Task Learning · Privacy-preserving

1 Introduction

Recommender systems constitute a fundamental component of modern digital platforms, providing users with personalized content. They are widely applied

H. Jung et al. (Eds.): DASFAA 2026, LNCS 16535, pp. 121–137, 2026.
https://doi.org/10.1007/978-981-92-0363-5_8

in domains such as e-commerce [4], streaming media [9], social networks [20], and online education [22]. Traditional recommender systems store users' interaction histories and personal attribute information on central servers to provide personalized recommendations and address the cold-start problem. This centralized storage directly exposes sensitive user information to potential unauthorized access, misuse, or data breaches, posing serious privacy risks.

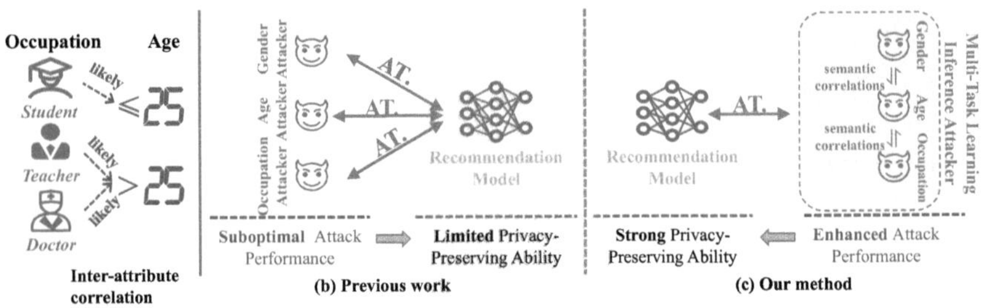

Fig. 1. (a): An example of inter-attribute correlation. (b-c): A model comparison: in contrast previous recommender systems using single-task attackers, our model employs a multi-task attacker to jointly infer multiple attributes. Here, AT. stands for Adversarial Training.

To prevent direct privacy leakage, federated recommender systems (FedRSs) [1,23] have become a widely adopted approach. FedRSs adopt a training method in which clients train the model locally, while a central server collects and aggregates the updates uploaded by clients. Such a training method allows users to retain their private data locally, and clients and the server exchange only model parameters or gradients.

Although FedRSs can effectively mitigate direct privacy leakage, they are still vulnerable to indirect privacy risks. Attackers leverage the model parameters or gradients uploaded by users to infer or reconstruct sensitive information. Attribute Inference Attacks (AIAs) are a typical form of privacy attack, and [3,24] demonstrate that attackers can leverage user embeddings learned from the recommendation model to infer user sensitive attributes (e.g., occupation and age as shown in Fig. 1(a)).

Recent studies [2,3,8,12,15] have introduced adversarial learning into recommender systems to defend against AIAs. In these methods, the framework typically consists of a recommendation model and a set of attackers. Each attacker is trained to predict a specific user attribute based on the user representations. The recommendation model is trained to remove privacy-related information from the user representations so that the attackers cannot accurately infer users' attributes. These two components are optimized alternately in an adversarial manner, where the attackers aim to maximize attribute inference accuracy, while the recommendation model strives to minimize privacy leakage. As shown in Fig. 1(b), current adversarial learning methods for defending against AIAs employ single-task learning attackers, where each attacker is responsible for predicting a single user attribute. These attackers mainly focus on capturing information relevant to their target attribute and largely ignore correlations among multiple attributes, which restricts the extraction of inter-attribute semantic information from user representations.

However, capturing inter-attribute semantic correlations is important for effective attacks. User representations may contain statistical dependencies among different attributes—for example, users who are students are more likely to be under 25 years old (as shown in Fig. 1(a)), and users who are nurses are more likely to be female. By leveraging such inter-attribute correlations, attackers can improve the precision of their attribute inference. Facing such stronger attackers in an adversarial learning manner, the recommendation model is consequently motivated to enhance privacy protection across all attributes.

To fully exploit inter-attribute semantic correlations and improve attacker performance, we propose the Multi-Attribute Collaborative Privacy-Preserving Federated Recommender System (MACPP-FedRS). As shown schematically in Fig. 1(c), MACPP-FedRS employs a multi-task learning attribute attacker that leverages a shared feature extractor to capture semantic correlations among attributes, thereby enhancing inference capabilities across all attributes. During the training of the multi-task learning attacker, imbalances often arise in the attack performance across different attributes. To address this issue, we introduce an uncertainty-weighted module that adaptively adjusts the training emphasis across attributes, preventing any single attribute attack from dominating the multi-task training process and maintaining balanced learning.

In summary, the main contributions of this work are as follows:

- To the best of our knowledge, our work is the first to employ multi-task learning for collaborative protection of multiple user attributes in FedRSs.
- We employ an uncertainty-weighted module to dynamically balance task weights during training, mitigating attribute protection imbalance.
- Experiments on three real-world datasets show that MACPP-FedRS effectively defends against AIAs while preserving recommendation performance.

2 Preliminaries

2.1 Federated Recommender Systems

In a federated recommender system, we consider M users and N items. Let $\mathcal{U} = \{u_1, \ldots, u_M\}$ denote the user set and $\mathcal{I} = \{i_1, \ldots, i_N\}$ the item set. To preserve privacy, each user u keeps the local interaction matrix $\mathbf{A}_u$ on the client side, which is composed of tuples (u, i, r_{ui}), where $r_{ui} = 1$ indicates that u has interacted with i, and $r_{ui} = 0$ otherwise. Using $\mathbf{A}_u$, client u trains a local recommendation model, which consists of an representation layer $f_\theta(\cdot)$ and a rating predictor $p_\phi(\cdot)$. $f_\theta(\cdot)$ maps users or items to representations, while $p_\phi(\cdot)$ predicts the user's preference for an item. The training objective for client u is then defined as follows:

$$\min_{\theta,\phi} \sum_{i \in \mathcal{I}} \mathcal{L}(p_\phi(f_\theta(u), f_\theta(i)), r_{ui}), \tag{1}$$

where $\mathcal{L}$ denotes the loss function. After local training, each client uploads its local parameters θ, ϕ to the server for aggregation. In each round, the server selects a subset $\mathcal{U}^* \in \mathcal{U}$ and aggregates their parameters $\{\Theta_u \mid u \in \mathcal{U}^*\}$ to construct a global model, which is then distributed to the clients for recommendation.

2.2 Attribute Inference Attacks

AIAs aim to infer user attributes from representations in FedRSs. Each user is associated with a set of attributes $\mathcal{T}$ (e.g., $\mathcal{T} = \{gender, age, occupation\}$), and the attribute value of user u on t is denoted as $y_{u,t}$, where $t \in \mathcal{T}$. We denote $\mathcal{T}_u \subseteq \mathcal{T}$ as the set of attributes of user u that require privacy protection (e.g., $\mathcal{T}_u = \{gender, age\}$), while $\mathcal{T} \setminus \mathcal{T}_u$ represents the non-sensitive attributes for u. We define γ as the probability that an attribute is regarded as private for a user. The presence of privacy-unaware users, denoted by $\mathcal{U}_{ins}$, creates opportunities for AIAs, as these users do not safeguard their attributes, making them potentially accessible together with model parameters. Formally, the publicly available information is represented as $D_{pub} = \{(\Theta_u, y_{u,t}, t) \mid u \in U_{ins}, t \in \mathcal{T}\}$. We assume the attacker has access to all users' local parameters Θ_u and the publicly revealed information D_{pub}, which is denoted as $\mathcal{K}$. A typical attacker trains an attack model A_t on the knowledge $\mathcal{K}$, so that the trained model takes Θ_u as input and outputs the attribute value $y_{u,t}$. This can be formalized as:

$$A_t : (\Theta_u, \mathcal{K}) \rightarrow y_{u,t}. \tag{2}$$

Therefore, FedRSs must design privacy-preserving algorithms to defend against AIAs while ensuring recommendation quality, with protection of sensitive attributes $\mathcal{T}_u$ essential to prevent leakage.

3 Methodology

In order to effectively leverage inter-attribute semantic correlations, we propose a new framework called Multi-Attribute Collaborative Privacy-Preserving Federated recommender System (MACPP-FedRS), as illustrated in Fig. 2. The framework adopts adversarial training, consisting of two modules: the recommendation model, which includes a representation layer f_θ and a rating predictor p_ϕ, and the multi-task learning attacker, which employs a shared feature extractor s_π and attribute towers g_ψ.

3.1 Multi-task Learning Attacker

The goal of the multi-task learning attacker is to infer multiple attributes of a user based on the user's representation. The attacker employs a shared feature extractor s_π at the bottom, while the top-level tower layers g_{ψ_t} handle the inference task for each individual attribute t. The shared feature extractor

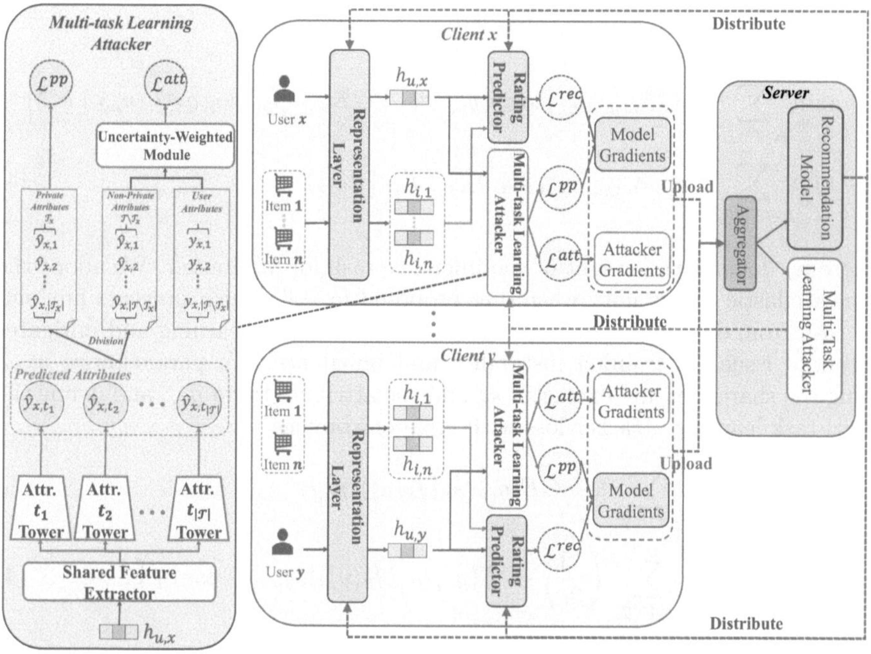

Fig. 2. Overall Framework of MACPP-FedRS, where Attr. stands for attribute.

allows multiple inference tasks to share the underlying feature representations, enabling collaborative learning between tasks. This design jointly identifies features that are prone to leaking multiple attributes, thereby fully leveraging the inter-attribute semantic correlation for inference.

To protect users' private attributes $\mathcal{T}_u$, the attacker is locally optimized with the objective of inferring the users' non-private attributes $\mathcal{T} \setminus \mathcal{T}_u$. Assume that the representation of user u obtained from the encoder f_θ is denoted as h_u, and the distribution of user u's attribute t is $p(y_{u,t} \mid h_u)$. To estimate the distribution $p(y_{u,t} \mid h_u)$, we introduce an auxiliary distribution $q_{\pi,\psi_t}(y_{u,t} \mid h_u)$ to approximate $p(y_{u,t} \mid h_u)$. We employ the Kullback–Leibler (KL) divergence to measure the discrepancy between the two distributions and minimize the KL divergence to make the two distributions similar. Owing to the varying magnitudes of KL divergences across different attributes, we introduce an attribute-specific weight ω. The ultimate objective is to minimize the weighted sum of the KL divergences over all non-private attributes:

$$\begin{aligned}
&\min_{\pi,\psi} \sum_{t\in\mathcal{T}\setminus\mathcal{T}_u} \omega_t \cdot KL(p(y_{u,t} \mid h_u) \parallel q_{\pi,\psi_t}(y_{u,t} \mid h_u)) \\
=&\min_{\pi,\psi} \sum_{t\in\mathcal{T}\setminus\mathcal{T}_u} \omega_t \cdot \left[\mathbb{E}_{p(y_{u,t}|h_u)} \log p(y_{u,t} \mid h_u) - \mathbb{E}_{p(y_{u,t}|h_u)} \log q_{\pi,\psi_t}(y_{u,t} \mid h_u)\right] \\
=&\min_{\pi,\psi} \sum_{t\in\mathcal{T}\setminus\mathcal{T}_u} -\omega_t \cdot \mathbb{E}_{p(y_{u,t}|h_u)} \left[\log q_{\pi,\psi_t}(y_{u,t} \mid h_u)\right], \qquad (3)
\end{aligned}$$

where w_t donates the weight of the inference task for attribute t. We adopt the homoscedastic uncertainty weighting proposed in [13], which prevents any single task from dominating the optimization in multi-task learning and mitigates potential issues of training instability and imbalance. We parameterize q_{π,ψ_t} using the shared feature extractor s_π and the attribute tower g_{ψ_t}, and define the multi-task learning attacker loss with cross-entropy as:

$$\begin{aligned}
\mathcal{L}^{att} &= \sum_{t\in\mathcal{T}\setminus\mathcal{T}_u} \omega_t \cdot CE(g_{\psi_t}(s_\pi(f_\theta(u)), y_{u,t}) \\
&= \sum_{t\in\mathcal{T}\setminus\mathcal{T}_u} \left(\frac{1}{2\sigma_t^2} \cdot CE\big(g_{\psi_t}(s_\pi(f_\theta(u))), y_{u,t}\big) + \log\sigma_t\right), \qquad (4)
\end{aligned}$$

where σ_t denotes the learnable uncertainty of attribute t, reflecting the inherent noise of its prediction task. By learning σ_t during training, the model automatically adjusts the relative weights of attribute-specific losses to promote balanced multi-task optimization. From Eq. (4), the attacker is trained on non-private attributes. Therefore, the privacy attribute probability γ, which denotes the likelihood that a user attribute is private, may influence the model performance.

3.2 Recommendation Model

The recommendation model consists of a representation layer f_θ and a rating predictor p_ϕ. The representation layer f_θ generates user and item embeddings through an embedding module and refines them through a propagation module yielding $h_u = f_\theta(u) \in \mathbb{R}^d$ and $h_i = f_\theta(i) \in \mathbb{R}^d$. The rating predictor p_ϕ maps the representations of the user and the item to a predicted preference score, indicating the user's affinity for the item.

The recommendation model has two objectives: protecting the private attributes of users and improving the precision of the recommendation. The recommendation model aims to protect each user's private attributes $\mathcal{T}_u$, preventing the attacker from correctly inferring these attributes. The privacy leakage risk in FedRSs mainly arises from the fact that user representations contain attribute information when clients train their local models. Therefore, during local model training, it is necessary to make each attribute t independent of the user representation $f_\theta(u)$. We use mutual information [18] to formally quantify the correlation between a user's representation $f_\theta(u)$ and attribute value $y_{u,t}$, and aim to minimize this mutual information:

$$\min_{\theta} \sum_{t \in \mathcal{T}_u} I(f_\theta(u); y_{u,t}). \tag{5}$$

We employ the contrastive log-ratio upper bound [7] to obtain an upper bound for the minimization objective described above:

$$\sum_{t \in \mathcal{T}_u} I(f_\theta(u); y_{u,t}) \leq \sum_{t \in \mathcal{T}_u} I_{vCLUB}(f_\theta(u); y_{u,t}), \tag{6}$$

with the condition:

$$\max_{\pi, \psi_t} \mathbb{E}_{p(f_\theta(u), y_{u,t})}[\log q_{\pi,\psi_t}(y_{u,t}|f_\theta(u))], \quad \forall t \in \mathcal{T}_u. \tag{7}$$

Therefore, the problem of minimizing mutual information can be equivalently reformulated as:

$$\begin{aligned} \min_{\theta} \sum_{t \in \mathcal{T}_u} I(f_\theta(u); y_{u,t}) &\Longleftrightarrow \min_{\theta} \sum_{t \in \mathcal{T}_u} I_{vCLUB}(f_\theta(u); y_{u,t}) \\ \Longleftrightarrow \min_{\theta} \sum_{t \in \mathcal{T}_u} \max_{\pi, \psi_t} \mathbb{E}_{p(f_\theta(u), y_{u,t})}[\log q_{\pi,\psi_t}(y_{u,t}|f_\theta(u))]. \end{aligned} \tag{8}$$

We parameterize q_{π,ψ_t} using the shared feature extractor s_π and the attribute tower g_{ψ_t}, and employ the KL divergence to approximate the above minimization objective:

$$\begin{aligned} &\min_{\theta} \sum_{t \in \mathcal{T}_u} \max_{\pi, \psi_t} \mathbb{E}_{p(f_\theta(u), y_{u,t})}[\log q_{\pi,\psi_t}(y_{u,t}|f_\theta(u))] \\ &\approx \max_{\theta} \sum_{t \in \mathcal{T}_u} KL\left(g_{\psi_t}\left(s_\pi\left(f_\theta(u)\right)\right) \| \, p(y_{u,t})\right). \end{aligned} \tag{9}$$

Since users do not use the values of private attributes during local training, we approximate the above KL divergence with a discrete uniform distribution:

$$\min_{\theta} \sum_{t \in \mathcal{T}_u} KL\left(g_{\psi_t}\left(s_\pi\left(f_\theta(u)\right)\right) \| \, \mathrm{Unif}(\mathcal{C}_t)\right), \tag{10}$$

where $\mathrm{Unif}(\mathcal{C}_t)$ represents the uniform distribution defined on the category set of attribute t. This implies that the recommendation model should eliminate as much attribute semantic information as possible from $f_\theta(u)$, so that the multi-task learning attacker tends to produce highly uncertain predictions that are close to a uniform distribution over the attribute categories. The privacy-preserving loss of the recommendation model can be formulated as:

$$\mathcal{L}^{pp} = \sum_{t \in \mathcal{T}_u} KL\left(g_{\psi_t}\left(s_\pi\left(f_\theta(u)\right)\right) \| \, \mathrm{Unif}(\mathcal{C}_t)\right). \tag{11}$$

Meanwhile, another optimization objective of the recommendation model is to enhance the accuracy of the recommendation information provided to users:

$$\min_{\theta,\phi} \sum_{i \in \mathcal{I}} \mathcal{L}\left(p_\phi\left(f_\theta(u), f_\theta(i)\right), r_{ui}\right). \tag{12}$$

We adopt BPR [19] as the recommendation loss. The training set $\mathcal{D}_u$ consists of pairs (i, i'), where i denotes a sampled positive item and i' denotes a sampled negative item. The recommendation loss is formulated as:

$$\mathcal{L}^{rec} = \sum_{(i,i') \in \mathcal{D}_u} - \log\left(\text{Sigmoid}(p_\phi(f_\theta(u), f_\theta(i)) - p_\phi(f_\theta(u), f_\theta(i')))\right). \tag{13}$$

Combining the two optimization objectives above, the recommendation model loss is formulated as:

$$\mathcal{L}^{total} = \beta \mathcal{L}^{rec} + (1 - \beta)\mathcal{L}^{pp}, \tag{14}$$

where $\beta \in [0, 1]$ is the trade-off that balances recommendation utility and privacy preservation. A higher value of β emphasizes the accuracy of the recommendation, while a lower value of β prioritizes privacy protection.

Algorithm 1: Procedures for Training MACPP-FedRS

```
Input: User set U, item set I, learning rates η1 for recommendation model, η2
       for attacker, trade-off β, initial parameters Θ0 = {θ0, φ0} for the
       recommendation model, Ψ0 = {π0, ψ0, ω0} for attacker, local epochs
       E_rec, E_att, global rounds R
Output: Trained recommendation model parameters Θ_R and attacker
        parameters Ψ_R
Initialize Θ^(0) ← Θ0, Ψ^(0) ← Ψ0;
for global round r = 1, ..., R do
    Server selects a subset of users U* ⊆ U;
    Server sends Θ^(r−1), Ψ^(r−1) to all users in U*;
    for each user u ∈ U* in parallel do
        Receive Θ^(r−1), Ψ^(r−1);
        Let T_u denote the private attribute subset of user u;
        for epoch e = 1, ..., E_rec do
            Sample local interaction data D_u;
            Compute recommendation loss L^rec on D_u;
            Compute privacy-preserving loss L^pp on private attributes T_u;
            // Private attributes are optimized without requiring true values
            L^total ← βL^rec + (1 − β)L^pp;
            Θ_u ← Θ_u − η1 ∇_{Θ_u} L^total;
        for epoch e = 1, ..., E_att do
            Compute attacker loss L^att on non-private attributes T \ T_u;
            // Non-private attributes are used with their true values
            Ψ_u ← Ψ_u − η2 ∇_{Ψ_u} L^att;
        Send Θ_u, Ψ_u to server;
    Aggregate {Θ_u | u ∈ U*} and {Ψ_u | u ∈ U*} to update Θ^(r), Ψ^(r);
return Θ^(R), Ψ^(R);
```

3.3 Training Procedures

The training procedure of MACPP-FedRS, summarized in Algorithm 1, proceeds in multiple federated rounds. In each round, a subset of clients is selected, and the current global recommendation model parameters Θ and attacker model parameters Ψ are distributed to them. Each client locally updates the recommendation model on its interaction data and private attributes, and the attacker on non-private attributes, then uploads both updated models to the server, which aggregates them to form the new global models. This process is repeated until convergence, yielding the final parameters Θ and Ψ.

3.4 Complexity Analysis

The computational complexity of the MACPP-FedRS is mainly affected by the embedding dimension d, the number of local samples per user in a round $|\mathcal{D}_u|$, local training epochs for recommendation model and attacker (E_{rec} and E_{att}), the number of attributes A, the number of users participating in each round $|\mathcal{U}^*|$, and global rounds R. For each user, local training consists of two parts: the recommendation model with complexity $\mathcal{O}(E_{\text{rec}} \cdot |\mathcal{D}_u| \cdot d)$, and the attacker complexity $\mathcal{O}(E_{\text{att}} \cdot A \cdot d)$. Therefore, the overall local training complexity for all users across R rounds is $\mathcal{O}(R \cdot |\mathcal{U}^*| \cdot d \cdot (E_{\text{rec}} \cdot |\mathcal{D}_u| + E_{\text{att}} \cdot A))$. Assuming E_{rec}, E_{att}, and A are constants, the total complexity simplifies to $\mathcal{O}(R \cdot d \cdot |\mathcal{U}^*| \cdot |\mathcal{D}_u|)$.

In summary, our MACPP-FedRS achieves competitive computational efficiency compared with other baselines (e.g., FedGNN [21] and UC-FedRec [12]).

4 Experiments

In this section, to validate the recommendation utility and privacy-preserving capability of MACPP-FedRS, we conducted a series of experiments. The experimental results address the following questions: (**RQ1**): How does MACPP-FedRS perform compared to other privacy-preserving federated recommendation methods? (**RQ2**): How does the hyperparameter β, which balances recommendation performance and privacy protection, affect the trade-off between these two objectives? (**RQ3**): How do the key components of MACPP-FedRS contribute to its recommendation performance and privacy protection? (**RQ4**): How does the factor γ, representing privacy attribute probability, affect the model's performance?

4.1 Experimental Settings

Datasets. We select three public datasets that contain user-side information for experiments, including MovieLens-100K[1], MovieLens-1M[2] [10], and Last.FM-

[1] https://grouplens.org/datasets/movielens/.
[2] https://grouplens.org/datasets/movielens/.

1K[3] [5]. Table 1 summarizes the statistics of the datasets. The user-side information of MovieLens-100K and MovieLens-1M includes gender, age, and occupation. The user-side information of Last.FM-1K includes gender, age, and location, but since the majority of users have missing age values, only gender and location are considered in the experiments. The age attribute in MovieLens is divided into 7 intervals, while the location attribute in Last.FM-1K is merged into 24 categories. For each user, each attribute is randomly marked as private information with a probability of γ. Users who have missing attribute values or insufficient interactions are excluded from the dataset. The implicit feedback of each user is treated as a positive instance, with the most recent positive instance reserved for testing and the remaining instances used for training.

Table 1. Statistics of the datasets.

Dataset	Number			Sparsity
	users	items	interactions	
MovieLens-100K	943	1,682	10,000	99.40%
MovieLens-1M	6,039	3,705	1,000,209	95.53%
Last.FM-1K	839	156,555	761,165	99.42%

Evaluation Metrics. We evaluate the model primarily from two aspects: recommendation performance and privacy protection capability. To evaluate the precision of the recommendation, we adopt the leave-one-out strategy [11]. The test set is constructed by combining each user's most recent positive instance with several negative instances that the user has not interacted with. *HR@K* and *NDCG@K* are employed to evaluate the quality of the recommendations in the top-K ranking lists. Higher *HR@K* and *NDCG@K* indicate better recommendation performance. To evaluate the resistance against AIAs, we construct and train a neural network as an attribute classifier. The classifier acts as an attacker to infer user attributes, thus assessing the risk of privacy leakage in user representations. Specifically, we use AUC and *F1 score* to evaluate binary and multi-class attributes respectively. Lower AUC or *F1 score* indicates stronger privacy protection.

Parameter Configurations. In our work, unless otherwise specified, we set the dimension of the representation layer output $d = 256$, the local batch size $|\mathcal{D}_u| = 256$, local epochs $E_{rec} = 10, E_{att} = 3$, and adopt the dot product as the rating predictor. For the multi-task learning attacker, we configure the shared feature extractor as a 3-layer perceptron. We perform grid search to find the best combination of learning rates, with η_1 for the recommendation model selected from $\{0.3, 0.5, 0.7, 1\}$ and η_2 for the attacker selected from $\{0.015, 0.025, 0.035, 0.05\}$. We set the trade-off $\beta = 0.5$ and the privacy attribute probability $\gamma = 0.2$.

[3] http://ocelma.net/MusicRecommendationDataset/lastfm-1K.html.

4.2 Baselines

Building on FedNCF [17] and FedGNN [21], we compare our proposed method with several baselines in terms of privacy protection and recommendation performance:

- **Early stopping**: By stopping the training of FedNCF or FedGNN early, the representations encode less user attribute information, sacrificing utility to enhance privacy protection.
- **Regularization**: L2 regularization is added to the parameters of the base model to restrict the representation of user attributes, thereby enhancing privacy protection at the cost of some recommendation utility.
- **Laplace Noise-based LDP** [16]: Laplace noise is added to client-side gradients before uploading in federated training to achieve local differential privacy. The privacy budget ϵ determines the strength of privacy protection.
- **UC-FedRec** [12]: A federated recommender system with compositional privacy protection, capable of meeting diverse user privacy demands.

Table 2. Comparison of MACPP-FedRS with other privacy-preserving methods w.r.t. recommendation performance and privacy protection. The best is marked in **bold** while the second best is in underline.

Base Model	Protection Methods	ML-100K					ML-1M				
		Utility(↑)		Privacy(↓)			Utility(↑)		Privacy(↓)		
		H@10	N@10	Gen	Age	Occu	H@10	N@10	Gen	Age	Occu
Fed GNN	No Protection	**0.8144**	**0.5146**	0.7249	0.4104	0.2351	**0.8513**	**0.5607**	0.8302	0.4922	0.1865
	Early Stopping	0.7922	<u>0.4993</u>	0.6962	0.3582	0.1978	0.8147	0.5111	0.7867	0.4492	0.1848
	Regularization	0.7964	0.4956	0.7218	0.4142	0.1828	0.8106	0.5040	0.8158	0.4707	0.1871
	LDP	0.7603	0.4652	0.7413	<u>0.3321</u>	<u>0.1642</u>	0.7396	0.4422	0.8007	0.4631	0.1732
	UC-FedRec	0.7869	0.4845	<u>0.6682</u>	<u>0.3321</u>	0.2052	0.8096	0.5084	<u>0.7548</u>	<u>0.4038</u>	<u>0.1546</u>
	MACPP-FedRS	<u>0.8006</u>	<u>0.4993</u>	**0.5537**	**0.3246**	**0.1567**	<u>0.8152</u>	<u>0.5312</u>	**0.6770**	**0.3887**	**0.1365**
Fed NCF	No Protection	**0.8123**	**0.5130**	0.7383	0.4030	0.1978	**0.8472**	**0.5585**	0.8102	0.4893	0.1825
	Early Stopping	0.7975	0.4921	0.7156	0.3769	<u>0.1791</u>	0.8070	0.5039	0.7857	0.4404	0.1610
	Regularization	0.7826	0.4712	<u>0.7072</u>	0.3955	0.1978	0.7846	0.4704	0.8127	0.4317	0.1830
	LDP	0.7784	0.4884	0.7642	0.4030	0.1866	0.7733	0.4786	0.7671	0.4428	0.1592
	UC-FedRec	0.7932	0.4890	0.7244	**0.3507**	0.2090	0.7661	0.4751	<u>0.7135</u>	<u>0.3969</u>	<u>0.1470</u>
	MACPP-FedRS	<u>0.7996</u>	<u>0.4926</u>	**0.6926**	<u>0.3545</u>	**0.1642**	<u>0.8084</u>	<u>0.5223</u>	**0.6871**	**0.3864**	**0.1429**

Base Model	Protection Methods	Last.FM-1K			
		Utility(↑)		Privacy(↓)	
		H@10	N@10	Gen	Loc
Fed GNN	No Protection	**0.7735**	**0.5716**	0.7172	0.3138
	Early Stopping	0.7569	0.5661	0.6869	0.2887
	Regularization	0.7449	0.5232	0.7306	0.2720
	LDP	0.7271	0.5187	<u>0.5682</u>	0.2887
	UC-FedRec	0.7426	0.5572	0.6104	<u>0.2679</u>
	MACPP-FedRS	<u>0.7581</u>	<u>0.5674</u>	**0.5547**	**0.2594**
Fed NCF	No Protection	**0.7616**	**0.5721**	0.6493	0.2971
	Early Stopping	0.7569	0.5628	0.6444	<u>0.2845</u>
	Regularization	0.7521	0.5376	0.6513	0.3096
	LDP	0.7390	0.5477	0.6481	<u>0.2845</u>
	UC-FedRec	0.7509	0.5601	<u>0.6433</u>	**0.2803**
	MACPP-FedRS	<u>0.7581</u>	<u>0.5654</u>	**0.6054**	**0.2803**

4.3 Performance Comparison (RQ1)

As shown in Table 2, MACPP-FedRS maintains top *HR@10* and *NDCG@10* scores while providing the strongest protection for nearly all user attributes, demonstrating effective privacy defense without sacrificing recommendation quality. The key advantage of MACPP-FedRS lies in its explicit modeling of inter-attribute semantic correlations through multi-task learning, which enables collaborative protection of multiple attributes and mitigates the risk of privacy leakage arising from correlations between them. In comparison, approaches based on local differential privacy or parameter regularization rely on noise injection or simple constraints, offering less precise and often inconsistent privacy control. UC-FedRec protects each attribute independently and does not exploit semantic dependencies, resulting in weaker overall privacy defense.

4.4 Hyperparameter Study (RQ2)

We vary the trade-off β in $\{0.1, 0.3, 0.5, 0.7\}$ to examine its effect on utility and privacy. As shown in Fig. 3, both recommendation performance and defense performance against AIAs are evaluated as percentage changes relative to the non-protected model. With increasing β, recommendation utility improves, whereas privacy protection gradually weakens. We observe that $\beta = 0.5$ achieves recommendation performance close to the non-protected model while providing effective privacy protection, and therefore we set it as the default value.

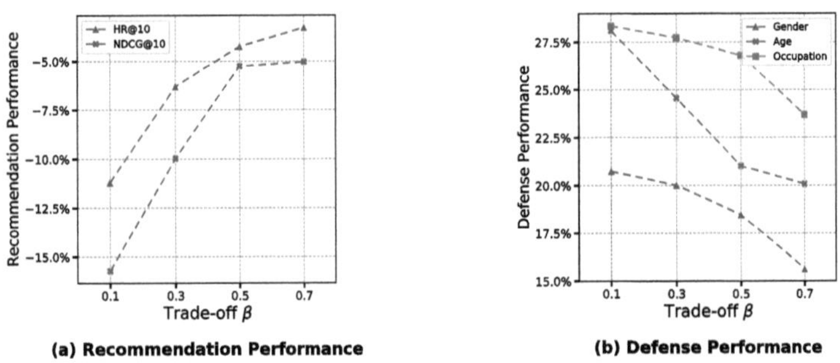

Fig. 3. Hyperparameter Study of MACPP-FedRS.

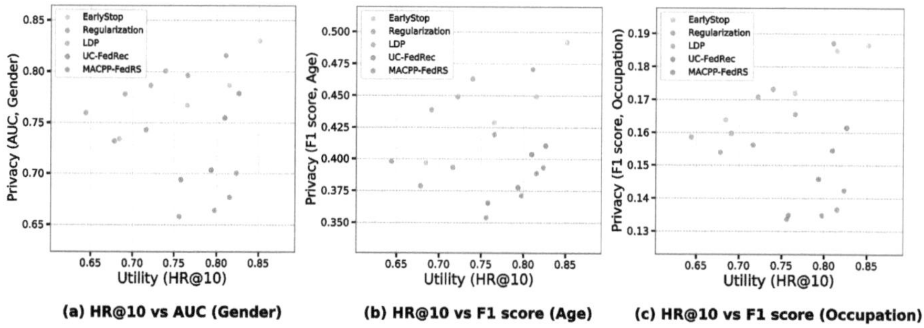

Fig. 4. Comparison of MACPP-FedRS and other privacy-preserving methods in balancing recommendation utility and privacy protection across different trade-off hyperparameter values.

Furthermore, we compare the trade-off capability of our method with other approaches under their respective hyperparameter settings. Figure 4 shows the utility–privacy trade-off for Regularization with weight $\lambda = \{0.0001, 0.0025, 0.0004, 0.0005\}$, LDP with privacy budget $\epsilon = \{1.5, 3, 4.5, 6\}$, UC-FedRec with trade-off $\beta = \{0.1, 0.3, 0.5, 0.7\}$, and MACPP-FedRS with trade-off $\beta = \{0.1, 0.3, 0.5, 0.7\}$. MACPP-FedRS consistently achieves the best balance between recommendation utility and multi-attribute privacy protection. For a similar level of utility, MACPP-FedRS delivers stronger privacy protection than competing methods.

4.5 Ablation Study (RQ3)

To study the effect of each component in MACPP-FedRS, we performed ablation experiments using several variants of the model:

- ***MACPP-w/o-W***: This variant removes the uncertainty-weighted module and calculates $\mathcal{L}^{att}$ by summing the attribute losses directly.
- ***MACPP-w/o-S***: In this version, the shared feature extractor is omitted, and each attribute tower directly takes the user representation as input.

Table 3 presents the ablation study of MACPP-FedRS. w/o-W slightly improves privacy for certain individual attributes under similar recommendation utility, but significantly weakens protection for most other attributes, indicating that the uncertainty-weighted module helps balance multi-task training and prevents dominance of specific attributes. w/o-S slightly reduces recommendation performance and weakens privacy protection across all attributes, as the model loses the ability to exploit inter-attribute semantic correlation, suggesting that the shared feature extractor facilitates improved multi-attribute privacy protection.

Table 3. Ablation Study of MACPP-FedRS. The best is marked in **bold**.

Protection Methods	ML-100K					ML-1M					Last.FM-1K			
	Utility(↑)		Privacy(↓)			Utility(↑)		Privacy(↓)			Utility(↑)		Privacy(↓)	
	H@10	N@10	Gen	Age	Occu	H@10	N@10	Gen	Age	Occu	H@10	N@10	Gen	Loc
w/o-W	0.7879	0.4838	0.6340	0.3582	**0.1455**	**0.8166**	0.5261	0.7270	**0.3841**	0.1517	0.7557	**0.5698**	0.6491	0.2678
w/o-S	0.7847	0.4823	0.6871	0.3321	0.2201	0.8098	0.5159	0.7349	0.4329	0.1505	**0.7581**	0.5550	0.5668	0.2636
MACPP	**0.8006**	**0.4993**	**0.5537**	**0.3246**	0.1567	0.8152	**0.5312**	**0.6770**	0.3864	**0.1365**	**0.7581**	0.5674	**0.5547**	**0.2594**

4.6 Factor Analysis (RQ4)

The privacy attribute probability γ denotes the proportion of user attributes designated as private. A larger γ means users are more likely to designate an attribute as private, whereas a smaller γ corresponds to a lower likelihood of being designated as private. To ensure fair comparison, we adjust the utility-privacy trade-off β so that models under different γ achieve similar recommendation performance. Figure 5 presents the results of the factor analysis. Across all datasets, increasing γ leads to weaker protection strength for all attributes. This is because a larger γ reduces the attacker's training data, constraining its learning of both single-attribute semantics and inter-attribute correlations, thereby weakening privacy protection.

5 Related Work

Federated Recommender Systems. Compared with the centralized recommender system, the federated recommender system is a distributed method, which avoids the situation that the server saves and processes all user data. It can guarantee the performance of the recommendations and protect the privacy of the users. Many researchers have studied the personalized modeling algorithm of the federated recommender system. FCF [1], FedMF [6] and FedNCF [17] integrate collaborative filtering into federated recommender systems, where user–user and item–item similarities are exploited to infer user preferences. Some studies [21] have explored the incorporation of graph neural networks into federated recommender systems to enhance recommendation performance. Some researchers [14] have applied meta-learning [14,27] to federated recommender systems and proposed a novel framework called MetaMF, which improves rating prediction performance in mobile environments. In addition, researchers have also investigated various privacy-preserving techniques, such as differential privacy [16], to ensure user privacy protection of federated recommender systems.

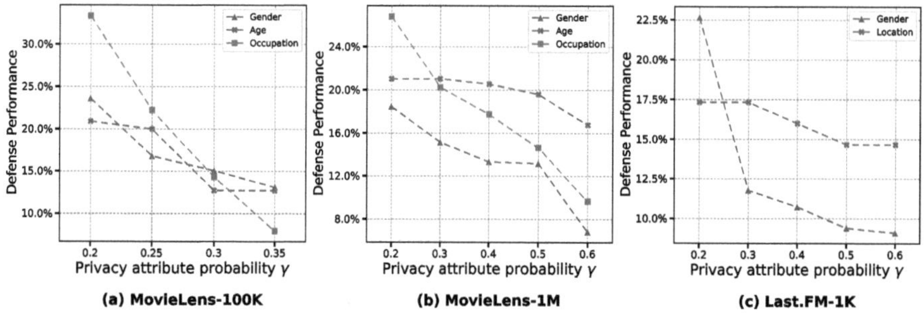

Fig. 5. Factor Analysis of MACPP-FedRS.

Defense Against Attribute Inference Attacks. Attribute inference attacks represent a typical category of indirect privacy threats [24,26]. By exploiting users' local model parameters and gradients, recommendation outputs, or behavioral traces, adversaries can infer sensitive attributes, including age, gender, occupation, and location. Some researchers [3] formulate the problem of recommendation and private attribute protection as an adversarial learning task. By introducing an attacker and training the recommendation model in an adversarial manner, they can effectively prevent the leakage of private attribute information while minimizing the recommendation loss. Some researchers have employed differential privacy to defend against attribute inference attacks. [24] incorporates differential privacy into graph convolutional networks by introducing perturbations at both the input and optimization stages. [25], building on a comprehensive privacy analysis, allocates different privacy budgets to the parameters of various modules in the recommendation model, thereby reducing the utility loss caused by defending against attribute inference attacks. [12] considers the heterogeneous privacy requirements of users and proposes a framework named UC-FedRec, which enables flexible and personalized protection of users' sensitive attributes.

6 Conclusion

In this paper, we propose MACPP-FedRS, a federated recommender framework against AIAs. It employs a multi-task learning attacker to capture inter-attribute correlations and an uncertainty-weighted module to balance optimization across attributes. Experiments show that MACPP-FedRS reduces inference risk and achieves a better trade-off between privacy protection and recommendation utility than existing defenses.

Acknowledgments. This research is supported in part by National Natural Science Foundation of China (No. 62472277, No. 62072304, No. 62572309, No. 62472176), and the Shanghai East Talents Program (2023-177).

Disclosure of Interests. The authors have no competing interests to declare that are relevant to the content of this article.

References

1. Ammad-Ud-Din, M., et al.: Federated collaborative filtering for privacy-preserving personalized recommendation system. arXiv preprint arXiv:1901.09888 (2019)
2. Arevalo, C.A., Noorbakhsh, S.L., Dong, Y., Hong, Y., Wang, B.: Task-agnostic privacy-preserving representation learning for federated learning against attribute inference attacks. In: Proceedings of the AAAI Conference on Artificial Intelligence, vol. 38, pp. 10909–10917 (2024)
3. Beigi, G., Mosallanezhad, A., Guo, R., Alvari, H., Nou, A., Liu, H.: Privacy-aware recommendation with private-attribute protection using adversarial learning. In: Proceedings of the 13th International Conference on Web Search and Data Mining, pp. 34–42 (2020)
4. Bodduluri, K.C., Palma, F., Kurti, A., Jusufi, I., Löwenadler, H.: Exploring the landscape of hybrid recommendation systems in e-commerce: a systematic literature review. IEEE Access **12**, 28273–28296 (2024)
5. Celma, Ò.: Music recommendation and discovery in the long tail.(phd). Universitat Pompeu Fabra, Barcelona, Spain (2008)
6. Chai, D., Wang, L., Chen, K., Yang, Q.: Secure federated matrix factorization. IEEE Intell. Syst. **36**(5), 11–20 (2020)
7. Cheng, P., Hao, W., Dai, S., Liu, J., Gan, Z., Carin, L.: Club: a contrastive log-ratio upper bound of mutual information. In: International Conference on Machine Learning, pp. 1779–1788. PMLR (2020)
8. Ganhör, C., Penz, D., Rekabsaz, N., Lesota, O., Schedl, M.: Unlearning protected user attributes in recommendations with adversarial training. In: Proceedings of the 45th International ACM SIGIR Conference on Research and Development in Information Retrieval, pp. 2142–2147 (2022)
9. Gomez-Uribe, C.A., Hunt, N.: The netflix recommender system: algorithms, business value, and innovation. ACM Trans. Manag. Inf. Syst. (TMIS) **6**(4), 1–19 (2015)
10. Harper, F.M., Konstan, J.A.: The movielens datasets: history and context. ACM Trans. Interact. Intell. Syst. (tiis) **5**(4), 1–19 (2015)
11. He, X., Zhang, H., Kan, M.Y., Chua, T.S.: Fast matrix factorization for online recommendation with implicit feedback. In: Proceedings of the 39th International ACM SIGIR Conference on Research and Development in Information Retrieval, pp. 549–558 (2016)
12. Hu, Q., Song, Y.: User consented federated recommender system against personalized attribute inference attack. In: Proceedings of the 17th ACM International Conference on Web Search and Data Mining, pp. 276–285 (2024)
13. Kendall, A., Gal, Y., Cipolla, R.: Multi-task learning using uncertainty to weigh losses for scene geometry and semantics. In: Proceedings of the IEEE Conference on Computer Vision and Pattern Recognition, pp. 7482–7491 (2018)
14. Lin, Y., et al.: Meta matrix factorization for federated rating predictions. In: Proceedings of the 43rd International ACM SIGIR Conference on Research and Development in Information Retrieval, pp. 981–990 (2020)
15. Liu, S., Du, J., Shrivastava, A., Zhong, L.: Privacy adversarial network: representation learning for mobile data privacy. Proc. ACM Interact. Mob. Wearable Ubiquitous Technol. **3**(4), 1–18 (2019)

16. Minto, L., Haller, M., Livshits, B., Haddadi, H.: Stronger privacy for federated collaborative filtering with implicit feedback. In: Proceedings of the 15th ACM Conference on Recommender Systems, pp. 342–350 (2021)
17. Perifanis, V., Efraimidis, P.S.: Federated neural collaborative filtering. Knowl.-Based Syst. **242**, 108441 (2022)
18. Poole, B., Ozair, S., Van Den Oord, A., Alemi, A., Tucker, G.: On variational bounds of mutual information. In: International Conference on Machine Learning, pp. 5171–5180. PMLR (2019)
19. Rendle, S., Freudenthaler, C., Gantner, Z., Schmidt-Thieme, L.: Bpr: bayesian personalized ranking from implicit feedback. arXiv preprint arXiv:1205.2618 (2012)
20. Sharma, K., et al.: A survey of graph neural networks for social recommender systems. ACM Comput. Surv. **56**(10), 1–34 (2024)
21. Wu, C., Wu, F., Lyu, L., Qi, T., Huang, Y., Xie, X.: A federated graph neural network framework for privacy-preserving personalization. Nat. Commun. **13**(1), 3091 (2022)
22. Wu, Z., Jiang, H., Cheng, X., Huang, H., Yang, Q., Zhang, J.: Metacognition-driven user-to-project recommendation for online education services. World Wide Web **26**(5), 2311–2327 (2023)
23. Yang, L., Tan, B., Zheng, V.W., Chen, K., Yang, Q.: Federated recommendation systems. In: Yang, Q., Fan, L., Yu, H. (eds) Federated Learning. LNCS, vol. 12500, pp. 225–239. Springer, Cham (2020). https://doi.org/10.1007/978-3-030-63076-8_16
24. Zhang, S., Yin, H., Chen, T., Huang, Z., Cui, L., Zhang, X.: Graph embedding for recommendation against attribute inference attacks. In: Proceedings of the Web Conference 2021, pp. 3002–3014 (2021)
25. Zhang, S., Yuan, W., Yin, H.: Comprehensive privacy analysis on federated recommender system against attribute inference attacks. IEEE Trans. Knowl. Data Eng. **36**(3), 987–999 (2023)
26. Zhao, X., et al.: Social relation-level privacy risks and preservation in social recommender systems. In: Proceedings of the 48th International ACM SIGIR Conference on Research and Development in Information Retrieval, pp. 1728–1737 (2025)
27. Zhao, X., Zhu, Y., Wang, C., Jing, M., Yu, J., Tang, F.: Task-difficulty-aware meta-learning with adaptive update strategies for user cold-start recommendation. In: Proceedings of the 32nd ACM International Conference on Information and Knowledge Management, pp. 3484–3493 (2023)

A Cognitive Distribution and Behavior-Consistent Framework for Black-Box Attacks on Recommender Systems

Hongyue Zhang[1,2], Mingming Li[1,2], Dongqin Liu[1,2](✉), Hui Wang[1,2], Yaning Zhang[3], Xi Zhou[1,2], Honglei Lv[1,2](✉), Jiao Dai[1,2], and Jizhong Han[1,2]

[1] Institute of Information Engineering, Chinese Academy of Sciences, Beijing, China
{zhanghongyue,limingming,liudongqin,wanghui4042,zhouxi,lvhonglei, daijiao,hanjizhong}@iie.ac.cn

[2] School of Cyber Security, University of Chinese Academy of Sciences, Beijing, China

[3] NSFOCUS, Beijing, China
zhangyaning2@nsfocus.com

Abstract. With the growing deployment of sequential recommender systems in e-commerce and other fields, their black-box interfaces raise security concerns: models are vulnerable to extraction and subsequent adversarial manipulation. Existing black-box extraction attacks primarily rely on hard labels or pairwise learning, often ignoring the importance of ranking positions, which results in incomplete knowledge transfer. Moreover, adversarial sequences generated via pure gradient methods lack semantic consistency with real user behavior, making them easily detectable. To overcome these limitations, this paper proposes a dual-enhanced attack framework. First, drawing on primacy effects and position bias, we introduce a cognitive distribution-driven extraction mechanism that maps discrete rankings into continuous value distributions with position-aware decay, thereby advancing from order alignment to cognitive distribution alignment. Second, we design a behavior-aware noisy item generation strategy that jointly optimizes collaborative signals and gradient signals. This ensures both semantic coherence and statistical stealth while effectively promoting target item rankings. Extensive experiments on multiple datasets demonstrate that our approach significantly outperforms existing methods in both attack success rate and evasion rate, validating the value of integrating cognitive modeling and behavioral consistency for secure recommender systems.

Keywords: Sequential recommendation · Model extraction · Profile pollution attack · Cognitive distribution alignment

H. Jung et al. (Eds.): DASFAA 2026, LNCS 16535, pp. 138–153, 2026.
https://doi.org/10.1007/978-981-92-0363-5_9

1 Introduction

Recommender systems have become a core component of modern information service platforms, widely deployed in applications such as short videos [30], social media [16], and e-commerce [10–12,20,23]. With the explosive growth of user behavior data, sequential recommendation models have significantly improved personalized recommendation accuracy by modeling the dynamic evolution of user interests [2]. Neural-structure-based models, including RNN and Transformers (e.g., BERT4Rec [18], SASRec [6], and NARM [9]), are particularly effective at capturing complex temporal dependencies and contextual semantic relationships, and are therefore widely adopted in industrial recommender systems.

However, the increasing model complexity and exposed interfaces introduce new security risks. Even without access to internal parameters or training data, attackers can replicate the behavior of a target model through limited Application Programming Interface (API) queries [28]. Once a high-fidelity surrogate model is constructed, it can be exploited to craft adversarial perturbations that mislead the original black-box system [3,8,26]. These findings reveal substantial vulnerabilities in modern recommender systems.

Despite recent progress, existing black-box model extraction and attack mechanisms still suffer from two key limitations. First, most black-box recommendation attacks rely on hard labels or pairwise ranking learning frameworks [21,24]. By modeling only discrete preference relations, these approaches overlook the continuous value information embedded in ranking positions. Such simplification leads to information loss and prevents the surrogate model from accurately capturing the fine-grained decision preferences of the black-box system. Moreover, it ignores attention decay characteristics inherent in human cognition, thereby constraining both the theoretical upper bound and practical effectiveness of model extraction.

Second, pollution items generated solely through gradient-based optimization often lack natural semantic relevance to the target item. As a result, the constructed user behavior sequences may exhibit statistical anomalies, making them vulnerable to detection by existing defense mechanisms [24].

To overcome these limitations, we propose a dual-enhanced black-box attack framework that improves both extraction fidelity and attack stealth. Inspired by the primacy effect [5] and position bias [1], our approach incorporates attention decay theory to reconstruct continuous value distributions from ranking signals, enabling cognitive distribution alignment between surrogate and target models. Meanwhile, we design a behavior-consistent pollution item generation mechanism that integrates collaborative relation analysis with gradient optimization, ensuring semantic coherence in user behavior sequences while maximizing attack impact.

The main contributions of this paper are summarized as follows:

- We advance black-box recommendation model extraction from discrete ranking imitation to cognitive distribution modeling, incorporating attention decay theory to enhance surrogate model fidelity.

- We propose a behavior-consistent pollution item generation paradigm that synergizes collaborative relations and gradient signals, improving attack stealth without sacrificing effectiveness.
- Extensive experiments on multiple public datasets demonstrate that our framework consistently outperforms existing baselines in attack performance and evasion rate, highlighting the importance of cognitive modeling and behavioral consistency in black-box recommendation attacks.

2 Related Work

2.1 Model Extraction in Recommendation

Model extraction attacks construct a surrogate model using limited API queries without access to the target model's parameters or training data. Early studies focused on image/text classification [19], including Knockoff nets [15] and MAZE [7]. For recommender systems, DFME [24] proposed a data-free extraction framework that distills surrogates from synthetic behavior sequences via pairwise ranking loss, and FSME [27] improved query efficiency by introducing limited real data. ME-MIA [31] further extended this line by combining BPR and Hinge losses for membership inference scenarios.

Despite progress, most black-box extraction methods still distill from discrete order constraints, under-utilizing the positional value signal and ignoring attention decay in user perception, which can yield sparse supervision and unstable surrogates. Our method addresses this by distilling a position-aware continuous value distribution while retaining local order preservation, enabling more faithful preference recovery and stronger transferability.

2.2 Attacks on Recommender Systems

Profile pollution (shilling) attacks [8] manipulate recommender outputs by injecting crafted items into user behavior sequences. Early heuristic strategies (e.g., mixing random or popular items) are simple but often ineffective [22]. More recent approaches leverage gradient guidance [24] to select poisoning items in the embedding space for stronger promotion, and INFAttack [3] incorporates influence functions to improve efficiency.

However, purely gradient-driven selection may ignore semantic and behavioral plausibility, producing abnormal sequences that are easier to detect. To enhance stealth without sacrificing effectiveness, our strategy fuses collaborative relation priors with gradient optimization to generate behavior-consistent injections that remain semantically coherent while reliably boosting the target item.

3 Methodology

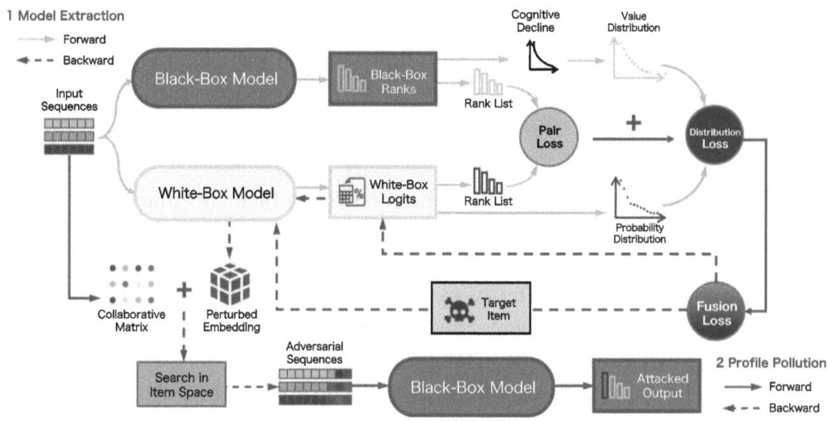

Fig. 1. Dual-enhanced black-box attack framework

We propose a dual-enhanced black-box attack framework, shown in Fig. 1, encompassing two core stages: (i) **cognitive distribution-aligned distillation** for learning a surrogate model that faithfully recovers ranking preferences from limited black-box queries; and (ii) **behavior-consistent pollution item generation**, achieved by a dual-signal fusion integrating gradient guidance with historical collaborative relations. This pipeline, from cognitive modeling and surrogate training to attack generation and promotion validation, ensures both stealthiness and efficacy in sequential recommendation systems.

3.1 Problem Definition and Notations

Let $f_b : \mathcal{X} \to \mathcal{I}^k$ be the target black-box recommender, where $\mathcal{X}$ is the user sequence space and $\mathcal{I}^k$ is an ordered top-k list. A user sequence $x = [i_1, i_2, \ldots, i_T]$ has length T, with $i_t \in \mathcal{I}$ over the item universe $\mathcal{I}$. The attacker, lacking f_b's parameters or training data, observes only truncated rankings via limited API queries. Our objective is learn a high-fidelity surrogate f_w and generate an injection sequence z effectively promote a target item i^* in f_b.

During distillation, we collect a query set $\mathcal{D} = \{(x_i, I_i^k)\}_{i=1}^B$, where $I_i^k = f_b(x_i)$. Surrogate f_w is then optimized by minimizing distillation loss $\mathcal{L}_{\text{dis}}$:

$$f_w^* = \arg\min_{f_w} \sum_{i=1}^{B} \mathcal{L}_{\text{dis}}\big(f_w(x_i), I_i^k\big). \quad (1)$$

For attack generation, given a collaborative matrix $\mathbf{S}$ (from historical logs) and surrogate gradients $\nabla_x \mathcal{L}_{\text{atk}}$, we seek an optimal injection sequence z to maximize target item i^*'s expected exposure under f_b:

$$z^* = \arg\max_z E_t\big(f_b([x; z])\big) \quad \text{s.t. } z \sim \text{Dual-Enhanced}(f_w, \mathbf{S}; w_g, w_s, K, L). \quad (2)$$

3.2 Data Generation and Model Distillation

Data Generation. We synthesize query sequences via an autoregressive sampling strategy. Starting with a random seed item, we repeatedly query the black-box API for a top-k list and sample the next item using a prescribed policy, e.g., position-decay or temperature sampling, until a preset length or budget is reached. This generates sequences reflecting real interaction temporal dependencies, within a limited query budget.

Model Distillation. Our surrogate aims to reproduce the black-box model's rankings and capture its implicit attention distribution pattern. Inspired by cognitive psychology's Primacy Effect and position bias, users' attention and perceived value for list items decay exponentially with ranking position. These insights drive our attention-decay mechanism, evolving distillation from order alignment to cognitive distribution alignment.

Specifically, for an input sequence x, the black-box model outputs an ordered top-k list: $I^k(x) = [i_1, i_2, \ldots, i_k]$. Assuming user attention intensity decays exponentially with position, we define this cognitive prior: $v(j) = \alpha^{j-1}$ with $\alpha \in (0,1)$ controlling decay rate (smaller α for stronger focus on top-ranked items). With temperature parameter $\tau_b > 0$, this prior converts to a soft black-box distribution:

$$p_b(i_j \mid x) = \frac{\exp(v(j)/\tau_b)}{\sum_{t=1}^{k} \exp(v(t)/\tau_b)}. \tag{3}$$

This reconstructs a continuous attention distribution from a discrete top-k ranking. Unlike traditional order-mimicking distillation, our approach better aligns with the user perception's gradient nature, providing a differentiable signal. Our attention-decay formulation is also theoretically consistent with NDCG-like position discounting. Specifically, while NDCG uses $d(i) = 1/\log_2(j+1)$, our cognitive value function is $v(j) = \alpha^{j-1}$. An $\alpha \in (0,1)$ ensures $v(j)$ and $d(j)$ are monotonic and rank-equivalent: $d(j_1) > d(j_2) \iff v(j_1) > v(j_2), \forall j_1 < j_2$. Thus, our attention-based cognitive distribution smoothly approximates the NDCG discounting mechanism, consistent with classical metrics.

For example, a black-box ranking of A, B, C (1st, 2nd, 3rd) is treated as a discrete ordering by traditional distillation (A > B > C). Our method, however, assigns attention weights (e.g., 0.6, 0.3, 0.1) to learn the attention intensity distribution. The surrogate model then produces scores $s_w(i_j; x)$ per position, normalized by a temperature τ_w into $p_w(\cdot \mid x; \tau_w)$. For ensuring cognitive-level consistency beyond rank imitation, we measure the distribution discrepancy using KullbackLeibler divergence:

$$\mathcal{L}_{\mathrm{KL}}(x) = \mathrm{KL}(p_b(\cdot \mid x) \,\|\, p_w(\cdot \mid x; \tau_w)) . \tag{4}$$

However, distribution alignment alone can compromise local ranking order. We thus incorporate a pairwise distillation loss, inspired by prior work, to maintain local structural stability:

$$\mathcal{L}_{\mathrm{pair}}(x) = \tfrac{1}{k-1} \textstyle\sum_{j=1}^{k-1} \max\{0,\, s_w(i_{j+1}) - s_w(i_j) + \delta_1\} + \tfrac{1}{k} \textstyle\sum_{j=1}^{k} \max\{0,\, s_w(i_j^{\mathrm{neg}}) - s_w(i_j) + \delta_2\}, \tag{5}$$

where i_j^{neg} denotes a negative sample paired with i_j, and $\delta_1, \delta_2 > 0$ are margin parameters. This loss ensures local ranking consistency and discriminability while enabling global cognitive alignment.

The final objective combines both global and local signals as follows. Here, λ_{dis} balances structure preservation: larger λ_{dis} prioritizes fine-grained ranking for precise distillation; smaller λ_{dis} emphasizes global cognitive imitation and user attention.

$$\mathcal{L}_{\text{distill}}(x) = \lambda_{\text{dis}}\, \mathcal{L}_{\text{pair}}(x) + (1 - \lambda_{\text{dis}})\, \mathcal{L}_{\text{KL}}(x), \quad \lambda_{\text{dis}} \in [0, 1]. \tag{6}$$

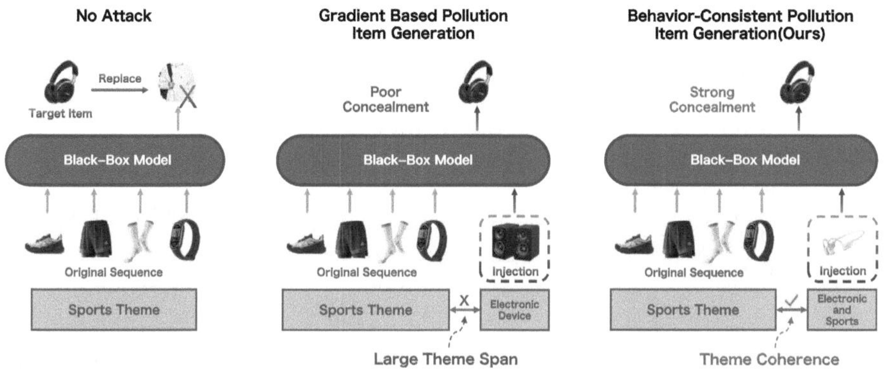

Fig. 2. Concealment comparison: gradient-based vs our behavior-consistent pollution. Pure gradient attacks select semantically unrelated items (e.g., an intelligent speaker for running headphones), improving ranking but violating user patterns and increasing detectability. Our method achieves higher concealment by introducing items (e.g., object motion headphones) aligning with both gradient effectiveness and historical collaborative behavior.

3.3 Behavior-Consistent Pollution Item Generation

We generate injection sequences using a behavior-consistent pollution item generation (Fig. 2) that integrates *gradient guidance* and *behavioral collaboration* to ensure both promotion strength and behavioral plausibility.

Candidate Pool via Collaborative Prior. From historical logs, we construct a collaborative matrix $\mathbf{S}$ and compute a conditional collaboration score $\tilde{s}(j \mid i^*)$ for candidate item j w.r.t. target item i^*. Items with high $\tilde{s}(j \mid i^*)$ constitute a semantically consistent candidate pool.

Gradient-Guided Scoring. We define an attack loss $\mathcal{L}_{\text{atk}}$ on the surrogate and compute the input-embedding gradient $g = \nabla_x \mathcal{L}_{\text{atk}}$. For candidate j with embedding $\mathbf{e}_j$, we measure its gradient alignment $\text{sim}_g(j)$, typically via cosine similarity with the negative averaged gradient direction.

Algorithm 1 Behavior-Consistent Profile Injection

Require: User sequence $\mathbf{x}$, target item i^*, total length L_{total}, surrogate model f_w, embedding layer E_w, gradient step ϵ, fusion weights w_g, w_s.
Ensure: Injected sequence $\mathbf{z}$
$\mathbf{z} \leftarrow \mathbf{x}$
$\mathcal{N}_{i^*} \leftarrow \text{Top-}K \ \arg\max_{v \in \mathcal{I}} \text{CoRel}(v, i^*)$ ▷ Candidate pool from collaborative prior
while $|\mathbf{z}| < L_{\text{total}}$ **do**
  $\mathbf{z}_{\text{curr}} \leftarrow [\mathbf{z}; i^*]$ ▷ Append placeholder target
  $\mathbf{E} \leftarrow E_w(\mathbf{z}_{\text{curr}})$ ▷ Get embeddings sequence $\in \mathbb{R}^{T \times d}$
  $\mathbf{g} \leftarrow \nabla_{\mathbf{E}_{\text{last}}} \mathcal{L}_{\text{ce}}(f_w(\mathbf{z}_{\text{curr}}), i^*)$
  $\mathbf{e}_{\text{adv}} \leftarrow \mathbf{E}_{\text{last}} - \epsilon \cdot \frac{\mathbf{g}}{\|\mathbf{g}\|_2}$ ▷ Gradient descent on embedding space
  $\mathcal{G} \leftarrow \text{Top-}K \ \arg\max_{v \in \mathcal{I}} \text{CosSim}(\mathbf{e}_{\text{adv}}, E_w(v))$
  $\mathcal{H} \leftarrow \mathcal{N}_{i^*} \cup \mathcal{G}$
  Compute scores for candidates:
  for $v \in \mathcal{H}$ **do**
    $S(v) \leftarrow w_g \cdot \text{CosSim}(\mathbf{e}_{\text{adv}}, E_w(v)) + w_s \cdot \text{CoRel}(v, i^*)$
  end for
  $v_{\text{best}} \leftarrow \arg\max_{v \in \mathcal{H}} S(v)$
  $\mathbf{z} \leftarrow [\mathbf{z}; v_{\text{best}}]$ ▷ Append the best candidate
end while
return $\mathbf{z}$

Dual-Signal Fusion. The final candidate score $S(i)$ for candidate item i linearly combines gradient alignment $\text{sim}_g(j)$ and collaborative score $\tilde{s}(j \mid i^*)$:

$$\text{S}(j) = w_g \cdot \text{sim}_g(j) + w_s \cdot \tilde{s}(j \mid i^*), \quad w_g, w_s \geq 0, \ w_g + w_s = 1. \tag{7}$$

Sequence Construction and Validation. Top-ranked candidates are selected, and a greedy or beam search assembles the injection sequence within budget. This sequence is then validated on f_b and limited refinements are made if necessary.

This mechanism jointly optimizes directional effectiveness and behavioral consistency, balancing stealthiness and promotion performance. Details are in Algorithm 1.

4 Experiments

In this section, we conduct extensive experiments to address the following research questions:

- **RQ1:** What does the proposed cognitive distribution-driven knowledge extraction mechanism outperform existing baseline methods in the general model extraction task?
- **RQ2:** How effective is the proposed behavioral pattern-consistent profile injection strategy in successfully promoting the rank of the target item?

- **RQ3:** What are the individual contributions of the cognitive distribution alignment mechanism and the behavioral pattern-consistent generation strategy to the overall performance of the attack framework, as revealed by ablation studies?
- **RQ4:** How sensitive are the model extraction and attack outcomes to the key hyper-parameters?
- **RQ5:** Do the generated injected sequences exhibit statistical indistinguishability from genuine user behavior sequences in terms of behavioral flow consistency and semantic coherence, thereby ensuring the stealthiness of the attack?

4.1 Experimental Setup

Dataset. For the experimental evaluation of our proposed framework, we utilize three publicly available datasets: Movielens-1M (ML-1M) [4], Steam [13], and Amazon Beauty [14]. As shown in Table 1, these datasets exhibit diverse characteristics, which allow for a comprehensive assessment of our framework's generality and robustness.

Baselines. To comprehensively evaluate our framework, we compare it against representative baselines in two aspects.

- **Model Extraction Attacks**
  - **DFME** [24]: Extracts black-box models via synthetic sequence generation and pairwise ranking distillation, aligning the order of top-k items through marginal ranking loss.
  - **ME-MIA** [31]: Optimizes extraction using a combined BPR ranking loss and hinge loss, achieving enhanced performance via engineering refinements.
- **Profile Pollution Attacks**
  - **RandAlter** [17]: Injects random items alternating with target items.
  - **DQN** [25,29]: Employs deep Q-learning with RNN to maximize target item ranking.
  - **WhiteBox SimAlter** [24]: Inserts items similar to the target based on white-box embeddings.
  - **DFME-based pollution** [24]: Leverages a distilled white-box model and gradient signals to craft adversarial sequences that promote target exposure.

All experiments are conducted using three widely used sequential recommendation models as surrogate architectures: **NARM** [9] (RNN-based with attention), **SASRec** [6] (self-attentive Transformer), and **BERT4Rec** [18] (bidirectional Transformer with masked modeling). These cover a spectrum from local dependency modeling to long-range contextual learning.

Table 1. Detailed statistics of experimental datasets

Datasets	Users	Items	Average length	Max length	Sparsity
ML-1M	6,040	3,416	166	2,277	95.16%
Steam	334,542	13,046	13	2,045	99.90%
Beauty	40,226	54,542	9	293	99.98%

Evaluation Metrics. To evaluate our framework, we employ standard ranking metrics (Recall@K and NDCG@K) to assess recommendation quality, along with Agreement@K (Agr@K) to measure consistency between the black-box target and extracted white-box models, following established protocols [6,18,24].

Implementation Details. Our experiments follow standard sequential recommendation protocols. For each user sequence $\boldsymbol{x}$ of length T, we use $\boldsymbol{x}_{[:T-2]}$ for training, the penultimate item for validation, and the last item for testing. To ensure fair comparison, we adhere to the hyperparameter settings from baseline methods [10,14,23], employing Adam optimizer (lr $= 0.001$, weight decay $= 0.01$, batch size $= 128$) with 100 warmup steps. Maximum sequence lengths are set to 200 (ML-1M) and 50 (Steam, Beauty).

Model architectures include: NARM (1 GRU layer), SASRec, and BERT4Rec (2 Transformer layers, 2 attention heads). Our framework uses $\alpha = 0.97$ for cognitive distribution attenuation and temperature $= 0.5$. Dataset-specific parameters (dropout: 0.1–0.5; BERT4Rec masking: 0.2–0.6) follow original publications. For attack evaluation, we set profile pollution sizes to 1% of sequence lengths (10 items for ML-1M, 2 for Steam/Beauty) with injected user counts of 60 (ML-1M), 3345 (Steam), and 402 (Beauty) [24].

4.2 Model Extraction (RQ1)

To rigorously evaluate our framework, we conducted comprehensive experiments across three benchmark datasets (ML-1M, Steam, and Beauty) from three critical perspectives. The specific results, as shown in Table 2, Table 3, and Table 4, systematically demonstrate the effectiveness of our method.

- **Within-Model Comparison**
  Using NARM on ML-1M as an example, our method achieves superior performance on both recommendation quality and ranking consistency, significantly outperforming DFME and MIA baselines. This validates that our cognitive distribution-driven distillation better captures the original model's ranking preferences, especially in critical top positions.
- **Cross-Architecture Transferability**
  Our method consistently achieves the best Agr@1 and Agr@10 metrics across all nine model-dataset combinations. The advantage is particularly pronounced for complex architectures like BERT4Rec, where Agr@1 improves by

over 20% compared to baselines. Notably, while traditional methods degrade when moving from RNN-based to Transformer-based architectures, our approach maintains stable performance, demonstrating strong adaptability to different model internal structures.

Table 2. Performance comparison of model extraction methods on ML-1M

Method	Model Extraction	N@10	R@10	Agr@1	Agr@10
NARM	BlackBox	0.626	0.819	–	–
	WhiteBox-DFME	0.615	0.812	0.571	0.747
	WhiteBox-ME-MIA	0.615	0.811	0.551	0.724
	WhiteBox-Ours	**0.617**	**0.813**	**0.601**	**0.757**
SASRec	BlackBox	0.605	0.827	–	–
	WhiteBox-DFME	**0.602**	0.802	0.454	0.662
	WhiteBox-ME-MIA	0.594	0.820	0.528	0.711
	WhiteBox-Ours	0.595	**0.818**	**0.573**	**0.724**
BERT4Rec	BlackBox	0.590	0.794	–	–
	WhiteBox-DFME	0.559	0.775	0.335	0.581
	WhiteBox-ME-MIA	0.560	0.773	0.308	0.569
	WhiteBox-Ours	**0.571**	**0.782**	**0.409**	**0.631**

Table 3. Performance comparison of model extraction methods on Steam

Method	Model Extraction	N@10	R@10	Agr@1	Agr@10
NARM	BlackBox	0.630	0.848	–	–
	WhiteBox-DFME	0.601	0.806	0.743	0.722
	WhiteBox-ME-MIA	0.612	0.821	0.728	0.752
	WhiteBox-Ours	**0.621**	**0.833**	**0.760**	**0.763**
SASRec	BlackBox	0.617	0.839	–	–
	WhiteBox-DFME	0.593	0.805	0.668	0.702
	WhiteBox-ME-MIA	0.602	0.826	0.639	0.767
	WhiteBox-Ours	**0.615**	**0.834**	**0.740**	**0.782**
BERT4Rec	BlackBox	0.627	0.851	–	–
	WhiteBox-DFME	0.585	0.793	0.708	0.607
	WhiteBox-ME-MIA	0.598	0.811	0.723	0.695
	WhiteBox-Ours	**0.613**	**0.827**	**0.744**	**0.721**

– **Cross-Dataset Robustness**
In data-rich scenarios (ML-1M and Steam), our method preserves recommendation quality while achieving the highest ranking consistency. In sparse

Table 4. Performance comparison of model extraction methods on Beauty

Method	Model Extraction	N@10	R@10	Agr@1	Agr@10
NARM	BlackBox	0.356	0.517	–	–
	WhiteBox-DFME	0.261	0.370	0.294	0.381
	WhiteBox-ME-MIA	0.271	0.378	0.267	0.369
	WhiteBox-Ours	**0.282**	**0.385**	**0.349**	**0.468**
SASRec	BlackBox	0.341	0.493	–	–
	WhiteBox-DFME	0.333	0.483	0.433	0.423
	WhiteBox-ME-MIA	0.334	0.486	0.413	0.424
	WhiteBox-Ours	**0.337**	**0.489**	**0.439**	**0.433**
BERT4Rec	BlackBox	0.321	0.493	–	–
	WhiteBox-DFME	0.191	0.269	0.351	0.342
	WhiteBox-ME-MIA	0.190	0.267	0.353	0.345
	WhiteBox-Ours	**0.191**	**0.278**	**0.365**	**0.362**

environments (Beauty), the relative advantage becomes more pronounced, with Agr@10 improving by 23% over DFME. These observations yield an important insight: in sparse data environments, conventional techniques relying on hard labels or pairwise learning exhibit heightened sensitivity to data noise, whereas our cognitive distribution modeling effectively dampens such interference through its attention decay mechanism, thereby demonstrating superior robustness and generalization capability.

4.3 Profile Pollution Attack (RQ2)

The proposed behavioral pattern-consistent poisoning strategy demonstrates significant effectiveness in promoting target item rankings across diverse experimental settings. The results are shown in Fig. 3. On the ML-1M dataset, our method achieves optimal performance under both white-box and black-box configurations, with the WhiteBox-Ours version improving target item hit rates by approximately 2.4% over the strongest baseline. This advantage proves particularly pronounced in complex architectures like BERT4Rec, while maintaining consistent superiority across different model types. The strategy also exhibits remarkable robustness in sparse data environments like the Beauty dataset, where it sustains leading performance despite overall metric degradation across all methods.

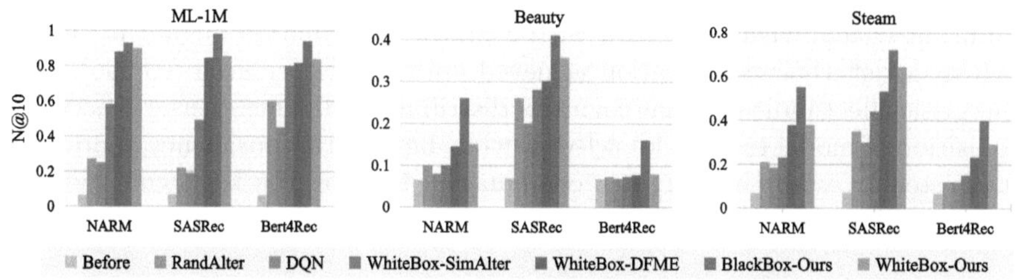

Fig. 3. Performance comparison of profile pollution attack methods

Comprehensive evaluation across three datasets confirms the method's stable and efficient attack capability under varying model architectures and data densities. The core innovation of synergistic collaboration analysis and gradient optimization enables the generated sequences to preserve semantic coherence with genuine user behavior while effectively elevating target item rankings. Particularly in black-box scenarios, attack success rates reach from 0.93 to 0.98, demonstrating an optimal balance between attack potency and behavioral stealth through precise gradient guidance.

4.4 Ablation Studies (RQ3)

The ablation experiment results as shown in Fig. 4 demonstrate that the cognitive distribution alignment mechanism is the core factor in improving attack effectiveness. In the BERT4Rec model, using the fusion loss alone significantly improved the attack success rate from 0.817 to 0.837, an improvement that far exceeds that achieved by adding the behavioral consistency strategy alone (0.817 to 0.8187). This proves that cognitive distribution alignment can more effectively capture ranking semantic information. Although the behavioral consistency strategy shows limited improvement when applied alone, it exhibits

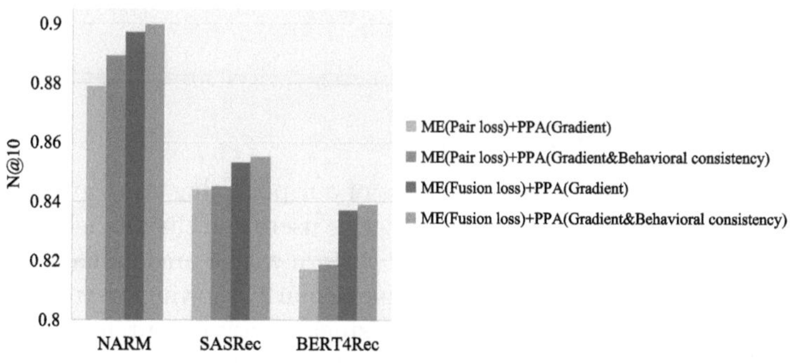

Note: ME stands for Model Extraction; PPA stands for Profile Pollution Attack

Fig. 4. Results of ablation experiments on ML-1M

significant effects when combined with cognitive distribution alignment. In the NARM model, their combination achieved optimal performance (0.8996). This synergistic effect indicates that cognitive distribution alignment ensures accurate acquisition of model ranking knowledge, while behavioral consistency guarantees attack stealth, with their organic combination forming the key foundation for our method's superiority.

4.5 Impact of Hyper-parameters (RQ4)

We conduct parameter sensitivity experiments on the attenuation factor α in the cognitive distribution mechanism. As shown in Fig. 5, as α increases from 0.7 to 0.97, the ranking consistency metrics Agr@1 and Agr@10 show a steady upward trend, peaking at 0.601 and 0.757, respectively, at $\alpha = 0.97$, representing approximately 18% and 10% improvements compared to $\alpha = 0.7$. This pattern confirms our theoretical hypothesis: higher α values (closer to 1) better align with the cognitive patterns of primacy effect and position bias, enabling more accurate capture of the essential characteristics of ranking distributions. Notably, when α exceeds 0.97, all metrics show slight degradation, indicating an optimal parameter range (0.95–0.97). This finding provides important guidance for parameter setting in practical applications while simultaneously verifying the robustness of the cognitive distribution alignment mechanism to parameter selection.

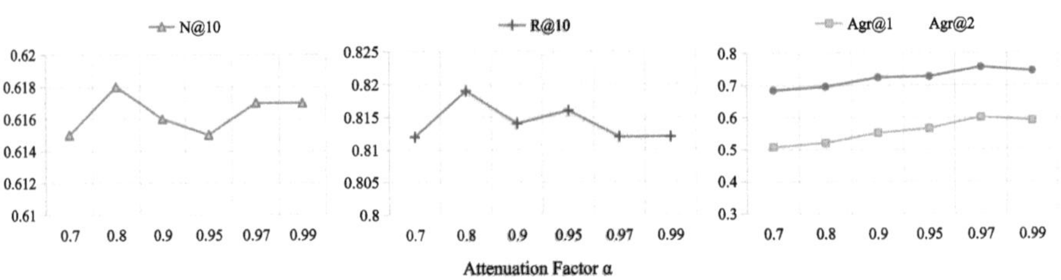

Fig. 5. Comparison of attenuation factors for NARM on ML-1M

4.6 Case Analysis (RQ5)

To quantitatively validate the stealthiness of our poisoning strategy, we conduct a fine-grained case analysis on a representative user from the MovieLens dataset. This user demonstrates a clear behavioral pattern with a predominant preference for 1990s films across multiple genres, particularly showing interest in sequel movies. We compare the behavioral consistency between our method and the baseline through genre coherence and stealthiness. As shown in Fig. 6, we can find that our method maintains superior behavioral consistency by preserving the user's temporal preferences and genre interests, while naturally integrating the target item. The baseline's temporal anomalies and excessive target repetition create detectable patterns that compromise stealthiness. This demonstrates that our behavior-aware generation strategy successfully balances attack effectiveness with behavioral authenticity.

User_ID: 1638

User History Sequence (Original):

["Perfect Murder, A (1998)", "EDtv (1999)", "This Is My Father (1998)", "Year of the Horse (1997)", "Children of the Corn III (1994)", "Modulations (1998)", "Mighty Joe Young (1949)", "Texas Chainsaw Massacre, The (1974)", "Slam (1998)", "Life (1999)", "Dead Ringers (1988)", "Liberty Heights (1999)", "Mumford (1999)", "Toy Story 2 (1999)", "What Dreams May Come (1998)", "Nashville (1975)", "Black Cauldron, The (1985)", "Emerald Forest, The (1985)", "Invasion of the Body Snatchers (1956)", "Mixed Nuts (1994)", "Welcome to the Dollhouse (1995)", "Sunchaser, The (1996)", "So Dear to My Heart (1949)"]

Attack Target:

["target_item_id": 3184, "target_item": "Montana (1998)"]

Comparison of Injected Sequences:

Method	Injected Items	Genre Coherence	Stealthiness
DFME	["Montana (1998)", "Jeremiah Johnson (1972)", "Montana (1998)", "Man Who Knew Too Much, The (1934)", "Montana (1998)", "Six-String Samurai (1998)", "Rosewood (1997)", "Lethal Weapon (1987)", "Mutiny on the Bounty (1935)", "Cavalcade (1933)", "Crimson Pirate, The (1952)"]	Abrupt shifts to westerns/musicals	**Low: Excessive repetition (3 times)**
Ours	["Glass Shield, The (1994)", "Montana (1998)", "King of New York (1990)", "Montana (1998)", "King of New York (1990)", "Second Jungle Book: Mowgli & Baloo, The (1997)", "Children of the Corn IV: The Gathering (1996)", "Music of the Heart (1999)", "Music of the Heart (1999)", "Safe Men (1998)", "Crimson Pirate, The (1952)"]	Natural alignment (crime/drama/sequels)	**High: Contextual embedding**

Fig. 6. Comparative analysis of injected sequence cases

5 Conclusion

This work establishes that the security of sequential recommender systems requires evaluation against adversaries mimicking both the statistical and semantic patterns of real user behavior. By formalizing the alignment of cognitive value distributions and integrating behavioral consistency directly into attack optimization, our framework introduces a paradigm shift from conspicuous manipulation to imperceptible influence. Our results confirm that such cognitively grounded and behaviorally stealthy attacks pose a significantly more potent and practical threat. This necessitates a fundamental rethinking of defensive strategies, moving beyond anomaly detection based on superficial statistical deviations towards auditing the semantic plausibility and cognitive consistency of user sequences. Ultimately, our study not only exposes a critical vulnerability vector but also provides a novel analytical lens, one that integrates cognitive psychology and behavioral analysis for building next-generation, robust recommender systems secure against advanced, semantically-aware adversaries.

Acknowledgments. This research was funded by the National Key Research and Development Program of China (Grant No. 2024YFC3307400), under the "Social Governance and Smart Society Technology Support" key special project.

Disclosure of Interests. The authors have no competing interests to declare that are relevant to the content of this article.

References

1. Atsusaka, Y.: Analyzing ballot order effects when voters rank candidates. Polit. Anal. **33**(1), 64–72 (2025)
2. Boka, T.F., Niu, Z., Neupane, R.B.: A survey of sequential recommendation systems: techniques, evaluation, and future directions. Inf. Syst. **125**, 102427 (2024)
3. Du, X., Chen, Y., Zhang, Y., Tang, J.: Precision profile pollution attack on sequential recommenders via influence function. arXiv preprint arXiv:2412.01127 (2024)
4. Harper, F.M., Konstan, J.A.: The movielens datasets: history and context. ACM Trans. Interact. Intell. Syst. (tiis) **5**(4), 1–19 (2015)
5. Hogarth, R.M., Einhorn, H.J.: Order effects in belief updating: the belief-adjustment model. Cogn. Psychol. **24**(1), 1–55 (1992)
6. Kang, W.C., McAuley, J.: Self-attentive sequential recommendation. In: 2018 IEEE International Conference on Data Mining (ICDM), pp. 197–206. IEEE (2018)
7. Kariyappa, S., Prakash, A., Qureshi, M.K.: Maze: data-free model stealing attack using zeroth-order gradient estimation. In: Proceedings of the IEEE/CVF Conference on Computer Vision and Pattern Recognition, pp. 13814–13823 (2021)
8. Lam, S.K., Riedl, J.: Shilling recommender systems for fun and profit. In: Proceedings of the 13th International Conference on World Wide Web, pp. 393–402 (2004)
9. Li, J., Ren, P., Chen, Z., Ren, Z., Lian, T., Ma, J.: Neural attentive session-based recommendation. In: Proceedings of the 2017 ACM on Conference on Information and Knowledge Management, pp. 1419–1428 (2017)
10. Li, M., et al.: Learning query-aware embedding index for improving e-commerce dense retrieval. In: Proceedings of the 46th International ACM SIGIR Conference on Research and Development in Information Retrieval, pp. 3265–3269. SIGIR '23, Association for Computing Machinery, New York, NY, USA (2023). https://doi.org/10.1145/3539618.3591834
11. Li, M., et al.: Adaptive hyper-parameter learning for deep semantic retrieval. In: Proceedings of the 2023 Conference on Empirical Methods in Natural Language Processing: Industry Track, pp. 775–782 (2023)
12. Liu, J., Chen, Q., Xu, J., Li, J., Li, B., Xu, S.: A unified search and recommendation framework based on multi-scenario learning for ranking in e-commerce. In: Proceedings of the 47th International ACM SIGIR Conference on Research and Development in Information Retrieval, pp. 2890–2894 (2024)
13. McAuley, J., Targett, C., Shi, Q., Van Den Hengel, A.: Image-based recommendations on styles and substitutes. In: Proceedings of the 38th international ACM SIGIR Conference on Research and Development in Information Retrieval, pp. 43–52 (2015)
14. Ni, J., Li, J., McAuley, J.: Justifying recommendations using distantly-labeled reviews and fine-grained aspects. In: Proceedings of the 2019 Conference on Empirical Methods in Natural Language Processing and the 9th International Joint Conference on Natural Language Processing (EMNLP-IJCNLP), pp. 188–197 (2019)
15. Orekondy, T., Schiele, B., Fritz, M.: Knockoff nets: stealing functionality of black-box models. In: Proceedings of the IEEE/CVF Conference on Computer Vision and Pattern Recognition, pp. 4954–4963 (2019)

16. Sharma, K., et al.: A survey of graph neural networks for social recommender systems. ACM Comput. Surv. **56**(10), 1–34 (2024)
17. Song, J., et al.: Poisonrec: an adaptive data poisoning framework for attacking black-box recommender systems. In: 2020 IEEE 36th International Conference on Data Engineering (ICDE), pp. 157–168. IEEE (2020)
18. Sun, F., et al.: Bert4rec: sequential recommendation with bidirectional encoder representations from transformer. In: Proceedings of the 28th ACM International Conference on Information and Knowledge Management, pp. 1441–1450 (2019)
19. Tramèr, F., Zhang, F., Juels, A., Reiter, M.K., Ristenpart, T.: Stealing machine learning models via prediction {APIs}. In: 25th USENIX Security Symposium (USENIX Security 16), pp. 601–618 (2016)
20. Wang, B., et al.: Learning multi-stage multi-grained semantic embeddings for e-commerce search. In: Companion Proceedings of the ACM Web Conference 2023, pp. 411–415. WWW '23 Companion, Association for Computing Machinery, New York, NY, USA (2023). https://doi.org/10.1145/3543873.3584638
21. Wang, Y., Su, J., Chen, C., Han, M., Zhang, C., Wang, J.: Sim4rec: data-free model extraction attack on sequential recommendation. In: Proceedings of the AAAI Conference on Artificial Intelligence, vol. 39, pp. 12766–12774 (2025)
22. Xing, X., Meng, W., Doozan, D., Snoeren, A.C., Feamster, N., Lee, W.: Take this personally: pollution attacks on personalized services. In: 22nd USENIX Security Symposium (USENIX Security 13), pp. 671–686 (2013)
23. Xu, E., et al.: Optimizing e-commerce search: toward a generalizable and rank-consistent pre-ranking model. In: Proceedings of the 47th International ACM SIGIR Conference on Research and Development in Information Retrieval, pp. 2875–2879. SIGIR '24, Association for Computing Machinery, New York, NY, USA (2024). https://doi.org/10.1145/3626772.3661343
24. Yue, Z., He, Z., Zeng, H., McAuley, J.: Black-box attacks on sequential recommenders via data-free model extraction. In: Proceedings of the 15th ACM Conference on Recommender Systems, pp. 44–54 (2021)
25. Zhang, H., Li, Y., Ding, B., Gao, J.: Practical data poisoning attack against next-item recommendation. In: Proceedings of the Web Conference 2020, pp. 2458–2464 (2020)
26. Zhang, H., et al.: Data poisoning attack against recommender system using incomplete and perturbed data. In: Proceedings of the 27th ACM SIGKDD Conference on Knowledge Discovery & Data Mining, pp. 2154–2164 (2021)
27. Zhang, H., Liu, F.: Few-shot model extraction attacks against sequential recommender systems. arXiv preprint arXiv:2411.11677 (2024)
28. Zhao, K., Li, L., Ding, K., Gong, N.Z., Zhao, Y., Dong, Y.: A survey on model extraction attacks and defenses for large language models. In: Proceedings of the 31st ACM SIGKDD Conference on Knowledge Discovery and Data Mining, vol. 2, pp. 6227–6236 (2025)
29. Zhao, X., Zhang, L., Xia, L., Ding, Z., Yin, D., Tang, J.: Deep reinforcement learning for list-wise recommendations. arXiv preprint arXiv:1801.00209 (2017)
30. Zhu, Y., Chen, J., Chen, L., Li, Y., Zhang, F., Liu, Z.: Interest clock: time perception in real-time streaming recommendation system. In: Proceedings of the 47th International ACM SIGIR Conference on Research and Development in Information Retrieval, pp. 2915–2919 (2024)
31. Zhu, Z., Wu, C., Fan, R., Lian, D., Chen, E.: Membership inference attacks against sequential recommender systems. In: Proceedings of the ACM Web Conference 2023, pp. 1208–1219 (2023)

Mitigating Popularity Bias for Two-Sided Fairness via Dual-Teacher Distillation in Recommendation

Chao Guo, Xuemin Wang(✉), Chuangying Zhu, Jialung Liang, Heng Yang, and Liang Chang

Guangxi Key Laboratory of Trusted Software, Guilin University of Electronic Technology, Guilin 541004, China
guoc4296@gmail.com, yangheng8748@126.com, {wangxuemin,zhucy,changl}@guet.edu.cn, leung@mails.guet.edu.cn

Abstract. Mitigating popularity bias is essential for two-sided fairness in recommendation systems. Existing methods usually require sensitive information such as demographic data and item attributes. However, due to policies and laws, such data are often inaccessible, limiting practical applications. To address this, we propose the Dual-Teacher Two-sided Fairness Distillation Framework (DTTFD), a teacher-student architecture that mitigates popularity bias without sensitive attributes. Specifically, we find that balancing user-item interactions or equalizing item exposures improves fairness, but inevitably sacrifices recommendation accuracy. Based on this insight, we utilize data augmentation to construct two unbiased datasets: one enforcing uniform user activities for user-side fairness, and the other ensuring equal item exposures for item-side fairness. We then train two fairness-specific teachers separately on these datasets and integrate them into a synthetic teacher, which distills fairness knowledge into the student model. Concurrently, an accuracy-focused teacher guides the student model to maintain high recommendation quality. Additionally, we introduce a momentum-based dynamic adjustment algorithm to adaptively balance fairness and accuracy throughout the distillation process. Experiments on three real-world datasets demonstrate significant fairness improvements over state-of-the-art methods while maintaining competitive recommendation accuracy.

Keywords: Recommendation systems · Popularity bias · Two-sided Fairness · Knowledge distillation

1 Introduction

In the era of information explosion, recommender systems (RS) serve as a core tool for filtering massive datasets and delivering personalized content, and have been widely deployed in e-commerce, social media, and online entertainment [9,10,38]. Nevertheless, a large body of research has uncovered an inherent

H. Jung et al. (Eds.): DASFAA 2026, LNCS 16535, pp. 154–170, 2026.
https://doi.org/10.1007/978-981-92-0363-5_10

popularity bias: the long-tailed training distribution causes the system to over-favor popular items and highly active users [1,2,28,39]. As illustrated in Fig. 1, only 5% of popular items attract the vast majority of interactions, whereas the remaining 95% of long-tail items—such as products from small local retailers—are almost "invisible" and receive negligible exposure [1,2]. Likewise, the active user group, though a minority, monopolizes most recommendation resources, while the overwhelming majority of inactive users suffer from a sharp drop in recommendation quality [11,26]. This two-sided unfairness not only inflicts systematic economic damage on small local businesses, but may also trigger broader social repercussions [7]. Consequently, achieving two-sided fairness in RS is of utmost importance.

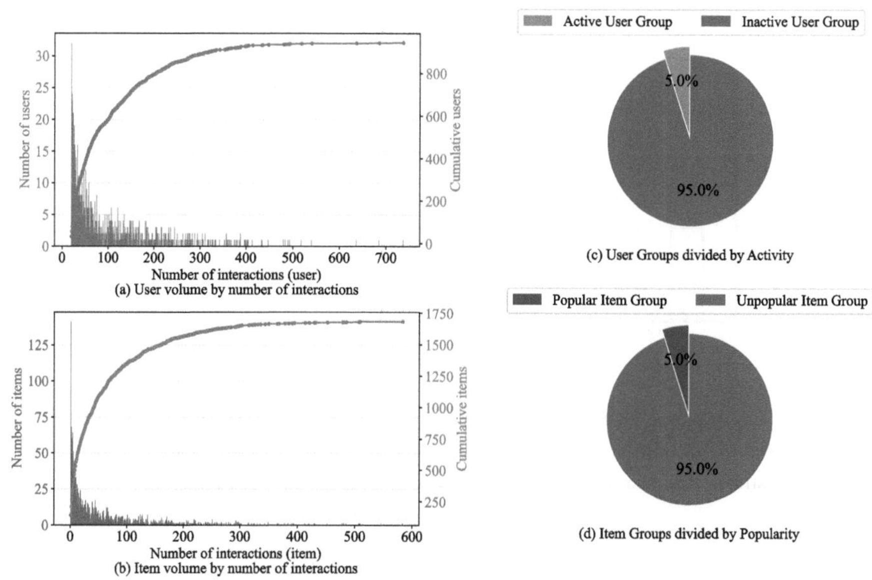

Fig. 1. Interaction Statistics & Grouping on the MovieLens-100k Dataset.

Over the past few years, numerous efforts have been devoted to enhancing fairness in recommender systems, such as propensity scoring [16], causal inference [39], and adversarial learning [28]. However, these approaches inherently depend on the availability of sensitive attributes (e.g., demographics), which significantly limits their real-world applicability due to privacy and legal constraints [5,15]. Recently, Wang et al. [35] explored fairness without sensitive attributes in graph neural networks via counterfactual learning, yet their method focuses on node classification rather than the two-sided fairness required in recommendation. Thus, developing a unified framework capable of simultaneously enhancing user and item fairness without relying on sensitive information remains a critical yet unresolved challenge.

Based on our investigation, the key to mitigating popularity bias and improving fairness in recommender systems lies in modifying the training data, with a particular focus on alleviating data sparsity by generating simulated user-item interactions from the existing data [4,8,28]. However, as mentioned earlier, most of these methods rely on sensitive information beyond the user-item interaction history to provide assistance, which limits their application scenarios. Recent studies have attempted to reduce recommendation disparities caused by long-tail distributions. For example, some works use interaction subsequences from popular items [21], while others apply random sampling methods [3]. These approaches have helped reduce differences in recommendation quality between highly active and less active users, as well as the exposure gap between popular and unpopular items. So we propose our hypothesis: *Can standardizing users' interaction history lengths and item exposure frequencies enhance fairness for both users and items?*

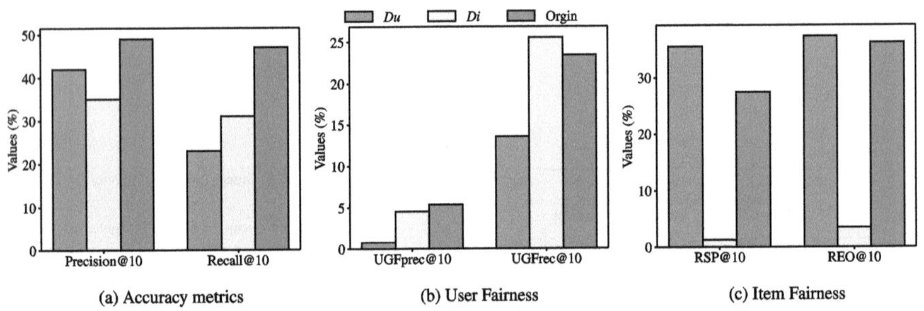

Fig. 2. The performance of NCF [19] on different datasets. UGF_{prec} and UGF_{rec} are widely adopted user-centric fairness metrics, whereas RSP and REO are standard item-centric measures. Smaller values indicate fairer model performance.

To verify this hypothesis, we standardize users' interaction-history lengths to generate D_u and equalize items' exposure frequencies to generate D_i. This is achieved by using data augmentation and random selection. As illustrated in Fig. 2, training the NCF model on D_u and D_i enhances fairness on either the user side or the item side, though this improvement comes at the expense of recommendation utility.

Informed by these findings, we propose a Dual-Teacher Two-sided Fairness Distillation (DTTFD) framework. Specifically, we first leverage multi-teacher distillation [24], incorporating an accuracy-oriented teacher trained on original data to retain recommendation utility. Further, inspired by [40], we integrate fairness-oriented teachers trained on unbiased datasets D_u and D_i via a synthetic-teacher strategy. To balance the fairness and accuracy objectives dynamically, we design a novel momentum-based algorithm that adaptively adjusts the teachers' influence during training. To sum up, we address the challenges through the following contributions:

- We demonstrate that adjusting the long-tail distribution via data augmentation, without sensitive attributes, enhances two-sided fairness.
- We propose DTTFD, a simple yet effective multi-teacher method to mitigate popularity bias, enhance two-sided fairness, and preserve accuracy. To our knowledge, this is the first work exploring two-sided fairness in recommendations via knowledge distillation without sensitive data.
- We introduce a momentum-based dynamic-adjustment algorithm that continually re-weights the fairness and accuracy teachers by monitoring the student model's real-time performance and fairness-metric fluctuations, thereby preventing either teacher from dominating the distillation process.
- We conduct experiments on three real-world datasets to verify the effectiveness of the proposed method in improving fairness while preserving utility.

2 Related Work

Fairness Recommendation for Popularity Bias. Fairness-aware learning in recommendation systems (RS) has attracted substantial interest in the machine learning community. Existing methods mainly tackle bias from sensitive attributes like demographics [5,36], emphasizing metrics such as statistical parity [13,36] or equal opportunity [17]. Yet, popularity bias—arising independently of user or item attributes—remains largely overlooked in fairness-focused RS studies. Adversarial learning, a prevalent technique, uses a generator-discriminator minimax game to debias recommendations. However, it suffers from (1) training instability from imbalanced dynamics, causing extended training and erratic results; (2) hyperparameter sensitivity (e.g., learning rate, batch size), risking performance drops or failures; and (3) high computational demands, necessitating GPUs/CPUs and hindering resource-limited researchers. Alternatives like pre-processing (e.g., data rebalancing [31,37]) and post-processing (e.g., ranking tweaks [25,33]) typically require sensitive data, curbing their use. In conclusion, a lightweight framework that does not require the use of sensitive data as an auxiliary factor is currently the optimal choice for achieving fairness in RS.

Data Augmentation. Data augmentation is widely used to tackle the long-tail distribution [21] in recommender systems. Recent work [8] improved user fairness through data augmentation but neglected item fairness. Since biases impact both users and items, single-sided mitigation is inadequate [21,28]. To address this, we extended data augmentation to achieve two-sided fairness by generating simulated data using latent factors from matrix decomposition with random sampling. By standardizing the length of user-item interactions and item exposure rates, we constructed two unbiased datasets to enhance both user-side and item-side fairness.

Knowledge Distillation. Knowledge distillation [12,20,27] compresses a large teacher model's knowledge into a compact student model for efficient RS. Soft labels—teacher predictions on user-item interactions—guide student training

[22,23,34], with their quality critical for effective distillation. While reducing inference latency, standard distillation intensifies popularity bias, favoring popular items and active users [6], undermining fair and accurate recommendations. However, like simple model training, existing debiasing methods, such as multi-teacher joint instruction [24] and unbiased teacher training, typically address either user or item fairness, rarely both.

3 Preliminaries

In this section, we formulate the problem statement aimed at mitigating popularity bias in RS. We define the recommendation task, and examine the fairness issues from both the user side and item side.

3.1 Recommendation Task

In a typical recommendation task setup, we are given a user set $\mathcal{U}$ and an item set $\mathcal{I}$. Because every user subjectively values an item, we encode this opinion as a real-valued rating $y_{u,i} \in \mathbb{R}$ that user $u \in \mathcal{U}$ assigns to item $i \in \mathcal{I}$. The available data are summarized by the triplet $\mathcal{D} = \{\mathcal{U}, \mathcal{I}, \mathbb{R}\}$: whenever the rating $y_{u,i}$ appears in $\mathcal{D}$ it is treated as ground-truth evidence of an actual useritem interaction, whereas any missing entry is taken to mean that no interaction has been observed. A recommendation system generates a *top-K* ranked list for each user$u$based on the predicted $\hat{y}_{u,i}$ for item i in the test set.

3.2 Definition of Fairness

In this paper, we investigate group-based fairness issues caused by popularity bias in both user-side and item-side perspectives in the recommendation task.

For user-side fairness, we aim to ensure user-oriented group fairness (UGF) by minimizing differential treatment across groups via the algorithm. we follow the definition of previous research [25]: Given m user groups $G_1^{\mathcal{U}}$, $G_2^{\mathcal{U}}$, $G_3^{\mathcal{U}}$, ..., $G_m^{\mathcal{U}}$, where all groups $G_j^{\mathcal{U}}$ are disjoint. The **UGF** requires

$$\mathbb{E}_{u \sim G_1^{\mathcal{U}}}[M(u)] = \mathbb{E}_{u \sim G_2^{\mathcal{U}}}[M(u)] = \cdots = \mathbb{E}_{u \sim G_m^{\mathcal{U}}}[M(u)], \tag{1}$$

where the subscript M denotes the accuracy measure of the recommendation. We divide users into two groups, $G_1^{\mathcal{U}}$ and $G_2^{\mathcal{U}}$, based on their activity level. The metric of **UGF** [25] is formulated as follows:

$$\text{UGF}_M = \left| \frac{1}{|G_1^{\mathcal{U}}|} \sum_{u \in G_1^{\mathcal{U}}} M(u) - \frac{1}{|G_2^{\mathcal{U}}|} \sum_{u \in G_2^{\mathcal{U}}} M(u) \right|. \tag{2}$$

A smaller UGF value indicates a smaller difference in accuracy M between groups, and hence better user-side fairness. For more than two groups, UGF can be extended by considering the average or maximum difference in M between group pairs.

For item-side fairness, two kinds of item-side fairness can be defined as below: Given n items groups $G_1^{\mathcal{I}}$, $G_2^{\mathcal{I}}$, $G_3^{\mathcal{I}}$,..., $G_n^{\mathcal{I}}$, the item **ranking-based statistical parity (RSP)** requires

$$P_1(K) = P_2(K) = P_3(K) = \cdots = P_n(K). \tag{3}$$

And the item **ranking-based equal opportunity (REO)** requires

$$P_1(K|y=1) = P_2(K|y=1) = P_3(K|y=1) \cdots = P_n(K|y=1). \tag{4}$$

Unlike user-side fairness, item-side fairness has two definitions: RSP, based on statistical parity [13], and REO, based on equal opportunity [17]. To comprehensively evaluate item-side fairness, we use both metrics, following prior work [41], assessing RSP and REO in practical models with non-binary fairness:

$$\text{RSP@}K = \frac{\text{std}(P_1(K), P_2(K), P_3(K), ..., P_n(K))}{\text{mean}(P_1(K), P_2(K), P_3(K), ..., P_n(K))}, \tag{5}$$

$$\text{REO@}K = \frac{\text{std}\left(P_1(K|y=1), P_2(K|y=1), \ldots, P_n(K|y=1)\right)}{\text{mean}\left(P_1(K|y=1), P_2(K|y=1), \ldots, P_n(K|y=1)\right)}, \tag{6}$$

where $P_j(K)$ is calculated as the ratio of top-K recommendations in the j-th item group to the total ranking candidates in that group, and $P_j(K|y=1)$ is the ratio of positive top-K recommendations in the j-th group to the total positive candidates in that group. We divide items into two groups, $G_1^{\mathcal{I}}$ and $G_2^{\mathcal{I}}$, based on their popularity. Similar to UGF, lower values in RSP and REO indicate better item-side fairness.

In summary, this paper addresses popularity bias in the distillation framework, enhancing fairness for users and items across popularity groups while maintaining competitive student performance with minimal impact on model efficacy.

4 Methodology

In order to address the fairness issues described in the previous section, we present the framework and the details of the proposed DTTFD framework. The whole framework is shown in Fig. 3.

4.1 Overall Framework

Motivated by empirical observations of training with randomly selected data, DTTFD aims to learn a fair Recommender System Model without accessing sensitive attributes. To achieve this, DTTFD addresses three key challenges: (1) mitigating popularity bias to ensure two-sided fairness without sensitive attributes;

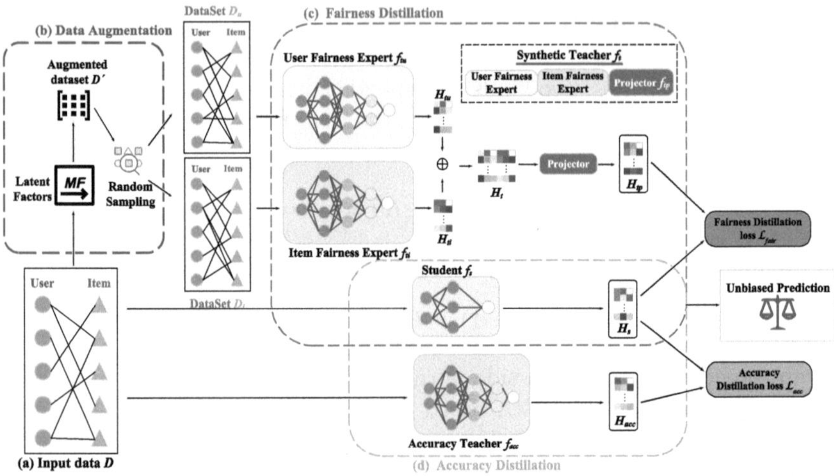

Fig. 3. Framework of the DTTFD. The framework consists of three main components: Data Augmentation, Fairness Distillation, and Accuracy Distillation.

(2) achieving user and item fairness simultaneously; and (3) balancing fairness and accuracy. To tackle the first and second challenges, DTTFD employs data augmentation to reduce popularity bias. Following the knowledge distillation paradigm [17], it utilizes two fairness experts—one focusing on user-side fairness and another on item-side fairness—to construct a synthetic teacher that outputs fair knowledge. Guided by this synthetic teacher and an accuracy-focused teacher, DTTFD extracts fair and informative knowledge. For the third challenge, a novel momentum-based dynamic adjustment algorithm balances fairness and accuracy losses. In comparative experiments, the two single-sided fairness experts and the accuracy teacher use the Neural Collaborative Filtering (NCF) [19] model as the backbone.

As shown in Fig. 3, DTTFD consists of the data augmentation module, a synthetic teacher model, an accuracy teacher model, and a student model, which are represented by g, f_t, f_{acc}, and f_s respectively. f_s is a two-layer MLP. The accuracy teacher conducts supervised training based on the original data and uses the early-stop mechanism to maintain the best accuracy state. We generate simulated data using data augmentation module without using sensitive information to supplement the dataset. We then form two unbiased datasets through random sampling. These two datasets have the characteristics of consistent user-item interaction lengths and the same number of item exposures. f_t aims to extract fair and detailed knowledge H_t for the student mode. Specifically, f_t consists of two fairness experts f_{tu} and f_{ti} and a projector f_{tp}. Here, f_{tu} and f_{ti} are trained on datasets with consistent lengths of user or item interactions, and without accessing sensitive attributes, they alleviate the popularity bias. Due to training on unbiased datasets, f_{tu} and f_{ti} may have information indicating better fairness on one side but poorer fairness on the other side, represented as H_{tu} and H_{ti}. To combine the fairness advantages, the projector f_{tp} is used to combine these information representations and perform mapping to generate the information representation H_t. H_t and H_s outputs of f_{acc} will be regarded as additional supervision that assists f_s in learning.

4.2 Data Augmentation

In this section, we delve into the specifics of the data-augmentation module. We employ the simplest matrix decomposition [32] combined with Singular Value Decomposition (SVD) to process the original dataset, thereby generating the complete prediction set $\hat{\mathcal{D}} = \{\mathcal{U}, \mathcal{I}, \hat{\mathbb{R}}\}$. We then randomly select useritem interactions that did not exist in the original data, combine them with the original dataset, and construct two new datasets: $\mathcal{D}_u$, in which the number of interactions for every user is set to the same constant value, and $\mathcal{D}_i$, in which the number of item exposures is set to the same constant value. For users or items whose interaction counts exceed the threshold, we retain only the threshold number of entries through random sampling.

4.3 Synthetic Teacher

In this section, we introduce the composition and optimization objective of the synthetic teacher f_t. The objective of f_t is to extract fair representations H_t for the student model. f_t consists of two fairness experts f_{tu}, f_{ti} and the projector f_{tp}. The MLP fairness experts f_{tu} and f_{ti} receive the user-item interaction history without attributes, $(u; i)$, as input and output their respective 8-dimensional intermediate-layer fairness representations H_{tu} and H_{ti}. The processing for obtaining the representation and is as follows:

$$H_{tu} = f_{tu}(u; i); \quad H_{ti} = f_{ti}(u; i), \tag{7}$$

where f_{tu} and f_{ti} are trained on Augmentation data, i.e., training set D_u and D_i, respectively. Based on the previous observations, f_{tu} and f_{ti} were trained on data where the interaction history lengths of users or items were the same, and the long-tail distribution was adjusted. Therefore, H_{tu} and H_{ti} are fairer representations because they alleviate the prevalence bias. However, since f_{tu} and f_{ti} were trained using different selected data, these two representations may not be able to achieve two-sided fairness simultaneously. To solve this problem, DTTFD uses the projector f_{tp} to combine these two representations to generate a fair and informative representation H_{tp}. Specifically, f_{tp} takes the connection of H_{tu} and H_{ti} as the input and outputs H_{tp}. The training objective is to ensure that the generated predictions achieve optimal fairness performance according to fairness metrics. This can be summarized as:

$$f_{tp}(x) = \text{ReLU}(Wx + b); \quad H_{tp} = f_{tp}(H_{tu} \oplus H_{ti}), \tag{8}$$

where $W \in \mathbb{R}^{8\times16}$ is the weight matrix, $b \in \mathbb{R}^8$ is the bias vector, and $\oplus$ denotes the concatenate operation. During the actual guidance process, the KL divergence (Kullback-Leibler Divergence) was employed to measure the difference between the soft label outputs of the student model and those of the synthesized teacher model.

4.4 Momentum-Based Dynamic Adjustment Algorithm

In order to balance the influence of the synthetic teacher and the accuracy teacher on the student model, and to prevent a single-sided expert from exerting excessive influence, we introduce a momentum-based dynamic adjustment algorithm. DTTFD uses the weighted sum of the fair distillation loss, the accuracy knowledge distillation loss, and the supervision learning loss of the student model as the total loss, which is expressed as:

$$\mathcal{L}_{\text{overall}}(r) = (1-\omega_s)\cdot(\omega_{acc}\cdot\mathcal{L}_{acc}+\omega_{fair}\cdot\mathcal{L}_{fair})+\omega_s\cdot\mathcal{L}_{\text{s}}, \tag{9}$$

where ω_{fair}, ω_{acc}, and ω_{s} represent the weights of the fairness distillation loss $\mathcal{L}_{\text{fair}}$, accuracy knowledge distillation loss $\mathcal{L}_{\text{acc}}$, and student model supervised learning $\mathcal{L}_{\text{s}}$. ω_s remains unchanged during the training process of the student model. The relevant formulas for each loss are as follows:

$$\mathcal{L}_s = \frac{1}{N}\sum_{i=1}^{N}(\hat{y}_{u,i}-y_{u,i})^2, \tag{10}$$

$$\mathcal{L}_{fair} = \text{KL}(Softmax(H_t)/T, Softmax(H_s)/T)\cdot T^2, \tag{11}$$

$$\mathcal{L}_{acc} = \text{KL}(Softmax(H_{acc})/T, Softmax(H_s)/T)\cdot T^2. \tag{12}$$

The changes in weights are dynamically adjusted based on the fairness performance and accuracy of the current student model. First, we quantify how much the student model has improved (or degraded) in both accuracy and fairness during the most recent epoch. To this end, we calculate the average changes $\Delta acc(r)$ in accuracy metrics (Precision@K and Recall@K) since the second epoch, as well as the average changes $\Delta fair$ in fairness measures (UGF$_{prec}$@K, UGF$_{rec}$@K, RSP@K and REO@K), as follows:

$$\begin{aligned}\Delta acc(r) =& (\text{Precision@}K(r)+\text{Recall@}K(r))/2-\\ &(\text{Precision@}K(r-1)+\text{Recall@}K(r-1))/2,\end{aligned} \tag{13}$$

$$\begin{aligned}\Delta fair(r) =& (\text{UGF}_{prec}\text{@}K(r)+\text{UGF}_{rec}\text{@}K(r)+\text{RSP@}K(r)+\text{REO@}K(r))/4\\ &-(\text{UGF}_{prec}\text{@}K(r-1)+\text{UGF}_{rec}\text{@}K(r-1)+\text{RSP@}K(r-1)\\ &+\text{REO@}K(r-1)))/4,\end{aligned} \tag{14}$$

where r is the number of iterations.

Next, we translate these raw changes into provisional weight updates. A momentum coefficient $m \in [0,1)$ is applied to $\Delta acc(r)$ so that the weights evolve smoothly rather than over-reacting to sudden fluctuations. This yields the unnormalized weights $\omega'_{acc}(r)$ and $\omega'_{fair}(r)$:

$$\omega'_{acc}(r) = m*\Delta acc(r)+\omega_{acc}(r);\ \omega'_{fair}(r) = m*\Delta fair(r)+\omega_{fair}(r). \tag{15}$$

Finally, we normalize the provisional weights to ensure they sum to one, obtaining the effective trade-off weights that will actually be used in the student's loss function during the upcoming epoch:

$$\omega_{acc}(r) = \frac{\omega'_{acc}(r)}{\omega'_{acc}(r) + \omega'_{fair}(r)}; \omega_{fair}(r) = \frac{\omega'_{fair}(r)}{\omega'_{acc}(r) + \omega'_{fair}(r)}. \tag{16}$$

With these normalized weights in hand, the training loop proceeds to the next iteration.

5 Experiments

In this section, we have conducted extensive experiments on three datasets to evaluate DTTFD[1], and address the following research questions: **RQ1**: How does DTTFD perform compared with existing fairness methods? **RQ2**: Can DTTFD adapt to baseline models with different architectures? **RQ3**: How does fairness distillation and accuracy distillation affect DTTFD? **RQ4**: How does the hyperparameter m affect the fairness and accuracy of DTTFD?

5.1 Experimental Settings

Table 1. Statistics of the Datasets

Dataset	#Users	#Items	#Pairs	#Sparsity
MovieLens-100K	943	1682	10,000	92.19%
MovieLens-1M	6040	3923	1,000,209	95.54%
BookCrossing	6358	6921	88,552	99.80%

Dataset. To comprehensively evaluate the effectiveness and fairness of DTTFD, we conduct experiments on three real-world datasets: MovieLens-100K[2], MovieLens-1M[3], and BookCrossing[4]. For user group construction, MovieLens-100K is split by gender to simulate scenarios with accessible sensitive attributes, while MovieLens-1M and BookCrossing adopt the top 5% most active users versus the rest to simulate scenarios without such access. For item groups, the top 5% most popular items form one group, and the rest form the other. Each dataset is split into training, validation, and testing sets in an 8:1:1 ratio. To construct the validation/test candidates for a user u, we first add all items in the validation/test set that have interacted with u as positive examples, and sample 10 non-interacting items as negative examples. Dataset statistics are summarized in Table 1.

[1] https://github.com/ChaGrown267/DTTFD.git.
[2] https://grouplens.org/datasets/movielens/100k/.
[3] https://grouplens.org/datasets/movielens/1m/.
[4] https://github.com/ashwanidv100/Recommendation-System---Book-Crossing-Dataset?tab=readme-ov-file.

Baselines. We compare DTTFD with various models in four broad categories, as follows:

(1) Conventional recommendation methods: **NCF** [19], **LightGCN** [18], **MF** [32].
(2) General fairness and diversity models: **FR(FairRec)** [29], **FULTR** [33].
(3) Fairness models for popularity bias: **FAiR** [28], **FDA** [8],**BIGFR** [14], **DC-GCN** [30].
(4) Compared distillation methods: **RD** [34], **UnKD** [6].

For all baselines, their associated parameters follow the original settings.

Metrics. For every user, we produce a Top-10 list and assess it along accuracy, user fairness, and item fairness. Accuracy adopts Prec@10(Precision@10) and Recall@10 (↑). User fairness is quantified by UGF_M, as defined in Eq. (2), computed on Prec@10 and Recall@10 across user groups. Item fairness is assessed with RSP@10 and REO@10, that measure exposure disparity across item segments, as defined in Eq. (5) and (6). For all fairness metrics, lower values indicate better fairness, representing reduced disparity across users or items.

Table 2. Comparison Between DTTFD and Baselines.

Model	Precision@10(↑)	Recall@10(↑)	UGF_{prec}@10(↓)	UGF_{rec}@10(↓)	RSP@10(↓)	REO@10(↓)	**Avg.(Rank)**
	MovieLens-100K						
NCF	49.32 ± 0.93%	46.51 ± 1.95%	5.33 ± 2.85%	21.65 ± 1.05%	23.61 ± 2.36%	31.24 ± 2.61%	7.17
LightGCN	46.29 ± 1.15%	44.57 ± 1.04%	4.85 ± 0.69%	19.98 ± 0.45%	11.23 ± 2.63%	16.79 ± 2.68%	5.33
MF	45.42 ± 0.33%	42.83 ± 0.29%	4.56 ± 0.66%	19.80 ± 0.33%	11.59 ± 1.17%	21.04 ± 1.19%	5.17
FR (FairRec)	48.01 ± 0.15%	45.81 ± 0.23%	4.19 ± 0.11%	21.08 ± 0.15%	16.69 ± 0.31%	17.31 ± 0.12%	5.50
FULTR	47.59 ± 1.06%	44.83 ± 0.95%	4.46 ± 0.47%	20.50 ± 0.45%	18.72 ± 2.12%	25.62 ± 1.40%	6.00
FAiR	45.34 ± 1.70%	42.95 ± 1.61%	3.59 ± 0.50%	19.76 ± 0.74%	9.57 ± 0.17%	**8.20 ± 2.83%**	3.33
FDA	40.55 ± 1.63%	**47.06 ± 1.01%**	3.88 ± 1.50%	20.53 ± 2.61%	11.57 ± 0.17%	9.55 ± 1.33%	4.17
BIGFR	43.51 ± 1.23%	41.21 ± 0.51%	4.34 ± 0.22%	**18.76 ± 0.33%**	5.21 ± 2.15%	8.67 ± 2.15%	3.83
DC-GCN	42.74 ± 1.03%	40.21 ± 0.43%	6.72 ± 0.35%	19.65 ± 0.20%	5.19 ± 1.07%	8.86 ± 0.37%	4.33
RD-NCF	**49.61 ± 0.23%**	46.78 ± 0.22%	5.75 ± 0.64%	21.92 ± 0.35%	26.09 ± 0.63%	32.86 ± 0.71%	7.83
UNKD	48.21 ± 0.17%	42.78 ± 0.22%	5.73 ± 0.12%	20.11 ± 0.19%	12.09 ± 0.51%	15.02 ± 0.23%	6.50
DTTFD-NCF	47.41 ± 0.85%	44.71 ± 0.85%	**2.71 ± 0.81%**	19.64 ± 1.84%	**5.11 ± 2.45%**	8.53 ± 3.91%	**2.50**
	MovieLens-1M						
NCF	**58.57% ± 0.36%**	34.27% ± 0.75%	4.53% ± 0.21%	16.94% ± 0.13%	34.63% ± 0.22%	45.21% ± 0.16%	8.33
LightGCN	57.47% ± 0.11%	34.19% ± 0.87%	44.03% ± 0.83%	29.09% ± 0.32%	9.99% ± 0.16%	15.20% ± 0.15%	5.67
MF	56.93% ± 0.45%	34.37% ± 0.27%	37.25% ± 1.30%	28.80% ± 0.32%	26.83% ± 1.58%	37.20% ± 1.29%	5.50
FR (FairRec)	57.47% ± 0.11%	34.23% ± 0.87%	6.06% ± 0.83%	17.16% ± 0.32%	40.84% ± 0.16%	41.14% ± 0.15%	6.17
FULTR	56.66% ± 0.56%	34.21% ± 0.36%	3.97% ± 0.48%	15.82% ± 0.21%	27.72% ± 2.09%	38.33% ± 2.07%	5.50
FAiR	52.54% ± 0.53%	31.72% ± 0.97%	3.67% ± 0.74%	15.67% ± 1.11%	4.44% ± 1.03%	7.91% ± 0.51%	2.83
FDA	53.66% ± 1.28%	32.55% ± 5.67%	3.22% ± 2.06%	16.67% ± 1.52%	7.62% ± 2.53%	15.91% ± 2.33%	4.00
BIGFR	51.23% ± 0.53%	31.25% ± 0.53%	2.88% ± 0.53%	15.37% ± 0.53%	4.53% ± 0.53%	9.12% ± 0.53%	3.17
DC-GCN	47.87% ± 0.23%	**35.65% ± 0.22%**	8.51% ± 0.33%	17.53% ± 0.47%	8.53% ± 0.63%	11.16% ± 3.52%	4.33
RD-NCF	57.57% ± 0.47%	34.31% ± 0.97%	36.20% ± 2.09%	29.21% ± 1.52%	32.57% ± 2.33%	43.11% ± 0.73%	7.83
UNKD	58.53% ± 0.21%	15.31% ± 0.97%	7.66% ± 0.25%	**14.37% ± 0.32%**	8.63% ± 0.32%	11.63% ± 0.53%	5.17
DTTFD-NCF	56.41% ± 4.35%	33.81% ± 2.29%	**2.21% ± 0.56%**	15.21% ± 1.16%	**4.06% ± 2.63%**	**6.61% ± 2.03%**	**2.17**
	BookCrossing						
NCF	18.83% ± 0.12%	67.71% ± 0.34%	37.66% ± 0.67%	48.13% ± 0.34%	2.34% ± 0.74%	4.87% ± 1.14%	6.50
LightGCN	**19.54% ± 0.09%**	**70.16% ± 0.35%**	37.57% ± 0.85%	50.62% ± 0.48%	4.28% ± 0.19%	9.20% ± 0.52%	5.50
MF	18.76% ± 0.11%	67.36% ± 0.42%	37.91% ± 0.95%	47.74% ± 0.45%	3.21% ± 0.59%	4.93% ± 0.75%	5.17
FR (FairRec)	18.55% ± 0.13%	66.71% ± 0.18%	37.27% ± 0.13%	47.56% ± 0.11%	2.93% ± 0.29%	4.23% ± 0.64%	4.83
FULTR	18.70% ± 0.13%	67.13% ± 0.38%	37.01% ± 0.81%	47.87% ± 0.31%	2.65% ± 0.86%	3.65% ± 0.49%	4.00
FAiR	18.58% ± 0.84%	66.73% ± 1.52%	37.32% ± 0.97%	47.38% ± 0.41%	2.35% ± 0.44%	2.93% ± 1.76%	2.83
FDA	15.03% ± 2.36%	43.73% ± 3.52%	36.99% ± 2.33%	49.38% ± 4.52%	4.35% ± 2.98%	8.84% ± 0.98%	8.33
BIGFR	18.21% ± 0.33%	65.73% ± 0.72%	36.89% ± 0.63%	47.22% ± 0.33%	3.22% ± 0.31%	2.85% ± 1.55%	3.33
DC-GCN	16.89% ± 0.87%	69.99% ± 0.87%	41.32% ± 0.30%	47.22% ± 0.61%	2.32% ± 1.33%	3.63% ± 2.07%	4.00
RD-NCF	19.01% ± 0.32%	68.37% ± 0.56%	37.82% ± 1.41%	49.22% ± 1.04%	2.83% ± 1.61%	4.51% ± 1.54%	7.17
UNKD	16.52% ± 0.12%	66.98% ± 0.22%	**35.82% ± 1.41%**	49.98% ± 0.62%	1.93% ± 1.61%	3.51% ± 1.54%	3.83
DTTFD-NCF	18.68% ± 0.24%	67.08% ± 0.84%	36.23% ± 1.41%	**47.21% ± 0.76%**	**1.22% ± 1.14%**	**2.85% ± 1.24%**	**2.17**

Implementation Details. We use an embedding dimension of 128 across all datasets. The model is optimized using Adam with a learning rate of $1e^{-3}$. We conduct experiments using NCF as our backbones. In the extension experiment, we also used MF and LightGCN as teachers' backbones to prove that our framework is not only effective for recommendation models based on MLP structure, but also adaptable to other structural models. For MF and LightGCN, we use the best user and item embeddings, while for NCF, we compute a weighted sum of embeddings from MF and MLP for recommendation tasks.

5.2 Performance Comparison

To answer RQ1, we conduct experiments on three datasets. Table 2 shows the performance of all models on the three datasets, where each model is repeated for ten runs and we report their mean ± std-dev. In the Table 2, the best result is bolded and the second-best result is underlined. From the tables, we make the following observations: Compared with traditional models (NCF, LightGCN, MF) and other methods, DTTFD performs better or at least as well in terms of fairness indicators, confirming its advantages in user fairness and item fairness. Furthermore, in terms of effectiveness, the accuracy of DTTFD is less than 2% away from the optimal baseline, confirming its advantages in terms of fairness and accuracy. It ranks first on average.

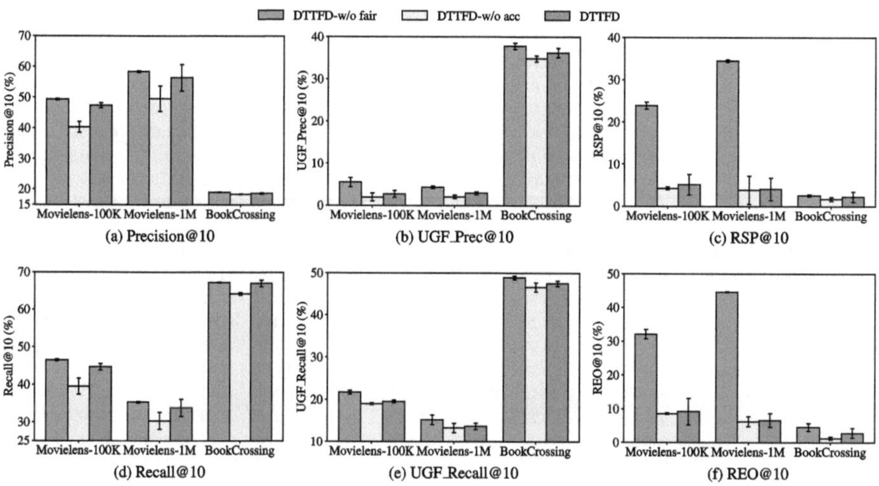

Fig. 4. Ablation study on the DTTFD

5.3 Ablation Study

To answer RQ2, we conducted an ablation study to explore the influence of each teacher. Specifically, we trained two variants, w/o fair and w/o acc, where w/o fair and w/o acc respectively represent the DTTFD variants that only have accuracy distillation and fairness distillation. For each variant, we conducted 5 experiments on each of the three datasets, and the comparison of the indicators is shown in Fig. 4. This enables us to observe the unique role of each teacher and the ability of the complete DTTFD to integrate the advantages of each teacher to achieve better results.

5.4 Extension Experiment

To address RQ3, we replaced the teacher model's backbone with LightGCN and MF, conducting comparative experiments across three datasets, including a simple KD framework for comparison. As shown in Table 3, DTTFD is effective across MLP, MF, and GCN structures in recommender systems, consistently enhancing two-sided fairness while preserving accuracy.

Table 3. Extension Experiment of DTTFD

				MovieLens-100K			
Model	Method	Precision@10(↑)	Recall@10(↑)	UGF_{prec}@10(↓)	UGF_{rec}@10(↓)	RSP@10(↓)	REO@10(↓)
NCF	Vanilla	49.32% ± 0.93%	46.51% ± 1.95%	5.33% ± 2.85%	21.65% ± 1.05%	23.61% ± 2.36%	31.24% ± 2.61%
	KD	49.22% ± 0.23%	46.43% ± 0.63%	6.17% ± 0.62%	20.65% ± 1.24%	26.61% ± 3.31%	30.24% ± 1.26%
	DTTFD	47.41% ± 0.85%	44.71% ± 0.85%	**2.71% ± 0.81%**	**19.64% ± 1.84%**	**5.11% ± 2.45%**	**8.53% ± 3.91%**
LightGCN	Vanilla	46.29% ± 1.15%	44.57% ± 1.04%	4.85% ± 0.69%	19.98% ± 0.45%	11.23% ± 2.63%	16.79% ± 2.68%
	KD	46.11% ± 0.53%	44.83% ± 0.52%	5.05% ± 0.51%	20.03% ± 0.61%	15.21% ± 2.23%	18.31% ± 3.08%
	DTTFD	46.23% ± 0.39%	43.81% ± 0.43%	**3.98% ± 0.47%**	**19.45% ± 0.41%**	**3.55% ± 1.62%**	**6.91% ± 2.11%**
MF	Vanilla	45.42% ± 0.33%	42.83% ± 0.29%	4.56% ± 0.66%	19.80% ± 0.33%	11.59% ± 1.17%	21.04% ± 1.19%
	KD	45.33% ± 0.33%	42.1% ± 0.33%	4.33% ± 0.33%	20.11% ± 0.33%	11.55% ± 0.33%	19.36% ± 0.33%
	DTTFD	45.63% ± 0.26%	43.03% ± 1.13%	**2.87% ± 0.77%**	**18.94% ± 0.36%**	**3.26% ± 2.57%**	**8.63% ± 1.56%**
				MovieLens-1M			
Model	Method	Precision@10(↑)	Recall@10(↑)	UGF_{prec}@10(↓)	UGF_{rec}@10(↓)	RSP@10(↓)	REO@10(↓)
NCF	Vanilla	58.57% ± 0.36%	34.27% ± 0.75%	4.53% ± 0.21%	18.94% ± 0.13%	34.63% ± 0.22%	45.21% ± 0.16%
	KD	59.31% ± 0.30%	34.07% ± 0.51%	6.23% ± 0.37%	18.37% ± 0.53%	34.42% ± 0.16%	45.57% ± 1.77%
	DTTFD	56.41% ± 4.35%	33.81% ± 2.29%	**2.21% ± 0.56%**	**15.21% ± 1.16%**	**4.06% ± 2.63%**	**6.61% ± 2.03%**
LightGCN	Vanilla	57.47% ± 0.11%	34.19% ± 0.87%	44.03% ± 0.83%	29.09% ± 0.32%	9.99% ± 0.16%	15.20% ± 0.15%
	KD	56.53% ± 0.41%	34.14% ± 0.33%	39.12% ± 0.87%	28.55% ± 0.44%	14.54% ± 0.12%	21.38% ± 1.63%
	DTTFD	55.10% ± 0.63%	33.28% ± 0.41%	**27.56% ± 1.17%**	**27.57% ± 0.50%**	**3.66% ± 2.11%**	**7.17% ± 1.32%**
MF	Vanilla	56.93% ± 0.45%	34.37% ± 0.27%	37.25% ± 1.30%	28.80% ± 0.32%	26.83% ± 1.58%	37.20% ± 1.29%
	KD	56.58% ± 0.32%	34.98% ± 0.22%	39.55% ± 0.41%	29.32% ± 0.07%	24.66% ± 0.31%	38.21% ± 1.42%
	DTTFD	57.91% ± 0.32%	34.97% ± 0.73%	**26.18% ± 0.86%**	**26.39% ± 1.51%**	**3.11% ± 2.67%**	**8.24% ± 1.33%**
				BookCrossing			
Model	Method	Precision@10(↑)	Recall@10(↑)	UGF_{prec}@10(↓)	UGF_{rec}@10(↓)	RSP@10(↓)	REO@10(↓)
NCF	Vanilla	18.83% ± 0.12%	67.71% ± 0.34%	37.66% ± 0.67%	48.13% ± 0.34%	2.34% ± 0.74%	4.87% ± 1.14%
	KD	18.53% ± 0.23%	67.63% ± 0.50%	38.67% ± 0.31%	50.11% ± 0.53%	5.14% ± 0.20%	6.87% ± 0.94%
	DTTFD	18.68% ± 0.24%	67.08% ± 0.84%	**36.23% ± 1.41%**	**47.21% ± 0.76%**	**2.22% ± 1.14%**	**2.85% ± 1.24%**
LightGCN	Vanilla	19.56% ± 0.15%	70.09% ± 1.04%	38.85% ± 0.69%	49.88% ± 0.45%	5.07% ± 1.63%	10.78% ± 2.68%
	KD	19.43% ± 0.66%	68.63% ± 0.42%	38.15% ± 0.23%	48.33% ± 0.62%	5.93% ± 0.53%	9.55% ± 1.53%
	DTTFD	19.06% ± 0.33%	68.42% ± 0.61%	**36.68% ± 0.92%**	**47.36% ± 0.47%**	**0.75% ± 0.66%**	**5.15% ± 1.15%**
MF	Vanilla	18.76% ± 0.11%	67.36% ± 0.42%	37.91% ± 0.95%	47.74% ± 0.45%	3.21% ± 0.59%	4.93% ± 0.75%
	KD	18.57% ± 0.23%	66.65% ± 0.31%	37.23% ± 0.72%	47.99% ± 0.35%	3.41% ± 0.36%	4.33% ± 1.11%
	DTTFD	18.23% ± 1.63%	65.45% ± 0.52%	**36.79% ± 1.66%**	**46.51% ± 0.53%**	**1.98% ± 0.11%**	**2.29% ± 0.79%**

5.5 Parameter Sensitivity

To address RQ4, we investigated the impact of hyperparameter m in Eq. (15) on performance, evaluating Prec@10, Recall@10, and fairness metrics using NCF over 25 epochs, with m ranging from 10^{-3} to 0.99. Due to space limitations, results are reported only for the MovieLens-100K dataset. As depicted in Fig. 5, accuracy metrics initially increase rapidly before declining with larger m, while fairness metrics rise and then stabilize, showing consistent trends. This suggests a balance, rather than a trade-off, between accuracy and fairness, with user- and item-side fairness mutually influencing each other. However, excessively large m delays achieving this balance. Thus, we recommend selecting m in the range (0.7, 0.9) for a stable and effective balance between accuracy and fairness. Furthermore, in Fig. 5(c), we can observe that the trend of the fair weight changes is completely opposite to that of the accuracy weight, eventually reaching a relatively stable balance.

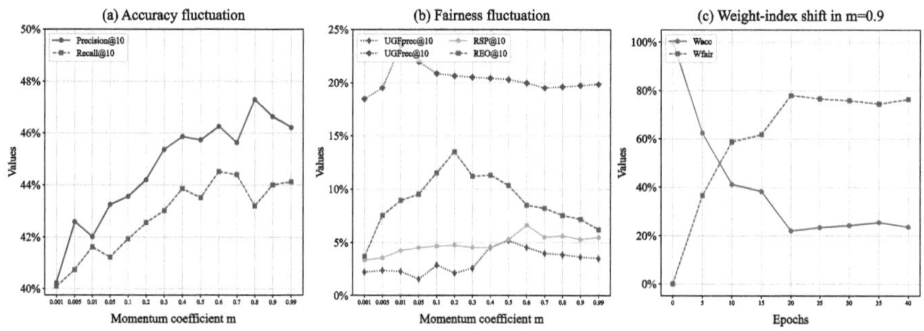

Fig. 5. Parameter Sensitivity on the DTTFD

6 Conclusion

In this paper, we propose a distillation framework capable of enhancing user fairness and item fairness, without sensitive attributes (e.g., demographic information and item properties) of the dataset. The core concept involves leveraging data augmentation techniques and randomly selecting data with consistent historical interaction lengths to train a fair teacher model. The student models to learn under the supervision of an accuracy-focused teacher, utilizing a momentum-based dynamic adjustment algorithm. Experimental results demonstrate that our framework effectively alleviates popularity bias while preserving model utility. In future work, we plan to explore the coexistence of group fairness and individual fairness at both the user-side and the item-side, as well as the ways to achieve two-sided fairness through knowledge distillation in large models.

Acknowledgements. This work is supported by the National Natural Science Foundation of China (Grant No. U22A2099 and No. 62336003) and Natural Science Foundation of Guangxi (Grant No. 2025GXNSFAA069551).

Conflict of Interest. The authors declare no conflict of interest.

References

1. Abdollahpouri, H., Mansoury, M.: Multi-sided exposure bias in recommendation. In: ACM KDD Workshop on Industrial Recommendation Systems (2020)
2. Abdollahpouri, H., Mansoury, M., Burke, R., Mobasher, B.: The connection between popularity bias, calibration, and fairness in recommendation. In: RecSys 2020, pp. 726–731. Association for Computing Machinery (2020)
3. Balasubramanian, K., Alshabanah, A., Markowitz, E., Ver Steeg, G., Annavaram, M.: Biased user history synthesis for personalized long-tail item recommendation. In: RecSys 2024, pp. 189–199. Association for Computing Machinery (2024)
4. Bin, C., Liu, W., Zhang, F., Chang, L., Gu, T.: Faircore: fairness-aware recommendation through counterfactual representation learning. IEEE Trans. Knowl. Data Eng. 1–14 (2025)
5. Bose, A., Hamilton, W.: Compositional fairness constraints for graph embeddings. In: Chaudhuri, K., Salakhutdinov, R. (eds.) ICML 2019, vol. 97, pp. 715–724. PMLR (2019)
6. Chen, G., Chen, J., Feng, F., Zhou, S., He, X.: Unbiased knowledge distillation for recommendation. In: WSDM 2023, pp. 976–984. Association for Computing Machinery (2023)
7. Chen, J., Dong, H., Wang, X., Feng, F., Wang, M., He, X.: Bias and debias in recommender system: a survey and future directions. ACM Trans. Inf. Syst. **41**, 1–39 (2023)
8. Chen, L., et al.: Improving recommendation fairness via data augmentation. In: WWW 2023, pp. 1012–1020. Association for Computing Machinery (2023)
9. Chen, W., et al.: Next-poi recommendation via spatial-temporal knowledge graph contrastive learning and trajectory prompt. IEEE Trans. Knowl. Data Eng. (2025)
10. Chen, W., et al.: Building and exploiting spatial-temporal knowledge graph for next poi recommendation. Knowl.-Based Syst. **258**, 109951 (2022)
11. Cheng, M., et al.: Empowering sequential recommendation from collaborative signals and semantic relatedness. In: DASFAA (2025)
12. Deng, X., Zhang, Z.: Comprehensive knowledge distillation with causal intervention. In: Advances in Neural Information Processing Systems, vol. 34, pp. 22158–22170. Curran Associates, Inc. (2021)
13. Dwork, C., Hardt, M., Pitassi, T., Reingold, O., Zemel, R.: Fairness through awareness. In: ITCS 2012, pp. 214–226. Association for Computing Machinery (2012)
14. Gan, Y., Wang, X., Liu, T., Chang, L., Gen, Q., Zeng, Y.: BIGFR: bridging individual and group fairness in recommendation systems. In: ICASSP 2025, pp. 1–5 (2025)
15. Goodfellow, I., et al.: Generative adversarial networks. Commun. ACM **63**, 139–144 (2020)
16. Gruson, A., et al.: Offline evaluation to make decisions about playlistrecommendation algorithms. In: WSDM 2019, pp. 420–428. Association for Computing Machinery (2019)

17. Hardt, M., Price, E., Price, E., Srebro, N.: Equality of opportunity in supervised learning. In: Advances in Neural Information Processing Systems, vol. 29. Curran Associates, Inc. (2016)
18. He, X., Deng, K., Wang, X., Li, Y., Zhang, Y., Wang, M.: Lightgcn: simplifying and powering graph convolution network for recommendation. In: SIGIR 2020, pp. 639–648. Association for Computing Machinery (2020)
19. He, X., Liao, L., Zhang, H., Nie, L., Hu, X., Chua, T.S.: Neural collaborative filtering. In: WWW 2017, pp. 173–182 (2017)
20. Hu, X., Tang, K., Miao, C., Hua, X.S., Zhang, H.: Distilling causal effect of data in class-incremental learning. In: CVPR, pp. 3957–3966 (2021)
21. Huang, Y., Yang, Z., Hu, W., Xu, B., Zhang, Z.: CSLP: collaborative solution to long-tail problem and popularity bias in sequential recommendation. In: SMC, pp. 4404–4411 (2024)
22. Kang, S., Hwang, J., Kweon, W., Yu, H.: DE-RRD: a knowledge distillation framework for recommender system. In: CIKM 2020, pp. 605–614. Association for Computing Machinery (2020)
23. Lee, J.W., Choi, M., Lee, J., Shim, H.: Collaborative distillation for top-n recommendation. In: ICDM 2019, pp. 369–378 (2019)
24. Li, J., Feng, X., Gu, T., Chang, L.: Dual-teacher de-biasing distillation framework for multi-domain fake news detection. In: ICDE 2024, pp. 3627–3639 (2024)
25. Li, Y., Chen, H., Fu, Z., Ge, Y., Zhang, Y.: User-oriented fairness in recommendation. In: WWW 2021, pp. 624–632. Association for Computing Machinery (2021)
26. Liu, Q., Feng, X., Gu, T., Liu, X.: Faircrs: towards user-oriented fairness in conversational recommendation systems. In: RecSys 2024, pp. 126–136. Association for Computing Machinery (2024)
27. Liu, Y., Chen, K., Liu, C., Qin, Z., Luo, Z., Wang, J.: Structured knowledge distillation for semantic segmentation. In: CVPR (2019)
28. Liu, Z., Fang, Y., Wu, M.: Mitigating popularity bias for users and items with fairness-centric adaptive recommendation. ACM Trans. Inf. Syst. **41**, 1–27 (2023)
29. Patro, G.K., Biswas, A., Ganguly, N., Gummadi, K.P., Chakraborty, A.: Fairrec: two-sided fairness for personalized recommendations in two-sided platforms. In: WWW 2020, pp. 1194–1204. Association for Computing Machinery (2020)
30. Peng, F., Liao, F., Lu, X., Zheng, J., Li, R.: Revisiting explicit recommendation with DC-GCN: divide-and-conquer graph convolution network. Inf. Syst. **130**, 102513 (2025)
31. Rahman, T., Surma, B., Backes, M., Zhang, Y.: Fairwalk: towards fair graph embedding. In: IJCAI (2019)
32. Rendle, S.: Factorization machines. In: ICDM 2010, pp. 995–1000 (2010)
33. Singh, A., Joachims, T.: Policy learning for fairness in ranking. In: Advances in Neural Information Processing Systems, vol. 32. Curran Associates, Inc. (2019)
34. Tang, J., Wang, K.: Ranking distillation: learning compact ranking models with high performance for recommender system. In: SIGIR 2018, pp. 2289–2298 (2018)
35. Wang, X., Gu, T., Bao, X., Chang, L.: Towards fair graph neural networks via graph counterfactual without sensitive attributes. In: ICDE, pp. 265–277. IEEE (2025)
36. Wu, L., Chen, L., Shao, P., Hong, R., Wang, X., Wang, M.: Learning fair representations for recommendation: a graph-based perspective. In: WWW 2021, pp. 2198–2208. Association for Computing Machinery (2021)
37. Zehlike, M., Bonchi, F., Castillo, C., Hajian, S., Megahed, M., Baeza-Yates, R.: FA*IR: a fair top-k ranking algorithm. In: CIKM 2017, pp. 1569–1578. Association for Computing Machinery (2017)

38. Zhang, S., Yao, L., Sun, A., Tay, Y.: Deep learning based recommender system: a survey and new perspectives. ACM Comput. Surv. (2019)
39. Zhang, Y., et al.: Causal intervention for leveraging popularity bias in recommendation. In: SIGIR 2021, pp. 11–20. Association for Computing Machinery (2021)
40. Zhu, Y., Li, J., Chen, L., Zheng, Z.: The devil is in the data: learning fair graph neural networks via partial knowledge distillation. In: WSDM 2024, pp. 1012–1021 (2024)
41. Zhu, Z., Wang, J., Caverlee, J.: Measuring and mitigating item under-recommendation bias in personalized ranking systems. In: SIGIR 2020, pp. 449–458. Association for Computing Machinery (2020)

CAMeRA: A Mamba-Based Context-Aware Adaptive Multimodal Architecture for Sequential Recommendation

Bohan Hu[1], Yuhang Li[1(✉)], and Beijia Cheng[2]

[1] Zhengzhou University, Zhengzhou, Henan, China
alchemist@gs.zzu.edu.cn
[2] Lanzhou University, Lanzhou, China

Abstract. Multimodal recommendation leverages auxiliary signals such as images and text to enhance personalization. However, two key challenges remain: semantic noise from auxiliary modalities, and the difficulty of modeling both long sequences efficiently and short or sparse behaviors robustly. We present **CAMeRA**, a structure-aware framework that repositions ID embeddings as semantic anchors to guide multimodal fusion. It introduces an ID-Guided Dual Modulation mechanism that aligns visual and textual features with user-specific behaviors while suppressing irrelevant noise. To preserve structural integrity, CAMeRA adopts a dual-phase optimization strategy that decouples ID-based sequential modeling from multimodal enhancement. Furthermore, we design a Context-Aware Adaptive Mamba (CA-Mamba) encoder, which extends Mamba with bidirectional context modeling and a GRU-based temporal compensation unit to address sparsity and instability in long-tail scenarios. Experiments on four Amazon benchmarks demonstrate that CAMeRA consistently outperforms strong baselines in both accuracy and robustness across diverse user behaviors.

Keywords: Multimodal learning · Recommendation system · Mamba

1 Introduction

In modern recommendation systems, useritem interactions extend beyond simple clicks or purchases, increasingly involving rich multimodal signals such as product images, reviews, and descriptions. These signals provide fine-grained cues for capturing user preferences. For instance, users may linger on visually appealing items or leave textual reviews that reflect personal intent. Such heterogeneous modalities carry complementary semantics and temporal patterns which, if effectively integrated, can significantly enhance personalization and prediction accuracy [22]. As illustrated in Fig. 1, hybrid multimodal sequences contain more

B. Hu and Y. Li—Contributed equally to this work.

H. Jung et al. (Eds.): DASFAA 2026, LNCS 16535, pp. 171–187, 2026.
https://doi.org/10.1007/978-981-92-0363-5_11

comprehensive behavioral information, and extracting the most salient signals from them can substantially benefit downstream preference modeling.

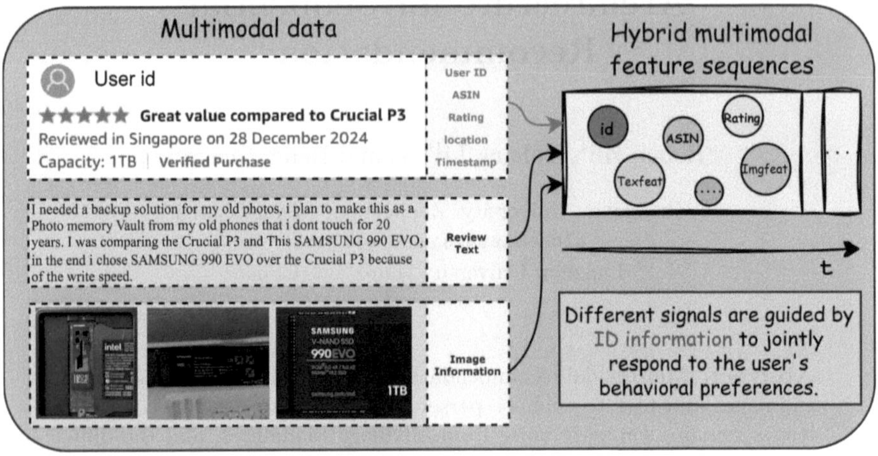

Fig. 1. Construction of hybrid multimodal input sequences. User ID serves as a guiding signal to align diverse modalities, facilitating downstream preference modeling.

With the rise of deep learning, sequential recommendation has advanced under an ID-centric paradigm [3]. Early CNN- and RNN-based models struggled to capture long-term dependencies due to limited temporal modeling [5,16, 35,37]. Transformer-based methods, such as SASRec [17] and BERT4Rec [31], improved the field by modeling global dependencies through self-attention. However, these methods primarily focus on unimodal ID sequences and neglect the rich multimodal context embedded in user behaviors.

Recent approaches [20,23,25] have attempted to incorporate auxiliary modalities into Transformer-based architectures to enhance personalization. While promising, this trend introduces two critical challenges. The first challenge is scalability: the quadratic complexity of self-attention [33] makes Transformers prohibitively expensive in long-sequence scenarios, a key bottleneck in high-throughput recommender systems. The second challenge lies in multimodal fusion: although ID sequences encode structured behavioral semantics, auxiliary modalities such as images and text are often noisy or redundant, which can interfere with preference modeling [13,27].

To mitigate the first challenge, Mamba [9] has recently emerged as an efficient alternative. As an enhanced Structured State Space Model (SSM), Mamba replaces computationally expensive self-attention with input-dependent selective scanning and hardware-aware parallelism, thereby achieving linear scalability with respect to sequence length. It has demonstrated strong sequential modeling capabilities in various recommendation tasks [11,19]. However, while Mamba alleviates the efficiency bottleneck, it leaves open a third challenge: its unidirectional design restricts contextual awareness, and its state estimation becomes unstable in short or sparse sequences—particularly common among long-tail users [26].

To address the second challenge, we propose ID-Guided Dual Modulation (ID-DM). Unlike conventional fusion strategies that treat ID, visual, and textual modalities equally, ID-DM leverages ID embeddings as semantic controllers rather than fusion targets. Specifically, ID embeddings modulate auxiliary features before sequence modeling, ensuring that multimodal representations align with user-specific interests while suppressing noise. This guided fusion offers two key benefits: (1) aligning multimodal features with structural behavioral semantics, and (2) preserving the integrity of the ID stream for independent and interpretable modeling. To further enhance this framework, we introduce a Dual-phase Optimization (D-Optimization) strategy. In the first phase, the model focuses on ID-only sequences to capture stable structural patterns. In the second phase, ID-guided fused representations are injected into a frozen encoder, introducing semantic signals without disturbing learned priors. This progressive design achieves coherent multimodal integration while avoiding the fragmentation caused by treating modalities as post hoc additions.

To overcome the third challenge, we design the Context-Aware Adaptive Mamba (CA-Mamba) encoder. CA-Mamba extends Mamba with two key components: a bidirectional backbone that captures both historical and anticipatory signals, and a lightweight Temporal Compensation Unit (TC-Unit) based on GRU [6], which enhances robustness for short or sparse sequences. Together, these designs preserve the efficiency of SSMs while improving contextual richness and stability, making CA-Mamba a practical choice for real-world multimodal recommendation.

Our key contributions are summarized below:

- We revisit the role of identity in multimodal recommendation by treating it as a structural anchor. Leveraging ID-guided modulation and dual-phase optimization, we establish a structure-first paradigm that enhances robustness, alignment, and training stability.
- We propose **CAMeRA**, an end-to-end architecture with ID-Guided Dual Modulation and a Context-Aware Adaptive Mamba encoder. The encoder combines bidirectional Mamba for global preference modeling and a GRU-based unit to improve robustness on sparse or short sequences.
- Extensive experiments on four Amazon benchmarks demonstrate the effectiveness and robustness of CAMeRA across diverse user behaviors and multimodal scenarios.

2 Related Work

2.1 Multimodal Sequential Recommendation

Sequential recommendation (SR) aims to predict users' next interactions by modeling temporal dependencies in behavior sequences. Early approaches leveraged CNNs [4] and RNNs [21,32] to capture local patterns and short-term dynamics, but struggled with long-range dependencies and efficiency. Transformer-based models [8,24,30,36,40] address this by using self-attention to model complex relations, though at the cost of quadratic complexity and high resource consumption. To enhance personalization and interpretability, recent work has introduced multimodal information—such as text, images, and audio—into sequential recommendation [22], giving rise to multimodal sequential recommendation. For example, BiVRec [14] jointly models user IDs and multimodal views to capture collaborative signals; [38] employs contrastive learning to align user-specific and cross-modal sequences; [34] leverages multi-level self-supervised objectives to reduce modality gaps.

While promising, this emerging field still faces challenges, including inefficient modality fusion and high training overhead, highlighting the need for more efficient and effective multimodal architectures.

2.2 State Space Models

State Space Models State Space Models (SSMs) [2] connect input and output sequences via latent states and have been widely adopted in sequence modeling. Structured State Space Sequence Models (S4) [10] extend SSMs with parameterized transitions to efficiently capture long-term dependencies. Mamba [9], a recent improvement over S4, introduces selective state spaces to dynamically filter inputs, achieving linear complexity and faster inference without attention mechanisms. Recent studies [18,19,26] have demonstrated the potential of Mamba in recommendation systems; however, its application to multimodal recommendation remains largely unexplored and underdeveloped. [29].

In this work, we propose CAMeRA, a structure-first Mamba-based framework that enhances contextual modeling and long-tail robustness through ID-guided modulation and dual-phase optimization.

3 Method

3.1 Overall Architecture

Given a user's interaction history $S_u = \{i_1, i_2, ..., i_n\}$, each item i is associated with a learnable ID embedding $\mathbf{e}_i \in \mathbb{R}^d$ and raw multimodal features $\mathbf{f}_i^m \in \mathbb{R}^{d_m}$, where $m \in \mathcal{M} = \{v, t\}$ denotes the visual and textual modalities.

We propose CAMeRA, as shown in Fig. 2, a context-aware multimodal encoder that models user preferences from sequential ID signals and filtered multimodal content. All modality features are projected into a shared d-dimensional space, where ID embeddings guide noise filtering and fusion before sequential modeling. Training occurs in two stages to decouple structural and multimodal learning, and the outputs are concatenated for final prediction.

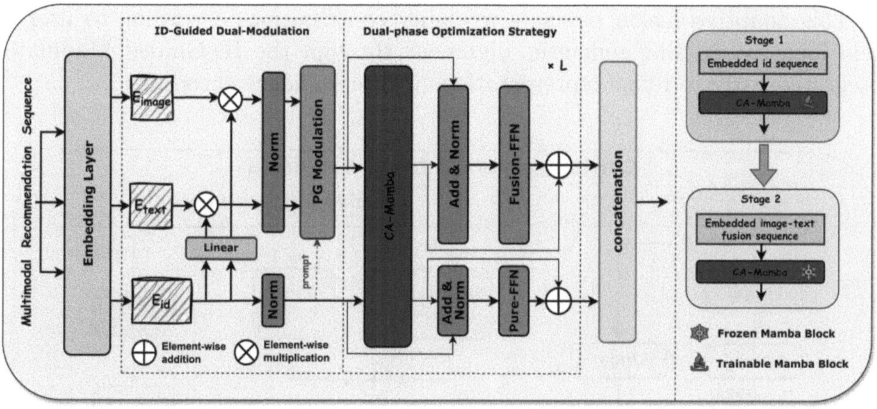

Fig. 2. CAMeRA framework overview. Identity-guided dual modulation (ID-DM) filters the noisy modalities through identity-conditional gating. The fused features are encoded by the CA-Mamba module, which combines bidirectional Mamba and GRU-based temporal compensation units (TCU). Structural and semantic information are modeled by two feedforward networks, while the two-stage optimization separates ID learning and multimodal fusion.

3.2 ID-Guided Dual Modulation

To effectively incorporate user-specific preferences into the multimodal representation, we propose an ID-Guided Dual Modulation (ID-DM) mechanism.

Given the visual feature $\mathbf{f}_i^v \in \mathbb{R}^{L\times d}$, textual feature $\mathbf{f}_i^t \in \mathbb{R}^{L\times d}$, and the corresponding item ID embedding $\mathbf{e}_i \in \mathbb{R}^{L\times d}$, the process consists of two steps:

Step 1: Intra-Modal ID-Guided Gating. We first apply element-wise modulation to each modality, where the gating factors are dynamically generated from the ID embedding. For the visual modality, we compute:

$$\mathbf{g}_i^v = \sigma\left(\mathbf{W}_v \cdot \mathrm{LayerNorm}(\mathbf{e}_i)\right) \tag{1}$$

$$\tilde{\mathbf{f}}_i^v = \mathrm{LayerNorm}\left(\mathbf{f}_i^v \odot \mathbf{g}_i^v\right) \tag{2}$$

A similar process is applied to the textual modality to obtain the gating factor $\mathbf{g}_i^t$ and the modulated feature $\tilde{\mathbf{f}}_i^t$.

Here, σ denotes the sigmoid activation, and $\odot$ represents element-wise multiplication. This step enables fine-grained feature filtering by emphasizing ID-relevant semantics within each modality.

Step 2: Prompt-Guided Modulation. To combine the filtered features, we introduce Prompt-Guided Modulation (PG Modulation), which employs a prompt-style fusion gate derived from the ID embedding to guide the integration process:

$$\mathbf{f}_i^{\mathrm{fused}} = \boldsymbol{\alpha}_i \odot \tilde{\mathbf{f}}_i^v + (1-\boldsymbol{\alpha}_i) \odot \tilde{\mathbf{f}}_i^t, \quad \boldsymbol{\alpha}_i = \sigma(\mathbf{W}_f \mathbf{e}_i) \tag{3}$$

This adaptive fusion balances modality contributions according to user or item identity, guiding semantic alignment through the ID-Guided Modulator integrated early to refine representations before sequence encoding.

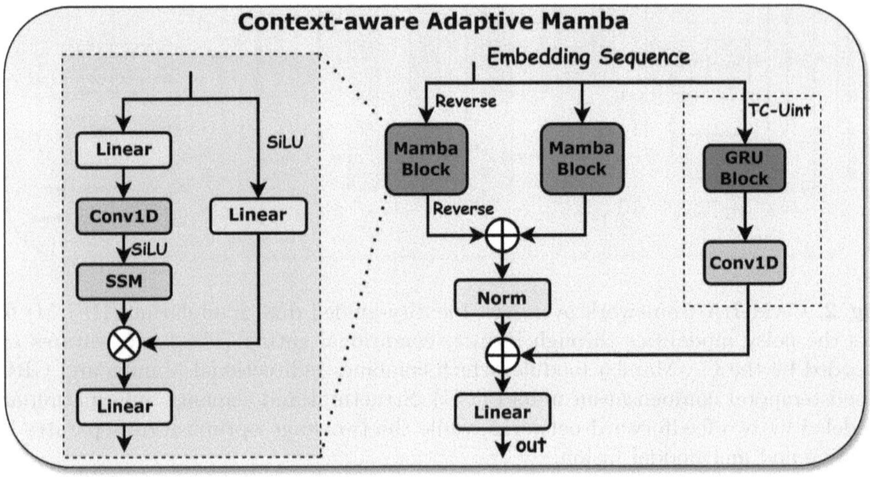

Fig. 3. Architecture of CA-Mamba. The encoder consists of two parallel branches: (1) a bidirectional Mamba block and (2) a Temporal Compensation Unit for short-term sequential refinement.

3.3 Context-Aware Adaptive Mamba

To effectively capture temporal dependencies in user interaction sequences, we propose **CA-Mamba**, as shown in Fig. 3, a context-aware and adaptive Mamba encoder that enhances structured state space modeling with bidirectional processing and Temporal Compensation Unit.

Both the original ID embeddings and fused multimodal features are processed by the CA-Mamba encoder. For clarity, we describe the following using the ID sequence as an example. Let $\mathbf{E} = [\mathbf{e}_1, \ldots, \mathbf{e}_T] \in \mathbb{R}^{T\times d}$ denote the input ID embedding sequence, The sequence is bidirectionally processed to construct context-enhanced representations.

Bidirectional Mamba Branch. To model temporal dependencies, we adopt a bidirectional Mamba encoder [39], as illustrated in Fig. 3, which processes inputs through parallel content and modulation streams to capture semantic dynamics and adaptive control. The bidirectional design integrates context from both directions, producing richer representations.

Given the input sequence $\mathbf{E} \in \mathbb{R}^{T\times d}$ to CA-Mamba, we project it into two latent streams:

$$\mathbf{x} = \text{Linear}_{\text{content}}(\mathbf{E}), \quad \mathbf{z} = \text{Linear}_{\text{mod}}(\mathbf{E}) \tag{4}$$

The content stream is refined via 1D convolution and SiLU activation:

$$\mathbf{x}' = \mathrm{SiLU}(\mathrm{Conv1D}(\mathbf{x})) \tag{5}$$

To capture long-range dynamics, $\mathbf{x}'$ is passed to a structured state space model:

$$\mathbf{v} = \mathrm{SSM}(\mathbf{x}') \tag{6}$$

The modulation stream is activated as $\mathbf{z}' = \mathrm{SiLU}(\mathbf{z})$ and used to reweight the SSM output:

$$\mathbf{y} = \mathrm{Linear}_{\mathrm{out}}(\mathbf{v} \odot \mathrm{SiLU}(\mathbf{z})) \tag{7}$$

For temporal consistency, both the input and output of the backward path are reversed. Final bidirectional representations are combined using a learnable gate and normalized:

$$\mathbf{H}_{\mathrm{bi}} = \mathrm{Norm}\left(\alpha \cdot \mathbf{y}^{\mathrm{fwd}} + (1-\alpha) \cdot \mathrm{Reverse}(\mathbf{y}^{\mathrm{bwd}})\right) \tag{8}$$

where $\alpha \in [0, 1]$ is a learnable scalar balancing directional outputs.

Temporal Compensation Unit (TC-Unit). To enhance short-term preference modeling—especially for long-tail users—we introduce a GRU-based TC-Unit. The normalized ID sequence $\mathbf{E}' = [\mathbf{e}'_1, \ldots, \mathbf{e}'_T] \in \mathbb{R}^{T \times d}$ serves as input.

The hidden state is updated at each time step as:

$$\mathbf{z}_t = \sigma(\mathbf{W}_z \mathbf{e}'_t + \mathbf{U}_z \mathbf{f}_{t-1} + \mathbf{b}_z) \tag{9}$$

$$\mathbf{r}_t = \sigma(\mathbf{W}_r \mathbf{e}'_t + \mathbf{U}_r \mathbf{f}_{t-1} + \mathbf{b}_r) \tag{10}$$

$$\tilde{\mathbf{f}}_t = \tanh(\mathbf{W}\mathbf{e}'_t + \mathbf{U}(\mathbf{r}_t \odot \mathbf{f}_{t-1}) + \mathbf{b}) \tag{11}$$

$$\mathbf{f}_t = \mathbf{z}_t \odot \mathbf{f}_{t-1} + (1-\mathbf{z}_t) \odot \tilde{\mathbf{f}}_t \tag{12}$$

We initialize $\mathbf{f}_0 = \mathbf{0}$ and collect all hidden states as $\mathbf{F} = [\mathbf{f}_1, \ldots, \mathbf{f}_T] \in \mathbb{R}^{T \times d}$. A 1D convolution is applied for smoothing:

$$\mathbf{H}_{\mathrm{TCU}} = \mathrm{Conv1D}(\mathbf{F}) \tag{13}$$

Final Output Aggregation. Outputs from both branches are combined using a learnable gate, followed by a linear transformation:

$$\mathbf{H}^{\mathrm{id}} = \mathrm{Linear}_{\mathrm{out}}(\alpha \cdot \mathbf{H}_{\mathrm{bi}} + (1-\alpha) \cdot \mathbf{H}_{\mathrm{gru}}) \tag{14}$$

where $\alpha \in [0, 1]$ balances global (Mamba) and local (GRU) signals. The result $\mathbf{H}^{\mathrm{id}} \in \mathbb{R}^{T \times d}$ is used in Stage 1 of our dual-phase optimization, while the same CA-Mamba encoder (frozen) is reused to process multimodal inputs in Stage 2.

3.4 Dual-Phase Optimization Strategy

Modeling user behavior in multimodal settings poses two key challenges: (1) core sequential patterns that capture user preferences are modality-agnostic, and (2) direct multimodal injection may introduce noisy or redundant signals that hinder intent learning. To address this, we introduce a Dual-phase Optimization Strategy (D-Optimization) that decouples structural modeling and semantic enhancement in a progressive and interpretable manner. Our approach is built upon a shared encoder CA-Mamba, as shown in Fig. 3, which has bidirectional Mamba blocks and parallel FFN branches for ID and multimodal inputs. The encoder is trained solely on ID sequences in the first phase to capture stable, modality-independent patterns, while in the second phase, multimodal features are injected with the encoder frozen to preserve structural priors and prevent semantic interference.

Stage 1: Structural Sequence Modeling with ID Only. We reuse the normalized ID embedding sequence $\mathbf{E}$ as input to an L-layer CA-Mamba encoder to extract direction-aware temporal representations. Each layer contains a bidirectional Mamba block and a dedicated feedforward branch for pure ID modeling:

$$\mathbf{H}_{\text{id}}^{\text{mamba}} = \text{CA-Mamba}(\mathbf{E}) \tag{15}$$

$$\mathbf{H}_{\text{pure}}^{\text{ffn}} = \text{FFN}_{\text{pure}}\left(\text{Norm}(\mathbf{H}_{\text{id}}^{\text{mamba}} + \mathbf{E})\right) \tag{16}$$

$$\mathbf{H}_{\text{pure}} = \mathbf{H}_{\text{id}}^{\text{mamba}} + \mathbf{H}_{\text{pure}}^{\text{ffn}} \tag{17}$$

During this phase, only the parameters of CA-Mamba and Pure-FFN are updated. The multimodal branch is kept frozen and optionally evaluated in parallel via `no_grad` to maintain consistency in architecture.

Stage 2: Multimodal Enhancement with CA-Mamba. After establishing stable temporal structures, we proceed to inject multimodal semantics by feeding the fused multimodal features $\mathbf{F} = [\mathbf{f}_1^{\text{fused}}, \ldots, \mathbf{f}_T^{\text{fused}}]$ into the same CA-Mamba encoder. Here, the Mamba block is frozen to preserve the learned structural priors, while the Fusion-FFN branch is independently updated:

$$\mathbf{H}_{\text{fused}}^{\text{mamba}} = \text{CA-Mamba}(\mathbf{F}) \quad \text{(frozen)} \tag{18}$$

$$\mathbf{H}_{\text{fused}}^{\text{ffn}} = \text{FFN}_{\text{fusion}}\left(\text{Norm}(\mathbf{H}_{\text{fused}}^{\text{mamba}} + \mathbf{F})\right) \tag{19}$$

$$\mathbf{H}_{\text{fused}} = \mathbf{H}_{\text{fused}}^{\text{mamba}} + \mathbf{H}_{\text{fused}}^{\text{ffn}} \tag{20}$$

This setup ensures that semantic information is injected into the representation space without altering the sequence modeling backbone.

Final Fusion and Prediction. To integrate both branches, we apply a gating mechanism that adaptively balances structural and semantic information. Let $\mathbf{H}_{\text{pure}}$ and $\mathbf{H}_{\text{fused}}$ be the outputs from Stage 1 and Stage 2 respectively. The final fused representation is:

$$\mathbf{H} = [\mathbf{H}_{\text{pure}};\ \boldsymbol{\beta} \cdot \mathbf{H}_{\text{fused}}] \tag{21}$$

where $\boldsymbol{\beta} \in [0, 1]^d$ is a learnable vector that modulates the contribution of multimodal signals across dimensions. The resulting representation $\mathbf{H}$ is used for downstream prediction.

3.5 Optimization

We optimize the model using a standard cross-entropy loss for next-item prediction. Following prior work, the fused visual and textual representations are adaptively weighted and concatenated with the ID embedding to form the final item representation. This representation is then used to compute prediction logits over candidate items.

The overall training objective is defined as:

$$\mathcal{L} = \mathcal{L}_{\text{rec}} \tag{22}$$

where $\mathcal{L}_{\text{rec}}$ denotes the cross-entropy loss for next-item recommendation.

4 Experimental

4.1 Experimental Setup

Datasets. We evaluate our method on four widely-used Amazon datasets [28]: Beauty, Sports, Toys, and Digital Music. Each dataset contains timestamped user-item interactions, which are preprocessed by filtering out users and items with less than five interactions. All interactions are chronologically sorted and split into training, validation, and test sets using the leave-one-out strategy (see Table 1).

Table 1. Statistics of the used datasets.

Dataset	# Users	# Items	Sparsity	Avg. length
Beauty	22,364	12,102	99.93%	8.88
Digital Music	5,542	3,569	99.67%	11.68
Sports	35,599	18,358	99.95%	8.32
Toys	19,413	11,925	99.93%	8.63

Evaluation Metrics. We evaluate top-k recommendation using Hit Rate (Hit@k) [1] and Normalized Discounted Cumulative Gain (NDCG@k) [15], with $k \in \{10, 20\}$. Hit@k checks whether the ground-truth item appears in the top-k list, while NDCG@k weights hits by their ranks. Results are averaged over all test users.

Implementation Details. All experiments are conducted on a single NVIDIA A100 80G GPU. We implement our model and all baselines using PyTorch. The embedding size is set to $d = 64$, and the batch size is 64. We adopt the Adam optimizer with an initial learning rate of 0.001. The maximum number of training epochs is 200 with early stopping based on validation NDCG@10. All models share the same data splits and evaluation protocol to ensure fair comparison.

Baselines. We compare our model with a set of representative baselines covering various architectural paradigms. STAMP [21] captures short-term user interests through attention-based memory networks, particularly effective in session-based scenarios. GRU4Rec [12] models sequential behavior via gated recurrent units and serves as a classic RNN-based baseline. BERT4Rec [31] leverages a bidirectional Transformer with masked item modeling to learn contextualized item representations. SASRec [17] introduces self-attention to capture dynamic user preferences in item sequences. FEARec [7] enhances traditional ID-based recommendation by incorporating multimodal features in a feature-aware manner. In addition, we include recent Mamba-based models for comparison, namely M3Rec [18], Mamba4Rec [19], and SIGMA [26], which leverage state space modeling or multimodal fusion on top of Mamba to improve sequential recommendation performance.

Table 2. Overall performance comparison between CAMeRA and other baselines. The best results are bold, and the second-best are underlined. All improvements demonstrate statistical significance ($p < 0.05$).

Datasets	Metrics	STAMP	GRU4Rec	BERT4Rec	FEARec	SASRec	SIGMA	Mamba4Rec	M3Rec	**CAMeRA**
Beauty	HR@10	0.0701	0.0550	0.0364	0.0846	0.0853	0.0831	0.0812	0.0853	**0.0905**
	HR@20	0.0942	0.0836	0.0543	0.1202	0.1208	0.1133	0.0854	0.1246	**0.1274**
	N@10	0.0431	0.0287	0.0202	0.0416	0.0424	0.0403	0.0402	0.0422	**0.0456**
	N@20	0.0492	0.0359	0.0247	0.0506	0.0513	0.0489	0.0499	0.0521	**0.0537**
Sports	HR@10	0.0369	0.0290	0.0173	0.0497	0.0483	0.0496	0.0525	0.0517	**0.0537**
	HR@20	0.0539	0.0471	0.0279	0.0728	0.0699	0.0731	0.0672	0.0765	**0.0781**
	N@10	0.0224	0.0148	0.0086	0.0235	0.0226	0.0231	0.0251	0.0240	**0.0259**
	N@20	0.0266	0.0193	0.0112	0.0293	0.0281	0.0293	0.0283	0.0302	**0.0313**
Toys	HR@10	0.0695	0.0464	0.0280	0.0917	0.0897	0.0927	0.0892	0.0952	**0.0969**
	HR@20	0.0919	0.0719	0.0419	0.1266	0.1211	0.1256	0.1013	0.1300	**0.1318**
	N@10	0.0453	0.0238	0.0154	0.0445	0.0438	0.0459	0.0457	0.0464	**0.0476**
	N@20	0.0509	0.0302	0.0189	0.0533	0.0517	0.0539	0.0498	0.0551	**0.0560**
Digital Music	HR@10	0.1301	0.1388	0.0942	0.1826	0.1881	0.1858	0.1592	0.1882	**0.1905**
	HR@20	0.1861	0.2077	0.1496	0.2550	0.2637	0.2563	0.2433	0.2633	**0.2651**
	N@10	0.0701	0.0728	0.0466	0.0862	0.0882	0.0879	0.0892	0.0913	**0.0925**
	N@20	0.0842	0.0901	0.0606	0.1045	0.1072	0.1059	0.1098	**0.1102**	0.1099

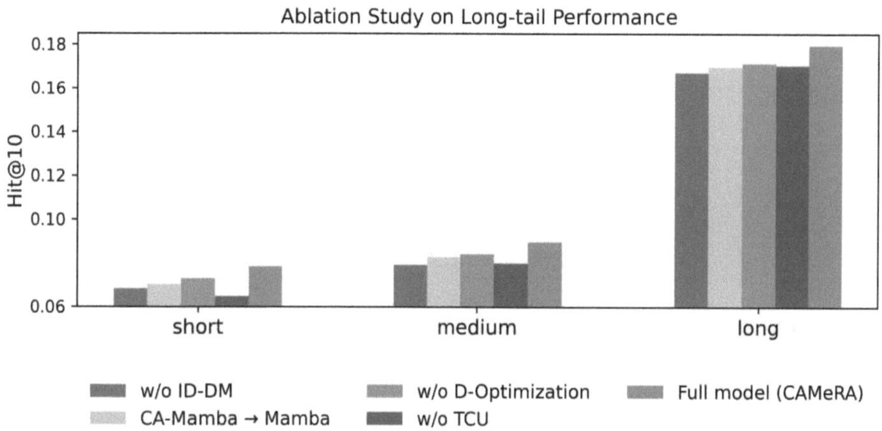

Fig. 4. Ablation study on Hit@10 performance across user groups with short, medium, and long interaction histories.

4.2 Overall Performance

As shown in Table 2, the proposed model demonstrates superior performance across all datasets and evaluation metrics. Compared with strong baselines, our method consistently achieves higher Hit@k and NDCG@k scores under both $k = 10$ and $k = 20$.

We summarize our findings as follows. Our model significantly outperforms traditional sequential models such as GRU4Rec and SASRec, demonstrating the advantage of combining bidirectional Mamba with the Temporal Compensation Unit (TCU) for robust preference modeling. Compared to multimodal baselines like FEARec and M3Rec, our method also achieves superior top-k accuracy. These improvements highlight the effectiveness of our ID-Guided Dual Modulation for early-stage filtering and the dual-phase optimization strategy for decoupling structural learning and semantic enhancement, thereby achieving more stable and generalizable recommendation performance.

4.3 Ablation Study

We conduct ablation studies on the Beauty dataset to assess the impact of each module in CAMeRA. As shown in Table 3, removing ID-guided dual modulation significantly degrades performance, validating its role in denoising and identity-aware preference modeling. Replacing the bidirectional CA-Mamba with a unidirectional variant weakens temporal modeling, while disabling dual-phase optimization leads to poorer convergence due to increased training interference. Excluding the temporal compensation unit slightly reduces accuracy, suggesting that short-term cues offer complementary benefits under sparse or long-tail conditions. These results confirm the necessity and complementarity of our structure-first, modulation-aware design.

Table 3. Ablation study on key modules in CAMeRA.

Variant	Hit@10	NDCG@10
w/o ID-DM	0.0802	0.0379
CA-Mamba → Mamba	0.0830	0.0401
w/o D-Optimization	0.0841	0.0415
w/o TCU	0.0813	0.0387
Full model (CAMeRA)	**0.0905**	**0.0456**

Table 4. Performance and efficiency comparison on the Beauty dataset.

Model	Hit@10	NDCG@10	Inf. Time	GPU Mem.
SASRec	0.0853	0.0424	233ms	14.48G
FEARec	0.0846	0.0416	245ms	14.91G
SIGMA	0.0831	0.0403	**132ms**	**4.30G**
M3Rec	0.0853	0.0422	139ms	4.53G
CAMeRA	**0.0905**	**0.0456**	148ms	4.74G

We also analyze performance across user behavior lengths based on Hit@10. As shown in Fig. 4, all ablated variants lag behind the full model, especially on long sequences. The absence of TCU particularly harms short-sequence users, highlighting its role in capturing short-term dynamics missed by global models.

4.4 Grouped Users Analysis

To gain deeper insights into model behavior under different user engagement patterns, we categorize users based on their interaction history length: short (1–5), medium (6–20), and long (21+). This grouping reflects the long-tail distribution in recommender systems, where most users have few interactions and a minority exhibit dense histories. Such imbalance challenges generalization—short-history users face data sparsity, while long-history users require strong long-range modeling. All interactions are ordered chronologically to ensure temporal consistency.

Figure 5 shows the Hit@10 performance for different user categories. CAMeRA consistently outperforms strong baseline models such as FEARec, SASRec, and M3Rec on both the Beauty and Sports datasets. Performance gains are even more significant for users with longer interaction histories, as our two-stage optimization and bidirectional CA-Mamba synergistically enhance long-range context modeling and representation stability. Notably, CAMeRA achieves consistent improvements even for users with short and medium interaction durations. This is due to the Temporal Compensation Unit (TCU), which captures transient behavioral cues often overlooked by traditional sequence models, and the adaptive multimodal fusion enhanced by ID-guided modulation.

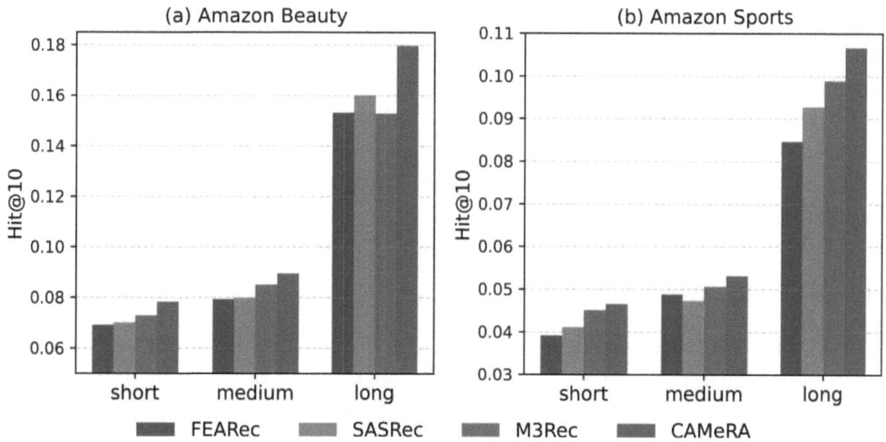

Fig. 5. User group analysis on Beauty and Sports datasets (Hit@10). Users are divided into short-, medium-, and long-history groups. CAMeRA consistently surpasses baselines across all types, showing strong adaptability to diverse behaviors.

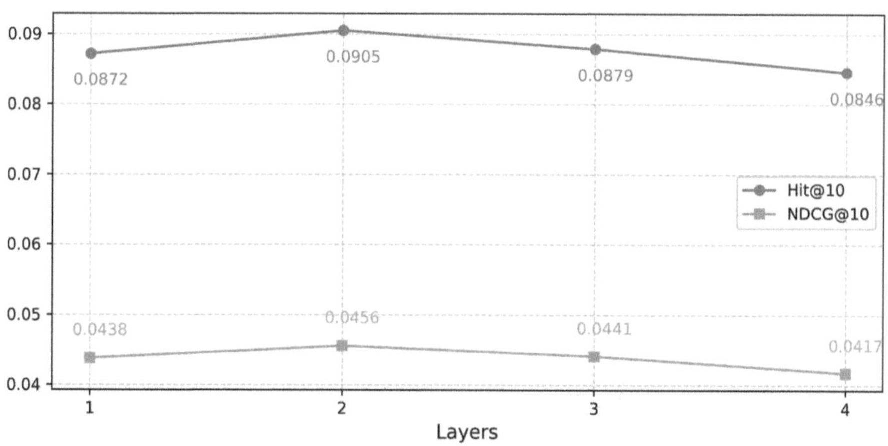

Fig. 6. Performance of CAMeRA with varying numbers of encoder layers on the Beauty dataset.

4.5 Model Efficiency Evaluation

To evaluate the performance-efficiency trade-off, we compare inference time and GPU memory usage on the Beauty dataset. As shown in Table 4, CAMeRA achieves the best Hit@10 and NDCG@10, outperforming both Mamba- and Transformer-based baselines. While slightly more resource-intensive than SIGMA and M3Rec, CAMeRA delivers significantly higher accuracy. This is understandable, as SIGMA only models ID sequences and ignores multimodal interactions, while M3Rec underutilizes the role of ID and lacks explicit mechanisms for context modeling and long-tail adaptation. Compared to Transformer-based models like SASRec and FEARec, CAMeRA achieves faster inference and requires significantly less GPU memory, demonstrating a strong balance between accuracy and efficiency.

Furthermore, as shown in Fig. 6, we analyze the performance of CAMeRA under different encoder depths. The results indicate that the optimal performance is achieved with two layers, while further increasing the depth yields diminishing returns.

4.6 Visualization

Figure 7 illustrates the progressive alignment of visual and textual features. ID-guided modulation enhances cross-modal coherence, while fusion and identity anchoring produce a compact, well-structured embedding space, confirming the effectiveness of our dual modulation design.

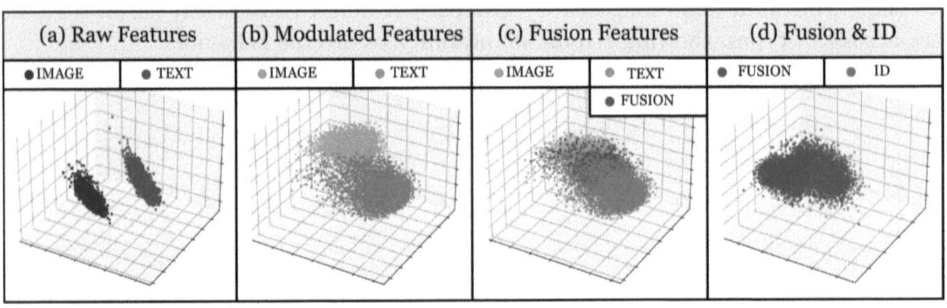

Fig. 7. Visualization of Feature Dynamics in ID-Guided Dual Modulation.

5 Conclusion

In this work, we propose CAMeRA, a multimodal recommendation framework that tackles semantic noise in fusion and inefficiencies in long- and short-sequence modeling. Using ID-guided dual modulation, our method achieves cleaner multimodal alignment via structural anchors. A dual-phase optimization strategy decouples structural learning from semantic enhancement, improving interpretability. For sequential modeling, we introduce CA-Mamba, a bidirectional Mamba encoder with a lightweight GRU-based compensation module for long-tail user behaviors. Experiments on four Amazon datasets show that CAMeRA outperforms strong baselines in accuracy and robustness. Future work will investigate dynamic fusion under incomplete modalities and develop lightweight versions for real-time inference.

References

1. Alsini, A., Huynh, D.Q., Datta, A.: Hit ratio: an evaluation metric for hashtag recommendation. arXiv preprint arXiv:2010.01258 (2020)
2. Aoki, M.: State Space Modeling of Time Series. Springer (2013)
3. Boka, T.F., Niu, Z., Neupane, R.B.: A survey of sequential recommendation systems: techniques, evaluation, and future directions. Inf. Syst. **125**, 102427 (2024)
4. Chen, M., Ma, T., Zhou, X.: CoCNN: co-occurrence CNN for recommendation. Expert Syst. Appl. **195**, 116595 (2022)
5. Cui, Q., Wu, S., Liu, Q., Zhong, W., Wang, L.: MV-RNN: a multi-view recurrent neural network for sequential recommendation. IEEE Trans. Knowl. Data Eng. **32**(2), 317–331 (2018)
6. Dey, R., Salem, F.M.: Gate-variants of gated recurrent unit (GRU) neural networks. In: 2017 IEEE 60th International Midwest Symposium on Circuits and Systems (MWSCAS), pp. 1597–1600. IEEE (2017)
7. Du, X., et al.: Frequency enhanced hybrid attention network for sequential recommendation. In: Proceedings of the 46th International ACM SIGIR Conference on Research and Development in Information Retrieval, pp. 78–88 (2023)
8. Fan, Z., Liu, Z., Zhang, J., Xiong, Y., Zheng, L., Yu, P.S.: Continuous-time sequential recommendation with temporal graph collaborative transformer. In: Proceedings of the 30th ACM International Conference on Information & Knowledge Management, pp. 433–442 (2021)
9. Gu, A., Dao, T.: Mamba: linear-time sequence modeling with selective state spaces. arXiv preprint arXiv:2312.00752 (2023)
10. Gu, A., Goel, K., Ré, C.: Efficiently modeling long sequences with structured state spaces. arXiv preprint arXiv:2111.00396 (2021)
11. Guo, X., Zhang, T., Xue, Y., Wang, C., Wang, F., Cui, Z.: M 3 rec: selective state space models with mixture-of-modality experts for multi-modal sequential recommendation. In: ICASSP 2025-2025 IEEE International Conference on Acoustics, Speech and Signal Processing (ICASSP), pp. 1–5. IEEE (2025)
12. Hidasi, B., Karatzoglou, A., Baltrunas, L., Tikk, D.: Session-based recommendations with recurrent neural networks. arXiv preprint arXiv:1511.06939 (2015)
13. Hu, H., Guo, W., Liu, Y., Kan, M.Y.: Adaptive multi-modalities fusion in sequential recommendation systems. In: Proceedings of the 32nd ACM International Conference on Information and Knowledge Management, pp. 843–853 (2023)
14. Hu, J., et al.: Bivrec: bidirectional view-based multimodal sequential recommendation. arXiv preprint arXiv:2402.17334 (2024)
15. Jayashree, R., Christy, A.: Improving the enhanced recommended system using Bayesian approximation method and normalized discounted cumulative gain. Procedia Comput. Sci. **50**, 216–222 (2015)
16. Jiang, J., et al.: Adamct: adaptive mixture of CNN-transformer for sequential recommendation. In: Proceedings of the 32nd ACM International Conference on Information and Knowledge Management, pp. 976–986 (2023)
17. Kang, W.C., McAuley, J.: Self-attentive sequential recommendation. In: 2018 IEEE International Conference on Data Mining (ICDM), pp. 197–206. IEEE (2018)
18. Lan, Z., Zhang, Y., Li, X.: M3REC: a meta-based multi-scenario multi-task recommendation framework. In: Proceedings of the 17th ACM Conference on Recommender Systems, pp. 771–776 (2023)
19. Liu, C., Lin, J., Wang, J., Liu, H., Caverlee, J.: Mamba4rec: towards efficient sequential recommendation with selective state space models. arXiv preprint arXiv:2403.03900 (2024)

20. Liu, H., Wei, Y., Song, X., Guan, W., Li, Y.F., Nie, L.: Mmgrec: multimodal generative recommendation with transformer model. arXiv preprint arXiv:2404.16555 (2024)
21. Liu, Q., Zeng, Y., Mokhosi, R., Zhang, H.: Stamp: short-term attention/memory priority model for session-based recommendation. In: Proceedings of the 24th ACM SIGKDD International Conference on Knowledge Discovery & Data Mining, pp. 1831–1839 (2018)
22. Liu, Q., et al.: Multimodal recommender systems: a survey. ACM Comput. Surv. **57**(2), 1–17 (2024)
23. Liu, Y., et al.: Pre-training graph transformer with multimodal side information for recommendation. In: Proceedings of the 29th ACM International Conference on Multimedia, pp. 2853–2861 (2021)
24. Liu, Z., Fan, Z., Wang, Y., Yu, P.S.: Augmenting sequential recommendation with pseudo-prior items via reversely pre-training transformer. In: Proceedings of the 44th International ACM SIGIR Conference on Research and Development in Information Retrieval, pp. 1608–1612 (2021)
25. Liu, Z., Ma, Y., Schubert, M., Ouyang, Y., Rong, W., Xiong, Z.: Multimodal contrastive transformer for explainable recommendation. IEEE Trans. Comput. Soc. Syst. **11**(2), 2632–2643 (2023)
26. Liu, Z., et al.: Sigma: selective gated mamba for sequential recommendation. In: Proceedings of the AAAI Conference on Artificial Intelligence, vol. 39, pp. 12264–12272 (2025)
27. Mo, F., Xiao, L., Song, Q., Gao, X., Liang, E.: Multimodal graph neural network for recommendation with dynamic de-redundancy and modality-guided feature denoisy. arXiv preprint arXiv:2411.01561 (2024)
28. Ni, J., Li, J., McAuley, J.: Justifying recommendations using distantly-labeled reviews and fine-grained aspects. In: Proceedings of the 2019 Conference on Empirical Methods in Natural Language Processing and the 9th International Joint Conference on Natural Language Processing (EMNLP-IJCNLP), pp. 188–197 (2019)
29. Qu, H., et al.: A survey of mamba. arXiv preprint arXiv:2408.01129 (2024)
30. de Souza Pereira Moreira, G., Rabhi, S., Lee, J.M., Ak, R., Oldridge, E.: Transformers4rec: bridging the gap between NLP and sequential/session-based recommendation. In: Proceedings of the 15th ACM Conference on Recommender Systems, pp. 143–153 (2021)
31. Sun, F., et al.: Bert4rec: sequential recommendation with bidirectional encoder representations from transformer. In: Proceedings of the 28th ACM International Conference on Information and Knowledge Management, pp. 1441–1450 (2019)
32. Tan, Y.K., Xu, X., Liu, Y.: Improved recurrent neural networks for session-based recommendations. In: Proceedings of the 1st Workshop on Deep Learning for Recommender Systems, pp. 17–22 (2016)
33. Vaswani, A., et al.: Attention is all you need. In: Advances in Neural Information Processing Systems, vol. 30 (2017)
34. Xu, J., Chen, Z., Yang, S., Li, J., Wang, H., Ngai, E.C.: Mentor: multi-level self-supervised learning for multimodal recommendation. In: Proceedings of the AAAI Conference on Artificial Intelligence, vol. 39, pp. 12908–12917 (2025)
35. Yan, A., Cheng, S., Kang, W.C., Wan, M., McAuley, J.: Cosrec: 2D convolutional neural networks for sequential recommendation. In: Proceedings of the 28th ACM International Conference on Information and Knowledge Management, pp. 2173–2176 (2019)

36. Yang, Y., Huang, C., Xia, L., Liang, Y., Yu, Y., Li, C.: Multi-behavior hypergraph-enhanced transformer for sequential recommendation. In: Proceedings of the 28th ACM SIGKDD Conference on Knowledge Discovery and Data Mining, pp. 2263–2274 (2022)
37. Yuan, L., Zhao, X.Y.: A hybrid MLP and CNN architecture for sequential recommendation. IEEE Access (2025)
38. Zhang, L., Zhou, X., Zeng, Z., Shen, Z.: Multimodal pre-training for sequential recommendation via contrastive learning. ACM Trans. Recomm. Syst. **3**(1), 1–23 (2024)
39. Zhang, X., et al.: Mamba in speech: towards an alternative to self-attention (2025). https://arxiv.org/abs/2405.12609
40. Zhang, Y., Wang, X., Chen, H., Zhu, W.: Adaptive disentangled transformer for sequential recommendation. In: Proceedings of the 29th ACM SIGKDD Conference on Knowledge Discovery and Data Mining, pp. 3434–3445 (2023)

Improving Question Recommendation Through Oracle Recommendation Imitation

Haipeng Liu[1], Ting Long[1(✉)], Yixiang Shan[1], Liang Yin[2], and Yi Chang[1]

[1] School of Artificial Intelligence, Jilin University, Changchun, China
{liuhp22,shanyx22}@mails.jlu.edu.cn, {longting,yichang}@jlu.edu.cn
[2] Computer Science Department, Shanghai Jiao Tong University, Shanghai, China
yinla@apex.sjtu.edu.cn

Abstract. Question recommendation aims to sequentially recommend personalized questions to improve student learning effects. Most existing question recommenders are built on online reinforcement learning pipelines, which simultaneously collect training data and update the model. While effective for exploration, the collected data is often low quality, limiting the recommender's ability to learn strategies that truly enhance learning. In this paper, we conduct an empirical study of this issue and propose a new pipeline named Oracle Recommendation Imitation (ORI). ORI first trains a preliminary recommender to generate high-quality samples and uses an oracle selection mechanism to efficiently extract reliable offline data enriched with collaborative information. This oracle dataset, combined with real-time online data, guides the recommender toward effective strategies. Extensive experiments show that ORI, despite its simplicity, can boost the performance of various recommenders by up to several tens of percentage points. Our code is available at https://github.com/wuming29/ORI.

Keywords: Question recommendation · Online education · Adaptive learning · reinforcement learning

1 Introduction

Question recommendation [12,16,22] has become an important task in the area of adaptive learning. Given the learning targets and question-answering (QA) history of an arbitrary student at each step, a question recommender is supposed to continuously recommend personalized questions for the student to practice, helping them gradually master their learning targets. The improvement in the student's mastery of the learning target is defined as the **learning effect**, and our goal is to train a recommender that can maximize the learning effect during the recommendation session. Considering the example illustrated in Fig. 1, assume that the learning targets of Bob are mastering *multiplication* and *division*. From his QA history, we can observe that he correctly answered questions related to *addition*, *subtraction*, and *multiplication*, which indicates that

H. Jung et al. (Eds.): DASFAA 2026, LNCS 16535, pp. 188–205, 2026.
https://doi.org/10.1007/978-981-92-0363-5_12

he may have already mastered these concepts. Therefore, the recommender recommends questions related to *division* to further enhance his learning effect. After the recommendation session, assuming that Bob has mastered only one of the two learning targets (*i.e.multiplication*), his learning effect can be recorded as $1/2 = 0.5$.

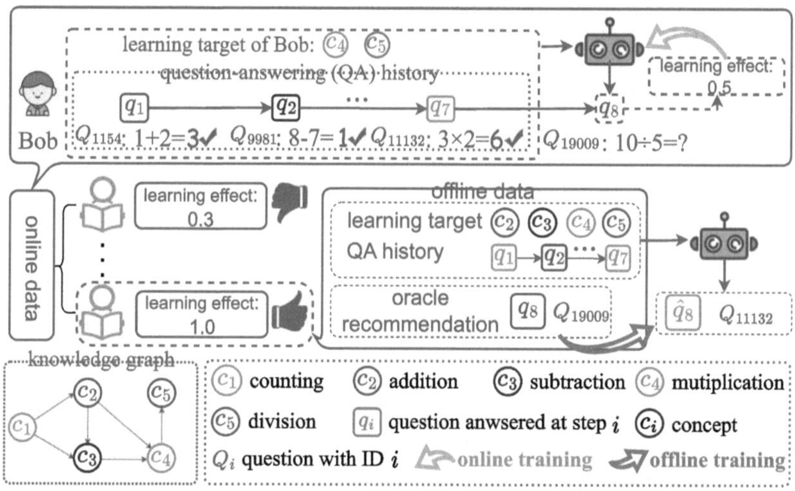

Fig. 1. Illustration of question recommendations and the core idea of ORI.

Since the goal of education is to enhance students' long-term learning effects rather than to cater to their immediate interests, previous methods used for predicting user click interests [3,7,28] are not suitable. Benefiting from the nature of reinforcement learning (RL) in maximizing long-term return, existing question recommender treat students' learning effects as the return and apply online RL to optimize the learning effects in long term [2,12,16,25]. The training pipeline of the online RL is shown in Fig. 1. Under this pipeline, the recommender dynamically updates each student's QA history during the recommendation process based on the recommended questions and the student's responses. After each session, the recommender then uses the updated QA history, along with the student's learning targets and learning effects, as training data to evaluate and refine the current recommendation strategy. The online RL pipeline enables the recommender to continuously explore and discover more effective strategies, making it the mainstream approach for training QA recommenders.

Although the online RL pipeline has been widely adopted and proven effective, we argue that it may not be the optimal approach for maximizing recommender performance. Under this pipeline, the training data is generated by a recommender that remains insufficiently trained and continuously explores untested strategies, making the collected data unlikely to yield high learning effects for students. We regard such data as **low-quality**, since it provides limited guidance for the recommender to learn how to produce effective recommendations. The continued use of such low-quality data can, in turn, significantly hinder the recommender's ability to improve its performance over time [9,13,24].

In this work, we first empirically analyze and demonstrate the aforementioned limitation of the online RL pipeline in detail (Sect. 4). To address this issue, we propose a novel training pipeline, termed Oracle Recommendation Imitation (ORI). As illustrated in Fig. 1, the core idea of ORI is quite simple: it extracts high-quality offline data from online interactions and uses this data to supplement the training process of the recommender. Since the offline data encapsulates informative patterns that guide the recommender toward effective decision-making, it offers stronger supervision and consequently enables further performance gains. Moreover, considering the enormous state-action space in question recommendation tasks, directly selecting high-quality data from limited online interactions is highly inefficient. Therefore, unlike traditional methods that filter high-quality samples, ORI introduces an oracle selection mechanism to efficiently extract reliable offline data enriched with collaborative information from online data.

Specifically, ORI takes a predefined RL-based recommender as its basic model and enhances it through a three-stage boosting strategy. In the first stage, ORI roughly trains the recommender to generate online data under the online RL pipeline. In the second stage, ORI collects interaction data between the recommender and students, identifies high-quality samples, and integrates collaborative information to construct a high-quality offline dataset, referred to as the oracle dataset. In the final stage, ORI jointly optimizes the recommender via online RL training and offline imitation learning using the oracle dataset. Since ORI can instantiate any RL-based recommender as its basic model, the proposed pipeline is broadly compatible with various existing question recommenders. Extensive experiments demonstrate that ORI's simple idea can nevertheless greatly enhance the performance of recommender systems.

In summary, the contributions of our paper are:

- Unlike previous works that only design recommender architectures within the online RL framework, we are the first, to the best of our knowledge, to propose the issue of the online RL pipeline and provide corresponding solutions in the educational domain.
- Our ORI is highly compatible. Theoretically, any RL-based question recommender can be utilized as the basic model and its performance can be enhanced by ORI.
- We conduct extensive experiments to evaluate ORI and compare it with numerous baselines. The results demonstrate the effectiveness of ORI.

2 Related Works

Previous question recommender typically defined the question recommendation task as a Markov Decision Process (MDP), optimizing the recommender using online RL algorithms. For example, CSEAL [16] encoded students' QA history and learning targets as the state, treated the recommended question as the action, and defined the learning effect as the reward. Following this, SRC [2] introduced self-attention mechanisms [27] to enhance recommender performance, while GEHRL [14] used a hierarchical reinforcement learning [20] architecture to improve performance. Building on these three works, HierLLM [17] further boosted the recommender's performance by leveraging the open-world knowledge of large language models.

Although previous works gain much achievement in question recommendation, all of them train the recommendation strategy within an online RL pipeline. Due to that most of the training data used in the online RL pipeline is of low quality, these works are unable to reach their full potential performance. To further improve the performance of question recommender, we introduce a novel pipeline that collecting high-quality data to train question recommenders.

3 Preliminaries

In this section, we will introduce the terminologies used in this paper and formulations of this task.

3.1 Terminologies

For a student u learning on an online education site, their **Question-answering (QA) history** at step t is $\mathcal{H}_t^u = \{(q_1, y_1), (q_2, y_2), \dots (q_i, y_i), \dots, (q_{t-1}, y_{t-1})\}$. Here (q_i, y_i) denotes the **QA record** at step i. q_i is the question they answers at step i, which is an element of question set, *i.e.*, $q_i \in \mathcal{Q}$. If the question ID of q_i is j, we use Q_j to denote q_i's ID. A question may cover one or more concepts. y_i is the correctness of their response to the q_i. If they answered q_i correctly, then $y_i = 1$, otherwise $y_i = 0$. For example, the QA record of Bob at step 1 is $(Q_{1154}, 1)$, the QA history of Bob at step 8 in Fig. 1 is $[(Q_{1154}, 1), (Q_{9981}, 1), \dots, (Q_{11132}, 1)]$, and the concept covered question Q_{1154} is *addition*.

The **learning target** is defined as the concepts students aim to master. Since concepts are covered by questions, *e.g.*, question $1 + 2 =?$ covers the concept *addition*, following previous works [2,16], we transform the learning target into the questions related to the target concepts for the ease of implementation, denoted as $\mathcal{T} = \{Q_j\}^{|\mathcal{T}|}$. For instance, if Bob in Fig. 1 aims to master *multiplication* and *division*, his learning target is composed of the questions related to *multiplication* and *division*, *i.e.*, $\mathcal{T} = \{Q_{11132}, Q_{19009}\}$.

The **learning effect** of a student is defined as the improvement of their mastering question under the learning target after the recommending session is complete [2,14,16]. We quantify this metric by:

$$\Delta_u = \frac{m_e - m_b}{m_{sup} - m_b}, \tag{1}$$

where m_b and m_e denote the mastery of the learning target at the beginning and end of the recommendation respectively, and m_{sup} denotes the maximum mastery of the learning target. Considering the example in the Fig. 1, if Bob's learning target is $\mathcal{T} = \{Q_{11132}, Q_{19009}\}$, the maximum number of questions he can master is 2, so $m_{sup} = 2$. If none of these questions are mastered at the beginning of the recommendation, then $m_b = 0$. After he exercises the recommended questions, if he correctly answers Q_{11132} but incorrectly answers Q_{19009}, we obtain $m_e = 1$. Therefore, Bob's learning effect is $\Delta_u = (1-0)/(2-0) = 0.5$.

3.2 Problem Formulation

In this paper, we aim to train a question recommender that continuously selects questions to maximize students' learning effects. Given a student u with QA history $\mathcal{H}_t^u$ and learning target $\mathcal{T}$, the recommender chooses the question $q_t \in \mathcal{Q}$ expected to yield the greatest long-term improvement in learning effect. After the student answers q_t, the correctness feedback updates $\mathcal{H}_{t+1}^u$, upon which the next recommendation is made. When a session ends, the student's overall learning effect Δ_u is evaluated. Since each recommendation step influences future learning, the task is naturally formulated as a Markov Decision Process (MDP) $< \mathcal{S}, \mathcal{A}, \mathcal{P}, R, \gamma >$ [2,14,16]. Here, $s_t \in \mathcal{S}$ denotes the state which are used for selecting questions from $\mathcal{Q}$ to recommend. $\mathcal{A}$ is a finite set of actions, and it denotes the question set $\mathcal{Q}$. R denotes the instant reward $r(\mathbf{s}_t, q_t)$ of recommending question q_t on state $\mathbf{s}_t$, which is only measured by the learning effect Δ_u.

As the recommender aims to maximize students' learning effect Δ_u, its primary optimization target could be formulated as:

$$\max_\pi J(\pi) = \max_\pi \Delta_u = \max_\pi \frac{1}{T} \sum_{t=0}^{T} \gamma^t r(\mathbf{s}_t, q_t), \tag{2}$$

where T denotes the total length of recommendation sessions. π is the recommender (see Sect. 5 for details).

4 Analysis of Online RL Pipeline

To maximize students' learning effects, prior works commonly adopt an online RL pipeline to train question recommenders. In this pipeline, a randomly initialized recommender sequentially selects questions for a student u, recording the QA history $\mathcal{H}_t^u$ and recommended question q_t at each step. After a session, the recommender associates the student's final learning effect Δ_u with the QA trajectory $\tau_u = \{(\mathcal{H}_t^u, q_t) \mid t \geq 0\}$ and learning target $\mathcal{T}$ to form a training sample $z_u = (\tau_u, \mathcal{T}, \Delta_u), z_u \in \mathcal{D}$, which is then used for model updating.

However, since the recommender generates data while still undertrained and continuously explores untested strategies, the resulting trajectories τ_u often lead to low learning effects, providing limited guidance for learning effective strategies. To quantify this issue, we define each sample's learning effect Δ_u as its **data quality**, and the average learning effect $\bar{\Delta}_u$ during testing as the recommender's **performance**. To validate our analysis, we conduct empirical studies showing that (1) higher-quality training data leads to better recommender performance, and (2) the performance of the recommender trained under the online RL pipeline remains far from optimal.

4.1 Quality vs Performance

To verify that higher-quality training data leads to better recommender performance, we trained three RL-based recommenders-DQN [18], SRC [2], and RAR [15]-on datasets with low, medium, and high average data quality. Each model was independently trained on the three datasets, and its performance was evaluated. As shown in Fig. 2(a), bar charts denote mean data quality and line plots denote recommender performance. Across all models, performance consistently improves with data quality, empirically confirming the positive correlation between data quality and model performance.

4.2 Potential of RAR

To further examine the online RL pipeline, we conducted an experiment using the state-of-the-art RAR [15]. We first trained RAR to convergence under the online RL pipeline, then retrained it with a manually constructed high-quality expert dataset under the offline RL pipeline. During training, we recorded the average data quality and corresponding performance in each episode. As shown in Fig. 2(b), online RL training produces data whose quality is constrained by the model's current performance, whereas retraining with higher-quality expert data yields further improvement even after convergence. This indicates that the online RL pipeline constrains the recommender's full potential and highlights its inherent limitations.

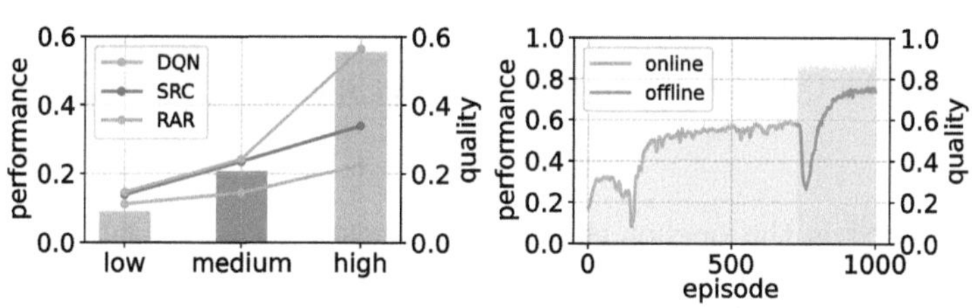

(a) quality vs performance (b) potential of RAR

Fig. 2. The impact of data quality on performance and the quality distribution of online data.

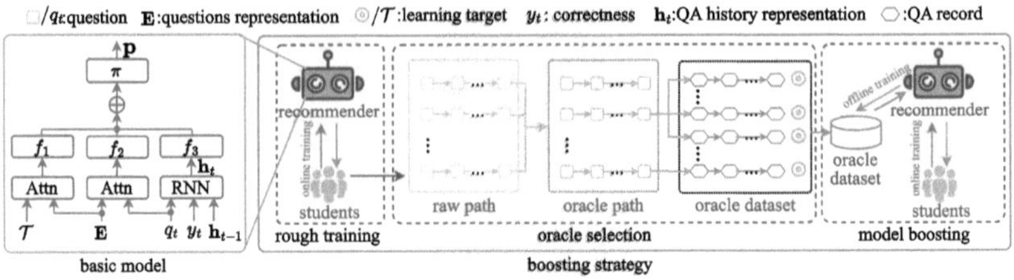

Fig. 3. The pipeline of our method.

5 Methodology

5.1 Overview

To overcome the limitations of online RL pipelines, we propose a novel training pipeline named Oracle Recommendation Imitation (ORI). As illustrated in Fig. 3, ORI aims to extract high-quality offline data from online interactions and enhance recommender performance through a hybrid of online and offline training. During data extraction, we introduce an oracle selection mechanism to ensure both the efficiency and reliability of the collected samples. Specifically, ORI consists of a basic model and a boosting strategy. The basic model is an arbitrary RL-based question recommender, while the boosting strategy-the core of ORI-trains it in three stages: rough train, oracle selection, and model boosting. Next, we will discuss the basic model and boosting strategy in detail.

5.2 Basic Model

The basic model is an arbitrary RL-based recommender forming the training foundation. To illustrate its training under our boosting strategy, we construct a sample basic model, which consists of a state encoder and a policy network.

State Encoder. The state encoder converts a student's QA history and learning target into a vector representing the MDP state. For each student u, we extract three types of information from the QA history $\mathcal{H}_t^u$ and the target $\mathcal{T}$: (1) the representation of $\mathcal{H}_t^u$, (2) the representation of $\mathcal{T}$, and (3) the representation of recent questions z_t, which are modeled separately due to their strong impact on future recommendations.

To encode the QA history, we first represent question q_i in $\mathcal{H}_t^u$ by:

$$\mathbf{e}_i = f_e(\mathbf{o}_i), \tag{3}$$

where $\mathbf{o}_i \in \mathbb{R}^{|\mathcal{Q}|}$ is the onehot encoding of question q_i. f_e denotes the linear function. $\mathbf{e}_i \in \mathbb{R}^{d_e}$ denotes the embedding of question q_i.

Next, we leverage a Recurrent Neural Network (RNN) to encode the student's QA history:

$$\begin{aligned} \mathbf{a}_i &= f_a\left(\mathbf{e}_i \oplus f_y(y_i)\right), \\ \mathbf{h}_{i+1} &= RNN(\mathbf{a}_i, \mathbf{h}_i), \text{for i = 0,..., t - 1,} \end{aligned} \tag{4}$$

where y_i is the correctness of response to q_i (see Sect. 3.1), $f_y(\cdot), f_a(\cdot)$ are linear functions. $\oplus$ denotes the concatenation. $\mathbf{a}_i \in \mathbb{R}^{d_a}$ denotes the representation of a QA record. $RNN(\cdot)$ denotes a Recurrent Neural Network, which is implemented with LSTM [8]. $\mathbf{h}_i \in \mathbb{R}^{d_h}$ denotes the representation of QA history.

To encode the learning target, we first encode the questions in the learning target by following [2,19]:

$$\mathbf{r}_i = Attn(f_r(\mathbf{e}_i), \mathbf{E}, \mathbf{E}) \oplus f_r(\mathbf{e}_i), \tag{5}$$

where $f_r(\cdot)$ is a linear function, $\mathbf{E} \in \mathbb{R}^{|\mathcal{Q}| \times d_e}$ denotes the embedding of questions in $\mathcal{Q}$, which is computed by feeding the questions in $\mathcal{Q}$ to Eq. (3). $\mathbf{r}_i \in \mathbb{R}^{d_r}$ represents the attentive representation of question q_i. $Attn(\cdot,\cdot,\cdot)$ denotes the multi-head attention [27], where key and value are $\mathbf{E}$ and query is $f_r(\mathbf{e}_i)$.

Then, we average the attentive representation of the questions in learning target $\mathcal{T}$ to obtain the representation of the learning target:

$$\mathbf{g} = Mean(\{\mathbf{r}_i | q_i \in \mathcal{T}\}). \tag{6}$$

It deserves to be mentioned here, we encode the questions in learning target via Eq. (5) but represent the questions in QA history via Eq. (3) due to the pressure of computation resource. It is feasible to replace $\mathbf{e}_i$ in Eq. (4) with $\mathbf{r}_i$ in theory.

To encode the recent questions, we apply the same encoding mechanism with the learning target, and we average the attentive representation of recent questions:

$$\mathbf{z}_t = Mean(\{\mathbf{r}_i | q_i \in \{q_{t-1}, ..., q_{t-n}\}\}), \tag{7}$$

where n is a hyperparameter, which denotes the number of recent questions we consider.

Finally, we sum the representation of the QA history, learning target, and recent questions to obtain the state $\mathbf{s}_t$:

$$\mathbf{s}_t = f_1(\mathbf{h}_t) + f_2(\mathbf{g}) + f_3(\mathbf{z}_t), \tag{8}$$

where $f_1(\cdot)$, $f_2(\cdot)$, $f_3(\cdot)$ are linear functions.

Policy Network. The policy network is responsible for selecting a suitable question from $\mathcal{Q}$. The recommended question q_t at step t is obtained by:

$$\begin{aligned} \mathbf{p}_t &= f_p(\mathbf{s}_t), \\ q_t &\sim \mathbf{p}_t, \end{aligned} \tag{9}$$

where $\mathbf{p}_t \in \mathbf{R}^{|\mathcal{Q}|}$, which represents the probability of the corresponding question being selected as the recommended question. f_p is a linear function. $q_t \sim \mathbf{p}_t$ represent sample question q_t according to $\mathbf{p}_t$.

5.3 Boosting Strategy

The boosting strategy is the core of ORI. It has three stages called rough training, oracle selection, and model boosting to train the basic model.

Rough Training. The goal of this stage is to roughly train the basic model to produce recommendations and generate the high-quality data we need subsequently. We design two optimization objectives to train the basic model: optimizing the learning effect and optimizing the representation of the QA history.

We first optimize the basic model by using the learning effect through the Policy Gradient [23]:

$$L_p = -\sum_{t=1}^{T} \Delta_u \ln p_t, \tag{10}$$

where p_t is the probability of question q_t is selected, obtained from $\mathbf{p}_t$ in Eq. (9).

In order to make full use of students' responses after each step of recommendation, we additionally use QA historical representation to predict students' performance on each recommended question. Specifically, we set this part of the loss function as follows:

$$\begin{aligned} \hat{y}_t &= f_k(\mathbf{h}_t), \\ L_k &= -\sum_{t=1}^{T} (y_t \ln \hat{y}_t + (1 - y_t) \ln (1 - \hat{y}_t)), \end{aligned} \tag{11}$$

where f_k is a linear function, y_t denotes the corresponding correctness of the question q_t.

Therefore, in this stage, the basic model is optimized by:

$$L_b = L_p + L_k. \tag{12}$$

Instead of fully training the basic model, we employ early stopping to obtain a roughly trained version. This speeds up ORI and helps generate higher-quality data: compared with an untrained model it performs better, while its incomplete convergence encourages more diverse recommendations [10,26], both of which increase the likelihood of obtaining high-quality data in the next stage.

Oracle Selection. In this stage, the basic model roughly trained by Eq. (12) interacts with students to collect data for building a high-quality oracle dataset that enhances model performance. However, due to the vast state-action space arising from numerous questions and diverse students, limited interactions make it hard to gather valuable samples. To address this, we propose a three-step Oracle Selection mechanism, which is based on the assumption that a recommendation failing for one student may be beneficial for another. For instance, a strategy emphasizing *addition* may be ineffective for students focusing on *multiplication* but beneficial for those targeting *addition*. As shown in Fig. 3, we take the following steps to obtain the oracle dataset:

Algorithm 1. The training of ORI

Randomly initialize the basic model.
// Rough training.
for episode $\leq \eta$ **do**
    Train basic model by Eq. (12).
end for
while not converge **do**
    // Oracle selection.
    Basic model interacts with students to obtain $\mathcal{D}_o$.
    Randomly initialize the basic model again for boosting.
    // Model boosting.
    while not converge **do**
        Train basic model by Eq. (14).
    end while
end while

First, the basic model interacts with students to obtain the **raw paths** ($\mathcal{P}'$). Specifically, for each student $u \in \mathcal{U}_1$, we let the basic model recommend questions for them. We denote the sequence of questions recommended to u as a path $\mathcal{P}_u = \{q_1, q_2, ..\}$. We collect the paths of all the students in $\mathcal{U}_1$, and form the set of paths $\mathcal{P}' = \{\mathcal{P}_u | u \in \mathcal{U}_1\}$.

Next, to ensure that the interaction data align with fundamental learning principles and improve generalizability and reliability, we generate **oracle paths** by recommending the questions from the raw paths to another group of students $\mathcal{U}_2$. Specifically, for each student $u' \in \mathcal{U}_2$, we take the questions from $\mathcal{P}_u$ ($\mathcal{P}_u \in \mathcal{P}'$) one by one, sequentially recommend them to student u' (Each path $\mathcal{P}_u$ is recommended to all students $u' \in \mathcal{U}_2$). We denote the learning effect after u' exercising the questions in $\mathcal{P}_u$ as $\Delta_{u'|u}$. Then, we compute the average learning effect of recommending the questions in $\mathcal{P}_u$ by:

$$\bar{\Delta}_u = Mean(\{\Delta_{u'|u} | u' \in \mathcal{U}_2\}), \tag{13}$$

where $\bar{\Delta}_u$ denotes the average learning effect students obtain after they exercise the questions in $\mathcal{P}_u$. If $\bar{\Delta}_u \geq \epsilon$, we put path $\mathcal{P}_u$ to the set of oracle paths $\mathcal{P}$.

Finally, for each oracle path, we match it with the most suitable student, thereby introducing collaborative information into the final **oracle dataset**. Specifically, for each student $u' \in \mathcal{U}_3$, we recommend the questions in $\mathcal{P}_u$ ($\mathcal{P}_u \in \mathcal{P}$) one by one to u'. If their learning effect $\Delta_{u'|u} \geq \zeta$, we put their QA history and learning target to the oracle dataset $\mathcal{D}_o$. The oracle dataset is then used in the stage of model boosting.

Model Boosting. In this stage, the basic model is further trained using the oracle dataset, which supplements the online RL pipeline with effective recommendation signals and enables model boosting. Specifically, the basic model is optimized in this stage via:

$$L_i = \sum_{t=1}^{T} BCE(\mathbf{p}_t, \mathbf{o}_t), \quad L = \alpha L_i + \beta L_b, \tag{14}$$

where $\mathbf{p}_t$ is obtained by Eq. (9). $\mathbf{o}_t$ represents the onehot encoding of a question at step t in the oracle dataset. L_i denotes the loss of imitating the recommendation in oracle dataset. α and β are hyperparameters.

Notably, the oracle selection and model boosting stages can be iterated until convergence. In our experiments, we perform this only once since it already produces sufficiently good results. The training procedure of our ORI is illustrated in Algorithm 1.

6 Experiment

Table 1. Learning effects of different methods on 5 simulators. **Bold** indicates the best performance among all methods. Underline indicates the second-best performance. * indicates p-value < 0.05 in the significance test.

		non-RL		classic RL			RL-based question recommender					ours
		GRU4REC	FMLP	DQN	SAC	TD3	CSEAL	SRC	GEHRL	HierLLM	RAR	ORI
KSS	$t=10$	0.0033	0.0023	0.0793	0.0482	0.0353	0.1558	0.1157	0.1528	0.2262	0.2227	**0.3145***
	$t=30$	0.0006	0.0024	0.1469	0.0788	0.0848	0.4835	0.2416	0.3173	0.4937	0.6160	**0.7182**
IEKT-JU	$t=10$	−0.0121	−0.0525	0.1593	−0.0048	−0.0162	−0.0106	0.0317	0.0611	0.2905	0.2855	**0.2949**
	$t=30$	0.0042	−0.0733	0.2487	0.0112	−0.0413	0.0245	0.0672	0.1357	0.4438	0.5106	**0.5125**
	$t=200$	0.1306	−0.0871	0.4795	0.1087	−0.0115	0.2687	0.2089	0.3130	0.6470	0.7162	**0.8350***
DKT-JU	$t=10$	0.0010	−0.0024	0.0077	0.0069	−0.0006	−0.0005	0.0009	0.0000	0.0182	0.0172	**0.0212***
	$t=30$	0.0000	−0.0030	0.0109	0.0063	−0.0003	−0.0004	0.0006	0.0018	0.0194	0.0167	**0.0214***
	$t=200$	0.0040	−0.0030	0.0096	0.0049	0.0040	0.0014	0.0028	0.0027	0.0176	0.0187	**0.0213***
IEKT-A09	$t=10$	0.0508	−0.0078	0.1574	0.2205	−0.0077	−0.0217	0.0618	0.0337	0.3533	0.4004	**0.4995**
	$t=30$	0.1003	−0.0504	0.2904	0.3358	−0.0329	−0.0606	0.1411	0.1269	0.5812	0.4882	**0.6505***
	$t=200$	0.0953	−0.1036	0.5176	0.5683	0.1108	0.1419	0.3041	0.2984	**0.7848**	0.6763	0.7750
DKT-A09	$t=10$	−0.0052	−0.0544	0.0150	0.0073	0.0066	−0.0360	0.0004	−0.0034	0.0498	0.0542	**0.0643**
	$t=30$	−0.0053	−0.0730	0.0175	0.0073	0.0076	−0.0436	0.0015	0.0058	0.0509	0.0547	**0.0648***
	$t=200$	−0.0020	−0.0743	0.0082	0.0140	0.0124	−0.0096	0.0021	0.0021	0.0510	0.0548	**0.0648**

In this section, we will discuss the details of our experiments.

6.1 Datasets and Simulators

To conduct our experiments, we selected two datasets: Assist09[1] [4] and Junyi[2] [1]. Since existing static datasets cannot provide real-time student responses to newly recommended questions or directly reflect students' final learning effects,

[1] https://sites.google.com/site/assistmentsdata/home/2009-2010-assistment-data.
[2] https://www.kaggle.com/datasets/junyiacademy/learning-activity-public-dataset-by-junyi-academy.

we followed prior works [2,15–17] and constructed four simulators based on two knowledge tracing (KT) models, DKT and IEKT. Specifically, we built DKT-A09, DKT-Ju, IEKT-A09, and IEKT-Ju on the Assist09 and Junyi datasets, respectively. Each simulator was created by training the corresponding KT model on one of the datasets. The trained KT models can predict students' responses to different exercises based on their QA histories, estimate their mastery levels of learning targets, and thereby compute the overall learning effects. These simulators were used to emulate student behavior when interacting with the recommender and to evaluate its performance. In addition, to further assess the effectiveness of the recommender, we incorporated the KSS simulator proposed by Liu et al. [16] for comprehensive evaluation.

6.2 Baseline

To evaluate the performance of ORI, we compare it with widely used baselines from previous works, which we roughly divide into three groups. The first group is the **non-RL** methods, including GRU4REC [7] and FMLP [28]. These methods are not specifically designed for the question recommendation task. We evaluate whether they are equally applicable to our task. The second group is the **classical RL** methods, including DQN [18], SAC [6], and TD3 [5], where we use a trained DKT [21] model to predict students' learning states as input states $\boldsymbol{s}_t$. This group is meant to validate the potential of classical RL algorithms for question recommendation. The third group consists of **RL-based question recommenders**, including CSEAL [16], SRC [2], GEHRL [14], HierLLM [17], and RAR [15]. These methods are specifically designed for the question recommendation task and are used to verify whether our method outperforms existing SOTA methods.

6.3 Experiment Setting

For our proposed method, we set $d_e = 48$, $d_r = 128$, $d_h = 64$, $d_a = 64$. The head of attention in Eq. (5) is 1. We use the Adam [11] to optimize the basic model in both stages of rough training and model boosting. We use one NVIDIA TITAN RTX GPUs (24 GB GPU memory) to train our model and the training time is about 15–20 hours. For the hyperparameters of baselines, we have carefully tuned them to ensure the baselines reach their optimal performance.

6.4 Evaluation Metrics

We evaluate model performance using the learning effect Δ_u, which measures a student's improvement after completing recommended questions. For each student, the simulator predicts their mastery of learning targets before and after the recommendation session. The learning effect is then computed via Eq. (1) for each target, and the results are averaged to obtain the overall model performance.

6.5 Experiment Result

We compare the performance of models after recommending t steps, where $t \in [10, 30, 200]$. Since the maximum steps of recommendation in KSS is 30, we only test the learning effect in $t = 10$ and $t = 30$ for KSS. The experimental results are presented in Table 1.

From Table 1, we can observe: (1) Our method achieves the best performance in almost all the cases. Also, its performance is several times higher than the baselines in most cases. That demonstrates the effectiveness and superiority of our method. (2) Models' performance increases as the step of recommendation increases in most cases. That is reasonable. Since students gain more exercise with more recommendations, they gain more improvement. However, in some cases, there are performance decreases with the step of recommendation increases. We assume this is because some harmful recommendations impede students' learning. (3) The second and third groups of baselines are all RL-based methods, and their performance is better than that of the first group on the whole. We believe this is because, compared to non-RL methods, RL-based methods can optimize recommendation strategies using rewards that reflect students' learning effects, rather than relying solely on QA history. This makes them more suitable for question recommendation tasks.

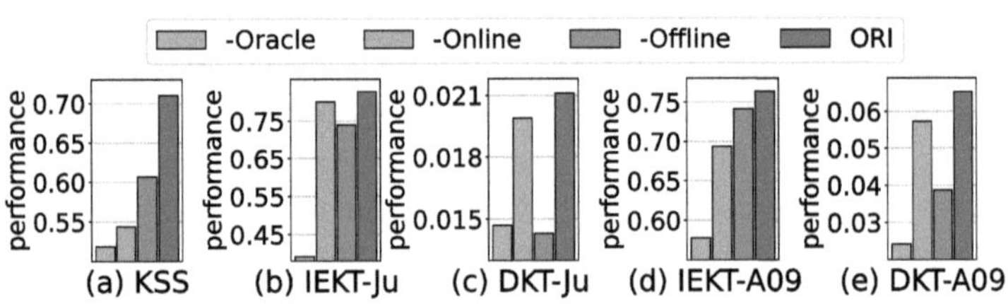

Fig. 4. The ablation study.

6.6 Ablation Study

To further investigate the effectiveness of our ORI pipeline, we conduct an ablation study. Specifically, we have three variants:

- **-Oracle** removes the selection of **oracle paths** $\mathcal{P}$ in Sect. 5.3. This means that we use the **raw path** $\mathcal{P}'$ directly to generate **oracle datasets** $\mathcal{D}_o$
- **-Online** removes online RL from boosting strategy. That means, the basic model only learns from the oracle dataset. Correspondingly, we remove L_b from Eq. (14)
- **-Offline** removes offline RL from boosting strategy. That means, the basic model only learns from the real-time data collected in the interaction with students. Correspondingly, we remove L_i from Eq. (14)

The experimental results are shown in the Fig. 4. We can see (1) No matter which one is removed, the performance of ORI will decrease. That demonstrates all of them has contribution to the performance. (2) In most cases, the -Oracle approach performs worse than both -Online and -Offline, highlighting the critical importance of selecting the oracle path carefully. Without this step, the oracle dataset may include many recommendations that are only suitable for specific students-such as those with exceptionally high learning ability or particularly simple learning targets-and do not follow general pedagogical principles. This can severely undermine the recommender's ability to learn and identify effective recommendation strategies. (3) -Online has worse performance than -Offline in KSS, IEKT-A09, but has better performance than -Offline in DKT-A09, DKT-Ju and IEKT-Ju. This indicates that the advantages of online and offline training vary across situations.: online RL pipeline allows the recommender to explore the effects of different recommendation strategies in real-time, while offline RL pipeline enables the recommender to leverage high-quality data for training.

6.7 Case Study

To investigate the differences in models' recommendations, we randomly select a student from KSS, and use different models to recommend questions for this student. As discussed previously, KSS defines a knowledge graph that indicates the prerequisite relationships between questions, which is assumed to fully adhere to the cognitive pattern. Therefore, in KSS, models are expected to produce recommendations that align with these cognitive patterns. The knowledge graph and the recommendations of models are illustrated in Fig. 5. Both the learning targets in the knowledge graph and the recommendation of models are presented via questions ID for the convenience of observation. The solid circles denote the questions that are answered correctly, the hollow circles denote the questions that are incorrectly answered. The dark blue nodes in the knowledge graph denotes the learning target, *i.e.*, question $0, 1, 2, 3, 4, 5$.

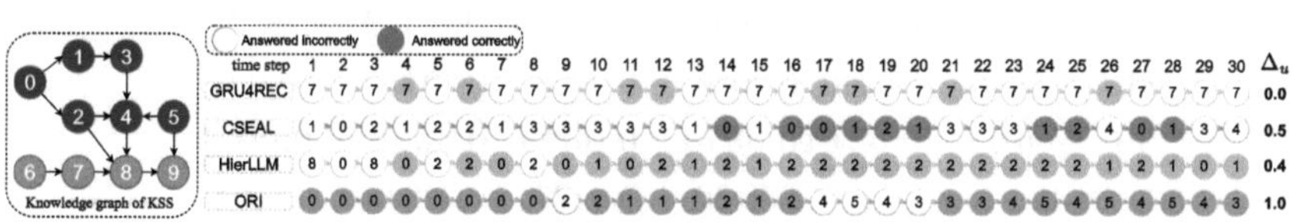

Fig. 5. Case study.

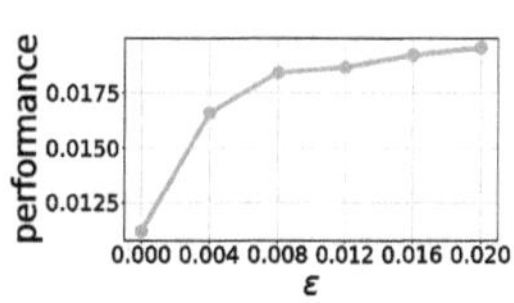

Fig. 6. Analysis of ϵ

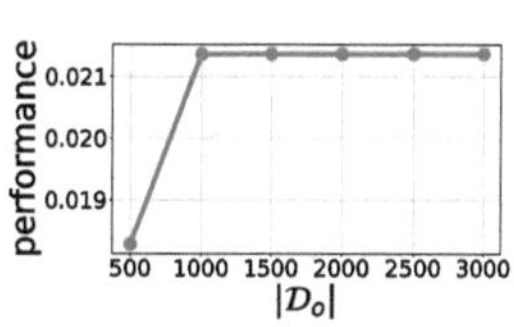

Fig. 7. Analysis of $|\mathcal{D}_o|$

From Fig. 5, we can observe: (1) GRU4REC consistently recommends question 7 to the student, even though it does not contribute to the student's mastery of learning targets. We suspect this is because GRU4REC is designed to cater to user interests, and it may assume that the student is only interested in question 7 and will continue to click on it for learning. As a result, it repeatedly recommends this question to the student without considering the student's learning effects. This further demonstrates that traditional non-RL methods are not suitable for question recommendation tasks. (2) CSEAL's recommendations are overly aggressive, while HierLLM's are too conservative. CSEAL continuously recommends subsequent questions $1, 2, 3$ even when the student has not yet mastered question 0, while HierLLM does not recommend further questions even after the student has mastered questions $0, 1, 2$. In comparison, ORI's recommendations are more reasonable. It not only consistently recommends question 0 to establish a solid foundation for the student at the beginning but also gradually guides the student master subsequent learning targets. We believe this is because the high-quality data used to train ORI effectively teaches it superior recommendation strategies. (3) All the recommendations given by both CSEAL and ORI adhere to the cognitive pattern in the knowledge graph. It is not surprising for CSEAL, since it references the same knowledge graph during recommendation. However, ORI also demonstrates this capability even without having access to the knowledge graph. We assume this is because the oracle dataset potentially contains the information of the knowledge graph. Training from the oracle dataset makes ORI learn the pattern of the knowledge graph.

6.8 Hyperparameter Analysis

We investigate the impact of the threshold ϵ for oracle path selection and the scale of the oracle dataset $|\mathcal{D}_o|$ on ORI performance on DKT-JU.

Table 2. Compatibility analysis.

	KSS	IEKT-JU	DKT-JU	IEKT-A09	DKT-A09
SRC	0.2416	0.2089	0.0028	0.3041	0.0021
SRC+ORI	**0.3214**	**0.3574**	**0.0099**	**0.3322**	**0.0031**
GEHRL	0.3173	0.3130	0.0027	0.2984	0.0021
GEHRL+ORI	**0.3486**	**0.3379**	**0.0094**	**0.3109**	**0.0045**

The Impact of Oracle Path Selection Threshold ϵ. To illustrate the impact of the threshold ϵ in Sect. 5.3 on ORI performance, we adjusted the threshold ϵ while keeping other hyperparameters unchanged to test the effect of ϵ on the performance of ORI. The experimental results are shown in Fig. 6. As shown in Fig. 6, the performance of ORI gradually improves and exhibits a converging trend as ϵ increases. This occurs because a higher ϵ more effectively prevents recommendation strategies that violate teaching principles from being included in the oracle dataset, thereby enhancing the generalizability and reliability of the data. As a result, the recommender is less affected by suboptimal strategies. When ϵ becomes sufficiently large, most non-compliant strategies are filtered out, so further increasing ϵ provides little additional improvement in recommender performance.

The Impact of the Oracle Dataset Scale $|\mathcal{D}_o|$. To verify how much oracle data is required to ensure ORI performance, we randomly select different amounts of oracle data to construct the oracle dataset for model boosting. The impact of the oracle dataset size $|\mathcal{D}_o|$ on model performance is shown in the Fig. 7. The results, shown in the Fig. 7, indicate that the model's performance peaks when $|\mathcal{D}_o|$ is 1000 or more. This experiment demonstrates that ORI can achieve maximum performance without requiring an excessive amount of oracle data, making our method easy to apply.

6.9 Compatibility Analysis

As we have emphasized, our pipeline can be adapted to any RL-based method. Therefore, to validate the compatibility of our pipeline, we adapt it to SRC and GEHRL. Specifically, we replace the basic model of ORI with them, such that they can be trained in a manner of both online and offline like ORI. The experiment results are shown in Table 2. We can observe that by applying our proposed pipeline, the performance of both SRC and GEHRL has been enhanced. This confirms the strong compatibility of our approach. Besides, although SRC and GEHRL have improved performance by replacing the basic model, their enhanced performance still does not surpass that of our proposed basic model. We speculate that this is because a basic model with superior performance can generate a higher quality oracle dataset, thereby learning more effective recommendation strategies and achieving higher performance.

7 Conclusion

In this paper, we identify the limitations of existing online RL pipelines and introduce ORI, a novel hybrid training framework specifically designed to address these challenges. Through extensive experiments, we demonstrate that although ORI adopts a relatively simple design, it significantly enhances the performance of question recommenders. Its strong compatibility allows it to be effectively integrated with a wide range of RL-based question recommenders, consistently yielding further performance improvements. As recommendation tasks and scenarios continue to grow in diversity, RL-based recommendation methods are expected to find even broader applications. Future research can focus on designing more tailored training pipelines and high-quality data extraction mechanisms to accommodate a wider range of real-world recommendation scenarios.

Acknowledgement. This work is supported by the National Natural Science Foundation of China (62307020).

References

1. Chang, H.S., Hsu, H.J., Chen, K.T.: Modeling exercise relationships in e-learning: a unified approach. In: EDM, pp. 532–535 (2015)
2. Chen, X., et al.: Set-to-sequence ranking-based concept-aware learning path recommendation. arXiv preprint arXiv:2306.04234 (2023)
3. Dai, S., et al.: Uncovering chatgpt's capabilities in recommender systems. In: Proceedings of the 17th ACM Conference on Recommender Systems, pp. 1126–1132 (2023)
4. Feng, M., Heffernan, N., Koedinger, K.: Addressing the assessment challenge with an online system that tutors as it assesses. User Model. User-Adap. Inter. **19**, 243–266 (2009)
5. Fujimoto, S., Hoof, H., Meger, D.: Addressing function approximation error in actor-critic methods. In: International Conference on Machine Learning, pp. 1587–1596. PMLR (2018)
6. Haarnoja, T., Zhou, A., Abbeel, P., Levine, S.: Soft actor-critic: off-policy maximum entropy deep reinforcement learning with a stochastic actor. In: International Conference on Machine Learning, pp. 1861–1870. PMLR (2018)
7. Hidasi, B., Karatzoglou, A., Baltrunas, L., Tikk, D.: Session-based recommendations with recurrent neural networks. arXiv preprint arXiv:1511.06939 (2015)
8. Hochreiter, S., Schmidhuber, J.: Long short-term memory. Neural Comput. **9**(8), 1735–1780 (1997)
9. Hong, Z.W., Agrawal, P., Combes, R.T.d., Laroche, R.: Harnessing mixed offline reinforcement learning datasets via trajectory weighting. arXiv preprint arXiv:2306.13085 (2023)
10. Huang, Z., Liu, Q.: Exploration and exploitation in hierarchical reinforcement learning with adaptive scheduling. In: ECAI 2023, pp. 1116–1123. IOS Press (2023)
11. Kingma, D.P., Ba, J.: Adam: a method for stochastic optimization. arXiv preprint arXiv:1412.6980 (2014)
12. Kubotani, Y., Fukuhara, Y., Morishima, S.: Rltutor: reinforcement learning based adaptive tutoring system by modeling virtual student with fewer interactions. arXiv preprint arXiv:2108.00268 (2021)
13. Levine, S., Kumar, A., Tucker, G., Fu, J.: Offline reinforcement learning: tutorial, review, and perspectives on open problems. arXiv preprint arXiv:2005.01643 (2020)
14. Li, Q., et al.: Graph enhanced hierarchical reinforcement learning for goal-oriented learning path recommendation. In: Proceedings of the 32nd ACM International Conference on Information and Knowledge Management, pp. 1318–1327 (2023)
15. Liu, H., Liu, Y., Long, T.: Personalized education with ranking alignment recommendation. In: International Conference on Database Systems for Advanced Applications, pp. 352–361. Springer, Cham (2025)
16. Liu, Q., et al.: Exploiting cognitive structure for adaptive learning. In: Proceedings of the 25th ACM SIGKDD International Conference on Knowledge Discovery & Data Mining, pp. 627–635 (2019)
17. Liu, Y., Liu, H., Long, T.: Hierllm: hierarchical large language model for question recommendation. In: International Conference on Database Systems for Advanced Applications, pp. 362–372. Springer, Cham (2025)
18. Mnih, V., et al.: Playing atari with deep reinforcement learning. arXiv preprint arXiv:1312.5602 (2013)
19. Pang, L., Xu, J., Ai, Q., Lan, Y., Cheng, X., Wen, J.: Setrank: learning a permutation-invariant ranking model for information retrieval. In: Proceedings of the 43rd International ACM SIGIR Conference on Research and Development in Information Retrieval, pp. 499–508 (2020)

20. Parr, R., Russell, S.: Reinforcement learning with hierarchies of machines. In: Advances in Neural Information Processing Systems, vol. 10 (1997)
21. Piech, C., et al.: Deep knowledge tracing. In: Advances in Neural Information Processing Systems, vol. 28 (2015)
22. Shi, D., Wang, T., Xing, H., Xu, H.: A learning path recommendation model based on a multidimensional knowledge graph framework for e-learning. Knowl.-Based Syst. **195**, 105618 (2020)
23. Silver, D., Lever, G., Heess, N., Degris, T., Wierstra, D., Riedmiller, M.: Deterministic policy gradient algorithms. In: International Conference on Machine Learning, pp. 387–395. PMLR (2014)
24. Swazinna, P., Udluft, S., Runkler, T.: Measuring data quality for dataset selection in offline reinforcement learning. In: 2021 IEEE Symposium Series on Computational Intelligence (SSCI), pp. 1–8. IEEE (2021)
25. Tang, X., Chen, Y., Li, X., Liu, J., Ying, Z.: A reinforcement learning approach to personalized learning recommendation systems. Br. J. Math. Stat. Psychol. **72**, 108–135 (2018)
26. Usama, M., Chang, D.E.: Learning-driven exploration for reinforcement learning. In: 2021 21st International Conference on Control, Automation and Systems (ICCAS), pp. 1146–1151. IEEE (2021)
27. Vaswani, A., et al.: Attention is all you need. In: Advances in Neural Information Processing Systems, vol. 30 (2017)
28. Zhou, K., Yu, H., Zhao, W.X., Wen, J.R.: Filter-enhanced MLP is all you need for sequential recommendation. In: Proceedings of the ACM Web Conference 2022, pp. 2388–2399 (2022)

FlowRec: Prior-Informed Flow Matching for Efficient Sequential Recommendation Generation

Li Li[1], Mingyue Cheng[1(✉)], Yuyang Ye[2], Zhiding Liu[1], and Enhong Chen[1]

[1] State Key Laboratory of Cognitive Intelligence, University of Science and Technology of China, Hefei, China
{lili0516,zhiding}@mail.ustc.edu.cn, {mycheng,cheneh}@ustc.edu.cn
[2] Rutgers University, New Brunswick, NJ, USA
yuyang.ye@rutgers.edu

Abstract. Sequential recommendation aims to predict each user's next preferred item based on their historical interactions. Recently, diffusion-based approaches have demonstrated strong generative capability in modeling complex user preferences. However, they still face two inherent limitations: (i) Gaussian priors are misaligned with user-specific interests, and curved noise schedules lead to error accumulation and unstable training; (ii) the stochastic denoising process introduces additional randomness and substantial computational overhead. To address these issues, we propose FlowRec, a flow-matching-based framework that formulates preference evolution as continuous flows between personalized priors and target items. Specifically, FlowRec constructs an informative behavior-based prior distribution derived from users' historical interactions, offering a distributionally closer initialization to the target distribution. It then learns a vector field to guide straight preference flows toward target interests. Moreover, a single-step alignment objective with positive and negative samples further enhances semantic consistency between generated representations and ground-truth items. Finally, FlowRec adopts deterministic ODE-based generation, achieving efficient and stable inference. Extensive experiments on multiple benchmark datasets demonstrate that FlowRec consistently outperforms state-of-the-art baselines in both recommendation accuracy and inference efficiency.

Keywords: Sequential recommendation · Flow matching

1 Introduction

Sequential recommendation (SR) aims to model users' dynamic preferences from chronologically ordered historical interactions [10,32]. It has been widely deployed in various online platforms, such as e-commerce and streaming media services. Early studies relied on Markov chains and matrix factorization to capture short-term sequential dependencies [28], while recent deep learning-based

H. Jung et al. (Eds.): DASFAA 2026, LNCS 16535, pp. 206–222, 2026.
https://doi.org/10.1007/978-981-92-0363-5_13

methods significantly improve SR by modeling both long- and short-term user interests [2,32]. More recently, diffusion models (DMs) [11,30] have emerged as a powerful generative paradigm for SR, owing to their ability to model complex preference distributions. Typically, these models adopt a standard Gaussian distribution as the prior and treat the ground-truth item distribution as the target. By predefining a noise schedule that perturbs item embeddings with Gaussian noise in the forward process and iteratively denoising them in the reverse process conditioned on user interaction sequences, diffusion-based recommenders [16,17,40] achieve expressive preference generation.

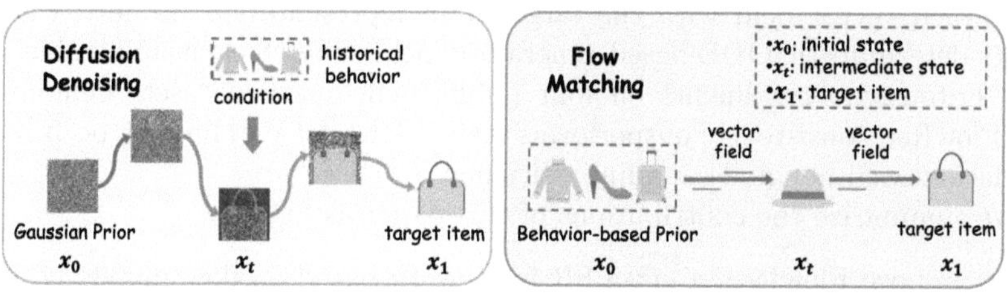

(a) Diffusion-based SR method (b) Flow-based SR method (ours)

Fig. 1. Comparison of our proposed method and diffusion-based method.

Despite their impressive effectiveness, diffusion-based approaches still suffer from two inherent limitations. First, their reliance on Gaussian priors and curved noise schedules limits their modeling capacity for SR. Gaussian priors introduce noise which is far from user-specific interests, forcing the model to reconstruct preferences from uninformative initializations. Although these models condition on users' historical interactions, biased or insufficient contextual information may cause them to generate unrealistic items or drift away from users' true preferences [20,37]. Meanwhile, the curved noise schedules exacerbate error accumulation and noise sensitivity, leading to unstable and inaccurate predictions. Moreover, such schedules typically require many sampling steps during generation, further increasing computational cost. Second, the stochastic SDE sampler used during inference introduces additional noise and requires diffusion term computation, resulting in substantial computational overhead. Overall, these challenges highlight the need for a more stable and efficient generative paradigm that mitigates noise sensitivity and reduces redundant sampling during inference.

To address the aforementioned challenges, we introduce the flow matching paradigm [19,24], a generalization of diffusion models that has shown strong generative capability across various domains, such as image [5], speech [26], and text generation [12]. Compared with diffusion models, flow matching offers greater flexibility in both the choice of prior distributions and the design of sampling schedules [7,33]. Leveraging these advantages, we propose FlowRec, a novel framework which explicitly models continuous preference flows from personalized priors to target items. The difference between FlowRec and diffusion-based methods is illustrated in Fig. 1. Specifically, FlowRec encodes users' historical behaviors into latent intent representations, which serve as behavior-based priors, replacing the conventional Gaussian priors. These personalized priors provide informative and user-specific initialization which better approximates target items, thus reducing the learning difficulty compared to Gaussian priors. A learnable vector field is then optimized to construct straight preference trajectories, refining the initial representations toward the ground-truth items. Furthermore, to better adapt flow matching to SR, we introduce a single-step alignment loss that incorporates both positive and negative samples. This loss complements the flow-matching objective by semantically aligning the single-step estimation from the learned vector field with the target item representation. Finally, FlowRec adopts deterministic ODE-based generation, achieving efficient and stable inference. Extensive experiments on four public benchmark datasets demonstrate that FlowRec consistently outperforms state-of-the-art baselines in both recommendation accuracy and sampling efficiency.

We summarize the contributions of this paper as follows:

- We propose FlowRec, a novel SR framework based on flow matching, which explicitly models straight preference trajectories from personalized priors to target items, enhancing recommendation accuracy and inference efficiency.
- We design two key components that tailor flow matching to sequential recommendation: a behavior-based prior for personalized and informative initialization, and a single-step alignment loss that complements the flow-matching objective and improves semantic alignment.
- We conduct extensive experiments on four public benchmark datasets, demonstrating the superiority of FlowRec over a range of state-of-the-art baselines.

2 Related Work

2.1 Sequential Recommendation

Sequential recommendation aims to capture users' dynamic interests from historical interactions. Early methods model users' dynamic interest transitions using Markov chains, such as FMPC [29] and Fossil [9]. In recent years, deep neural networks have become a dominant approach for SR, leading to notable performance improvements [13,22,31]. GRU4Rec [10] employs the gated recurrent unit [3] to predict the next item. Caser [32] treats recent item sequence as an "image" and utilizes convolutional filters to capture sequential patterns. Self-attention-based methods further enhance the effectiveness of SR. SASRec [13] introduces uni-directional self-attention, while BERT4Rec [31] employs a bidirectional Transformer to encode user behavior sequences.

Beyond conventional methods, generative approaches such as VAEs [15] and GANs [6] have also been explored in SR. Methods like ACVAE [39] and MFGAN [27] aim to model the uncertainty of user preferences and alleviate exposure bias. However, they often suffer from limited representation capacity and unstable training [18]. Recently, diffusion models have emerged as a new promising paradigm in SR, achieving state-of-the-art performance [4,14,17,36]. DiffuRec [17] and DiffRec [35] directly apply diffusion models to SR. Specifically, these methods corrupt the next-item representation with Gaussian noise and iteratively reconstruct it through a denoising process. DreamRec [40] utilizes a Transformer encoder to encode historical interactions as conditional guidance for the denoising procedure. CaDiRec [4] employs a context-aware diffusion model to generate augmented views for SR via contrastive learning.

2.2 Flow Matching

Flow Matching [19,24] (FM) is a recently proposed generative modeling paradigm based on continuous normalizing flows [1]. While diffusion models are formulated as stochastic differential equation (SDE), FM models are based on ordinary differential equation (ODE). Unlike diffusion models starting from a Gaussian prior, FM allows more flexible prior distributions [7]. Instead of iterative denoising, FM learns a probability flow by regressing a vector field between the prior and target distribution, significantly accelerating the sampling process and improving stability. FM has demonstrated superior performance across various generative tasks, such as image generation [5,34], speech generation [26], and reinforcement learning [14]. Recent studies have begun adapting flow matching to recommendation tasks. FlowCF [20] introduces a discrete flow framework to address sparsity and implicit feedback in collaborative filtering, while FMRec [21] incorporates straight flow trajectory and task-specific objectives for sequential recommendation. Nevertheless, the potential of flow matching to leverage behavior-informed priors for SR remains underexplored, offering new opportunities for research.

3 Preliminaries

3.1 Problem Statement

Given an item set $\mathcal{I} = \{i_1, i_2, \dots, i_{|\mathcal{I}|}\}$ and a user set $\mathcal{U} = \{u_1, u_2, \dots, u_{|\mathcal{U}|}\}$, we denote each behavior sequence of user $u \in \mathcal{U}$ as $\mathcal{S} = [i_1^u, i_2^u, \dots, i_{n-1}^u]$, where $i_j^u \in \mathcal{I}$ is the j-th item interacted with by user u and n is the sequence length. The goal of SR is to predict the next item i_n^u that the user is likely to interact with, based on historical behavior sequence $\mathcal{S}$. This can be formulated as:

$$\arg\max_{i_j \in \mathcal{I}} P(i_n^u = i_j | \mathcal{S}), \tag{1}$$

where P denotes the likelihood of item i_j being the next item, based on the user's historical interaction sequence $\mathcal{S}$.

3.2 Flow Matching

Flow matching [19,24] is a class of generative models that learn a continuous transformation from a prior distribution p_0 to a target distribution p_1. Unlike other generative paradigms, flow matching allows the prior distribution p_0 to be arbitrary. It aims to learn matching flows, represented by vector field, between p_0 and p_1 via an ODE: $d\boldsymbol{x}_t = u(\boldsymbol{x}_t, t)dt$, where $u(\boldsymbol{x}_t, t)$ denotes a time-dependent vector field describing the probability flow over time.

Forward Process. Given data samples $\boldsymbol{x}_1 \sim p_1(\boldsymbol{x})$ and prior samples $\boldsymbol{x}_0 \sim p_0(\boldsymbol{x})$, an interpolated flow path can be constructed as

$$\boldsymbol{x}_t = a_t \boldsymbol{x}_0 + b_t \boldsymbol{x}_1, \quad t \in [0, 1], \tag{2}$$

where a_t and b_t are predefined interpolation schedules. The corresponding vector field is given by the time derivative:

$$u_t(\boldsymbol{x}_t|\boldsymbol{x}_1) = \frac{d\boldsymbol{x}_t}{dt} = a'_t \boldsymbol{x}_0 + b'_t \boldsymbol{x}_1, \tag{3}$$

where $u_t(\boldsymbol{x}_t|\boldsymbol{x}_1)$ is the conditional vector field. The marginal vector field is defined as $u(\boldsymbol{x}_t, t) \triangleq \mathbb{E}_{p_t(u_t|x_t)}[u_t]$. A neural network v_θ parameterized by θ is trained to fit the marginal vector field: $\mathcal{L}_{\mathrm{FM}}(\theta) = \mathbb{E}_{t,p_t(\boldsymbol{x}_t)} \|v_\theta(\boldsymbol{x}_t, t) - u(\boldsymbol{x}_t, t)\|^2$. However, directly computing this loss function is intractable due to the marginalization, so the conditional flow matching loss is used instead:

$$\mathcal{L}_{\mathrm{CFM}}(\theta) = \mathbb{E}_{t,\boldsymbol{x}_0,\boldsymbol{x}_1} \|v_\theta(\boldsymbol{x}_t, t) - u_t(\boldsymbol{x}_t|\boldsymbol{x}_1)\|^2 . \tag{4}$$

Minimizing $\mathcal{L}_{\mathrm{CFM}}$ is equivalent to minimizing $\mathcal{L}_{\mathrm{FM}}$ with respect to θ.

Reverse Process. After the vector field $v_\theta$is learned, new samples can be generated by integrating the ODE backward from the source distribution p_0:

$$d\boldsymbol{x}_t = v_\theta(\boldsymbol{x}_t, t)dt, \quad \boldsymbol{x}_0 \sim p_0. \tag{5}$$

Integrating from $t = 0$ to $t = 1$ yields deterministic trajectories toward the target distribution p_1. This reverse process defines a stable and efficient generative mechanism, in contrast to the stochastic sampling used in diffusion models.

4 The Proposed FlowRec

4.1 Overview of FlowRec

As illustrated in Fig. 2, FlowRec learns a vector field to construct a straight flow path from a behavior-based prior to the target item. A user's interaction sequence is first encoded into a latent representation $\boldsymbol{x}_0 \sim p_0$, which serves as the closest accessible approximation to the target item. The embedding of the ground-truth target item is denoted as $\boldsymbol{x}_1 \sim p_1$. We train a flow model

f_θ to estimate the vector field that guides the initial representation $\boldsymbol{x}_0$ toward the true target item $\boldsymbol{x}_1$. In addition to the flow matching objective, we tailor a single-step alignment loss that leverages the estimated vector field to obtain approximation $\tilde{\boldsymbol{x}}_1$ and semantically align it with the target item. FlowRec jointly optimizes the behavior-based prior loss, flow matching loss, and alignment loss to generate representations consistent with user preferences. During inference, FlowRec performs deterministic generation by solving the learned ODE from $\boldsymbol{x}_0$, yielding a refined item embedding for efficient candidate ranking.

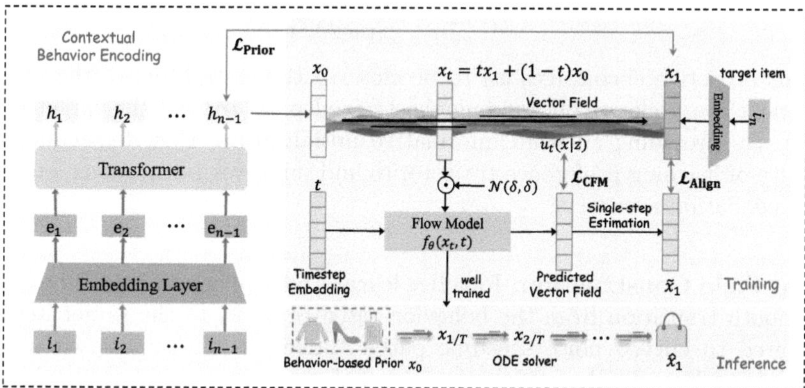

Fig. 2. Overview of FlowRec. The left figure shows a Transformer-based behavior encoder, while the right figures illustrate the training and inference phases.

4.2 Preference Flows with Flow Matching

Contextual Behavior Encoding. Building upon the flow matching formulation, we construct a personalized and informative prior that captures users' historical preferences. Unlike diffusion-based models that adopt a simple Gaussian prior, FlowRec leverages flow matching's flexibility to define a behavior-based prior distribution. This prior serves as the initial state of the preference flow, providing a closer and more informative initialization for generation.

Formally, each item $i \in \mathcal{I}$ is represented by an embedding vector $\boldsymbol{e} \in \mathbb{R}^d$, where d denotes the embedding dimension. Given a user's historical interaction sequence $[i_1, i_2, \ldots, i_{n-1}]$, we obtain its embedding representation $\boldsymbol{E}_{1:n-1} = [\boldsymbol{e}_1, \boldsymbol{e}_2, \ldots, \boldsymbol{e}_{n-1}] \in \mathbb{R}^{(n-1)\times d}$, while the ground-truth next item is represented by $\boldsymbol{e}_n$. Following previous works [37,40], we employ a Transformer encoder to capture complex sequential dependencies, formulated as:

$$\boldsymbol{H} = \mathrm{Trm}(\boldsymbol{e}_1, \boldsymbol{e}_2, \ldots, \boldsymbol{e}_{n-1}), \tag{6}$$

where $\boldsymbol{H} \in \mathbb{R}^{(n-1)\times d}$ denotes the hidden states of the Transformer encoder and the last state $\boldsymbol{h}_{n-1} = \boldsymbol{H}[-1] \in \mathbb{R}^d$ is used as the behavior-based prior $\boldsymbol{x}_0$. We regard the distribution of these learned representations as the prior distribution p_0. Compared to a Gaussian noise prior, this prior preserves contextual information and user-specific intent. Notably, while we employ a Transformer encoder to model sequential dependencies, FlowRec is model-agnostic and can be integrated with other sequential encoders. It further optimizes the encoder's output representations, consistently improving recommendation accuracy.

To ensure that $\boldsymbol{x}_0$ faithfully reflects users' true preferences, the behavior encoder is trained with a cross-entropy loss function:

$$\mathcal{L}_{\text{Prior}} = -\frac{1}{|\mathcal{U}|}\sum_{u\in\mathcal{U}} \log \frac{\exp(\boldsymbol{x}_0 \cdot \boldsymbol{e}_n^u)}{\sum_{i\in\mathcal{I}} \exp(\boldsymbol{x}_0 \cdot \boldsymbol{e}_i^u)}. \tag{7}$$

This objective encourages $\boldsymbol{x}_0$ to be close to the embedding of the ground-truth next item, effectively guiding the learned prior toward the target distribution. By providing such an informative initialization, FlowRec reduces the difficulty of learning preference trajectories and improves both convergence and recommendation accuracy.

Vector Field Construction. FlowRec learns a continuous vector field to model the smooth transition from the behavior-based prior $\boldsymbol{x}_0$ to the target item $\boldsymbol{x}_1$. Compared to curved noise schedule paths applied in diffusion-based models, our framework explicitly models straight preference flows, enabling stable and efficient preference evolution.

To this end, we train a flow model f_θ to estimate the vector field that guides the transformation between the prior and target states. The straight trajectory, characterized by interpolation coefficients $a_t = 1-t$ and $b_t = t$, is widely adopted in various studies [19,33] for its simplicity and computational efficiency. We randomly sample a time step $t \in [0, 1]$ from a uniform distribution. Following Eq. (2), the flow path $\boldsymbol{x}_t$ is obtained by linear interpolation between $\boldsymbol{x}_0$ and $\boldsymbol{x}_1$:

$$\boldsymbol{x}_t = (1-t)\boldsymbol{x}_0 + t\boldsymbol{x}_1, \tag{8}$$

which defines a straight preference trajectory between the prior and target states.

Following [17], to improve robustness and avoid direct exposure of $\boldsymbol{x}_0$ and $\boldsymbol{x}_1$, we introduce a stochastic modulation term $\boldsymbol{\lambda} \in \mathbb{R}^d$, where each component $\lambda_i \sim \mathcal{N}(\delta, \delta)$. The modulated input $(\boldsymbol{\lambda} \odot \boldsymbol{x}_t)$ and the embedded time step $\boldsymbol{t}$ are then fed into a two-layer MLP parameterized by θ:

$$v_\theta(\boldsymbol{x}_t, \boldsymbol{t}) = f_\theta(\boldsymbol{x}_t, t) = MLP((\boldsymbol{\lambda} \odot \boldsymbol{x}_t) \oplus \boldsymbol{t}), \tag{9}$$

where $\odot$ and $\oplus$ denote element-wise multiplication and concatenation, respectively. The predicted vector field $f_\theta(\boldsymbol{x}_t, t)$ captures the instantaneous direction of preference evolution, which will later be integrated through an ODE solver during inference to generate deterministic and smooth trajectories.

4.3 Model Optimization

Conditional Flow Matching Loss. Following the straight trajectory defined in Eq. (8), the corresponding conditional vector field can be expressed as:

$$u_t(\boldsymbol{x}_t|\boldsymbol{x}_1) = \frac{d\boldsymbol{x}_t}{dt} = \boldsymbol{x}_1 - \boldsymbol{x}_0, \tag{10}$$

which is constant along the flow path and independent of t. According to Eq. (4), the vector field neural network v_θ (implemented as f_θ in FlowRec) is trained to approximate the true field u_t via the Conditional Flow Matching (CFM) loss:

$$\mathcal{L}_{\text{CFM}} = \mathbb{E}_{t,\boldsymbol{x}_0,\boldsymbol{x}_1}||f_\theta(\boldsymbol{x}_t, t) - (\boldsymbol{x}_1 - \boldsymbol{x}_0)||^2. \tag{11}$$

This objective enforces geometric consistency between the predicted and true vector fields, guiding the model to learn a vector field that drives the straight flow from the behavior-based prior $\boldsymbol{x}_0$ toward the target item $\boldsymbol{x}_1$.

However, minimizing this Euclidean distance alone only ensures that the predicted embedding is close to its corresponding target, without enforcing that it is farther from other negative items. In recommendation, such absolute proximity is insufficient – the model must learn relative distances that reflect ranking preferences. Without this constraint, the learned representation may remain ambiguous with respect to other items, leading to suboptimal ranking accuracy.

Single-Step Alignment Loss. To address the above limitation, we introduce a *single-step alignment loss* that provides a ranking-aware contrastive signal to complement the flow-matching objective. Starting from the middle state in Eq. (8), the terminal state of the trajectory can be expressed by integrating the vector field over time: $\boldsymbol{x}_1 = \boldsymbol{x}_t + \int_t^1 u_t(\boldsymbol{x}_t|\boldsymbol{x}_1)dt$. In practice, since the conditional vector field $u_t(\boldsymbol{x}_t|\boldsymbol{x}_1)$ is constant and time-invariant under the straight trajectory assumption, we can approximate the integration using a first-order Euler step:

$$\tilde{\boldsymbol{x}}_1 = \boldsymbol{x}_t + (1 - t) \times f_\theta(\boldsymbol{x}_t, t). \tag{12}$$

where f_θ denotes the learned approximation of the true vector field u_t. Then, we align the estimated embedding $\tilde{\boldsymbol{x}}_1$ with the ground-truth target embedding $\boldsymbol{e}_n$ using a contrastive cross-entropy loss:

$$\mathcal{L}_{\text{Align}} = -\frac{1}{|\mathcal{U}|}\sum_{u\in\mathcal{U}} \log \frac{\exp(\tilde{\boldsymbol{x}}_1 \cdot \boldsymbol{e}_n^u)}{\sum_{i\in\mathcal{I}} \exp(\tilde{\boldsymbol{x}}_1 \cdot \boldsymbol{e}_i^u)}. \tag{13}$$

This auxiliary loss explicitly incorporates both positive and negative items, thereby enforcing that the predicted representation is not only close to its target but also discriminative from irrelevant candidates. By combining the geometric supervision from $\mathcal{L}_{\text{CFM}}$ with the ranking-aware contrastive signal from $\mathcal{L}_{\text{Align}}$, FlowRec achieves semantically aligned and discriminative preference trajectories, enabling accurate ranking with minimal sampling steps during inference.

Joint Optimization. Finally, the overall objective of FlowRec integrates all learning signals into a unified ed-to-end optimization framework:

$$\mathcal{L} = \mathcal{L}_{\text{Prior}} + \alpha\mathcal{L}_{\text{CFM}} + \beta\mathcal{L}_{\text{Align}}, \tag{14}$$

where α and β are hyperparameters that control the relative contributions of the flow-matching and alignment loss.

4.4 Model Inference

During inference, FlowRec generates the next-item representation $\hat{\boldsymbol{x}}_1$ through deterministic integration of the learned flow. Given a user's historical interaction sequence, the behavior encoder first produces an initial state $\boldsymbol{x}_0$, which serves as the starting point of the learned flow. Once the flow model f_θ has been trained, inference proceeds by solving the ODE that defines the flow trajectory:

$$d\boldsymbol{x}_t = f_\theta(\boldsymbol{x}_t, t)dt, \quad \boldsymbol{x}_0 \sim p_0. \tag{15}$$

Unlike diffusion-based methods that adopts stochastic denoising with additional noise injection and diffusion term computation, this ODE-based formulation only involves simple numerical integration, ensuring both stability and efficiency. To numerically solve the ODE, we adopt the Euler method. Given a predefined number of integration steps T, the flow is computed iteratively as:

$$\boldsymbol{x}_{t+\frac{1}{T}} = \boldsymbol{x}_t + \frac{1}{T} f_\theta(\boldsymbol{x}_t, t), \tag{16}$$

where $t = 0, \frac{1}{T}, \frac{2}{T}, \ldots, 1 - \frac{1}{T}$. After T steps, we obtain the predicted embedding $\hat{\boldsymbol{x}}_1$ which approximates the user's next item embedding. Finally, FlowRec computes relevance scores between $\hat{\boldsymbol{x}}_1$ and all candidate item embeddings via an inner product, and recommends the top-K items with the highest scores.

The deterministic inference process not only eliminates the stochastic uncertainty of diffusion-based sampling but also substantially reduces computational overhead, enabling fast and stable recommendation generation.

5 Experiments

5.1 Experimental Settings

Datasets. We evaluate the proposed method on four public real-world datasets. All datasets have been widely used in sequential recommendation.

- *Amazon Beauty* and *Amazon Toys* are two subsets of the Amazon dataset [25], containing user reviews and ratings for beauty products and toys.
- *ML-1M* [8] contains one million movie ratings from approximately 6000 users on 4000 movies, provided by the MovieLens platform.
- *Yelp* is a well-known business review dataset collected by Yelp.

Following previous studies [13,17], we remove inactive users and unpopular items with fewer than five interaction records. The detailed statistics of all datasets are summarized in Table 1.

Table 1. The detailed description and statistics of datasets.

Datasets	#Sequence	#Items	#Actions	Avg_len	Sparsity
Beauty	22,363	12,101	198,502	8.53	99.93%
Toys	19,412	11,924	167,597	8.63	99.93%
ML-1M	6,040	3,416	999,611	165.50	95.16%
Yelp	30,431	20,033	316,354	10.4	99.95%

Evaluation Settings. Following previous work [31,35], we set the max sequence length to 50 for all datasets. Recommendation performance is evaluated with two widely recognized metrics: HR@k (Hit Rate) and NDCG@k (Normalized Discounted Cumulative Gain), with results reported for $k = 5, 10$. All datasets are split into training, validation, and test sets following the leave-one-out strategy. To ensure a fair comparison, we analyze ranking results on the full item set.

Baseline Methods. To ensure a comprehensive assessment, we compare FlowRec with nine baseline methods in SR, divided into three categories: (i) **classical methods**: GRU4Rec [10], SASRec [13], and Bert4Rec [31], which model sequential dependencies through recurrent and self-attentive architectures; (ii) **contrastive learning-based methods:** CL4SRec [38] and CaDiRec [4], which enhance representation robustness via jointly self-supervised contrastive objectives; (iii) **generative-based methods:** ACVAE [39], DiffuRec [35], and PreferDiff [23], which formulate recommendation as a generative modeling problem to capture the uncertainty of user preferences.

Implementation Details. All experiments are conducted on a single NVIDIA GeForce RTX 4090 GPU. Baselines are implemented based on their official open-source codes. We employ a 4-layer bidirectional Transformer with 4 attention heads as the behavior-based prior encoder, and a 2-layer MLP as the flow model. The embedding size and hidden state size are both set to 128, with a batch size of 512. Dropout rates for the embedding and hidden layers are 0.1 and 0.3, respectively. The mean and variance δ of the Gaussian distribution are set to 0.001. We explore the sampling steps in $[1, 5, 10, 15, 20, 25, 30, 35]$. The model is optimized using Adam optimizer with an initial learning rate of 0.005. The loss function coefficients α and β are tuned from $[5, 10, 15, 20]$ and $[1, 2, 3, 4, 5]$, respectively. Early stopping is applied with a patience of 10 epochs.

Table 2. Experimental results (%) on the four datasets. We use "H" and "N" to represent HR and NDCG, respectively. The highest score is typed in bold, while the second-best score is underlined.

Datasets	Metrics	GRU4Rec	SASRec	BERT4Rec	CL4SRec	CaDiRec	ACVAE	PreferDiff	DiffuRec	FlowRec	Improv.
Beauty	H@5	3.1910	3.4623	2.4427	4.3438	4.9221	5.0912	4.6147	5.3812	**5.9123**	9.87%
	N@5	2.1850	2.4802	1.5684	2.8437	3.1462	4.0230	3.4595	3.9230	**4.2166**	7.48%
	H@10	4.7054	4.8909	4.0865	6.5313	7.2445	7.1074	5.8087	7.5912	**8.2760**	9.02%
	N@10	2.6735	2.9359	2.9080	3.5489	3.9183	4.6807	3.8435	4.6324	**4.9804**	7.51%
Toys	H@5	1.7780	4.7127	2.2170	5.4039	5.3263	5.5878	5.0421	5.7039	**6.2539**	9.64%
	N@5	1.2350	3.4509	1.3413	3.7329	3.6209	4.2412	3.8068	4.2871	**4.6209**	7.79%
	H@10	2.7398	6.4063	3.5463	7.4232	7.6914	7.3213	6.0890	7.3512	**8.2261**	6.95%
	N@10	1.5402	3.9944	1.7706	4.3782	4.3842	4.9891	4.1855	4.8195	**5.2578**	5.39%
Yelp	H@5	1.6976	2.6009	2.6602	1.9684	2.3183	1.3223	2.0505	3.8268	**3.9683**	3.70%
	N@5	1.1749	1.8038	1.8953	1.2151	1.4708	0.9135	1.4777	2.6579	**2.7929**	5.08%
	H@10	2.6390	3.9924	4.2884	3.4274	3.8524	2.3951	2.7800	5.7047	**5.9408**	4.14%
	N@10	1.4750	2.2499	2.4194	1.6846	1.9625	1.2754	1.7135	3.2619	**3.4253**	5.01%
ML-1M	H@5	5.8405	9.0894	8.222	1.7219	12.7745	9.5543	7.7716	15.2918	**15.5848**	1.92%
	N@5	3.8281	5.5502	5.1276	1.0221	8.3723	6.5349	5.0609	10.0919	**10.3634**	2.69%
	H@10	9.2965	15.5286	13.7086	3.3609	20.2898	16.1316	11.9801	23.7949	**24.1482**	1.48%
	N@10	4.9321	7.6131	6.8960	1.5433	10.7855	7.8832	6.2955	12.8233	**13.1228**	2.34%

5.2 Overall Performance

The overall performance of FlowRec and all baselines across four benchmark datasets is presented in Table 2. We can make the following observations: (i) Generative models, especially diffusion models, outperform traditional methods, demonstrating their strong capability in modeling complex data distributions and preference uncertainty; (ii) FlowRec achieves state-of-the-art results on all datasets. For instance, on the Beauty dataset, it improves HR@5 and HR@10 by 9.87% and 9.02% over the strongest baseline DiffuRec. These results confirm the effectiveness and generalizability of our framework; (iii) FlowRec benefits from its behavior-based prior and learnable vector field. Unlike models such as SASRec that treat next-item prediction as a one-step classification, FlowRec explicitly models continuous preference transitions from a personalized prior to the target item. This formulation enables it to refine both the initial representation and the generated trajectory iteratively. Compared with diffusion-based methods (e.g., DiffuRec, PreferDiff), FlowRec generates straight and stable preference flows from informative priors, avoiding stochastic sampling noise and achieving deterministic, efficient inference.

5.3 Generalizability on Insufficient Context

Head and Long-Tail Items. Following previous work [2], we split the top 20% most frequent items as head items and the rest as long-tail items. Users are grouped based on their most recent interaction i_{n-1} before the target item. As shown in Fig. 3, long-tail items are typically more difficult to recommend due to their sparse interaction contexts. DiffuRec, which reverses a diffusion process from pure Gaussian noise, is sensitive to such weak context conditions, often producing unstable predictions. In contrast, FlowRec leverages informative and personalized priors to generate deterministic preference trajectories, ensuring stable inference even under sparse supervision. Consequently, FlowRec achieves significant gains on long-tail items. For example, on the Toys dataset, FlowRec achieves only an 8.85% improvement over DiffuRec on head items but a substantial 21.28% gain on long-tail items. A similar trend is observed across other dataset, demonstrating FlowRec's strong generalization ability.

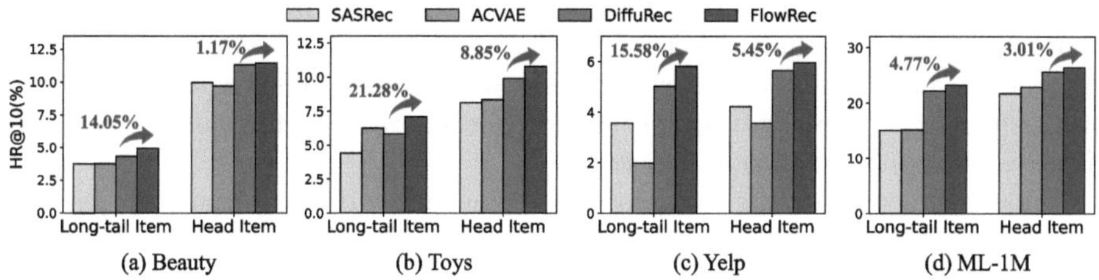

Fig. 3. Performance on the long-tail and head items.

Different Sequence Length. Based on the percentile of sequence length, we divide all datasets into three groups (short, middle, and long) in a 1:2:1 ratio. The results are shown in Fig. 4. Short sequences often lack sufficient information to capture users' current preferences for a precise recommendation, whereas it is also challenging to handle very long sequences. Both DiffuRec and FlowRec perform better on long sequences than SASRec and ACVAE, showing their superior ability to capture complex historical patterns. Notably, FlowRec achieves significantly higher accuracy on short sequences, benefiting from its deterministic ODE formulation that avoids noise injection and yields more stable inference.

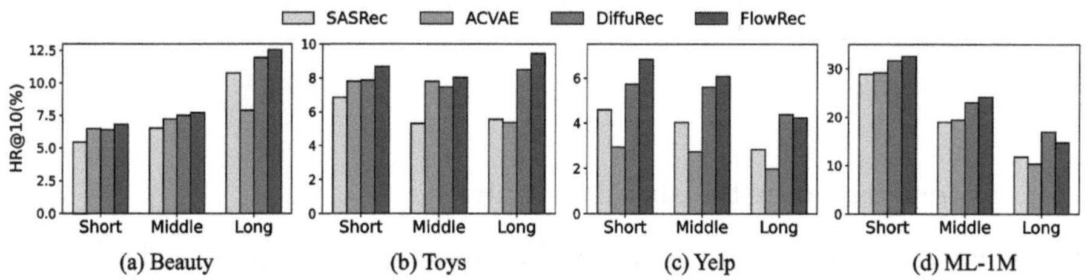

Fig. 4. Performance on different length of sequence.

5.4 Ablation Study

We conduct an ablation experiments to verify the effectiveness of each component in FlowRec. The experimental results are presented in Table 3.

Removing any of the three objectives – $\mathcal{L}_{\text{Prior}}$, $\mathcal{L}_{\text{CFM}}$, or $\mathcal{L}_{\text{Align}}$ – leads to a clear performance degradation, verifying that all components are indispensable. In particular, eliminating $\mathcal{L}_{\text{Align}}$ causes a notable drop, highlighting the necessity of contrastive alignment for distinguishing relevant items from negatives. The variant without $\mathcal{L}_{\text{Prior}}$ performs worst, confirming that the behavior-based prior provides an informative and personalized initialization that simplifies the generative process.

We further replace the Transformer encoder with a GRU ("w GRU"). The performance decrease indicates that the quality of prior representations strongly influences overall generation. Nevertheless, this variant still surpasses GRU4Rec by a large margin, demonstrating that FlowRec's flow-based formulation remains effective regardless of encoder choice.

Finally, we compare the recommendation accuracy of the initial state $\boldsymbol{x}_0$ and the generated state $\hat{\boldsymbol{x}}_1$. The strong performance of $\boldsymbol{x}_0$ verify that the learned priors are already well-aligned with user intent, while the full FlowRec further refines these representations through continuous flow modeling, achieving the highest accuracy and most stable convergence.

Table 3. Results(%) of ablation experiments.

Datasets	Metrics	w/o $\mathcal{L}_{\text{Prior}}$	w/o $\mathcal{L}_{\text{CFM}}$	w/o $\mathcal{L}_{\text{Align}}$	w GRU	Initial	FlowRec
Beauty	HR@5	2.9476	5.3070	3.7949	4.3986	5.6843	**5.9125**
	NDCG@5	1.9851	3.7331	2.4130	3.0822	4.0517	**4.2166**
	HR@10	4.4039	7.7036	5.8232	6.4790	8.0083	**8.2760**
	NDCG@10	2.4586	4.5058	3.0668	3.7474	4.7952	**4.9804**
Toys	HR@5	4.0813	5.9959	4.0246	3.9605	6.0529	**6.2539**
	NDCG@5	2.8858	4.0643	2.5603	2.8067	4.4397	**4.6209**
	HR@10	5.6610	8.1911	5.9105	5.5336	8.1090	**8.2261**
	NDCG@10	3.3927	4.7712	3.1662	3.3144	5.0672	**5.2578**

5.5 Hyperparameter Sensitivity Analysis

We analyze the impact of two important hyperparameters α and β, which control the relative weight of the conditional flow matching loss $\mathcal{L}_{\text{CFM}}$ and the single-step alignment loss $\mathcal{L}_{\text{Align}}$, respectively. The results on the Toys and Beauty datasets are shown in Fig. 5. When α increases from small values, performance improves steadily, indicating that moderate trajectory-level supervision facilitates the learning of smooth preference flow. However, excessive α weakens the influence of $\mathcal{L}_{\text{Align}}$, leading to performance degradation. A similar pattern is observed for β: proper alignment regularization enhances ranking discrimination.

Overall, FlowRec maintains stable performance across a wide range of hyperparameter values, demonstrating its robustness to the balance between trajectory guidance and alignment supervision.

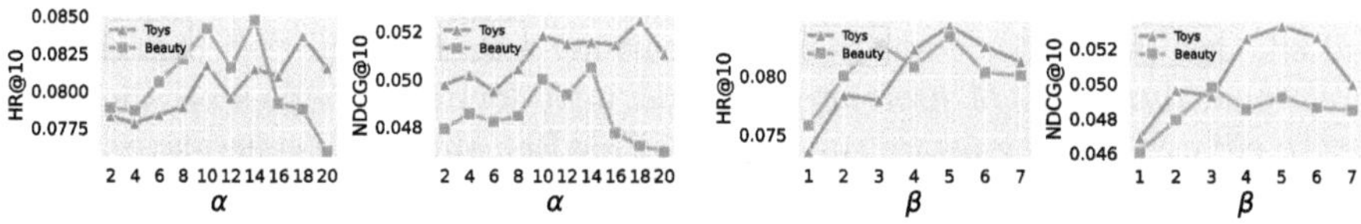

Fig. 5. The sensitivity of FlowRec to α and β.

5.6 Inference Efficiency Analysis

Sampling Steps. The number of sampling steps is a key factor influencing the inference speed. To evaluate FlowRec's inference efficiency, we varied the number of sampling steps on the Toys dataset. As illustrated in Fig. 6, FlowRec achieves optimal performance with only 8–10 sampling steps, whereas DiffuRec requires 30–35 steps. Even with a single step, FlowRec remains competitive, benefiting from its straight flow formulation and the simple, efficient ODE solver that avoids iterative denoising.

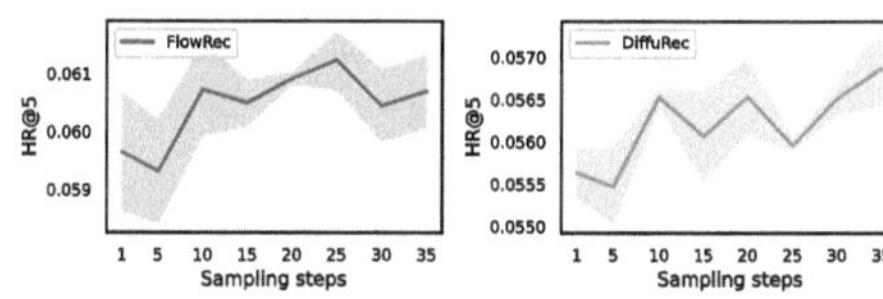

Fig. 6. The impact of sampling steps on FlowRec and DiffuRec.

Inference Time. We further compare the training and inference time of different models on the Toys and Yelp datasets, as summarized in Table 4. For fairness, both DiffuRec and FlowRec are evaluated with 10 sampling steps. FlowRec exhibits comparable training efficiency to DiffuRec but achieves substantially faster inference, with runtime close to SASRec. This improvement arises from the use of a simple Euler ODE solver, which less computationally intensive than DiffuRec's stochastic denoising process.

Table 4. Training and inference time (s/epoch).

Datasets	Phase	SASRec	ACVAE	PreferDiff	DiffuRec	FlowRec
Toys	Training	5.33	8.687	31.99	10.11	9.98
	Inference	0.88	18.00	5.25	8.51	1.27
Yelp	Training	11.67	13.59	58.18	19.10	17.62
	Inference	2.93	28.99	8.30	14.77	3.67

6 Conclusion

This paper introduced FlowRec, a simple yet effective framework for SR based on the flow matching paradigm. FlowRec explicitly models user preference trajectories from personalized behavior-based priors to target items, enabling stable and efficient preference transitions. By learning a vector field that captures the directional dynamics of user interests and introducing a single-step alignment loss to enforce ranking-aware supervision, FlowRec aligns the generative process with the recommendation objective. Extensive experiments on four benchmark datasets demonstrate that FlowRec consistently surpasses state-of-the-art methods in terms of recommendation accuracy and inference efficiency.

Acknowledgments. This research was supported by grants from the National Natural Science Foundation of China (No. 62502486), the grants of Provincial Natural Science Foundation of Anhui Province (No. 2408085QF193), USTC ResearchFunds of the DoubleFirst-Class Initiative (No. YD2150002501), the Fundamental Research Funds for the Central Universities of China (No. WK2150110032).

References

1. Chen, R.T., Rubanova, Y., Bettencourt, J., Duvenaud, D.K.: Neural ordinary differential equations. In: Advances in Neural Information Processing Systems, vol. 31 (2018)
2. Cheng, M., Liu, Q., Zhang, W., Liu, Z., Zhao, H., Chen, E.: A general tail item representation enhancement framework for sequential recommendation. Front. Comput. Sci. **18**, 186333 (2024)
3. Cho, K., Van Merriënboer, B., Bahdanau, D., Bengio, Y.: On the properties of neural machine translation: encoder-decoder approaches. arXiv preprint arXiv:1409.1259 (2014)
4. Cui, Z., Wu, H., He, B., Cheng, J., Ma, C.: Context matters: enhancing sequential recommendation with context-aware diffusion-based contrastive learning. In: Proceedings of the 33rd ACM International Conference on Information and Knowledge Management, pp. 404–414 (2024)
5. Esser, P., et al.: Scaling rectified flow transformers for high-resolution image synthesis. In: Forty-First International Conference on Machine Learning (2024)
6. Goodfellow, I.J., et al.: Generative adversarial nets. In: Advances in Neural Information Processing Systems, vol. 27 (2014)
7. Guerreiro, J.J.A., Inoue, N., Masui, K., Otani, M., Nakayama, H.: Layoutflow: flow matching for layout generation. In: European Conference on Computer Vision, pp. 56–72. Springer, Cham (2024)
8. Harper, F.M., Konstan, J.A.: The movielens datasets: history and context. ACM Trans. Interact. Intell. Syst. (TIIS) **5**(4), 1–19 (2015)
9. He, R., McAuley, J.: Fusing similarity models with Markov chains for sparse sequential recommendation. In: 2016 IEEE 16th International Conference on Data Mining (ICDM), pp. 191–200. IEEE (2016)
10. Hidasi, B., Karatzoglou, A., Baltrunas, L., Tikk, D.: Session-based recommendations with recurrent neural networks. arXiv preprint arXiv:1511.06939 (2015)

11. Ho, J., Jain, A., Abbeel, P.: Denoising diffusion probabilistic models. Adv. Neural. Inf. Process. Syst. **33**, 6840–6851 (2020)
12. Hu, V., et al.: Flow matching for conditional text generation in a few sampling steps. In: Proceedings of the 18th Conference of the European Chapter of the Association for Computational Linguistics (Volume 2: Short Papers), pp. 380–392 (2024)
13. Kang, W.C., McAuley, J.: Self-attentive sequential recommendation. In: 2018 IEEE International Conference on Data Mining (ICDM), pp. 197–206. IEEE (2018)
14. Kim, M., Lee, Y., Kang, S., Oh, J., Chong, S., Yun, S.Y.: Preference alignment with flow matching. Adv. Neural. Inf. Process. Syst. **37**, 35140–35164 (2024)
15. Kingma, D.P., Welling, M.: Auto-encoding variational bayes. arXiv e-prints pp. arXiv–1312 (2013)
16. Li, W., et al.: Dimerec: a unified framework for enhanced sequential recommendation via generative diffusion models. In: Proceedings of the Eighteenth ACM International Conference on Web Search and Data Mining, pp. 726–734 (2025)
17. Li, Z., Sun, A., Li, C.: Diffurec: a diffusion model for sequential recommendation. ACM Trans. Inf. Syst. **42**(3), 1–28 (2023)
18. Liang, D., Krishnan, R.G., Hoffman, M.D., Jebara, T.: Variational autoencoders for collaborative filtering. In: Proceedings of the 2018 World Wide Web Conference, pp. 689–698 (2018)
19. Lipman, Y., Chen, R.T., Ben-Hamu, H., Nickel, M., Le, M.: Flow matching for generative modeling. arXiv preprint arXiv:2210.02747 (2022)
20. Liu, C., Zhang, Y., Wang, J., Ying, R., Caverlee, J.: Flow matching for collaborative filtering. arXiv preprint arXiv:2502.07303 (2025)
21. Liu, F., et al.: Flow matching based sequential recommender model. arXiv preprint arXiv:2505.16298 (2025)
22. Liu, Q., et al.: LLM-ESR: large language models enhancement for long-tailed sequential recommendation. Adv. Neural. Inf. Process. Syst. **37**, 26701–26727 (2024)
23. Liu, S., Zhang, A., Hu, G., Qian, H., Chua, T.: Preference diffusion for recommendation. arXiv preprint arXiv:2410.13117 (2024)
24. Liu, X., Gong, C., Liu, Q.: Flow straight and fast: learning to generate and transfer data with rectified flow. arXiv preprint arXiv:2209.03003 (2022)
25. McAuley, J., Targett, C., Shi, Q., Van Den Hengel, A.: Image-based recommendations on styles and substitutes. In: Proceedings of the 38th International ACM SIGIR Conference on Research and Development in Information Retrieval, pp. 43–52 (2015)
26. Mehta, S., Tu, R., Beskow, J., Székely, É., Henter, G.E.: Matcha-TTS: a fast TTS architecture with conditional flow matching. In: ICASSP 2024-2024 IEEE International Conference on Acoustics, Speech and Signal Processing (ICASSP), pp. 11341–11345. IEEE (2024)
27. Ren, R., et al.: Sequential recommendation with self-attentive multi-adversarial network. In: Proceedings of the 43rd International ACM SIGIR Conference on Research and Development in Information Retrieval, pp. 89–98 (2020)
28. Rendle, S., Freudenthaler, C., Gantner, Z., Schmidt-Thieme, L.: BPR: Bayesian personalized ranking from implicit feedback. arXiv preprint arXiv:1205.2618 (2012)
29. Rendle, S., Freudenthaler, C., Schmidt-Thieme, L.: Factorizing personalized Markov chains for next-basket recommendation. In: Proceedings of the 19th International Conference on World Wide Web, pp. 811–820 (2010)

30. Song, Y., Sohl-Dickstein, J., Kingma, D.P., Kumar, A., Ermon, S., Poole, B.: Score-based generative modeling through stochastic differential equations. arXiv preprint arXiv:2011.13456 (2020)
31. Sun, F., et al.: Bert4rec: sequential recommendation with bidirectional encoder representations from transformer. In: Proceedings of the 28th ACM International Conference on Information and Knowledge Management, pp. 1441–1450 (2019)
32. Tang, J., Wang, K.: Personalized top-n sequential recommendation via convolutional sequence embedding. In: Proceedings of the Eleventh ACM International Conference on Web Search and Data Mining, pp. 565–573 (2018)
33. Tong, A., et al.: Improving and generalizing flow-based generative models with minibatch optimal transport. Trans. Mach. Learn. Res. 1–34 (2024)
34. Tong, A., et al.: Conditional flow matching: simulation-free dynamic optimal transport. arXiv preprint arXiv:2302.00482 (2023)
35. Wang, W., Xu, Y., Feng, F., Lin, X., He, X., Chua, T.S.: Diffusion recommender model. In: Proceedings of the 46th International ACM SIGIR Conference on Research and Development in Information Retrieval, pp. 832–841 (2023)
36. Xie, W., Wang, H., Zhang, L., Zhou, R., Lian, D., Chen, E.: Breaking determinism: fuzzy modeling of sequential recommendation using discrete state space diffusion model. Adv. Neural. Inf. Process. Syst. **37**, 22720–22744 (2024)
37. Xie, W., Zhou, R., Wang, H., Shen, T., Chen, E.: Bridging user dynamics: transforming sequential recommendations with Schrödinger bridge and diffusion models. In: Proceedings of the 33rd ACM International Conference on Information and Knowledge Management, pp. 2618–2628 (2024)
38. Xie, X., et al.: Contrastive learning for sequential recommendation. In: 2022 IEEE 38th International Conference on Data Engineering (ICDE), pp. 1259–1273. IEEE (2022)
39. Xie, Z., Liu, C., Zhang, Y., Lu, H., Wang, D., Ding, Y.: Adversarial and contrastive variational autoencoder for sequential recommendation. In: Proceedings of the Web Conference 2021, p. 449–459 (2021)
40. Yang, Z., Wu, J., Wang, Z., Wang, X., Yuan, Y., He, X.: Generate what you prefer: reshaping sequential recommendation via guided diffusion. Adv. Neural. Inf. Process. Syst. **36**, 24247–24261 (2023)

HISIR: End2End Hierarchical Intent-Based Sequential Recommendation

Yan Jiang and Chao Li(✉)

Zhejiang University, Hangzhou, China
{yan_jiang,chaoli}@zju.edu.cn

Abstract. Sequential recommendation is pivotal for modern web services, yet accurately capturing the multifaceted nature of user needs remains challenging. Existing intent-aware methods attempt to model user intent, but they often suffer from critical limitations: (1) a Monolithic Interest Representation that fails to capture concurrent user interests, and (2) a Flattened Intent Modeling approach that overlooks the inherent hierarchy of user goals. To address these issues, we propose HISIR, an End2End Hierarchical Intent-based Sequential Recommendation framework. It first employs a Multi-Expert Interest Extraction Layer to disentangle a user's diverse interests. Subsequently, a Hierarchical Intent Modeling Layer maps these interests into a structured, multi-level representation. The entire framework is optimized by a Dual-Decoupling Loss Optimization Layer to ensure the independence and logical consistency of the learned representations. Extensive experiments on four public benchmarks demonstrate that HISIR consistently and significantly outperforms state-of-the-art baselines. The code repository is available at https://github.com/JessieJiangyan/HISIR.

Keywords: Sequential Recommendation · Intent-based Recommendation · Hierarchical Intent

1 Introduction

With the exponential growth of online information, recommender systems have become an indispensable component of modern web services, playing a pivotal role in alleviating information overload and enhancing user experience [24,26]. Among various recommendation paradigms, sequential recommendation has emerged as a critical research direction [22]. By modeling the temporal dynamics of user behaviors, sequential recommendation aims to predict the user's next action based on their historical interaction sequence, such as purchases, clicks, or views.

The evolution of sequential recommendation has witnessed a progression of techniques. Early approaches often relied on Markov chains to capture transitional patterns [5]. Subsequently, deep learning models, particularly those based on Recurrent Neural Networks (RNNs) [6] and Transformers [8], have demonstrated remarkable success by learning sophisticated representations of user

H. Jung et al. (Eds.): DASFAA 2026, LNCS 16535, pp. 223–239, 2026.
https://doi.org/10.1007/978-981-92-0363-5_14

behavior sequences. More recently, a promising avenue of research has focused on intent-based sequential recommendation [4,12]. These methods infer latent user intents from literal user-item interactions to gain a deeper understanding of user needs. By modeling these latent intents, such approaches can achieve a deeper understanding of user needs, leading to more accurate and explainable recommendations, thus marking it as a promising research direction.

Despite the progress made, existing sequential recommendation methods, including those that model user intent, face several critical challenges in accurately capturing the complex and multifaceted nature of user needs.

- **Monolithic Interest Representation:** Most existing models generate a single, holistic user representation from their behavior sequence [8]. This monolithic approach fails to capture the concurrent, diverse interests a user may possess. For instance, a user might simultaneously be interested in "outdoor camping gear" and "professional work attire," and a single vector representation struggles to distinctly encapsulate these parallel and often unrelated interests [2].
- **Flattened Intent Modeling:** Current intent-aware models typically adopt a flat structure, treating each user intention as an independent entity (e.g., "buying a wireless charger") [12]. This overlooks the inherently hierarchical nature of user goals, where specific, low-level intents (e.g., "buy a smart wristband") are often manifestations of broader, high-level objectives (e.g., "digital health management"). This flattened perspective misses the crucial contextual dependencies between different levels of intent.
- **Entangled Interest and Intent Representations:** There is a lack of explicit mechanisms to disentangle the representation of diverse interests from that of hierarchical intents. This entanglement leads to ambiguous and confounded user profiles, where the boundaries between distinct interests are blurred and the logical structure of hierarchical intents is obscured [10]. For example, the model might erroneously associate an "outdoor backpack" interest with a high-level "smart living upgrade" intent, corrupting the clarity and precision of the user representation.

Figure 1 provides a conceptual illustration of these challenges and contrasts the traditional approach with our proposed HISIR framework.

To address the aforementioned challenges, we propose a novel framework, the End2End **H**ierarchical **I**ntent-based **S**equent**i**al **R**ecommendation (**HISIR**). HISIR is designed to learn decoupled representations of multiple user interests and their hierarchical intents for precise sequential recommendation.

Our proposed HISIR model is comprised of three core components. First, a Multi-Expert Interest Extraction Layer disentangles the diverse interests of a user from their behavior sequence. Next, Inspired by the principles of vector quantization, the Hierarchical Intent Modeling Layer organizes these varied interests into a structured, multi-level intent hierarchy, ranging from high-level abstract intents to more specific, low-level intents. Finally, a Dual Disentanglement Optimization Layer enforces the independence of these interest represen-

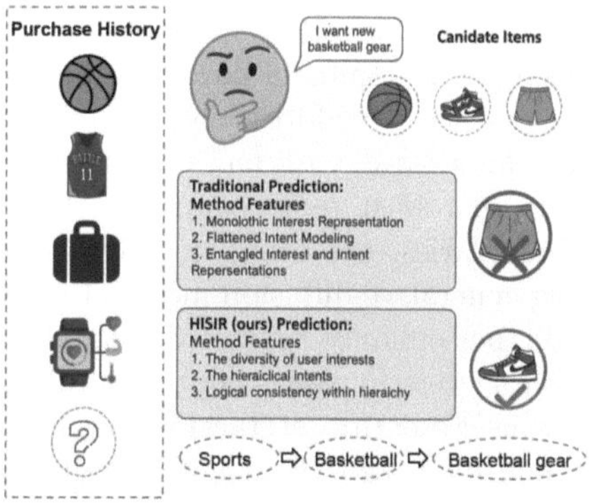

Fig. 1. An illustration of challenges in sequential recommendation. While traditional methods struggle with diverse user interests, HISIR leverages hierarchical intents (e.g., Sports → Basketball → Basketball Gear)) for precise prediction.

tations and the logical consistency within the intent hierarchy through a novel dual-loss objective.

The main contributions of this paper are summarized as follows:

- We propose HISIR, a novel sequential recommendation framework that explicitly models the diversity of user interests and the hierarchical structure of user intents in a disentangled manner.
- We introduce a Multi-Expert Interest Extraction Layer to effectively identify and represent a user's multiple, concurrent interests, and a Hierarchical Intent Modeling Layer that uses residual quantizers and a high-level guidance mechanism to build a structured and interpretable intent hierarchy, which distinguishes among high-level goal intents, mid-level domain intents, and low-level specific intents.
- We design a Dual Disentanglement Optimization Layer, incorporating a Mutual Information-based loss and a Hierarchical Contrastive loss, to effectively decouple interest representations from each other and to enforce logical consistency within the intent hierarchy.
- We conduct extensive experiments on several real-world benchmark datasets. The results demonstrate the superiority of HISIR over state-of-the-art methods, and a series of ablation studies validate the effectiveness of each component within our proposed model.

2 Related Work

2.1 Sequential Recommendation

Sequential recommendation aims to predict a user's next action based on their historical behavior. Early approaches evolved from Markov chains [5] to deep learning models like Recurrent Neural Networks (RNNs) [6]. More recently, self-attention mechanisms, popularized by models such as SASRec [8] and BERT4Rec [21], have become the dominant paradigm for their superior ability to capture long-range dependencies. To address data sparsity, self-supervised learning (SSL) has been widely integrated, with methods like CL4SRec [25] leveraging contrastive learning to learn robust representations. The latest advancements continue to push the boundaries, exploring iterative sparse attention mechanisms [11] and pioneering generative diffusion models to directly synthesize item representations [14] to further enhance predictive accuracy.

However, despite their success in capturing temporal dynamics, these methods typically compress a user's entire interaction history into a single, holistic representation. This monolithic approach struggles to disentangle the concurrent and diverse interests a user may have, motivating our design of a Multi-Expert Interest Extraction Layer to capture these multifaceted signals explicitly.

2.2 Intent-Aware Recommendation

To improve performance and interpretability, another line of research focuses on modeling latent user intents. Early works often relied on auxiliary information, but recent methods infer intents directly from user behavior sequences. These approaches commonly employ latent variable models like DSSRec [13], or leverage contrastive learning to align sequence representations with intent prototypes derived from clustering, as seen in ICLRec [4]. To address the scalability limitations of such clustering-based methods, ELCRec [12] proposes an end-to-end learnable clustering framework, treating intent centers as learnable parameters optimized within mini-batches. Another recent direction involves leveraging generative models; for instance, InDiRec [18] employs an intent-guided diffusion model to generate semantically consistent augmented views for contrastive learning. Other works have focused on designing sophisticated contrastive tasks on subsequences to capture stable and dynamic intents [17,23] or have successfully integrated external knowledge, such as knowledge graphs, to create more structured intent representations [28].

Nevertheless, these intent-aware models predominantly operate on a flat structure, treating different intents as independent entities. They overlook the inherent hierarchical nature of user goals, where specific actions are often manifestations of broader objectives. This limitation leads to entangled representations and motivates our work on a Hierarchical Intent Modeling Layer and a Dual-Decoupling Loss, designed to learn structured, multi-level, and disentangled intent representations.

3 Methodology

In this section, we introduce our proposed model, **HISIR**, which stands for End2End **H**ierarchical **I**ntent-based **S**equential **R**ecommendation. The core idea of HISIR is to explicitly model user intents in a structured, multi-level hierarchy while simultaneously capturing diverse user interests through a disentangled

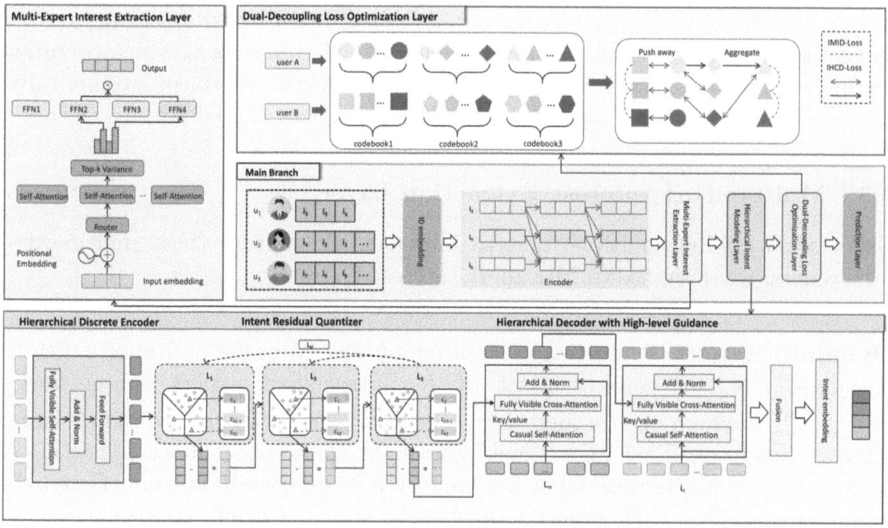

Fig. 2. The overall architecture of HISIR.

representation learning framework. As illustrated in Fig. 2, the architecture comprises three main components: (1) a Multi-Expert Interest Extraction Layer; (2) a Hierarchical Intent Modeling Layer that learns structured and discrete user intents; and (3) a Dual-Decoupling Loss Optimization Layer that ensures representational independence.

3.1 Problem Definition

Let $\mathcal{U} = \{u_1, u_2, \ldots, u_{|\mathcal{U}|}\}$ be the set of users and $\mathcal{I} = \{i_1, i_2, \ldots, i_{|\mathcal{I}|}\}$ be the set of items. For each user $u \in \mathcal{U}$, we have a chronologically ordered interaction sequence $S_u = [i_1^{(u)}, i_2^{(u)}, \ldots, i_{|S_u|}^{(u)}]$, where $i_t^{(u)} \in \mathcal{I}$ is the item that user u interacted with at time step t. Given the historical interaction sequence S_u, the task of sequential recommendation is to predict the next item $i_{|S_u|+1}^{(u)}$ that the user is most likely to interact with.

3.2 Input Layer

To process the raw item sequences, we first transform them into dense vector representations. Each item $i \in \mathcal{I}$ is associated with a learnable embedding vector $\mathbf{e}_i \in \mathbb{R}^d$, where d is the embedding dimension. To capture the temporal dynamics and emphasize more recent interactions, we incorporate a learnable temporal decay factor. Specifically, for a sequence S_u truncated or padded to a fixed length N, the input representation $\mathbf{X}_u \in \mathbb{R}^{N \times d}$ is formed by combining item embeddings with positional embeddings, modulated by a temporal scaling factor:

$$\mathbf{x}_t = \alpha_t(\mathbf{e}_{i_t} + \mathbf{p}_t), \quad t \in [1, N] \tag{1}$$

where $\mathbf{p}_t \in \mathbb{R}^d$ is the learnable positional embedding for position t, and α_t is a learnable scalar that models the importance of the t-th position, typically assigning higher weights to more recent interactions. This layer produces a temporally-aware behavior sequence representation $\mathbf{X}_u = [\mathbf{x}_1, \mathbf{x}_2, \ldots, \mathbf{x}_N]$ [8].

3.3 Multi-expert Interest Extraction Layer

Users often exhibit multiple, diverse interests. To capture these multifaceted preferences, we design a Multi-Expert Interest Extraction Layer.

Dynamic Expert Encoding. Inspired by the Mixture of Experts (MoE) paradigm [20], we employ a set of K independent expert sub-networks, $\{f_{\text{expert}}^{(k)}\}_{k=1}^{K}$, where each expert is designed to specialize in a distinct latent interest dimension. Each expert is a lightweight Transformer block consisting of a multi-head self-attention layer followed by a feed-forward network. Given the input sequence $\mathbf{X}_u$, each expert k produces a specialized representation:

$$\mathbf{H}_u^{(k)} = f_{\text{expert}}^{(k)}(\mathbf{X}_u) \in \mathbb{R}^{N \times d} \tag{2}$$

To obtain a user-level interest representation from each expert, we apply mean pooling over the time dimension, resulting in K interest vectors $\{\mathbf{h}_u^{(k)}\}_{k=1}^{K}$, where $\mathbf{h}_u^{(k)} \in \mathbb{R}^d$.

Variance-Gated Interest Filtering. To ensure interest diversity and prevent expert redundancy, we introduce a novel interest filtering mechanism. The intuition is that an expert effectively capturing a meaningful interest dimension should produce distinct representations for users with varied behaviors, leading to a higher representation variance across a mini-batch. We quantify this by calculating the variance of each expert's representations across a mini-batch $\mathcal{B}$ of users:

$$\text{Var}^{(k)} = \frac{1}{|\mathcal{B}|} \sum_{u \in \mathcal{B}} \|\mathbf{h}_u^{(k)} - \bar{\mathbf{h}}^{(k)}\|_2^2, \quad \text{where} \quad \bar{\mathbf{h}}^{(k)} = \frac{1}{|\mathcal{B}|} \sum_{u \in \mathcal{B}} \mathbf{h}_u^{(k)} \tag{3}$$

We then select the Top-K' experts with the highest variance. Here, K' is a fixed hyperparameter, but the set of active experts is determined adaptively for each batch, allowing the model to focus on the most salient interest facets for the current group of users. This dynamic selection is stabilized by our decoupling objectives, which encourage experts to learn distinct and consistent roles. The final multi-interest representation for user u is the concatenation of the outputs from these selected experts:

$$\mathbf{h}_u^{\text{multi-interest}} = \text{Concat}(\mathbf{h}_u^{(k_1)}, \mathbf{h}_u^{(k_2)}, \ldots, \mathbf{h}_u^{(k_{K'})}) \tag{4}$$

where $\{k_1, \ldots, k_{K'}\}$ are the indices of the experts with the top K' variances. This filtering step enhances both performance and efficiency.

3.4 Hierarchical Intent Modeling Layer

While interests reflect what a user likes, intents represent the underlying purpose of their actions. We model these intents in a structured, hierarchical, and interpretable manner.

Hierarchical Discrete Encoder. We employ a stack of L Residual Quantizers (RQs), a technique effective for structured vector quantization [7], to map the continuous interest representation into discrete intent codes. Each layer $l \in [1, L]$ has a learnable codebook $\mathcal{C}_l \in \mathbb{R}^{M_l \times d_l}$, where M_l is the number of codes and d_l is the code dimension. Each vector in this codebook serves as a learnable prototype representing a latent user intent at level l. These codebooks are structured hierarchically: high-level codebooks (e.g., $l = L$) represent abstract goals like "digital health management," while low-level ones (e.g., $l = 1$) represent concrete intents like "purchase a smart bracelet".

Given the user's interest representation $\mathbf{z}_0 = \mathbf{h}_u^{\text{multi-interest}}$, the encoder at level l quantizes the input residual from the previous level, $\mathbf{r}_{l-1}$ (with the initial residual defined as $\mathbf{r}_0 = \mathbf{z}_0$), by finding the closest code vector $\mathbf{c}_{l,j} \in \mathcal{C}_l$:

$$\hat{\mathbf{c}}_l = \arg\min_{\mathbf{c}\in\mathcal{C}_l} \|\mathbf{r}_{l-1} - \mathbf{c}\|_2^2 \tag{5}$$

The residual for the next layer is then computed as $\mathbf{r}_l = \mathbf{r}_{l-1} - \hat{\mathbf{c}}_l$. This process is optimized by minimizing the quantization error, also known as the codebook loss: $\mathcal{L}_{\text{RQ}} = \sum_{l=1}^{L} \|\mathbf{r}_{l-1} - \hat{\mathbf{c}}_l\|_2^2$.

Hierarchical Decoder with High-Level Guidance. To ensure the learned intents are logically consistent across levels, we design a hierarchical decoder. The decoder at level l refines the intent representation $\hat{\mathbf{c}}_l$ using guidance from the level above, $\hat{\mathbf{c}}_{l+1}$. This is achieved through a "high-level guidance attention" mechanism, where the higher-level intent acts as the query:

$$\tilde{\mathbf{c}}_l = \text{Attention}(\mathbf{Q} = \hat{\mathbf{c}}_{l+1}, \mathbf{K} = \hat{\mathbf{c}}_l, \mathbf{V} = \hat{\mathbf{c}}_l) \tag{6}$$

This allows abstract intents (e.g., "home renovation") to guide the representation of more specific intents (e.g., "replace sofa"), enforcing a top-down dependency.

Hierarchical Consistency Constraint. To further enforce logical consistency, we introduce a hierarchical KL divergence loss, which is widely used to align probability distributions in generative modeling [9]. This loss encourages the distribution of lower-level intent representations to align with the conditional distribution guided by their corresponding higher-level intents.

We achieve this by projecting the intent vectors from adjacent levels into a shared probabilistic space. Specifically, for a user u, we define two probability distributions. The first, $q_{u,l}$, represents the distribution derived from the refined lower-level intent $\tilde{\mathbf{c}}_{u,l}$. The second, $p_{u,l}$, is a target distribution derived from the higher-level intent $\tilde{\mathbf{c}}_{u,l+1}$. These distributions are obtained by passing the intent vectors through shared, learnable projection matrices $(\mathbf{W}_q, \mathbf{W}_p)$ followed by a softmax function:

$$\mathbf{q}_{u,l} = \mathrm{Softmax}(\mathbf{W}_q \tilde{\mathbf{c}}_{u,l}) \tag{7}$$

$$\mathbf{p}_{u,l} = \mathrm{Softmax}(\mathbf{W}_p \tilde{\mathbf{c}}_{u,l+1}) \tag{8}$$

Here, $\mathbf{q}_{u,l}$ and $\mathbf{p}_{u,l}$ are probability vectors over a latent categorical space. The matrix $\mathbf{W}_p$ can be interpreted as a learnable transformation that maps a higher-level intent to a prediction about the lower-level intent's distribution.

We then enforce consistency by minimizing the KL divergence between these two distributions. This pushes the lower-level intent's distribution $\mathbf{q}_{u,l}$ to match the target distribution $\mathbf{p}_{u,l}$ predicted by the level above. The consistency loss is summed over all adjacent levels and all users in a mini-batch $\mathcal{B}$:

$$\mathcal{L}_{\text{Consistency}} = \sum_{u \in \mathcal{B}} \sum_{l=1}^{L-1} D_{KL}\left(\mathbf{q}_{u,l} \parallel \text{stop\_gradient}(\mathbf{p}_{u,l})\right) \tag{9}$$

where $D_{KL}(\mathbf{q} \parallel \mathbf{p}) = \sum_i \mathbf{q}_i \log(\frac{\mathbf{q}_i}{\mathbf{p}_i})$. We apply a stop-gradient operation to the target distribution $\mathbf{p}_{u,l}$ to ensure a stable training signal, where the lower-level representation learns to match the prediction from the fixed higher-level representation. This mechanism ensures that the hierarchy is semantically and probabilistically coherent. The final user intent representation $\mathbf{h}_u^{\text{intent}}$ is an aggregation of the refined intent vectors from all levels $\{\tilde{\mathbf{c}}_{u,l}\}_{l=1}^{L}$, fused via a self-attention mechanism.

3.5 Dual-Decoupling Loss Optimization Layer

A key challenge is to ensure that the learned representations are both diverse and independent. We address this with a dual-decoupling loss.

Multi-interest Mutual Information Decoupling Loss (IMID-Loss). To ensure that different experts capture distinct interests, we minimize the mutual information (MI) between their output representations. We leverage the Mutual Information Neural Estimator (MINE), which maximizes a lower bound on MI based on the Donsker-Varadhan representation [1]. We minimize the negative of this bound, which is equivalent to minimizing the MI. The loss is implemented via a discriminator network T_ϕ with parameters ϕ:

$$\mathcal{L}_{\text{IMID}} = -\sum_{j \neq k} \left(\mathbb{E}_{\mathcal{P}_{jk}}[T_\phi(\mathbf{h}^{(j)}, \mathbf{h}^{(k)})] - \log(\mathbb{E}_{\mathcal{P}_j \otimes \mathcal{P}_k}[e^{T_\phi(\mathbf{h}^{(j)}, \mathbf{h}^{(k)})}]) \right) \tag{10}$$

In practice, the joint distribution $\mathcal{P}_{jk}$ is approximated by pairing interest vectors $(\mathbf{h}_u^{(j)}, \mathbf{h}_u^{(k)})$ from the same user u within a mini-batch $\mathcal{B}$. The product of marginals $\mathcal{P}_j \otimes \mathcal{P}_k$ is approximated by pairing $\mathbf{h}_u^{(j)}$ with $\mathbf{h}_{u'}^{(k)}$ from a different, randomly shuffled user u' in the same batch.

Hierarchical Intent Contrastive Decoupling Loss (IHCD-Loss). To disentangle the hierarchical intent representations, we propose a contrastive loss composed of two parts. Based on the principles of contrastive representation learning [3], we define a similarity function $\text{sim}(\mathbf{a}, \mathbf{b}) = \mathbf{a}^\top \mathbf{b}/(\|\mathbf{a}\| \|\mathbf{b}\|)$ and use a temperature parameter τ.

Cross-Level Contrastive Loss. This loss enforces logical alignment between adjacent intent levels. For a user u, the refined low-level intent $\tilde{\mathbf{c}}_{u,l}$ and its corresponding high-level intent $\tilde{\mathbf{c}}_{u,l+1}$ form a positive pair. High-level intents from other users $u' \in \mathcal{B}$ serve as negative examples. The loss for a user u at level l is:

$$\mathcal{L}_{\text{cross}}^{(u,l)} = -\log \frac{\exp(\text{sim}(\tilde{\mathbf{c}}_{u,l}, \tilde{\mathbf{c}}_{u,l+1})/\tau)}{\sum_{u' \in \mathcal{B}} \exp(\text{sim}(\tilde{\mathbf{c}}_{u,l}, \tilde{\mathbf{c}}_{u',l+1})/\tau)} \tag{11}$$

Intra-Level Distinction Loss. This loss ensures that intents at the same level are discriminative. We generate two augmented views of the user's input sequence, including mask, crop, and reorder, resulting in two distinct refined intent representations at each level, denoted as $\tilde{\mathbf{c}}_{u,l}^{(1)}$ and $\tilde{\mathbf{c}}_{u,l}^{(2)}$. These form a positive pair. The representations of the same level from other users $u' \in \mathcal{B}$ are treated as negative examples. The loss is:

$$\mathcal{L}_{\text{intra}}^{(u,l)} = -\log \frac{\exp(\text{sim}(\tilde{\mathbf{c}}_{u,l}^{(1)}, \tilde{\mathbf{c}}_{u,l}^{(2)})/\tau)}{\sum_{u' \in \mathcal{B}} \exp(\text{sim}(\tilde{\mathbf{c}}_{u,l}^{(1)}, \tilde{\mathbf{c}}_{u',l}^{(2)})/\tau)} \tag{12}$$

The total IHCD-Loss is a weighted sum of these two components across all applicable levels:

$$\mathcal{L}_{\text{IHCD}} = \sum_{u \in \mathcal{B}} \sum_{l=1}^{L-1} \left(\mathcal{L}_{\text{cross}}^{(u,l)} + \beta \mathcal{L}_{\text{intra}}^{(u,l)}\right) \tag{13}$$

where β is a balancing hyperparameter.

3.6 Overall Objective

The final objective function of HISIR combines the primary recommendation loss (e.g., cross-entropy) with the auxiliary losses for quantization, consistency, and decoupling:

$$\mathcal{L}_{\text{total}} = \mathcal{L}_{\text{rec}} + \lambda_1 \mathcal{L}_{\text{RQ}} + \lambda_2 \mathcal{L}_{\text{IMID}} + \lambda_3 \mathcal{L}_{\text{IHCD}} + \lambda_4 \mathcal{L}_{\text{Consistency}} \tag{14}$$

where $\lambda_1, \lambda_2, \lambda_3, \lambda_4$ are hyperparameters that balance the contribution of each term. This multi-task learning framework enables HISIR to learn rich, structured, and decoupled representations for superior recommendation performance.

4 Experiments

In this section, we conduct a series of experiments to comprehensively evaluate the performance of our proposed HISIR model. Our primary goal is to answer the following research questions:

- **RQ1:** Does HISIR outperform state-of-the-art baselines?
- **RQ2:** How do key components of HISIR contribute to overall performance?
- **RQ3:** How sensitive is HISIR to its crucial hyperparameters, such as the number of experts and the structure of the intent levels?
- **RQ4:** What is the computational efficiency of HISIR in terms of training time, inference speed, and parameter count compared to a strong baseline?
- **RQ5:** Can HISIR learn meaningful and interpretable hierarchical intent representations, as demonstrated through a qualitative case study?

4.1 Experimental Settings

Datasets. We evaluate our model on four widely-used public benchmark datasets: Beauty, Toys, Sports, and Yelp. These datasets have been established as standard benchmarks in sequential recommendation research, frequently utilized to assess and compare model performance [4,12,21]. The first three are subcategories from the Amazon Review dataset [15], known for their sparsity. The Yelp dataset contains user reviews for businesses[1]. Following standard preprocessing procedures, we filter out users and items with fewer than 5 interactions to ensure data quality [27]. The statistics of the datasets are summarized in Table 1.

Table 1. Statistics of the experimental datasets after preprocessing.

Dataset	#Users	#Items	#Actions	Avg. Len.	Sparsity
Beauty	22,363	12,101	198,502	8.9	99.95%
Toys	19,412	11,924	167,597	8.6	99.93%
Sports	35,598	18,357	296,337	8.3	99.95%
Yelp	30,431	20,033	316,354	8.3	99.95%

Evaluation Metrics. To evaluate the performance of all models for the next-item prediction task, we adopt two standard top-K ranking metrics: Hit Ratio (HR@k) and Normalized Discounted Cumulative Gain (NDCG@k). These metrics are the de-facto standard for evaluating ranking quality in sequential recommendation [22]. HR@k measures whether the ground-truth item is present in the top-k recommendation list, while NDCG@k is a position-aware metric that assigns higher scores to hits at higher ranks. We report results for k = 5 and k = 20. For evaluation, we rank all items in the item set that the user has not interacted with previously.

[1] https://www.yelp.com/dataset.

Baselines. We compare HISIR against several competitive baselines, categorized as follows:

- **Classic Sequential Models:**
  - BPR-MF [19]: a classic matrix factorization method for implicit feedback.
  - GRU4Rec [6]: An RNN-based model for session-based recommendation.
  - SASRec [8]: A state-of-the-art model using self-attention to capture sequential patterns.
- **SSL-based Models:**
  - BERT4Rec [21]: Employs a deep bi-directional self-attention model using the Cloze task.
  - CL4SRec [25]: A contrastive learning framework for sequential recommendation.
- **Intent-based Models:**
  - DSSRec [13]: A model that disentangles user intents in latent spaces using a seq2seq strategy.
  - ICLRec [4]: A strong baseline that uses contrastive learning to model latent user intents.
  - ECLRec [12]: An end-to-end framework for intent learning that jointly optimizes behavior representation and learnable clustering.
  - ICMA [16]: An intent contrastive learning framework that creates augmented views via relevance-based item insertion and substitution.
  - InDiRec [18]: An intent-aware diffusion model that generates contrastive views guided by learned intent prototypes.

Implementation Details. We implemented all models in PyTorch, using an embedding dimension of $d = 64$ and a maximum sequence length of 50. All models were trained with the Adam optimizer. For HISIR, key architectural hyperparameters like the number of experts K and intent layers L were tuned as detailed in Sect. 4.4. Similarly, the loss balancing hyperparameters were optimized via grid search on a validation set. Based on this tuning, we adopted the configuration of $\lambda_1 = 1.0$ (Quantization), $\lambda_2 = 0.1$ (IMID), $\lambda_3 = 0.1$ (IHCD), $\lambda_4 = 0.2$ (Consistency) for all experiments, as it yielded the best overall performance. For baselines, we adopted settings from their original papers or tuned them on our validation set. We report the average results of three runs with different seeds, all conducted on a single NVIDIA RTX 3090 GPU.

4.2 Performance Comparison (RQ1)

To comprehensively evaluate the effectiveness of our proposed model, we compare the performance of HISIR against all baseline models. The main results are presented in Table 2. From the table, we can draw several key conclusions.

HISIR Consistently Achieves State-of-the-Art Performance. Across all four datasets, HISIR significantly outperforms all competitive baselines on both

HR and NDCG metrics. This demonstrates the clear superiority of our proposed architecture. Notably, on the challenging Beauty dataset, HISIR achieves an NDCG@5 of 0.0516, improving upon the strongest baseline (InDiRec) by 5.74%. A similar significant gain is observed on the Toys dataset, where HISIR improves NDCG@5 by 6.65% over InDiRec. These substantial and consistent improvements validate the overall effectiveness of our end-to-end hierarchical intent-based framework.

Intent-Aware Models are Superior to Generic Sequential Models. The results reveal a distinct trend where intent-aware models (e.g., ICLRec, ELCRec, ICMA, InDiRec) generally show a strong advantage over classic sequential models like SASRec and general self-supervised methods like CL4SRec. This performance gap underscores the importance of moving beyond mere sequential patterns to explicitly model the underlying user intent. By inferring the "why" behind user actions, intent-aware approaches gain a deeper understanding of user needs, which translates into more accurate predictions of their future behavior.

HISIR's Hierarchical and Disentangled Modeling Surpasses Other Intent-Based Methods. Most importantly, HISIR surpasses even the most competitive and recent intent-based baselines. This superiority stems from HISIR's unique design that directly addresses the core limitations of existing intent-aware methods—namely, the Monolithic Interest Representation and Flattened Intent Modeling. While strong baselines like InDiRec and ICLRec model latent intents, they typically do so within a flat structure, failing to capture the inherent hierarchy of user goals. In contrast, HISIR's Hierarchical Intent Modeling Layer explicitly constructs a multi-level intent representation. Furthermore, its Multi-Expert Interest Extraction Layer effectively disentangles a user's diverse interests, leading to a more precise and robust user profile. The significant performance gains strongly support the benefits of modeling user interests and intents in a hierarchical and decoupled manner.

Table 2. Overall performance comparison on four datasets. Best results are in **bold**, and the second best are underlined. "Improv." denotes the relative improvement of HISIR over the best baseline.

Dataset	Beauty				Toys				Sports				Yelp			
Metric	HR@5	NDCG@5	HR@20	NDCG@20	HR@5	NDCG@5	HR@20	NDCG@20	HR@5	NDCG@5	HR@20	NDCG@20	HR@5	NDCG@5	HR@20	NDCG@20
BPR-MF	0.0212	0.0130	0.0589	0.0236	0.0120	0.0082	0.0312	0.0136	0.0141	0.0091	0.0323	0.0142	0.0127	0.0082	0.0346	0.0143
GRU4Rec	0.0111	0.0058	0.0478	0.0104	0.0097	0.0059	0.0301	0.0116	0.0162	0.0103	0.0421	0.0186	0.0152	0.0091	0.0371	0.0145
SASRec	0.0374	0.0241	0.0901	0.0387	0.0463	0.0306	0.0941	0.0441	0.0206	0.0135	0.0497	0.0216	0.0160	0.0101	0.0443	0.0179
BERT4Rec	0.0360	0.0216	0.0984	0.0391	0.0274	0.0174	0.0688	0.0291	0.0217	0.0143	0.0604	0.0251	0.0196	0.0121	0.0564	0.0223
CL4SRec	0.0423	0.0281	0.0994	0.0441	0.0526	0.0362	0.1038	0.0506	0.0217	0.0137	0.0540	0.0227	0.0229	0.0144	0.0630	0.0256
DSSRec	0.0410	0.0261	0.0914	0.0403	0.0502	0.0337	0.0975	0.0471	0.0214	0.0142	0.0495	0.0220	0.0171	0.0112	0.0464	0.0193
ICLRec	0.0495	0.0326	0.1072	0.0491	0.0586	0.0397	0.1130	0.0550	0.0263	0.0173	0.0630	0.0276	0.0233	0.0146	0.0645	0.0261
ELCRec	0.0529	0.0355	0.1079	0.0509	0.0585	0.0403	0.1138	0.0560	0.0286	0.0185	0.0648	0.0286	0.0236	0.0150	0.0653	0.0266
ICMA	0.0568	0.0386	0.1155	0.0551	0.0683	0.0467	0.1242	0.0625	0.0306	0.0208	0.0663	0.0308	0.0245	0.0161	0.0684	0.0278
InDiRec	0.0686	0.0488	0.1303	0.0662	0.0747	0.0541	0.1339	0.0707	0.0378	0.0268	0.0787	0.0383	0.0261	0.0169	0.0705	0.0288
HISIR	**0.0722**	**0.0516**	**0.1349**	**0.0684**	**0.0780**	**0.0577**	**0.1402**	**0.0754**	**0.0400**	**0.0279**	**0.0798**	**0.0390**	**0.0287**	**0.0184**	**0.0750**	**0.0313**
Improvement	5.25%	5.74%	3.53%	3.32%	4.41%	6.65%	4.71%	6.64%	5.82%	4.10%	1.40%	1.83%	9.96%	8.87%	6.38%	8.68%

4.3 Ablation Study (RQ2)

To validate the contribution of each key component in HISIR, we conduct a comprehensive ablation study. We create several variants by individually removing the core architectural modules (w/o Multi-Expert, w/o Hierarchical Intent) and the auxiliary objective functions (w/o IMID-Loss, w/o IHCD-Loss). The results, presented in Table 3, reveal a clear and consistent hierarchy of importance.

Removing the core architectural modules causes the most significant performance degradation. The severe drop in the "w/o Multi-Expert" variant (e.g., a 26.8% decrease in HR@20 on Beauty) highlights the fundamental importance of disentangling diverse user interests, while the substantial decline in the "w/o Hierarchical Intent" variant confirms that our structured modeling is a primary driver of performance. Furthermore, ablating the objective functions also leads to a consistent performance drop. For instance, removing the IHCD-Loss causes a notable 12.3% drop in NDCG@20 on Sports, which demonstrates that our contrastive objective is vital for shaping a semantically coherent intent hierarchy.

In summary, this study validates that HISIR's superiority stems not from any single component, but from the powerful synergy between its disentangled architecture and the specialized objectives that govern it.

Table 3. Ablation study of key components on different datasets.

Model Variant	Beauty		Toys		Sports		Yelp	
Metric	HR@20	NDCG@20	HR@20	NDCG@20	HR@20	NDCG@20	HR@20	NDCG@20
HISIR (full model)	**0.1349**	**0.0684**	**0.1402**	**0.0754**	**0.0798**	**0.0390**	**0.0750**	**0.0313**
w/o Multi-Expert	0.0989	0.0500	0.1135	0.0569	0.0618	0.0283	0.0636	0.0262
w/o Hierarchical Intent	0.1015	0.0604	0.1302	0.0696	0.0710	0.0306	0.0653	0.0261
w/o IMID-Loss	0.1225	0.0650	0.1355	0.0711	0.0736	0.0336	0.0698	0.0281
w/o IHCD-Loss	0.1248	0.0673	0.1368	0.0723	0.0746	0.0342	0.0721	0.0293

4.4 Parameter Analysis (RQ3)

We analyze the sensitivity of HISIR to its key hyperparameters: the number of experts K and hierarchical intent layers L. The results are shown in Fig. 3.

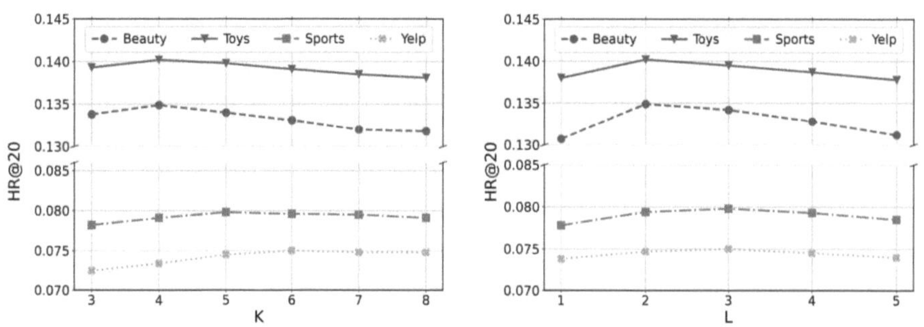

(a) Number of experts K. (b) Number of hierarchical intent layers L.

Fig. 3. Hyperparameter sensitivity analysis of HR@20 on all four datasets.

Impact of the Number of Experts (K). As shown in Fig. 3a, performance consistently improves with more experts before declining due to redundancy and overfitting. The model achieves peak performance on the Beauty and Toys datasets at $K = 4$, while the more diverse Sports and Yelp datasets show a flatter curve, peaking around $K = 5$ and $K = 6$. This pattern suggests the optimal number of experts correlates with the diversity of user behaviors in the dataset, confirming that multiple experts are crucial for disentangling interests.

Impact of the Number of Hierarchical Layers (L). Figure 3b strongly validates our core hypothesis: moving from a flat structure ($L = 1$) to a hierarchical one ($L \geq 2$) brings a significant performance improvement across all datasets. The optimal depth depends on the dataset's complexity, peaking at $L = 2$ for the more focused Beauty and Toys datasets. In contrast, the more diverse Sports and Yelp datasets benefit from a deeper hierarchy up to $L = 3$. This suggests that a two-level hierarchy is sufficient for focused domains, while a deeper structure is necessary to model the progression from concrete actions to abstract goals in more complex scenarios. Performance tends to decline when $L > 3$, likely due to increased model complexity.

Based on this comprehensive analysis, we select the optimal dataset-specific values for K and L in our experiments to ensure a fair and robust comparison.

4.5 Efficiency Analysis (RQ4)

We analyze the efficiency of HISIR by comparing its training time, inference time, and parameter count against strong baselines on the Beauty dataset. As shown in Table 4, HISIR demonstrates a compelling balance of efficiency and performance.

Notably, HISIR is the most lightweight model with only 0.99M parameters and achieves the fastest inference speed (17.5 s), making it highly suitable for online deployment. Its training time (265.3 s/epoch) is also competitive, being the most efficient among the compared models. This efficiency analysis confirms that HISIR's significant performance gains do not come at the cost of excessive computational resources, highlighting its practical value for real-world systems.

Table 4. Efficiency comparison on the Beauty dataset.

Model	Training Time (s/epoch)	Inference Time (s)	#Parameters (M)
ICMA	276.6	18.9	1.03
InDiRec	284.5	20.8	1.20
HISIR	265.3	15.5	0.99

4.6 Case Study: Visualization of Hierarchical Intents (RQ5)

To qualitatively assess the learned representations (RQ5), we visualize user intents from the Beauty test set using t-SNE. As shown in Fig. 4, each intent is colored by the category of the ground-truth next item.

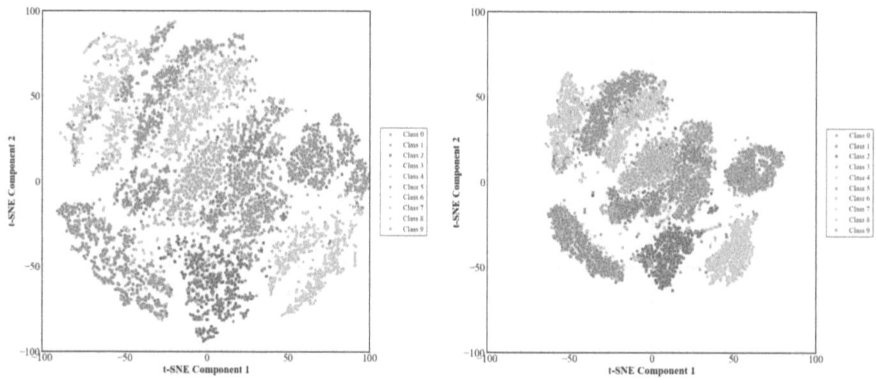

(a) Low-level Intent Representations (b) High-level Intent Representations

Fig. 4. t-SNE visualization of user intents on the Beauty dataset.

The visualizations reveal distinct, well-separated clusters that strongly align with item categories. This indicates that HISIR successfully captures the semantic purpose behind user behaviors, learning an interpretable intent space. Furthermore, a clear hierarchical structure is evident. The high-level intent clusters (Fig. 4b) are more compact and aggregated compared to the more granular low-level intents (Fig. 4a). This suggests the model effectively abstracts specific, fine-grained intents into broader, category-level ones.

In conclusion, this study provides strong qualitative evidence that HISIR learns meaningful and hierarchically organized user intents.

5 Conclusion

In this paper, we proposed HISIR, a novel End2End Hierarchical Intent-based Sequential Recommendation framework designed to capture the multifaceted and structured nature of user needs. By employing a Multi-Expert Interest Extraction Layer, HISIR effectively disentangles diverse user interests from behavior sequences. Subsequently, the Hierarchical Intent Modeling Layer maps these interests into a structured, multi-level intent representation. The independence and logical consistency of these representations are explicitly enforced by a Dual-Decoupling Loss Optimization Layer. Extensive experiments on four public benchmarks demonstrate that HISIR significantly outperforms state-of-the-art baselines, yielding improvements of up to 9.96% in HR@5. Ablation studies further validate the indispensable contribution of each proposed component. For future work, exploring the application of our hierarchical intent modeling approach in cross-domain recommendation scenarios presents a promising direction.

Acknowledgement. This work was supported in part by NSFC under Grant No. U23A20326, and in part by the Research and Development Program of Zhejiang under Grant No. 2026C02A1235.

References

1. Belghazi, M.I., et al.: Mutual information neural estimation. In: Proceedings of ICML, pp. 531–540 (2018)
2. Cen, Y., Zhang, J., Zou, X., Zhou, C., Yang, H., Tang, J.: Controllable multi-interest framework for recommendation. In: Proceedings of KDD, pp. 2942–2951 (2020)
3. Chen, T., Kornblith, S., Norouzi, M., Hinton, G.: A simple framework for contrastive learning of visual representations. In: Proceedings of ICML, pp. 1597–1607 (2020)
4. Chen, Y., Liu, Z., Li, J., McAuley, J., Xiong, C.: Intent contrastive learning for sequential recommendation. In: Proceedings of WWW, pp. 2172–2182 (2022)
5. He, R., McAuley, J.: Fusing similarity models with Markov chains for sparse sequential recommendation. In: Proceedings of ICDM, pp. 191–200 (2016)
6. Hidasi, B., Karatzoglou, A., Baltrunas, L., Tikk, D.: Session-based recommendations with recurrent neural networks. In: Proceedings of ICLR (2016)
7. Jégou, H., Douze, M., Schmid, C.: Product quantization for nearest neighbor search. IEEE Trans. Pattern Anal. Mach. Intell. **33**(1), 117–128 (2011)
8. Kang, W.C., McAuley, J.: Self-attentive sequential recommendation. In: Proceedings of ICDM, pp. 197–206 (2018)
9. Kingma, D.P., Welling, M.: Auto-encoding variational bayes. In: Proceedings of ICLR (2014)
10. Li, X., et al.: Multi-intention oriented contrastive learning for sequential recommendation. In: Proceedings of WSDM, pp. 411–419 (2023)
11. Lin, G., Luo, J., Li, Y., Gao, C., Luo, Q., Jin, D.: Iterative sparse attention for long-sequence recommendation. In: Proceedings of AAAI, vol. 39, pp. 12147–12155 (2025)
12. Liu, Y., et al.: End-to-end learnable clustering for intent learning in recommendation. In: Proceedings of NeurIPS, vol. 37, pp. 5913–5949 (2024)
13. Ma, J., Zhou, C., Yang, H., Cui, P., Wang, X., Zhu, W.: Disentangled self-supervision in sequential recommenders. In: Proceedings of KDD, pp. 483–491 (2020)
14. Mao, W., Liu, S., Liu, H., Liu, H., Li, X., Hu, L.: Distinguished quantized guidance for diffusion-based sequence recommendation. In: Proceedings of WWW (2025)
15. McAuley, J., Targett, C., Shi, Q., Van Den Hengel, A.: Image-based recommendations on styles and substitutes. In: Proceedings of SIGIR, pp. 43–52 (2015)
16. Pei, B., et al.: Intent contrastive learning based on multi-view augmentation for sequential recommendation. In: Proceedings of COLING, pp. 3300–3309 (2025)

17. Qin, X., Yuan, H., Zhao, P., Liu, G., Zhuang, F., Sheng, V.S.: Intent contrastive learning with cross subsequences for sequential recommendation. In: Proceedings of WSDM, pp. 548–556 (2024)
18. Qu, Y., Nobuhara, H.: Intent-aware diffusion with contrastive learning for sequential recommendation. In: Proceedings of SIGIR, pp. 1552–1561 (2025)
19. Rendle, S., Freudenthaler, C., Gantner, Z., Schmidt-Thieme, L.: BPR: Bayesian personalized ranking from implicit feedback. In: Proceedings of UAI (2009)
20. Shazeer, N., et al.: Outrageously large neural networks: the sparsely-gated mixture-of-experts layer. In: Proceedings of ICLR (2017)
21. Sun, F., et al.: Bert4rec: sequential recommendation with bidirectional encoder representations from transformer. In: Proceedings of CIKM, pp. 1441–1450 (2019)
22. Wang, S., Hu, L., Wang, Y., Cao, L., Sheng, Q.Z., Orgun, M.: Sequential recommender systems: challenges, progress and prospects. In: Proceedings of IJCAI, pp. 6332–6338 (2019)
23. Wang, W., et al.: Intent oriented contrastive learning for sequential recommendation. In: Proceedings of AAAI, vol. 39, pp. 12748–12756 (2025)
24. Wu, C., Wu, F., Huang, Y., Xie, X.: Personalized news recommendation: methods and challenges. ACM Trans. Inf. Syst. **41**(1), 1–50 (2023)
25. Xie, X., et al.: Contrastive learning for sequential recommendation. In: Proceedings of ICDE, pp. 1259–1273 (2022)
26. Zhang, S., Yao, L., Sun, A., Tay, Y.: Deep learning based recommender system: a survey and new perspectives. ACM Comput. Surv. **52**(1), 1–38 (2020)
27. Zhou, K., et al.: S3-rec: self-supervised learning for sequential recommendation with mutual information maximization. In: Proceedings of CIKM, pp. 1893–1902 (2020)
28. Zou, D., Wei, W., Zhu, F., Xu, C., Zhang, T., Huo, C.: Knowledge enhanced multi-intent transformer network for recommendation. In: Proceedings of WWW, pp. 1–9 (2024)

Low-Rank Guided Attention with Wavelet Augmentation for Sequential Recommendation

Hao Li[1], Mingxing Shao[1], Tiancheng Zhang[1(✉)], Minghe Yu[1], Xue Geng[2], and Ge Yu[1]

[1] School of Computer Science and Engineering, Northeastern University, Shenyang 110819, China
{2401854,2301935}@stu.neu.edu.cn, {tczhang,yuminghe,yuge}@mail.neu.edu.cn
[2] Institute for Infocomm Research, A*STAR, Singapore, Singapore
geng_xue@i2r.a-star.edu.sg

Abstract. Transformer-based deep learning models have achieved remarkable success in sequential recommendation. However, their attention mechanism still faces two inherent challenges: its high representational capacity tends to overfit noise; and it suffers from the oversmoothing problem, making it insufficiently sensitive to high-frequency signals in user behavior sequences. To address these issues, we propose LoWARec, which introduces structural priors into the Transformer through a low-rank attention to alleviate the impact of noise and captures multi-scale user behavior patterns in the frequency domain to mitigate the oversmoothing problem. Specifically, LoWARec first employs a low-rank attention to compress long interaction sequences into compact latent interests, thereby introducing structural priors. It then fuses these latent interests with the original sequences, seamlessly integrating low-rank representations into the Transformer to alleviate overfitting. Finally, it applies Fourier and wavelet transforms to extract low- and high-frequency components, thereby refining user preference modeling and mitigating the oversmoothing problem. Extensive experiments on five public datasets show that LoWARec consistently outperforms eight baselines, achieving up to 4.48% and 7.07% HR@5 improvement on Sports and LastFM, respectively. The code is released at https://github.com/juexuan/LoWARec.

Keywords: Sequential Recommendation · Attention Bias · Low-Rank Model

1 Introduction

As a subfield of recommendation systems, sequential recommendation plays a crucial role in applications such as e-commerce and video platforms [5,14,32]. The goal of sequential recommendation is to predict oracle items based on a

H. Jung et al. (Eds.): DASFAA 2026, LNCS 16535, pp. 240–257, 2026.
https://doi.org/10.1007/978-981-92-0363-5_15

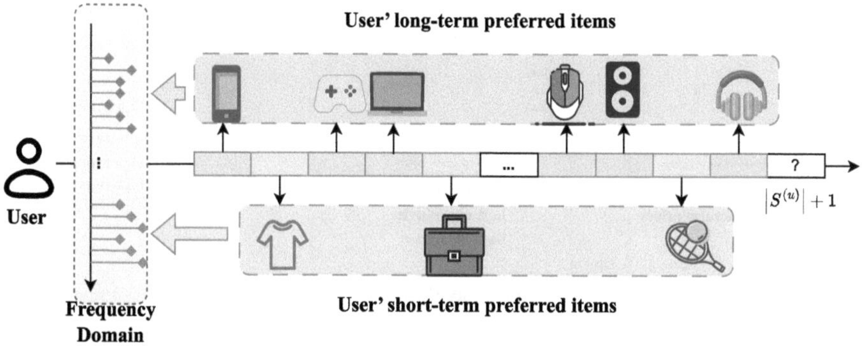

Fig. 1. Illustration of high- and low-frequency signals in sequential recommendation. User's long-term preferred items correspond to the low-frequency component, while short-term preferred items correspond to the high-frequency component.

user's historical interaction sequence [27,29]. User interaction data are inherently dynamic, making it essential for recommendation models to effectively capture both long-term and short-term user preferences for accurate predictions.

Transformer-based deep learning models have recently achieved considerable success in sequential recommendation tasks due to their powerful representation capabilities [11,23]. However, they still face two primary challenges. First, user behavior data often contains noise, as user interactions can be influenced by factors such as item popularity rather than their genuine interests [28,33]. The strong modeling capacity of Transformers can lead to overfitting to this noise. Second, these models tend to have an oversmoothing problem. This issue stems from the insufficient inductive bias for sequential modeling in self-attention architectures, which limits their ability to capture fine-grained temporal dependencies and avoid overfitting [2]. Additionally, the global dependency modeling inherent in self-attention exhibits a low-pass filtering nature, which tends to average out local variations and results in over-smoothed embeddings with less discriminative feature representations [21]. As shown in Fig. 1, high-frequency signals play an important role in sequential data by reflecting users' short-term interests. However, the oversmoothing problem smooths out high-frequency signals, leading to less accurate modeling of short-term preferences.

Another line of research focuses on low-rank models, which aim to extract low-dimensional latent interest representations from high-dimensional interaction sequences [4,20]. These representations are subsequently used to predict the item that the user is most likely to interact with next. This approach can be viewed as incorporating structural priors, which helps mitigate the influence of noisy data and is particularly effective for modeling long-term user preferences. However, due to their limited representational capacity, low-rank models often struggle to make accurate predictions for long-tail users [15], that is, those with sparse interaction histories.

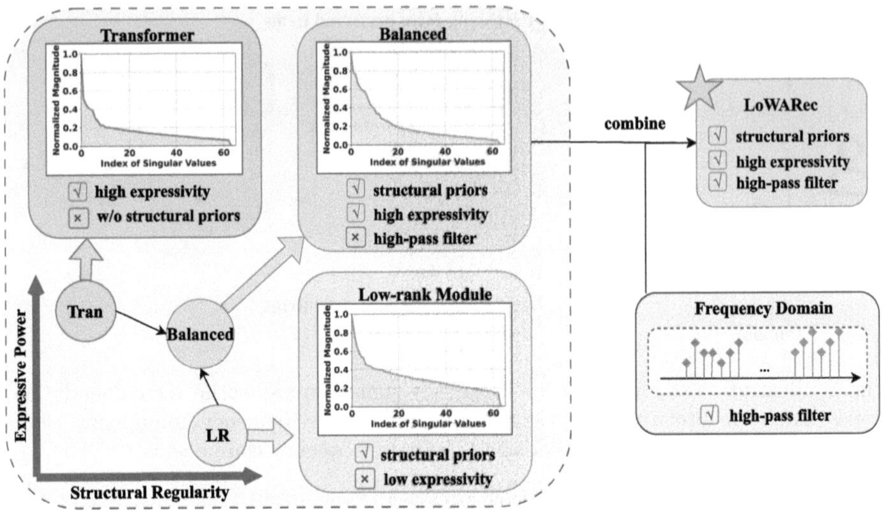

Fig. 2. Illustration of our motivations. The Transformer (the purple area) exhibits high expressivity but lacks structural priors, causing a rapid singular value decay where a few dominant components prevail, thereby leading to overfitting noise. In contrast, the low-rank module (the orange area) incorporates structural priors and yields a smoother singular value decay, but its representational capacity is limited. Combining both and introducing a high-pass filter in the frequency domain alleviates oversmoothing, enabling more accurate user interest modeling. (Color figure online)

As shown in Fig. 2, our goal is to combine the strengths of low-rank models and Transformer-based deep learning models by introducing structural priors while maintaining strong representation capability. Furthermore, we employ the wavelet transform to extract high-frequency components in the frequency domain to alleviate the over-smoothing problem.

To be more specific, **LoWARec** mainly consists of three components: Item-to-Interest Aggregation Attention, Interest-Context Interaction Attention, and Frequency-Aware Bias. First, Item-to-Interest Aggregation Attention compresses long interaction sequences into compact latent interests, introducing low-rank information that helps reduce noise and capture long-term preferences. Second, Interest-Context Interaction Attention fuses these latent interests with the original sequence, seamlessly integrating low-rank representations into the Transformer to enhance expressive power while preserving temporal dynamics. Finally, the Frequency-Aware Bias module extracts both low- and high-frequency components through Fourier and wavelet transforms, refining user preference modeling and alleviating the oversmoothing problem. This design mirrors a natural learning journey: it begins with establishing a foundational understanding, progresses to deeper and more complex modeling, and concludes with meticulous fine-tuning. The contributions of this work are as follows:

- We integrate low-rank modeling into Transformer architectures, introducing structural priors while preserving expressive capacity, which facilitates learning long-term user preferences and reduces the impact of noise.
- We incorporate frequency-domain modeling that uses the Fourier transform to capture global low-frequency components and the wavelet transform to extract localized high-frequency variations, thereby enhancing user preference modeling and mitigating the oversmoothing issue.
- Extensive experiments on public datasets demonstrate the effectiveness of our method compared to several representative baselines.

2 Related Work

Sequential recommendation uses the historical interaction sequences of users to predict the next likely item. Early studies primarily adopted Markov Chains [19] and KNN [7,9] to capture short-term user preferences. While these methods have shown effectiveness in modeling immediate interactions, they struggle to capture long-range dependencies. To address this limitation, researchers have introduced deep learning models such as CNNs [30] and RNNs [13] to better model sequential patterns and complex user behaviors. Such as GRU4Rec [8] models short session-based interactions using RNNs with a ranking loss tailored for recommendation. Caser [24] treats embedded item sequences as images and uses convolutional filters to capture user preferences.

More recently, inspired by the success of Transformers [25] in NLP, researchers have increasingly applied them to sequential recommendation. SASRec [11] is a representative model that uses self-attention to model user preferences. BERT4Rec [23] employs deep bidirectional self-attention with a masked prediction objective. DuoRec [18] addresses representation degeneration by applying contrastive regularization with model-level augmentation. Recently, the over-smoothing issue of Transformer-based models has attracted increasing attention in the research community. Methods such as FEARec [3] and BSARec [21] address this issue by applying the Fourier transform to separate signals into high- and low-frequency components in the frequency domain, followed by the inverse Fourier transform to reconstruct the processed signal in the original domain. Although these approaches have shown promising results, their reliance on global Fourier filtering is mismatched to the inherently local nature of high-frequency signals, which typically occur only within short time intervals. Consequently, although the Fourier transform followed by its inverse can capture the overall high-frequency components, it yields only global frequency characteristics and lacks timefrequency localization, thereby limiting its ability to accurately identify the temporal occurrence of these components in the signal.

Another line of research focuses on leveraging low-rank structures to reduce the computational complexity of Transformer [26]. Low-rank methods effectively reduce model complexity, improve generalization, and mitigate overfitting. In the context of sequential recommendation, LightSANs [4] is a representative low-rank self-attention model that addresses the high complexity and weak positional encoding of standard attention-based recommenders. FedSeqRec [12] leverages low-rank tensor projection to model long-term preferences in federated sequential recommendation. Although these methods effectively reduce model complexity, their limited expressiveness often leads to suboptimal performance.

3 Preliminaries

3.1 Problem Definition

In sequential recommendation, the objective is to predict the next item $v^{(u)}_{|S^u|}$ that a user u might be interested in based on their historical interaction sequence $\mathcal{S}^{(u)} = [v^{(u)}_1, v^{(u)}_2, ..., v^{(u)}_{|S^u|}]$. Specifically, the task is to select an item $v_j \in \mathcal{V}$ that maximizes the probability of being the user's next interaction, conditioned on their previous behavior.

$$\arg\max_{v_j \in \mathcal{V}} P(v^{(u)}_{|S^u|} = v_j \mid \mathcal{S}^{(u)}). \tag{1}$$

3.2 Discrete Wavelet Transform

User behavior sequences can be viewed as temporal signals. The Discrete Wavelet Transform (DWT) is a time-frequency analysis technique that decomposes a signal into components of different frequency bands while preserving temporal information. This makes it particularly suitable for analyzing non-stationary signals [16]. DWT enables the model to separate low-frequency trends from high-frequency local variations, thereby enhancing its representation capacity [31]. The Haar wavelet is the simplest orthogonal wavelet [6]. It defines the scaling and wavelet functions as:

$$\phi(t) = \begin{cases} 1, & 0 \leq t < 1, \\ 0, & \text{otherwise}, \end{cases} \qquad \psi(t) = \begin{cases} 1, & 0 \leq t < \frac{1}{2}, \\ -1, & \frac{1}{2} \leq t < 1, \\ 0, & \text{otherwise}. \end{cases} \tag{2}$$

In discrete form, DWT can be implemented using convolution followed by downsampling. For Haar wavelets, the associated low-pass and high-pass filters are:

$$h = \left[\frac{1}{\sqrt{2}}, \frac{1}{\sqrt{2}}\right], \quad g = \left[\frac{1}{\sqrt{2}}, -\frac{1}{\sqrt{2}}\right]. \tag{3}$$

Given a discrete signal $x = \{x_t\}_{t=1}^{T}$, the approximation and detail coefficients are computed by:

$$A = \text{downsample}(x * h), \quad D = \text{downsample}(x * g), \tag{4}$$

where $*$ denotes convolution and downsample reduces the sequence by a factor of 2. The approximation coefficients A retain the overall structure, while the detail coefficients D capture local variations. Together, they allow the signal to be accurately reconstructed.

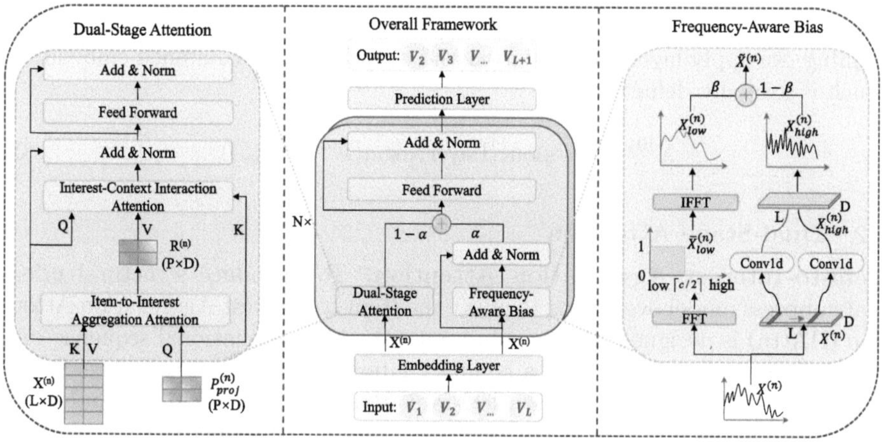

Fig. 3. The architecture of LoWARec.

4 Proposed Method

To address the overfitting and oversmoothing issues inherent in Transformer-based recommenders, we propose LoWARec, which introduces structural priors through low-rank attention and leverages Fourierwavelet decomposition to incorporate attention bias into the Transformer, effectively balancing expressive power and regularization. The overall architecture of LoWARec is illustrated in Fig. 3. LoWARec primarily consists of two core components: the Dual-Stage Attention mechanism and the Frequency-Aware Bias module. The Dual-Stage Attention mechanism first employs Item-to-Interest Aggregation Attention to extract low-rank representations, followed by Interest-Context Interaction Attention to effectively integrate these low-rank features into the Transformer, thereby leveraging the complementary strengths of low-rank modeling and the Transformer. Meanwhile, the Frequency-Aware Bias module alleviates the oversmoothing problem by decomposing user-item interaction signals into high- and low-frequency components using Fourier and wavelet transforms, respectively, and incorporating them as adaptive bias terms to enhance representation learning.

4.1 Embedding Layer

First, the historical interaction sequence of a user is transformed into a fixed-length sequence $\mathcal{S}^{(u)} = [v_1^{(u)}, v_2^{(u)}, ..., v_L^{(u)}]$. Specifically, if the original sequence length exceeds L, it is truncated from the most recent interactions. If it is shorter than L, the remaining positions are padded with zeros. The final sequence $\mathcal{S}^{(\sqcap)}$ is then used to retrieve embeddings from the item embedding matrix $M \in \mathbb{R}^{|\mathcal{I}| \times D}$, resulting in the input embedding $E^u \in \mathbb{R}^{L \times D}$, where each row $E_j = M_{s_j}$.

Positional information is crucial for capturing the order of interactions and improving recommendation performance. Therefore, a learnable positional embedding matrix $P_{\text{pos}} \in \mathbb{R}^{L \times D}$ is added to the input embedding. To stabilize training, we apply layer normalization and dropout to the combined embedding, which is formally defined as:

$$X^{(0)} = \text{Dropout}(\text{LayerNorm}(E^u + P_{\text{pos}})). \tag{5}$$

4.2 Dual-Stage Attention

Item-to-Interest Aggregation Attention. To introduce structural priors that suppress model overfitting to noise, the Item-to-Interest Aggregation Attention (IIAttn) is designed. IIAttn compresses the high-dimensional sequence representation $X^{(n)} \in \mathbb{R}^{L \times D}$ into a compact latent interest matrix $R^{(n)} \in \mathbb{R}^{P \times D}$ with $P < L$. By introducing the low-rank structure, this module emphasizes long-term interest and effectively alleviates noise in the sequence.

To be more specific, we introduce a learnable projection matrix $P_{\text{proj}}^{(n)} \in \mathbb{R}^{P \times D}$, which serves as a query matrix to handle the input sequence $X^{(n)}$. The attention mechanism is defined as follows:

$$\begin{aligned} R^{(n)} &= \text{SelfAtten}(P_{\text{proj}}^{(n)}, X^{(n)}, X^{(n)}) \\ &= \text{Softmax}\left(\frac{P_{\text{proj}}^{(n)} W_Q (X^{(n)} W_K)^T}{\sqrt{D'}}\right) X^{(n)} W_V. \end{aligned} \tag{6}$$

Here, $R^{(n)} \in \mathbb{R}^{P \times D}$ denotes the latent interest matrix. $W_Q, W_K, W_V \in \mathbb{R}^{D \times D}$ are learnable linear weights. The matrix $P_{\text{proj}}^{(n)} \in \mathbb{R}^{P \times D}$ acts as a compact query matrix to summarize the sequence representations in $X^{(n)}$, facilitating the extraction of diverse latent interests. It is worth noting that, unlike the Item-to-Interest Aggregation (IIAggr) in LightSANs, which performs a direct linear projection to map items into latent interests, the IIAttn incorporates attention-based interactions to capture dynamic dependencies within the latent interest space.

Interest-Context Interaction Attention. The Interest-Context Interaction Attention (ICAttn) module integrates latent interest representations into the Transformer framework, introducing low-rank information while preserving the model's expressive capacity. Specifically, we treat the original sequence embedding $X^{(n)} \in \mathbb{R}^{L \times D}$ as the query, the projection matrix $P_{\text{proj}}^{(n)} \in \mathbb{R}^{P \times D}$ as the key, and the latent interest matrix $R^{(n)}$ as the value in a cross-attention mechanism:

$$\begin{aligned} S^{(n)} &= \text{SelfAtten}(X^{(n)}, P_{\text{proj}}^{(n)}, R^{(n)}) \\ &= \text{Softmax}\left(\frac{X^{(n)} W_Q (P_{\text{proj}}^{(n)} W_K)^T}{\sqrt{D'}}\right) R^{(n)} W_V. \end{aligned} \tag{7}$$

ICAttn enables each item embedding to attend to the most relevant latent interests, bridging detailed sequence representations with compact latent interests. Then, to enhance model stability and enable deeper representation learning, we follow the standard Transformer encoder architecture. A residual connection and layer normalization are applied to the interaction output $S^{(n)}$:

$$\hat{S}^{(n)} = \text{LayerNorm}(S^{(n)} + X^{(n)}). \tag{8}$$

Then, a feed-forward network with a residual layer yields the output of the module:

$$\hat{Y}^{(n)} = \text{LayerNorm}(\hat{S}^{(n)} + \text{FeedForward}(\hat{S}^{(n)})). \tag{9}$$

4.3 Frequency-Aware Bias

To alleviate the oversmoothing problem, we design Frequency-Aware Bias (FAB) module, which integrates Fourier-based low-frequency extraction with wavelet-based high-frequency extraction. Unlike traditional approaches that rely on Fourier transforms to extract high-frequency components, FAB employs the Haar wavelet for high-frequency analysis due to its superior time-frequency locality, which enables more precise detection of local discontinuities and abrupt variations in sequences [10].

To reduce computational complexity while preserving the ability to capture high-frequency details, we employ a depthwise separable 1D convolution [1,22] to approximate the high-pass filtering operation of the Haar wavelet transform. This operation effectively performs local differencing, allowing the model to focus on short-term variations. Formally, given a sequence representation $X^{(n)} \in \mathbb{R}^{L \times D}$, we apply a 1D convolution independently to each feature channel to extract its local high-frequency components:

$$X_{\text{high}}^{(n)} = \text{Conv}_{\text{wavelet}}(X^{(n)}). \tag{10}$$

The convolution kernel is initialized as $K_h = [1, -1]$, which serves as a simple first-order difference operator, effective for identifying local trends and abrupt changes within the sequence. Notably, each channel utilizes an independent kernel, and the convolution weights are learnable and differentiable, enabling the model to adaptively extract distinct high-frequency features across channels.

Following previous work [21], we utilize the Discrete Fourier transform(DFT) and the Inverse Discrete Fourier transform(IDFT) to obtain low-frequency information. More specifically, we leverage the DFT to project input sequences into the frequency domain, enabling explicit separation of low-frequency (long-term interest) and high-frequency (short-term preference) components. Specifically, The frequency-domain representation is computed as $\bar{X}^{(n)} = F(X^{(n)})$, where F is the DFT. The low-frequency component $\bar{X}_{\text{low}}^{(n)}$ is extracted by selecting the $\lceil c/2 \rceil$ lowest-magnitude frequency coefficients, and is then transformed back into the time domain using IDFT:

$$X_{\text{low}}^{(n)} = F^{-1}(\bar{X}_{\text{low}}^{(n)}). \tag{11}$$

Then, to balance long-term and short-term user preferences, we fuse the extracted low- and high-frequency components via a learnable parameter β:

$$\tilde{X}^{(n)} = \beta X_{\text{low}}^{(n)} + (1-\beta) X_{\text{high}}^{(n)}. \tag{12}$$

Finally, to preserve the original semantic information while incorporating frequency-enhanced features, we apply dropout, residual connection, and layer normalization to the fused representation $\tilde{X}^n$ as follows:

$$\hat{X}^{(n)} = \text{LayerNorm}\left(X^{(n)} + \text{Dropout}(\tilde{X}^{(n)})\right). \tag{13}$$

4.4 Post-processing Module

Inspired by previous work [21], we integrate frequency-enhanced representation $\hat{X}^{(n)}$ with the output of the self-attention module $\hat{Y}^{(n)}$ to alleviate the over-smoothing issue commonly observed in Transformer models. Specifically, the final representation of the n-th layer is obtained by linearly interpolating these two components:

$$Z^{(n)} = \alpha \hat{X}^{(n)} + (1-\alpha)\hat{Y}^{(n)}, \tag{14}$$

where $\alpha \in [0, 1]$ is a hyperparameter that controls the trade-off between the frequency rescaler layer and the attention-derived features.

The fused output $Z^{(n)}$ is then processed by a feed-forward network, followed by a residual connection and layer normalization to produce the input for the next layer:

$$X^{(n+1)} = \text{LayerNorm}(Z^{(n)} + \text{FeedForward}(Z^{(n)})). \tag{15}$$

4.5 Prediction Layer

After processing through N layers, the final representation $X^{(N)}$ is obtained. The user's preference score for a candidate item v_j is computed via the inner product:

$$\Pr(v_j \mid S^{(u)}) = \langle X^{(N)}, E_j \rangle. \tag{16}$$

LoWARec is trained using a cross-entropy loss function:

$$\mathcal{L} = -\log\left(\frac{\exp(\langle X^{(N)}, E_g \rangle)}{\sum_{j=1}^{|\mathcal{I}|} \exp(\langle X^{(N)}, E_j \rangle)}\right), \tag{17}$$

where g denotes the ground truth item, and $|\mathcal{I}|$ is the total number of items.

5 Experiments

We evaluate the performance of our proposed model on five real-world datasets against eight competitive baselines. Overall, our experiments are designed to answer the following research questions:

- **RQ1:** How does LoWARec perform overall, and how well does it handle both short and long user behavior sequences?
- **RQ2:** Are the individual components of our proposed model effective?
- **RQ3:** How do hyperparameters influence the performance of LoWARec?
- **RQ4:** How do the low-rank properties and frequency spectrum modeling capabilities of LoWARec contribute to its performance improvement?

5.1 Experiments Settings

Datasets. To evaluate the effectiveness of our proposed model, we conduct experiments on five datasets spanning different domains. Specifically, the *Beauty* and *Sports* datasets are derived from the Amazon datasets [17], while the other datasets include the Yelp business review dataset, the ML-1M movie rating dataset, and the LastFM music dataset. We follow the data preprocessing procedures described in [34,35], and the statistics of the processed datasets are summarized in Table 1.

Evaluation Metrics. We evaluate model performance using two widely adopted metrics: Hit Rate (HR@K) and Normalized Discounted Cumulative Gain (NDCG@K), where $K \in \{5, 10, 20\}$. HR@K reflects the model's ability to retrieve relevant items within the top-K recommendations, while NDCG@K further assesses the ranking quality by considering the positions of the correctly recommended items.

Table 1. Statistics of the processed datasets.

Dataset Name	# Users	# Items	# Interactions	Avg.Len	Sparsity
Sports	25,598	18,357	296,337	8.3	99.95%
Beauty	22,363	12,101	198,502	8.9	99.93%
Yelp	30,431	20,033	316,354	10.4	99.95%
LastFM	1,090	3,646	52,551	48.2	98.68%
ML-1M	6,041	3,417	999,611	165.5	95.16%

Baselines. To validate the effectiveness of our proposed model, we compare it with four categories of baseline methods:

- **RNN/CNN-based sequential models**: GRU4Rec [8] and Caser [24].
- **Transformer-based sequential models**: SASRec [11], BERT4Rec [23], and DuoRec [18].
- **Transformer-based models enhanced with frequency adjustment techniques**: FEARec [3] and BSARec [21].
- **Low-rank attention based sequential models**: LightSANs [4].

Implementation Details. All experiments are implemented using PyTorch and conducted on an NVIDIA RTX A4000 GPU with 16 GB of memory. Hyperparameter tuning is performed based on the recommendations from the baselines. For our proposed model, the following hyperparameters are used: the coefficient α is selected from $\{0.1, 0.3, 0.5, 0.7, 0.9\}$, c is chosen from $\{1, 3, 5, 7, 9\}$, and the projection dimension P is selected from $\{2, 4, 6, 8, 10\}$. The number of layers N is set to 2, and the number of attention heads is selected from $\{1, 2, 4\}$. The hidden layer dimension D is set to 64, the maximum sequence length N is set to 50, and the batch size is set to 256. The learning rate is selected from the range $\{3 \times 10^{-4}, 10^{-3}\}$. For the dropout rate, a value of 0.2 is used for the ML-1M dataset, and 0.5 for all other datasets.

5.2 Overall Performance (RQ1)

Overall Performance. Table 2 presents the performance of LoWARec and the baselines. Overall, LoWARec achieves the best performance across all datasets. Specifically, RNN- and CNN-based models show weaker performance compared to Transformer-based models. The LightSANs model, which utilizes low-rank module, significantly outperforms simpler Transformer models like SASRec and BERT4Rec on datasets with longer average sequence lengths. This demonstrates the advantage of low-rank module in incorporating structural priors and modeling long-term interests. FEARec and BSARec outperform standard Transformer models, indicating the importance of frequency-domain adjustments. Additionally, LoWARec shows a greater improvement in HR compared to NDCG.

Performance on Long and Short Sequences. To evaluate model performance on sequences of varying lengths, we conduct experiments on the ML-1M dataset with a maximum sequence length $N = 200$. We define *short* and *long* sequences based on the 15th and 85th percentiles of sequence lengths, respectively. The learning rate is set to 0.003, and the batch size is fixed at 1024. The results are presented in Fig. 4. We observe that models tend to perform worse on long sequences compared to short ones, which suggests that the rich information in long sequences may lead to overfitting. The low-rank-based model LightSANs outperforms the traditional Transformer-based model SASRec, especially in the long-sequence scenario, demonstrating the advantage of low-rank structures in capturing users' long-term preferences. Overall, our proposed model achieves consistently competitive performance on both short and long sequences.

Table 2. Overall performance comparison on the five datasets. The best performing method in each row is boldfaced, and the second best method in each row is underlined. Improvements are shown in the last column.

Dataset	Metric	Caser	GRU4Rec	SASRec	BERT4Rec	LightSans	DuoRec	FEARec	BSARec	LoWARec	Improv.
Beauty	HR@5	0.0118	0.0174	0.0346	0.0419	0.0579	0.0690	0.0700	0.0712	**0.0715**	0.42%
	HR@10	0.0212	0.0302	0.0549	0.0662	0.0742	0.0956	0.0969	0.0976	**0.0995**	1.95%
	HR@20	0.0358	0.0467	0.0855	0.0991	0.0980	0.1298	0.1327	0.1337	**0.1359**	1.65%
	NDCG@5	0.0070	0.0105	0.0226	0.0278	0.0430	0.0492	0.0500	0.0512	**0.0514**	0.39%
	NDCG@10	0.0100	0.0147	0.0291	0.0356	0.0482	0.0577	0.0587	0.0597	**0.0604**	1.17%
	NDCG@20	0.0136	0.0188	0.0368	0.0439	0.0542	0.0663	0.0676	0.0688	**0.0696**	1.16%
ML-1M	HR@5	0.0839	0.1063	0.1394	0.1530	0.1914	0.2083	0.2061	0.2101	**0.2180**	3.76%
	HR@10	0.1387	0.1745	0.2146	0.2321	0.2737	0.2932	0.2917	0.2945	**0.3131**	6.32%
	HR@20	0.2222	0.2707	0.3157	0.3373	0.3675	0.4013	0.3897	0.4106	**0.4171**	1.58%
	NDCG@5	0.0520	0.0667	0.0908	0.1004	0.1318	0.1434	0.1408	0.1436	**0.1486**	3.48%
	NDCG@10	0.0695	0.0887	0.1150	0.1258	0.1583	0.1709	0.1684	0.1708	**0.1792**	4.92%
	NDCG@20	0.0905	0.1129	0.1405	0.1524	0.1820	0.1982	0.1931	0.2000	**0.2055**	2.75%
Yelp	HR@5	0.0119	0.0130	0.0146	0.0246	0.0168	0.0260	0.0269	0.0265	**0.0279**	5.28%
	HR@10	0.0206	0.0226	0.0243	0.0425	0.0278	0.0431	0.0436	0.0439	**0.0462**	5.24%
	HR@20	0.0325	0.0391	0.0398	0.0688	0.0475	0.0706	0.0678	0.0724	**0.0754**	4.14%
	NDCG@5	0.0073	0.0078	0.0094	0.0154	0.0104	0.0167	0.0168	0.0167	**0.0174**	4.19%
	NDCG@10	0.0100	0.0109	0.0125	0.0211	0.0139	0.0222	0.0222	0.0223	**0.0233**	4.48%
	NDCG@20	0.0130	0.0150	0.0164	0.0277	0.0188	0.0291	0.0283	0.0294	**0.0306**	4.08%
LastFM	HR@5	0.0229	0.0248	0.0450	0.0312	0.0339	0.0431	0.0413	0.0523	**0.0560**	7.07%
	HR@10	0.0358	0.0376	0.0633	0.0459	0.0495	0.0596	0.0578	0.0716	**0.0881**	23.04%
	HR@20	0.0560	0.0606	0.0881	0.0752	0.0734	0.0881	0.0872	0.1083	**0.1229**	13.48%
	NDCG@5	0.0176	0.0165	0.0320	0.0188	0.0235	0.0307	0.0303	0.0358	**0.0362**	1.04%
	NDCG@10	0.0216	0.0207	0.0379	0.0234	0.0285	0.0360	0.0356	0.0443	**0.0465**	4.97%
	NDCG@20	0.0268	0.0265	0.0440	0.0308	0.0345	0.0432	0.0431	0.0536	**0.0552**	2.99%
Sports	HR@5	0.0085	0.0120	0.0195	0.0262	0.0298	0.0390	0.0398	0.0424	**0.0443**	4.48%
	HR@10	0.0138	0.0195	0.0314	0.0410	0.0389	0.0553	0.0575	0.0596	**0.0625**	4.87%
	HR@20	0.0222	0.0318	0.0467	0.0608	0.0517	0.0784	0.0811	0.0842	**0.0880**	4.51%
	NDCG@5	0.0054	0.0078	0.0132	0.0174	0.0221	0.0276	0.0278	0.0299	**0.0304**	1.67%
	NDCG@10	0.0071	0.0102	0.0170	0.0221	0.0250	0.0328	0.0335	0.0354	**0.0362**	2.26%
	NDCG@20	0.0092	0.0133	0.0208	0.0271	0.0282	0.0386	0.0394	0.0416	**0.0426**	2.40%

5.3 Ablation Studies (RQ2)

To verify the effectiveness of each component in LoWARec, we conducted ablation studies on the ML-1M and Yelp dataset, as shown in Table 3. Specifically, we individually removed the Item-to-Interest Aggregation Attention (IIAttn) and Interest-Context Interaction Attention (ICAttn) modules. The results show that the performance of LoWARec degrades in both cases, with a more significant decline observed when IIAttn is removed. In addition, by adjusting the parameter α in Eq. (14), we separately eliminated the Dual-Stage Attention (DSA) module and Frequency-Aware Bias (FAB) module, both of which led to performance drops. Furthermore, replacing the IIAttn module with the IIAggr in LightSANs, substituting the wavelet transform with a Fourier transform, or replacing the DSA module with a multihead attention layer all led to performance drops, confirming the effectiveness of the proposed architectural components.

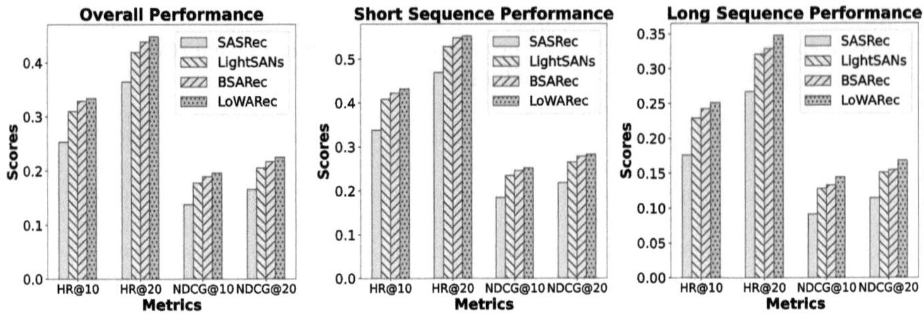

Fig. 4. Performance comparison on short and long sequences.

Table 3. Ablation Study on ML-1M and Yelp.

Dataset	Metric	LoWARec	*w/o* IIAttn	*w/o* ICAttn	*w/o* DSA	*w/o* FAB	IIAttn→IIAggr	Wavelet→FFT	DSA→Multi-Head Attn
ML-1M	HR@20	**0.4171**	0.3985	0.4124	0.4030	0.4045	0.4076	0.4078	0.4139
	NDCG@20	**0.2055**	0.1977	0.2021	0.1934	0.1973	0.2001	0.2024	0.2046
Yelp	HR@20	**0.0754**	0.0734	0.0740	0.0727	0.0674	0.0730	0.0707	0.0743
	NDCG@20	**0.0306**	0.0297	0.0300	0.0298	0.0270	0.0291	0.0282	0.0305

5.4 Impact of Hyperparameter (RQ3)

We conducted a comprehensive analysis of LoWARec's hyperparameters on the ML-1M and Sports datasets. As shown in Fig. 5, the top and bottom rows report results on ML-1M and Sports, respectively, with columns corresponding to α, c, and P. LoWARec shows clear sensitivity to α: on the relatively sparse Sports dataset, higher α values lead to better performance, whereas on the denser ML-1M dataset, the best accuracy is achieved when $\alpha = 0.3$. These findings highlight the importance of frequency-domain adjustment, particularly for sparse data. For the hyperparameter c, LoWARec achieves optimal performance when $c = 5$ on Sports and $c = 9$ on ML-1M, reflecting dataset-dependent characteristics. Regarding the projection dimension P, performance first increases and then slightly declines as P grows, suggesting that a too-small dimension fails to capture sufficient user representations, while an overly large one introduces redundant features and weakens the model's generalization ability.

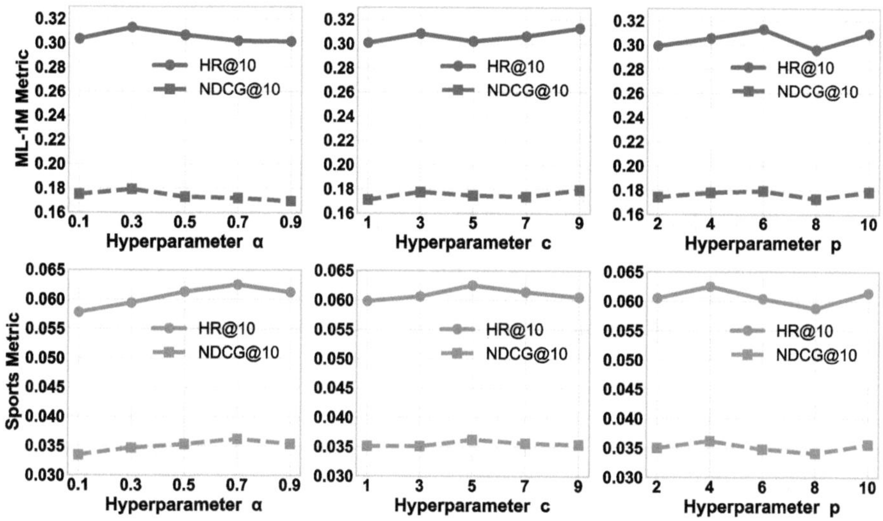

Fig. 5. Impact of Hyper-parameter.

5.5 Visualization (RQ4)

Figure 6 presents the visualization results of singular value decomposition (SVD) and frequency spectrum analysis for different models.

In the singular value spectrum analysis, LightSANs, as a representative low-rank model, exhibits a relatively slow singular value decay throughout the training process, indicating a more balanced energy distribution across embedding dimensions. This slow decay reflects the structural regularization imposed by its low-rank design, which helps maintain stable representations but restricts the model's ability to capture complex and dynamic user interests. In contrast, Transformer-based models such as SASRec show a much faster singular value decay as training proceeds, suggesting that a few dominant components prevail, leading to higher expressivity but increased risk of overfitting.

Figure 6(a) shows the singular value distributions when LoWARec and BSARec have their frequency adjustment modules removed (without FreqAdjust), while the other models remain in the original configurations. It can be observed that LoWARec presents a moderate singular value decay, lying between the rapid decay of deep models and the flat spectrum of low-rank models. This indicates that the Dual-Stage Attention effectively balances structural regularization and expressive capacity, enabling LoWARec to capture diverse user interests without suffering from information redundancy.

Figure 6(b) illustrates the singular value distributions with the frequency adjustment module retained. The introduction of frequency adjustment improves the singular value spectra of both LoWARec and BSARec, indicating that the frequency adjustment module contributes to capturing critical information across different frequency bands. Notably, under the same settings, LoWARec retains

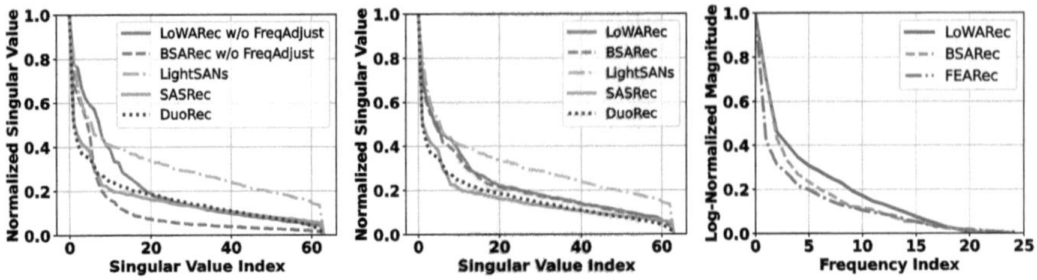

(a) (b) (c)

Fig. 6. Low-Rank and Frequency Spectrum Analysis of Models on the LastFM Dataset.

higher singular values and exhibits a slower decay compared to BSARec, suggesting that LoWARec can preserve more potentially useful information and has stronger representational capacity.

Figure 6(c) presents the log-normalized frequency spectra of different models. Compared with BSARec and FEARec, LoWARec retains higher spectral energy not only in the low-frequency region but also across the high-frequency range. This suggests that LoWARec preserves both the global temporal structures and finer-grained variations, enabling it to capture richer sequential patterns. The relatively slower spectral decay of LoWARec further indicates its enhanced ability to model detailed signal components while maintaining robustness against noise, contributing to its superior performance.

6 Conclusion

Based on existing approaches, this paper further optimizes Transformer-based architectures for sequential recommendation. While Transformers possess strong representational capacity, they are prone to overfitting noise in the data and often suffer from over-smoothing issues. To address these challenges, we propose LoWARec, a novel model that combines the strengths of low-rank models and the Transformer to enhance generalization. Moreover, it integrates both Fourier and wavelet transforms to extract low- and high-frequency information, effectively mitigating the oversmoothing problem. Experimental results on five benchmark datasets demonstrate the effectiveness of LoWARec. In future work, we plan to further explore the integration of low-rank modeling with deep architectures, as well as the potential of frequency-domain enhancements.

Acknowledgments. This research was funded by the National Natural Science Foundation of China (Nos. 62272093, 62137001).

References

1. Chollet, F.: Xception: deep learning with depthwise separable convolutions. In: IEEE Conference on Computer Vision and Pattern Recognition, pp. 1800–1807 (2017)
2. Dosovitskiy, A., et al.: An image is worth 16x16 words: transformers for image recognition at scale. In: Proceedings of the 9th International Conference on Learning Representations (2021)
3. Du, X., et al.: Frequency enhanced hybrid attention network for sequential recommendation. In: Proceedings of the 46th International ACM SIGIR Conference on Research and Development in Information Retrieval, pp. 78–88 (2023)
4. Fan, X., Liu, Z., Lian, J., Zhao, W.X., Xie, X., Wen, J.R.: Lighter and better: low-rank decomposed self-attention networks for next-item recommendation. In: Proceedings of the 44th International ACM SIGIR Conference on Research and Development in Information Retrieval, pp. 1733–1737 (2021)
5. Guo, S., Zhang, S., Sun, W., Ren, P., Chen, Z., Ren, Z.: Towards explainable conversational recommender systems. In: Proceedings of the 46th International ACM SIGIR Conference on Research and Development in Information Retrieval, pp. 2786–2795 (2023)
6. Haar, A.: Zur theorie der orthogonalen funktionensysteme - erste mitteilung. Math. Ann. **69**(3), 331–371 (1910)
7. He, R., McAuley, J.J.: Fusing similarity models with Markov chains for sparse sequential recommendation. In: IEEE International Conference on Data Mining, pp. 191–200 (2016)
8. Hidasi, B., Karatzoglou, A., Baltrunas, L., Tikk, D.: Session-based recommendations with recurrent neural networks. In: International Conference on Learning Representations (2016)
9. Hu, H., He, X., Gao, J., Zhang, Z.L.: Modeling personalized item frequency information for next-basket recommendation. In: Proceedings of the 43rd International ACM SIGIR Conference on Research and Development in Information Retrieval, pp. 1071–1080 (2020)
10. Huang, Y., et al.: Freqwave-trandud: a multivariate time series anomaly detection method based on wavelet and fourier transforms. IEEE Access **13**, 68384–68397 (2025)
11. Kang, W.C., McAuley, J.: Self-attentive sequential recommendation. In: IEEE International Conference on Data Mining, pp. 197–206 (2018)
12. Li, L., Lin, F., Xiahou, J., Lin, Y., Wu, P., Liu, Y.: Federated low-rank tensor projections for sequential recommendation. Knowl.-Based Syst. **255**, 109483 (2022)
13. Liu, Q., Wu, S., Wang, D., Li, Z., Wang, L.: Context-aware sequential recommendation. In: IEEE International Conference on Data Mining, pp. 1053–1058 (2016)
14. Liu, Q., et al.: LLMEmb: large language model can be a good embedding generator for sequential recommendation. In: Proceedings of the AAAI Conference on Artificial Intelligence, vol. 39, no. 11, pp. 12183–12191 (2025)
15. Liu, Q., et al.: LLM-ESR: large language models enhancement for long-tailed sequential recommendation. In: Advances in Neural Information Processing Systems (2024)
16. Ma, H., Liu, D., Xiong, R., Wu, F.: iWave: CNN-based wavelet-like transform for image compression. IEEE Trans. Multimedia **22**(7), 1667–1679 (2020)

17. Ni, J., Li, J., McAuley, J.: Justifying recommendations using distantly-labeled reviews and fine-grained aspects. In: Proceedings of the Conference on Empirical Methods in Natural Language Processing and the 9th International Joint Conference on Natural Language Processing, pp. 188–197 (2019)
18. Qiu, R., Huang, Z., Yin, H., Wang, Z.: Contrastive learning for representation degeneration problem in sequential recommendation. In: Proceedings of the 15th ACM International Conference on Web Search and Data Mining, pp. 813–823 (2022)
19. Rendle, S., Freudenthaler, C., Schmidt-Thieme, L.: Factorizing personalized Markov chains for next-basket recommendation. In: Proceedings of the 19th International Conference on World Wide Web, pp. 811–820 (2010)
20. Sedhain, S., Menon, A., Sanner, S., Xie, L., Braziunas, D.: Low-rank linear cold-start recommendation from social data. In: Proceedings of the AAAI Conference on Artificial Intelligence, vol. 31, no. 1 (2017)
21. Shin, Y., Choi, J., Wi, H., Park, N.: An attentive inductive bias for sequential recommendation beyond the self-attention. In: Proceedings of the AAAI Conference on Artificial Intelligence, pp. 8984–8992 (2024)
22. Sifre, L., Mallat, S.: Rigid-Motion Scattering for Texture Classification. arXiv preprint arXiv:1403.1687 (2014)
23. Sun, F., et al.: BERT4Rec: sequential recommendation with bidirectional encoder representations from transformer. In: Proceedings of the 28th ACM International Conference on Information and Knowledge Management, pp. 1441–1450 (2019)
24. Tang, J., Wang, K.: Personalized top-N sequential recommendation via convolutional sequence embedding. In: Proceedings of the Eleventh ACM International Conference on Web Search and Data Mining, pp. 565–573 (2018)
25. Vaswani, A., et al.: Attention is all you need. In: Advances in Neural Information Processing Systems, pp. 5998–6008 (2017)
26. Wang, S., Li, B.Z., Khabsa, M., Fang, H., Ma, H.: Linformer: Self-Attention with Linear Complexity. arXiv preprint arXiv:2006.04768 (2020)
27. Wu, W., Wang, C., Shen, D., Qin, C., Chen, L., Xiong, H.: AFDGCF: adaptive feature de-correlation graph collaborative filtering for recommendations. In: Proceedings of the 47th International ACM SIGIR Conference on Research and Development in Information Retrieval, pp. 1242–1252 (2024)
28. Yang, Z., Wu, J., Wang, Z., Wang, X., Yuan, Y., He, X.: Generate what you prefer: reshaping sequential recommendation via guided diffusion. In: Advances in Neural Information Processing Systems (2023)
29. Ye, Y., et al.: Harnessing multimodal large language models for multimodal sequential recommendation. In: Proceedings of the AAAI Conference on Artificial Intelligence, vol. 39, no. 12, pp. 13069–13077 (2025)
30. Yuan, F., Karatzoglou, A., Arapakis, I., Jose, J.M., He, X.: A simple convolutional generative network for next item recommendation. In: Proceedings of the 12th ACM International Conference on Web Search and Data Mining, pp. 582–590 (2019)
31. Zhang, J., Zeng, Z., Sharma, P.K., Alfarraj, O., Tolba, A., Wang, J.: A dual encoder crack segmentation network with Haar wavelet-based high-low frequency attention. Expert Syst. Appl. **256**, 124950 (2024)
32. Zhao, J., Wang, W., Lin, X., Qu, L., Zhang, J., Chua, T.S.: Popularity-aware distributionally robust optimization for recommendation system. In: Proceedings of the 32nd ACM International Conference on Information and Knowledge Management, pp. 4967–4973 (2023)

33. Zhao, J., Wenjie, W., Xu, Y., Sun, T., Feng, F., Chua, T.S.: Denoising diffusion recommender model. In: Proceedings of the 47th International ACM SIGIR Conference on Research and Development in Information Retrieval, pp. 1370–1379 (2024)
34. Zhou, K., et al.: S3-rec: self-supervised learning for sequential recommendation with mutual information maximization. In: Proceedings of the 29th ACM International Conference on Information & Knowledge Management, pp. 1893–1902 (2020)
35. Zhou, K., Yu, H., Zhao, W.X., Wen, J.R.: Filter-enhanced MLP is all you need for sequential recommendation. In: Proceedings of the ACM Web Conference, pp. 2388–2399 (2022)

Beyond Static Representation: Coarse-to-Fine Dynamic Latent Reasoning for Sequential Recommendation

Xiaochen Yang[1], Yin Xu[1(✉)], Xiaobei Wang[1], Hao Wang[1], Zhida Jiang[2], Zhen Chen[2], and Zhaolong Xing[2]

[1] School of Computer Science and Technology, University of Science and Technology of China, Hefei, China
stephen8@mail.ustc.edu.cn, {yinxu,wanghao3}@ustc.edu.cn
[2] JD.com, Beijing, China
{jiangzhida5,chenzhen48,xingzhaolong1}@jd.com

Abstract. Sequential recommendation is crucial for predicting user intent from historical behaviors. However, existing methods are constrained by static representations that compress a user's history into a single vector, lacking the reasoning capability to capture dynamic user intents. To address this limitation, we introduce **CARARec**, a unified framework that embodies a **Dynamic Reasoning** paradigm. This paradigm models user intent as a sequential trajectory of "thoughts", where each step progressively refines the user representation within a continuous latent hyper action space. The framework is built on three core components to solve key challenges: (1) Latent Concept Synthesis (LCS) anchors the reasoning process within the item vocabulary, ensuring semantic coherence; (2) Coarse-to-Fine Concept Selection (CFCS) makes this process computationally tractable; and (3) Reasoning Path Alignment (RPA) imposes a coherent and generalizable logic. Extensive experiments on real-world benchmarks demonstrate that CARARec substantially outperforms state-of-the-art baselines. Moreover, ablation studies confirm the necessity of each component, highlighting the principled and effective design of the framework. Notably, the benefits of CARARec are architecture-agnostic, enabling seamless integration with various backbone models. The specific implementation of CARARec is available in the code repository: https://github.com/stephen863256/CARARec.

Keywords: Sequential Recommendation · Representation Learning · Latent Reasoning

1 Introduction

Recommender Systems (RS) have become key parts of modern digital services to reduce information overload and provide personalized content for different

H. Jung et al. (Eds.): DASFAA 2026, LNCS 16535, pp. 258–273, 2026.
https://doi.org/10.1007/978-981-92-0363-5_16

users. Among various approaches, Sequential Recommendation (SR), such as SASRec [11], BERT4Rec [23], stands out as a core task: it aims to capture dynamic interest patterns and predict users' next action based on their historical behavior, as illustrated in Fig. 1(a). In recent years, inspired by the remarkable reasoning capabilities of Large Language Models (LLMs), particularly through mechanisms like Chain-of-Thought (COT), latent reasoning for sequential recommendation (LRSR) has emerged as a transformative paradigm for advancing SR, as illustrated in Fig. 1(b). LRSR addresses the limitations of traditional "direct inference" methods that lack the ability to iteratively refine user representations. Methods like LARES [13] and ReaRec [24] demonstrate that multi-step implicit reasoning can significantly enhance model capacity to capture intricate user intent, outperforming non-reasoning baselines by simulating human-like "deliberation" before generating recommendations.

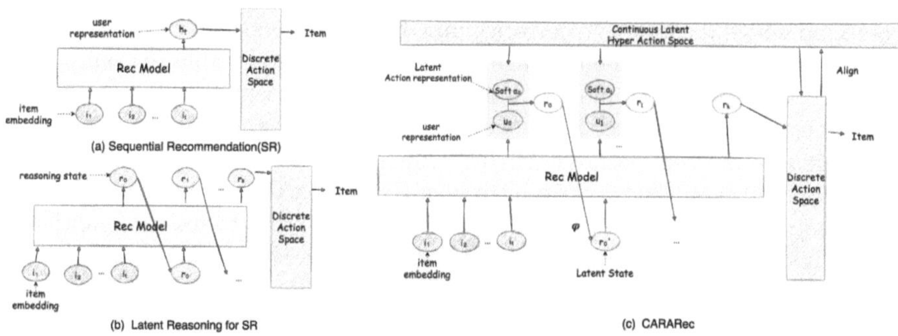

Fig. 1. Illustration of different SR paradigms

This formulation inherently faces challenges due to the large dynamic discrete action space. Unlike LLMs which operate in a dense and semantically continuous natural language space, RS actions lie within a sparse item-ID space. This fundamental divergence makes it difficult to define and explore meaningful reasoning processes in recommendation tasks. Specifically, reasoning in LLMs is often realized through COT processes that model potential output pathways, while LRSR methods autoregressively generate user representations without the capacity to "think the action" at each step—a process that is computationally suboptimal and prone to unstable reasoning trajectories. To overcome these limitations, we propose a novel paradigm named *Dynamic Reasoning*, as illustrated in Fig. 1(c), where models form a trajectory of evolving representations within a continuous latent hyper action space (hereafter, **Hyperspace**), enabling fine-grained exploration of user intent. While this paradigm is inspired by multi-step inference in LLMs [4,21,30], a direct translation of this paradigm to recommendation is non-trivial, primarily due to the fundamental disparity in action spaces.

To address the challenge of performing dynamic reasoning within the Hyperspace (i.e., the **Anchoring Problem**) [1,5], we introduce Latent Concept Synthesis **(LCS)**, which anchors the Hyperspace through connections to the discrete item world, thereby improving semantic grounding and making intermediate states easier to interpret. This mechanism constrains each reasoning step to be a weighted synthesis of actual item embeddings, helping keep the reasoning trajectory close to tangible concepts and reducing the risk of semantic drift.

However, the recommendation action space encompasses the complete item set, whose large cardinality introduces substantial complexity in the mapping process from the discrete action space to the Hyperspace (i.e., the **Tractability Problem**). A practical solution should navigate this space without an exhaustive search. We propose Coarse-to-Fine Concept Selection **(CFCS)** strategy, which reduces the computation by using a retrieve-and-rank style funnel. Concretely, CFCS adopts an annealed Top-K sampling, starting with a broad exploration of concepts and progressively narrowing its focus, which improves scalability in large item sets.

Additionally, we consider the **Structuring Problem** in reasoning paths, where numerous disparate trajectories may hinder the learning of precise representations. Our goal is to encourage coherent and generalizable reasoning patterns. To this end, we employ a dual optimization strategy that integrates two complementary techniques. First, we adopt dense supervision to provide step-wise guidance in deep reasoning chains, which facilitates iterative refinement of user representations. Second, we introduce Reasoning Path Alignment **(RPA)**, a contrastive learning objective that aligns reasoning trajectories of semantically similar user sequences, serving as a regularizer of the Hyperspace geometry.

In this paper, we introduce **CARARec**, a unified framework to instantiate the dynamic reasoning paradigm for sequential recommendation. It targets anchoring, tractability, and structuring via three corresponding components. Our main contributions are summarized as follows:

- **New dynamic paradigm.** We propose Latent Concept Synthesis to improve grounding of the reasoning process within the Hyperspace, moving beyond purely static representations.
- **Hyperspace alignment.** We employ contrastive learning to align reasoning trajectories for semantically similar intents, encouraging coherent reasoning patterns.
- **Model and validation.** We conduct experiments on four benchmarks to demonstrate the effectiveness of the proposed framework under our settings.

2 Related Work

2.1 The Paradigm of Static Representation in Sequential Recommendation

Sequential recommendation, which aims to predict a user's next action from historical interactions [15,29], has evolved from early Markov Chain-based methods [20] to deep learning models. This transition included RNNs in GRU4Rec [6], CNNs in Caser [25], and more recently, the Transformer architecture in models like SASRec [11] and BERT4Rec [23], which has become standard due to its self-attention mechanism. Despite architectural advances, these models operate on a static representation paradigm, mapping item IDs to an embedding space. Subsequent research has focused on enhancing this single representation with side information [8,32] or through Self-Supervised Learning (SSL) [16,17,31]. However, these methods still compress a user's history into one static vector in a single forward pass. This single-step encoding process restricts the model's capacity for dynamic reasoning, imposes a persistent performance ceiling, and calls for a more flexible, reasoning-oriented approach.

2.2 The Emergence of Dynamic Reasoning

The concept of multi-step reasoning, illustrated by Large Reasoning Models (LRMs) such as DeepSeek-R1 [3] and Kimi-k2 [26], has evolved from the resource-heavy Chain-of-Thought (CoT) prompting [30] to more efficient latent reasoning [12,36]. This latent paradigm encodes the reasoning process as a sequence of hidden states, formed within either continuous latent spaces [2,5] or discrete abstract tokens [34]. Inspired by this shift, recent research in sequential recommendation has begun exploring latent reasoning to overcome the limitations of static models. Early studies such as ReaRec [24], STREAM-Rec [33], and LARES [13] generate latent tokens autoregressively to refine user representations. While these works demonstrate the potential of iterative reasoning, they leave two key issues open: their intermediate "thoughts" remain disconnected from the real item space, failing to address the **Anchoring Problem**, and they lack consistent logical constraints, leaving the **Structuring Problem** unresolved. This raises a central question: how can we build a framework that is not only iterative but also anchored, tractable, and structured?

3 Overview

3.1 Revisiting the Static Representation Paradigm

In sequential recommendation, we let $\mathcal{U}$ and $\mathcal{I}$ denote a set of users and items, respectively. Given a user's historical behavior sequence $S_{u,t} = (i_1, \ldots, i_t)$, the sequential recommendation system aims to accurately predict the next item i_{t+1}, where t is the length of the behavior sequence. In order to achieve this goal, the conventional static representation paradigm uses item embeddings $\mathbf{E} \in \mathbb{R}^{N \times d}$ and positional embeddings $\mathbf{P} \in \mathbb{R}^{L_{max} \times d}$, and then converts the item sequence into an input matrix $\mathbf{X}^{(0)} = [\mathbf{e}_{i_1} + \mathbf{p}_1, \ldots, \mathbf{e}_{i_t} + \mathbf{p}_t]$. A sequential encoder, such as a transformer model, processes this matrix in a single forward pass to produce hidden states $(\mathbf{h}_1, \ldots, \mathbf{h}_t) = \text{Encoder}(\mathbf{X}^{(0)})$. The final hidden state $\mathbf{h}_t$ serves as the user representation $\mathbf{h}_{u,t}$, which can be applied to compute recommendation scores via an inner product (i.e., $\mathbf{s} = \mathbf{h}_{u,t}\mathbf{E}^\top$). The core limitation of this approach is compressing a user's entire history into a single static point in the embedding space, which inevitably leads to oversimplification of complex preference dynamics, failing to capture the step-by-step reasoning process behind how user interests evolve or shift over time, as illustrated in Fig. 1(a).

3.2 The Dynamic Reasoning Paradigm

In this paper, our proposed framework introduces a dynamic reasoning process. Instead of a single pass, we treat the initial encoder output as the starting point for a reasoning chain, defined as $\mathbf{r}_0 \triangleq \mathbf{h}_{u,t}$. This state is iteratively refined over K steps, where the model augments the input sequence with generated thoughts. The refinement process at each step k (from 0 to $K-1$) can be defined as follows:

Definition 1 (Dynamic Reasoning Trajectory). *Given an initial reasoning state* $\mathbf{r}_0$ *and input sequence* $\mathbf{X}^{(0)}$*, a reasoning trajectory of length* K*, denoted as* $(\mathbf{r}_0, \mathbf{r}_1, \ldots, \mathbf{r}_K)$*, is constructed by iteratively refining the state for each step* $k \in [0, K-1]$ *as follows:*

1. ***Thought Generation:*** *The current state* $\mathbf{r}_k$ *is transformed into a latent state (i.e., a* ***"thought"****)* $\mathbf{r}'_k = \varphi(\mathbf{r}_k)$.
2. ***Sequence Augmentation:*** *The thought is appended to the input sequence:* $\mathbf{X}^{(k+1)} = Concat\left(\mathbf{X}^{(k)}, \mathbf{r}'_k + \mathbf{p}_{t+k+1}\right)$.
3. ***State Refinement:*** *The augmented sequence is processed by a shared encoder to yield the next state:* $\mathbf{r}_{k+1} = Encoder\left(\mathbf{X}^{(k+1)}\right)_{t+k+1}$.

The last state $\mathbf{r}_K$ is used for computing recommendation scores: $\mathbf{s} = \mathbf{r}_K \mathbf{E}^\top$, while each preceding state $\mathbf{r}_k$ marks a deliberative step where a refined user representation and a latent action representation jointly advance the reasoning within the Hyperspace to progressively align its abstract trajectory with the tangible discrete action space.

As depicted in Fig. 1(c), compared to the static paradigm that condenses the entire user history into a single fixed representation in one pass, our dynamic reasoning paradigm iteratively refines the user state through K steps of thought generation and sequence augmentation. This enables explicit modeling of the evolution of user intentions, captures multi-step logical dependencies in behaviors, and avoids oversimplifying complex preferences into a single static embedding. Furthermore, while previous latent reasoning paradigms (as illustrated in Fig. 1(b)) lack φ for aligning the Hyperspace and discrete action space, the innovation of our **dynamic reasoning paradigm** lies in the design of the thought generation function φ and reasoning trajectory rules. Crucially, the function φ perpetually grounds the abstract reasoning process in the tangible item space, and these components are detailed next.

4 Methodology

To address the limitations of single-step inference, we propose a dynamic reasoning framework for sequential recommendation, called CARARec. It uses a multi-step reasoning process, iteratively refining user representations for more accurate and robust recommendations. Specifically, CARARec incorporates three key components: (1) Latent Concept Synthesis (LCS) module is designed to anchor the Hyperspace to the discrete action space, solving the anchoring problem; (2)

Coarse-to-Fine Concept Selection (CFCS) module is presented to reduce computational complexity, addressing the tractability problem; and (3) model optimization module achieves a dual optimization objective with dense supervision for predictive accuracy and Reasoning Path Alignment (RPA) for Hyperspace structuring, tackling the structuring problem.

4.1 Latent Concept Synthesis Module

To solve the key **Anchoring Problem**, we introduce **Latent Concept Synthesis (LCS)** module to link our Hyperspace with the real world of discrete items. Existing reasoning frameworks often suffer from instability, as their reasoning states $\mathbf{r}_k$ tend to deviate from the manifold of valid item embeddings. Thus, the encoder receives inputs that lie outside the distribution of its training data, causing the reasoning process to become disentangled from the semantics of real items. To avoid this issue, LCS proposes a simple yet effective strategy: every thought is constructed as a weighted mixture of item embeddings (i.e., **hybrid item embedding**), ensuring that all reasoning states remain anchored within the valid embedding space. This design not only stabilizes the reasoning dynamics but also enhances interpretability, as each intermediate reasoning step can be explicitly expressed as a soft combination of real items. For brevity, this design can be defined as the "thought generation function" φ, where $\mathbf{r}'_k = \varphi(\mathbf{r}_k)$. At each reasoning step k, the module performs the following operations:

1. **Projection:** The reasoning state $\mathbf{r}_k$ is projected onto the item space to get logits, which are normalized by Softmax with temperature T_{soft} as π_k.

$$\mathbf{l}_k = \mathbf{r}_k \mathbf{E}^\top, \quad \pi_k = \text{Softmax}(\mathbf{l}_k / T_{soft}) \tag{1}$$

2. **Synthesis:** This distribution provides weights to synthesize the hybrid item embeddings $\mathbf{r}'_k$ as a convex combination of all item embeddings.

$$\mathbf{r}'_k = \sum\nolimits_{i \in \mathcal{I}} \pi_{k,i} \cdot \mathbf{e}_i = \pi_k \mathbf{E} \tag{2}$$

4.2 Coarse-to-Fine Concept Selection Module

To address the axiomatic **Tractability Problem**, we introduce **Coarse-to-Fine Concept Selection (CFCS)**. The full-vocabulary synthesis in LCS, with its $O(Nd)$ complexity, is computationally intractable for large-scale systems where the number of items N can be in the millions [18]. CFCS transforms this challenge into a manageable exploration of high-relevance subspaces without sacrificing expressiveness, by employing an **annealed Top-K selection** mechanism. This coarse-to-fine strategy narrows the focus from a broad set of concepts in early steps to a smaller, refined set in later ones, mirroring the coarse-ranking, fine-ranking, and re-ranking paradigm commonly adopted in traditional recommendation systems. The candidate set size at step k is determined by:

$$K_k = \max(\lfloor K_{initial} \cdot \gamma^k \rfloor, K_{min}) \tag{3}$$

with initial size $K_{initial}$, decay rate $\gamma \in (0, 1]$, and minimum K_{min}. This annealed approach operationalizes the coarse-to-fine principle by initially exploring a wider conceptual space with a larger candidate set, ensuring broad coverage. As the process iterates, the shrinking candidate size allows for a more focused and precise refinement on the most promising concepts, effectively balancing computational efficiency with the depth of semantic exploration. This strategic narrowing of the search space at each step is crucial for making the reasoning process both tractable and effective in large-scale environments. The procedure has two stages, reducing synthesis cost to $O(K_k d)$, as illustrated in Fig. 2:

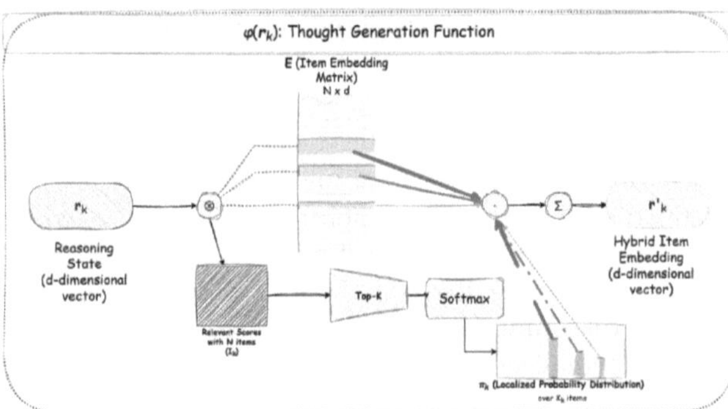

Fig. 2. Illustration of the Thought Generation Function $\varphi(\mathbf{r}_k)$

1. **Candidate Retrieval:** We calculate logits over the vocabulary, $\mathbf{l}_k = \mathbf{r}_k \mathbf{E}^\top$, and select the Top-$K_k$ items to form a high-relevance subset, $\mathcal{I}_k^{\text{top}}$.
2. **Localized Synthesis:** We leverage the Softmax function from the subset, i.e., $\pi_k^{\text{top}} = \text{Softmax}(\mathcal{I}_k^{\text{top}} / T_{soft})$, and synthesize the hybrid item embedding using the following equation:

$$\mathbf{r}'_k = \sum\nolimits_{i \in \mathcal{I}_k^{\text{top}}} \pi_{k,i}^{\text{top}} \cdot \mathbf{e}_i \tag{4}$$

4.3 Model Optimization Module

The optimization objective of our model needs to consider two aspects: dense supervision for predictive accuracy and a contrastive objective to solve the fundamental **Structuring Problem**. First, in order to address the credit assignment problem in the deep reasoning chain, we apply dense supervision at each reasoning step to provide a more direct and stable gradient path. This ensures that each intermediate state is itself guided towards a predictively meaningful representation. The loss is calculated as follows:

$$\mathcal{L}_{REC} = \sum_{k=0}^{K} \mathcal{L}_{CE}(\mathbf{r}_k, i_{t+1}) \tag{5}$$

where $\mathcal{L}_{CE}$ is the standard cross-entropy function. This guides the model to learn a path of iterative refinement, where each state improves upon the last.

Then, we introduce **Reasoning Path Alignment (RPA)** to address the **Structuring Problem** of path degeneracy, as illustrated in Fig. 3. This contrastive objective enforces a structural prior by aligning the reasoning trajectories ($[\mathbf{r}_k]_{k=0}^{K}$ and $[\tilde{\mathbf{r}}_k]_{k=0}^{K}$) from sequences sharing the same target item [17], which forces the model to distill a generalizable reasoning logic for semantically similar intents. We generate these positive pairs, each comprising an original user sequence S_u and its semantically similar augmented counterpart $\tilde{S}_u$, and use an InfoNCE-based loss [27] $\ell(\cdot,\cdot)$ for alignment at each step, where negative samples are drawn from other user sequences in the mini-batch:

$$\mathcal{L}_{CL} = \sum_{k=1}^{K} \frac{1}{2} \left[\ell(\mathbf{r}_k, \tilde{\mathbf{r}}_k) + \ell(\tilde{\mathbf{r}}_k, \mathbf{r}_k)\right] \tag{6}$$

This objective regularizes the geometry of the Hyperspace, forcing the model to learn a generalizable and coherent logic. The final loss is a weighted combination:

$$\mathcal{L}_{final} = \mathcal{L}_{REC} + \lambda \mathcal{L}_{CL} \tag{7}$$

where λ is a balancing hyperparameter.

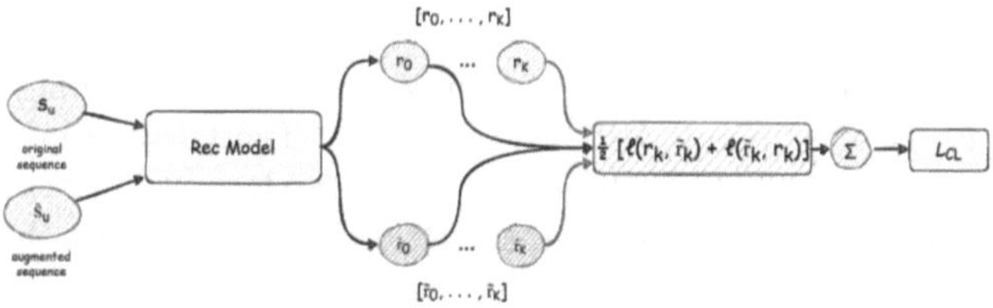

Fig. 3. Illustration of the RPA

4.4 Complexity Analysis

Time Complexity. The naive inference has a time complexity of approximately $\mathcal{O}((K+1)Nd)$, which is impractical for real-world applications due to the linear scan over the large item vocabulary N at each reasoning step. However, this is easily optimized with standard Approximate Nearest Neighbor (ANN) search techniques [10,14,28], reducing the vocabulary-dependent cost from linear to $\mathcal{O}(\log N)$. This shifts the computational bottleneck to the Transformer's self-attention mechanism, making our approach feasible for large-scale systems.

Space Complexity. The space complexity is dominated by the $O(Nd)$ item embedding matrix, while our multi-step reasoning process adds only a negligible $O(Kd)$ overhead for extra positional embeddings. Therefore, CARARec is parameter-efficient and imposes no significant memory burden.

5 Experiment

We evaluate our proposed dynamic reasoning paradigm by addressing the following three Research Questions (RQs):

- **RQ1:** How does CARARec demonstrate superior performance over SOTA models in dynamic reasoning tasks, and can its framework be effectively generalized across different backbone architectures?
- **RQ2:** How do the core components of CARARec address the key challenges of *Anchoring*, *Tractability*, and *Structuring* in dynamic reasoning?
- **RQ3:** How do key hyperparameters and computational efficiency impact the practical deployment of CARARec?

5.1 Experiment Setup

Datasets. We conduct offline experiments on four Amazon datasets [7]: **Scientific**, **Game**, **Baby**, and **Instrument.** For dataset construction, following standard preprocessing [18], user interaction sequences are chronologically ordered, and only the most recent 50 items are retained. We adopt a leave-one-out strategy: the last item for testing, the second last for validation, and the rest for training. To ensure data quality and alleviate sparsity, we apply 5-core filtering by removing users and items with fewer than five interactions (statistics in Table 1).

Baselines. We compare our method with representative baselines across four categories: **(1) RNN-based**: GRU4Rec [6]. **(2) Transformer-based**: SASRec [11], BERT4Rec [23], FMLPRec [35], and BSARec [22]. **(3) SSL-based:** DuoRec [17]. **(4) Reasoning-enhanced**: ReaRec-ERL [24], ReaRec-PRL [24], and LARES [13].

Evaluation and Implementation Details. Following prior work [8,9,19], we report Recall@K and NDCG@K as metrics for $K \in \{10, 20\}$, averaged over all users. To ensure a rigorous evaluation, we assess all models by ranking the ground-truth item against the entire item pool. Our base Transformer encoder uses 2 layers, 2 heads, 64-d embeddings, 256 FFN size, dropout 0.5, and GELU activation. We use the AdamW optimizer (lr=0.001) with a batch size of 1024, training for up to 300 epochs with early stopping (patience 10). We tune reasoning steps $K \in \{1, 2, 3, 4\}$ for CARARec and ReaRec, and follow the two-stage training for LARES. CARARec uses a single-stage scheme with $K_{\text{initial}} \in \{100, 150, 200, 250, 300\}$, decay rate $\gamma \in \{0.8, 0.85, 0.9, 0.95, 0.99\}$, and balancing hyperparameter $\lambda \in \{0.001, 0.01, 0.1, 0.5, 1.0\}$. Other parameters are fixed: $T_{\text{soft}} = 1.0$, $\tau = 0.07$, and cosine similarity.

5.2 Main Results (RQ1)

We evaluate the effectiveness, generalizability, and efficiency of CARARec with state-of-the-art baselines. As shown in Table 2, **CARARec consistently achieves the best performance across all datasets and metrics**, especially in terms of NDCG. Specifically, it surpasses LARES (the sota baseline) on NDCG@10 by **4.8%** on Scientific, **3.7%** on Game, **3.5%** on Baby, and **6.4%** on Instrument, demonstrating the superiority of its dynamic reasoning mechanism.

Table 1. Statistics of the proposed datasets

Dataset	#Users	#Items	#Actions	Avg.L	Sparsity
Scientific	50,985	25,848	412,947	8.10	99.969%
Game	94,762	25,612	814,586	8.60	99.966%
Baby	150,777	36,013	1,241,083	8.23	99.977%
Instrument	57,439	24,587	511,836	8.91	99.964%

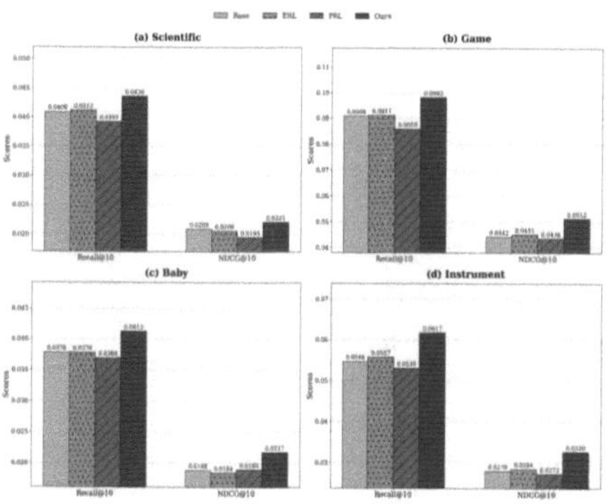

Fig. 4. Performance comparison with FMLPRec as the backbone. 'Base' refers to the original FMLPRec model, while 'ERL', 'PRL' (from ReaRec), and 'Ours' denote CARARec.

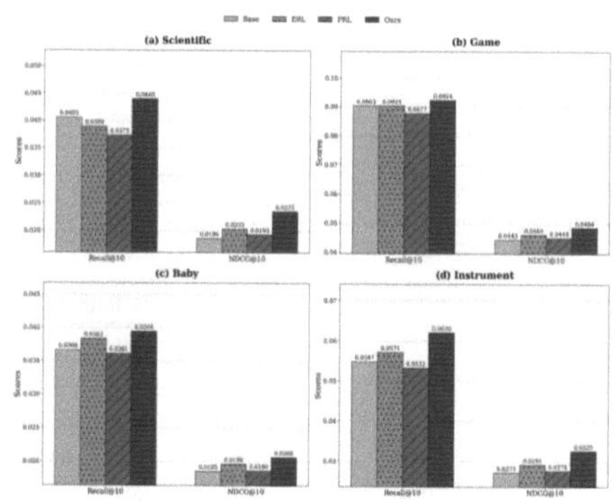

Fig. 5. Performance comparison with BSARec as the backbone

To further examine its generalizability and computational efficiency, CARARec is integrated with two representative backbones, FMLPRec and BSARec. As illustrated in Fig. 4 and Fig. 5, CARARec consistently outperforms both the original backbone models and their ReaRec variants, indicating that the benefits of dynamic reasoning are **architecture-agnostic** and can be seamlessly incorporated into diverse model structures.

5.3 Ablation Study

The Components of CARARec (RQ2). To answer **RQ2**, we perform an additive ablation study to evaluate each CARARec component—LCS, CFCS, and RPA. Starting from a **Base** model, we incrementally add each module under the same settings. Results are shown in Table 4.

- **LCS (Anchoring).** Adding Latent Concept Synthesis (*+LCS*) improves NDCG across all datasets (e.g., 0.0185→0.0200 on Scientific). This confirms our idea: by making the model's middle "thoughts" use real item concepts, LCS by its design inherently ties abstract reasoning to actual items, prevents semantic drift off the item manifold, and strengthens the anchoring effect.
- **CFCS (Tractability).** Introducing Coarse-to-Fine Concept Selection (*+CFCS*) further enhances both Recall and NDCG (e.g., 0.0200→0.0215 on Scientific). This surprising gain shows that progressively refining the search focus into manageable subspaces not only secures tractability, but also proves to be a more effective strategy for guiding the model to relevant concepts.
- **RPA (Structuring).** Adding Reasoning Path Alignment (*+RPA*) yields the largest gains (e.g., Recall@10: 0.0544→0.0624 on Instrument), demonstrating the benefit of consistent, structured reasoning trajectories. By aligning reasoning trajectories for semantically similar intents, RPA enforces a structural prior to combat path degeneracy, forcing the model to distill a more generalizable and coherent logic beyond what dense supervision offers.

Table 2. Performance comparison of different methods on the four datasets. R stands for Recall and N stands for NDCG. The **best results are in bold with a star** (*), and the second-best results are underlined with a dagger (†).

Model	Scientific				Game			
	R@10	R@20	N@10	N@20	R@10	R@20	N@10	N@20
GRU4Rec	0.0319	0.0504	0.0164	0.0210	0.0800	0.1230	0.0410	0.0518
SASRec	0.0383	0.0581	0.0185	0.0235	0.0913	0.1373	0.0429	0.0545
BERT4Rec	0.0287	0.0451	0.0148	0.0190	0.0767	0.1191	0.0394	0.0501
FMLPRec	0.0396	0.0601	0.0198	0.0250	0.0908	0.1374	0.0442	0.0559
BSARec	0.0405	0.0594	0.0186	0.0234	0.0903	0.1355	0.0443	0.0556
DuoRec	0.0409	0.0614	0.0209	0.0261	0.0917	0.1360	0.0464	0.0575
ReaRec-ERL	0.0394	0.0594	0.0195	0.0245	0.0890	0.1399	0.0446	0.0559
ReaRec-PRL	0.0395	0.0586	0.0202	0.0250	0.0897	0.1344	0.0458	0.0570
LARES w/o RPT	0.0430	0.0643	0.0222	0.0273	0.0947†	**0.1416***	0.0470	0.0588
LARES	0.0444†	0.0657†	0.0230†	0.0283†	0.0944	0.1406†	0.0482†	0.0598†
Ours	**0.0459***	**0.0670***	**0.0241***	**0.0294***	**0.0948***	0.1401	**0.0500***	**0.0614***
Model	Baby				Instrument			
	R@10	R@20	N@10	N@20	R@10	R@20	N@10	N@20
GRU4Rec	0.0340	0.0546	0.0175	0.0227	0.0511	0.0768	0.0266	0.0331
SASRec	0.0365	0.0568	0.0183	0.0234	0.0546	0.0826	0.0264	0.0334
BERT4Rec	0.0266	0.0440	0.0134	0.0177	0.0443	0.0674	0.0232	0.0290
FMLPRec	0.0378	0.0589	0.0188	0.0242	0.0546	0.0807	0.0279	0.0345
BSARec	0.0366	0.0578	0.0185	0.0239	0.0547	0.0824	0.0271	0.0341
DuoRec	0.0369	0.0570	0.0182	0.0233	0.0582	0.0875	0.0299	0.0372
ReaRec-ERL	0.0370	0.0581	0.0187	0.0240	0.0546	0.0825	0.0270	0.0340
ReaRec-PRL	0.0365	0.0578	0.0187	0.0241	0.0548	0.0825	0.0282	0.0352
LARES w/o RPT	0.0390†	0.0605†	0.0193	0.0246	0.0599†	0.0908†	0.0302	0.0380
LARES	0.0387	0.0596	0.0201†	0.0253†	0.0596	0.0885	0.0312†	0.0384†
Ours	**0.0399***	**0.0611***	**0.0208***	**0.0263***	**0.0624***	**0.0920***	**0.0332***	**0.0406***

Practical Deployment (RQ3). For practical deployment, we examine CARARec's sensitivity to key hyperparameters, verify its robustness, and test its efficiency.

- **Reasoning Steps** (K). As shown in Fig. 6, performance first increases and then declines with more steps. Optimal K is 3 for most datasets, while Scientific peaks at $K = 4$. We hypothesize this is because Scientific's complex, multi-aspect intents demand more steps of iterative refinement via our Coarse-to-Fine Synthesis to converge on a specific target.
- **Decay Rate** (γ). Figure 7 shows a concave trend as γ varies from 0.80 to 0.95, peaking at $\gamma = 0.90$. A small γ (e.g., 0.80) anneals the search subspace too rapidly, risking the loss of vital concepts. A large γ (e.g., 0.95), however, results in a slow annealing, retaining too much noise, which fails to adequately focus the synthesis on the most promising candidates. The optimal $\gamma = 0.90$ achieves the ideal annealing trajectory for reasoning.

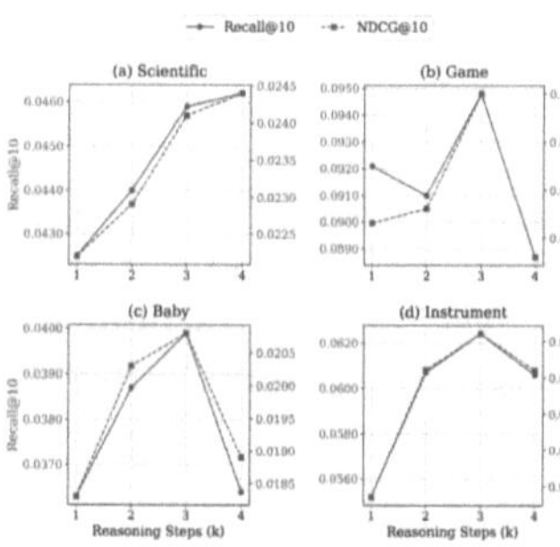

Fig. 6. Impact of the number of reasoning steps (K) on model performance across four datasets

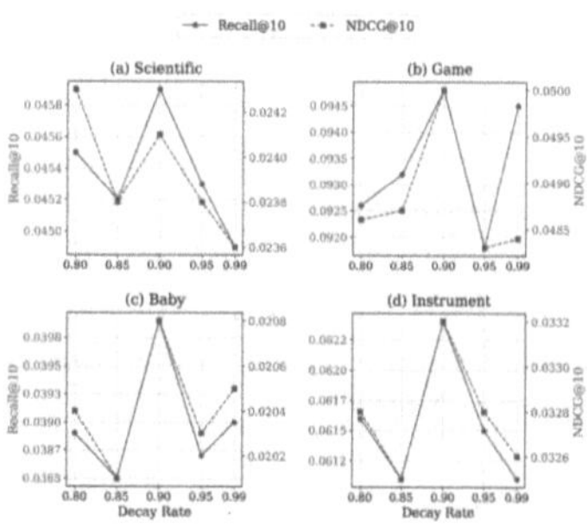

Fig. 7. Impact of the decay rate (γ) on model performance

- **Initial Candidate Size** (K_{initial}). As shown in Fig. 8, performance peaks at $K_{\text{initial}} = 200$. This parameter is crucial as it defines the scope of the initial "coarse" exploration in our CFCS framework. A small set (e.g., 100) excessively restricts the high-relevance subspace, risking the exclusion of concepts vital for an accurate initial synthesis. Conversely, a large set (e.g., 300) pollutes the candidate pool with noise, which degrades the synthesis quality and impedes the subsequent fine-grained refinement process.
- **Balancing Hyperparameter** (λ). In Fig. 9, $\lambda = 0.1$ achieves the best balance between recommendation accuracy and path alignment. A small λ fails to impose a strong enough structural prior, leading to path degeneracy, while a large λ over-constrains the reasoning geometry, sacrificing predictive accuracy for structural uniformity. This balance is particularly sensitive in complex domains like Scientific, which require both coherent logic and diverse reasoning trajectories to accurately model nuanced user intents.

- **Efficiency analysis.** Table 3 shows that the time cost per reasoning step decreases after the initial iteration ($k = 1$), due to the annealed Coarse-to-Fine mechanism reducing candidate pool size. The initial step is the most computationally intensive because it operates on the largest set of candidates, which is then progressively narrowed by the annealed selection strategy in subsequent steps. This confirms that CARARec's deliberative reasoning is both effective and computationally efficient, making it practically viable.

Table 3. Time taken (ms) at each step k for a total of 3 reasoning steps. We use k = 0 for the initial encode phase. The final column shows the percentage decrease from k = 1 to k = 3.

Dataset/k	0	1	2	3	Trend (k = 1 to 3)
Scientific	1.21	1.31	1.12	1.10	↓ 16.0%
Game	1.14	1.24	1.13	1.11	↓ 10.5%
Baby	1.29	1.43	1.36	1.34	↓ 6.3%
Instrument	1.46	1.56	1.33	1.30	↓ 16.7%

Table 4. Additive ablation study on the key components of CARARec. We start with a **Base** model (a standard SASRec Model) and progressively add LCS, CFCS and RPA. The **Full Model** includes all components. The **best result is in bold with a star** (*), and the second-best is underlined with a dagger (†). R stands for Recall and N stands for NDCG.

Dataset/Metric		CARARec Variants (Cumulative)			
		Base	+ LCS	+ LCS + CFCS	Full Model (+ RPA)
Scientific	R@10	0.0383	0.0385	<u>0.0414</u>†	**0.0459***
	R@20	0.0581	0.0600	<u>0.0632</u>†	**0.0670***
	N@10	0.0185	0.0200	<u>0.0215</u>†	**0.0241***
	N@20	0.0235	0.0254	<u>0.0270</u>†	**0.0294***
Game	R@10	<u>0.0913</u>†	0.0884	0.0907	**0.0948***
	R@20	<u>0.1373</u>†	0.1343	0.1372	**0.1401***
	N@10	0.0429	0.0459	<u>0.0466</u>†	**0.0500***
	N@20	0.0545	0.0574	<u>0.0583</u>†	**0.0614***
Baby	R@10	0.0365	0.0358	<u>0.0382</u>†	**0.0399***
	R@20	0.0568	0.0568	<u>0.0595</u>†	**0.0611***
	N@10	0.0183	0.0185	<u>0.0200</u>†	**0.0208***
	N@20	0.0234	0.0238	<u>0.0253</u>†	**0.0263***
Instrument	**R@10**	<u>0.0546</u>†	0.0542	0.0544	**0.0624***
	R@20	0.0826	0.0821	<u>0.0834</u>†	**0.0920***
	N@10	0.0264	0.0284	<u>0.0287</u>†	**0.0332***
	N@20	0.0334	0.0354	<u>0.0360</u>†	**0.0406***

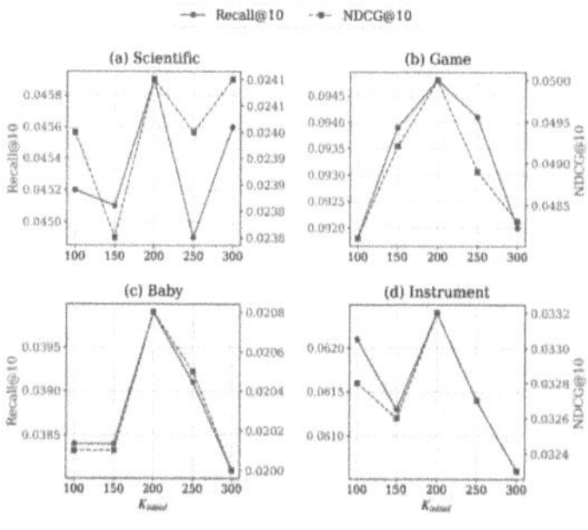

Fig. 8. Impact of the initial candidate size (K_{initial}) on model performance

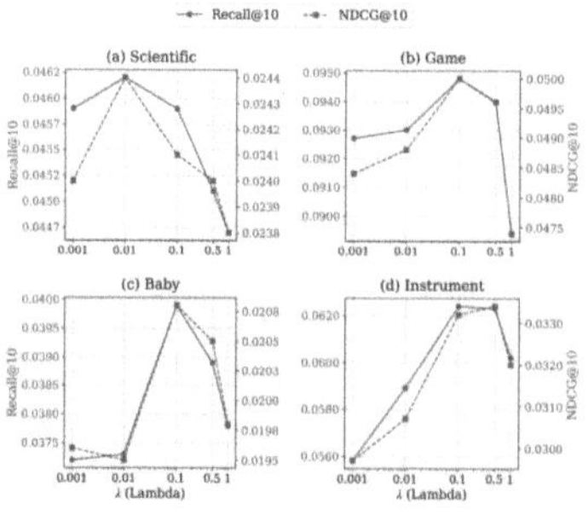

Fig. 9. Impact of the balancing hyperparameter (λ) for the contrastive loss on model performance

6 Conclusion

In this work, we introduce **Dynamic Reasoning** for sequential recommendation and propose **CARARec**, a unified framework with three components: Latent Concept Synthesis (LCS), Coarse-to-Fine Concept Selection (CFCS), and Reasoning Path Alignment (RPA). Experiments on four benchmarks show that CARARec yields consistent improvements over competitive baselines under our evaluation settings, and the gains transfer to multiple backbone models.

Limitations. Our study is evaluated on four Amazon sub-categories. We do not fully examine generalization to broader domains, edge cases with extremely short/sparse histories, robustness to key hyperparameters (e.g., reasoning steps and candidate sizes) across different item-set scales, or end-to-end inference overhead against lightweight baselines.

Future Work. Promising directions include:

- **Anchoring Enhancement:** Investigating stronger grounding strategies to further improve stability and semantic coherence in latent concept synthesis.
- **Tractability Adaptation:** Designing adaptive mechanisms to balance reasoning depth and computation for changing item sets.
- **Structuring Optimization:** Exploring alternative trajectory regularization objectives beyond the current contrastive formulation.

Acknowledgments. This work was supported in part by the National Natural Science Foundation of China (NSFC) under Grant 62502483 and in part by the Fundamental Research Funds for the Central Universities under Grant WK2150250040.

Disclosure of Interests. The authors have no competing interests to declare that are relevant to the content of this article.

References

1. Butt, N., Kwiatkowski, A., Labiad, I., Kempe, J., Ollivier, Y.: Soft Tokens, Hard Truths (2025). https://doi.org/10.48550/arXiv.2509.19170
2. Geiping, J., et al.: Scaling up test-time compute with latent reasoning: a recurrent depth approach. arXiv preprint arXiv:2502.05171 (2025)
3. Guo, D., et al.: Deepseek-r1: incentivizing reasoning capability in LLMs via reinforcement learning. arXiv preprint arXiv:2501.12948 (2025)
4. Guo, D., et al.: Deepseek-coder: when the large language model meets programming–the rise of code intelligence. arXiv preprint arXiv:2401.14196 (2024)
5. Hao, S., et al.: Training large language models to reason in a continuous latent space. arXiv preprint arXiv:2412.06769 (2024)
6. Hidasi, B., Karatzoglou, A., Baltrunas, L., Tikk, D.: Session-based recommendations with recurrent neural networks. arXiv preprint arXiv:1511.06939 (2015)
7. Hou, Y., Li, J., He, Z., Yan, A., Chen, X., McAuley, J.: Bridging language and items for retrieval and recommendation. arXiv preprint arXiv:2403.03952 (2024)
8. Hou, Y., Mu, S., Zhao, W.X., Li, Y., Ding, B., Wen, J.R.: Towards universal sequence representation learning for recommender systems. In: Proceedings of the 28th ACM SIGKDD Conference on Knowledge Discovery and Data Mining, pp. 585–593 (2022)
9. Jia, J., et al.: LEARN: Knowledge Adaptation from Large Language Model to Recommendation for Practical Industrial Application (2024). https://doi.org/10.48550/arXiv.2405.03988
10. Johnson, J., Douze, M., Jégou, H.: Billion-scale similarity search with GPUs. IEEE Trans. Big Data **7**(3), 535–547 (2019)
11. Kang, W.C., McAuley, J.: Self-attentive sequential recommendation. In: 2018 IEEE International Conference on Data Mining (ICDM), pp. 197–206. IEEE (2018)
12. Li, J., et al.: Implicit reasoning in large language models: a comprehensive survey. arXiv preprint arXiv:2509.02350 (2025)
13. Liu, E., et al.: Lares: latent reasoning for sequential recommendation. arXiv preprint arXiv:2505.16865 (2025)
14. Malkov, Y.A., Yashunin, D.A.: Efficient and robust approximate nearest neighbor search using hierarchical navigable small world graphs. IEEE Trans. Pattern Anal. Mach. Intell. **42**(4), 824–836 (2018)
15. Pan, L., Pan, W., Wei, M., Yin, H., Ming, Z.: A survey on sequential recommendation (2025). https://arxiv.org/abs/2412.12770
16. Qin, X., et al.: Meta-optimized contrastive learning for sequential recommendation. In: Proceedings of the 46th International ACM SIGIR Conference on Research and Development in Information Retrieval, pp. 89–98 (2023)
17. Qiu, R., Huang, Z., Yin, H., Wang, Z.: Contrastive learning for representation degeneration problem in sequential recommendation. In: Proceedings of the Fifteenth ACM International Conference on Web Search and Data Mining, pp. 813–823 (2022)
18. Rajput, S., Mehta, N., Singh, A., Hulikal Keshavan, R., Vu, T., Heldt, L., et al.: Recommender systems with generative retrieval. Adv. Neural. Inf. Process. Syst. **36**, 10299–10315 (2023)
19. Ren, X., et al.: Representation learning with large language models for recommendation. In: Proceedings of the ACM Web Conference 2024, pp. 3464–3475 (2024). https://doi.org/10.1145/3589334.3645458

20. Rendle, S., Freudenthaler, C., Schmidt-Thieme, L.: Factorizing personalized Markov chains for next-basket recommendation. In: Proceedings of the 19th International Conference on World Wide Web, WWW 2010, pp. 811–820. Association for Computing Machinery, New York, NY, USA (2010). https://doi.org/10.1145/1772690.1772773
21. Shao, Z., et al.: Deepseekmath: pushing the limits of mathematical reasoning in open language models. arXiv preprint arXiv:2402.03300 (2024)
22. Shin, Y., Choi, J., Wi, H., Park, N.: An attentive inductive bias for sequential recommendation beyond the self-attention. In: Proceedings of the AAAI Conference on Artificial Intelligence, vol. 38, pp. 8984–8992 (2024)
23. Sun, F., et al.: Bert4rec: sequential recommendation with bidirectional encoder representations from transformer. In: Proceedings of the 28th ACM International Conference on Information and Knowledge Management, pp. 1441–1450 (2019)
24. Tang, J., et al.: Think before recommend: unleashing the latent reasoning power for sequential recommendation. arXiv preprint arXiv:2503.22675 (2025)
25. Tang, J., Wang, K.: Personalized top-n sequential recommendation via convolutional sequence embedding. In: Proceedings of the Eleventh ACM International Conference on Web Search and Data Mining, pp. 565–573 (2018)
26. Team, K., et al.: Kimi k2: open agentic intelligence (2026). https://arxiv.org/abs/2507.20534
27. Tschannen, M., Djolonga, J., Rubenstein, P.K., Gelly, S., Lucic, M.: On mutual information maximization for representation learning. arXiv preprint arXiv:1907.13625 (2019)
28. Wang, M., Xu, X., Yue, Q., Wang, Y.: A comprehensive survey and experimental comparison of graph-based approximate nearest neighbor search. arXiv preprint arXiv:2101.12631 (2021)
29. Wang, S., Hu, L., Wang, Y., Cao, L., Sheng, Q.Z., Orgun, M.: Sequential recommender systems: challenges, progress and prospects. In: Proceedings of the Twenty-Eighth International Joint Conference on Artificial Intelligence, pp. 6332–6338. IJCAI-2019, International Joint Conferences on Artificial Intelligence Organization (2019). https://doi.org/10.24963/ijcai.2019/883
30. Wei, J., Wang, X., Schuurmans, D., Bosma, M., Xia, F., Chi, E., et al.: Chain-of-thought prompting elicits reasoning in large language models. Adv. Neural. Inf. Process. Syst. **35**, 24824–24837 (2022)
31. Xie, X., et al.: Contrastive learning for sequential recommendation. In: 2022 IEEE 38th International Conference on Data Engineering (ICDE), pp. 1259–1273. IEEE (2022)
32. Xu, L., et al.: Sequence-level semantic representation fusion for recommender systems. In: Proceedings of the 33rd ACM International Conference on Information and Knowledge Management, pp. 5015–5022 (2024)
33. Zhang, J., et al.: Slow thinking for sequential recommendation. arXiv preprint arXiv:2504.09627 (2025)
34. Zhang, Z., et al.: Soft thinking: unlocking the reasoning potential of LLMs in continuous concept space. arXiv preprint arXiv:2505.15778 (2025)
35. Zhou, K., Yu, H., Zhao, W.X., Wen, J.R.: Filter-enhanced MLP is all you need for sequential recommendation. In: Proceedings of the ACM Web Conference 2022, pp. 2388–2399 (2022)
36. Zhu, R.J., et al.: A survey on latent reasoning. arXiv preprint arXiv:2507.06203 (2025)

DiffKGR: Diffusion-Based Virtual Edge Generation for Knowledge Graph Recommendation

Lyuwen Wu[1,2], Xiaoying Gan[1,2(✉)], Luoyi Fu[1,2], Lei Zhou[1,2], Xinbing Wang[1,2], and Chenghu Zhou[1,2]

[1] Shanghai Jiao Tong University, Shanghai, China
{wlw2016,ganxiaoying,yiluofu,zhoulei1588,xwang8}@sjtu.edu.cn
[2] Chinese Academy of Sciences, Beijing, China

Abstract. Knowledge graphs (KGs) have been increasingly integrated in recommendation systems. They capture relations among items and attributes, offering supplementary knowledge beyond observed user-item interactions. However, KG-enhanced recommendation still faces the challenge of sparsity, given that most entities in real-world KGs are associated with only a limited number of attributes. This scarcity impedes information propagation across KG and undermines the quality of item representations. To this end, we propose DiffKGR, a diffusion-based framework that generates virtual edges to alleviate sparsity in KGs and enhance downstream recommendation performance. DiffKGR consists of two key components: diffusion-based virtual edge generation and relevance-aware KG aggregation. The generation module employs a structure-guided diffusion process to produce structurally consistent virtual edges. This effectively mitigates the sparsity of the knowledge graph. The KG aggregation module introduces relevance scores to evaluate task-specific relevance of edges, which are then used to adjust edges' weights during aggregation. This mechanism suppresses the influence of irrelevant information during message passing. To evaluate the effectiveness of DiffKGR, we conduct experiments on three public benchmark datasets. Experimental results show that our method outperforms state-of-the-art recommendation models, achieving up to 4.74% improvement. The implementations of DiffKGR are available at https://github.com/WLW8991/DiffKGR.

Keywords: Recommendation · Knowledge Graph · Diffusion Model · Sparsity Issue

1 Introduction

In the Internet era, recommendation systems have been widely deployed across various domains, including video streaming [39], e-commerce [26], online advertising [44] and social media [18]. These systems play a crucial role in alleviating information overload and facilitating content discovery for users [11,20].

H. Jung et al. (Eds.): DASFAA 2026, LNCS 16535, pp. 274–291, 2026.
https://doi.org/10.1007/978-981-92-0363-5_17

Recently, knowledge graphs (KGs) have gained increasing attention in recommendation systems, due to their ability to associate between items and attributes [32]. For example, embedding-based methods [1,30,45] leverage KG embedding techniques to learn representations of users and items, and reveal semantic relatedness between them. Path-based methods [10,37] capture long-range user–item dependencies with predefined relational patterns. Propagation-based methods [23,33,41] automatically aggregate information from multi-hop neighbors, facilitating data-driven discovery of high-order relational patterns.

Despite their effectiveness, KG-enhanced recommendation methods still face the challenge of sparsity. In real-world knowledge graphs, most entities lack comprehensive attribute information, resulting in missing edges and sparse KG structures. This is illustrated in Fig. 1(a), which presents the degree distribution of entities in the Amazon-Book KG. The distribution reveals that only a small fraction of entities exhibit high connectivity - up to 80% attributes - while over 67% are associated with fewer than 20 attributes. A specific example is provided in Fig. 1(b). Consider a user who has purchased the suspense fiction *The Last Precinct*. With a well-connected KG, the system could leverage genre information to infer the user's preference and recommend fictions in the same genre, such as *Children of the Night*. However, due to missing connections in the KG, the system fails to capture the book's genre and cannot recognize the user's underlying interest in suspense fiction. As a result, *Children of the Night* is excluded from the recommendation list for this user.

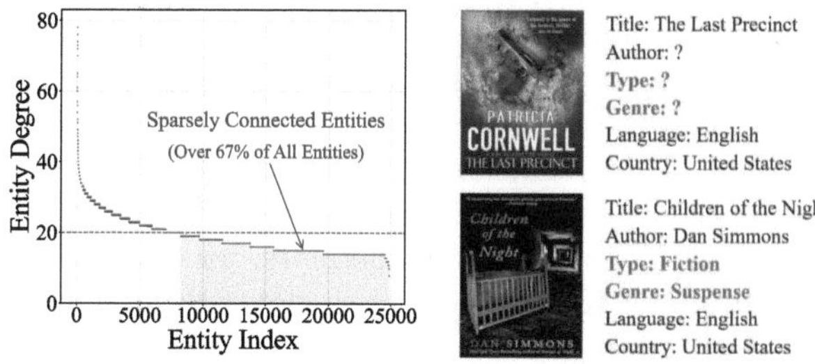

(a) Distribution of entity degrees in KG. (b) Missing attribute connections.

Fig. 1. Sparsity in Amazon-book KG.

Therefore, it is essential to alleviate sparsity issue for KG-enhanced recommendation. Recent studies have explored KG augmentation to construct additional edges, enriching item attributes and semantic relations. However, prior methods mainly rely on local structures, such as co-occurring attributes [28] and reweighted relations [17,19], to generate new edges. While effective in well-connected regions, such reliance can be problematic for sparsely connected entities. As a result, these methods struggle to preserve global graph consistency and can introduce irrelevant or misleading connections. Consequently, two major challenges remain:

– **Overfitting to Local Structure.** Entities with sparse connections aggregate little information from their neighborhoods, which obstructs propagation of global information. As a result, the model tends to overfit to local struc-

tural patterns, generating edges that deviate from the global topology. These edges introduce structural inconsistency and distort item representations.
- **Representation Drift in Knowledge Aggregation.** Knowledge graphs gather heterogeneous data from diverse sources, inevitably incorporating numerous task-irrelevant edges. New edges generated from structural patterns introduce additional irrelevant information. The indiscriminate aggregation of these irrelevant connections leads to representation drift.

To address these challenges, we propose a diffusion-based framework, which integrates a diffusion-based virtual edge generation module with a relevance-aware KG aggregation mechanism. The generation module employs a structure-guided diffusion model where the *head - relation* pair serves as conditional guidance. This conditional generation aligns generated edges with the global distribution, preserving its consistency while enriching entity connectivity and alleviating KG sparsity. Furthermore, we design a relevance-aware aggregation mechanism for KG message passing over the enriched graph. For each edge associated with item entities, a relevance score is computed to evaluate its relevance to the recommendation task. By adaptively weighting edges according to their relevance scores, this mechanism suppresses task-irrelevant connections, reduces representation drift, and ultimately improves the quality of recommendations.

Overall, our contributions are summarized as follows:

- We propose DiffKGR, a novel diffusion-based framework that performs virtual edge generation on knowledge graphs to mitigate sparsity issue and improve recommendation performance.
- We employ a structure-guided diffusion model for virtual edge generation. This module generates structurally consistent edges and alleviates KG sparsity, and enhances information propagation for downstream recommendation.
- We develop a relevance-aware aggregation mechanism that incorporates relevance score to guide message passing. This suppresses task-irrelevant edges and lead to more discriminative representations.

We conduct comprehensive experiments on three benchmark datasets. Results demonstrate that DiffKGR outperforms state-of-the-art KG-enhanced recommendation methods.

2 Problem Formulation

User-Item Interaction Graph. Let $\mathcal{U}$ denote the set of users and $\mathcal{I}$ denote the set of items. Users $u \in \mathcal{U}$ and items $i \in \mathcal{I}$ form a bipartite interaction graph, which is defined as $\mathcal{G}_u = (u, y_{u,i}, i)$. Here $y_{u,i} = 1$ indicates user u has interacted with item i, and $y_{u,i} = 0$ means no interaction. We denote the embedding of item i and user u as $\mathbf{e}_i$ and $\mathbf{e}_u$, respectively.

Knowledge Graph. We denote a knowledge graph as $\mathcal{G} = \{(h, r, t) \mid h, t \in \mathcal{E}, r \in \mathcal{R}\}$. Each triplet (h, r, t) denotes a directed relation r from a head entity h to a tail entity t. Here, $\mathcal{E}$, and $\mathcal{R}$ represent the sets of entities and relations. We denote the embedding of head h, relation r, and tail t as $\mathbf{e}_h$, $\mathbf{e}_r$ and $\mathbf{e}_t$, respectively. Specifically, we denote the primary KG as $\mathcal{G}_k$ and the virtual-enhanced knowledge graph as $\mathcal{G}_v$.

Task Formulation. We define the knowledge graph-enhanced recommendation task as follows: given a user-item interaction graph $\mathcal{G}_u$ and an associated knowledge graph $\mathcal{G}_k$, the goal is to learn a recommendation model $\mathcal{F}(u, i \mid \mathcal{G}_u, \mathcal{G}_k, \Theta)$, parameterized by Θ. This model aims to estimate the likelihood $\hat{y}_{u,i}$ that a user u will interact with an item i.

3 Methodology

We provide an overview of the DiffKGR in Fig. 2.

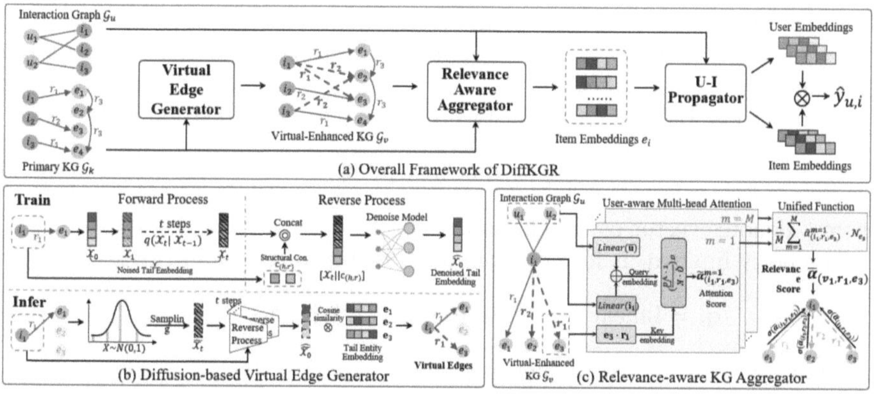

Fig. 2. (a) Overall framework of DiffKGR. (b) Diffusion-based Virtual Edge Generator that enriches connectivity by generating virtual edges. (c) Relevance-aware KG Aggregator that integrates information with relevance scores.

3.1 Diffusion-Based Virtual Edge Generation

Forward Diffusion Process. In the forward process, the tail entity embedding is gradually corrupted by the addition of Gaussian noise. Let the initial state of a tail entity denote as $\chi_0 \in \mathbb{R}^d$. The perturbed embedding χ_t at step t is:

$$q(\chi_t|\chi_{t-1}) = \mathcal{N}(\chi_t; \sqrt{1-\beta_t}\chi_{t-1}, \beta_t \boldsymbol{I}), \tag{1}$$

where $\mathcal{N}$ refers to the Gaussian distribution, and $\beta_t \in (0, 1)$ determines the magnitude of Gaussian noise at step t.

As $T \to \infty$, χ_T converges to a standard Gaussian distribution. By applying the reparameterization trick and leveraging the additive property of independent Gaussian noise, χ_t can be directly derived from χ_0:

$$q(\chi_t|\chi_0) = \mathcal{N}(\chi_t; \sqrt{\bar{\alpha}_t}\chi_0, (1-\bar{\alpha}_t)\boldsymbol{I}), \quad \alpha_t = 1-\beta_t, \quad \bar{\alpha}_t = \prod_{s=1}^{t} \alpha_s. \tag{2}$$

where a linear noise scheduler is applied to control the variance schedule with hyperparameters $s \in [0,1]$, α_{low}, and α_{up}:

$$1-\bar{\alpha}_t = s \cdot \left[\alpha_{low} + \frac{t-1}{T-1}(\alpha_{up} - \alpha_{low})\right], t \in \{1, \cdots, T\}. \tag{3}$$

χ_t then can be reparameterized as $\chi_t = \sqrt{\bar{\alpha}_t}\chi_0 + \sqrt{1-\bar{\alpha}_t}\epsilon, \quad \epsilon \sim \mathcal{N}(0, \boldsymbol{I})$.

Reverse Diffusion with Structural Guidance. In the reverse phase, the model gradually denoises noisy embedding χ_t to reconstruct clean embeddings χ_0 under structural condiction.

Structural guidance $\mathbf{c}_{h,r}$ is constructed from head and relation embedding:

$$\mathbf{c}_{h,r} = \text{Linear}([\mathbf{e}_h \parallel \mathbf{e}_r]), \tag{4}$$

where $\parallel$ denotes vector concatenation.

The noisy input $\mathbf{q}_t$ is formed from noisy embedding χ_t and timestep t:

$$\mathbf{q}_t = \text{Linear}(\chi_t + \text{Linear}([\cos(t \cdot \boldsymbol{\omega}) \parallel \sin(t \cdot \boldsymbol{\omega})])), \tag{5}$$

where $\boldsymbol{\omega} \in \mathbb{R}^{d/2}$ is a fixed frequency vector.

Then noisy input $\mathbf{q}_t$ and structural condition $\mathbf{c}_{h,r}$ are fused via multi-head attention, and the reconstructed embedding $\widehat{\chi}_0$ is produced through an MLP:

$$\widehat{\chi}_0 = \text{MLP}(\text{MultiheadAttention}(\mathbf{c}_{h,r}, \mathbf{q}_t, \mathbf{q}_t)). \tag{6}$$

Virtual Edges Generation. Given structure-guided denoise model and noisy samples, the generation module produces virtual edges in alignment of global distribution of $\mathcal{G}_k$. This generation begins with recovering a tail embedding $\widehat{\widetilde{\chi}}_0$ from a noisy sample $\widetilde{\chi}_T \sim \mathcal{N}(\mathbf{0}, \boldsymbol{I})$:

$$p_\theta(\widetilde{\chi}_{t-1} \mid \widetilde{\chi}_t, \mathbf{e}_h, \mathbf{e}_r) = \mathcal{N}\big(\widetilde{\chi}_{t-1}; \boldsymbol{\mu}_\theta(\widetilde{\chi}_t, t, \mathbf{e}_h, \mathbf{e}_r), \boldsymbol{\Sigma}_t\big), \tag{7}$$

where $\boldsymbol{\Sigma}_t$ and $\boldsymbol{\mu}_\theta$ denote the variance and mean of the posterior distribution at timestep t, which is computed by:

$$\boldsymbol{\mu}_\theta(\widetilde{\chi}_t, t, \mathbf{e}_h, \mathbf{e}_r) = \frac{\beta_t\sqrt{\alpha_{t-1}}}{1-\bar{\alpha}_t} \cdot \mathbf{f}_\theta(\widetilde{\chi}_t, t, \mathbf{e}_h, \mathbf{e}_r) + \frac{\sqrt{\alpha_t}(1-\bar{\alpha}_{t-1})}{1-\bar{\alpha}_t} \cdot \widetilde{\chi}_t,$$

where $\mathbf{f}_\theta(\widetilde{\chi}_t, t, \mathbf{e}_h, \mathbf{e}_r)$ is the structure-guided denoising function at timestep t, and the scalars $\beta_t, \alpha_t, \bar{\alpha}_t$ are diffusion coefficients defined in Eq. 1.

In practice, the next embedding is deterministically set to $\widetilde{\chi}_{t-1} = \widetilde{\boldsymbol{\mu}}_\theta(\widetilde{\chi}_t, t, \mathbf{e}_h, \mathbf{e}_r)$, and this process is iterated for $t = T, T-1, \ldots, 1$. After the generation process, we compute the cosine similarity between the recovered tail embedding $\widehat{\widetilde{\chi}}_0$ and each candidate tail embedding $\mathbf{e}_t$:

$$\text{sim}(\widehat{\widetilde{\chi}}_0, \mathbf{e}_{t\prime}) = \frac{\widehat{\widetilde{\chi}}_0 \cdot \mathbf{e}_{t\prime}}{\|\widehat{\widetilde{\chi}}_0\| \, \|\mathbf{e}_{t\prime}\|}, \quad \forall t\prime \in \mathcal{E}_r, \tag{8}$$

where $\mathcal{E}_r$ denotes the set of candidate tail entities associated with the relation r.

We select the top-k tail entities with $\text{sim}(\widehat{\widetilde{\chi}}_0, \mathbf{e}_{t\prime}) \geq \tau_r$. Based on selected tail entities $\mathcal{J}_r$, we construct virtual edges of the form $(h, r, \hat{t})$. The virtual-enhanced graph $\mathcal{G}_v$ is then constructed from primary KG $\mathcal{G}_k$ and virtual edges:

$$\mathcal{G}_v = \mathcal{G}_k \cup \{(h, r, \hat{t}) \mid \hat{t} \in \mathcal{J}_r, \, r \in \mathcal{R}, \, h \in \mathcal{H}\}. \tag{9}$$

3.2 Relevance-Aware Knowledge Aggregation

Computation of Task-Specific Relevance Score. Based on virtual-enhanced KG $\mathcal{G}_v$ and user-item interactions, we evaluate task-specific relevance.

For each item i, we aggregate user-context embeddings $\mathbf{e}_{\bar{u}_i}$ from its interacted user set $\mathcal{U}_i$:

$$\mathbf{e}_{\bar{u}_i} = \frac{1}{|\mathcal{U}_i|} \sum_{u \in \mathcal{U}_i} \mathbf{e}_u. \tag{10}$$

We use a multi-head attention to incorporate both item entity $\mathbf{e}_h$ and user context $\mathbf{e}_{\bar{u}_i}$. For each head, the query and key vectors are obtained by:

$$\mathbf{k}_{(h,r,t)}^{(m)} = \mathbf{e}_t^{(m)} \odot \mathbf{e}_r^{(m)}, \quad \mathbf{q}_h^{(m)} = \mathbf{W}_Q^{(m)} \mathbf{e}_h^{(m)} + \mathbf{W}_{QU}^{(m)} \mathbf{e}_{\bar{u}_h}^{(m)}, \tag{11}$$

where $m = 1, \ldots, M$ is attention head index, $\mathbf{W}_Q^{(m)}$ and $\mathbf{W}_{QU}^{(m)}$ are the learnable projection matrices, $\mathbf{e}_h^{(m)}$, $\mathbf{e}_t^{(m)}$ and $\mathbf{e}_r^{(m)}$ represent the corresponding slices of the head, tail and relation, respectively. $\odot$ is element-wise multiplication.

The normalize attention score $\tilde{\alpha}_{(h,r,t)}^{(m)}$ of attention head m is calculated as:

$$\tilde{\alpha}_{(h,r,t)}^{(m)} = softmax\left(\frac{\langle \mathbf{q}_h^{(m)}, \mathbf{k}_{(h,r,t)}^{(m)} \rangle}{\tau \cdot \sqrt{d_k}}\right), \tag{12}$$

where d_k is the dimension of each head, and τ is a temperature hyperparameter.

To decouple the attention score from varying number of neighbors, we rescale the normalized attention $\tilde{\alpha}_{(h,r,t)}^{(m)}$ to get a unified relevance score $\bar{\alpha}_{(h,r,t)}$:

$$\bar{\alpha}_{(h,r,t)} = \frac{1}{M} \sum_{m=1}^{M} (\tilde{\alpha}_{(h,r,t)}^{(m)} \cdot |\mathcal{N}_h|). \tag{13}$$

Relevance-Aware KG Aggregation. Relevance-aware KG aggregation performs a weighted aggregation guided by relevance score $\bar{\alpha}_{(h,r,t)}$.

We perform relevance-aware aggregation for item entities, and relevance-based weight $\beta_{(h,r,t)}$ is calculated from relevance score $\bar{\alpha}_{(h,r,t)}$:

$$\beta_{(h,r,t)} = softmax(\bar{\alpha}_{(h,r,t)}), \tag{14}$$

The item embedding $\mathbf{e}_h^{l(\text{agg})}$ is obtained by a residual connection with the entity embedding from last layer:

$$\mathbf{e}_h^{l(\text{agg})} = \mathbf{e}_h^{l-1(\text{agg})} + \mathbf{e}_h^l, \quad \mathbf{e}_h^l = \sum_{(h,r,t)\in\mathcal{N}_h} \beta_{(h,r,t)}(\mathbf{e}_t \odot \mathbf{e}_r), \tag{15}$$

where $\mathcal{N}_h$ is neighbors of entity h.

For attribute entities, we apply a lightweight, relevance-agnostic aggregation:

$$\mathbf{e}_h^{l(\text{agg})} = \mathbf{e}_h^{l-1(\text{agg})} + \frac{1}{|\mathcal{N}_h|} \sum_{(h,r,t)\in\mathcal{N}_h} (\mathbf{e}_t \odot \mathbf{e}_r). \tag{16}$$

3.3 U-I Propagation and Model Prediction

User-Item Interaction Propagation. We concatenate item embeddings from primary KG and virtual-enhanced KG to preserve both perspectives:

$$\mathbf{e}_i^{l(\text{concat})} = [\mathbf{e}_i^{l(\text{agg-prim})} \,\|\, \mathbf{e}_i^{l(\text{agg-virt})}], \tag{17}$$

where $\mathbf{e}_i^{\text{prim}}$ and $\mathbf{e}_i^{\text{virt}}$ are item embeddings from the primary KG $\mathcal{G}_k$ and virtual-enhanced KG $\mathcal{G}_v$, and l denotes the l-th layer.

Let $\mathbf{A}_{UI}$ and $\mathbf{A}_{IU}$ be the adjacency matrices for user-item and item-user interactions, respectively. Mutual propagation for user and item embeddings is:

$$\hat{\mathbf{e}}_u^l = \mathbf{A}_{UI}\mathbf{e}_i^{l(\text{concat})}, \quad \hat{\mathbf{e}}_i^l = \mathbf{A}_{IU}\mathbf{e}_u^l. \tag{18}$$

Interaction Prediction. We sum the multi-layer output as final representation for users and items, and use cosine similarity to forecast how likely the user u would interact with item i:

$$\hat{y}_{u,i} = \cos(\hat{\mathbf{e}}_u, \hat{\mathbf{e}}_i), \quad \hat{\mathbf{e}}_u = \sum_{l=0}^{L} \hat{\mathbf{e}}_u^l, \quad \hat{\mathbf{e}}_i = \sum_{l=0}^{L} \hat{\mathbf{e}}_i^l. \tag{19}$$

3.4 Model Optimization

Optimization of KG Diffusion Process. We use a joint loss combining a contrastive objective and a regularization term for diffusion model:

$$\mathcal{L}_{\text{diff}} = \mathcal{L}_{\text{contrast}} + \lambda_{\text{reg}} \cdot \mathcal{L}_{\text{reg}}, \tag{20}$$

where λ_{reg} is a weighting coefficient balancing $\mathcal{L}_{\text{contrast}}$ and $\mathcal{L}_{\text{reg}}$.

The contrastive loss and regularization loss are defined as:

$$\mathcal{L}_{\text{contrast}} = -\frac{1}{N}\sum_{i=1}^{N}\log\frac{\exp(\text{sim}(z_{1,i}, z_{2,i}))}{\sum_{j=1}^{N}\exp(\text{sim}(z_{1,i}, z_{2,j}))}, \quad \mathcal{L}_{\text{reg}} = \frac{1}{N}\sum_{i=1}^{N}||z_{2,i}||, \tag{21}$$

where $\text{sim}(\cdot,\cdot)$ denotes cosine similarity, z_1 and z_2 denote ground-truth embeddings and generated embeddings, respectively.

Optimization for the Recommendation Model. Let $\mathbf{S} \in \mathbb{R}^{N\times M}$ denote the predicted user-item score matrix, where $\mathbf{S}_{ij}$ is the predicted score between user i and item j. The cross entropy loss for recommendation is defined as:

$$\mathcal{L}_{\text{rec}} = -\frac{1}{N}\sum_{i=1}^{N}\sum_{j=1}^{M} q_{ij}\log\frac{\exp(\mathbf{S}_{ij})}{\sum_{k=1}^{M}\exp(\mathbf{S}_{ik})}, \tag{22}$$

where q_{ij} denotes the smoothed label distribution over items.

3.5 Time Complexity Analysis

Let N be the number of entities, d the embedding dimension, T the number of diffusion steps, and $E_a \subseteq E$ the number of item-related edges.

Diffusion-Based Virtual Edge Generation. Each forward pass processes one candidate pair with $O(d^2)$ complexity, mainly from the denoising model. During inference, one reverse diffusion step costs $O(Td^2)$, computing similarity adds $O(Nd)$, and top-k selection adds $O(N)$. The total inference complexity per pair is therefore $O(Td^2 + Nd)$. In practice, diffusion training takes an average of 12.72 s/epoch, and virtual edge generation costs 9.42 s on average across datasets.

Relevance-Aware KG Aggregation. The attention branch costs $O(E_a d)$, sparse user–item interaction costs $O(Kd)$, and the mean branch costs $O((E - E_a)d)$. Final aggregation adds $O(Nd)$, resulting in overall complexity $O(Kd + Ed + Nd)$ per forward pass. With virtual edges added, the average training time per epoch is 280.07 s (Last-FM), 1444.88 s (Yelp2018), and 1592.18 s (Amazon-Book), while average testing times are 33.67 s, 87.13 s, and 62.29 s, respectively.

4 Experiments

To evaluate the effectiveness of DiffKGR, we design a series of experiments to address the following research questions:

- **RQ1**: How does DiffKGR perform compared to state-of-the-art recommendation approaches?
- **RQ2**: How do the two modules respectively contribute to the performance?
- **RQ3**: How does DiffKGR address the sparsity problem in knowledge graphs?
- **RQ4**: Are the virtual edges structurally consistent with the primary KG?
- **RQ5**: How effective is the relevance-aware aggregation in suppressing task-irrelevant information?
- **RQ6:** What are the effects of different hyperparameters of DiffKGR?

4.1 Experimental Settings

Dataset. We evaluate on three real-world datasets: Last-FM, Amazon-Book, and Yelp2018. Following previous works [28, 42], items are aligned with Freebase [4] entities, and primary KG is constructed with two-hop neighbors. Dataset statistics are shown in Table 1.

Table 1. Statistics of the experimental datasets.

Statistics	Last-FM	Amazon-book	Yelp2018
Users	23,566	70,679	45,919
Items	48,123	24,915	45,538
Interactions	3,034,796	846,434	1,183,610
Density	2.7e-3	4.8e-4	5.7e-4
Knowledge Graph			
Entities	58,266	29,714	47,472
Relations	9	39	42
Triplets	464,567	686,516	869,603

Evaluation Metrics. Following [35,36,42], we evaluate the recommendation performance with Recall@20 and NDCG@20 for DiffKGR and all baseline methods. We report the average results over all users in the test set.

Compared Baseline Models. We compare DiffKGR with various baseline models, which can be grouped into three categories: Collaborative Filtering (CF) methods (BPR [25], NeuMF [9], GC-MC [29], LightGCN [8], and SGL [40].), Embedding-based knowledge-enhanced methods (CKE [45] and KTUP [6]), GNN-based knowledge-enhanced methods (KGNN-LS [31], KGCN [33], KGAT [35], KGIN [36], MCCLK [47], KGCL [43], KGRec [42], KRDN [46], and EditKG [28]).

Parameter Settings. DiffKGR is implemented in PyTorch. We set the ID embedding dimension $d = 128$, batch size $= 4096$, and learning rate $= 1 \times 10^{-3}$. Model parameters are initialized with Xavier initialization and optimized using Adam. For diffusion-based generation, we discuss the selection of top$k$ and τ_r in Sect. 4.7. For fair comparison, we report each baseline's best results from the original papers or reproduce them using released code with default settings.

4.2 RQ1: Overall Performance Comparison

Table 2 presents the performance across all baselines and DiffKGR.

We observe that DiffKGR achieves state-of-the-art performance across all datasets. In terms of Recall@*20*, DiffKGR outperforms the best baselines by 4.70% on Last-FM, 4.57% on Amazon-book, and 2.46% on Yelp2018. Similar trends are observed for NDCG *@20*, with improvements of 4.74%, 4.69%, and 1.60%, respectively. The performance gains mainly stem from two components. By injecting structurally consistent virtual edges, the diffusion-based generation module compensates for missing links, thereby alleviating sparsity in KG. Meanwhile, the relevance-aware aggregation reduces task-irrelevant noise by weighting item edges according to their relevance to the recommendation task.

Table 2. Comparison on real-world datasets in terms of Recall@*20* and NDCG@*20*. Bold indicates the best and underline the second best.

Model	Last-FM		Amazon-book		Yelp2018	
	Recall	NDCG	Recall	NDCG	Recall	NDCG
BPR	0.0690	0.0585	0.0822	0.0501	0.0384	0.0253
NeuMF	0.0699	0.0615	0.0506	0.0276	0.0308	0.0237
GC-MC	0.0709	0.0631	0.0845	0.0502	0.0386	0.0261
LightGCN	0.0738	0.0647	0.1058	0.0652	0.0419	0.0253
SGL	0.0879	0.0775	0.1141	0.0713	0.0429	0.0275
CKE	0.0845	0.0718	0.0835	0.0512	0.0387	0.0247
KTUP	0.0865	0.0671	0.0976	0.0634	0.0362	0.0302
KGNN-LS	0.0881	0.0690	0.0983	0.0633	0.0395	0.0302
KGCN	0.0879	0.0694	0.0983	0.0633	0.0396	0.0302
KGAT	0.0870	0.0743	0.0957	0.0577	0.0340	0.0287
KGIN	0.0900	0.0779	0.1144	0.0723	0.0357	0.0225
MCCLK	0.0671	0.0603	0.1089	0.0707	0.0327	0.0194
KGCL	0.0905	0.0769	0.1146	0.0719	0.0399	0.0247
KGRec	0.0943	0.0810	0.1443	0.0782	0.0745	0.0478
KRDN	0.1023	0.0946	0.1574	0.0886	0.0842	0.0544
EditKG	<u>0.1212</u>	<u>0.1095</u>	<u>0.1770</u>	<u>0.1003</u>	<u>0.0854</u>	<u>0.0564</u>
DiffKGR	**0.1269**	**0.1147**	**0.1851**	**0.1050**	**0.0875**	**0.0573**

Dataset-wise, Table 1 shows that Last-FM has lower triple-to-entity ratio than other two datasets. This indicates that KG from Last-fm is sparser than those of Amazon-Book and Yelp2018. Correspondingly, DiffKGR demonstrates the highest relative improvement on Last-FM among three datasets.

Finally, we note that KG-enhanced methods generally outperform CF baselines. Among them, GNN-based and contrastive models provide stronger representations through structural modeling and regularization.

4.3 RQ2: Ablation Study

This study aims to evaluate the effectiveness of the key modules incorporated in DiffKGR. We develop three distinct model variants:

- "w/o V-KG": This variant excludes the diffusion-based virtual edge generation module and performs message passing solely over the primary KG.
- "w/o R-Agg": We remove the relevance-aware computation and assign uniform weights to edges of knowledge graphs during message passing.
- "w/o both": We disable both the diffusion-based generation module and the relevance-aware aggregation. Message passing is conducted over the primary KG with uniform edge weights.

Table 3 show that removing either module reduces performance. Removing the virtual edge generation causes a larger drop, underscoring the importance of virtual edge in mitigating sparsity and improving information propagation. Disabling relevance-aware aggregation also degrades performance by introducing task-irrelevant noise. The largest decline occurs when both modules are removed, demonstrating their complementary roles in enhancing recommendation quality.

Table 3. Ablation study on key components.

Models	Last-FM		Amazon-book		Yelp2018	
	Recall	NDCG	Recall	NDCG	Recall	NDCG
w/o V-KG	0.1212	0.1094	0.1819	0.1022	0.0856	0.0561
w/o R-Agg	0.1264	0.1142	0.1841	0.1042	0.0837	0.0548
w/o both	0.1211	0.1092	0.1801	0.1018	0.0840	0.0552
DiffKGR	**0.1269**	**0.1147**	**0.1851**	**0.1050**	**0.0875**	**0.0573**

4.4 RQ3: Model Benefits Investigation in Alleviating KG Sparsity

To examine how our framework alleviates knowledge graph sparsity, we plot each entity's degree in the primary KG (blue) and the virtual-enhanced KG (green).

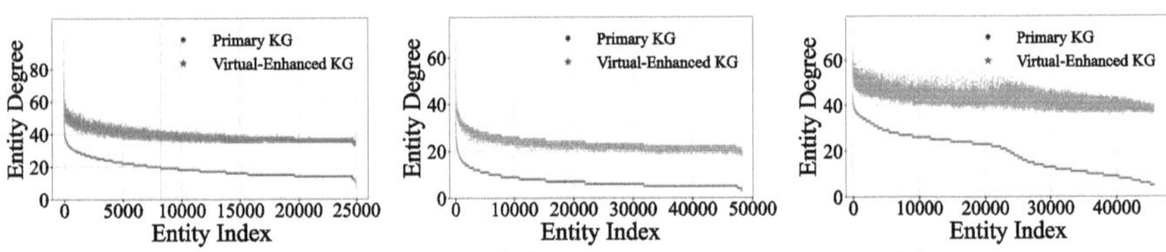

(a) Amazon-book. (b) Last-FM (c) Yelp2018

Fig. 3. Distribution of entity degrees in different datasets. (Color figure online)

As shown in Fig. 3, the superior position of the green curve over the blue one confirms that most entities acquire additional neighbors through virtual edge generation. Taking the Amazon-Book dataset as an example, approximately 67% of entities originally have fewer than 20 edges. This number increases to around 40 after applying our generation module, demonstrating the method's effectiveness in mitigating KG sparsity. Similar trends can be observed in other datasets, as shown in Fig. 3(b) and (c).

4.5 RQ4: Consistency of Virtual Edge Generation

To assess the structural consistency of virtual edges, we first examine the embedding distributions of translated triples ($\mathbf{e}_h + \mathbf{e}_r - \mathbf{e}_t$), following the TransE [38] paradigm. For clarity, we randomly sample 100,000 edges from each dataset and reduce their dimensionality with SVD for visualization.

As shown in Fig. 4, virtual edges (blue) largely overlap with primary ones (red) across all datasets, suggesting that the diffusion-based generator preserves the overall structural distribution of the primary KG.

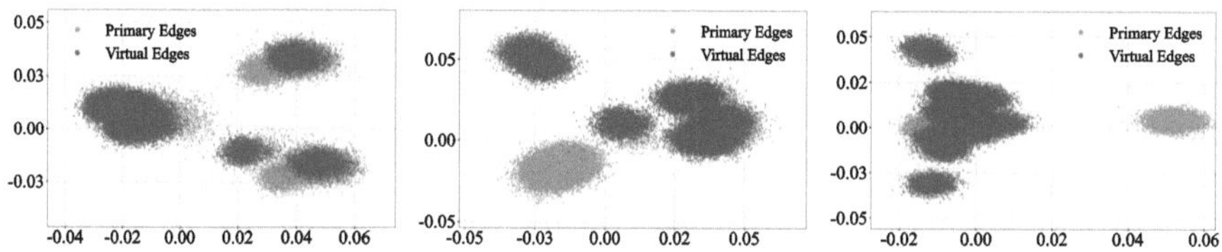

(a) Amazon-book. (b) Last-FM (c) Yelp2018

Fig. 4. Visualization of edge embeddings in different datasets. (Color figure online)

To further validate consistency of generation, we evaluate diffusion model through a link prediction task. Edges in primary KG are divided into 80% training and 20% evaluation, and results are reported in Table 4.

Table 4. Results of Link prediction.

Models	Last-FM		Amazon-book		Yelp2018	
	MR	hit@5	MR	hit@5	MR	hit@5
TransE [5]	1563.31	0.2518	7542.86	0.3954	550.79	0.3907
CooccurrenceFiltered [2]	964.45	0.8567	248.86	0.5232	22.47	0.6076
Diffusion	186.72	0.6700	42.13	0.7337	24.44	0.7757

Compared with classical KG embedding models, the strong MR (24.44-186.72) and Hit@5 (0.6700–0.7757) demonstrate that generated edges are consistently ranked top among all candidates in link prediction. This confirms the module's effectiveness in preserving global structure distribution of primary KG.

4.6 RQ5: Effect of Relevance-Aware Aggregation

We examine how relevance score influences the KG aggregation. As shown in Fig. 5, x-axis denotes the relevance score (log-scaled), y-axis denotes aggregation weight, and the color intensity reflects the value of weights during aggregation.

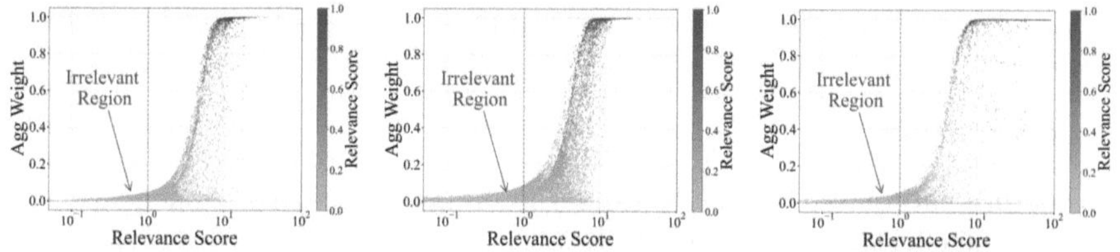

(a) Amazon-book. (b) Last-FM (c) Yelp2018

Fig. 5. Relevance score vs. Aggregation weight in different datasets.

We observe that a high concentration of edges have low relevance scores (score < 1), as shown by the irrelevant region left of the red dashed line. These edges receive near-zero aggregation weights, indicating that relevance-aware aggregation effectively suppresses noise while emphasizing informative edges.

4.7 RQ6: Hyperparameter Analysis

We analyze parameter sensitivity by varying the top-k candidate and threshold τ_r in diffusion-based virtual edge generation. Specifically, we test top-1, top-3, and top-5 settings for k, and τ_r values from 0.8 to 0.95, as shown in Fig. 6.

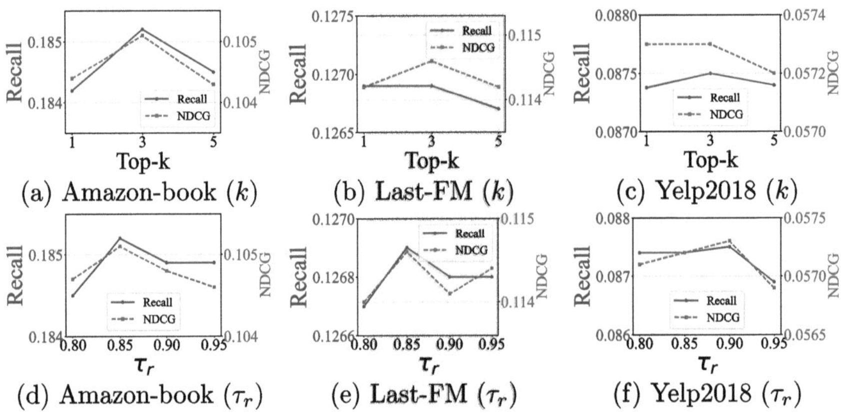

Fig. 6. Hyperparameter sensitivity analysis across datasets. (a–c) Results with different top-k virtual edges; (d–f) Results with different thresholds τ_r.

Figure 6(a–c) shows that top-3 achieves the best or comparable performance across datasets. Top-1 provides too few connections to reduce KG sparsity effectively, while top-5 adds noise and computational cost without clear gains. Thus, we adopt top-3 as the default setting for virtual edge generation.

Similar to the top-k setting, lower τ_r values generate more virtual edges with noise. We select the threshold that yields the best performance for each dataset: $\tau_r = 0.85$ for Amazon-Book and Last-FM, and $\tau_r = 0.9$ for Yelp2018.

4.8 Case Study

Figure 7 presents a real example from the Amazon-book recommendation to illustrate the effectiveness of DiffKGR.

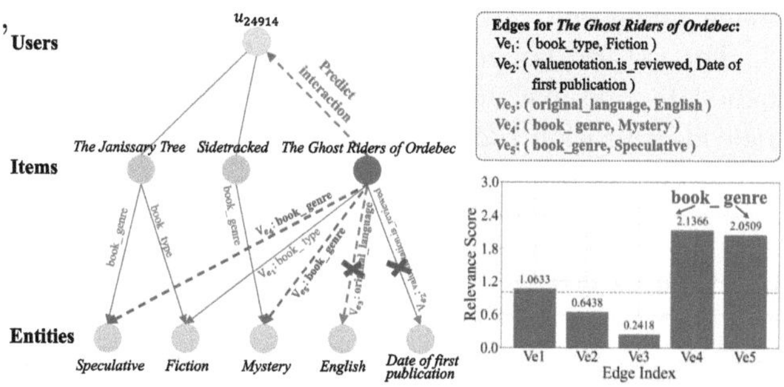

Fig. 7. Case study from Amazon-Book dataset.

Consider user u_{24914}, who has previously read *The Janissary Tree* and *Sidetracked*, both labeled as *Fiction* with genres *Speculative* and *Mystery*. When evaluating the candidate book *The Ghost Riders of Ordebec* for recommendation, its sparse attribute connections in the knowledge graph make it difficult to assess its alignment with the user's preferences. To address this, our diffusion-based module generates virtual edges V_{e_3} (*original_language*), V_{e_4} (*book_genre*), and V_{e_5} (*book_genre*) for this candidate. These inferred attributes enrich the book's relational context, enabling more effective information propagation and improving the modeling of user preferences, thereby enhancing recommendation quality.

Subsequently, we compute task-specific relevance scores for edges V_{e_1} to V_{e_5}, obtaining values of 1.0633, 0.6438, 0.2418, 2.1366, and 2.0509, respectively. Among these, V_{e_4} and V_{e_5} receive the highest scores and are accordingly assigned greater weights during knowledge graph aggregation. Based on the enriched item representation, the virtual-enhanced book *The Ghost Riders of Ordebec* is predicted to exhibit a high interaction likelihood with user u_{24914}, and is therefore recommended to this user.

5 Related Work

KG-Enhanced Recommendation. KG-enhanced recommendation methods can be grouped into three types. Embedding-based approaches [1,6,30,45] encode entities and relations with KG embeddings techniques. Path-based methods capture semantic connections along KG paths [37]. Propagation-based methods employ GNNs to aggregate multi-hop neighbors, with KGAT [35] using attention and KGIN [36] applying relation-aware disentanglement.

Knowledge Graph Augmentation for Recommendation. To address KG sparsity, recent studies have proposed various augmentation. EditKG [28] adds attribute links under the assumption that similar items share common attributes. SDK [17] employs a self-supervised hypergraph framework to generate high-order relational links. VRKG4Rec [19] builds virtual relational graphs to reveal implicit relations. KGRec [42] focuses on task-relevant links by aligning KG structures with collaborative filtering signals.

Diffusion Models. Diffusion probabilistic models (DPMs) have demonstrated strong performance across domains. In computer vision, they are applied to image inpainting [21], super-resolution [14], denoising [27], and segmentation [3]. For natural language processing, DPMs improve text generation [22], dialogue [16], and machine translation [7]. DPMs also support graph tasks such as node classification [12] and link prediction [24], and have recently been applied to recommendation systems [13,15,34].

6 Conclusion

In this paper, we proposed DiffKGR, a novel diffusion-based framework mitigating the sparsity issue for knowledge graph-enhanced recommendation. DiffKGR consists of two key components: (i) a diffusion-based module that generates structurally consistent virtual edges through a structure-guided diffusion process, and (ii) a relevance-aware KG aggregation mechanism that leverages task-specific relevance scores to suppress irrelevant message passing. Extensive experiments on real-world datasets demonstrate that DiffKGR achieves superior performance over state-of-the-art baselines.

Acknowledgments. This study was funded by National Natural Science Foundation of China (No. 92579211, 62272301, 62525209, T2421002, 62432002) and Natural Science Foundation of Shanghai (No. 21TQ1400214).

References

1. Ai, Q., Azizi, V., Chen, X., Zhang, Y.: Learning heterogeneous knowledge base embeddings for explainable recommendation. Algorithms **11**(9), 137 (2018)
2. Ali, M., et al.: PyKEEN 1.0: a python library for training and evaluating knowledge graph embeddings. J. Mach. Learn. Res. **22**(82), 1–6 (2021). http://jmlr.org/papers/v22/20-825.html
3. Amit, T., Shaharbany, T., Nachmani, E., Wolf, L.: Segdiff: image segmentation with diffusion probabilistic models. arXiv preprint arXiv:2112.00390 (2021)
4. Bollacker, K., Evans, C., Paritosh, P., Sturge, T., Taylor, J.: Freebase: a collaboratively created graph database for structuring human knowledge. In: Proceedings of the 2008 ACM SIGMOD International Conference on Management of Data, pp. 1247–1250 (2008)
5. Bordes, A., Usunier, N., Garcia-Duran, A., Weston, J., Yakhnenko, O.: Translating embeddings for modeling multi-relational data. In: Advances in Neural Information Processing Systems, vol. 26 (2013)
6. Cao, Y., Wang, X., He, X., Hu, Z., Chua, T.S.: Unifying knowledge graph learning and recommendation: towards a better understanding of user preferences. In: The World Wide Web Conference, pp. 151–161 (2019)
7. Chen, L., Feng, A., Yang, B., Li, Z.: XDLM: cross-lingual diffusion language model for machine translation. arXiv preprint arXiv:2307.13560 (2023)
8. He, X., Deng, K., Wang, X., Li, Y., Zhang, Y., Wang, M.: Lightgcn: simplifying and powering graph convolution network for recommendation. In: Proceedings of the 43rd International ACM SIGIR Conference on Research and Development in Information Retrieval, pp. 639–648 (2020)

9. He, X., Liao, L., Zhang, H., Nie, L., Hu, X., Chua, T.S.: Neural collaborative filtering. In: Proceedings of the 26th International Conference on World Wide Web, pp. 173–182 (2017)
10. Hu, B., Shi, C., Zhao, W.X., Yu, P.S.: Leveraging meta-path based context for top-n recommendation with a neural co-attention model. In: Proceedings of the 24th ACM SIGKDD International Conference on Knowledge Discovery & Data Mining, pp. 1531–1540 (2018)
11. Isinkaye, F.O., Folajimi, Y.O., Ojokoh, B.A.: Recommendation systems: principles, methods and evaluation. Egypt. Inform. J. **16**(3), 261–273 (2015)
12. Jang, H., Park, S., Mo, S., Ahn, S.: Diffusion probabilistic models for structured node classification. Adv. Neural. Inf. Process. Syst. **36**, 32075–32101 (2023)
13. Jiang, Y., Xia, L., Wei, W., Luo, D., Lin, K., Huang, C.: Diffmm: multi-modal diffusion model for recommendation. In: Proceedings of the 32nd ACM International Conference on Multimedia, pp. 7591–7599 (2024)
14. Li, H., et al.: Srdiff: single image super-resolution with diffusion probabilistic models. Neurocomputing **479**, 47–59 (2022)
15. Li, Z., Sun, A., Li, C.: Diffurec: a diffusion model for sequential recommendation. ACM Trans. Inf. Syst. **42**(3), 1–28 (2023)
16. Liu, S., Chen, H., Ren, Z., Feng, Y., Liu, Q., Yin, D.: Knowledge diffusion for neural dialogue generation. In: Proceedings of the 56th Annual Meeting of the Association for Computational Linguistics (Volume 1: Long Papers), pp. 1489–1498 (2018)
17. Liu, Y., Xuan, H., Li, B., Wang, M., Chen, T., Yin, H.: Self-supervised dynamic hypergraph recommendation based on hyper-relational knowledge graph. In: Proceedings of the 32nd ACM International Conference on Information and Knowledge Management, pp. 1617–1626 (2023)
18. Lu, J., Wu, D., Mao, M., Wang, W., Zhang, G.: Recommender system application developments: a survey. Decis. Support Syst. **74**, 12–32 (2015)
19. Lu, L., Wang, B., Zhang, Z., Liu, S., Xu, H.: Vrkg4rec: virtual relational knowledge graph for recommendation. In: Proceedings of the Sixteenth ACM International Conference on Web Search and Data Mining, pp. 526–534 (2023)
20. Lü, L., Medo, M., Yeung, C.H., Zhang, Y.C., Zhang, Z.K., Zhou, T.: Recommender systems. Phys. Rep. **519**(1), 1–49 (2012)
21. Lugmayr, A., Danelljan, M., Romero, A., Yu, F., Timofte, R., Van Gool, L.: Repaint: inpainting using denoising diffusion probabilistic models. In: Proceedings of the IEEE/CVF Conference on Computer Vision and Pattern Recognition, pp. 11461–11471 (2022)
22. Luu, A.T., Buntine, W.L., et al.: Discrete diffusion language model for long text summarization. CoRR (2024)
23. Ma, T., Huang, L., Lu, Q., Hu, S.: KR-GCN: knowledge-aware reasoning with graph convolution network for explainable recommendation. ACM Trans. Inf. Syst. **41**(1), 1–27 (2023)
24. Nguyen, T.K., Fang, Y.: Diffusion-based negative sampling on graphs for link prediction. In: Proceedings of the ACM Web Conference 2024, pp. 948–958 (2024)
25. Rendle, S., Freudenthaler, C., Gantner, Z., Schmidt-Thieme, L.: BPR: Bayesian personalized ranking from implicit feedback. In: Proceedings of the Twenty-Fifth Conference on Uncertainty in Artificial Intelligence, pp. 452–461 (2009)
26. Schafer, J.B., Konstan, J.A., Riedl, J.: E-commerce recommendation applications. Data Min. Knowl. Disc. **5**, 115–153 (2001)
27. Shen, C., Yang, Z., Zhang, Y.: Pet image denoising with score-based diffusion probabilistic models. In: International Conference on Medical Image Computing and Computer-Assisted Intervention, pp. 270–278. Springer (2023)

28. Tang, G., et al.: Editkg: editing knowledge graph for recommendation. In: Proceedings of the 47th International ACM SIGIR Conference on Research and Development in Information Retrieval, pp. 112–122 (2024)
29. Van Den Berg, R., Thomas, N.K., Welling, M.: Graph convolutional matrix completion. arXiv preprint arXiv:1706.02263, vol. 2, no. 8, p. 9 (2017)
30. Wang, H., Zhang, F., Xie, X., Guo, M.: DKN: deep knowledge-aware network for news recommendation. In: Proceedings of the 2018 World Wide Web Conference, pp. 1835–1844 (2018)
31. Wang, H., et al.: Knowledge-aware graph neural networks with label smoothness regularization for recommender systems. In: Proceedings of the 25th ACM SIGKDD International Conference on Knowledge Discovery & Data Mining, pp. 968–977 (2019)
32. Wang, H., Zhang, F., Zhao, M., Li, W., Xie, X., Guo, M.: Multi-task feature learning for knowledge graph enhanced recommendation. In: The World Wide Web Conference, pp. 2000–2010 (2019)
33. Wang, H., Zhao, M., Xie, X., Li, W., Guo, M.: Knowledge graph convolutional networks for recommender systems. In: The World Wide Web Conference, pp. 3307–3313 (2019)
34. Wang, W., Xu, Y., Feng, F., Lin, X., He, X., Chua, T.S.: Diffusion recommender model. In: Proceedings of the 46th International ACM SIGIR Conference on Research and Development in Information Retrieval, pp. 832–841 (2023)
35. Wang, X., He, X., Cao, Y., Liu, M., Chua, T.S.: KGAT: knowledge graph attention network for recommendation. In: Proceedings of the 25th ACM SIGKDD International Conference on Knowledge Discovery & Data Mining, pp. 950–958 (2019)
36. Wang, X., et al.: Learning intents behind interactions with knowledge graph for recommendation. In: Proceedings of the Web Conference 2021, pp. 878–887 (2021)
37. Wang, X., Wang, D., Xu, C., He, X., Cao, Y., Chua, T.S.: Explainable reasoning over knowledge graphs for recommendation. In: Proceedings of the AAAI Conference on Artificial Intelligence, vol. 33, pp. 5329–5336 (2019)
38. Wang, Z., Zhang, J., Feng, J., Chen, Z.: Knowledge graph embedding by translating on hyperplanes. In: Proceedings of the AAAI Conference on Artificial Intelligence, vol. 28 (2014)
39. Wei, Y., Wang, X., Nie, L., He, X., Hong, R., Chua, T.S.: MMGCN: multi-modal graph convolution network for personalized recommendation of micro-video. In: Proceedings of the 27th ACM International Conference on Multimedia, pp. 1437–1445 (2019)
40. Wu, J., et al.: Self-supervised graph learning for recommendation. In: Proceedings of the 44th International ACM SIGIR Conference on Research and Development in Information Retrieval, pp. 726–735 (2021)
41. Xian, Y., Fu, Z., Muthukrishnan, S., De Melo, G., Zhang, Y.: Reinforcement knowledge graph reasoning for explainable recommendation. In: Proceedings of the 42nd International ACM SIGIR Conference on Research and Development in Information Retrieval, pp. 285–294 (2019)
42. Yang, Y., Huang, C., Xia, L., Huang, C.: Knowledge graph self-supervised rationalization for recommendation. In: Proceedings of the 29th ACM SIGKDD Conference on Knowledge Discovery and Data Mining, pp. 3046–3056 (2023)
43. Yang, Y., Huang, C., Xia, L., Li, C.: Knowledge graph contrastive learning for recommendation. In: Proceedings of the 45th International ACM SIGIR Conference on Research and Development in Information Retrieval, pp. 1434–1443 (2022)
44. Yuan, S.T., Tsao, Y.W.: A recommendation mechanism for contextualized mobile advertising. Expert Syst. Appl. **24**(4), 399–414 (2003)

45. Zhang, F., Yuan, N.J., Lian, D., Xie, X., Ma, W.Y.: Collaborative knowledge base embedding for recommender systems. In: Proceedings of the 22nd ACM SIGKDD International Conference on Knowledge Discovery and Data Mining, pp. 353–362 (2016)
46. Zhu, X., Du, Y., Mao, Y., Chen, L., Hu, Y., Gao, Y.: Knowledge-refined denoising network for robust recommendation. In: Proceedings of the 46th International ACM SIGIR Conference on Research and Development in Information Retrieval, pp. 362–371 (2023)
47. Zou, D., et al.: Multi-level cross-view contrastive learning for knowledge-aware recommender system. In: Proceedings of the 45th International ACM SIGIR Conference on Research and Development in Information Retrieval, pp. 1358–1368 (2022)

PLRF: A Personalized Learning Recommendation Framework Based on Federated Knowledge Graphs

Yiping Teng[1], Tiantian Yu[1], Gang Wang[1](✉), Zhen Song[2], Jiajia Li[1], and Bing Hu[3]

[1] School of Computer, Shenyang Aerospace University, Shenyang, China
{typ,gangwang,lijiajia}@sau.edu.cn, yutiantian@stu.sau.edu.cn
[2] School of Software, Shandong University, Jinan, China
songzhen@sdu.edu.cn
[3] Faculty of Computer Science, Dalhousie University, Halifax, Canada
binghu@dal.ca

Abstract. The rapid advancement of artificial intelligence and the proliferation of online education platforms have made accurate and personalized recommendation services a critical component of modern learning systems. Traditional recommendation models often exhibit low accuracy due to the cold-start problem and the limited availability of user–item interaction data. Existing solutions have introduced knowledge graphs (KGs) to incorporate semantic relations into recommendation models. However, the improvement in accuracy remains limited. Moreover, KG-based methods commonly rely on locally constructed graphs that are incomplete and unable to exploit information distributed across other educational platforms. In this work, we propose PLRF, a federated knowledge-graph-based recommendation framework that enables collaborative learning across multiple platforms while preserving data privacy. To handle data heterogeneity, PLRF introduces a regularization mechanism during local training to align heterogeneous representations. Extensive evaluations on two real-world datasets demonstrate that PLRF achieves improvements in NDCG, Recall, Precision, and F1 score compared with both non-KG and single-source KG approaches, highlighting the potential of federated knowledge graph learning in cross-platform personalized recommendation.

Keywords: Knowledge graph · Federated learning · Personalized Recommendation

1 Introduction

The rapid advancement of artificial intelligence has opened new possibilities for personalized education. Modern digital learning environments, including online

Supplementary Information The online version contains supplementary material available at https://doi.org/10.1007/978-981-92-0363-5_18.

H. Jung et al. (Eds.): DASFAA 2026, LNCS 16535, pp. 292–307, 2026.
https://doi.org/10.1007/978-981-92-0363-5_18

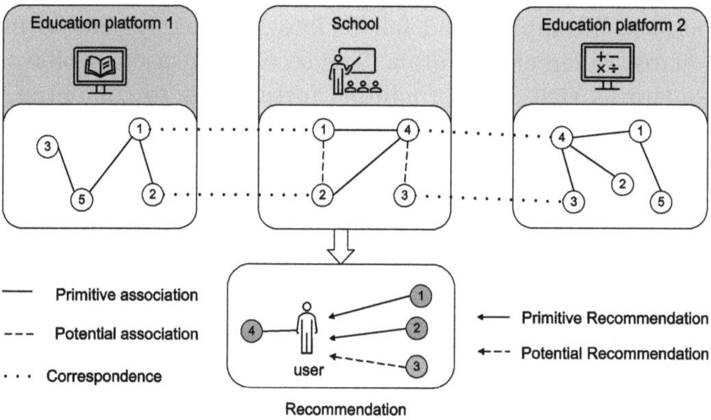

Fig. 1. Example of personalized recommendation based on primitive and potential associations.

courses, learning management systems, and institutional repositories, continuously generate massive amounts of learner-related data. These data originate from diverse sources and exhibit strong heterogeneity, leading to complex multi-source educational datasets. In a typical educational environment, students continuously generate learning traces across multiple platforms, including the school's internal system and several external online education services. Each platform maintains its own KG that encodes problem entities and their associations. Figure 1 illustrates an example of cross-platform personalized recommendation based on such heterogeneous KGs. When a student answers Problem ④ incorrectly in the school system, the recommendation module retrieves related problems within the school KG through primitive associations such as prerequisite or concept similarity relations. At the same time, the system exploits potential associations inferred from external platforms. On Education Platform 2, Problem ④ is related to Problem ③ through latent or learned connections. By aligning corresponding entities between the school and platform KGs, the system expands the candidate recommendation set to include Problem ③, although it is not explicitly connected to Problem ④ in the school KG. However, such educational data are often isolated within individual institutions or platforms, forming information silos that impede cross-platform integration. Moreover, educational data contain highly sensitive personal information and are subject to strict privacy protections, which are often even more stringent than those in commercial domains [8]. These characteristics create substantial challenges for developing accurate and privacy-preserving educational recommendation systems.

Existing approaches have primarily evolved in two directions to improve recommendation quality in educational contexts. The first focuses on enhancing semantic reasoning through knowledge-graph-based recommendation models. These methods [18,23] exploit rich relational structures among entities to alleviate data sparsity and cold-start problems [14,18]. While effective at capturing conceptual associations, they typically rely on locally constructed knowledge graphs, which are incomplete and fail to integrate heterogeneous information distributed across different platforms. The second direction explores federated learning frameworks that enable multiple institutions to collaboratively train recommendation models without exposing raw data [9,11]. This paradigm helps address data-sharing restrictions and privacy concerns, allowing broader participation across organizations. However, current federated systems [22] still depend on structurally inconsistent local knowledge graphs that remain disconnected, leading to poor semantic alignment and limited information integration. Despite these advances, neither knowledge-graph enhancement nor federated collaboration alone can fully overcome the challenges of data heterogeneity and cross-platform isolation.

Data heterogeneity, privacy constraints, and data isolation significantly limit the performance of recommendation frameworks, with heterogeneity posing a major challenge due to the divergent data distributions among participating clients. Such heterogeneity often leads to slow global convergence, reduced training efficiency, and client drift, whereby local models overfit to their own data and fail to benefit from global aggregation. To solve these limitations, we propose PLRF, a federated learning–based joint optimization framework for knowledge graph completion and recommendation. PLRF integrates knowledge graph completion and recommendation into a bidirectional synergistic learning process that shares a semantic embedding layer across tasks, enabling unified representation learning over both graph structures and user–item interactions. During local training, a regularization term mitigates client drift in FKG aggregation, while in global aggregation, the server combines entity embeddings from all clients to refine the shared model. This design leverages the richer cross-client knowledge encoded in the FKG to enhance recommendation performance. In contrast, recommendation signals simultaneously improve knowledge graph completion quality, achieving mutual reinforcement under heterogeneous and privacy-preserving conditions. This paper's main contributions are as follows:

- We propose PLRF, a federated recommendation framework that leverages a federated knowledge graph (FKG) to enrich item representations and incorporate cross-client knowledge. This substantially alleviates the cold-start problem by enabling recommendations for items and users with limited interaction history.
- To address data heterogeneity in FKG aggregation, the PLRF presents a regularization-integrated local training strategy. This strategy mitigates client drift, stabilizes global model convergence, and thereby enhances both robustness and efficiency in updating knowledge graph embeddings within heterogeneous federated environments.
- We conduct comprehensive experiments on two real-world datasets, jointly evaluating PLRF on Top-N recommendation and federated KG completion tasks. Results demonstrate performance obtains over state-of-the-art baselines, validating the mutual enhancement between the recommendation model and the knowledge graph completion model within the proposed framework.

The remainder of the paper is organized as follows: Sect. 2 reviews related work, Sect. 3 introduces the background and preliminary. Section 4 elaborates on our proposed methodology, Sect. 5 presents the experimental results, and Sect. 6 concludes the paper. Finally, acknowledgments are provided.

2 Related Work

This section provides a comprehensive review of existing recommendation research that is closely related to our study. The relevant works can be grouped into two main categories: knowledge graph-based recommendation and federated learning recommendation systems. Table 1 summarizes representative studies on recommender systems and presents a comparative analysis.

Table 1. Comparison of Recommendation Systems in Cross-Domain, Heterogeneity, and Privacy Settings.

References	Data Silo	Privacy	Heterogeneity
BPRMF [12]	Centralized	None	None
CKE [23]	Centralized	None	None
CFKG [1]	Centralized	None	None
MKR [16]	Centralized	None	None
KTUP [4]	Centralized	None	None
FCF [2]	Distributed	Sending Gradient Updates	None
FedRec [11]	Distributed	Anonymization	None
FedMF [5]	Distributed	Homomorphic Encryption	None
FedGNN [20]	Distributed	Local Differential Privacy	None
Ours	Distributed	Mask	Regularization term

Knowledge Graph-Based Recommendation. Conventional collaborative filtering (CF) models, such as BPRMF [12], rely solely on user–item interaction data. These methods do not exploit external contextual information (e.g., item attributes or domain knowledge), which severely limits their performance under sparse data conditions. In particular, they struggle with the cold-start problem, where new users or items have insufficient interaction history to support effective preference learning [13]. To overcome the limitations of traditional CF, recent studies have integrated structured semantic information from knowledge graphs into the recommendation process. By linking items to entities and relationships in a KG, these models can propagate and enrich user and item representations with auxiliary semantic knowledge. The literature broadly categorizes KG-based recommendation methods into three paradigms: &1 Path-based methods infer user–item relationships, by exploiting connectivity patterns in the knowledge graph, typically computing preference scores via inner product. They are simple to implement and can be seamlessly integrated with traditional collaborative filtering approaches; however, they primarily capture shallow structural features and fail to model deeper semantic correlations between entities. &2 Propagation-based methods enhance user and item representations [17,18]. They start from user-interacted items and propagate preferences through multi-hop neighbors in the knowledge graph. This approach leverages semantic neighborhoods to capture richer contextual signals; however, it is susceptible to noise propagation and necessitates careful selection of neighboring entities to ensure recommendation accuracy. &3 Embedding-based methods [1,4,16,23] represent entities and relations in a continuous vector space, often jointly optimizing knowledge graph embeddings and recommendation objectives within a unified learning framework. This paradigm enables integrated representation learning for heterogeneous data sources; however, its effectiveness is highly dependent on the coverage and correctness of the underlying knowledge graph.

Federated Learning Recommendation System. To address the problem of collaborative recommendation in multi-agency and multi-client data quarantine, federated learning is applied to the recommendation system because of its data immobility and model mobility characteristics. Recommendations based on federated learning methods are broadly categorized in the literature into three paradigms: &1 Federated recommendation algorithms based on collaborative filtering [2,5]. This category of algorithms preserves the core rationale of collaborative filtering, which centers on modeling user-item interaction patterns for preference inference, and achieves model collaboration by transmitting either the item feature matrix or the gradients of item feature vectors to the server for global updates. &2 Federated recommendation algorithms based on deep learning [20]. Adopting a client-server architecture, these algorithms enable each client to receive the global model from the server and then fine-tune it using local data to obtain a personalized federated recommendation model. &3 Federated recommendation algorithms based on meta-learning [10]. These methods leverage a small number of samples to train personalized models even when the amount of client-side data is limited.

3 Background and Preliminaries

3.1 Problem Definition

We consider a federated recommendation scenario with K clients $\mathcal{C} = \{C_1, C_2, \ldots, C_K\}$, each representing an independent organization (e.g., a school) with its own local dataset. For client C_k, let $\mathcal{S}_k$ denote the set of students, $\mathcal{Q}_k$ the set of questions, and $\mathbf{Y}_k \in \mathbb{R}^{|\mathcal{S}_k| \times |\mathcal{Q}_k|}$ the student–question interaction matrix, where $y_{s,q}^k$ encodes the implicit feedback between student $s \in \mathcal{S}_k$ and question $q \in \mathcal{Q}_k$ (e.g., number of incorrect attempts).

Each client C_k also maintains a local knowledge graph where E_k is the set of entities (e.g., students, questions, concepts), R_k is the set of relation types, and $T_k \subseteq E_k \times R_k \times E_k$ is the set of triples (h, r, t) describing the semantic relationships among entities. In the context of question recommendation, the objective is to predict the likelihood that a student s will interact with or correctly answer an unseen question q. Formally, the recommendation model aims to learn a scoring function:

$$\hat{y}_{s,q} = f(s, q \mid \Theta),$$

where $\hat{y}_{s,q}$ represents the predicted interaction score between s and q, and Θ denotes the parameters of the model. Finally, for each student s, the system ranks candidate questions according to $\hat{y}_{s,q}$ and recommends the top-N questions with the highest predicted scores.

3.2 Federated Knowledge Graph Embedding

We consider a federated knowledge graph setting with C isolated clients, where each client i holds a private knowledge graph $G_i = (E_i, R_i, T_i)$ containing entities E_i, relations R_i, and triples $T_i \subseteq E_i \times R_i \times E_i$. These local graphs are disjoint in triple and relation sets, but may share a subset of aligned entities across clients.

A knowledge graph embedding (KGE) model assigns low-dimensional vectors to entities and relations, enabling triple scoring via functions $f_r(h, t)$.

We denote by T the server-maintained *global entity table* containing all aligned entities, by $P_c \in \{0,1\}^{n \times n_c}$ the permutation matrix mapping local entity indices to global positions, and by $v_c \in \{0,1\}^n$, the existence vector indicating which global entities occur in client c.

The learning process consists of iterative local KGE training on each G_i, secure aggregation of entity embeddings at the server, and redistribution of updated global embeddings to clients.

4 Methodology

In this section, we give a detailed introduction to our PLRF. The overview of PLRF is demonstrated in Fig. 2. By and large, the PLRF consists of two parts: a central server and a set of clients. The workflow is designed to ensure privacy preservation during client-server communication through masking techniques.

Initially, the server instantiates a global entity embedding matrix and disseminates it to all clients. Each client updates the received embeddings using its local data, producing local entity embeddings. Before transmission, clients apply the masking mechanism to protect sensitive information, and then upload the masked updates to the server. The server aggregates the received embeddings to refine the global matrix. This process is iteratively repeated until the global entity embeddings converge. The whole process of EMFD is summarized in Algorithm 1.

Central Server. The server receives masked entity embeddings from clients, ensuring privacy preservation during transmission. It performs secure aggregation over the received updates and refines the global entity embedding matrix accordingly.

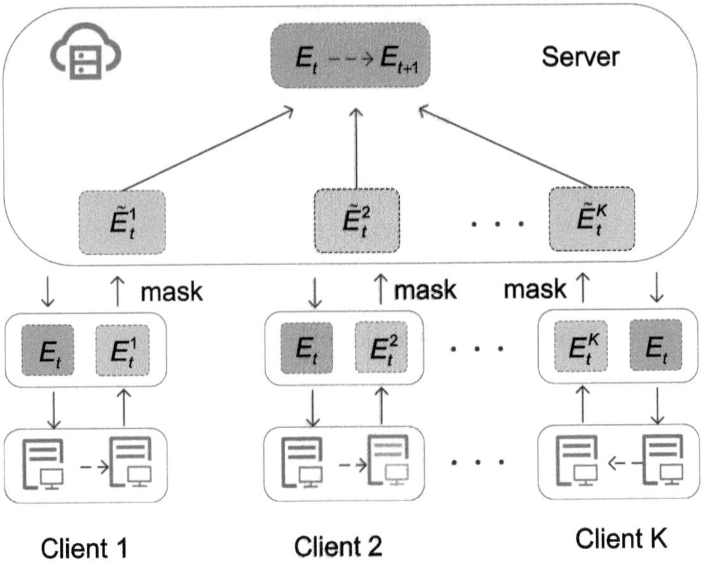

Fig. 2. Overview of the PLRF.

Clients. Each client maintains its local knowledge graph and student–question interaction records. Upon receiving the global entity embeddings from the server, the client performs local updates by jointly training a knowledge graph completion model and a recommendation model. The updated entity embeddings are then masked to prevent information leakage before being transmitted back to the server for aggregation.

4.1 Global Aggregation

Global aggregation enables the collaborative optimization of multi-source entity embeddings. The central server is responsible for aggregating entity embeddings received from multiple clients and distributing the aggregated results back to them. To align heterogeneous client embeddings with the global embedding space, the server maintains a set of *permutation matrices* P_s, used to map each client's entity embedding matrix to the global entity embedding matrix.

Initially, the server constructs the global embedding matrix E_0^s and generates each client's initial embedding via permutation transformation:

$$E_0^c \leftarrow P_s^T E_0^s. \tag{1}$$

At communication round t, the server sends the permutation-transformed global embeddings to all clients. Each client performs local updates and returns masked entity embeddings to the server. The server then applies the *Federated Averaging* (FedAvg) strategy to aggregate the embeddings:

Algorithm 1: The procedure of the PLRF framework

Input: Clients K, epochs P, batch size B, lr η

Key Setup
- **for** $c = 1$ **to** K **do**
  - Generate (pk_c, sk_c), broadcast pk_c;
  - **for** $j \neq c$ **do**
    - $K_{c,j} \leftarrow \text{KeyAgree}(sk_c, pk_j)$;
    - Generate mask $M_{c,j}$ from $K_{c,j}$;

Server
- Initialize E_0^0;
- **for** $t = 0, 1, \ldots$ **do**
  - **for** $c = 1$ **to** K **do**
    - Send E_t^c to all clients;
  - **for** $c = 1$ **to** K **do**
    - $E_t^c \leftarrow \text{LocalTrain}(c, E_t^c)$;
  - $E_{t+1}^0 \leftarrow \frac{1}{K} \sum_{c=1}^{K} P_c \hat{E}_t^c$;

LocalTrain c, E_t^c
- **for** $e = 1$ **to** P **do**
  - **for** *batch b to B* **do**
    - $\mathcal{L} \leftarrow \mathcal{L}_{\text{rec}} + \mathcal{L}_{\text{TUP}}$;
    - Update $E_t^c, P_i^c, S_i^c, Q_i^c$ with $-\eta \nabla \mathcal{L}$;
- $\hat{E}_t^c \leftarrow E_t^c + \sum_{j=1, j \neq c}^{K} M_{c,j}$;
- $\hat{E}_{t+1}^c$ to server;
- **return** E_t^c;

$$E_{t+1}^s \leftarrow \frac{1}{K} \sum_{c=1}^{K} P_s \tilde{E}_t^c, \tag{2}$$

where K denotes the total number of clients, $\tilde{E}_t^c$ is the masked embedding matrix uploaded by client c at round t, and E_{t+1}^s is the global embedding matrix for round $t+1$. The procedure consists of four steps: distribution, local update, secure upload, and aggregation. These steps are repeated until the global entity embeddings converge.

4.2 Local Joint Training

To achieve bidirectional representation enhancement between the recommendation model and the completion model, a local joint training approach is adopted. In the local phase, jointly training the KG completion model and the recommendation model is a core component of the PLRF. The rich entity and relation embeddings derived from the KG complement the item and preference embeddings in the recommendation model, leading to mutually enhanced performance. This joint representation effectively improves recommendation performances while providing interpretability by associating recommended items with explicit KG semantics. The overall workflow is illustrated in Fig. 3.

During joint training, the recommendation model enhances item embeddings q from the recommendation task by directly adding the corresponding entity embeddings e from the KG via the mapping table $\mathcal{M}$:

$$\hat{q} = q + e, \quad (q, e) \in \mathcal{M}. \tag{3}$$

Similarly, preference embeddings p are augmented with the associated relation embeddings r:

$$\hat{p} = p + r. \tag{4}$$

This process enriches implicit user preferences with explicit KG semantics that facilitate interpretable recommendation outcomes. The scoring function used in joint training is:

$$g(s, q; p) = \|\mathbf{s}^{\perp} + \hat{\mathbf{p}} - \hat{\mathbf{q}}^{\perp}\|. \tag{5}$$

In the backward pass, the KG completion model computes gradients for all parameters with respect to the total loss. Owing to the embedding alignments (*item–entity* and *knowledge-point–relation*) established in the forward pass, gradients propagate across both tasks, enabling bidirectional knowledge transfer. All parameters are updated synchronously via a unified optimizer, ensuring coherent training dynamics and avoiding parameter decoupling.

The total training objective is defined as follows:

$$\mathcal{L}_{\text{total}} = \lambda\mathcal{L}_t + (1 - \lambda)\mathcal{L}_{\text{rec}}, \tag{6}$$

where λ controls the trade-off between the KG completion loss $\mathcal{L}_t$ and the recommendation loss $\mathcal{L}_{\text{rec}}$.

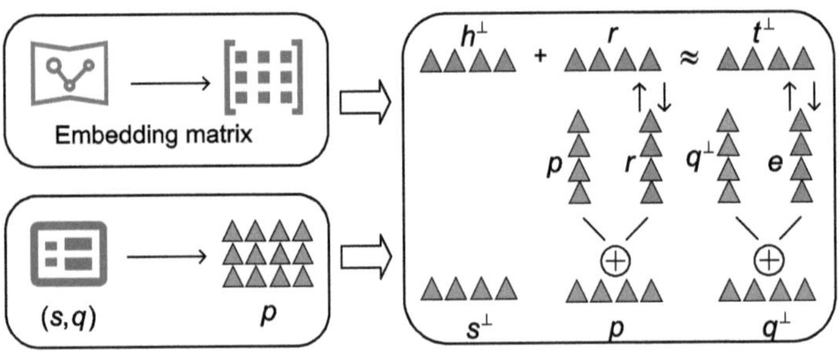

Fig. 3. Local Embedding Enhancement Representation

4.3 Regularizing Local Embeddings

To improve generalization and stabilize training, we introduce a regularization term that constrains local embeddings over successive rounds. For each local client, the knowledge graph completion task consists of two components: knowledge graph embedding and a regularization term. We adopt the TransH model [19], which represents entities and relations as vectors and projects entity vectors onto a hyperplane defined by the relation vector.

Formally, for a triple (e^h, r, e^t), the head and tail entity projections are:

$$\mathbf{e^h}_{\perp} = \mathbf{e^h} - \mathbf{w}_r^{\top}\mathbf{e^h}\mathbf{w}_r, \quad \mathbf{e^t}_{\perp} = \mathbf{e^t} - \mathbf{w}_r^{\top}\mathbf{e^t}\mathbf{w}_r, \tag{7}$$

where $\mathbf{w}_r$ denotes the normal vector of the hyperplane associated with relation r. The scoring function is:

$$f_r(\mathbf{h}, \mathbf{t}) = \left\| \left(\mathbf{h} - \mathbf{w}_r^\top \mathbf{h} \mathbf{w}_r\right) + \mathbf{d}_r - \left(\mathbf{t} - \mathbf{w}_r^\top \mathbf{t} \mathbf{w}_r\right) \right\|_2 . \tag{8}$$

TransH is trained using a margin-based ranking loss, which facilitates effective discrimination between positive and negative triples by maintaining a reasonable score difference between these two categories of triples:

$$\mathcal{L}_{\text{kge}} = \sum_{\substack{(e^h, r, e^t) \in \Delta \\ (e^{h'}, r, e^{t'}) \in \Delta'}} \max\left(0,\ f_r(e^h, e^t) + \gamma - f_r(e^{h'}, e^{t'})\right), \tag{9}$$

where γ is the margin hyperparameter, Δ denotes the set of positive triples, and Δ' denotes the set of negative triples generated by corrupting one entity in a positive triple such that the resulting triple does not exist in the KG.

Let client c at round t receive the global entity embeddings E_t^c from the server. We penalize deviations from the entity embeddings E_{t-1}^c obtained in the previous local round, enforcing smooth updates:

$$\mathcal{L}_{\text{p}} = \frac{1}{2} \| E_{t-1}^c - E_t^c \|^2. \tag{10}$$

The final local KG training objective is:

$$\mathcal{L}_t = \mathcal{L}_{\text{kge}} + \mu \mathcal{L}_{\text{p}}, \tag{11}$$

where μ is a hyperparameter controlling the trade-off between KG completion and regularization.

4.4 Security Analysis

Direct exposure of entity embeddings to the server introduces severe privacy vulnerabilities. Specifically, when the server colludes with a client—motivated by shared interests—it can exploit the embeddings and triple structures provided by the colluding client to infer the private triple data of other clients. To preserve client data privacy during server aggregation without compromising the accuracy of the aggregated model, we employ a secure masking scheme. In this method, the server never receives raw entity embeddings; instead, it obtains masked embeddings that conceal individual client contributions.

Preparation Phase. Each pair of clients establishes a shared secret via a key agreement protocol (e.g., Diffie–Hellman). Let client c and client j generate private keys, exchange public keys, and compute an identical shared key $K_{c,j}$. The shared key serves as a seed for a pseudo-random generator (PRG) to produce the mask matrix:

$$M_{c,j} = \Delta_{c,j} \cdot \text{PRG}(K_{c,j}), \tag{12}$$

where $\Delta_{c,j} \in \{-1, +1\}$ is an indicator for ordering: $\Delta_{c,j} = 1$ if $c \geq j$ and $\Delta_{c,j} = -1$ otherwise.

Data Upload. In training round t, c sends the masked entity embedding matrix:

$$\tilde{E}_t^c = E_t^c + \sum_{j=1, j \neq c}^{K} M_{c,j}, \tag{13}$$

where E_t^c is the local entity embedding matrix before masking.

Server Aggregation. Upon receiving masked embeddings from all K clients, the server performs standard *FedAvg* aggregation:

$$E_{t+1}^s = \frac{1}{K} \sum_{c=1}^{K} \left(E_t^c + \sum_{j=1, j \neq c}^{K} M_{c,j} \right). \tag{14}$$

Due to the symmetric design of the mask matrices $M_{c,j}$, their summation equals zero. This ensures that E_{t+1}^s is equivalent to the average of the original unmasked embeddings. Meanwhile, the server only receives a cryptographically masked aggregated result, making the reconstruction of any individual client's original contribution computationally infeasible. Consequently, even if the server colludes with a subset of clients, the private data of non-colluding clients cannot be reliably inferred from locally observable information, thereby providing a formal guarantee against client-level privacy leakage.

5 Experiments

In this section, we demonstrate the experiment results on the two datasets. We first give the experimental settings, and then show the main experimental results compared with baseline methods.

Table 2. Statistics of the KDD CUP 2010 and ASSIST2009 datasets after preprocessing.

Dataset	#Users	#Hierarchies	#Questions	#Triplets	#Ratings
KDD CUP 2010	574	140	1,079	187,054	28,884
ASSIST2009	4,165	125	17,753	338,002	55,756

5.1 Experimental Setup

Datasets. We employ two publicly available datasets from the education domain: KDD CUP 2010 and ASSIST2009. Both datasets comprise student records, problem-solving activities, and corresponding question attributes. We model interactions using the number of incorrect responses for each student–question pair, and represent the data in the form of triplets ⟨student, question,

Table 3. Experimental results on KDD CUP 2010 and ASSIST2009 datasets.

KGE	Setting	KDD CUP 2010				ASSIST2009			
		MRR	Hits@1	Hits@5	Hits@10	MRR	Hits@1	Hits@5	Hits@10
TransE	Single	0.3812	0.2645	0.5218	0.6135	0.3247	0.2156	0.4532	0.5618
	FedE	0.4025	0.2876	0.5534	0.6421	0.3512	0.2431	0.4876	0.5987
	FedEC	0.4158	0.3012	0.5678	0.6543	0.3689	0.2587	0.5012	0.6123
TransH	Single	0.3945	0.2789	0.5367	0.6254	0.3378	0.2289	0.4678	0.5765
	FedE	0.4213	0.3023	0.5712	0.6634	0.3745	0.2654	0.5123	0.6234
	FedEC	0.4389	0.3178	0.5890	0.6789	0.3912	0.2813	0.5345	0.6456
DistMult	Single	0.3765	0.2570	0.5134	0.6021	0.3189	0.2091	0.4456	0.5543
	FedE	0.4087	0.2912	0.5456	0.6345	0.3567	0.2489	0.4912	0.6012
	FedEC	0.4234	0.3056	0.5612	0.6415	0.3734	0.2634	0.5056	0.6156
ComplEx	Single	0.3876	0.2670	0.5278	0.6189	0.3312	0.2212	0.4589	0.5618
	FedE	0.4123	0.2934	0.5578	0.6456	0.3623	0.2534	0.4934	0.6034
	FedEC	0.4312	0.3134	0.5812	0.6678	0.3876	0.2789	0.5278	0.6312
KTUP		0.4522	0.3228	0.6115	0.6923	0.3998	0.2808	0.5450	0.6419
PLRF(Ours)		**0.4534**	**0.3278**	**0.6064**	**0.6930**	**0.4047**	**0.2986**	**0.5569**	**0.6597**

question hierarchy⟩. After retaining relevant columns and filtering out entities with frequencies lower than five. For triplets, we randomly assign relationships to clients, then partition the triplets according to the client where each relationship is located. Finally, the knowledge graph of a client consists of the triplets associated with the relationships owned by that client. The rating data is partitioned into different clients based on questions, the resulting dataset statistics are summarized in Table 2. For federated training, data of each client are split into training, validation, and test sets in a 70%/20%/10% ratio.

Baseline. For the knowledge graph completion task, we compare PLRF against a diverse set of representative models, including TransE [3], TransH [19], DistMult [21], ComplEx [15], and KTUP [4]. The comparisons are conducted under three distinct training configurations: individual training, the FedE [6] setting, and the FedEC [7] setting, to assess performance across centralized and federated environments.

For the recommendation task, PLRF is compared with a broad range of baseline recommendation methods, including CFKG [1], CKE [23], and again KTUP [4]. These baselines encompass both traditional collaborative filtering approaches and knowledge-graph-enhanced recommendation models, enabling a comprehensive evaluation of PLRF across different methodological paradigms.

Evaluation Metrics. We adopt two widely used metrics for evaluating link prediction performance: **Mean Reciprocal Rank (MRR)**, **Hits@n**, and **Normalized Discounted Cumulative Gain (NDCG)**.

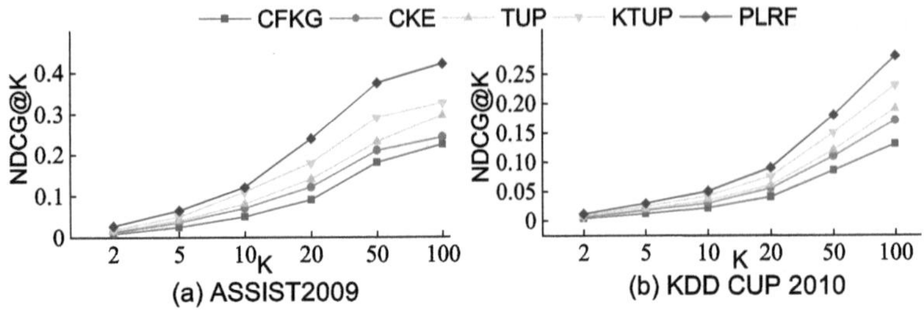

Fig. 4. NDCG on two datasets

Mean Reciprocal Rank (MRR). Given a test triple (h, r, t), the model assigns scores to all candidate entities and ranks them accordingly. MRR measures the average of the reciprocal ranks of the correct entity in these ranked lists:

$$\mathrm{MRR} = \frac{1}{|T|} \sum_{(h,r,t)\in T} \frac{1}{\mathrm{rank}(t)} \tag{15}$$

where T denotes the set of test triples and $\mathrm{rank}(t)$ is the position of the correct entity in the ranking.

Hits@n. This metric calculates the proportion of test triples in which the correct entity appears within the top-N ranked candidates:

$$\mathrm{Hits@}n = \frac{|\{(h, r, t) \in T \mid \mathrm{rank}(t) \leq n\}|}{|T|} \tag{16}$$

Normalized Discounted Cumulative Gain (NDCG). NDCG is an evaluation metric that jointly considers the relevance and positional ranking of recommended items. It assigns higher scores when highly relevant items appear in top ranks, reflecting their increased likelihood of being examined by users. Relevance contributions are logarithmically discounted according to item positions, with later ranks receiving stronger penalties due to reduced user attention. The final score is normalized by the ideal DCG, ensuring that NDCG values lie within the range $[0, 1]$ and enabling consistent comparison across different recommendation lists (Fig. 4).

5.2 Knowledge Graph Completion Performance

We evaluate PLRF on the knowledge graph completion task using MRR and Hits@K (K = 1, 5, 10) across KDD CUP 2010 and ASSIST2009 datasets (Table 3). On the KDD CUP 2010 dataset under the FedE setting, PLRF yields relative MRR improvements of 12.64% over TransE, 10.94% over TransH, 7.62% over DistMult, and 9.97% over ComplEx; gains in Hits@K reach up to 8.45% for Hits@1, 6.38% for Hits@5, and 5.12% for Hits@10. On the ASSIST20092009 dataset, the corresponding MRR gains are15.23%, 8.06%, 13.46%, and 11.70%,

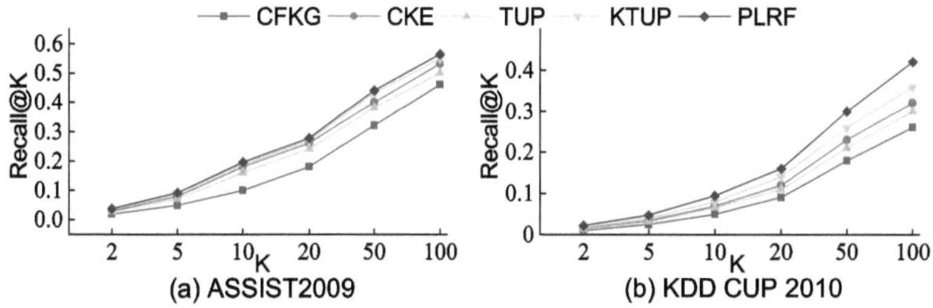

Fig. 5. Recall on two datasets

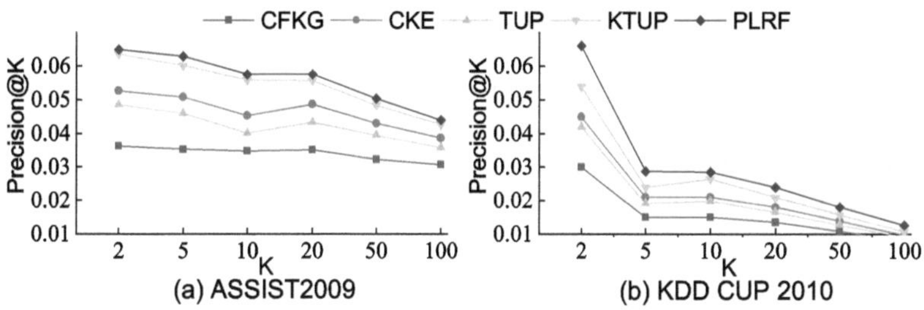

Fig. 6. Precision on two datasets

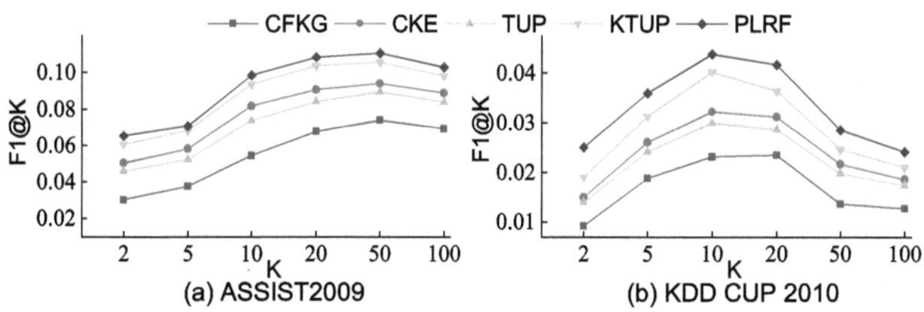

Fig. 7. F1 score on two datasets

with Hits@K improvements peaking at 9.21% for Hits@1, 7.85% for Hits@5, and 6.94% for Hits@10. Under the FedEC setting, PLRF maintains consistent advantages over all baselines across both datasets. This may be caused by the local regularization term, which effectively mitigates entity ranking bias in clients with sparse data and further boosts completion accuracy. Unlike KTUP [4], PLRF's federated design integrates local student–question interactions with cross-client knowledge graph triples, resulting in systematic gains across all evaluation metrics (Figs. 5, 6 and 7).

5.3 Recommendation Performance

We assess recommendation quality using NDCG@K, Recall@K, Precision@K, and F1 score@K with $K \in \{2, 5, 10, 20, 50, 100\}$, as reported in Fig. 4–7. Across both KDD CUP 2010 and ASSIST2009 datasets, PLRF consistently surpasses KTUP [4] on NDCG@K and Recall@K for $K \in \{2, 5, 10, 20, 50, 100\}$. At $K = 10$, on the ASSIST2009 dataset, PLRF improves over KTUP [4] by 9.52% in NDCG, 2.25% in Recall, 4.32% in Precision, and 5.12% in F1 score. On the KDD CUP 2010 dataset at $K = 10$, the gains are 19.11% in NDCG, 18.75% in Recall, 7.55% in Precision, and 8.93% in F1 score. The performance advantages arise from architectural differences. CKE combines collaborative filtering item factors with knowledge-base representations in a unidirectional manner, limiting cross-source interaction and underutilizing knowledge graph signals. CFKG enforces similarity between entity and item embeddings to transfer information, but this constraint can overwrite informative structure in the embeddings. KTUP is restricted to local knowledge graphs and cannot exploit cross-client triples, which weakens representation learning under data sparsity and heterogeneity. In contrast, PLRF uses federated knowledge graphs to enrich local triples without raw data sharing, and jointly trains knowledge graph completion and recommendation, allowing entities and relations to directly inform question and preference representations. The design improves ranking quality and consistently boosts NDCG, Recall, Precision, and F1 score.

6 Conclusion

This paper presents PLRF, a personalized learning recommendation framework based on federated knowledge graphs. The method leverages triplet information from distributed clients to enrich local knowledge graphs while preserving privacy. PLRF jointly trains a knowledge graph completion module and a recommendation module, enabling mutual performance improvement. To address non-IID data across clients, a regularization term is introduced to reduce the impact of data heterogeneity. Experiments on multiple benchmarks show that PLRF supports effective cross-platform collaboration and consistently achieves state-of-the-art results compared with federated and centralized baselines.

Acknowledgments. The work was partially supported by the Natural Science Foundation of Liaoning Province (2025-MS-126), the Young and Middle-aged Science and Technology Innovation Talent Support Plan of Shenyang (RC230832), the Key Research and Development Program of Liaoning Province under Grant 2023JH26/10300022, and the Shandong Provincial Natural Science Foundation of China (ZR2025QC643).

References

1. Ai, Q., Azizi, V., Chen, X., Zhang, Y.: Learning heterogeneous knowledge base embeddings for explainable recommendation. Algorithms **11**(9), 137 (2018)
2. Ammad-ud-din, M., et al.: Federated collaborative filtering for privacy-preserving personalized recommendation system. CoRR abs/1901.09888 (2019)

3. Bordes, A., Usunier, N., García-Durán, A., Weston, J., Yakhnenko, O.: Translating embeddings for modeling multi-relational data. In: NIPS, pp. 2787–2795 (2013)
4. Cao, Y., Wang, X., He, X., Hu, Z., Chua, T.: Unifying knowledge graph learning and recommendation: towards a better understanding of user preferences. In: WWW, pp. 151–161. ACM (2019)
5. Chai, D., Wang, L., Chen, K., Yang, Q.: Secure federated matrix factorization. IEEE Intell. Syst. **36**(5), 11–20 (2021)
6. Chen, M., Zhang, W., Yuan, Z., Jia, Y., Chen, H.: Fede: embedding knowledge graphs in federated setting. In: IJCKG, pp. 80–88. ACM (2021)
7. Chen, M., Zhang, W., Yuan, Z., Jia, Y., Chen, H.: Federated knowledge graph completion via embedding-contrastive learning. Knowl. Based Syst. **252**, 109459 (2022)
8. Contreras, M.R., Jaimes, J.O.P.: AI ethics in the fields of education and research: a systematic literature review. In: ICACIT, pp. 1–6. IEEE (2024)
9. Huang, W., Liu, J., Li, T., Ji, S., Wang, D., Huang, T.: Fedcke: cross-domain knowledge graph embedding in federated learning. IEEE Trans. Big Data **9**(3), 792–804 (2023)
10. Jalalirad, A., Scavuzzo, M., Capota, C., Sprague, M.R.: A simple and efficient federated recommender system. In: BDCAT, pp. 53–58. ACM (2019)
11. Lin, G., Liang, F., Pan, W., Ming, Z.: Fedrec: federated recommendation with explicit feedback. IEEE Intell. Syst. **36**(5), 21–30 (2021)
12. Rendle, S., Freudenthaler, C., Gantner, Z., Schmidt-Thieme, L.: BPR: Bayesian personalized ranking from implicit feedback. CoRR abs/1205.2618 (2012)
13. Schafer, J.B., Frankowski, D., Herlocker, J.L., Sen, S.: Collaborative filtering recommender systems. In: The Adaptive Web. Lecture Notes in Computer Science, vol. 4321, pp. 291–324. Springer (2007)
14. Sun, Z., Deng, Z., Nie, J., Tang, J.: Rotate: knowledge graph embedding by relational rotation in complex space. In: ICLR (Poster). OpenReview.net (2019)
15. Trouillon, T., Welbl, J., Riedel, S., Gaussier, É., Bouchard, G.: Complex embeddings for simple link prediction. In: ICML. JMLR Workshop and Conference Proceedings, vol. 48, pp. 2071–2080. JMLR.org (2016)
16. Wang, H., Zhang, F., Zhao, M., Li, W., Xie, X., Guo, M.: Multi-task feature learning for knowledge graph enhanced recommendation. In: WWW, pp. 2000–2010. ACM (2019)
17. Wang, H., Zhao, M., Xie, X., Li, W., Guo, M.: Knowledge graph convolutional networks for recommender systems. In: WWW, pp. 3307–3313. ACM (2019)
18. Wang, X., He, X., Cao, Y., Liu, M., Chua, T.: KGAT: knowledge graph attention network for recommendation. In: KDD, pp. 950–958. ACM (2019)
19. Wang, Z., Zhang, J., Feng, J., Chen, Z.: Knowledge graph embedding by translating on hyperplanes. In: AAAI, pp. 1112–1119. AAAI Press (2014)
20. Wu, C., Wu, F., Cao, Y., Huang, Y., Xie, X.: Fedgnn: federated graph neural network for privacy-preserving recommendation. CoRR abs/2102.04925 (2021)
21. Yang, B., Yih, W., He, X., Gao, J., Deng, L.: Embedding entities and relations for learning and inference in knowledge bases. In: ICLR (Poster) (2015)
22. Yang, N., Yuan, X., Lin, H., Zhang, H., Lv, P., Wang, J.: Feddm: federated learning incorporating dissimilarity measure for mobile edge computing systems. IEEE Trans. Cogn. Commun. Netw. **12**, 981–993 (2026)
23. Zhang, F., Yuan, N.J., Lian, D., Xie, X., Ma, W.: Collaborative knowledge base embedding for recommender systems. In: KDD, pp. 353–362. ACM (2016)

SeeKRec: Toward Semantic-Empowered Knowledge-Aware Recommendation

Xiangming Li[1], Qingshan Li[1], Hua Chu[1], Yangtao Zhou[1(✉)], Jianan Li[1], and Wanqiang Yang[2]

[1] Xidian University, Xian 710071, China
{xiangmli,zhou_yt}@stu.xidian.edu.cn, {qshli,hchu}@mail.xidian.edu.cn, lijianan@xidian.edu.cn

[2] Shanghai Fairyland Software Co., Ltd., Shanghai 201103, China
yangwan@fulan.com.cn

Abstract. Knowledge-aware recommendation aims to enhance recommendation performance by leveraging item-side information in knowledge graphs (KGs). Existing methods typically assign identifiers (IDs) to items and entities in a KG and model their structural relations. However, these methods overlook rich semantic information and therefore fail to capture the latent semantic associations between users and items, leading to suboptimal performance. To address this, we propose a **Se**mantic-**e**mpowered **K**nowledge-aware **Rec**ommendation approach called **SeeKRec**, which captures the semantic-collaborative features of users and items in an extract-refine-integrate manner. Specifically, to extract semantic information, we design an adaptive semantic preference extractor that hierarchically integrates intra-user preference discovery and inter-user semantic alignment, achieving accurate preference understanding and uncovering latent cross-user relations. To achieve task-driven semantic refinement, we design a sample-then-enhance framework that first samples task-relevant semantics through meticulously designed semantic sampling and then enhances them via a task-aware semantic calibrator, yielding high-quality semantic representations aligned with recommendation objectives. To seamlessly integrate refined semantic representations with conventional collaborative signals, we propose a semantics-aware mixture-of-experts model with a distill-then-fuse strategy, which distills semantics to bridge the dimensional gap, and fuses them with ID embeddings via dynamic expert routing to capture diverse semantic-collaborative relations. Extensive experiments show SeekRec achieves state-of-the-art results.

Keywords: Knowledge-aware Recommendation · Large Language Models · Knowledge Graphs · Mixture of Experts

1 Introduction

Knowledge-aware recommendation has emerged as an effective approach for improving recommendation systems by integrating knowledge graphs as external prior knowledge. Leveraging rich relational information from knowledge graphs,

H. Jung et al. (Eds.): DASFAA 2026, LNCS 16535, pp. 308–323, 2026.
https://doi.org/10.1007/978-981-92-0363-5_19

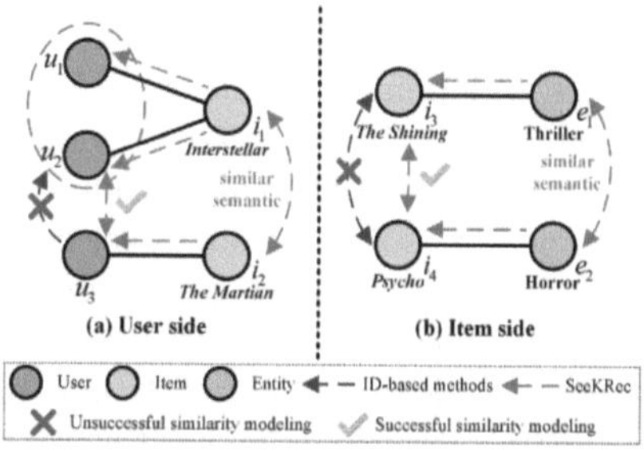

Fig. 1. A toy example illustrating the potential issues of ID-based methods.

these approaches enhance user and item representations, thereby mitigating the long-standing data sparsity and cold-start issues that hinder traditional collaborative filtering [18]. Typically, knowledge-aware recommendation assigns unique IDs to items and entities in a knowledge graph and exploits their adjacency relations to capture diverse high-order associations, thereby enhancing collaborative filtering [4]. For instance, KGRec [18] proposes a self-supervised rationalization strategy to identify the most contributive ID connections within the knowledge graph, thereby enhancing recommendation accuracy.

Although knowledge-aware recommendation has advanced considerably, existing ID-based methods characterize different users and items solely through ID data, overlooking the potential of rich semantic information. Consequently, these methods fail to capture the semantic features of users and items, as well as their latent semantic similarities, resulting in deficiencies on both the user and item sides of the recommendation process. In terms of users, existing methods overlook users' semantic preferences, resulting in inadequate preference modeling and hindering the discovery of latent users with similar preferences. As illustrated in Fig. 1(a), u_1 and u_2 have watched the science fiction movie *Interstellar*, while u_3 has engaged with another science fiction film, *The Martian*. As a result, these three users share similar semantic preferences for movies. However, due to the absence of explicit ID connections between u_3 and the other two users, ID-based methods fail to recognize the similarity between them. In terms of items, existing methods struggle to capture intrinsic semantic relationships between them. As shown in Fig. 1(b), *The Shining* and *Psycho* have the attributes "Thriller" and "Horror" respectively in the knowledge graph, which are semantically related. However, due to their distinct IDs and the absence of explicit connections, ID-based methods fail to recognize the semantic relationship between them.

To address the omission of semantic information in current approaches, we intend to extract semantic information and incorporate them into the knowledge-aware recommendation system. In extracting semantic information, we encounter the challenge that user semantics are implicit. Meanwhile, item semantics in the knowledge graph, although abundant, contain substantial task-irrelevant content, impeding the extraction of recommendation-aligned semantics. Motivated by the strong capabilities of large language models (LLMs) in knowledge extraction and semantic understanding [4], we aim to leverage them to explicitly extract user semantics and refine both user and item semantics to align with recommendation objectives. In integrating the semantics, we observe a significant dimensional gap between the high-dimensional semantic features and the low-dimensional ID embeddings [7], the latter serving as the foundation for user and item representations in recommendation systems. Most existing methods employ trainable adapters such as MLPs to project semantic features into the ID space, but this often leads to substantial semantic loss [7]. Moreover, mapping heterogeneous features into a single latent space constrains the model's ability to capture the diverse relational patterns between them. Motivated by the low-rank approximation capacity of semantic decomposition [7] and the demonstrated ability of mixture-of-experts (MoE) to fuse diverse information sources, we aim to compress high-dimensional semantic features via low-rank approximation, such as singular value decomposition (SVD), to retain the principal semantic information, and introduce an MoE to seamlessly integrate them with collaborative signals, thereby enhancing recommendation performance.

Building on these insights, we propose a **Se**mantic-**e**mpowered **K**nowledge-aware **Rec**ommendation method called **SeeKRec**, which captures the semantic-collaborative features of users and items in an extract-refine-integrate manner. Specifically, to extract semantic information, we develop an adaptive semantic preference extractor that combines intra-user preference discovery with inter-user semantic alignment. The extractor dynamically adjusts the quantity of semantic preferences for each user based on the diversity of their historical behaviors, and autonomously aligns these preferences across users, enabling precise preference modeling and uncovers latent cross-user semantic relations. We then abstract the extracted user preferences into semantic preference nodes and merge them with the knowledge graph to build a unified semantic-aware heterogeneous graph, thereby structurally modeling user-item semantics. To achieve recommendation-oriented semantic refinement, we propose a sample-then-enhance framework that first dynamically explores the semantic-aware heterogeneous graph via a carefully designed search-prune-evaluate subgraph sampling strategy. This strategy identifies semantic subgraphs that are most relevant to the recommendation objectives, enabling adaptive discovery of task-driven semantic structures. A task-aware semantic calibrator is then introduced to enhance the sampled semantics, aligning them with the recommendation objectives. To integrate refined semantic representations with conventional collaborative signals, we propose a semantic-aware MoE network that seamlessly integrates them through a distill-then-fuse strategy. The network bridges the dimensional gap between semantic features and ID embeddings by applying a semantic distillation process, leveraging the low-rank approximation capability of singular value decomposition (SVD) to identify informative semantic components, reducing the dimensionality of semantic features while preserving semantic integrity. The network then fuses distilled semantics and collaborative signals by employing multiple experts to capture diverse association patterns between them, effectively harnessing semantic information to achieve superior recommendation performance. Extensive experiments show that our method outperforms state-of-the-art baselines. The code, along with further theoretical and experimental analyses, can be found at https://github.com/xiangmli/SeKGRec. Our contributions are summarized as follows:

- We propose a novel semantic-empowered knowledge-aware recommendation method, which captures the semantic-collaborative features of users and items in an extract-refine-integrate manner.
- We design an adaptive semantic preference extractor that adaptively discovers intra-user preferences and performs inter-user semantic alignment, yielding precise user semantics and unveiling latent cross-user relations.
- We refine semantics through a sample-then-enhance framework, enhancing their effectiveness in the recommendation through the proposed search-prune-evaluate semantic sampling strategy and semantic calibrator.
- We propose a semantic-aware MoE network that seamlessly integrates semantics with collaborative signals through a distill-then-fuse strategy.
- Extensive evaluation on three real-world datasets shows that our method consistently outperforms existing approaches across all benchmarks.

2 Related Work

Knowledge-Aware Recommendation. Knowledge-aware recommendation utilizes knowledge graphs to enhance user and item representation modeling. Existing methods fall into three categories: embedding-based methods, path-based methods, and propagation-based methods [4]. Embedding-based methods learn embeddings of entities and relations via knowledge graph embedding techniques to augment user and item representations [22]. Path-based methods exploit meta-paths to uncover long-range dependencies between users and items [18]. Propagation-based methods employ graph neural networks (GNNs) to aggregate information over the knowledge graph, capturing higher-order connectivity [15]. However, most of these methods represent entities and relations solely by IDs, neglecting their rich textual semantics, which hinders the discovery of latent semantic associations between users and items. Currently, only CoLaKG [4] incorporates semantic information from knowledge graphs. However, this approach fails to adequately consider user-side semantic preferences and indiscriminately uses item-side knowledge, which may introduce irrelevant information. Furthermore, it does not fully consider how semantic information can enhance recommendations, resulting in significant semantic loss and severely affecting the model's ability to leverage semantic information for improved recommendation performance. In contrast, our proposed SeeKRec explicitly extracts user semantic preferences, refines semantics in a task-driven manner, and employs a carefully designed semantic-aware MoE network that mitigates semantic loss and achieves seamless integration of semantic and collaborative information.

LLMs for Recommendation Systems. Recent advancements in LLMs have spurred increasing interest in LLM-based recommendation systems. Current approaches can be broadly classified into two categories: LLMs as recommenders and LLMs as enhancers [8]. In the LLMs as recommenders paradigm, early studies applied LLMs to zero-shot and few-shot recommendation scenarios [2], while other approaches fine-tune LLMs on recommendation datasets to customize them for recommendation tasks [23]. However, the essence of LLMs lies in predicting the probability of the next token, which inherently limits their ability to understand recommendation tasks. Therefore, these methods often struggle to capture collaborative signals [9]. To address the limitations of these methods, researchers have shifted their focus to the study of LLMs as enhancers, which leverage LLM-derived information to enhance recommendations. For instance, some studies directly extract representations from LLMs based on item textual features [21], while others leverage LLMs to enhance the collaborative knowledge in pre-trained collaborative filtering recommender systems [9]. However, most existing LLM-based recommendation methods are designed for knowledge-free recommendation and face significant challenges when applied to knowledge-aware recommendation. Although knowledge graphs provide rich domain knowledge, they also contain substantial task-irrelevant information, which risks distracting LLMs from capturing task-relevant semantics. Moreover, effectively leveraging the complementary strengths of LLMs' world knowledge, the structured domain knowledge from knowledge graphs, and collaborative signals in a unified manner remains an unsolved problem. In contrast, our proposed SeeKRec addresses these challenges by performing recommendation-oriented semantic refinement to filter out irrelevant information and employing a semantic-aware MoE network to achieve seamless integration of semantics and collaborative signals.

3 Problem Definition

User-Item Interaction Matrix. In recommendation systems, historical user-item interactions (e.g., clicks or purchases), denoted as $\mathcal{Y}^+$, can be modeled as a user-item bipartite graph $\mathcal{G} = (\mathcal{V}, \mathcal{Y}^+)$, where $\mathcal{V} = (U \cup I)$ consists of the user set U and the item set I.

Knowledge Graphs. A knowledge graph is a structured organization of external information, composed of triplets (h, r, t), where h and t denote the head and tail entities, respectively, and r specifies the relation between them. Formally, a knowledge graph is defined as $G_K = \{(h, r, t)|h, t \in \mathcal{E}, r \in \mathcal{R}\}$, where $\mathcal{E}$ and $\mathcal{R}$ are the sets of entities and relations. Items form an entity subset, i.e., $I \subseteq \mathcal{E}$, and their connected entities serve as attributes.

Knowledge-Aware Recommendation. Knowledge-aware recommendation seeks to learn a parameterized function $\hat{y}_{uv} = F(u, v|G, G_K, \Theta)$ that predicts the interaction probability $\hat{y}_{uv}$ for user u and item v, where Θ represents the learnable parameters of the model.

4 Methodology

Our approach captures the semantic-collaborative features of users and items in an extract-refine-integrate manner. The overall framework is shown in Fig. 2.

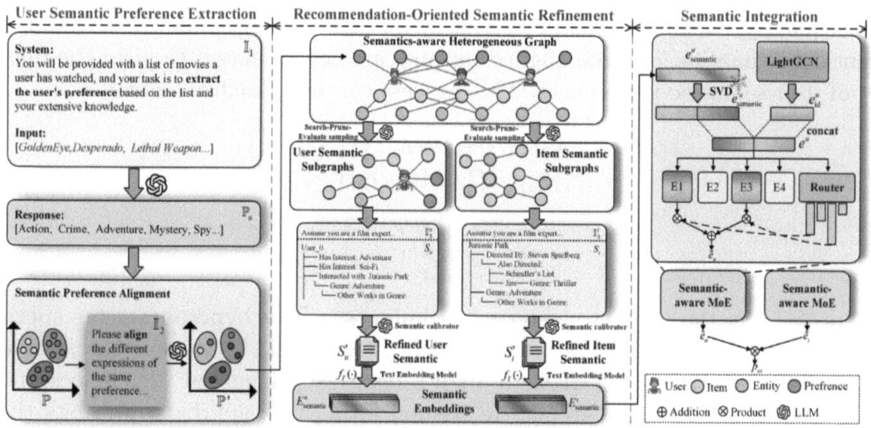

Fig. 2. The overall framewrok of SeeKRec.

The overall framework, as shown in Fig. 2, consists of three key components: the user semantic preference extraction module, the recommendation-oriented semantic refinement module, and the semantic integration module.

4.1 Adaptive User Semantic Preference Extraction

Given the lack of explicit semantic knowledge on user side and the crucial role of modeling user preferences in recommendation systems, we propose an adaptive semantic preference extractor that derives intra-user semantic preferences and perform inter-user semantic alignment. Specifically, the extractor adaptively determines the number of semantic preferences for each user according to the diversity of their historical behaviors, and further autonomously aligns semantically equivalent preferences across users. In this way, the extractor can precisely extract individual preferences and uncover latent cross-user semantic relations.

Inspired by the superior knowledge extraction and user behavior modeling capabilities of LLMs, the extractor first employs an LLM to derive users' semantic preferences from their historical interactions. LLMs can infer highly distilled, precise preferences that both summarize users' observed interactions and reveal potential unobserved interests [8]. This process can be formally defined as:

$$\mathbb{P} = \{\mathbb{P}_1, \mathbb{P}_2, \ldots, \mathbb{P}_{|U|}\}, \quad \mathbb{P}_u = \mathrm{LLM}(\mathbb{I}_1, T_u), \tag{1}$$

where $\mathbb{I}_1$ is the system prompt used to extract semantic preferences (all prompts are available in our publicly accessible code). T_u is the set of item texts in user u's interaction history. $\mathbb{P}_u$ represents the semantic preferences extracted for user u in textual form, and $\mathbb{P}$ aggregates these preferences across all users. Since LLMs may express the same preference in varied ways, exaggerating preference counts and hindering the discovery of cross-user similarity, we rigorously align semantically equivalent preferences into unified expressions. To achieve this, semantic preference are encoded and clustered using text clustering techniques [13].

Subsequently, each cluster is processed by an LLM to align preferences sharing identical semantics, and the aligned outputs are concatenated to yield the final set of user semantic preferences. This process can be formalized as follows:

$$\begin{aligned} \mathcal{C} = \{c_1, c_2, \ldots, c_k\} &= f_C(\mathbb{P}, \kappa), \\ \{c_i'\}_{i=1}^{\kappa} &= \{\text{LLM}(\mathbb{I}_2, c_i)\}_{i=1}^{\kappa}, \\ \mathbb{P}' &= \bigcup_{i=1}^{\kappa} c_i', \end{aligned} \tag{2}$$

where $f_C(\cdot)$ is the textual clustering techniques, κ is a hyperparameter specifying the number of clusters, and c_i is the i-th cluster. $\mathbb{I}_2$ denotes the system prompt that instructs the LLM to align semantically identical preferences and $\mathbb{P}'$ is the aligned user preferences in textual form. This approach prevents the LLM's input from becoming excessively long while exploiting the intra-cluster semantic similarity to effectively recognize identical preferences. Notably, our method unifies semantically similar preferences into coherent semantic expressions, ensuring that the merged preferences remain semantically meaningful. In contrast, CIKGRec [8] collapse preferences into ID-level representations that rely solely on structural information, limiting their ability to capture latent cross-user semantic relations. Furthermore, while such ID-based approaches, exemplified by CIKGRec, typically assume a fixed number of user preferences, our clustering parameter κ serves only to facilitate preference alignment and does not determine the final preference count. The total number of semantic preferences is adaptively determined through the semantic preference alignment.

4.2 Recommendation-Oriented Semantic Refinement

To uncover high-order semantic relations of users and items, we merge the extracted user semantic preferences with item semantics in the knowledge graph, forming a unified semantics-aware heterogeneous graph. However, knowledge graphs may contain task-irrelevant information [18], weakening the ability of semantic to enhance recommendation performance. Therefore, we propose a recommendation-oriented semantic refinement to mitigate task-irrelevant information and enhance semantic representations toward better alignment with the recommendation objective. Inspired by the powerful semantic comprehension capabilities of LLMs, we first leverage an LLM to sample task-relevant semantic subgraphs from the unified heterogeneous graph, and subsequently employ an task-aware semantic calibrator to refine these sampled semantics, aligning them more closely with recommendation objectives. We propose that these semantic subgraphs must meet two criteria: noise independence (the subgraphs should exclude noise irrelevant to the recommendation task) and semantic sufficiency (the subgraphs should adequately represent the semantics of the center node).

Based on the two criteria, we propose the search-prune-evaluate semantic subgraph sampling strategy. Specifically, let $\mathcal{T}_v^{\text{sub}}$ denote the set of triplets in the sampled subgraph centered at node $v \in \{u, i\}$, and $\mathcal{V}_v^{sub}$ be the corresponding node set. Initially, the triplet set $\mathcal{T}_v^{\text{sub}}$ is empty, and the node set $\mathcal{V}_v^{sub}$

contains only the center node v. Our sampling strategy consists of three phases: search, prune and evaluation. 1) In the **search phase**, we retrieve the neighboring triplet set $\mathcal{T}^{\text{adj}}$ for each node in the current node set $\mathcal{V}_v^{sub}$ in the unified heterogeneous graph, while excluding any triplets already included in $\mathcal{T}_v^{\text{sub}}$. 2) In the **prune phase**, we feed the adjacent triplet set $\mathcal{T}^{\text{adj}}$ into our LLM-based subgraph pruner. The subgraph pruner discards triplets irrelevant to the recommendation task, yielding the pruned set $\mathcal{T}^{\text{pru}}$, which is subsequently merged into $\mathcal{T}_v^{\text{sub}}$. 3) In the **evaluate phase**, the updated subgraph is fed into the LLM-based subgraph evaluator to determine whether it adequately represents the center node's features. This process is repeated until the subgraph sufficiently captures the center node's features. The search-prune-evaluate strategy yields compact subgraphs that capture the most informative semantics for the recommendation task while containing as few triplets as possible. This allows LLMs to identify task-relevant semantic structures with only a few iterations, ensuring that the overall computational cost of the sampling process remains limited.

After sampling semantic subgraphs for users and items, we employ a semantic calibrator that leverages the semantic understanding capability of LLMs to perform task-aware refinement, aligning the sampled semantics with recommendation objectives. To further enhance the LLM's comprehension of graph data, we serialize the sampled subgraphs into a tree-structured textual format that can be easily parsed by the LLM [24]. Moreover, we design a system prompt to guide the calibrator in enhancing subgraph semantics toward better alignment with the recommendation task.

$$\begin{aligned} S_v &= f_P(\mathcal{T}_v^{\text{sub}}), \\ S_v' &= \text{LLM}(\mathbb{I}_3^v, S_v) \end{aligned} \tag{3}$$

where $f_P(\cdot)$ denotes the projection of the semantic subgraph of node v to the tree-structured text S_v, and $\mathbb{I}_3^v$ is the system prompt for the semantic enhancement of node v. S_v' represents the output of the semantic calibrator. Our design empowers LLMs to gain a deep understanding of graph data, effectively refining the semantics within the subgraphs to better align with the recommendation objectives. Moreover, since the sampled subgraphs are highly condensed and contain only task-relevant semantics, the refinement process incurs limited computational cost, as the LLM can efficiently enhance the concise semantics without extensive inference. The refined semantics are then encoded as continuous vectors for utilization in downstream recommendation models.

$$E_{\text{semantic}}^v = f_T(A_v), \tag{4}$$

where $f_T(\cdot)$ is a pre-trained text embedding model, such as SimCSE, which can be substituted with other embedding models. E_{semantic}^v is the semantic embedding of node v. Our recommendation-oriented semantic refinement module alleviates the influence of task-irrelevant semantics and aligns semantic information with recommendation objectives, thereby enabling the semantics to better support the recommendation task. This distinguishes our approach from CoLaKG

[4], the only existing work incorporating semantics into knowledge-aware recommendation. CoLaKG integrates all knowledge without discrimination, leading to task-irrelevant semantics that undermine the contribution of semantic information to recommendation performance.

4.3 Semantic-Collaborative Feature Integration

Traditional recommendation models represent users and items using ID embeddings $e_{\mathrm{id}}^{v} \in \mathbb{R}^{d}$, which primarily capture collaborative relations but lack semantic awareness. Building upon the extracted and refined semantic representations, we integrate them into ID embeddings, effectively leveraging semantic information to enhance recommendation performance. To achieve this, we propose a semantic-aware MoE network, which integrates semantic information with collaborative signals through a distill-then-fuse strategy.

A major challenge in integrating high-dimensional semantic features with low-dimensional ID embeddings is the substantial dimensional gap between them. Most existing methods, exemplified by CoLaKG [4], employ trainable adapters to project semantic features into the ID space. However, existing studies have demonstrated that such projections often lead to significant semantic loss [7]. To address this issue, the semantic-aware MoE applies semantic distillation to the semantic features, leveraging the low-rank approximation capability of SVD. Formally, this process can be expressed as follows:

$$\mathbf{U}\mathbf{S}\mathbf{U}^{\mathbf{T}} = \mathrm{SVD}(\Sigma), \tag{5}$$

where each column in $\mathbf{U}$ is an eigenvector of Σ, defining a basis direction in the feature space. The diagonal entries of S are the corresponding eigenvalues that weight these basis vectors. Larger eigenvalues indicate greater preservation of semantics from E_{semantic} in the associated subspace. By selecting the top d eigenvalues and their corresponding subspaces, we construct a compressed semantic embedding $E_{\mathrm{semantic}}^{'}$ with the same dimension as the ID embedding while retaining the majority of the semantic information.

Traditional dense fusion architectures, such as MLPs or attention-based models [11], often compress heterogeneous features into a single representation space, limiting their ability to model the heterogeneous and diverse relational patterns between semantic embeddings and collaborative signals. Inspired by the proven ability of mixture-of-experts (MoE) networks to integrate diverse information sources, the semantic-aware MoE then fuse the distilled semantic features with ID embeddings through dynamic expert routing, leveraging multiple experts to capture diverse association patterns between semantic knowledge and collaborative signals. Specifically, we employ N expert networks $E_{i}{}_{i=1}^{N}$ to perform the fusion. Notably, the value of N is relatively small, and our experiments demonstrate that choosing $N \in [2, 8]$ suffices to achieve strong performance. To further improve efficiency, we adopt a sparse gating mechanism [14] in our MoE design, significantly reducing the computational cost by activating only a subset of experts each time. This can be formally expressed as:

$$h(e^v) = \text{Top-S}(e^v \cdot W),$$
$$G(e^v)_i = \frac{\exp(h(e^v)_i)}{\sum_{j=1}^{S} \exp(h(e^v)_j)} \quad (6)$$

where $e^v = \left[e'^v_{\text{semantic}} \parallel e^v_{\text{id}}\right]$, and $e'^v_{\text{semantic}} \in \mathbb{R}^d$ is the compressed semantic embedding of node v. $W \in \mathbb{R}^{2d \times N}$ is the learnable parameters of the gating network. $h(e^v)$ denotes the logit score for each expert with respect to node v, and $G(e^v)_i$ is the weight of expert i obtained by applying a softmax to those logits. The final output of the expert network $\tilde{e}_v$ is a weighted sum of the individual expert outputs:

$$\tilde{e}_v = \sum_{i=1}^{N} G(e^v)_i E_i(e^v). \quad (7)$$

The outputs of the semantic-aware MoE network serve as the final user and item embeddings, and their dot product is used to predict the probability of interaction p_{ui}, i.e. $p_{ui} = \tilde{e}_u^\top \tilde{e}_i$. We optimize the semantic-aware MoE using contrastive loss [10] with an expanded set of negative samples to promote consistent and stable expert learning. In addition, the margin-based formulation adaptively filters out uninformative negative samples, encouraging the model to focus on hard negatives, thereby improving learning stability and accelerating convergence. The loss function can be formally written as:

$$\mathcal{L}_{\text{MoE}} = \sum_{u,i \in \mathcal{D}} \left[(1 - p_{ui})_+ + \frac{1}{|\mathcal{M}|} \sum_{j \in \mathcal{M}} (p_{uj} - \beta)_+ \right] \quad (8)$$

where (u, i) is an observed interaction from training set $\mathcal{D}$, and $\mathcal{M}$ is the negative item set sampled from unobserved items for user u. $(\cdot)_+$ is the ramp function $max(\cdot, 0)$. The objective of this loss is to maximize the similarity of positive pairs while decreasing the similarity of negative pairs with a margin β.

5 Experiments

5.1 Experimental Settings

Datasets. We conduct extensive experiments on three real-world datasets: DBbook2014 [3], Book-Crossing [5], and MovieLens-1M [12]. We follow previous work [8] to filter out users with fewer than 10 interactions in MovieLens-1M and fewer than 5 interactions in DBbook2014 and Book-Crossing. The statistics are presented in Table 1.

Baselines. We compare SeeKRec with a wide range of representative baselines, grouped into three categories: (1) Knowledge-free recommenders, including LightGCN [6], SGL [17], and SimGCL [20]; (2) ID-based knowledge-aware recommenders, comprising embedding-based methods such as CKE [22] and CFKG [1], propagation-based methods such as KGAT [15], KGIN [16], KGCL [19], MCCLK [25], KGRec [18], and CIKGRec [8]; and (3) Semantic-based knowledge-aware recommenders, exemplified by CoLaKG [4].

Table 1. Statistics of the datasets.

Dataset	DBbook2014	Book-Crossing	MovieLens-1M
#Users	5,576	6,616	6,040
#Items	2,680	8,853	3,260
#Ratings	65,961	110,662	998,539
#Relations	13	4	20
#Entities	8,762	1,404	14,377
#Triplets	134,223	1,137	415,104

Table 2. The overall performance of SeeKRec and baseline models. The best and second-best performances are highlighted in **bold** and underline, respectively.

MODEL	DBbook2014				Book-Crossing				MovieLens-1M			
	R@10	R@20	N@10	N@20	R@10	R@20	N@10	N@20	R@10	R@20	N@10	N@20
LightGCN	0.2142	0.2894	0.1684	0.1944	0.0667	0.0915	0.0535	0.0613	0.1559	0.2442	0.3687	0.3804
SGL	0.2165	0.2897	0.1774	0.2011	0.0672	0.0911	0.0562	0.0660	0.1564	0.2411	0.3713	0.3826
SimGCL	0.2172	0.2902	0.1791	0.2052	0.0686	0.0921	0.0601	0.0664	0.1576	0.2459	0.3786	0.3871
CKE	0.1504	0.2120	0.1429	0.1609	0.0565	0.0788	0.0456	0.0508	0.1517	0.2350	0.3008	0.3369
CFKG	0.1560	0.2216	0.1292	0.1426	0.0529	0.0751	0.0399	0.0441	0.1509	0.2428	0.2915	0.3219
KGAT	0.2119	0.2880	0.1595	0.1843	0.0651	0.0890	0.0606	0.0684	0.1555	0.2422	0.3668	0.3765
KGIN	0.2138	0.2907	0.1664	0.1919	0.0611	0.0859	0.0478	0.0545	0.1565	0.2445	0.3708	0.3829
KGCL	0.2372	0.3157	0.1890	0.2146	0.0651	0.0848	0.0604	0.0659	0.1560	0.2429	0.3701	0.3822
MCCLK	0.2406	0.3191	0.1910	0.2182	0.0615	0.0844	0.0461	0.0528	0.1595	0.2500	0.3588	0.3736
KGRec	0.2405	0.3198	0.1912	0.2187	0.0614	0.0859	0.0491	0.0570	0.1579	0.2491	0.3780	0.3876
CIKGRec	0.2597	0.3351	0.2032	0.2295	0.0747	0.1023	0.0624	0.0715	0.1623	0.2528	0.3864	0.3952
CoLaKG	0.2614	0.3381	0.2047	0.2311	0.0749	0.1035	0.0633	0.0734	0.1630	0.2574	0.3877	0.4007
SeeKRec	**0.2839**	**0.3633**	**0.2169**	**0.2464**	**0.0838**	**0.1132**	**0.0684**	**0.0786**	**0.1704**	**0.2696**	**0.3988**	**0.4091**
%Imp.	**8.61%↑**	**7.45%↑**	**5.96%↑**	**6.61%↑**	**11.88%↑**	**9.37%↑**	**8.06%↑**	**7.08%↑**	**4.54%↑**	**4.75%↑**	**2.85%↑**	**2.10%↑**

Implementation Details and Evaluation Metrics. We split each user's interaction history into training, validation, and test sets in a 7:1:2 ratio. The LLM used in this paper is the 'gpt-3.5-turbo-0125' model. For all baseline models, we use the official implementations released by the authors on GitHub. Hyperparameters are either adopted as recommended by the authors or carefully tuned on the validation sets. To assess our model's performance, we employ two widely used metrics: Recall@K and NDCG@K, with K set to 10 and 20. Recall@K evaluates the model's ability to retrieve relevant items, while NDCG@K measures ranking quality. Higher values for both metrics indicate better performance.

5.2 Overall Performance

The overall performance of SeeKRec and the baseline models is presented in Table 2. Three key observations can be drawn from the results. (1) Our method outperforms all baselines across the three datasets, demonstrating its effectiveness in leveraging semantic information to enhance recommendation performance. Specifically, the results indicate that SeeKRec effectively extracts users' semantic preferences, refines them in a recommendation-oriented manner and

Table 3. Comparison of computational costs.

Dataset	Method	Input Token	Output Token	Total Token	Time (h)
DBbook2014	CoLaKG	5,171,558	1,852,646	7,024,204	2.24
	SeeKRec	**4,574,756**	**1,516,504**	**6,091,260**	**1.33**
Book-Crossing	CoLaKG	3,686,461	1,740,118	5,426,579	3.81
	SeeKRec	**2,817,560**	**1,043,787**	**3,861,347**	**1.84**
MovieLens-1M	CoLaKG	3,980,404	2,220,847	6,201,251	15.13
	SeeKRec	**3,253,983**	**1,735,012**	**4,988,995**	**8.92**

Table 4. Ablation study of SeeKRec.

	DBbook2014		Book-Crossing		MovieLens-1M	
	R@10	N@10	R@10	N@10	R@10	N@10
Ours	**0.2839**	**0.2169**	**0.0838**	**0.0684**	**0.1704**	**0.3988**
w/o USE	0.2748	0.2072	0.0774	0.0652	0.1627	0.3901
w/o USA	0.2763	0.2102	0.0782	0.0658	0.1618	0.3859
w/o SPE	0.2684	0.2077	0.0739	0.0628	0.1583	0.3806
w/o SD	0.2907	0.2092	0.0750	0.0633	0.1605	0.3846
w/o MoE	0.2624	0.2051	0.0744	0.0621	0.1597	0.3822

integrates them with collaborative signals. (2) In scenarios with rich auxiliary knowledge (e.g., DBbook2014 and MovieLens-1M), the ID-based knowledge-aware recommenders outperform knowledge-free recommenders, suggesting that ID-based auxiliary knowledge can enhance recommendation performance. However, its performance deteriorates in scenarios with sparse auxiliary knowledge (e.g., Book-Crossing) due to the limited ID-based structural information and insufficient exploration of latent semantic information, resulting in inadequate modeling of user preferences and item features. (3) The semantic-based methods (e.g., SeeKRec and CoLaKG) outperforms ID-based methods in most scenarios. This underscores the efficacy of semantic information in capturing user preferences and uncovering latent relationships among items.

5.3 Efficiency Study

We evaluate the efficiency of SeekRec and CoLaKG, the only existing knowledge-aware recommendation method that incorporates semantic information, in terms of token consumption and total time required for complete recommendation. Results are shown in Table 3. Across all datasets, SeekRec demonstrates clear advantages over CoLaKG on all metrics. In terms of token consumption, SeekRec shows a significant advantage in both input and output token usage. This highlights SeekRec's efficiency in reducing token consumption by leveraging only recommendation-relevant semantic information, whereas CoLaKG indiscriminately utilizes all semantics within the knowledge graphs. Regarding total time cost, SeekRec achieves significant speedups across all datasets, demonstrating its high efficiency. This efficiency results from SeekRec's ability to quickly capture recommendation-relevant semantic information and efficiently integrate it with collaborative filtering signals through a sparsely activated MoE network. In summary, SeekRec not only enhances recommendation performance but also offers substantial efficiency advantages, highlighting the superiority of our proposed method.

5.4 Ablation Study

To evaluate the contribution of each innovative component in SeeKRec, we created five ablation variants.

- **w/o USE**: Exclude the **U**ser **S**emantic Preference **E**xtraction module and use only the item semantics from the knowledge graph.
- **w/o USA**: Omit the inter-**U**ser **S**emantic **A**lignment process.
- **w/o SPE**: Replace the **S**earch-**P**rune-**E**valuate semantic subgraph sampling strategy with full first-order neighbor sampling and random second-order neighbor sampling, as in CoLaKG [4].
- **w/o SD**: Exclude the **S**emantic **D**istillation process and directly fuse high-dimensional semantic features with low-dimensional ID embeddings.
- **w/o MoE**: Omit the MoE fusion of semantic and ID embeddings. Instead, project the semantic embeddings into the ID embedding space and combine them via element-wise addition for the final recommendation, similar to traditional methods and CoLaKG [4].

We also evaluated the complete model, referred to as "Ours". The results for all variants are presented in Table 4, from which the following conclusions can be drawn: First, removing any component results in a decline in overall performance, highlighting the indispensability of each part of SeeKRec and further validating the effectiveness of each innovation. Moreover, the performance decline in the w/o USE variant highlights the importance of extracting user-side semantics, while the decline in the w/o USA variant indicate the critical role of semantic preference alignment in identifying potential users with similar preferences. Furthermore, the results of w/o SPE show that our carefully designed subgraph sampling strategy samples high-quality subgraphs that ensure noise independence and semantic sufficiency. Additionally, the performance degradation of the w/o SD variant shows the importance of bridging the dimension gap between semantic and ID embeddings. Finally, the performance decline in the w/o MoE variant indicates that our method captures diverse association patterns between semantic and collaborative signals and effectively fuses them.

5.5 Hyperparameter Sensitivity

To assess the impact of different hyperparameters on model performance, we conducted a series of experiments focusing on three key parameters: the number of semantic alignment clusters κ, the number of experts N, and the value of K when selecting Top-K experts. The results are shown in Fig. 3. Based on the

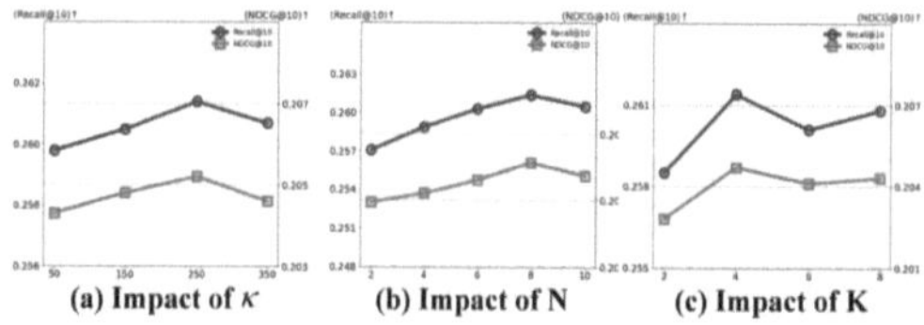

Fig. 3. Hyperparameter experiments on DBbook2014.

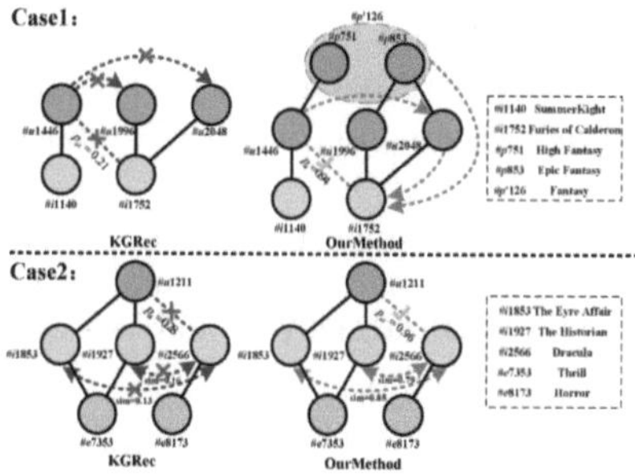

Fig. 4. Two real cases in DBbook2014 dataset.

results, we derive the following observations: **(1)** Since clustering merely serves to facilitate semantic preference alignment, and κ does not correspond to the actual number of preferences, its impact on the final results is negligible. **(2)** We observe that an insufficient number of experts N impairs the MoE's ability to capture diverse associations between semantic and ID embeddings, thereby degrading performance. At $N = 8$, however, these associations are adequately learned, and further increasing N yields no additional benefit. **(3)** In experiments with N fixed at 8 and K varied, a too small K leads to insufficient expert information being selected, thereby degrading fusion performance. However, once K exceeds a certain threshold, performance stagnates while resource overhead increases.

5.6 Case Study

This section presents two real cases from DBbook2014 to illustrate the explainability and rationality of SeeKRec. As depicted in Fig. 4, in the first scenario, user $\#u1446$ interacted with book $\#i1140$, a fantasy novel. Meanwhile, book $\#i1752$, another fantasy work, was engaged by $\#u1996$ and $\#u2048$, and it also appears in $\#u1446$'s test set. However, ID-based methods cannot infer their similarity due to the lack of ID linkage between $\#u1446$ and these two preference-similar users and thus fail to recommend $\#i1752$ to $\#u1446$. In contrast, our approach captures $\#u1446$'s semantic preference as "High Fantasy" and extracts "Epic Fantasy" for $\#u1996$ and $\#u2048$, then aligns them under a unified "Fantasy" preference, capturing user similarity based on the aligned preference and successfully recommending the target item. In the second case, user $\#u1211$ interacted in the training set with $\#i1927$ and $\#i1853$, both featuring the "Thrill" attribute, and in the test set with #*i*2566, which features the "Horror" attribute. Although these attributes are semantically proximate, ID-based methods fail to detect such similarity and thus cannot recommend #*i*2566. In contrast, our approach leverages semantic information to capture latent semantic associations between items, significantly improving recommendation performance.

6 Conclusion

In this paper, we propose SeeKRec to extract and refine semantics of users and items, and integrate them with collaborative signals adaptively, yielding a substantial improvement in recommendation performance.

Acknowledgments. This work was supported by the National Natural Science Foundation of China (62472341, 62302373, 62506282, and 62376203), Natural Science Foundation of Jiangsu Province (BK20232028), and the STI 2030—Major Projects (2022ZD0208500).

References

1. Ai, Q., Azizi, V., Chen, X., Zhang, Y.: Learning heterogeneous knowledge base embeddings for explainable recommendation. Algorithms **11**(9), 137 (2018)
2. Bao, K., Zhang, J., Zhang, Y., Wang, W., Feng, F., He, X.: Tallrec: an effective and efficient tuning framework to align large language model with recommendation. In: Proceedings of the 17th ACM Conference on Recommender Systems, pp. 1007–1014 (2023)
3. Cao, Y., Wang, X., He, X., Hu, Z., Chua, T.S.: Unifying knowledge graph learning and recommendation: towards a better understanding of user preferences. In: The World Wide Web Conference, pp. 151–161 (2019)
4. Cui, Z., et al.: Comprehending knowledge graphs with large language models for recommender systems. In: Proceedings of the 48th International ACM SIGIR Conference on Research and Development in Information Retrieval, pp. 1229–1239 (2025)
5. Dong, X., Yu, L., Wu, Z., Sun, Y., Yuan, L., Zhang, F.: A hybrid collaborative filtering model with deep structure for recommender systems. In: Proceedings of the AAAI Conference on Artificial Intelligence, vol. 31 (2017)
6. He, X., Deng, K., Wang, X., Li, Y., Zhang, Y., Wang, M.: Lightgcn: simplifying and powering graph convolution network for recommendation. In: Proceedings of the 43rd International ACM SIGIR Conference on Research and Development in Information Retrieval, pp. 639–648 (2020)
7. Hu, G., Zhang, A., Liu, S., Cai, Z., Yang, X., Wang, X.: Alphafuse: learn ID embeddings for sequential recommendation in null space of language embeddings. In: Proceedings of the 48th International ACM SIGIR Conference on Research and Development in Information Retrieval, pp. 1614–1623 (2025)
8. Hu, Z., et al.: Bridging the user-side knowledge gap in knowledge-aware recommendations with large language models. In: Proceedings of the AAAI Conference on Artificial Intelligence, pp. 11799–11807 (2025)
9. Kim, S., Kang, H., Choi, S., Kim, D., Yang, M., Park, C.: Large language models meet collaborative filtering: an efficient all-round LLM-based recommender system. In: Proceedings of the 30th ACM SIGKDD Conference on Knowledge Discovery and Data Mining, pp. 1395–1406 (2024)

10. Mao, K., et al.: Simplex: a simple and strong baseline for collaborative filtering. In: Proceedings of the 30th ACM International Conference on Information & Knowledge Management, pp. 1243–1252 (2021)
11. Nagrani, A., Yang, S., Arnab, A., Jansen, A., Schmid, C., Sun, C.: Attention bottlenecks for multimodal fusion. Adv. Neural. Inf. Process. Syst. **34**, 14200–14213 (2021)
12. Noia, T.D., Ostuni, V.C., Tomeo, P., Sciascio, E.D.: Sprank: semantic path-based ranking for top-n recommendations using linked open data. ACM Trans. Intell. Syst. Technol. (TIST) **8**(1), 1–34 (2016)
13. Reimers, N., Gurevych, I.: Sentence-bert: sentence embeddings using siamese bert-networks. arXiv preprint arXiv:1908.10084 (2019)
14. Shazeer, N., et al.: Outrageously large neural networks: the sparsely-gated mixture-of-experts layer. arXiv preprint arXiv:1701.06538 (2017)
15. Wang, X., He, X., Cao, Y., Liu, M., Chua, T.S.: KGAT: knowledge graph attention network for recommendation. In: Proceedings of the 25th ACM SIGKDD International Conference on Knowledge Discovery & Data Mining, pp. 950–958 (2019)
16. Wang, X., et al.: Learning intents behind interactions with knowledge graph for recommendation. In: Proceedings of the Web Conference 2021, pp. 878–887 (2021)
17. Wu, J., et al.: Self-supervised graph learning for recommendation. In: Proceedings of the 44th International ACM SIGIR Conference on Research and Development in Information Retrieval, pp. 726–735 (2021)
18. Yang, Y., Huang, C., Xia, L., Huang, C.: Knowledge graph self-supervised rationalization for recommendation. In: Proceedings of the 29th ACM SIGKDD Conference on Knowledge Discovery and Data Mining, pp. 3046–3056 (2023)
19. Yang, Y., Huang, C., Xia, L., Li, C.: Knowledge graph contrastive learning for recommendation. In: Proceedings of the 45th International ACM SIGIR Conference on Research and Development in Information Retrieval, pp. 1434–1443 (2022)
20. Yu, J., Yin, H., Xia, X., Chen, T., Cui, L., Nguyen, Q.V.H.: Are graph augmentations necessary? Simple graph contrastive learning for recommendation. In: Proceedings of the 45th International ACM SIGIR Conference on Research and Development in Information Retrieval, pp. 1294–1303 (2022)
21. Yuan, Z., et al.: Where to go next for recommender systems? ID-vs. modality-based recommender models revisited. In: Proceedings of the 46th International ACM SIGIR Conference on Research and Development in Information Retrieval, pp. 2639–2649 (2023)
22. Zhang, F., Yuan, N.J., Lian, D., Xie, X., Ma, W.Y.: Collaborative knowledge base embedding for recommender systems. In: Proceedings of the 22nd ACM SIGKDD International Conference on Knowledge Discovery and Data Mining, pp. 353–362 (2016)
23. Zhang, Y., Bao, K., Yan, M., Wang, W., Feng, F., He, X.: Text-like encoding of collaborative information in large language models for recommendation. arXiv preprint arXiv:2406.03210 (2024)
24. Zhao, J., et al.: Graphtext: graph reasoning in text space. arXiv preprint arXiv:2310.01089 (2023)
25. Zou, D., et al.: Multi-level cross-view contrastive learning for knowledge-aware recommender system. In: Proceedings of the 45th International ACM SIGIR Conference on Research and Development in Information Retrieval, pp. 1358–1368 (2022)

PreFact: Knowledge Propagation Regulating Network Toward Preferred Facts for Knowledge-Aware Recommendation

Chengyu Feng[1,2], Hua Chu[1,2], Yangtao Zhou[1,2(✉)], Zhenjiang Ding[1,2], Jianan Li[1,2], Qingshan Li[1,2], Xiangming Li[1,2], Zhongqi Lu[1,2], and Wanqiang Yang[2,3]

[1] School of Computer Science and Technology, Xidian University, Xi'an 710071, China
{cyfeng,zhou_yt,dzj,xiangmli,luzhongqi}@stu.xidian.edu.cn, {hchu,qshli}@mail.xidian.edu.cn, lijianan@xidian.edu.cn

[2] Intelligent Financial Software Engineering New Technology Joint Laboratory, Xidian University, Xi'an 710071, China

[3] Shanghai Fairyland Software Corp., Ltd., Shanghai 200233, China
yangwan@fulan.com.cn

Abstract. Knowledge-aware recommendation (KR) systems can enhance recommendation performance by utilizing factual knowledge in knowledge graphs (KGs). Existing methods typically adopt graph connectivity as the only propagation principle, spreading all factual knowledge into the recommendation process. However, this injects non-preferred facts into recommendation signals, driving the results away from user preferences and severely limiting knowledge-aware recommendation performance. To address this, we propose a knowledge-aware recommendation framework called PreFact, which introduces fact-preference relevance as an additional propagation principle to regulate knowledge propagation toward preferred facts. Specifically, to evaluate the fact-preference relevance, we design a relevance evaluation mixture-of-experts that simultaneously specializes in fact semantic encoding and user preference mining, which establishes cross-space associations in a heterogeneity-coordinated manner. To propagate knowledge around preferred facts, we design a self-adaptive switch network that continuously maintains a relevance-proportionate knowledge propagation flow and delicately cuts off the spread of non-preferred facts, which integrates factual knowledge into recommendation in a preferred-fact-centric manner. Extensive experiments on three public datasets demonstrate that PreFact consistently surpasses state-of-the-art models. The implementation of PreFact is available at https://github.com/fonzoin/PreFact.

Keywords: Knowledge-aware Recommendation · Knowledge Graphs · Knowledge Propagation Regulating · Mixture of Experts · Switch Networks

H. Jung et al. (Eds.): DASFAA 2026, LNCS 16535, pp. 324–339, 2026.
https://doi.org/10.1007/978-981-92-0363-5_20

1 Introduction

In the age of information, information overload has emerged as a critical barrier to users' efficient access to valuable content, thereby elevating the importance of recommendation systems across diverse domains [15,24]. Traditional recommendation systems, predominantly based on collaborative filtering (CF) techniques [3,6,10], are heavily dependent on historical user-item interaction data, which inevitably suffers from data sparsity and the cold-start issue. To address these issues, emerging knowledge-aware recommendation (KR) systems [8,17,23] incorporate knowledge graphs (KGs) as a powerful auxiliary information source. They employ techniques such as masking mechanisms [27], contrastive learning [25], and graph invariant learning [20] to extract rich factual knowledge from KGs, and then propagate it into the recommendation based on graph connectivity via graph neural network (GNN) [4,14,19]. By leveraging the abundant facts within KGs, such methods significantly enrich user and item representations and mitigate the issues of data sparsity and cold start, which has consequently attracted considerable scholarly attention [7,21].

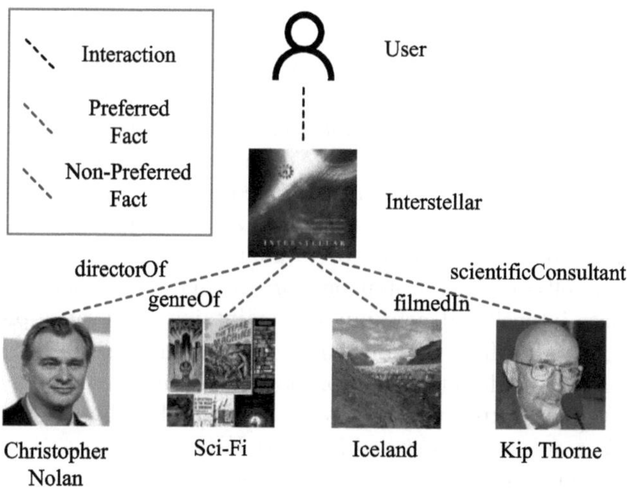

Fig. 1. An illustrative example in the movie scenario. The movie *Interstellar* has four associated facts in the KG, some of which are non-preferred facts to the user. Allowing such facts to propagate into recommendation drives the recommendation results away from the user's preferences.

However, existing methods overlook that due to the vast scale and high degree of generalization of KGs [11,13,25], some of their facts are semantically irrelevant to user preferences. For example, as shown in Fig. 1, the movie *Interstellar* that a user interacted with has four related facts in the knowledge graph. The facts that *Interstellar* was filmed in Iceland and the theoretical physicist Kip Thorne was the scientific consultant of *Interstellar* are semantically irrelevant to the user's preferences. Such non-preferred facts are inevitably introduced into downstream recommendation tasks when knowledge is propagated solely according to graph connectivity, as in existing methods. This disrupts the accurate modeling of user preferences and consequently drives the recommendation results away from the user's preferences. Therefore, regulating knowledge propagation toward preferred facts is crucial for knowledge-aware recommendation tasks.

Despite its necessity, regulating knowledge propagation will encounter two main challenges. First, user preferences are reflected in user-item interactions, while facts reside in knowledge graphs. These two types of information exist in heterogeneous semantic spaces, resulting in a semantic gap that hinders the alignment of facts with user preferences. It is necessary to bridge this gap and enable a direct evaluation of fact-preference relevance. Second, how to set the regulating criterion that best balances information purity and diversity remains a critical concern. There exists an inherent trade-off between information purity and diversity [2,18]. Overly regulating knowledge propagation inevitably diminishes informational diversity which is crucial for robust recommendation [1]. Hence, carefully regulating knowledge propagation without compromising information diversity is essential to ensure best recommendation performance.

To address the aforementioned challenges, we propose a knowledge-aware recommendation framework called PreFact, which introduces fact-preference relevance as an additional propagation principle to regulate knowledge propagation toward preferred facts. PreFact introduces two key modules to realize knowledge propagation regulating. Specifically, a relevance evaluation mixture-of-experts bridges the semantic gap between heterogeneous semantic spaces. It simultaneously specializes in fact semantic encoding and user preference mining, and establishes cross-space associations to evaluate the fact-preference relevance. Furthermore, a self-adaptive switch network regulates knowledge propagation with appropriate intensity. It continuously maintains a relevance-proportionate knowledge propagation flow and delicately cuts off the spread of non-preferred facts, allowing preferred-fact-centric propagation while balancing information purity and diversity. Experimental results on three benchmark datasets show that our PreFact outperforms state-of-the-art KR methods.

In summary, our contributions are highlighted as follows:

- We develop an end-to-end KR framework named PreFact to prevent recommendation results from drifting away from user preferences.
- We design a relevance evaluation mixture-of-experts that evaluates the fact-preference relevance in a heterogeneity-coordinated manner.
- We design a self-adaptive switch network that carefully regulates knowledge propagation while balancing information purity and diversity.
- Extensive experiments conducted on three public benchmark datasets demonstrate the superiority of PreFact.

2 Related Work

In this section, we briefly introduce the knowledge-aware recommendation and its related works.

Knowledge-aware recommendation methods focus on the effective extraction of KG information and its integration into recommendation systems and have boosted recommendation performance. They extract knowledge from KGs in various ways, ranging from traditional embedding learning to self-supervised graph learning and emerging diffusion models. For instance, CKE [26] integrates TransR [11] embeddings with traditional collaborative filtering, enabling the model to leverage both structural and semantic information of entities. KGAT [21] advances this by explicitly modeling high-order relations through attentive embedding propagation layers over the combined knowledge and user-item graphs, thus refining the representation of entities and users. Similarly, KGIN [22] employs a GNN-based architecture that models user intents as combinations of KG relations and captures relational path semantics via a dual aggregation mechanism. KRDN [27] adopts an adaptive knowledge modeling strategy with parameterized binary masks that learn deep knowledge associations. KGRec [23] introduces a rationale-enhanced framework that discovers the knowledge connection rationale through the masking-reconstructing process. DiffKG [8] leverages a diffusion-based process to learn robust representations from the KG and then integrates them through a collaborative graph convolution. EditKG [17] simultaneously generates item attributes and relations to address knowledge imbalance. These approaches effectively extract rich factual knowledge from KGs, significantly enhancing recommendation quality.

3 Problem Formulation

We begin by introducing the symbolic concepts relevant to this paper, subsequently presenting a formal definition of knowledge-aware recommendation task.

User-Item Interaction. In a typical recommendation scenario, user-item interactions are continuously generated as behavioral data from users. Let $\mathcal{U} = \{u_1, u_2, \ldots, u_{|\mathcal{U}|}\}$ and $\mathcal{I} = \{i_1, i_2, \ldots, i_{|\mathcal{I}|}\}$ denote the sets of users and items, respectively, where $|\mathcal{U}|$ and $|\mathcal{I}|$ are the total number of users and items. The interaction between users and items is represented by a binary matrix $\mathrm{Y} \in \mathbb{R}^{|\mathcal{U}| \times |\mathcal{I}|}$, where $y_{ui} = 1$ indicates that user u has interacted with item i, such as by clicking, rating, or purchasing. Conversely, $y_{ui} = 0$ means no interaction has occurred. The user-item interaction graph can be further constructed based on Y.

Knowledge Graph. In addition to behavioral data, knowledge graphs provide rich structured information that describes semantic relationships among entities, including items. A knowledge graph is typically formulated as a set of triplets $\mathcal{G} = \{(h, r, t) | h, t \in \mathcal{E}, r \in \mathcal{R}\}$, where h, r, and t refer to the head entity, relation, and tail entity, respectively. $\mathcal{E}$ and $\mathcal{R}$ denote the sets of entities and relations,

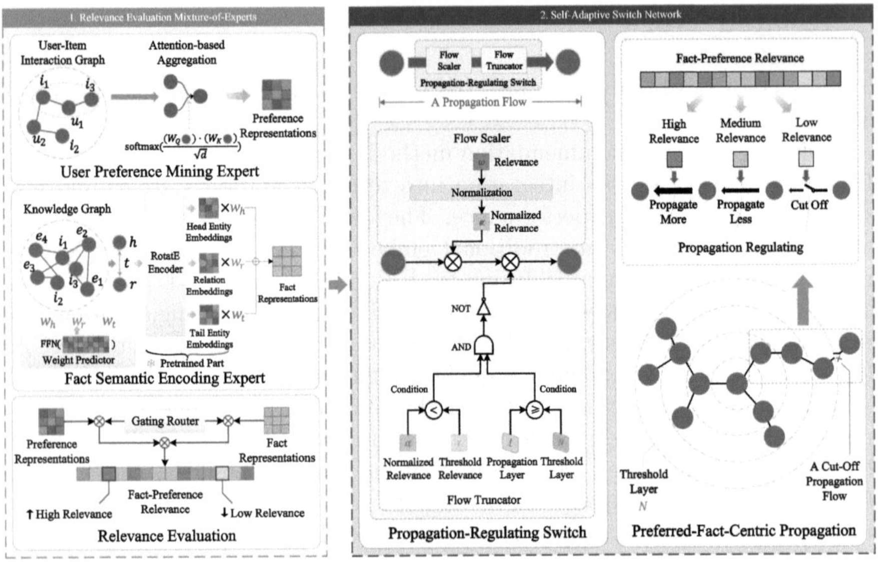

Fig. 2. The overall framework of the PreFact model (best viewed in color).

respectively. For example, the triplet (*Christopher Nolan*, *directorOf*, *Inception*) conveys the fact that "Christopher Nolan is the director of Inception." The user-item interaction graph and the knowledge graph can be connected through items, jointly forming the topology of knowledge-aware recommendation. By leveraging the factual knowledge in the KG, we can better capture complex features of the items.

Task Formulation. Given a user-item interaction graph Y and a knowledge graph $\mathcal{G}$, the goal of knowledge-aware recommendation task is to develop a model capable of precisely predicting the likelihood of interaction between a user and an item, thereby enabling more accurate personalized recommendations.

4 Methodology

In this section, we present the proposed PreFact. As demonstrated in Fig. 2, it consists of two main modules: 1) Relevance Evaluation Mixture-of-Experts: Two mini expert networks are built for user preference mining and fact semantic encoding. After obtaining the representations from each expert, we route them into a shared semantic space and evaluate fact-preference relevance by a gating router. 2) Self-Adaptive Switch Network: This switch network employs a flow scaler and a flow truncator to regulate each propagation flow and achieve preferred-fact-centric propagation. The flow scaler continuously maintains a relevance-proportionate knowledge propagation flow. The flow truncator delicately cuts off the propagation of non-preferred facts. With the collaborative effort of these two modules, PreFact is able to bridge the semantic gap between user-item interactions and KGs and perform knowledge propagation regulating without compromising information diversity.

4.1 Relevance Evaluation Mixture-of-Experts

In this module, we adopt a mixture-of-experts (MoE) architecture to evaluate fact-preference relevance. Since user preferences and fact semantics are very complex, the MoE architecture can effectively decompose diverse features among user preferences and fact semantics, thus leading to an improved modeling capacity in both semantic spaces. After capturing the features, the gating router of the MoE architecture can bridge the semantic gap heterogeneous semantic space and evaluate fact-preference relevance directly.

User Preference Mining Expert. The user preference mining expert aims to accurately model users' personalized preferences. We implement the user preference mining expert by employing attention-based aggregation on the user-item interaction graph to extract complex user preference features. Specifically, we treat users as queries and items as keys and values, using the items a user has interacted with to represent her preference. The aggregation process is as follows:

$$e_u^t = \sum_{i \in \mathcal{N}_u} \text{softmax}\left(\frac{\mathbf{W}_Q e_u \cdot (\mathbf{W}_K e_i)}{\sqrt{d}}\right) \cdot e_i, \tag{1}$$

where e_u^t denotes the representation of user u's preference, and $\mathcal{N}_u$ represents the neighborhood set of user u in the user-item interaction graph. The embeddings e_u and e_i are randomly initialized representations of user u and item i, respectively. W_Q and W_K are learnable weight matrices, and d is the embedding dimension.

Fact Semantic Encoding Expert. To capture the semantics of facts, the fact semantic encoding expert first encodes entities and relations into pretrained embeddings, then combines them to form unified fact representations. We adopt the Knowledge Graph Embedding (KGE) technique as the encoder of entities and relations:

$$e_h, e_r, e_t = \text{KGE}(h, r, t), \tag{2}$$

where e_h, e_r, e_t denote the embeddings of head entity h, relation r, and tail entity t, respectively. In practice, we adopt RotatE [16] as the specific KGE implementation.

To further obtain fact representations, we need to deeply fuse the embeddings of the head entity, relation, and tail entity to interconnect their semantics. Considering that the contributions of the head entity, relation, and tail entity may vary across facts, we design an affine feed forward network (FFN) as a weight predictor to self-adaptively learn their contribution weights for each fact:

$$\text{FFN}(x) = \text{softmax}(\mathbf{W}_{out}(\sigma(\mathbf{W}_{in} x + \mathbf{b}_{in})) + \mathbf{b}_{out}), \tag{3}$$

$$w_h, w_r, w_t = \mathbf{FFN}(e_h||e_r||e_t), \tag{4}$$

$$e_{(h,r,t)} = w_h e_h + w_r e_r + w_t e_t, \tag{5}$$

where we assign $\mathbf{W}_{in} \in \mathbb{R}^{3d \times d}$, $\mathbf{b}_{in} \in \mathbb{R}^{d}$ to handle the individual characteristics of the concatenated embeddings and $\mathbf{W}_{out} \in \mathbb{R}^{d \times 3}$, $\mathbf{b}_{out} \in \mathbb{R}^{3}$ act as a prediction head to generate weights w_h, w_r, w_t for fact (h, r, t). $\sigma(\cdot)$ is the sigmoid function. $e_{(h,r,t)}$ denotes the representation of fact (h, r, t).

Fact-Preference Relevance Evaluation. After obtaining two kinds of representations, we evaluate fact-preference relevance. We first route them into a shared semantic space to establish cross-space associations by a multilayer perceptron (MLP) architecture gating router:

$$\tilde{e}_u^t = \mathbf{W}^{(2)}\sigma(\mathbf{W}^{(1)} e_u^t + \mathbf{b}^{(1)}) + \mathbf{b}^{(2)}, \tag{6}$$

$$\tilde{e}_{(h,r,t)} = \mathbf{W}^{(2)}\sigma(\mathbf{W}^{(1)} e_{(h,r,t)} + \mathbf{b}^{(1)}) + \mathbf{b}^{(2)}, \tag{7}$$

where $\tilde{e}_u^t$, $\tilde{e}_{(h,r,t)}$ denote the mapped preference and fact representations, $\mathbf{W}^{(1)}$, $\mathbf{W}^{(2)}$, $\mathbf{b}^{(1)}$, and $\mathbf{b}^{(2)}$ denote the weights and biases of the gating router. After mapping, we evaluate the fact-preference relevance. Here, the relevance is quantified by computing the similarity between the representations:

$$\omega_{(h,r,t)} = \sum_{u \in \mathcal{U}} \mathrm{sim}(\tilde{e}_u^t, \tilde{e}_{(h,r,t)}), \tag{8}$$

where $\mathrm{sim}(\cdot)$ denotes the dot product operation, $\omega_{(h,r,t)}$ is the fact-preference relevance of fact (h, r, t).

4.2 Self-adaptive Switch Network

To regulate knowledge propagation toward preferred facts based on fact-preference relevance, we propose a self-adaptive switch network. Specifically, we model the knowledge propagation regulating as assigning a appropriate propagation weight for each propagation flow. We integrate both a flow scaler and a flow truncator to assign propagation weights controlling the extent of knowledge propagation, and then perform preferred-fact-centric propagation based on those weights.

Flow Scaler. The flow scaler continuously maintains a relevance-proportionate knowledge propagation flow by scaling propagation weights based on fact-preference relevance. It normalizes the fact-preference relevance of each fact as its propagation weight:

$$\alpha_{(h,r,t)} = \frac{\omega_{(h,r,t)}}{\sum_{(h,r,t) \in \mathcal{N}_h} \omega_{(h,r,t)}}, \tag{9}$$

where $\alpha_{(h,r,t)}$ denotes the propagation weight of fact (h, r, t), and $\mathcal{N}_h$ represents the neighborhood set of head entity h.

Flow Truncator. The propagation flow then proceeds to the flow truncator. Specifically, it truncates the propagation of facts with low factpreference relevance and those located in deeper propagation layers. This ensures that all facts are propagated with appropriate weights in the early stages while preventing non-preferred facts from spreading as propagation deepens, preserving both information purity and diversity. The regulating process of the flow truncator can be formalized as:

$$\rho_{(h,r,t)} = (1 - \mathbb{1}(\alpha_{(h,r,t)} < \gamma \wedge l \geq N)) \cdot \alpha_{(h,r,t)}, \tag{10}$$

where $\rho_{(h,r,t)}$ denotes the final propagation weight of fact (h, r, t), $\mathbb{1}(\cdot)$ is a logic function that returns 1 when the condition is met and 0 otherwise, l denotes the current propagation layer. γ, N are the threshold relevance and the threshold layer, which are both model hyperparameters.

Preferred-Fact-Centric Propagation. We then perform graph propagation on the user-item interaction graph and the knowledge graph based on the obtained propagation weights:

$$x_h^{(l)} = \frac{1}{|\mathcal{N}_h|} \sum_{(h,r,t)\in\mathcal{N}_h} \rho_{(h,r,t)} x_r \odot x_t^{(l-1)}, \tag{11}$$

$$x_u^{(l)} = \frac{1}{|\mathcal{N}_u|} \sum_{i\in\mathcal{N}_u} x_i^{(l-1)}, \tag{12}$$

where $x_h^{(l)}$ and $x_t^{(l)}$ denote the representations of head entity h and tail entity t at the l-th propagation layer, while x_r is the representation of relation r, all randomly initialized at $l = 0$. $x_i^{(l)}$ is the representation of item i at layer l, derived from its corresponding entity, and $x_u^{(l)}$ is the representation of user u, also randomly initialized at the beginning, and $\odot$ is the element-wise product. This propagation can be iterated for L layers to form a multi-layer GNN. The final user and item representations are obtained by fusing outputs from all layers:

$$\tilde{x}_u = \sum_{l=0}^{L} x_u^{(l)}, \tag{13}$$

$$\tilde{x}_i = \sum_{l=0}^{L} x_i^{(l)}, \tag{14}$$

where $\tilde{x}_u$ and $\tilde{x}_i$ represent the final representations of user u and item i, respectively. The entire model is optimized using the cosine contrastive loss (CCL) [12,27] function:

$$\hat{y}_{u,i} = \mathrm{s}(\tilde{x}_u, \tilde{x}_i), \tag{15}$$

$$\mathcal{L} = \sum_{u,i\in\mathcal{D}} [(1 - \hat{y}_{u,i})_+ + \frac{1}{|\mathcal{N}|} \sum_{j\in\mathcal{N}} (\hat{y}_{u,j} - \delta)_+], \tag{16}$$

where $\hat{y}_{u,i}$ denotes the predicted score between user u and item i, s($\cdot$) is the cosine similarity function, $\mathcal{D}$ and $\mathcal{N}$ denote the positive and negative sample sets, respectively, and δ is a margin controlling the CCL loss. The operator $(\cdot)_+$ represents the ramp function, *i.e.*, $\max(\cdot, 0)$.

4.3 Model Analysis

Model Size. The trainable parameters of PreFact consist of: (1) embeddings of users, items $\{e_u, e_i | u \in \mathcal{U}, i \in \mathcal{I}\}$ and weight matrices $\{\mathbf{W}_Q, \mathbf{W}_K\}$ in the user preference mining expert; (2) weight matrices $\{\mathbf{W}_{in}, \mathbf{W}_{out}\}$ and bias terms $\{\mathbf{b}_{in}, \mathbf{b}_{out}\}$ used in the fact semantic encoding expert; (3) weight matrices $\{\mathbf{W}^{(1)}, \mathbf{W}^{(2)}\}$ and bias terms $\{\mathbf{b}^{(1)}, \mathbf{b}^{(2)}\}$ used in the fact-preference relevance evaluation; (4) representations of head entities, relations, and tail entities $\{x_h^{(0)}, x_r, x_t^{(0)} | h, t \in \mathcal{E}, r \in \mathcal{R}\}$ and representations of users, items $\{x_u^{(0)}, x_i^{(0)} | u \in \mathcal{U}, i \in \mathcal{I}\}$ in the preferred-fact-centric propagation.

Time Complexity. The time complexity of our model mainly arises from two parts: (1) the attention-based graph aggregation over the user-item interaction graph, with complexity $O(|C|d)$, where $|C|$ denotes the number of user-item interactions, and d is the embedding dimension; (2) the preferred-fact-centric propagation, with complexity $O(L|C|d + L|\mathcal{G}|d)$, where L is the number of aggregation layers, and $|\mathcal{G}|$ is the number of knowledge graph facts. Overall, the total time complexity is $O((L+1)|C|d + L|\mathcal{G}|d)$, comparable to mainstream knowledge-aware recommendation methods. Note that pretraining the KGE embeddings can be completed offline in advance, thus excluded from both the time cost and the model size cost.

5 Experiment

In this section, we present a series of experiments to empirically validate the effectiveness of our proposed PreFact framework. Specifically, we aim to answer the following research questions:

- **RQ1**: Can PreFact achieve superior recommendation performance compared to state-of-the-art KR models?
- **RQ2**: How do the key components of PreFact individually contribute to its overall performance?
- **RQ3**: How sensitive is PreFact to changes in key hyperparameters (*i.e.* the threshold layer, and the threshold relevance)?

5.1 Experimental Settings

Datasets. To ensure a thorough and diverse evaluation, we utilize three publicly available datasets from different real-world domains: Last-FM for music

Table 1. Statistics of the datasets.

Statistics	Last-FM	MIND	Alibaba-iFashion
#Users	23,566	100,000	114,737
#Items	48,123	30,577	30,040
#Interactions	3,034,796	2,975,319	1,781,093
Density	2.7e-3	9.7e-4	5.2e-4
Knowledge Graph			
#Entities	58,266	24,733	59,156
#Relations	9	512	51
#Triplets	464,567	148,568	279,155

Table 2. The overall performance evaluation results. R is an abbreviation for Recall and N is an abbreviation for NDCG. The best and second-best performances are highlighted in bold and with underline, respectively.

Models	Last-FM				MIND				Alibaba-iFashion			
	R@10	R@20	N@10	N@20	R@10	R@20	N@10	N@20	R@10	R@20	N@10	N@20
CKE	0.0571	0.0825	0.0633	0.0706	0.0246	0.0379	0.0207	0.0250	0.0520	0.0811	0.0390	0.0495
KGAT	0.0571	0.0830	0.0629	0.0704	0.0260	0.0326	0.0228	0.0282	0.0524	0.0837	0.0385	0.0498
KGIN	0.0686	0.0950	0.0764	0.0836	0.0196	0.0376	0.0173	0.0241	0.0756	0.1146	0.0577	0.0717
KGCL	0.0422	0.0647	0.0523	0.0579	0.0226	0.0395	0.0175	0.0239	0.0755	0.1149	0.0576	0.0717
KRDN	<u>0.0732</u>	<u>0.1027</u>	0.0848	<u>0.0937</u>	0.0283	0.0435	0.0245	0.0294	<u>0.0920</u>	<u>0.1369</u>	<u>0.0717</u>	<u>0.0878</u>
KGRec	0.0647	0.0915	0.0722	0.0796	0.0334	0.0422	<u>0.0368</u>	0.0373	0.0764	0.1163	0.0586	0.0729
DiffKG	0.0731	0.0974	<u>0.0852</u>	0.0911	<u>0.0382</u>	<u>0.0615</u>	0.0299	<u>0.0387</u>	0.0836	0.1265	0.0635	0.0789
KGIL	0.0725	0.0945	0.0783	0.0840	0.0135	0.0250	0.0110	0.0150	0.0774	0.1151	0.0592	0.0712
Ours	**0.0795**	**0.1074**	**0.0913**	**0.0984**	**0.0396**	**0.0634**	**0.0377**	**0.0392**	**0.0937**	**0.1395**	**0.0729**	**0.0890**

recommendation, MIND for news recommendation, and Alibaba-iFashion for e-commerce scenarios. To preprocess the data, we adopt a 10-core filtering strategy by removing users and items that appear fewer than 10 times, ensuring sufficient interaction density for reliable model training. For Last-FM, we follow the previous work [21] and [17] to construct the knowledge graph by mapping items to the corresponding Freebase entities and extracting relevant knowledge triplets accordingly. For MIND, we construct the knowledge graph by collecting relevant facts from Wikidata based on the representative entities provided in the original data, following the approach proposed in [23]. For Alibaba-iFashion, the knowledge graph is manually constructed by utilizing the category information, in line with the method described in [22]. Table 1 presents a summary for the statistics of the three datasets.

Evaluation Protocols. To ensure a fair and consistent evaluation, we adopt a full-ranking protocol and split each dataset into a training set (*i.e.* 80%) and a test set (*i.e.* 20%). To assess the performance of PreFact and all baseline models, we adopt widely used Recall@K and NDCG@K as evaluation metrics, with K set to 20 for top-K recommendations. We initialize all trainable parameters with Xavier [5] and adopt Adam [9] as the optimizer. To optimize the performance of PreFact, we conduct a grid search for optimal hyperparameter settings. We implement PreFact using PyTorch and compare its performance with various baseline models using their official code.

Baseline Models. To conduct a comprehensive evaluation, we perform an extensive comparative analysis of our proposed PreFact against a wide range of baseline methods.

- **CKE** [26]: A recommendation framework that integrates TransR [11] with traditional collaborative filtering.
- **KGAT** [21]: A knowledge-aware recommendation model that explicitly captures high-order relations in the hybrid structure of knowledge graph and user-item graph through attentive embedding propagation layers.
- **KGIN** [22]: A GNN-based recommendation model that explicitly models user intents as combinations of KG relations and captures relational path semantics via a dual aggregation scheme.
- **KGCL** [25]: A knowledge graph contrastive learning framework for recommendation systems that integrates knowledge graph augmentation and cross-view contrastive learning to mine entity relationships and enhance robust item representations.
- **KRDN** [27]: A knowledge-aware recommendation framework that employs adaptive knowledge modeling with parameterized binary masks to learn deep knowledge associations.
- **KGRec** [23]: A rationale-enhanced framework utilizing rationale discovery with rationale-aware contrasting to extract knowledge connections in knowledge graphs in a rationale-aware manner.
- **DiffKG** [8]: A diffusion-based model for knowledge-aware recommendation systems that employs a forward process to corrupt the knowledge graph with Gaussian noise and a collaborative knowledge graph convolution mechanism to integrate user-item interaction signals.
- **KGIL** [20]: An invariant learning framework for recommendation systems that learns factual features across diverse environments using environment generators and an invariant learning objective.

5.2 Performance Comparison (RQ1)

Table 2 presents the performance results of all evaluated models. Based on these results, we derive the following observations.

The experimental results demonstrate that our proposed model PreFact consistently outperforms all baseline models on both metrics and all three datasets.

These enhancements account for two reasons: First, the proposed relevance evaluation mixture-of-experts specializes in fact semantic encoding and user preference mining, which establishes cross-space associations in a heterogeneity-coordinated manner. Second, the self-adaptive switch network continuously maintains a relevance-proportionate knowledge propagation flow and delicately cuts off the spread of non-preferred facts, which integrates factual knowledge into recommendation in a preferred-fact-centric manner.

KRDN exhibits strong competitiveness, ranking as the suboptimal model across most settings. This advantage can be attributed to its adaptive knowledge refining and contrastive denoising learning, which refining knowledge associations and denoising implicit interactions simultaneously. However, its knowledge refining operations still lack user-specific adaptivity and diversity-preserving ability. As a result, KRDN is less capable of maintaining preferred facts and preserving both the diversity and purity of recommendation.

5.3 Ablation Study of PreFact (RQ2)

In this section, we conduct ablation experiments to investigate the contribution of each key module in the proposed PreFact framework. Specifically, we analyze the effectiveness of the relevance evaluation mixture-of-experts and the self-adaptive switch network (including the flow scaler and flow truncator). To this end, we design the following model variants for comparison:

- **PreFact w/o. MoE:** This variant removes the entire relevance evaluation mixture-of-experts module. It simply assigns a learnable parameter to each fact as its factpreference relevance.
- **PreFact w/o. Scal:** This variant removes the flow scaler from the self-adaptive switch network. The propagation intensity is no longer proportional to the relevance score, resulting in uniform propagation intensity across all facts.
- **PreFact w/o. Trun:** This variant discards the flow truncator. All facts continue to propagate throughout the entire network regardless of their relevance.
- **PreFact w/o. Switch:** This variant removes the entire self-adaptive switch network, meaning the knowledge propagation is performed by a normal GNN without being dynamically regulated.

The experimental results are summarized in Table 3. We observe consistent performance drops across both datasets when any key module is removed, highlighting their indispensable roles. The key findings can be summarized as follows: (i) Removing the relevance evaluation mixture-of-experts (*i.e.*, PreFact w/o. MoE) causes the most significant performance degradation, underscoring its crucial role in simultaneously specializing in fact semantic encoding and user preference mining, which establishes cross-space associations in a heterogeneity-coordinated manner. Assigning a learnable parameter to each fact as its fact-preference relevance is overly simplistic, failing to accurately model the nuances of user preferences for specific facts. (ii) Removing the flow scaler (PreFact w/o.

Table 3. Ablation study results on Alibaba-iFashion and Last-FM datasets. R denotes Recall and N denotes NDCG. The best and second-best performances are highlighted in bold and with underline, respectively.

Models	Alibaba-iFashion				Last-FM			
	R@20	N@20	R@40	N@40	R@20	N@20	R@40	N@40
w/o. MoE	0.0832	0.1237	0.0648	0.0794	0.0703	0.0965	0.0827	0.0864
w/o. Scal	0.0911	0.1370	0.0719	0.0871	0.0778	0.1049	0.0885	0.0957
w/o. Trun	0.0927	0.1384	0.0714	0.0885	0.0788	0.1065	0.0901	0.0974
w/o. Switch	0.0893	0.1320	0.0682	0.0844	0.0759	0.1024	0.0879	0.0928
PreFact	**0.0937**	**0.1395**	**0.0729**	**0.0890**	**0.0795**	**0.1074**	**0.0913**	**0.0984**

Scal) results in a noticeable performance decline, indicating that proportionally scaling propagation intensity according to fact relevance is essential for regulating knowledge propagation toward preferred facts. (iii) Discarding the flow truncator (PreFact w/o. Trun) weakens model performance. Without cutting off propagation of non-preferred facts, the spreading of them limits recommendation performance. (iv) Removing the entire self-adaptive switch network (PreFact w/o. Switch) leads to an overall decline across all metrics, confirming that maintaining a relevance-proportionate knowledge propagation flow and delicately cutting off the spread of non-preferred facts is essential to prevent semantic drift and recommendation performance degradation.

Overall, these results verify that each component of PreFact contributes to its superior performance, and their collaborative functioning enables effective, relevance-aware knowledge propagation.

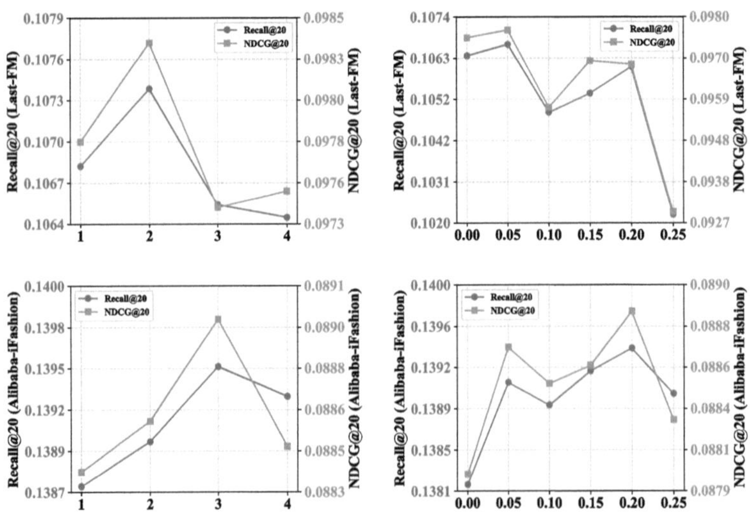

(a) The impact of the threshold layer (b) The impact of the threshold relevance

Fig. 3. Impact of hyperparameters.

5.4 Hyperparameter Analysis (RQ3)

In this subsection, we focus on examining the effects of different hyperparameters on our method, including the threshold layer, and the threshold relevance. Due to space limitations, we report results only on Last-FM and Alibaba-iFashion.

Impact of the Threshold Layer. We tune the threshold layer in {1,2,3,4}, and report the results in Fig. 3a. The results reveal that truncating propagation too early (*e.g.*, at layer 1) prevents the model from fully exploiting the relational knowledge in the early propagation stage, leading to insufficient information diversity. In contrast, applying truncation too late (*e.g.*, at layer 4) allows non-preferred facts to spread excessively, causing semantic drift and degrading recommendation performance. A moderate switching point (layer 2 or 3) achieves the best trade-off between purity and diversity, allowing PreFact to reach maximum performance.

Impact of the Threshold Relevance. We tune the threshold relevance in {0.0,0.05,0.1,0.15,0.2,0.25}, and record the results in Fig. 3b. A too-low threshold allows low-relevance facts to continue propagating, introducing non-preferred facts into the recommendation process. Conversely, an overly strict threshold blocks potentially valuable but moderately relevant facts. A moderate threshold achieves the best balance between suppressing non-preferred facts and preserving potentially useful factual knowledge.

6 Conclusion

In this paper, we propose PreFact, a knowledge-aware recommendation framework designed to regulate knowledge propagation toward preferred facts by introducing fact-preference relevance as an additional propagation principle. Firstly, we design a relevance evaluation mixture-of-experts that simultaneously specializes in fact semantic encoding and user preference mining, and establishes cross-space associations to evaluate the fact-preference relevance. Secondly, we design a self-adaptive switch network, consisting of a flow scaler and a flow truncator, which regulates knowledge propagation intensity to maintain high information purity without compromising diversity. Extensive experiments conducted on three real-world datasets demonstrate that PreFact consistently surpasses state-of-the-art knowledge-aware recommendation methods across all evaluated metrics.

Acknowledgments. This work was supported by the National Natural Science Foundation of China (62472341, 62302373, 62506282, and 62376203), Natural Science Foundation of Jiangsu Province (BK20232028), and the STI 2030—Major Projects (2022ZD0208500).

References

1. Abdollahpouri, H., Mansoury, M., Burke, R., Mobasher, B.: The unfairness of popularity bias in recommendation. arXiv preprint arXiv:1907.13286 (2019)
2. Chen, L., Yuan, F., Yang, J., He, X., Li, C., Yang, M.: User-specific adaptive fine-tuning for cross-domain recommendations. IEEE Trans. Knowl. Data Eng. **35**(3), 3239–3252 (2021)
3. Du, Y., et al.: Towards explainable collaborative filtering with taste clusters learning. In: Proceedings of the ACM Web Conference 2023, pp. 3712–3722 (2023)
4. Dwivedi, V.P., Joshi, C.K., Luu, A.T., Laurent, T., Bengio, Y., Bresson, X.: Benchmarking graph neural networks. J. Mach. Learn. Res. **24**(43), 1–48 (2023)
5. Glorot, X., Bengio, Y.: Understanding the difficulty of training deep feedforward neural networks. In: Proceedings of the Thirteenth International Conference on Artificial Intelligence and Statistics, pp. 249–256. JMLR Workshop and Conference Proceedings (2010)
6. Hu, Y., Koren, Y., Volinsky, C.: Collaborative filtering for implicit feedback datasets. In: 2008 Eighth IEEE International Conference on Data Mining, pp. 263–272. IEEE (2008)
7. Huang, C., et al.: Knowledge-aware coupled graph neural network for social recommendation. In: Proceedings of the AAAI Conference on Artificial Intelligence, vol. 35, pp. 4115–4122 (2021)
8. Jiang, Y., Yang, Y., Xia, L., Huang, C.: Diffkg: knowledge graph diffusion model for recommendation. In: Proceedings of the 17th ACM International Conference on Web Search and Data Mining, pp. 313–321 (2024)
9. Kingma, D.P., Ba, J.: Adam: a method for stochastic optimization. CoRR abs/1412.6980 (2014). https://api.semanticscholar.org/CorpusID:6628106
10. Koren, Y.: Factorization meets the neighborhood: a multifaceted collaborative filtering model. In: Proceedings of the 14th ACM SIGKDD International Conference on Knowledge Discovery and Data Mining, pp. 426–434 (2008)
11. Lin, Y., Liu, Z., Sun, M., Liu, Y., Zhu, X.: Learning entity and relation embeddings for knowledge graph completion. In: Proceedings of the AAAI Conference on Artificial Intelligence, vol. 29 (2015)
12. Mao, K., et al.: Simplex: a simple and strong baseline for collaborative filtering. In: Proceedings of the 30th ACM International Conference on Information & Knowledge Management, pp. 1243–1252 (2021)
13. Pujara, J., Augustine, E., Getoor, L.: Sparsity and noise: where knowledge graph embeddings fall short. In: Proceedings of the 2017 Conference on Empirical Methods in Natural Language Processing, pp. 1751–1756 (2017)
14. Scarselli, F., Gori, M., Tsoi, A.C., Hagenbuchner, M., Monfardini, G.: The graph neural network model. IEEE Trans. Neural Networks **20**(1), 61–80 (2008)
15. Sun, F., et al.: Bert4rec: sequential recommendation with bidirectional encoder representations from transformer. In: Proceedings of the 28th ACM International Conference on Information and Knowledge Management, pp. 1441–1450 (2019)
16. Sun, Z., Deng, Z.H., Nie, J.Y., Tang, J.: Rotate: knowledge graph embedding by relational rotation in complex space. arXiv preprint arXiv:1902.10197 (2019)
17. Tang, G., et al.: Editkg: editing knowledge graph for recommendation. In: Proceedings of the 47th International ACM SIGIR Conference on Research and Development in Information Retrieval, pp. 112–122 (2024)

18. Tian, C., Xie, Y., Li, Y., Yang, N., Zhao, W.X.: Learning to denoise unreliable interactions for graph collaborative filtering. In: Proceedings of the 45th International ACM SIGIR Conference on Research and Development in Information Retrieval, pp. 122–132 (2022)
19. Wang, H., Zhao, M., Xie, X., Li, W., Guo, M.: Knowledge graph convolutional networks for recommender systems. In: The World Wide Web Conference, pp. 3307–3313 (2019)
20. Wang, S., Sui, Y., Wang, C., Xiong, H.: Unleashing the power of knowledge graph for recommendation via invariant learning. In: Proceedings of the ACM Web Conference 2024, pp. 3745–3755 (2024)
21. Wang, X., He, X., Cao, Y., Liu, M., Chua, T.S.: KGAT: knowledge graph attention network for recommendation. In: Proceedings of the 25th ACM SIGKDD International Conference on Knowledge Discovery & Data Mining, pp. 950–958 (2019)
22. Wang, X., et al.: Learning intents behind interactions with knowledge graph for recommendation. In: Proceedings of the Web Conference 2021, pp. 878–887 (2021)
23. Yang, Y., Huang, C., Xia, L., Huang, C.: Knowledge graph self-supervised rationalization for recommendation. In: Proceedings of the 29th ACM SIGKDD Conference on Knowledge Discovery and Data Mining, pp. 3046–3056 (2023)
24. Yang, Y., Huang, C., Xia, L., Huang, C., Luo, D., Lin, K.: Debiased contrastive learning for sequential recommendation. In: Proceedings of the ACM Web Conference 2023, pp. 1063–1073 (2023)
25. Yang, Y., Huang, C., Xia, L., Li, C.: Knowledge graph contrastive learning for recommendation. In: Proceedings of the 45th International ACM SIGIR Conference on Research and Development in Information Retrieval, pp. 1434–1443 (2022)
26. Zhang, F., Yuan, N.J., Lian, D., Xie, X., Ma, W.Y.: Collaborative knowledge base embedding for recommender systems. In: Proceedings of the 22nd ACM SIGKDD International Conference on Knowledge Discovery and Data Mining, pp. 353–362 (2016)
27. Zhu, X., Du, Y., Mao, Y., Chen, L., Hu, Y., Gao, Y.: Knowledge-refined denoising network for robust recommendation. In: Proceedings of the 46th International ACM SIGIR Conference on Research and Development in Information Retrieval, pp. 362–371 (2023)

Enhancing Item Understanding in Knowledge Graph-Based Recommendation via Large Language Models

Xinjie Lin and Yitong Wang(✉)

School of Computer Science, Fudan University, Shanghai, China
{xjlin23,yitongw}@m.fudan.edu.cn

Abstract. In recent years, recommender systems based on knowledge graphs (KGs) have gained significant attention. By effectively leveraging the rich factual information within KGs and capturing the relationships between items, these methods are able to provide more accurate recommendations. However, KG-based recommender systems still face challenges including missing facts and limited coverage, which restrict the in-depth understanding of items. Given the rich open-world knowledge and strong common-sense reasoning capabilities of large language models (LLMs), in this paper, we propose a novel method named EKLRec to address the above challenges. Specifically, EKLRec employs LLMs to understand and analyze items from the perspectives of item profiling and user preferences, and construct a dual-view recommendation framework. We also devise a knowledge fusion network to enhance the interaction and fusion of item features from different semantic spaces. Furthermore, an adaptive weighting multi-level contrastive learning module is proposed to promote the alignment of representations by maximizing the mutual information between different views. Extensive experiments conducted on three real-world datasets demonstrate the superior performance of our method over the state-of-the-arts.

Keywords: Large Language Models · Knowledge Graphs · Recommender Systems

1 Introduction

With the rapid development of online services, recommender systems have become increasingly important in matching user information needs and alleviating information overload. Traditional collaborative filtering (CF) methods [7,13] are typically built on discrete ID features, despite their effectiveness, they still struggle to address data sparsity and cold start problems [14]. Therefore, KG-based recommendation methods [18,27,28] have been proposed to incorporate KGs as auxiliary information into recommender systems. But manually designed KGs usually suffer from missing triples and limited coverage problem [4,6] and

H. Jung et al. (Eds.): DASFAA 2026, LNCS 16535, pp. 340–355, 2026.
https://doi.org/10.1007/978-981-92-0363-5_21

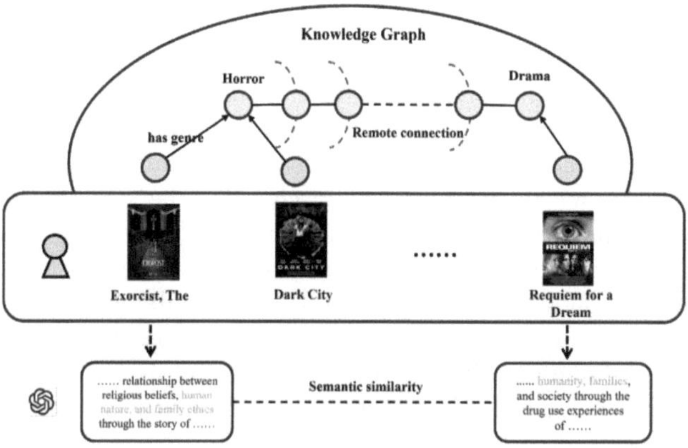

Fig. 1. Example of LLMs addressing limited item understanding in KG-based methods.

they mainly consists of basic attributes of items [6], such as the genre or director of a movie, which lacks in-depth interpretation of items.

Currently, LLMs have demonstrated great potential in enhancing recommender systems due to their rich open-world knowledge and strong commonsense reasoning capabilities. While some studies use LLMs to generate semantic representations of item profiles [2,15,24] or use LLMs to determine complementary relationships between items and users [8,21,29], these methods ignore the accurate and rich semantic associations provided by entities and relations in KGs. Moreover, large language models are extensively pre-trained for natural language tasks such as text understanding and generation. The semantic representations encoded by language models and the ID representations trained end-to-end in knowledge graph-based recommender systems originate from distinct semantic spaces. Due to the discrepancy in their adapted downstream tasks and vector dimensions, the semantic gap [12,15] between different semantic spaces would lead to negative transfer in recommendation performance.

We attempt to tackle the problem of missing facts and semantic gap by effectively integrating KGs and LLMs into the recommender system. For example, in Fig. 1, existing KG-based methods could only capture the similarity between the movies "The Exorcist" and "Dark City" in a user's viewing history through the relation of genre, but fail to capture connections of the movie "Requiem for a Dream" with other two movies due to long range relationships. However, through the analysis of LLMs, we can extract the semantic similarity between "The Exorcist" and "Requiem for a Dream" from external knowledge and help the recommender system to understand items and capture user preferences.

To fully integrate the factual knowledge from KGs and external knowledge from LLMs, we have to address two challenges. i) How can we effectively introduce external knowledge from LLMs to enhance item understanding? LLMs can not only provide in-depth analyses of items by knowledge reservation but also infer potential target users who may be interested in these items by reasoning ability, which can serve as a complement to item semantics and user interests in recommendation. ii) Due to the existence of semantic gap, how can we integrate the semantic understanding from language models with collaborative signals of ID format used in recommender systems to obtain a more comprehensive item representation? Existing methods typically use simple addition or average pooling operations, which not only fail to consider the semantic gap between language models and recommender systems but also ignore the structural relationships within the semantic space. Therefore, it is necessary to fully explore integration and fusion of feature representations from different semantic spaces.

To address the above issues, in this paper, we propose a new method named **E**nhancing Item Understanding in **K**nowledge Graph-based Recommendation via **L**arge Language Models (**EKLRec**). The core idea is to integrate external item knowledge extracted by LLMs to enhance the item understanding of recommender systems. Firstly, based on the textual attributes of items, we guide the LLMs to generate item-profiling texts and user-preference texts of items. We then construct a knowledge view and a collaborative view with the KG and the user-item interaction graph, achieving detailed dual-view knowledge fusion. Specifically, in the knowledge fusion phase, we propose an item-item graph construction module and a knowledge fusion network, allowing sufficient interaction and fusion of representations from different semantic spaces. Finally, we introduce a multi-level contrastive learning module to enhance the alignment of different representations within each view and maximize the mutual information between different views through additional self-supervision signals.

Our main contributions are summarized as follows:

- We propose a novel method that constructs a dual-view recommendation framework, which incorporates external semantic information from LLMs and empowers the KG-based recommender models with enhanced item understanding.
- We propose a knowledge fusion network and employ contrastive learning, so as to realize fusion of semantic representations and achieve alignment between different views.
- Extensive experiments on three real-world datasets demonstrate the superiority of our method compared to various baseline methods. Further analysis verifies the ability of our method to address the data sparsity issue.

2 Related Work

2.1 KG-Based Recommendation

Existing KG-based recommendation methods can be categorized into three types: embedding-based methods, path-based methods, and propagation-based methods [6]. Embedding-based methods leverage abundant facts from KGs to enhance the representations of users and items. Path-based methods utilize KGs to explore distant connections. KPRN [20] extracts meta-paths and aggregates user preference along each path by fully-connected layers. Despite effectiveness of path-based methods, they rely on domain knowledge and manual effort in meta-path design. Recently, propagation-based methods have been proposed to enhance entity and relationship representations by aggregating embeddings from multi-hop neighbors using Graph Neural Network (GNN). For example, KGAT [18] proposes the graph attention mechanism to propagate user and item embeddings on the KG. KGRec [25] integrates generative and contrastive self-supervised tasks to highlight significant knowledge and align signals from both the knowledge and user-item interaction views. Despite the effectiveness of existing KG-based recommender systems, their side information is limited to the existing KG and do not consider the external knowledge from LLMs.

2.2 LLM-Enhanced Recommendation

Due to the open world knowledge and common sense reasoning ability of LLMs, it exhibits extraordinary general intelligence in various natural language processing tasks. By combining LLMs with traditional recommender systems, recommendation tasks can benefit from LLMs. LLM-KERec [29] determines complementary relationships in each item pair, and constructs a complementary KG to enhance the recommender system. LLMRec [22] uses LLMs for various graph augmentation strategies in recommendation. Besides the LLM-enhanced methods, in generative recommender systems, LLMs are expected to generate reasonable recommendation results directly. For example, P5 [5] converts multiple recommendation tasks into natural language processing tasks, capturing deeper semantics through the rich information of natural language for personalized recommendations. While generative recommender systems suffer from the high cost of large-scale deployment and unsatisfactory performance, existing LLM-enhanced recommender systems use cached representations generated by language models but still struggle to utilize semantic information from LLMs effectively.

3 Preliminaries

In this section, We introduce some preliminaries and the problem formulation.

User-Item Interaction Graph. $\mathcal{G}_u = \{(u_i, y_{u_i v_j}, v_j) | u_i \in \mathcal{U}, v_j \in \mathcal{V}\}$ is constructed in the form of bipartite graph, where $\mathcal{U}$ denotes the user set and $\mathcal{V}$ denotes the item set. Here, $y_{u_i v_j} = 1$ if user u_i interacted with item v_j, and vice versa. Moreover, each item $v_j \in \mathcal{V}$ is attached with a text feature p_j.

Knowledge Graph. Knowledge graphs (KGs) store item-related attributes in the form of triples, which is formally represented as $\mathcal{G}_k = \{(h, r, t) | h, t \in \mathcal{E}, r \in \mathcal{R}\}$. Specifically, head entity h and tail entity t belong to the entity set $\mathcal{E}$, while r is a relation from the relation set $\mathcal{R}$. For instance, the triple (Titanic, Film.Genre, Love) characterizes the fact that the movie "Titanic" has the genre of "Love". Notably, the item set is a subset of the entity set, denoted as $\mathcal{V} \subset \mathcal{E}$.

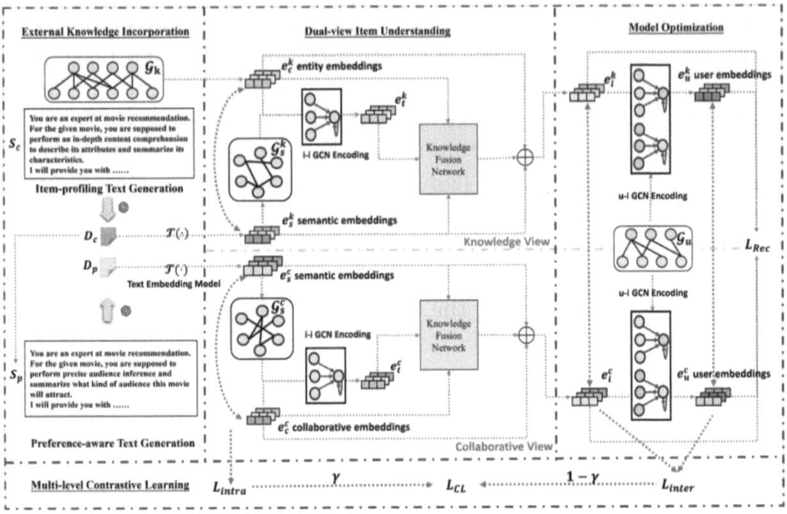

Fig. 2. The framework of our proposed EKLRec.

Task Formulation. We define the KG-aware recommendation task as follows: given the user-item interaction graph $\mathcal{G}_u$ and the corresponding KG $\mathcal{G}_k$, the task objective is to train a recommender model that predicts the probability of user u interacting with item v.

4 Methodology

In this section, we elaborate on the proposed approach, EKLRec, which improves item understanding in recommender systems by semantic augmentation with LLMs and further combination with KG-based methods. An overview of EKLRec is illustrated in Fig. 2. It consists of three components: external knowledge incorporation, dual-view item understanding, and multi-level contrastive learning, which are described in details in following sections respectively. We enhance item semantics by in-depth understanding of both item content and user preference with LLMs. Through the knowledge fusion module and multi-level contrastive learning, external semantic information from LLMs could effectively combined with collaborative filtering information from interaction data and factual knowledge provided by KGs.

4.1 External Knowledge Incorporation

In the KG-based recommender systems, KGs help to solve data sparsity, but they still suffer from the problem of missing facts and limited coverage. Furthermore, they fail to enable in-depth understanding of items such as the core theme or emotional tone of a movie. Due to LLMs' rich external knowledge and common sense reasoning abilities, LLMs can analyze and summarize the essence of items so as to enhance item understanding significantly. Therefore, we designed a prompt to guide LLMs to deeply understand items in the scene of recommendation. We use textual features as item profiles p and combine it with our designed item profiling system prompt S_c, then we leverage the LLMs to generate the item-profiling text D_c as follows:

$$D_c = LLMs\left(S_c, p\right) \tag{1}$$

The text features p here consist of titles and other dataset-specific attributes of items, and user reviews are added if available. Since the item-profiling text of an item is already a high generalization of the essence of the item, we use it as a basis to guide the LLM to infer potential target users interested in the items from the perspective of user preference. We combine the preference reasoning system prompt S_p with previously generated item-profiling text D_c as input and leverage the LLM to generate the preference-aware text D_p, which can be formulated as:

$$D_p = LLMs\left(S_p, D_c\right) \tag{2}$$

The preference-aware text D_P implies the coincidence degree of potential users between different items in the scene of recommendation task. The contents of two prompts S_c and S_p are shown in Figure 2.

We use pre-trained text embedding models [11] to convert the above D_c, D_p into embedded vectors to retain items' inherent meaning. The embedding process can be defined as follows:

$$e_s^k = \mathcal{T}\left(D_c\right), \quad e_s^c = \mathcal{T}\left(D_p\right) \tag{3}$$

By caching the text embeddings generated by language models for subsequent recommendation task, the inference process of language models does not affect the computational cost and inference efficiency of the recommender model.

4.2 Dual-View Item Understanding

We propose a dual-view recommendation framework consisting of knowledge view and collaborative view, which aligns semantic embeddings with factual knowledge in KGs and collaborative signals in user-item interaction respectively.

In the knowledge view, we integrate the factual knowledge from the KG with the item-profiling knowledge from LLMs to accurately capture the semantic information of items. We use the attention weighted heterogeneous information aggregation [25] on the KG to obtain corresponding entity embeddings of items. In addition, to fully learn the entity objective dependency inherent in the KG, we alternately implement the heterogeneous information aggregation and TransE [1] during model training phase.

As indicated in existing researches [15], simply adding or average pooling the embeddings from different semantic spaces ignores the inherent semantic gap and prevents the effective utilization of external knowledge. We propose to construct an item-item graph based on semantic embeddings to distill the implicit structural relationships from semantic space to enhance the entity embeddings. Specifically, the item-item graph is constructed using cosine similarity, which can be formulated as:

$$S_{ij} = cos(e_s^i, e_s^j). \tag{4}$$

where cos denotes the cosine similarity and e_s^i, e_s^j are semantic embeddings of two items v_i, v_j generated from pre-trained models. Subsequently, the KNN sparsification is employed on the fully connected item-item graph to extract relatively important semantic connection relationships and finally obtain the constructed item-item semantic graph $\mathcal{G}_s$ in both views.

We use the entity embeddings corresponding to the items and perform the widely applied LightGCN [7] operation on the item-item graph to inject the semantic relationships from external knowledge into entity embeddings. Outputs at different convolution layers are summed up to obtain the augmented item embedding e_t.

For different items, factual knowledge from KGs and external knowledge from LLMs have different effects, while augmented embeddings achieve the interaction between them. Therefore, a knowledge fusion network is proposed to realize the customized fusion of semantic embeddings, entity embeddings and augmented embeddings. In the knowledge fusion network, we first generate two customized transformation matrices. We use the customized transformation matrices and nonlinear mapping function to build our knowledge fusion network as follows:

$$e_m^k = \sigma(\psi^1(e_z)\psi^2(e_z)e_s^k), \quad e_z = e_s^k \| e_c^k \| e_t^k \tag{5}$$

where e_s^k, e_c^k, e_t^k denote semantic embedding, entity embedding and augmented embedding in knowledge view respectively. ψ^1, ψ^2 are two fully-connected layers, which takes $e_z \in \mathbb{R}^{n \times 3d}$ as input and output the customized transformation matrices $\psi^1(e_z) \in \mathbb{R}^{n \times d \times k}$, $\psi^2(e_z) \in \mathbb{R}^{n \times k \times d}$ for items. $\sigma(\cdot)$ refers to nonlinear activation. By the transformation matrices and the non-linear mapping function σ, we can obtained customized embedding e_m^k and it is then used to obtain the fused item representation e_i^k. The process of item knowledge fusion is carried out through the following weighted summation:

$$e_i^k = \alpha * e_c^k + (1 - \alpha) * (e_m^k + e_s^k) \tag{6}$$

where α controls the weight between the entity embedding and embedded auxiliary information, e_i^k is obtained through fine knowledge fusion and then serve as the fused item embeddings of the recommendation task in knowledge view.

We also carry out the same process of item understanding enhancement in the collaborative view. In the collaborative view, collaborative embeddings of items are initialized to capture the pure collaborative signals on the user-item interaction graph. After the construction of item-item semantic graph, collaborative embeddings are used to generate the augmented embeddings. We use the preference-aware semantic embeddings e_s^c, collaborative embeddings e_c^c and augmented embeddings e_t^c as the input of knowledge fusion network to achieve enhanced understanding of items in the collaborative view. After the dual-view fine-grained knowledge fusion, we finally obtain the fused item embeddings e_i^k in knowledge view and e_i^c in collaborative view respectively.

4.3 Multi-level Contrastive Learning

After successfully fusing knowledge in both views to obtain item embeddings, we model user interests on the user-item interaction graph. Without loss of generality, we use LightGCN [7] as the graph encoder to generate the refined embeddings e_i, e_u of items and users in both views for recommendation.

In order to facilitate the adaptation of semantic embeddings to recommendation tasks and the alignment of recommendation signals between knowledge view and collaborative view, we design a multi-level contrastive learning module, including contrastive learning within and between views. To be exact, in the intra-view contrastive learning, we regard the trained ID embeddings $e_{c,v}$ and semantic embeddings $e_{s,v}$ of the same item v as positive pairs, and pushing the positive pairs closer to each other while treating the rest of the samples as negative. The ID embeddings $e_{c,v}$ of item v is entity embeddings in the knowledge view, and it is collaborative embeddings in the collaborative view. The semantic embeddings $e_{s,v}$ represent for embeddings generated by language models. Based on the InfoNCE loss [3], the contrastive learning objective can be defined as:

$$\mathcal{L}_{cl}(v) = -\log \frac{\exp(cos(e_{s,v}, e_{c,v})/\tau)}{\sum_{j \in \mathbb{N} \cup \{v\}} \exp(cos(e_{s,v}, e_{c,j})/\tau) + \exp(cos(e_{s,j}, e_{c,v})/\tau)} \tag{7}$$

where $cos(\cdot)$ denotes the cosine similarity of two vectors, $\mathbb{N}$ is the set of negative samples and τ is the temperature hyper-parameter. At the same time, in the inter-view contrastive learning, we take the embeddings belonging to the same item or user from knowledge view and collaborative view as positive pairs to maximize the mutual information and achieve the alignment of recommendation signals between two different views.

Furthermore, given the progressive relationship between the objectives of intra-view contrastive learning and inter-view contrastive learning, the effectiveness of the final recommendation task is based on well refined item embeddings. We propose an adaptive weighting mechanism [9] to coordinate the weights between different contrastive learning tasks. The proposed adaptive weighting mechanism can be formulated as follows:

$$\gamma^{(t)} = \tanh\left(\frac{1}{\epsilon^{t/T}}\right), \quad t: 0 \rightarrow T, \tag{8}$$

where t, T denote the current epoch and the maximum epoch, ϵ is the hyper-parameter controlling the process of focus shifting and $tanh(\cdot)$ denotes the non-linear activation. The obtained adaptive weights are further combined into the contrastive learning objective as follows:

$$\mathcal{L}_{cl} = \gamma^{(t)} \mathcal{L}_{intra} + (1 - \gamma^{(t)}) \mathcal{L}_{inter}. \tag{9}$$

Table 1. Statistics of the datasets.

Dataset	#Users	#Items	#Interactions	#Entities	#Relations	#Triples
MovieLens	6,033	3,099	832,207	67,748	47	943,813
Amazon-book	6,802	5,472	832,207	74,250	20	652,745
Yelp	10,330	8,917	246,643	83,124	42	1,466,998

At the beginning, the contrastive learning task focus on the adaptation from semantic embeddings to recommender system to promote the knowledge fusion process, and then gradually shift the focus to the main recommendation tasks to align the recommendation signals between different views.

4.4 Model Prediction and Optimization

In both knowledge view and collaborative view, we use cosine similarity to predict user interest in candidate items. Finally, the model's prediction score $\hat{y}_{u,i}$ is determined by the sum of similarities in the two views:

$$\hat{y}_{u,i} = \cos(\mathbf{e}_u^k, \mathbf{e}_i^k) + \cos(\mathbf{e}_u^c, \mathbf{e}_i^c) \tag{10}$$

To alleviate the issue of convergence in the dual-view recommender model, we use the cosine contrastive loss [10] to introduce more negative samples and filter out those with insufficient information.

$$\mathcal{L}_{rec} = (1 - \hat{y}_{ui}) + \frac{1}{|\mathcal{N}|} \sum_{j \in \mathcal{N}} \max(0, \hat{y}_{uj} - m) \tag{11}$$

By combining the loss of contrastive learning and recommendation, we minimize the following loss function to optimize model parameters:

$$\mathcal{L}_{main} = \mathcal{L}_{rec} + \lambda_1 \mathcal{L}_{cl} + \lambda_2 \|\Theta\|_2^2, \tag{12}$$

where λ_1, λ_2 are hyperparameters controlling the measures of loss functions and $\|\Theta\|$ in the L_2 regularization term denotes the model parameter set. It is worth noting that we conduct alternating training between the main loss $\mathcal{L}_{main}$ and TransE loss $\mathcal{L}_t$.

5 Experiments

5.1 Experimental Setup

Datasets. We conducted experiments on three public datasets (MovieLens[1], Amazon book[2], Yelp[3]). The datasets cover a wide range of application scenarios.

[1] https://grouplens.org/datasets/movielens/.
[2] https://jmcauley.ucsd.edu/data/amazon/.
[3] https://www.yelp.com/dataset.

Table 2. Performance comparison of different methods, where R denotes Recall and N denotes NDCG. The best results are bolded, and the second best results are underlined.

Model	MovieLens				Amazon-book				Yelp			
	R@20	R@40	N@20	N@40	R@20	R@40	N@20	N@40	R@20	R@40	N@20	N@40
BPR-MF	0.2454	0.3644	0.3274	0.3461	0.1918	0.2705	0.1220	0.1454	0.1148	0.1817	0.0692	0.0890
LightGCN	0.2633	0.3843	0.3506	0.3677	0.2169	0.3014	0.1401	0.1651	0.1211	0.1911	0.0737	0.0944
SGL	0.2606	0.3820	0.3487	0.3665	0.2298	0.3185	0.1508	0.1771	0.1409	0.2173	0.0875	0.1101
CKE	0.2446	0.3614	0.3270	0.3444	0.1912	0.2727	0.1217	0.1459	0.1143	0.1788	0.0691	0.0884
KGAT	0.2494	0.3690	0.3404	0.3572	0.2089	0.2938	0.1351	0.1605	0.1166	0.1854	0.0706	0.0907
KGIN	0.2515	0.3713	0.3357	0.3540	0.2126	0.2919	0.1405	0.1642	0.1172	0.1852	0.0724	0.0924
MCCLK	0.2569	0.3791	0.3430	0.3612	0.2178	0.3032	0.1433	0.1687	0.1191	0.1920	0.0726	0.0942
KGRec	0.2756	0.3996	0.3672	0.3842	0.2361	0.3238	0.1553	0.1815	0.1398	0.2189	0.0872	0.1102
RLMRec	0.2514	0.3705	0.3439	0.3589	0.2266	0.3101	0.1512	0.1758	0.1366	0.2128	0.0849	0.1075
CIKGRec	0.2739	0.3979	0.3653	0.3824	0.2443	0.3328	0.1587	0.1850	0.1429	0.2202	0.0876	0.1104
CoLaKG	0.2881	0.4145	0.3872	0.4033	0.2354	0.3260	0.1521	0.1790	0.1434	0.2193	0.0890	0.1117
EKLRec	**0.2936**	**0.4175**	**0.3963**	**0.4097**	**0.2598**	**0.3470**	**0.1807**	**0.2064**	**0.1442**	**0.2219**	**0.0911**	**0.1142**

We adopt settings similar to those in previous researches [18,31]. For explicit feedback data, we reserve interactive records with scores no less than 3, and use the commonly used 10-core method to filter out items and users with less than 10 interactive records. For each dataset, we randomly assign the historical behavior of each user in a ratio of 7:1:2 to construct the training set, validation set and test set. For the construction of KG, we follow the work of [30,31] to build item KG for each dataset while filtering out uncommon entities and relationships. The statistics of these processed datasets are shown in Table 1.

Baselines. For comprehensive performance evaluation, We compare EKLRec with various representative baselines, including CF-based methods (BPR-MF, LightGCN, SGL), KG-based methods (CKE, KGAT, KGIN, MCCLK, KGRec), and LLM-enhanced methods (RLMRec, CIKGRec, CoLaKG). Some methods [16,17,24,26] that have been compared in previous research and shown to have inferior performances are not included in our evaluation.

BPR-MF [13] is based on matrix factorization, it uses the pairwise Bayesian personalized sorting (BPR) loss as the optimization goal.

LightGCN [7] improves efficiency by simplifying the structure of graph convolution network for recommendation.

SGL [23] constructs different views for contrastive learning by data augmention and incorporates self-supervision signals as a supplement to recommendation.

CKE [27] proposes to integrate entity representations from the KGs with item representations for better performance.

KGAT [18] uses attention mechanism to process the information in KG and improves recommendation by learning the relationship between different entities.

KGIN [19] considers user intentions over different relations in the KG and injects relational embedding in the aggregation layer to improve performance.

MCCLK [32] employs the contrastive learning in a hierarchical way and mining comprehensive graph features and structure information.

KGRec [25] improve the recommender systems by generative and contrastive self-supervised learning and emphasize informative knowledge connections.

RLMRec [12] proposes to combine representational learning with LLMs to capture the complex semantic aspects of user preferences.

CIKGRec [8] leverages LLMs to construct user-side structural knowledge and integrate explicit user interests with the recommender system.

ColaKG [4] leverages LLMs to transforming KG data into semantic embeddings and generate the retrieval-augmented representations for recommendation.

Evaluation Protocols and Implementation Details. We implement all baseline methods based on their published code. For fair comparison, the embedding size d of all the recommended methods is set to 64, the batchsize is set to 4096, the Adam optimizer is used. Meanwhile, we perform the early stop technology according to the performance of the model on the validation set. We utilize DeepSeek-V2[4] by invoking its API[5] for text generation. For the text embedding model, we choose to use the BGE-M3 [11]. For fair comparison, the same large language model and text embedding model are used in all LLM-enhanced methods. We use Recall@N and NDCG@N as the top-K recommendation metrics, where N is set to 20 and 40.

5.2 Overall Performance Comparison

The overall performance of EKLRec and baseline models is shown in Table 2. Based on the results, we make several observations. Firstly, EKLRec outperforms all baseline models in all metrics and datasets, proving the effectiveness of our method, and emphasizes the efficiency of dual-view item understanding enhancement and multi-level contrastive learning. Although the performance of the traditional method BPRMF is limited, LightGCN and SGL surpass some KG-based methods in performance, which indicates that effective utilization of KG remains a challenging task. Secondly, among the KG-based baselines, MCCLK and KGRec are notable for introduce self-supervised learning into KG-based recommendation framework. However, they are facing the challenge of limited understanding of items. In contrast, our method leverages LLMs to incorporate external knowledge, enables the integration of external knowledge and structural knowledge from KG. Thirdly, for LLM-based recommendation methods, RLM-Rec incorporates external knowledge from LLMs, CIKGRec and ColaKG further

[4] https://github.com/deepseek-ai/DeepSeek-V2.

[5] https://api-docs.deepseek.com/.

considers the auxiliary information from LLMs and KG simultaneously. However, these methods only exhibit marginal improvements over traditional techniques. In contrast, our method shows superior performance, which further verifies the effectiveness of the semantic fusion module and multi-level contrastive learning in enhancing item understanding for recommender system.

5.3 Ablation Study

In this section, we demonstrate the effectiveness of the model by comparing its performance with four different versions across three datasets. The results are shown in Table 3, "w/o WA" means removing the weight adaptation in multi-level contrastive learning, "w/o CL" means removing the multi-level contrastive learning module, "w/o KF" means replacing the knowledge fusion module with average pooling and "w/o EK" means removing the incorporation of external knowledge from LLMs, using auxiliary information from knowledge graphs only. Firstly, we observed the performance degradation of the model on all three datasets without the incorporation of external knowledge, which emphasized the importance of external knowledge from LLMs for item understanding in recommender system. It is noteworthy that we observe a significant performance decline when using average pooling, which indicates that simply combining representations from different semantic spaces will lead to the contamination of noise. When the multi-level contrastive learning module is removed, due to the lack of self-supervision signals, the performance of the model decreases. Moreover, the role of weight adaptation between different levels of contrastive learning tasks is non-negligible.

Table 3. Ablation study on all three datasets.

Dataset	MovieLens		Amazon-book		Yelp	
Method	R@20	N@20	R@20	N@20	R@20	N@20
EKLRec	**0.2936**	**0.3963**	**0.2598**	**0.1807**	**0.1442**	**0.0911**
w/o WA	0.2902	0.3908	0.2573	0.1795	0.1430	0.0885
w/o CL	0.2869	0.3870	0.2533	0.1763	0.1417	0.0879
w/o KF	0.2741	0.3721	0.2370	0.1638	0.1349	0.0861
w/o EK	0.2790	0.3778	0.2417	0.1692	0.1370	0.0866

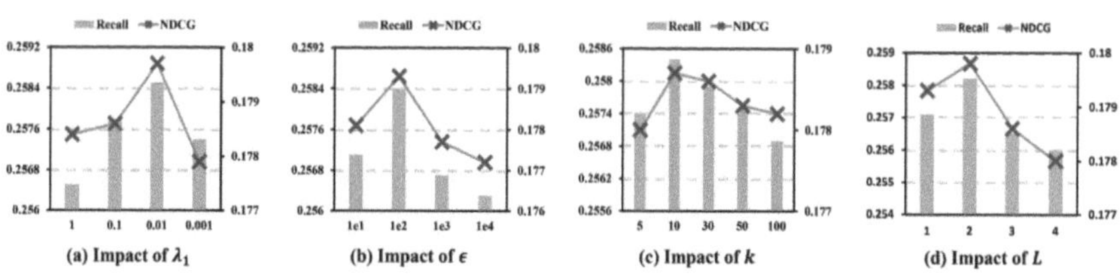

Fig. 3. Hyperparameter sensitive experiments on Amazon-book.

5.4 Sensitivity Analysis

In this section, we study the effects of four critical hyperparameters, including the weight λ of contrastive learning tasks, ϵ which controls the adaptive weighting, the number of neighbors k when constructing the item-item graph, and the number of layers L when propagating on the item-item graph. The results of the hyperparameter sensitivity experiment on Amazon-Book dataset are illustrated in Fig. 3. The primary vertical axis on the left side presents Recall@20 results as a bar chart, while the secondary vertical axis on the right side presents the NDCG@20 Results as a line chart. As shown in Fig. 3, we can draw several conclusions. First, λ adjusts the weight between the contrastive learning loss and the recommendation loss and λ=0.01 brings the best model performance while too extreme values may interfere with each other between different tasks. Second, ϵ controls the speed of training focus shifting from intra-view to inter-view contrastive learning. When ϵ is too large, the intra-view contrastive learning task cannot be fully trained, and the model performance achieves the best when it is set to 100. Third, K represents the number of neighbors reserved for each item in the process of KNN sparsification, and L represents the layers of information propagation on the item-item graph, too large K or L may cause contamination or over-smoothing problems which lead to worse model performance.

5.5 Robustness to Data Sparsity

One of the motivations of KG-based and LLM-enhanced methods is to introduce auxiliary information so as to alleviate the effect of data sparsity. Most existing recommender models are sensitive to data sparsity. Following the previous work [18], We divide users into four groups according to the interaction frequency in the dataset. We evaluate EKLRec with several baselines in terms of NDCG@20 on each user group in two datasets. As depicted in Fig. 4, the horizontal axis represents user interaction from sparse to dense and the results demonstrate superior performance of EKLRec on sparse user groups, which highlights the positive effect of our model in alleviating data sparsity problem. We attribute the advantage to our proposed external knowledge incorporation and knowledge fusion strategy.

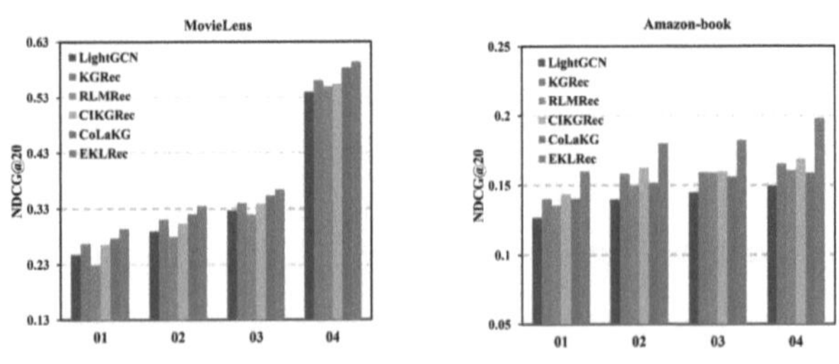

Fig. 4. Performance comparison on different user groups.

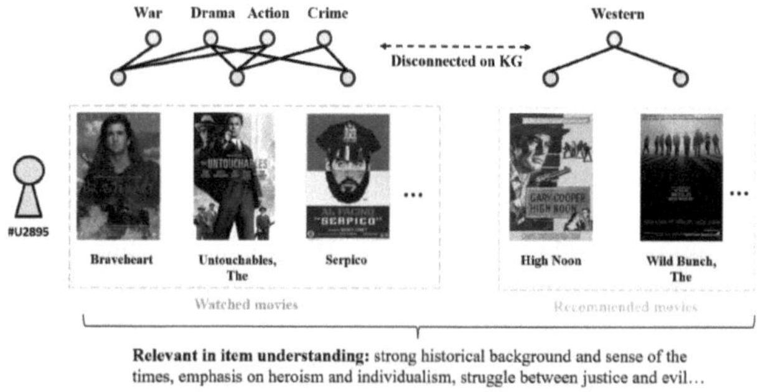

Fig. 5. Case study of semantic similarity brought by LLMs on MovieLens dataset.

5.6 Case Study

In order to understand the effectiveness of EKLRec more intuitively, a case study from the MovieLens dataset is shown in Fig. 5. As is depicted in the figure, the left side is the user's historical interaction in the training set, and the right side is the recommendation result of EKLRec for the user in the test set. The adjacent entities of the item with the relationship "has genre" in the KG are displayed at the top of each item. The watched movies of the user are connected through entities including "drama", "action" and "crime" in the KG. However, for the movies "High Noon" and "Wild Bunch" recommended for the user, their connected entity in the KG is only "western". Obviously, the connection between the user's different interests cannot be fully captured in the KG-based methods. Although it is difficult to observe the correlation between "High Noon" and "Wild Bunch" with the user's watched movies in KG, they are semantically related under the analysis of LLMs. Therefore, our method leverages LLMs to introduce external knowledge to enhance the item understanding, successfully captures high-level semantic similarity between items, and addresses the limitations of KG-based methods.

6 Conclusion

We analyze the limitations of existing KG-based recommender systems and propose a novel method, EKLRec, to address these issues. EKLRec enhance the item understanding of recommender systems by incorporating external knowledge and achieving sufficient fusion of knowledge from different semantic space. Moreover, the multi-level contrastive learning module can introduces additional self-supervised signals. We conducted extensive experiments on three real datasets to validate the superiority of our method. In future research, we will further explore the potential of prompt engineering and fine-tuning techniques in enhancing recommender systems, as well as more advanced methods for leveraging external knowledge.

Disclosure of Interests. The authors have no competing interests to declare that are relevant to the content of this article.

References

1. Bordes, A., Usunier, N., Garcia-Duran, A., Weston, J., Yakhnenko, O.: Translating embeddings for modeling multi-relational data. In: Advances in Neural Information Processing Systems, vol. 26 (2013)
2. Chen, H., et al.: LKPNR: large language models and knowledge graph for personalized news recommendation framework. Comput. Mater. Continua **79**(3) (2024)
3. Chen, T., Kornblith, S., Norouzi, M., Hinton, G.: A simple framework for contrastive learning of visual representations. In: International Conference on Machine Learning, pp. 1597–1607. PmLR (2020)
4. Cui, Z., et al.: Comprehending knowledge graphs with large language models for recommender systems. In: Proceedings of the 48th International ACM SIGIR Conference on Research and Development in Information Retrieval, pp. 1229–1239 (2025)
5. Geng, S., Liu, S., Fu, Z., Ge, Y., Zhang, Y.: Recommendation as language processing (RLP): a unified pretrain, personalized prompt & predict paradigm (P5). In: Proceedings of the 16th ACM Conference on Recommender Systems, pp. 299–315 (2022)
6. Guo, Q., et al.: A survey on knowledge graph-based recommender systems. IEEE Trans. Knowl. Data Eng. **34**(8), 3549–3568 (2020)
7. He, X., Deng, K., Wang, X., Li, Y., Zhang, Y., Wang, M.: Lightgcn: simplifying and powering graph convolution network for recommendation. In: Proceedings of the 43rd International ACM SIGIR Conference on Research and Development in Information Retrieval, pp. 639–648 (2020)
8. Hu, Z., et al.: Bridging the user-side knowledge gap in knowledge-aware recommendations with large language models. In: Proceedings of the AAAI Conference on Artificial Intelligence, vol. 39, pp. 11799–11807 (2025)
9. Jeon, H., Lee, J.E., Yun, J., Kang, U.: Cold-start bundle recommendation via popularity-based coalescence and curriculum heating. In: Proceedings of the ACM Web Conference 2024, pp. 3277–3286 (2024)
10. Mao, K., et al.: Simplex: a simple and strong baseline for collaborative filtering. In: Proceedings of the 30th ACM International Conference on Information & Knowledge Management, pp. 1243–1252 (2021)
11. M3-embedding: Multi-linguality, multi-functionality, multi-granularity text embeddings through self-knowledge distillation (2024)
12. Ren, X., et al.: Representation learning with large language models for recommendation. In: Proceedings of the ACM Web Conference 2024, pp. 3464–3475 (2024)
13. Rendle, S., Freudenthaler, C., Gantner, Z., Schmidt-Thieme, L.: BPR: Bayesian personalized ranking from implicit feedback. arXiv preprint arXiv:1205.2618 (2012)
14. Togashi, R., Otani, M., Satoh, S.: Alleviating cold-start problems in recommendation through pseudo-labelling over knowledge graph. In: Proceedings of the 14th ACM International Conference on Web Search and Data Mining, pp. 931–939 (2021)
15. Wang, C., et al.: Collaborative alignment for recommendation. In: Proceedings of the 33rd ACM International Conference on Information and Knowledge Management, pp. 2315–2325 (2024)

16. Wang, H., et al.: Ripplenet: propagating user preferences on the knowledge graph for recommender systems. In: Proceedings of the 27th ACM International Conference on Information and Knowledge Management, pp. 417–426 (2018)
17. Wang, H., Zhao, M., Xie, X., Li, W., Guo, M.: Knowledge graph convolutional networks for recommender systems. In: The World Wide Web Conference, pp. 3307–3313 (2019)
18. Wang, X., He, X., Cao, Y., Liu, M., Chua, T.S.: KGAT: knowledge graph attention network for recommendation. In: Proceedings of the 25th ACM SIGKDD International Conference on Knowledge Discovery & Data Mining, pp. 950–958 (2019)
19. Wang, X., et al.: Learning intents behind interactions with knowledge graph for recommendation. In: Proceedings of the Web Conference 2021, pp. 878–887 (2021)
20. Wang, X., Wang, D., Xu, C., He, X., Cao, Y., Chua, T.S.: Explainable reasoning over knowledge graphs for recommendation. In: Proceedings of the AAAI Conference on Artificial Intelligence, vol. 33, pp. 5329–5336 (2019)
21. Wang, Y., et al.: Enhancing recommender systems with large language model reasoning graphs. arXiv preprint arXiv:2308.10835 (2023)
22. Wei, W., et al.: LLMRec: large language models with graph augmentation for recommendation. In: Proceedings of the 17th ACM International Conference on Web Search and Data Mining, pp. 806–815 (2024)
23. Wu, J., et al.: Self-supervised graph learning for recommendation. In: Proceedings of the 44th International ACM SIGIR Conference on Research and Development in Information Retrieval, pp. 726–735 (2021)
24. Xi, Y., et al.: Towards open-world recommendation with knowledge augmentation from large language models. In: Proceedings of the 18th ACM Conference on Recommender Systems, pp. 12–22 (2024)
25. Yang, Y., Huang, C., Xia, L., Huang, C.: Knowledge graph self-supervised rationalization for recommendation. In: Proceedings of the 29th ACM SIGKDD Conference on Knowledge Discovery and Data Mining, pp. 3046–3056 (2023)
26. Yang, Y., Huang, C., Xia, L., Li, C.: Knowledge graph contrastive learning for recommendation. In: Proceedings of the 45th International ACM SIGIR Conference on Research and Development in Information Retrieval, pp. 1434–1443 (2022)
27. Zhang, F., Yuan, N.J., Lian, D., Xie, X., Ma, W.Y.: Collaborative knowledge base embedding for recommender systems. In: Proceedings of the 22nd ACM SIGKDD International Conference on Knowledge Discovery and Data Mining, pp. 353–362 (2016)
28. Zhang, Y., Ai, Q., Chen, X., Wang, P.: Learning over knowledge-base embeddings for recommendation. arXiv preprint arXiv:1803.06540 (2018)
29. Zhao, Q., Qian, H., Liu, Z., Zhang, G.D., Gu, L.: Breaking the barrier: utilizing large language models for industrial recommendation systems through an inferential knowledge graph. In: Proceedings of the 33rd ACM International Conference on Information and Knowledge Management, pp. 5086–5093 (2024)
30. Zhao, W.X., et al.: KB4Rec: a data set for linking knowledge bases with recommender systems. Data Intell. **1**(2), 121–136 (2019)
31. Zhu, X., Du, Y., Mao, Y., Chen, L., Hu, Y., Gao, Y.: Knowledge-refined denoising network for robust recommendation. In: Proceedings of the 46th International ACM SIGIR Conference on Research and Development in Information Retrieval, pp. 362–371 (2023)
32. Zou, D., et al.: Multi-level cross-view contrastive learning for knowledge-aware recommender system. In: Proceedings of the 45th International ACM SIGIR Conference on Research and Development in Information Retrieval, pp. 1358–1368 (2022)

Spurious Correlation Knowledge Graph Disentanglement for Multi-behavior Recommendation

Tongxin Xu[1], Chenzhong Bin[1(✉)], Cihan Xiao[2], Zhixin Zeng[1], and Yunhui Li[1]

[1] School of Computer Science and Information Security, Guilin University of Electronic Technology, Guilin, China
{binchenzhong,zxzeng,liyh}@guet.edu.cn

[2] Center for Language and Speech Processing, Johns Hopkins University, Baltimore, USA
cxiao7@jhu.edu

Abstract. Multi-behavior recommendation systems have been shown to effectively mitigate data sparsity in target behavior by leveraging various auxiliary behaviors. Existing multi-behavior recommendation methods primarily focus on modeling auxiliary behaviors and applying them to target behavior recommendations. However, the interaction correlations between item features and user intents in the auxiliary behaviors contain both genuine and spurious correlations. Indiscriminately modeling these correlations can easily lead to a degradation in recommendation performance. Therefore, we propose a novel SCKGD framework for multi-behavior recommendation. In particular, we construct a spurious correlation knowledge graph (SCKG) from the item perspective to capture the spurious correlations between items. Based on the generated SCKG, a time-projected knowledge encoder is designed to dynamically weight the strength of spurious correlations between items and learn the corresponding item embeddings. To disentangle the spurious correlations, we propose a disentangling contrastive learning method that preserves the genuine correlated semantics aligning with the user's target intent. Concurrently, to enrich the supervision signal in the target behavior, we perform inter-behavior contrastive learning to transfer the genuine correlated semantics from auxiliary behaviors into the target behavior. Experiments on three datasets demonstrate that our framework outperforms state-of-the-art (SOTA) methods and disentangles spurious correlations. Our implementation is available (https://github.com/LokHsu/SCKGD).

Keywords: Multi-Behavior Recommendation · Spurious Correlation Disentanglement · Dynamic Knowledge Embedding

1 Introduction

Recently, multi-behavior recommendation systems have received widespread attention, due to their ability to effectively integrate various auxiliary behav-

H. Jung et al. (Eds.): DASFAA 2026, LNCS 16535, pp. 356–371, 2026.
https://doi.org/10.1007/978-981-92-0363-5_22

iors to alleviate the sparsity of target behavior data. Specifically, some studies leverage the cascading effects among behaviors to model embeddings for each behavior [4,26], while other works adopt data augmentation strategies to compensate for the lack of userâĂŞitem interaction supervision in the target behavior [18,22,25,29]. In addition, to address the inherent noise in auxiliary behaviors, recent research has proposed a series of denoising methods to prevent auxiliary behaviors from directly affecting the final recommendation results [2,6,23,24,27].

Although recent multi-behavior recommendation methods have demonstrated their effectiveness in modeling auxiliary behavior intents and applying them to target behavior recommendations, they still face a challenge in practical applications: during the process of modeling auxiliary behaviors, it is difficult to discriminate the genuine and spurious interaction correlations between user intents and item features. Neglecting the existence of spurious correlations may disrupt the model's ability to learn user target intent. In single-behavior recommendation, spurious correlations are caused by item features that are not the causal features of user behaviors but frequently co-occur with user interactions [11]. However, in multi-behavior recommendation, since users exhibit varying behavioral intents across different behaviors [17], their interaction intents are more complex than those in single-behavior scenarios. Therefore, we especially define **spurious correlation** in multi-behavior scenarios as the inconsistency between dynamic user intents and static item features. Figure 1 shows an example of spurious correlations in multi-behavior recommendation scenario: Jane viewed a dress, added a teddy bear to her cart, and eventually purchased a pair of summer overalls. Since her target intent is to buy clothing, the dress poses a genuine correlation with her. In contrast, other items interacted by Jane under auxiliary behaviors (e.g., the teddy bear) do not reflect the actual purchase intent, resulting in a spurious correlation. As for Henry, he viewed a pair of summer overalls, added a sweater to the cart, but ended up buying a scarf. As the potential purchasing intent is driven by seasonal factors, the overalls pose a spurious correlation with him. Empirically, user intents are revealed by the items they have previously interacted with. Due to the dynamic and multifaceted user intents, directly measuring spurious correlations between users and items in auxiliary behaviors is intractable. Thus, we turn to reckoning the spurious correlations between items by calculating the cumulative item difference between users' auxiliary and target behaviors.

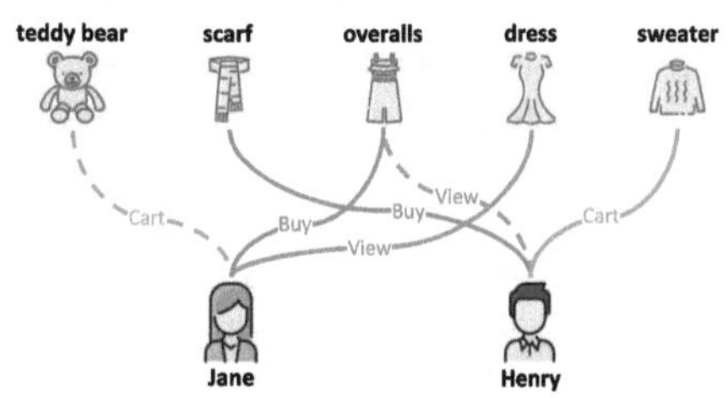

Fig. 1. An example showing spurious correlations in multi-behavior recommendations, where dashed and solid lines represent spurious and genuine correlations between users and items, respectively.

Contribution. In this work, we focus on exploring the disentanglement of spurious correlations in multi-behavior recommendation. Though being prevalent in natural language processing (NLP) [14] and computer vision (CV) [1], spurious correlation disentanglement is relatively less explored in multi-behavior recommendation. Recently, many studies have introduced KG into recommendation to explore fine-grained differences between items [16,17,20,28]. Inspired by these works, we propose a framework–**S**purious **C**orrelation **K**nowledge **G**raph **D**isentanglement (SCKGD) to tackle the inconsistency between dynamic user intents and static item features, a phenomenon we formally define as spurious correlation. However, unlike most existing knowledge-aware recommendation methods [20,25,28] that utilize KG to capture the relationships between users and items for enhancing recommendations, our approach constructs a spurious correlation-oriented knowledge graph for capturing the fine-grained differences between item features and user intents, and modeling the irrelevance between items. Specifically, we first propose a spurious correlation measurement method for generating spurious correlation knowledge graph (SCKG). Subsequently, based on the generated SCKG, a time-projected knowledge encoder is designed to dynamically weight the strength of spurious correlations between items and learn the corresponding SCKG item embeddings. To disentangle the spurious correlations in multi-behavior recommendation, we propose a disentangling contrastive learning method that preserves the semantics consistent with the user's target intent by minimizing the mutual information (MI) between multi-behavior embeddings and SCKG embeddings. Concurrently, to enrich the supervision signal in the target behavior, we perform inter-behavior contrastive learning to transfer the genuine correlated semantics from auxiliary behaviors into the target behavior. In summary, our main contributions are as follows:

1. We propose a new framework SCKGD to measure and disentangle the spurious correlations existing in multi-behavior recommendations. To the best of our knowledge, this is the first attempt to identify and address the spurious correlations in multi-behavior recommendations from a KG perspective.
2. In SCKGD framework, we construct a spurious correlation knowledge graph (SCKG) and design a time-projected knowledge encoder to dynamically measure the spurious correlations between items and learn the corresponding SCKG item embeddings. Furthermore, two contrastive learning method are proposed to preserve and transfer the semantics aligned with the user's target intent from auxiliary behaviors to the target behavior.
3. Experiments on three real-world datasets demonstrate that our framework outperforms SOTA methods and disentangles spurious correlations.

2 Related Work

2.1 Knowledge-Aware Recommendation

Recently, KG has become an important data source for enhancing the recommendation system frameworks, leading to the emergence of a series of knowledge-aware recommendation methods. For instance, KHGT [20] attempts to explore the fine-grained differences between items by integrating the learned item-side

KG embeddings into the multi-behavior interactive item representations. Moreover, other methods have also achieved some results in utilizing KG information to enhance the target semantic. For example, KGCL [28] and KMCLR [25] transfer KG semantics into user-item graph to alleviate the issue of data sparsity.

2.2 Multi-behavior Recommendation

Recent studies have attempted to explore effective methods for addressing behavioral diversity. MBGMN [21] employs meta-learning networks to capture interaction diversity and behavioral heterogeneity. CRGCN [26] and MB-CGCN [4] leverage the cascading effects in multi-behavior to facilitate the embedding learning between different behaviors. To alleviate data sparsity of the target behavior, some studies transfer the user intents from auxiliary behaviors into the target behavior by introducing different data augmentation schemes [15,18,22,25,29,30]. Moreover, recent research has recognized the negative impact of the noise within auxiliary behaviors. BCIPM [27] utilize auxiliary behavior data solely to learn prior knowledge and network parameters, reducing auxiliary behavior influence on recommendations. MPC [6] treats items in earlier behaviors but absent later as negatives, learns their embeddings, and mitigates their influence by subtracting them from the final representations. MBID [23] decomposes the auxiliary behavior embeddings into genuine and spurious correlation parts through a projection mechanism, and disentangles the spurious correlation intent using the information bottleneck principle.

3 Preliminaries

3.1 Concepts and Notations

Multi-behavior Interaction Graph. We let $\mathcal{U}$ represent the set of users and $\mathcal{I}$ represent the set of items. Then, we use $\mathcal{G} = (\mathcal{U} \cup \mathcal{I}, \mathcal{E})$ to denote the user-item interaction bipartite graph, and the edge set $\mathcal{E}$ represents observed interactions. In the multi-behavior recommendation scenario with K types of behaviors, the bipartite graph $\mathcal{G}$ can be divided into K behavior subgraphs $\{\mathcal{G}_1, \mathcal{G}_2, \cdots, \mathcal{G}_K\}$, and user-item interaction matrices are represented as $\{\mathbf{R}_1, \mathbf{R}_2, \cdots, \mathbf{R}_K\}$.

Spurious Correlation Knowledge Graph (Item-Item Relation). The knowledge graph is typically considered as a static graph composed of triples in the form of $(\mathcal{H}, \mathcal{R}, \mathcal{T})$, where $\mathcal{H}$ represents the head-entity set, $\mathcal{R}$ is the interaction relation set (a.k.a the multi-behavior set in the context of multiple behaviors) and $\mathcal{T}$ denotes the tail-entity set. Adding an independent temporal dimension to the triples constructs a dynamic KG. Therefore, we define the knowledge graph using the quadruple $\mathbb{G} = (\mathcal{H}, \mathcal{R}, \mathcal{T}, \mathcal{D})$, where $\mathcal{D}$ is the discrete-time interval set. Given $\{\tau_1, \tau_2, \cdots, \tau_D\} \in \mathcal{D}$, the graph can be decomposed into several static subgraphs, i.e., $\mathbb{G} = \mathbb{G}_{\tau_1} \cup \mathbb{G}_{\tau_2} \cup \cdots \cup \mathbb{G}_{\tau_D}$, each consisting of triples valid at their respective time intervals.

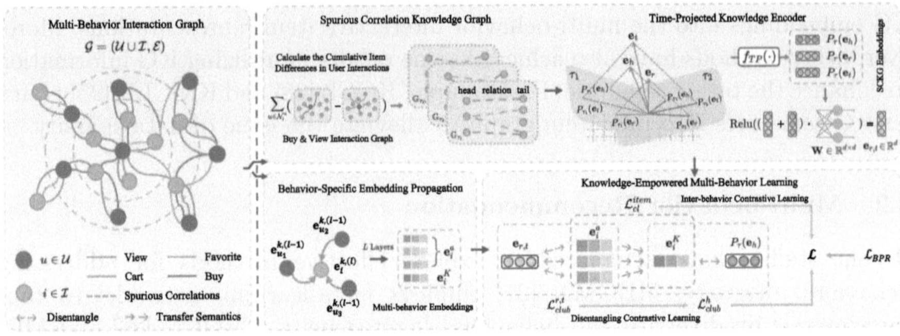

Fig. 2. SCKGD framework overview, where $f_{TP}(\cdot)$ denotes TP-TransR or TP-TATEC.

3.2 Problem Definition

Based on the above notations, the studied task is formally stated as:

Input: the user-item multi-behavior interaction graph $\mathcal{G}$ and the spurious correlation knowledge graph $\mathbb{G}$.

Output: a prediction function predicts the probability $\hat{y}_{ui}$ of user u interacting with item i under target behavior K.

4 Methodology

In this section, we present our SCKGD framework, which encapsulates two key components: spurious correlation-oriented knowledge graph and knowledge-empowered multi-behavior learning. The overall framework is shown in Fig. 2.

4.1 Spurious Correlation-Oriented Knowledge Graph

The presence of spurious correlations can lead to significant biases in representation learning. To disentangle the spurious correlations in multi-behavior recommendation, we propose a spurious correlation knowledge graph (SCKG) from the perspective of items. Building on the SCKG, we propose a time-projected knowledge encoder that dynamically learns the SCKG embeddings of items based on temporal factors. The SCKG embeddings are further used for disentangled representation learning at the item-level.

Spurious Correlation Knowledge Graph Construction. Since measuring the consistency between user intents and item features across multiple behaviors is intractable, genuine correlations are prevalent for most items in users' rich interactions, whereas spurious correlations are relatively rare. Therefore, an intuitive idea is to sample the head entity set by calculating the cumulative item difference between users' multi-behavior interactions and identify the N most uncorrelated items as their corresponding spurious correlated entities. In this way, we can disentangle the interference of these spurious correlated items, preserving and enhancing the genuine correlations in the user interactions.

Formally, we first measure the spurious correlation score through the item difference in user interactions between the target behavior K and the auxiliary behavior a:

$$\rho(i, a) = 1 - \sum_{u \in \mathcal{N}_i^a} \frac{|\mathcal{N}_u^K \cap \mathcal{N}_u^a|}{|\mathcal{N}_u^K \cup \mathcal{N}_u^a|}, \tag{1}$$

where $\mathcal{N}_i^a$ denotes the user set that interacted with item i under auxiliary behavior a, while $\mathcal{N}_u^K$ and $\mathcal{N}_u^a$ represent the item set that user u interacted with under target behavior K and auxiliary behavior a, respectively. After obtaining the spurious correlation score of each item, we use the following equation to sample items as the head-entity set corresponding to relation a in the SCKG:

$$\mathcal{H}_a = \{h \mid \rho(h, a) \geq \beta : h \in \mathcal{I}\}, \tag{2}$$

where β is the threshold for controlling the spurious correlation score. Subsequently, we define the spurious correlation entity set (i.e., the tail entity set) corresponding to the head entity $h \in \mathcal{H}_a$ under relation a using the following equations:

$$\begin{aligned} \mathbf{A} &= \mathbf{R}_a - \mathbf{R}_a \cap \mathbf{R}_K, \\ \mathcal{T}_a^h &= \{t \mid t \in TopIndex(\textstyle\sum_{u \in \mathcal{N}_h^K} \mathbf{A}_{u,:}, N)\}, \end{aligned} \tag{3}$$

where $TopIndex(\cdot, N)$ is the function that returns the indices of the N largest elements in the input vector, and $\mathbf{A}_{u,:}$ represents the row of the interaction matrix $\mathbf{A}$ corresponding to user u. In this way, we can identify the N items that are least related to the head entity h as the tail entities, based on the cumulative item interaction differences of the user.

Furthermore, two items with shorter interaction intervals are expected to exhibit weaker spurious correlations, whereas those with longer interaction intervals should exhibit stronger spurious correlations. Therefore, we introduce a temporal factor to measure the strength of spurious correlations between items. Specifically, we first calculate the average time span between the users' interactions with h and t. Then, we adopt a percentile-based discretization strategy to divide the time span into D discrete time intervals with approximately balanced samples, forming the interval set $\mathcal{D}$. As a result, the final SCKG is the union of the quadruples for all auxiliary behavior a: $\mathbb{G} = \{(h, a, t, \tau) | t \in \mathcal{T}_a^h, \tau \in \mathcal{D}\}$.

Time-Projected Knowledge Encoder. Since the spurious correlations between dynamic user intents and static item features typically evolve over time, we propose a time-projected knowledge encoder to disentangle time-related factors from the entity semantics in the SCKG, aiming to accurately learn the spurious correlation embeddings of items at different time points.

Specifically, we treat discrete-time intervals as a series of hyperplanes and project the embeddings of entities and relations onto these time-specific hyperplanes. This enables the original semantic space to be segmented along the temporal dimension, thereby extracting spurious correlation representations corresponding to each specific time interval. Mathematically, for the SCKG with D

time intervals, there are D different hyperplanes represented by normal vectors $w_{\tau_1}, w_{\tau_2}, \cdots, w_{\tau_D}$. Therefore, based on these hyperplanes, the SCKG can be decomposed into several static subgraphs $\mathbb{G}_{\tau_1}, \mathbb{G}_{\tau_2}, \cdots, \mathbb{G}_{\tau_D}$. Consequently, we can project triples of $\mathbb{G}_\tau = \{(h, r, t)\}$ onto the specific time hyperplane w_τ. The projected representation of SCKG triples can be formulated as:

$$P_\tau(\mathbf{e}) = \mathbf{e} - (w_\tau^\top \mathbf{e}) w_\tau, \quad \forall \mathbf{e} \in \{\mathbf{e}_h, \mathbf{e}_r, \mathbf{e}_t\}. \tag{4}$$

where $\|w_\tau\|_2 = 1$, and we learn $\{w_\tau\}_{\tau=1}^D$ for each time interval τ, along with the entity and relation embeddings. Furthermore, to accurately capture the spurious correlations between items, inspired by the translation distance method TransR [9] and the semantic matching method TATEC [5], we model the spurious correlations between items from multiple perspectives.

To be specific, we propose time-projected versions of TransR and TATEC, denoted as TP-TransR and TP-TATEC. TP-TransR can dynamically learn entity representations in the relation space, allowing the head (tail) entities to reflect the spurious correlation features of their corresponding tail (head) entities in the SCKG for different relations and times. Similarly, TP-TATEC can capture the latent semantics of entities over time, thus modeling the spurious correlation between entities. Accordingly, for each quadruple (h, r, t, τ), the scoring functions are defined as follows:

$$\begin{aligned} f_{TP-TransR}(h, r, t, \tau) &= -\left\|\mathbf{M}_r P_\tau(\mathbf{e}_h) + P_\tau(\mathbf{e}_r) - \mathbf{M}_r P_\tau(\mathbf{e}_t)\right\|_2^2, \\ f_{TP-TATEC}(h, r, t, \tau) &= P_\tau(\mathbf{e}_h)^\top \mathbf{M}_r P_\tau(\mathbf{e}_t) + P_\tau(\mathbf{e}_h)^\top P_\tau(\mathbf{e}_r) \\ &\quad + P_\tau(\mathbf{e}_t)^\top P_\tau(\mathbf{e}_r) + P_\tau(\mathbf{e}_h)^\top \mathbf{D} P_\tau(\mathbf{e}_t), \end{aligned} \tag{5}$$

where $\mathbf{M}_r \in \mathbb{R}^{d \times d}$ is a projection matrix of relation r, which projects entities from the entity space into the relation space. $\mathbf{D}$ is a diagonal matrix shared across relations. Optimized by the two scoring functions, ($\mathbf{e}_h$,$\mathbf{e}_r$,$\mathbf{e}_t$) capture the temporal dynamics of spurious correlations, and effectively model the spurious correlation representations of entities and relations in the SCKG.

4.2 Knowledge-Empowered Multi-behavior Learning

Based on the spurious correlation embeddings ($\mathbf{e}_h$,$\mathbf{e}_r$,$\mathbf{e}_t$) learned by the above steps, the disentangling contrastive learning method is proposed for disentangling the spurious correlations in the multi-behavior representations. Additionally, to further transfer the genuine correlated semantics from auxiliary behaviors into the target behavior, an inter-behavior contrastive learning paradigm is introduced to provide additional supervisory signals for the target behavior.

Disentangling Contrastive Learning. We use MI to quantify and minimize the correlation between multi-behavior embeddings and the SCKG embeddings, thereby disentangling the spurious correlations in multi-behavior scenarios.

To be precise, we leverage the Contrastive Log-ratio Upper Bound (CLUB) [3], a method for estimating MI by calculating the difference in conditional probabilities, to model the irrelevance between multi-behavior and SCKG representations. Therefore, having obtained the user embeddings $\mathbf{e}_u^k$ and item embeddings

$\mathbf{e}_i^k$ for each specific behavior k learned by the LightGCN, we apply CLUB on the head-entity set $\mathcal{H}$ to refine the item embeddings by disentangling the spurious correlations:

$$\mathcal{L}_{club}^{h} = \frac{1}{|\mathcal{H}|}\sum_{h\in\mathcal{H}}\left[\log q_\phi(P_\tau(\mathbf{e}_h)|\mathbf{e}_h^K) - \frac{1}{|\mathcal{H}|}\sum_{h'\in\mathcal{H}}\log q_\phi(P_\tau(\mathbf{e}_{h'})|\mathbf{e}_h^K)\right], \tag{6}$$

where $q_\phi(\cdot|\cdot)$ is a neural network for fitting the conditional distribution $p(\cdot|\cdot)$. The positive item h and the negative item h', both belonging to the head entity set $\mathcal{H}$, are sampled as described in Eq. (2), while $\mathbf{e}_h^K$ is the embedding of item h under the target behavior K, learned by LightGCN. Subsequently, given that tail entities are defined by specific relations in the SCKG, we use GCN Aggregator [8] to aggregate the tail entity embeddings $\mathbf{e}_t$ and its associated relation embeddings $\mathbf{e}_r$ as its relation-specific embeddings $\mathbf{e}_{r,t}$:

$$\mathbf{e}_{r,t} = \text{LeakyReLU}\left(\mathbf{W}(\mathbf{e}_r + \mathbf{e}_t)\right), \tag{7}$$

where we set the activation function to LeakyReLU [10], and $\mathbf{W} \in \mathbb{R}^{d\times d}$ is the trainable weight matrix to distill useful information for graph propagation. Similarly, we apply CLUB on the tail-entity set $\mathcal{T}_a$ to learn the item embeddings disentangled from spurious correlations under the auxiliary behavior a:

$$\mathcal{L}_{club}^{a,t} = \frac{1}{|\mathcal{T}_a|}\sum_{t\in\mathcal{T}_a}\left[\log q_\phi(\mathbf{e}_{a,t}|\mathbf{e}_t^a) - \frac{1}{|\mathcal{T}_a|}\sum_{t'\in\mathcal{T}_a}\log q_\phi(\mathbf{e}_{a,t'}|\mathbf{e}_t^a)\right]. \tag{8}$$

As a result, the final objective function of spurious correlation disentanglement is: $\mathcal{L}_{club} = \mathcal{L}_{club}^{h} + \sum_{a=1}^{K-1}\mathcal{L}_{club}^{a,t}$.

Inter-behavior Contrastive Learning. After disentangling the spurious correlations existing in multi-behavior scenarios, the semantic information in the auxiliary behaviors that aligns with the target intent is sufficiently preserved. Therefore, to enrich the supervision signal in the target behavior, we perform inter-behavior contrastive learning to transfer the genuine corrrelated auxiliary behavior semantics into the target behavior. A widely used idea [19] is to treat the same node embeddings from different behaviors as positive pairs and the different node embeddings from different behaviors as negative pairs. Thus, the item-side optimization objective based on InfoNCE [12] is formulated as:

$$\mathcal{L}_{cl}^{item} = \sum_{a=1}^{K-1}\sum_{i\in\mathcal{I}} -\log\frac{\exp(s(\mathbf{e}_i^a, \mathbf{e}_i^K)/\mathbf{t})}{\sum_{i'\in\mathcal{I}, i'\neq i}\exp(s(\mathbf{e}_{i'}^a, \mathbf{e}_i^K)/\mathbf{t})}, \tag{9}$$

where $s(\cdot)$ measures the cosine similarity between two vectors; $\mathbf{t}$ is the temperature hyperparameter in softmax. Incorporating a similarly designed user-side loss, the final objective function of inter-behavior contrastive learning is: $\mathcal{L}_{cl} = \mathcal{L}_{cl}^{user} + \mathcal{L}_{cl}^{item}$.

4.3 Model Training

In this section, we introduce the two-stage learning method for our SCKGD, and provide the analysis on the computational complexity.

Two-Stage Learning Method. Based on the aforementioned modules, we obtain embeddings generated in two different ways: one using multi-behavior information and the other using SCKG information. However, since multi-behavior embeddings rely on SCKG embeddings to disentangle the spurious correlations, the training of our model is composed of two stages within an optimization loop. In particular: i) In the first stage, we utilize the SCKG to learn the items' spurious correlation embeddings from both TP-TransR and TP-TATEC perspectives, thereby guiding the second stage to disentangle spurious correlations. ii) In the second stage, we leverage the two types of the SCKG embeddings to disentangle the spurious correlations in multi-behavior representations and further transfer the genuine correlation semantics from auxiliary behaviors into the target behavior. Therefore, we adopt a multi-task training strategy in the second stage to jointly optimize the main supervised task loss, the disentangling contrastive learning loss, and the inter-behavior contrastive learning loss.

Specifically, to optimize the model parameters, we first adopt the Bayesian Personalized Ranking (BPR) [13] as the main supervised task loss:

$$\mathcal{L}_{BPR}^{k} = -\sum_{u \in \mathcal{U}} \sum_{i \in \mathcal{N}_u^k} \sum_{j \notin \mathcal{N}_u^k} \ln \text{Softplus}(\hat{y}_{ui}^{k} - \hat{y}_{uj}^{k}), \tag{10}$$

where $\hat{y}_{ui}^{k}$ represents the inner product between the final embeddings of user u and item i under behavior k. Then, the overall optimization objective in the second stage is formulated as follows:

$$\mathcal{L} = \sum_{k=1}^{K} \mathcal{L}_{BPR}^{k} + \gamma \mathcal{L}_{club} + \mu \mathcal{L}_{cl} + \lambda \|\Theta\|^2, \tag{11}$$

where Θ represents the learnable parameters and the L_2 regularization is applied to alleviate the overfitting issue, while γ and μ are used to control the combination weights for disentangling contrastive learning and inter-behavior contrastive learning, respectively.

5 Experiments

In this section, we evaluate the performance of SCKGD through extensive experiments conducted on three real-world recommendation datasets and answer the following research questions:

RQ1: How does the performance of SCKGD in multi-behavior recommendation scenarios compare to existing methods?
RQ2: What are the impacts of key components in SCKGD on performance?

RQ3: How robust is SCKGD to data sparsity and spurious correlations in recommendation tasks?
RQ4: What is the influence of different hyperparameter settings on the recommendation performance of SCKGD?

5.1 Experimental Settings

Implementation and Metrics. We use LightGCN as the backbone model. Therefore, the embedding size is fixed at 64 for all models, and the number of propagation layers is 3. Our model is optimized with a learning rate of $1e^{-3}$ and a mini-batch size of 2048. The combination weights γ and μ are selected from the range $\{1e^{-3}, 5e^{-3}, 1e^{-2}, 1e^{-1}\}$.

To ensure a fair and effective evaluation, we strictly follow the evaluation experiment settings of the MBID [23], and report the average HR (Hit Ratio) and NDCG (Normalized Discounted Cumulative Gain) @K where $K = 10$, 20.

Datasets. We evaluate our SCKGD on three real-world datasets, whose statistics are shown in Table 1: **Tmall** from Alibaba, **Retail** from Retail Rocket, and **IJCAI-Contest** from the IJCAI competition.

Table 1. Statistics of multi-behavior datasets.

Dataset	Users	Items	Interactions	Density
Tmall	31,882	31,232	1,451,219	0.00145
Retail	147,894	99,037	1,584,238	0.00011
IJCAI-Contest	17,435	35,920	799,368	0.00127

Baselines. We compare our SCKGD with two types of SOTA baselines: single-behavior methods MF-BPR [13], LightGCN [7], KGCL [28] and multi-behavior methods KHGT [20], CML [18], KMCLR [25], StarRec [29], BCIPM [27], MPC [6], and MBID [23]. It is worth noting that knowledge-aware recommendation methods are also included in both types of methods.

5.2 Performance Comparison (RQ1)

We present the detailed evaluation results of all methods on different datasets in Table 2, where the best-performing results are highlighted in **bold**, and the sub-optimal results are underlined. Our SCKGD consistently outperforms all baselines on all datasets, with the improvements over the best baseline ranging from 2.17% to 10.82% in terms of HR and NDCG. On this basis, we can find that:

i) The knowledge-aware recommendation methods (e.g., KGCL, KHGT) and the multi-behavior recommendation methods (e.g., BCIPM, MPC) always outperform SOTA single-behavior recommendation methods (e.g., LightGCN), indicating that introducing additional auxiliary information helps mitigate data sparsity in single behavior and enhances recommendation performance.

Table 2. Performance comparison on different datasets, where $*$ indicates statistical significance with $p < 0.05$ based on a two-tailed paired t-test.

Methods	From'Year	Tmall		Retail		IJCAI-Contest	
		H@10	N@10	H@10	N@10	H@10	N@10
MF-BPR	UAI'9	0.2443	0.1501	0.2608	0.1653	0.1641	0.0917
LightGCN	SIGIR'20	0.3418	0.2048	0.3065	0.1847	0.2571	0.1231
KGCL	SIGIR'22	0.3751	0.2058	0.3098	0.1863	0.2635	0.1373
KHGT	SIGIR'21	0.4032	0.2351	0.4199	0.2471	0.3173	0.1824
CML	WSDM'22	0.5185	0.3055	0.4256	0.2503	0.4773	0.2831
KMCLR	WSDM'23	0.5671	0.3540	0.4557	0.2735	0.5255	0.3174
BCIPM	SIGIR'24	0.5498	0.3459	0.4549	0.2719	0.5187	0.3115
StarRec	SDM'25	0.5625	0.3493	0.4571	0.2759	0.5242	0.3205
MPC	AAAI'25	0.5774	0.3561	0.4626	0.2826	0.5281	0.3205
MBID	CIKM'25	0.6006	0.3731	0.4895	0.3021	0.5483	0.3332
SCKGD	-	**0.6174***	**0.3985***	**0.5235***	**0.3348***	**0.5602***	**0.3554***
Impr.	-	2.80%	6.81%	6.95%	10.82%	2.17%	6.66%

ii) Both SCKGD, MPC and MBID significantly outperform CML, KMCLR, and StarRec, as the latter ones primarily focus on modeling auxiliary behaviors and treat the intent of auxiliary behaviors as equivalent to the target intent, without considering the inherent noise in auxiliary behaviors. In contrast, SCKGD, MPC and MBID recognize the existence of noise in auxiliary behaviors and mitigate the negative transfer caused by such noise on modeling users' target intents, thereby achieving superior performance.

iii) Compared with MBID, our proposed SCKGD achieves significant performance improvements. This is primarily because MBID rigidly enforces a projection mechanism to dichotomize auxiliary behaviors into genuine and spurious intents, and further disentangles the spurious ones. Such a binary division is overly idealized and oversimplifies the intricate correlations among behaviors. In contrast, our method constructs a spurious correlation-oriented knowledge graph to explicitly quantify and model the noise in auxiliary behaviors from the item perspective. By leveraging MI, we disentangle spurious correlations from auxiliary behaviors and transfer the genuine correlation intent into the target behavior, enabling our model to achieve the best overall performance.

5.3 Ablation and Effectiveness Analyses (RQ2)

To show the effectiveness of the key components, we conducted an ablation study on SCKGD as shown in Table 3. The analysis is summarized as follows:

Effect of Spurious Correlation Knowledge Graph. We first aim to clarify whether the SCKG is beneficial for disentangling the spurious correlations in

Table 3. Ablation study in terms of HR@10 and NDCG@10.

Data	Tmall		Retail		IJCAI-Contest	
Metrics	HR	NDCG	HR	NDCG	HR	NDCG
w/o-SCKG	0.5486	0.3221	0.4486	0.2708	0.5141	0.3042
w/o-TPKE	0.5702	0.3544	0.4702	0.2846	0.5306	0.3178
w/o-IBCL	0.4251	0.2096	0.3724	0.2301	0.4572	0.2702
SCKGD	**0.6174**	**0.3985**	**0.5235**	**0.3348**	**0.5602**	**0.3554**

multi-behavior recommendation. To this end, we design a variant w/o-SCKG, which disables the SCKG and the modules related to disentangled contrastive learning. As depicted in Table 3, compared with the full SCKGD, the performance of w/o-SCKG is significantly degraded in both indicators. This suggests that the SCKG can accurately capture the spurious correlations between items and supervise the subsequent spurious correlation disentanglement.

Effect of Time-Projected Knowledge Encoder. To investigate the role of the time-projected knowledge encoder, we propose a variant w/o-TPKE by disabling the time-projected knowledge encoder. Instead, we only use the standard versions of TransR and TATEC as objective functions to learn the SCKG embeddings. Clearly, SCKGD obtains better performance than w/o-TPKE. It suggests that by employing the time-projected knowledge encoder, SCKGD can dynamically weight the strength of spurious correlations between items, thereby accurately modeling the spurious correlations of entities over time.

Effect of Inter-behavior Contrastive Learning. To verify the role of the inter-behavior contrastive learning in the model, we propose a variant w/o-IBCL, which disables the inter-behavior contrastive learning. Due to the inability to transfer the auxiliary behavior supervision signal into the target behavior, this leads w/o-IBCL to completely degenerate into a common single-behavior recommendation model, which further results in the failure of the disentanglement module to function properly. The experimental results indicate that inter-behavior contrastive learning is essential for our model to transfer the disentangled auxiliary behavior semantics into the target behavior.

5.4 Robustness Analysis (RQ3)

Due to the common existence of sparse interaction data in real-world scenarios [31], and given that the intensity of spurious correlations may vary with different degrees of sparsity, our goal is to demonstrate the robustness of our method under different levels of data sparsity and different degrees of spurious correlations. To this end, we conduct robustness experiments on the Retail dataset, as it features over 100K users with severely sparse user-item interactions (i.e., an interaction density of around 0.01%), as detailed in Table 1. Following the setup in [20,25], we divide each user's interaction counts in Retail into five levels of data sparsity based on the total number of user interactions across all behaviors. The recommendation accuracy of the models at different sparsity degrees is shown in Fig. 3, where the x-axis represents the user's interaction density, the left side of the y-axis uses bars to quantify the cumulative item difference (abbr., Cum. Item Diff.) between target and auxiliary behaviors for each user group, and the right side of the y-axis presents the average metric value using line charts.

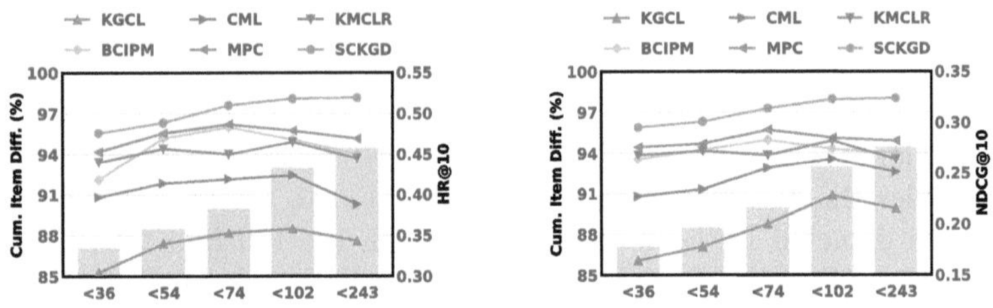

(a) Retail HR@10

(b) Retail NDCG@10

Fig. 3. Performance of SCKGD and baseline methods w.r.t different data sparsity degrees on Retail.

We have the following findings: The performance gap between our method and other competitors becomes more evident as user interactions become denser. Particularly when the number of interactions ranges from 74 to 243 (a period characterized by increasingly frequent user interactions but growing disparities in interaction items between behaviors), all other methods gradually suffer from interference caused by the spurious correlations, leading to a significant performance degradation. This demonstrates that SCKGD can effectively disentangle the spurious correlations and transfer the genuine correlated semantics into the target behavior, thereby maintaining robust performance across varying degrees of data sparsity and spurious correlation intensity.

5.5 Hyperparameter Analysis (RQ4)

We conduct extensive experiments to investigate the sensitivity of SCKGD to several key hyperparameters, including the spurious correlation threshold β, the top N spurious items for head entities, the discrete-time intervals D, and the combination weight γ. Figure 4 illustrates the performance variations relative to the optimal setting.

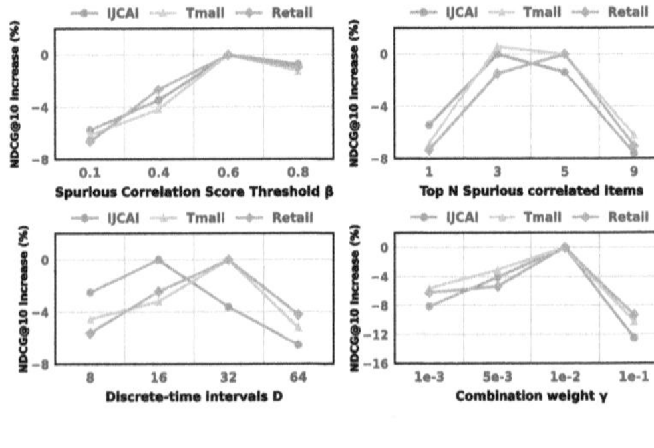

Fig. 4. Hyperparameter analysis of SCKGD.

Spurious Correlation Score Threshold β**.** The results show that a too-low threshold (<0.6) may misclassify genuinely correlated items as spurious, hurting performance, while a too-high threshold limits spurious correlation capture. A moderate threshold allows SCKG to effectively capture spurious correlations, yielding stable improvements across datasets with different sparsity levels.

Top N **Spurious Correlated Items for the Head Entity.** Through experimental results, it is found that as the number of Top N uncorrelated items associated with each head entity increases, SCKGD can accurately model and disentangle spurious correlations. However, an too large number of spurious correlated items (i.e., >5) may lead to some genuine correlated items being involved in the modeling of spurious correlations, resulting in a performance drop.

Discrete-Time Intervals D**.** From Fig. 4, we find that the optimal selection of D varies with the dataset. Specifically, for Tmall and Retail, the distribution of user interaction times is more dispersed, requiring a greater number of discrete-time intervals to meet the model's requirements. In contrast, the distribution of user interaction times in IJCAI-Contest is more concentrated, allowing for fewer discrete-time intervals to satisfy the model's needs.

Combination Weight γ**.** We vary the combination weight γ while keeping other hyperparameters fixed. As the weight increases from $1e^{-3}$ to $1e^{-2}$, the model further effectively disentangles the spurious correlations, resulting in an improved performance. However, a larger weight may cause the model to overemphasize disentangling spurious correlations, which could lead to overfitting issues.

6 Conclusion

In this work, we design a novel multi-behavior recommendation framework that leverages a spurious correlation-oriented knowledge graph to disentangle the spurious correlations in multi-behavior scenarios. Specifically, the SCKG and the time-projected knowledge encoder indirectly reckon the spurious correlation between user intent and item features through the spurious correlations between items. On this basis, the disentangling contrastive learning method can disentangle the spurious correlations from auxiliary behaviors, and then transfer the genuine semantics from auxiliary behaviors to the target behaviors through the inter-behavior contrastive learning, thereby alleviating the issue of sparse supervision signals in the target behavior. Extensive experimental results on different real-world datasets prove the effectiveness of SCKGD. In the future, we will further improve the disentangling process by introducing a counterfactual generator to capture users' latent semantics.

Acknowledgments. This work was supported in part by the National Natural Science Foundation of China (Grant Nos. 62566015 and U22A2099), and by the Innovation Project of GUET Graduate Education (Grant No. 2025YCXS058).

References

1. Agarwal, V., Shetty, R., Fritz, M.: Towards causal VQA: revealing and reducing spurious correlations by invariant and covariant semantic editing. In: Proceedings of the IEEE/CVF Conference on Computer Vision and Pattern Recognition, pp. 9690–9698 (2020)
2. Bin, C., Xu, T., Zhang, F.: User intent disentanglement for multi-behavior recommendation via information bottleneck principle. Neurocomputing **656**, 131454 (2025)
3. Cheng, P., Hao, W., Dai, S., Liu, J., Gan, Z., Carin, L.: Club: a contrastive log-ratio upper bound of mutual information. In: International Conference on Machine Learning, pp. 1779–1788. PMLR (2020)
4. Cheng, Z., Han, S., Liu, F., Zhu, L., Gao, Z., Peng, Y.: Multi-behavior recommendation with cascading graph convolution networks. In: Proceedings of the ACM Web Conference 2023, pp. 1181–1189 (2023)
5. García-Durán, A., Bordes, A., Usunier, N.: Effective blending of two and three-way interactions for modeling multi-relational data. In: Calders, T., Esposito, F., Hüllermeier, E., Meo, R. (eds.) ECML PKDD 2014. LNCS (LNAI), vol. 8724, pp. 434–449. Springer, Heidelberg (2014). https://doi.org/10.1007/978-3-662-44848-9_28
6. Gong, S., Liu, Y., Dang, Y., Guo, G., Zhao, J., Wang, X.: Multiple purchase chains with negative transfer elimination for multi-behavior recommendation. In: Proceedings of the AAAI Conference on Artificial Intelligence, vol. 39, pp. 11717–11725 (2025)
7. He, X., Deng, K., Wang, X., Li, Y., Zhang, Y., Wang, M.: Lightgcn: simplifying and powering graph convolution network for recommendation. In: Proceedings of the 43rd International ACM SIGIR Conference on Research and Development in Information Retrieval, pp. 639–648 (2020)
8. Kipf, T.N., Welling, M.: Semi-supervised classification with graph convolutional networks. arXiv preprint arXiv:1609.02907 (2016)
9. Lin, Y., Liu, Z., Sun, M., Liu, Y., Zhu, X.: Learning entity and relation embeddings for knowledge graph completion. In: Proceedings of the AAAI Conference on Artificial Intelligence, vol. 29 (2015)
10. Maas, A.L., Hannun, A.Y., Ng, A.Y., et al.: Rectifier nonlinearities improve neural network acoustic models. In: Proceedings of the 30th International Conference on Machine Learning, Atlanta, GA, vol. 30, p. 3 (2013)
11. Mu, S., Li, Y., Zhao, W.X., Wang, J., Ding, B., Wen, J.R.: Alleviating spurious correlations in knowledge-aware recommendations through counterfactual generator. In: Proceedings of the 45th International ACM SIGIR Conference on Research and Development in Information Retrieval, pp. 1401–1411 (2022)
12. van den Oord, A., Li, Y., Vinyals, O.: Representation learning with contrastive predictive coding. arXiv preprint arXiv:1807.03748 (2018)
13. Rendle, S., Freudenthaler, C., Gantner, Z., Schmidt-Thieme, L.: BPR: Bayesian personalized ranking from implicit feedback. In: Proceedings of the Twenty-Fifth Conference on Uncertainty in Artificial Intelligence, pp. 452–461 (2009)
14. Srivastava, M., Hashimoto, T., Liang, P.: Robustness to spurious correlations via human annotations. In: International Conference on Machine Learning, pp. 9109–9119. PMLR (2020)
15. Wang, X., Wang, W., Feng, F., Rong, W., Yin, C., Xiong, Z.: Causal intervention for fairness in multibehavior recommendation. IEEE Trans. Comput. Soc. Syst. **11**(5), 6320–6332 (2024)

16. Wang, X., He, X., Cao, Y., Liu, M., Chua, T.S.: Kgat: knowledge graph attention network for recommendation. In: Proceedings of the 25th ACM SIGKDD International Conference on Knowledge Discovery & Data Mining, pp. 950–958 (2019)
17. Wang, X., et al.: Learning intents behind interactions with knowledge graph for recommendation. In: Proceedings of the Web Conference 2021, pp. 878–887 (2021)
18. Wei, W., Huang, C., Xia, L., Xu, Y., Zhao, J., Yin, D.: Contrastive meta learning with behavior multiplicity for recommendation. In: Proceedings of the Fifteenth ACM International Conference on Web Search and Data Mining, pp. 1120–1128 (2022)
19. Wu, J., et al.: Self-supervised graph learning for recommendation. In: Proceedings of the 44th International ACM SIGIR Conference on Research and Development in Information Retrieval, pp. 726–735 (2021)
20. Xia, L., et al.: Knowledge-enhanced hierarchical graph transformer network for multi-behavior recommendation. In: Proceedings of the AAAI Conference on Artificial Intelligence, vol. 35, pp. 4486–4493 (2021)
21. Xia, L., Xu, Y., Huang, C., Dai, P., Bo, L.: Graph meta network for multi-behavior recommendation. In: Proceedings of the 44th International ACM SIGIR Conference on Research and Development in Information Retrieval, pp. 757–766 (2021)
22. Xu, J., et al.: Multi-behavior self-supervised learning for recommendation. In: Proceedings of the 46th International ACM SIGIR Conference on Research and Development in Information Retrieval, pp. 496–505 (2023)
23. Xu, T., Bin, C., Xiao, C., Li, Y., Gu, T.: Multi-behavior intent disentanglement for recommendation via information bottleneck principle. In: Proceedings of the 34th ACM International Conference on Information and Knowledge Management, pp. 5391–5395 (2025)
24. Xu, T., Bin, C., Xiao, C., Zeng, Z., Gu, T.: Disc: disentangling spurious correlations for multi-behavior recommendation. IEEE Trans. Comput. Soc. Syst. (2026)
25. Xuan, H., Liu, Y., Li, B., Yin, H.: Knowledge enhancement for contrastive multi-behavior recommendation. In: Proceedings of the Sixteenth ACM International Conference on Web Search and Data Mining, pp. 195–203 (2023)
26. Yan, M., et al.: Cascading residual graph convolutional network for multi-behavior recommendation. ACM Trans. Inf. Syst. **42**(1), 1–26 (2023)
27. Yan, M., Liu, F., Sun, J., Sun, F., Cheng, Z., Han, Y.: Behavior-contextualized item preference modeling for multi-behavior recommendation. In: Proceedings of the 47th International ACM SIGIR Conference on Research and Development in Information Retrieval, pp. 946–955 (2024)
28. Yang, Y., Huang, C., Xia, L., Li, C.: Knowledge graph contrastive learning for recommendation. In: Proceedings of the 45th International ACM SIGIR Conference on Research and Development in Information Retrieval, pp. 1434–1443 (2022)
29. Zhang, W., et al.: Starrec: a hypergraph-based framework with star-expansion for multi-behavior recommendation. In: Proceedings of the 2025 SIAM International Conference on Data Mining (SDM), pp. 181–191. SIAM (2025)
30. Zhang, X., Cheng, X., Xiao, Y., Zheng, W.: Advancing multibehavior recommendation with dual-mode augmented contrastive learning. IEEE Trans. Comput. Soc. Syst. **12**(5), 3116–3130 (2025)
31. Zhu, Y., et al.: Personalized transfer of user preferences for cross-domain recommendation. In: Proceedings of the Fifteenth ACM International Conference on Web Search and Data Mining, pp. 1507–1515 (2022)

DynamicPO: Dynamic Preference Optimization for Recommendation

Xingyu Hu[1,2(✉)], Kai Zhang[1], Jiancan Wu[1(✉)], Shuli Wang[3], Chi Wang[3], Wenshuai Chen[3], Yinhua Zhu[3], Haitao Wang[3], Xingxing Wang[3], and Xiang Wang[1]

[1] University of Science and Technology of China, Hefei, China
huxy@mail.ustc.edu.cn, kaizhang1215@gmail.com, wujcan@gmail.com, xiangwang1223@gmail.com
[2] Shanghai Innovation Institute, Shanghai, China
[3] Meituan, Chengdu, China
{wangshuli03,wangchi06,chenwenshuai,zhuyinhua,wanghaitao13, wangxingxing04}@meituan.com

Abstract. In large language model (LLM)-based recommendation systems, direct preference optimization (DPO) effectively aligns recommendations with user preferences, necessitating multi-negative objective functions to leverage abundant implicit-feedback negatives and to sharpen preference boundaries. However, our empirical analyses reveal a counterintuitive phenomenon–*preference optimization collapse*–where increasing the number of negative samples can lead to performance degradation despite a continuously decreasing training loss. We further theoretically demonstrate that this collapse arises from *gradient suppression*, caused by the dominance of easily discriminable negatives over boundary-critical negatives that truly define user preference boundaries, leaving boundary-relevant signals under-optimized and ultimately weakening the model's decision boundary. Motivated by these observations, we propose **DynamicPO** (**Dynamic P**reference **O**ptimization), a lightweight and plug-and-play framework comprising two adaptive mechanisms: (1) *Dynamic Boundary Negative Selection*, which identifies and prioritizes informative negatives near the model's decision boundary, and (2) *Dual-Margin Dynamic β Adjustment*, which calibrates optimization strength per sample according to boundary ambiguity. Extensive experiments on three public datasets show that DynamicPO effectively prevents optimization collapse and improves recommendation accuracy on multi-negative preference optimization methods, with negligible computational overhead. Our code and datasets are available at https://github.com/xingyuHuxingyu/DynamicPO.

Keywords: Large Language Models · Sequential Recommendation · Large Language Model-based Recommendation · Direct Preference Optimization

X. Hu—Work done at Meituan.

H. Jung et al. (Eds.): DASFAA 2026, LNCS 16535, pp. 372–387, 2026.
https://doi.org/10.1007/978-981-92-0363-5_23

1 Introduction

Sequential recommendation systems aim to predict the next item a user is likely to interact with by modeling their historical behaviors [3]. With the advent of large language models (LLMs), which possess extensive world knowledge and powerful reasoning capabilities, the landscape of recommender systems is rapidly evolving. Recent studies have shown that LLMs can serve as a strong backbone for next-generation recommendation algorithms, outperforming traditional architectures in various domains [2,4,16,21].

Despite recent progress in leveraging large language models (LLMs) for recommendation tasks, the majority of current approaches predominantly employ supervised fine-tuning (SFT) based on language modeling objectives. Specifically, these methods train models to predict the next token or to continue sequences of text, which primarily encourages the generation of coherent and contextually appropriate language. However, this training paradigm does not explicitly guide the model to discern or rank user preferences, as it focuses on sequence continuation rather than on preference discrimination. Consequently, such objectives are inherently misaligned with the goal of recommendation, which is to accurately rank items according to users' likes and dislikes. This misalignment may limit the effectiveness of LLM-based recommendation models in capturing nuanced user preferences and delivering truly personalized recommendations [1].

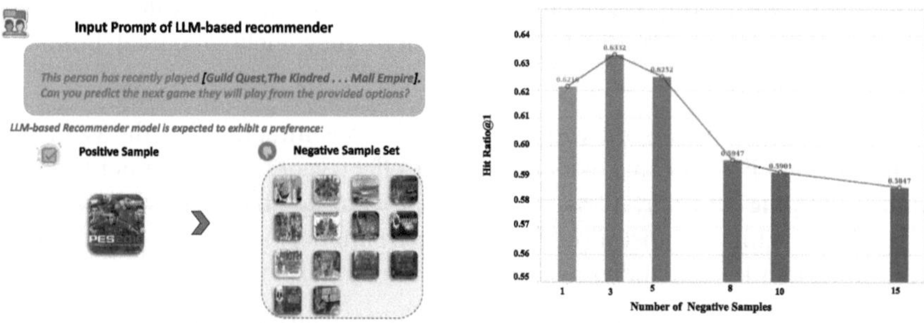

(a) Preference optimization of LLM-based Recommenders.

(b) The preference optimization collapse phenomenon observed in DMPO.

Fig. 1. Demonstration of preference optimization and collapse phenomenon in LLM-based recommenders.

To bridge this gap, preference optimization techniques–originally developed for dialogue and instruction tuning–have been adapted to the recommendation domain. These methods, especially in two-stage pipelines (first distilling item-level knowledge via SFT, then refining subtle user preferences through preference optimization), have demonstrated substantial and consistent improvements over SFT alone [1]. The process trains models to prioritize the positive sample while deprioritizing negatives (see Fig. 1a). A key insight underlying these methods is that implicit-feedback datasets inherently provide a wealth of negative samples—namely, items that are skipped or not interacted with by users. This abundance of negatives facilitates multi-negative objective functions, which sharpen preference boundaries and enhance the robustness of recommendation models [1,6,22].

Previous research has shown that increasing negative samples in preference optimization initially benefits model performance [1]. However, as demonstrated in Fig. 1b, our experimental analysis finds that beyond a critical threshold, this trend reverses: performance not only plateaus but also drops sharply, even as the training loss continues to decline. This unexpected degradation suggests that multi-negative aggregation may introduce non-trivial negative optimization biases. We identify this performance degradation as preference optimization collapse and attribute it, both theoretically and empirically, to gradient suppression caused by an imbalance between model-discriminative negatives and boundary-critical negatives (see Sect. 3.1).

To mitigate this collapse, we propose **Dynamic Preference Optimization (DynamicPO)**–a plug-and-play method that maintains effective boundary learning through two adaptive mechanisms:

- **Dynamic Boundary Negative Selection:** We intelligently identify the most informative boundary samples through real-time clustering, ensuring optimization focuses precisely on regions where preference discrimination remains ambiguous.
- **Dual-Margin Dynamic β Adjustment:** We assign a dynamic, sample-specific β parameter based on dual-margin discrimination difficulty, adaptively concentrating gradient updates where they yield maximal boundary refinement.

Extensive experiments on three widely used recommendation datasets (i.e., Goodreads, LastFM, Steam) demonstrate the effectiveness of DynamicPO. Additionally, our plug-and-play strategy preserves training efficiency with negligible overhead, yet achieves substantial performance enhancements. The contributions of this work can be concluded as follows:

- We experimentally identify the preference optimization collapse phenomenon and theoretically characterize it as gradient suppression caused by imbalance between model-discriminative ($\mathcal{S}$) and boundary-critical ($\mathcal{B}$) negatives for LLM-based recommendation systems.
- We propose DynamicPO, a preprocessing-free and plug-and-play method that adaptively selects boundary negatives and dynamically assigns negative sample specific β during training, thereby enabling precise preference boundary refinement and robust optimization in recommendation tasks.
- Extensive experiments on multiple LLM backbones and datasets demonstrate that DynamicPO consistently improves recommendation performance, effectively alleviates preference optimization collapse, and generalizes across diverse multi-negative DPO objectives without incurring additional computational cost.

2 Preliminary

In this section, we briefly review the foundation of preference optimization and its adaptation for LLM-based recommender systems.

2.1 Direct Preference Optimization

Large language models (LLMs) are usually trained via supervised fine-tuning (SFT), which enhances linguistic coherence but not user alignment. Reinforcement Learning from Human Feedback (RLHF) improves alignment through reward modeling, yet training instability and cost remain high. DPO [17] provides a closed-form alternative derived from RLHF, directly aligning the model with human preference pairs without reinforcement learning:

$$\mathcal{L}_{\text{DPO}}(\pi_\theta; \pi_{\text{ref}}) = -\mathbb{E}_{(x,y_w,y_l)}\left[\log \sigma\left(\beta \log \frac{\pi_\theta(y_w \mid x)}{\pi_{\text{ref}}(y_w \mid x)} - \beta \log \frac{\pi_\theta(y_l \mid x)}{\pi_{\text{ref}}(y_l \mid x)}\right)\right]. \quad (1)$$

Here, π_θ denotes the optimized policy, π_{ref} is a fixed reference model, and β controls the regularization strength. DPO thus achieves stable preference alignment more efficiently than RLHF-based methods.

2.2 Preference Optimization for LLM-Based Recommenders

Supervised fine-tuning (SFT) adapts LLMs to recommendation tasks but only maximizes the generation probability of the next item, failing to utilize negative feedback. Direct Multi-Preference Optimization (DMPO) [1] extends DPO by introducing multiple negatives to refine user preference boundaries:

$$\mathcal{L}_{\text{DMPO}} = -\mathbb{E}_{(x_u,y_w,y_l)}\left[\log \sigma\left(\beta \log \frac{\pi_\theta(y_w|x_u)}{\pi_{\text{ref}}(y_w|x_u)} - \frac{1}{k}\sum_{i=1}^{k} \beta \log \frac{\pi_\theta(y_{l_i}|x_u)}{\pi_{\text{ref}}(y_{l_i}|x_u)}\right)\right]. \quad (2)$$

This formulation guides LLMs to increase the likelihood of positive items (y_w) while suppressing negatives $\{y_{l_i}\}_{i=1}^{k}$, enabling explicit learning of user preference boundaries and improving recommendation accuracy.

3 Method

In this section, we investigate the preference optimization collapse phenomenon in Direct Multi-Preference Optimization [1] (DMPO), where model performance stagnates despite declining training loss. We attribute this collapse to gradient suppression caused by the imbalance between model-discriminative negatives and boundary-critical negatives. To resolve this, we propose DynamicPO, a dual-mechanism framework introducing: (1) adaptive boundary negative selection and (2) fine-grained β-differentiation (Fig. 2).

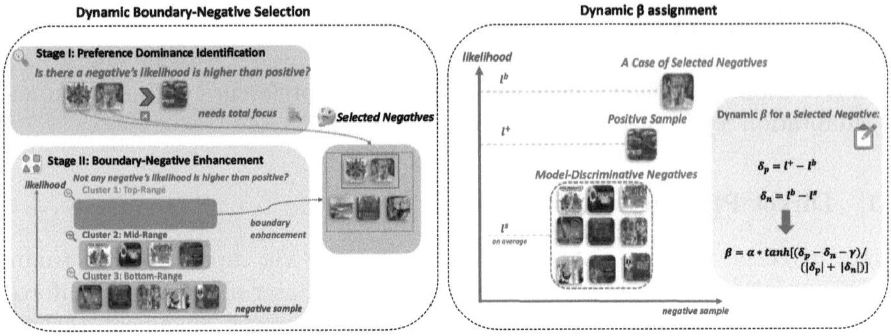

Fig. 2. Overview of DynamicPO: dynamic boundary negative selection and dynamic β-adjustment

3.1 Motivation: Gradient Suppression of Boundary-Critical Negatives in Multi-negative DPO

In DMPO, we observe a paradoxical phenomenon: model performance stagnates or even degrades, while training loss continues to decrease. This suggests the optimization process becomes decoupled from the true objective of refining preference boundaries. To investigate this, we leverage the insight that the likelihood assigned to a sequence reflects the model's confidence in that response [10]. When two responses exhibit similar likelihoods, they typically share semantic or logical coherence under the model's internal representation [7,11]. Consequently, the likelihood gap $\Delta = \log \pi_\theta(y_w \mid x_u) - \log \pi_\theta(y_{l_i} \mid x_u)$ serves as a quantitative measure of the model's preference resolution. By monitoring the evolution of these gaps during training, we find that negative samples can be partitioned into two distinct categories based on their proximity to the decision boundary:

- **Model-Discriminative Negatives** ($\mathcal{S}$): negatives with large likelihood gaps ($\log \pi_\theta(y_w \mid x_u) - \log \pi_\theta(y_{l_i} \mid x_u) \gg 0$), which the model already distinguishes well.
- **Boundary-Critical Negatives** ($\mathcal{B}$): negative samples on the model's decision boundary, with likelihood gaps ($\log \pi_\theta(y_w \mid x_u) - \log \pi_\theta(y_{l_i} \mid x_u) \leq 0$, or $\log \pi_\theta(y_w \mid x_u) - \log \pi_\theta(y_{l_i} \mid x_u) \approx 0$), which indicates unresolved or ambiguous preference discrimination.

To empirically investigate the distribution of these two categories, we adopt zero as a coarse-grained threshold for a first-order approximation. As illustrated in Fig. 3, this reveals a stark imbalance where $\mathcal{S}$ vastly outnumbers $\mathcal{B}$ throughout the training process. Ideally, effective optimization should resolve ambiguities, reducing the proportion of $\mathcal{B}$ over time. However, we observe a counter-intuitive trend: despite continuous loss reduction, the proportion of $\mathcal{B}$ actually **increases** from 6.3% to 7.6%. This signals a **boundary deterioration**: the dominance of easy negatives ($\mathcal{S}$) in the gradient creates a shortcut for loss minimization,

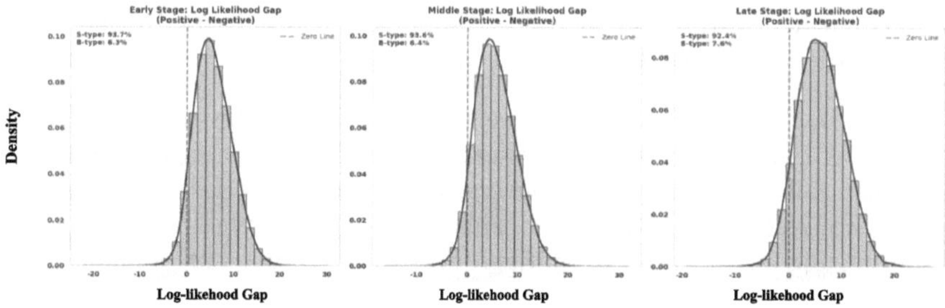

Fig. 3. Proportions of $\mathcal{S}$ and $\mathcal{B}$ through three training stages of DMPO.

causing the model to over-optimize trivial distinctions while neglecting–or even worsening–the resolution of critical boundary cases.

We attribute this deterioration to the **initialization bias introduced by** the SFT stage. Specifically, SFT equips the model with a coarse-grained discrimination capability, resulting in an abundance of already well-separated negatives ($\mathcal{S}$). However, standard multi-negative DPO naively aggregates these samples, inadvertently forcing the model to further amplify these trivial distinctions rather than focusing on the sparse samples near the boundary. Consequently, the optimization suffers from a **distributional imbalance**–where the overwhelming gradient contribution from $\mathcal{S}$ suppresses the critical updates required by the informative $\mathcal{B}$ set. In the following, we formally analyze the mechanism by which multi-negative DPO over-optimizes model-discriminative negatives.

Mechanism of Gradient Suppression. Formally, during optimization the aggregated gradient over k negatives can be decomposed as:

$$\begin{aligned}\frac{1}{k}\sum_{i=1}^{k}\nabla\log\pi_\theta(y_l^i|x_u) = \frac{|\mathcal{S}|}{k}&\underbrace{\left(\frac{1}{|\mathcal{S}|}\sum_{i\in\mathcal{S}}\nabla\log\pi_\theta(y_l^i|x_u)\right)}_{\text{average gradient of }\mathcal{S}}\\ &+\frac{|\mathcal{B}|}{k}\underbrace{\left(\frac{1}{|\mathcal{B}|}\sum_{j\in\mathcal{B}}\nabla\log\pi_\theta(y_l^j|x_u)\right)}_{\text{average gradient of }\mathcal{B}},\end{aligned} \tag{3}$$

Since $|\mathcal{S}| \gg |\mathcal{B}|$, the gradient is dominated by $\mathcal{S}$, forming a self-reinforcing loop where the model keeps further lowering the likelihood of already well-separated $\mathcal{S}$ negatives, even though they have been clearly distinguished from positives. This unnecessary amplification of easy distinctions consumes gradient capacity, leaving boundary-critical samples $\mathcal{B}$ under-optimized. The resulting imbalance suppresses boundary learning and drives the observed preference optimization collapse.

3.2 Dynamic Boundary Negative Selection: Capturing Informative Samples for User Preference Modeling

Guided by our theoretical insights, we prioritize **Boundary-Critical Negatives** ($\mathcal{B}$) as the primary source of meaningful optimization signals. Standard DMPO can fall into a deceptive convergence, whereby the loss decreases even as preference resolution deteriorates. This occurs because the model focuses on over-optimizing trivial gaps. We mitigate this through two core principles:

- **Principle 1:** Focus on boundary negatives resembling positives, as they best capture unresolved preferences and refine the decision boundary.
- **Principle 2:** Select boundary negatives adaptively according to sample informativeness, avoiding fixed thresholds.

These principles are operationalized via a **fully online and preprocessing-free** selection strategy consisting of two hierarchical stages:

Stage 1: Preference Dominance Identification. We first identify critical preference violations where the model's current policy contradicts the intended ordering. Let L_p and $L(n)$ denote the log-likelihoods of the positive and negative samples $n \in \mathcal{N}$, respectively. We define the violation set as:

$$\mathcal{N}_{vio} = \{n \in \mathcal{N} \mid L(n) > L_p\} \tag{4}$$

If $\mathcal{N}_{vio} \neq \emptyset$, these samples directly expose boundary deficiencies and are prioritized for optimization by setting the selected boundary negative set $\mathcal{B} = \mathcal{N}_{vio}$.

Stage 2: Boundary Negative Enhancement via Likelihood Clustering. If no such violation exists ($\mathcal{N}_{vio} = \emptyset$), we focus on negatives near the decision boundary to resolve preference ambiguities. We adaptively partition the negative samples into three clusters $\{\mathcal{C}_{top}, \mathcal{C}_{mid}, \mathcal{C}_{bot}\}$ using k-means clustering ($k = 3$) based on their log-likelihoods $\{L(n)\}_{n \in \mathcal{N}}$. The selection is then defined as $\mathcal{B} = \mathcal{C}_{top}$, where $\mathcal{C}_{top}$ denotes the cluster with the highest centroid. This adaptive grouping allows the model to flexibly capture informative boundary negatives across varying likelihood distributions without relying on rigid static thresholds.

By adaptively balancing major violation correction with boundary refinement, this mechanism effectively mitigates signal dilution and enhances the precision of preference modeling.

3.3 Dynamic β-Adjustment: Fine-Grained Boundary Regularization

In DPO-based frameworks, the hyperparameter β modulates the Kullback-Leibler (KL) regularization between the learned policy π_θ and the reference policy π_{ref} [17]. A smaller β facilitates aggressive adaptation to preference data (plasticity), whereas a larger β enforces conservative alignment to preserve the reference model's prior knowledge (stability). While β-DPO [20] introduces a dynamic β for natural language generation, we observe that its direct application to our *Dynamic Boundary Negative Selection* unexpectedly yields a 0.32%

performance drop. This suggests that existing adjustment strategies–primarily designed for open-ended generation–fail to account for the *multi-negative* and *implicit-feedback* nature of recommendation tasks.

To this end, we propose a fine-grained dynamic β-adjustment strategy to adaptively amplify optimization signals from each selected boundary negative and prevent boundary dilution through two principles:

- **Principle 1:** Assign β dynamically to each boundary negative based on its own informativeness.
- **Principle 2:** Modulate β using both positiveâĂŞnegative ambiguity and distance from easy negatives to emphasize harder samples.

To operationalize these principles, we introduce a dual-margin mechanism for each selected boundary negative sample:

- **Positive-to-Boundary Margin** ($\delta_p = L^+ - L^b$): measures *boundary ambiguity* using the likelihoods of the positive sample (L^+) and the boundary negative (L^b).
- **Boundary-to-Easy Margin** ($\delta_n = L^b - L^e$): quantifies *informativeness contrast* relative to the average log-likelihood of normal negatives (L^e).

We then compute the dynamic β for each sample as follows:

$$\beta = \beta_0 \cdot (1 + \alpha \cdot \tanh\left(\frac{\delta_p - \delta_n - \gamma}{|\delta_p| + |\delta_n|}\right)), \tag{5}$$

In this formula, β_0 is the base regularization weight and α scales the adjustment intensity. Crucially, γ denotes the *intrinsic preference margin*, representing the natural likelihood superiority that a positive sample is expected to maintain over a negative one. The normalized denominator ($|\delta_p|+|\delta_n|$) ensures the adjustment is sensitive to the *relative* likelihood distribution. Notably, the $\tanh(\cdot)$ function constrains β within a stable range, yielding updates that are both expressive for challenging cases and robust against outliers.

The Intuition Behind this Design is to Adaptively Balance Plasticity and Stability Based on Sample-Specific Difficulty. Specifically, for a challenging and informative negative that exhibits high ambiguity (small δ_p) and significant contrast from discriminative negatives (large δ_n), the numerator ($\delta_p - \delta_n - \gamma$) becomes negative. This leads to a decreased β, thereby reducing KL-regularization and allowing the model more *plasticity* to aggressively refine the decision boundary. Conversely, for trivial samples that are easily discriminable, β increases to enforce *stability* and prevent the model from over-adapting to non-critical signals. By precisely customizing β, DynamicPO provides fine-grained gradients that focus the model's capacity on refining the most critical preference regions.

4 Experiments

In this section, we conduct extensive experiments to evaluate the effectiveness and robustness of **DynamicPO**. We first outline the experimental setup, encompassing diverse LLM backbones, benchmark datasets, and representative baselines. Subsequently, we present a comparative analysis of experimental results, focusing on DynamicPO's ability to mitigate preference optimization collapse and its generalization across various multi-negative objectives. Finally, we provide in-depth analyses of reward boundary evolution and computational efficiency to further validate the superiority of our approach.

4.1 Experimental Settings

Base Model. Our approach is evaluated using three distinct base models: Llama2-7b-hf for the main experiments, supplemented by Llama3-8B-Instruct and Qwen2.5-7B-Instruct for exploratory studies. The diversity of these base models provides a robust foundation for our experimental analysis.

Datasets. We evaluate our approach on three widely used recommendation datasets from diverse domains: LastFM [5] (music recommendation), Goodreads[1] (book ratings and reviews), and Steam[2] (video game ratings and reviews). We follow prior work [16] for preprocessing and chronological 8:1:1 splitting. Detailed statistics for each dataset are summarized in Table 1.

Evaluation. Following prior work [6,16], we adopt HitRatio@1 to evaluate recommendation accuracy via re-ranking: for each sequence, models identify the correct item from a candidate set containing 20 negative samples plus the ground-truth item. Valid Ratio measures LLM-specific behaviors (e.g., instruction-following), and consistently approaches 1.0 (>0.95) across all experiments–indicating near-perfect instruction compliance. Consequently, we omit Valid Ratio in exploration studies as only HitRatio@1 reflects recommendation capability.

Table 1. Statistics of Datasets.

Dataset	LastFM	Steam	Goodreads
# Sequence	1,220	11,938	6,031
# Item	4,606	3,581	4,500
# Interaction	73,510	274,726	220,100

[1] https://www.goodreads.com.
[2] https://store.steampowered.com/.

Baselines. We compare our approach with several representative methods, covering both traditional and LLM-based paradigms. Traditional baselines include GRU4Rec [12] (GRU for sequential modeling), Caser [18] (convolutional filters for sequence patterns), and SASRec [14] (self-attention for long-term dependencies). LLM-based models include MoRec [24] (BERT for text encoding and SASRec for sequences), TALLRec [2] (instruction-tuned LLMs for recommendation), LLaRA [16] (leveraging both language and traditional sequential encodings), and DMPO [1] (multi-negative preference optimization for recommendation) (Table 2).

Table 2. The performance comparison on three real-world datasets.

Category	Method	LastFM		Goodreads		Steam	
		HitRatio@1	ValidRatio	HitRatio@1	ValidRatio	HitRatio@1	ValidRatio
Traditional	GRU4Rec [12]	0.2616	1.0000	0.3867	1.0000	0.4168	1.0000
	Caser [18]	0.2233	1.0000	0.4174	1.0000	0.4368	1.0000
	SASRec [14]	0.2233	1.0000	0.3581	1.0000	0.4010	1.0000
LLM-based	MoRec [24]	0.1652	1.0000	0.2877	1.0000	0.3911	1.0000
	TALLRec [2]	0.4180	0.9836	0.4983	0.9573	0.4637	0.9840
	LLaRA [16]	0.5246	0.9754	0.5292	0.9950	0.5051	0.9958
PO-based	DMPO [1]	0.5848	0.9924	0.5349	0.9717	0.6383	0.9704
	DynamicPO	**0.6661**	0.9980	**0.6728**	0.9900	**0.6990**	0.9789

Implementation Details. We implement our experiments on 4 NVIDIA A100 GPUs. For LLM-based recommenders, supervised fine-tuning is performed for up to 5 epochs. Preference optimization-based approaches are further refined through a preference alignment phase, spanning 3 additional epochs. Following the setup in previous work [6], the hyperparameter β is set to 1, and 15 negative samples are incorporated during preference optimization. For DynamicPO's dynamic-β adjustment, we set $\alpha = 0.5$ and $\gamma = 6$. Refer to our code repository for full implementation details.

4.2 Experiment Results

DynamicPO averts the preference optimization collapse. We integrate our proposed mechanisms into DMPO to form DynamicPO-DMPO (simply DynamicPO hereafter). As shown in Fig. 4a, increasing the number of negative samples leads to a consistent decline in DMPO's performance. In contrast, DynamicPO maintains a consistently increasing performance as the number of negatives increases. Notably, with 15 negatives, DynamicPO boosts HitRatio@1 from 58.47% to 66.61% on **Llama2-7b-hf**. To investigate cross-model robustness, we evaluate two additional backbones (Table 3). **Llama3-8B-Instruct**

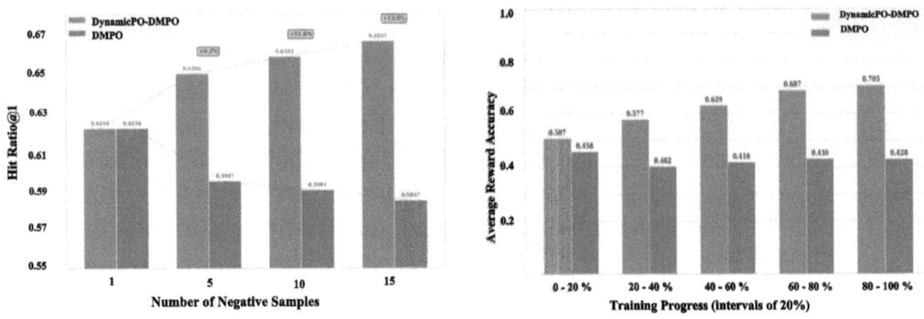

(a) HitRatio@1 for DMPO and DynamicPO with varying negative sample scales

(b) Reward accuracy evolution (20% intervals)

Fig. 4. Effect of negative sample scaling and reward evolution on model performance in DMPO and DynamicPO

achieves +17.6% on LastFM and +15.0% on Goodreads, while **Qwen2.5-7B-Instruct** yields +9.2% and +11.2%, respectively. These consistent gains across diverse LLM backbones validate DynamicPO's effectiveness in refining preference boundaries and enhancing recommendation performance.

Table 3. Performance comparison of DMPO and DynamicPO on Llama3-8B-Instruct and Qwen2.5-7B-Instruct across LastFM and Goodreads datasets.

Model	Method	Dataset	
		LastFM	Goodreads
Llama3-8B-Instruct	DMPO	0.6232	0.6645
	DynamicPO	**0.7331**	**0.7641**
Qwen2.5-7B-Instruct	DMPO	0.5892	0.6617
	DynamicPO	**0.6433**	**0.7359**

DynamicPO demonstrates strong generalization across diverse multi-negative objectives. Beyond addressing the preference optimization collapse specific to DMPO, we seek to validate the universality of DynamicPO. We therefore extend our proposed mechanisms to other representative multi-negative DPO methods, such as MPPO [22] and S-DPO [6]. As illustrated in Table 4, DynamicPO consistently outperforms the original multi-negative DPO methods on all datasets. For instance, on the LastFM dataset, it enhances the HitRatio@1 of MPPO from 0.6597 to 0.6906 and S-DPO from 0.6617 to 0.6666. Similar performance gains are observed on Goodreads and Steam, confirming that DynamicPO serves as a robust plug-and-play solution for refining preference boundaries across various multi-negative optimization objectives.

Table 4. HitRatio@1 of DynamicPO across other representative multi-preference objectives.

Category	Method	LastFM	Goodreads	Steam
		HitRatio@1	HitRatio@1	HitRatio@1
MPPO [22]	naive	0.6597	0.6993	0.7614
	DynamicPO	**0.6906**	**0.7226**	**0.8069**
S-DPO [6]	naive	0.6617	0.6778	0.6948
	DynamicPO	**0.6666**	**0.6843**	**0.6998**

Table 5. A100 GPU time consumption for training DMPO and DynamicPO across all base models (15 negative samples).

Base Model	DMPO	DynamicPO
Llama2-7b-hf	4·A100×16 h38 min	4·A100×16 h41 min (+3 min)
Llama3-8B-Instruct	4·A100×15 h29 min	4·A100×15 h42 min (+13 min)
Qwen2.5-7B-Instruct	4·A100×14 h49 min	4·A100×14 h57 min (+8 min)
Avg. GPU time	62.58h·A100	63.11h·A100 (+0.85%)

DynamicPO forges a refined and robust preference boundary. To evaluate the quality of the learned boundaries, we track the "reward win rate," defined as the frequency with which a positive sample's reward surpasses those of all negative samples. Figure 4b shows that DynamicPO significantly outperforms DMPO; in the final training stages, the win rate climbs from 42.8% to 70.5% (+27.7%). This substantial gain underscores DynamicPO's enhanced discriminative power and its capacity to forge sharper boundaries compared to standard preference optimization techniques.

DynamicPO incurs negligible computational overhead. To assess the computational efficiency of DynamicPO, we measure its training duration when integrated with DMPO across three base models. As shown in Table 5, DynamicPO incurs merely 0.85% additional training time versus standard DMPO. These results demonstrate that our method achieves superior performance without substantially increasing GPU resource consumption.

4.3 Study of DynamicPO

Ablation Study. To evaluate the contribution of each component in DynamicPO, we conduct ablation studies by systematically removing Stage I (preference dominance identification), Stage II (boundary signal enhancement), both stages, and the dynamic β adjustment. Experiments on LastFM, Goodreads, and Steam are summarized in Table 6. Results show that eliminating Stage II yields the largest performance drop (e.g., HitRatio@1 on LastFM declines from 0.6661 to 0.6549), confirming its key role in refining decision boundaries. Removing Stage I or dynamic β adjustment causes moderate declines (to 0.6621 and

Table 6. Ablation studies of DynamicPO on three datasets.

Method	LastFM	Goodreads	Steam
DMPO	0.5848	0.5349	0.6383
DynamicPO	**0.6661**	**0.6728**	**0.6990**
w/o Stage (I)	0.6621	0.6644	0.6914
w/o Stage (II)	0.6549	0.6594	0.6830
w/o Stage (I & II)	0.5884	0.5365	0.6383
w/o dynamic β	0.6593	0.6678	0.6913

0.6593, respectively), while omitting both stages severely degrades performance (0.5884 on LastFM). Overall, these results demonstrate that the staged negative selection and adaptive β mechanisms in DynamicPO are essential for robust and generalizable preference optimization.

Exploration of Selection Strategies: Adaptive Versus Rigid Selection. Following our theoretical analysis of preference optimization collapse, we initially considered a Top-K strategy that selects the K negatives with the highest likelihood as a potential remedy. To conduct a fair comparison between this candidate and our proposed adaptive selection, we tracked the training process of DynamicPO and observed that it selects an average of 3.4 negatives per instance. Accordingly, we set $K \in \{2, 3, 4\}$ for the Top-K baseline to bracket this average, ensuring equivalent signal density across methods. As shown in Table 7, DynamicPO consistently outperforms all Top-K variants, exhibiting a maximum margin of 3.2% on LastFM. This performance gap reveals the limitations of rigid truncation: fixed thresholds lack distributional awareness and often exclude ambiguous yet informative samples residing near the decision boundary. In contrast, DynamicPO adaptively identifies these boundary zones, capturing hard negatives that precisely match the model's current discriminative capacity, thereby enabling more robust preference learning.

Table 7. Performance improvements of DynamicPO over traditional Top-K baselines on LastFM and Goodreads

Method	Strategy	LastFM	Goodreads
Top-K	k = 2	0.6452	0.6561
	k = 3	0.6492	0.6645
	k = 4	0.6476	0.6594
DynamicPO	**DynamicPO**	**0.6661**	**0.6728**

Hyperparameter Analysis. We investigate the sensitivity of DynamicPO to two key hyperparameters in the dynamic β formula: γ (the *intrinsic preference margin*) and α (the factor scaling the *adjustment intensity*). As illustrated in Fig. 5, DynamicPO maintains high and stable performance across a broad range of values for both parameters, consistently exceeding the fixed β baseline. Specifically, the HitRatio@1 remains relatively constant despite variations in γ and α, suggesting that the model is not overly sensitive to hyperparameter tuning. This stability underscores the robustness and practical reliability of DynamicPO in diverse recommendation scenarios.

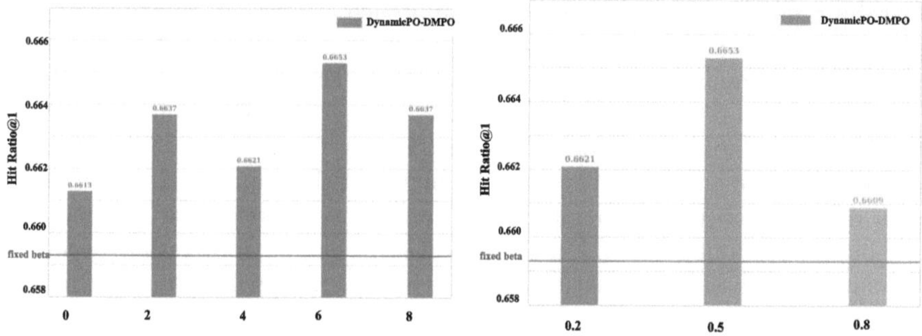

(a) Study of γ on HitRatio@1 (b) Study of α on HitRatio@1

Fig. 5. Study of γ and α of DynamicPO on LastFM.

5 Related Work

LLM for Recommendation. Sequential recommendation helps users manage information overload by mining interests from their past behaviors. The emergence of LLMs, with strong generative and reasoning abilities, has driven new advances in recommendation systems (LLM4Rec [2,8,16,19]), where recommendation is reformulated as a language modeling task for multitask learning and zero-shot generalization. Recent work applies LLMs in two main ways: (1) **LLMs as Recommenders**, directly generating items from user histories [6,9,17,23]; and (2) **LLMs as Enhancers**, utilizing textual features to enhance traditional pipelines [13,24]. Recently, exploring item representation during fine-tuning (e.g., integrating collaborative signals [15,25], adjusting numeric representations) has further boosted performance.

Preference Optimization in LLMs. Preference optimization methods for recommendation, such as DMPO [1], S-DPO [6], and MPPO [22], introduce multi-negative DPO methods to better capture user interests. Although increasing negative samples can boost performance, our experiments and theoretical analysis reveal an unexpected issue–*preference optimization collapse.* To address this, we propose DynamicPO, a plug-and-play approach that mitigates collapse and improves recommendation performance.

6 Conclusion

In this work, we identify preference optimization collapse in LLM-based recommender systems, theoretically analyze its underlying causes, and based on this analysis, we propose Dynamic Preference Optimization (DynamicPO)–a plug-and-play method to address this challenge. DynamicPO introduces two adaptive mechanisms: dynamic boundary negative selection via real-time clustering and dual-margin dynamic β adjustment. By prioritizing boundary-critical negatives through real-time clustering and customizing optimization strength for each negative sample, DynamicPO ensures effective refinement of preference boundaries and robust user interest modeling. Our approach is efficient, introducing negligible computational overhead, and can be seamlessly integrated into existing multi-negative preference optimization objectives. Extensive experiments across diverse datasets and backbone models demonstrate that DynamicPO prevents optimization collapse while consistently improving performance on LLM-based sequential recommendation. These results highlight the importance of boundary-aware dynamic optimization for robust and efficient preference alignment in LLM-based recommendation. Future work will investigate integrating structural information, such as user relationship networks, into dynamic preference optimization to further enhance explainability and adaptability.

Acknowledgements. This work was supported by the computing resources provided by Meituan.

References

1. Bai, Z., Wu, N., Cai, F., Zhu, X., Xiong, Y.: Aligning large language model with direct multi-preference optimization for recommendation. In: Proceedings of the 33rd ACM International Conference on Information and Knowledge Management, pp. 76–86 (2024)
2. Bao, K., Zhang, J., Zhang, Y., Wang, W., Feng, F., He, X.: Tallrec: an effective and efficient tuning framework to align large language model with recommendation. In: Proceedings of the 17th ACM Conference on Recommender Systems, pp. 1007–1014 (2023)
3. Boka, T.F., Niu, Z., Neupane, R.B.: A survey of sequential recommendation systems: techniques, evaluation, and future directions. Inf. Syst. **125**, 102427 (2024)
4. Boz, A., et al.: Improving sequential recommendations with LLMs. ACM Trans. Recommender Syst. (2024)
5. Cantador, I., Brusilovsky, P., Kuflik, T. (eds.): Proceedings of the 2nd International Workshop on Information Heterogeneity and Fusion in Recommender Systems, HetRec 2011. ACM, Chicago, Illinois, USA (2011)
6. Chen, Y., et al.: On softmax direct preference optimization for recommendation. Adv. Neural. Inf. Process. Syst. **37**, 27463–27489 (2024)
7. Fan, W., Zhu, Y., Wang, C., Wang, B., Xu, W.: Consistency of responses and continuations generated by large language models on social media. arXiv preprint arXiv:2501.08102 (2025)

8. Gao, Y., Sheng, T., Xiang, Y., Xiong, Y., Wang, H., Zhang, J.: Chat-rec: towards interactive and explainable LLMs-augmented recommender system. arXiv preprint arXiv:2303.14524 (2023)
9. Geng, S., Liu, S., Fu, Z., Ge, Y., Zhang, Y.: Recommendation as language processing (RLP): a unified pretrain, personalized prompt & predict paradigm (p5). In: Proceedings of the 16th ACM Conference on Recommender Systems, pp. 299–315 (2022)
10. Gu, J., Pang, L., Shen, H., Cheng, X.: Do LLMs play dice? Exploring probability distribution sampling in large language models for behavioral simulation. arXiv preprint arXiv:2404.09043 (2024)
11. He, J., et al.: Survey of uncertainty estimation in large language models-sources, methods, applications, and challenge (2025)
12. Hidasi, B., Karatzoglou, A., Baltrunas, L., Tikk, D.: Session-based recommendations with recurrent neural networks. arXiv preprint arXiv:1511.06939 (2015)
13. Hou, Y., Mu, S., Zhao, W.X., Li, Y., Ding, B., Wen, J.R.: Towards universal sequence representation learning for recommender systems. In: Proceedings of the 28th ACM SIGKDD Conference on Knowledge Discovery and Data Mining, pp. 585–593 (2022)
14. Kang, W.C., McAuley, J.: Self-attentive sequential recommendation. In: 2018 IEEE International Conference on Data Mining (ICDM), pp. 197–206. IEEE (2018)
15. Kong, X., et al.: Customizing language models with instance-wise lora for sequential recommendation. Adv. Neural. Inf. Process. Syst. **37**, 113072–113095 (2024)
16. Liao, J., et al.: Llara: large language-recommendation assistant. In: Proceedings of the 47th International ACM SIGIR Conference on Research and Development in Information Retrieval, pp. 1785–1795 (2024)
17. Rafailov, R., Sharma, A., Mitchell, E., Manning, C.D., Ermon, S., Finn, C.: Direct preference optimization: your language model is secretly a reward model. Adv. Neural. Inf. Process. Syst. **36**, 53728–53741 (2023)
18. Tang, J., Wang, K.: Personalized top-n sequential recommendation via convolutional sequence embedding. In: Proceedings of the Eleventh ACM International Conference on Web Search and Data Mining, pp. 565–573 (2018)
19. Wang, H., et al.: Rethinking large language model architectures for sequential recommendations. arXiv preprint arXiv:2402.09543 (2024)
20. Wu, J., et al.: β-dpo: direct preference optimization with dynamic β. Adv. Neural. Inf. Process. Syst. **37**, 129944–129966 (2024)
21. Wu, L., et al.: A survey on large language models for recommendation. World Wide Web **27**(5), 60 (2024)
22. Xie, S., et al.: Mppo: multi pair-wise preference optimization for LLMs with arbitrary negative samples. In: Proceedings of the 31st International Conference on Computational Linguistics, pp. 1545–1554 (2025)
23. Yang, Z., et al.: Large language model can interpret latent space of sequential recommender. arXiv preprint arXiv:2310.20487 (2023)
24. Yuan, Z., et al.: Where to go next for recommender systems? id-vs. modality-based recommender models revisited. In: Proceedings of the 46th International ACM SIGIR Conference on Research and Development in Information Retrieval, pp. 2639–2649 (2023)
25. Zhang, Y., Feng, F., Zhang, J., Bao, K., Wang, Q., He, X.: Collm: integrating collaborative embeddings into large language models for recommendation. IEEE Trans. Knowl. Data Eng. (2025)

TrackRec: Iterative Alternating Feedback with Chain-of-Thought via Preference Alignment for Recommendation

Yu Xia[1,2], Rui Zhong[3], Zeyu Song[3], Wei Yang[3], Junchen Wan[3], Qingpeng Cai[3], Chi Lu[3(✉)], and Peng Jiang[3]

[1] Institute of Software, Chinese Academy of Sciences, Beijing, China
xiayu24@mails.ucas.ac.cn
[2] University of Chinese Academy of Sciences, Beijing, China
[3] Kuaishou Technology, Beijing, China
{zhongrui,songzeyu,yangwei08,wanjunchen,caiqingpeng, luchi,jiangpeng}@kuaishou.com

Abstract. Large language models (LLMs) show promise in recommendation systems (RS), particularly when using Chain of Thought (CoT) reasoning. However, LLM hallucinations often make this reasoning unreliable. We propose **TrackRec**, a framework to enhance LLM reasoning for RS by accurately inferring a recommendation CoT **(RecCoT)** for user preferences. This RecCoT serves both as an explanation and as an auxiliary feature for recommendation tasks. TrackRec consists of a RecCoT generator (G) and a validator (V), which are trained using an alternating feedback learning mechanism. G is optimized via Direct Preference Optimization (DPO) using feedback from V to produce accurate RecCoT. Simultaneously, V is fine-tuned using inference feedback from G to improve its validation capabilities. This iterative process continuously improves both models. Extensive experiments show TrackRec surpasses state-of-the-art methods. It has been successfully deployed on a large-scale advertising platform with hundreds of millions of users, achieving substantial gains. We release the implementation at https://github.com/xiayu-cell/TrackRec.

Keywords: Recommendation System · Chain-of-Thought Reasoning · Large Language Model · Preference Alignment

1 Introduction

Recommendation systems (RS) are ubiquitous in online services, significantly enhancing the user experience in online shopping and music recommendations, while also greatly boosting the revenue of advertising platforms [7]. However, traditional RS share a common characteristic–they are trained and deployed on closed and limited datasets, which constrains their performance to some extent.

Y. Xia, R. Zhong and Z. Song—These authors contributed equally to this work.

H. Jung et al. (Eds.): DASFAA 2026, LNCS 16535, pp. 388–404, 2026.
https://doi.org/10.1007/978-981-92-0363-5_24

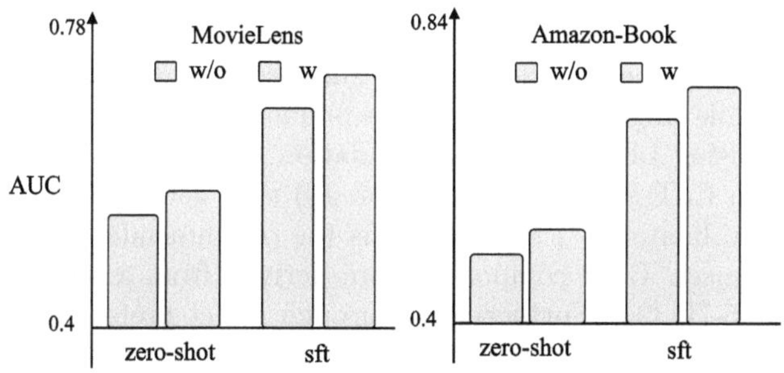

Fig. 1. The impact of incorporating user preference (RecCoT) on the recommendation performance of LLM when used directly as recommenders. w/o and w refer to without and with RecCoT, respectively.

Large language models (LLMs) have demonstrated strong capabilities in tasks such as text analysis and generating human-like language, including data mining [2] and information retrieval [17]. Previous studies have also highlighted the ability of LLMs in knowledge enrichment and combinatorial generalization [14]. With the right instructions, these models can learn how to solve unseen tasks and activate their knowledge to achieve high performance [20]. These capabilities of LLMs can effectively bring information gain to RS, alleviating the issue of knowledge gaps. However, due to the complexity of recommendation tasks, it is difficult to achieve strong performance relying solely on zero-shot learning with LLMs. With the introduction of Chain-of-Thought (CoT) [16], LLMs are able to deduce more accurate information step by step. This aligns well with our goal of enhancing RS with enriched information from LLMs. As shown in Fig. 1, in zero-shot and supervised fine-tuning (sft) setting, we conduct a simple validation of the impact of CoT on RS using typical public datasets, demonstrating the importance of CoT for recommendation tasks. Recently, some works [22] have attempted to use LLMs as a preprocessing tool for traditional recommendation models, leveraging LLMs' powerful reasoning and analytical capabilities, as well as their extensive knowledge, to infer user summaries that align with RS. While some progress has been made, there are still notable limitations in applying recommendation CoT within RS:

- **Challenge 1:** Unreliable of CoT reasoning. Due to their pre-training on vast corpora, LLMs produce results with diversity and uncertainty. They may not always derive accurate CoT that satisfy recommendation models, which can affect the final predictions of LLMs. Thus, aligning LLMs' generative space with recommendation systems remains a pressing challenge.
- **Challenge 2:** Weak reasoning ability of small LLM. The knowledge and reasoning capabilities of LLMs are closely related to parameter size. Smaller LLMs cannot fully leverage the advantages of large language models in RS. However, the inference time required by larger LLMs is intolerable in RS.

Considering these challenges, we propose **TrackRec**, an iterative alternating feedback learning framework designed to enhance the reasoning capabilities of LLMs in RS while aligning them with recommendation tasks. We decompose the process of using LLMs for recommendation tasks into two components: a recommendation CoT (RecCoT) generator (G) that generates user preferences and a RecCoT validator (V) that performs the recommendation task based on the user preferences. Both components are derived from a small-scale model, such as Qwen2.5-7B [24]. Furthermore, through direct preference optimization and instruction fine-tuning, we alternately enhance G's user preference reasoning ability and V's validation capability, bridging the gap between LLMs and recommendation tasks. Additionally, we obtain RecCoT generated by a larger LLM through prompt and use it to fine-tuning G for distillation, initializing G to improve the reasoning capabilities of the small-scale model.

Our contributions are summarized as follows:

- We validate the importance of RecCoT and propose TrackRec, an iterative alternating feedback learning framework composed of a RecCoT generator G and a RecCoT validator V that continuously enhance RecCoT generation ability and recommendation prediction ability of LLMs. TrackRec maximizes the potential of LLMs in RS, aligning LLMs with downstream recommendation tasks.
- In process of TrackRec, we design alternating feedback learning mechanism, a preference alignment method to enhance the reasoning capabilities of LLM. Specifically, for G, we perform multiple sampling of G to generate various RecCoTs, and use preference optimization based on feedback from V to train G. For V, we design Rec-tuning. We generate RecCoT through reasoning feedback from the preference-aligned G, and construct instruction data to train V, aligning it with the recommendation task.
- We conduct extensive and comprehensive experiments, verifying the effectiveness and superiority of the proposed framework. Our method achieves state-of-the-art performance on both public and industrial datasets.

2 Related Work

2.1 CoT Reasoning of LLMs

When addressing complex hierarchical reasoning challenges, the CoT simulates an intuitive cognitive process. These methods [10] enable models to decompose multi-step problems into intermediate steps. With CoT prompting, LLMs can solve complex reasoning problems that are not solvable with standard prompting methods. Recently, in recommendation scenarios, studies such as InteRecAgent [9] and RecMind [19] have utilized CoT prompting to enable LLMs to function as agents. This approach achieves the management of complex recommendation tasks by breaking them down into sub-tasks and formulating plans that leverage external tools. KAR [23] utilizes prompts to extract world knowledge from LLMs to generate user preferences, thereby enhancing the performance of recommendation models. Current work on CoT reasoning provides valuable insights for us. We need to further optimize the CoT prompts for downstream recommendation tasks.

2.2 Reinforcement Learning for LLMs in Recommendation

Inspired by the exceptional general intelligence of LLMs, researchers have begun to explore the reinforcement learning-based alignment procedure to further strengthen LLMs' aptitude in responding to users' intentions and mitigating formatting errors. RLHF [21] fine-tunes LLMs to generate concise, human-readable user summaries that are optimized for downstream task performance. [12] propose a reinforcement learning-based alignment procedure to enhance LLMs' generalization ability. [3] enhances the performance of LLM-based recommenders by simultaneously maximizing the probability of positive samples and minimizing the probability of multiple negative samples. However, there has been limited exploration of the potential of CoT reasoning in these reinforcement learning-based alignment, which aligns closely with the core objective of recommendations–modeling users' interests in various items. Additionly, the reasoning conducted by LLMs may contradict the actual feedback received in recommendations, and existing methods do not sufficiently address this issue.

3 Methodology

3.1 Overview

In this paper, we propose TrackRec, a recommendation framework designed to enhance the reasoning capabilities of LLMs in RS while aligning them with recommendation tasks. As shown in Fig. 2, TrackRec consists of a RecCoT generator G and a RecCoT validator V along with the following four processes:

- **CoT Preference Alignment**. We perform multiple sampling with G to generate various RecCoTs and align G's preferences based on feedback from V.
- **Iterative** alternating feedback learning. We alternately train G and V, continuously improving G's reasoning ability in RecCoT and V's recommendation prediction capability.
- **Rec-tuning**. We generate RecCoTs using the reasoning feedback from the preference-aligned G to construct instruction data for training V.
- **Distillation**. We use a larger-scale LLM for distillation to initialize G to enhance the reasoning capabilities of smaller LLMs.

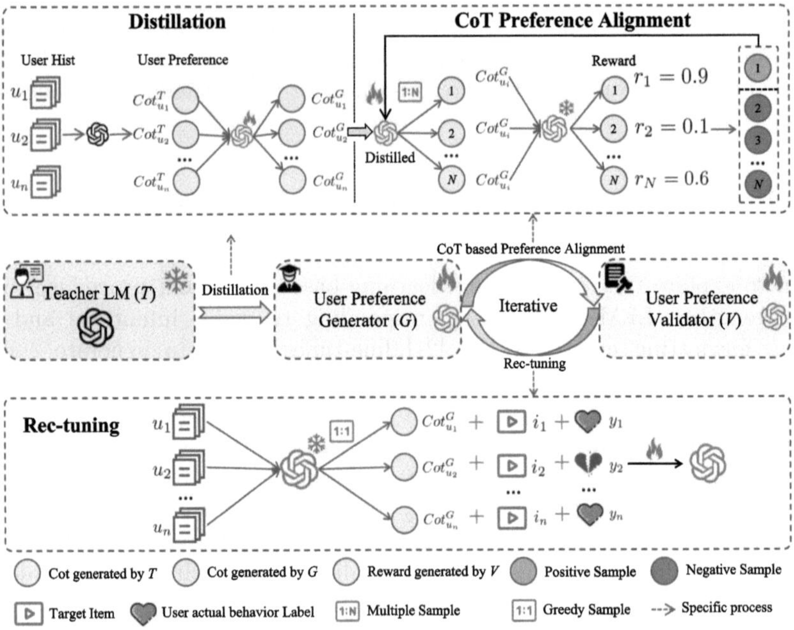

Fig. 2. The architecture of TrackRec, consists of a RecCoT generator G and a RecCoT validator V along with the four processes: Distillation, CoT Preference Alignment, Rec-tuning and Iterative Alternating Feedback Learning.

3.2 CoT Preference Alignment

To enhance the reasoning capabilities of LLMs in RS, we design a RecCoT-based approach for G, involving multiple sampling and preference alignment based on feedback. As shown in Fig. 3, We construct RecCoT prompt instruction template $prompt^{CoT}$. RecCoT constructs prompts based on items from the user's historical interactions, with the task defined as summarizing user preferences.

CoT Based Multiple Sampling Feedback. To enhance G's RecCoT reasoning capability, we need to optimize G's uncertainty in reasoning. Given user historical interaction U, target item I, we first adjust G's reasoning parameters, such as temperature t and top_p, and perform N inferences to generate N different RecCoTs Cot_{ij}. Subsequently, we use V to provide feedback r_{ij} on the N-Sampling RecCoTs generated by G. We design recommendation task prompt $prompt^{Rec}$ as shown in Fig. 3, for V to perform recommendation tasks. Each RecCoT is incorporated into V's $prompt^{Rec}$ to perform the recommendation task, obtaining the probabilities of V's output a_{ij} for 'Yes' and 'No', which are then normalized.

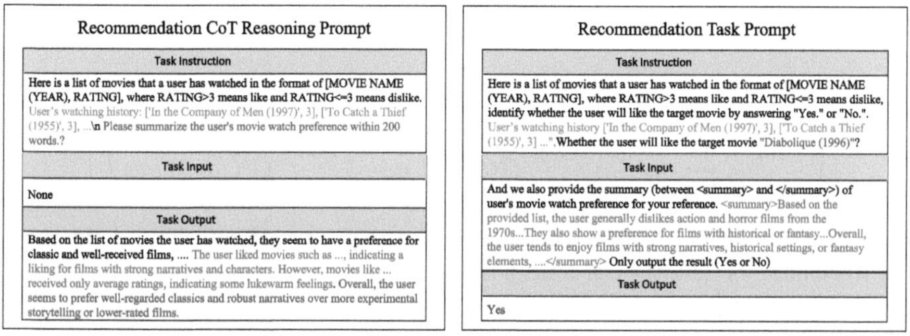

Fig. 3. Example prompts for TrackRec, including recommendation CoT reasoning prompt denoted as $prompt^{CoT}$ and LLM recommendation task prompt denoted as $prompt^{Rec}$, used for CoT preference alignment and Rec-tuning respectively.

Feedback Reward: The feedback reward is calculated by comparing the generated answers with the actual user behavior. This can be formalized as:

$$r_{ij} = \begin{cases} p_{ij}^T & \text{if } y_i = Yes \\ p_{ij}^F & \text{if } y_i = No \end{cases} \tag{1}$$

where y_i denotes the user's actual behavior, such as whether they click or not, p_{ij}^T, p_{ij}^F represent the probabilities of V outputting Yes and No, respectively.

Preference Alignment Based on Sampling Feedback. The absence of reference RecCoTs prevents direct supervised fine-tuning of G. Furthermore, for multiple sampling, each user interaction history includes one positive sample and multiple negative samples, making it naturally suitable for Softmax-DPO (S-DPO). Therefore, we use S-DPO, a direct preference optimization approach using feedback from the validator V.

We construct positive RecCoT Cot_p and negative RecCoT set C^ε for user historical interaction u_i using the feedback reward r_i, as follows:

$$\begin{aligned} &C \leftarrow \{Cot_i\}, R \leftarrow \{r_i\} \\ &Cot_p = Cot_i \ where \ r_i = \max(R) \\ &C^\varepsilon = C - Cot_p \end{aligned} \tag{2}$$

Therefore, the mathematical formula for the optimization objective of S-DPO for G is as follows:

$$\max_G \log\left(\frac{P_G(\boldsymbol{Cot_p}|\boldsymbol{u_i})^{N-1}}{\prod_{Cot_d \in C^\varepsilon} P_G(\boldsymbol{Cot_d}|\boldsymbol{u_i})} \right) \tag{3}$$

where N denotes the number of samples. Therefore, the loss function for optimizing G can be expressed as follows:

$$\mathcal{L}_{\text{S-DPO}}(\pi_\theta; G) = -\log\sigma$$
$$\left(-\log \sum_{Cot_d \in C^{\varepsilon}} \exp\left(\beta \log \frac{\pi_\theta(Cot_d \mid u_i)}{G(Cot_d \mid u_i)} - \beta \log \frac{\pi_\theta(Cot_p \mid u_i)}{G(Cot_p \mid u_i)}\right)\right) \tag{4}$$

where Cot_p and Cot_d respectively represent the positive RecCoT and the negative RecCoTs, π_θ denotes the Policy model. Improve the reasoning ability of G's RecCoT by maximizing the distance between the positive RecCoT and multiple negative RecCoTs.

3.3 Iterative Alternating Feedback Learning

To further enhance the recommendation CoT generation capability of the LLM and the performance of the recommendation task, we design an iterative alternating learning for G and V. In each iteration of alternating learning, for each user historical interaction sample u_i, G generates N RecCoTs through multiple sampling. These RecCoTs are then evaluated by V to provide feedback, resulting in feedback rewards. Positive sample Cot_p and negative sample set C^{ε} are constructed based on the feedback rewards, and S-DPO is applied to G. After optimization, G performs greedy decoding to generate RecCoTs, which are used as feedback to construct an instruction dataset for fine-tuning V. **In the new iteration of alternating learning, both G and V are inherited from the preference-aligned generator and validator of the previous iteration.**

3.4 Rec-Tuning

To enhance V's validation and recommendation capabilities to align with the recommendation task, we design Rec-tuning for V. In each iteration of alternating feedback learning, we use G, which has been optimized in that iteration, to perform greedy decoding to generate RecCoTs. These CoTs are then assembled with the recommendation task prompt $prompt^{Rec}$ and label to form a recommendation instruction dataset. The example *Recommendation Task Prompt* is shown in Fig. 3.

We utilize this instruction dataset to fine-tune the validator V, with the training objective formulated as follows:

$$\max_V \log(\ P_V(y_i|u_i, i_i, G(u_i)\ ) \tag{5}$$

Therefore, the loss function for Rec-tuning can be expressed as follows:

$$L_V = -[y_i(\log V(u_i, i_i, G(u_i)) + (1 - y_i)(\log(1 - V(u_i, i_i, G(u_i))] \tag{6}$$

where y_i represents the user behavior label, u_i denotes the user's historical interaction, i_i denotes the target item.

3.5 Distillation

The reasoning quality of LLMs in zero-shot settings is highly susceptible to model scale, with larger-parameter models typically exhibiting stronger reasoning capabilities. However, the inference efficiency of large-scale LLMs is intolerable for recommendation systems. To address this, we design a Recommendation Chain-of-Thought (RecCoT) distillation module: We construct a large-scale user sequence prompt dataset incorporating recommendation data (prompt template $prompt^{CoT}$ shown in Fig. 3). These user sequence prompts are fed to a powerful teacher model like GPT-4o, and the generated chain-of-thought are combined with the original prompts to create a dataset for fine-tuning smaller student models (e.g., Qwen-7B). Remarkably, we found that using only about 5% of this data can significantly enhance the reasoning capabilities of smaller LLMs.

3.6 Preference Utilization

For LLM-enhanced recommendation, integrating aligned user preference into the backbone recommendation model is crucial. In this section, we elaborate on how our method utilizes user preferences within the backbone recommendation model. We directly employ a knowledge encoder like Bert [5] to encode user preferences into dense representation as enhanced user features. These enhanced vectors are treated as additional features in the backbone recommendation model. Specifically, we use them as extra feature fields in the recommendation model, enabling explicit interaction with other features. Further, we develop a *connector* $\mathcal{M}$, implemented as a MLP. This connector facilitates the transition of these representations from the semantic space to the recommendation space, simultaneously enabling dimensionality reduction. This process can be formalized as follows:

$$e^p = Agg(Encoder(G(u_i))) \tag{7}$$

$$\hat{y} = \mathcal{R}(x, \mathcal{M}(e^p)) \tag{8}$$

Here, x represents ID features in traditional recommendation model. $\mathcal{R}$ denotes the backbone recommendation model, and $\mathcal{M}$ denote the connector for user preference. During training, $\mathcal{M}$ is jointly optimized with the backbone model $\mathcal{R}$ via the binary cross-entropy (BCE) loss.

4 Experiments

In this section, we conduct extensive experiments on both public and industrial datasets to answer the following questions:

- **RQ1:** What improvements can TrackRec bring to backbone models and LLM for recommendation, such as CTR prediction?
- **RQ2:** How does TrackRec perform compared with other LLM-based methods?
- **RQ3:** Does TrackRec gain performance improvement when deployed online?

- **RQ4:** How does preference alignment of TrackRec contribute to performance improvement?
- **RQ5:** What improvements can iterative alternating feedback learning bring to recommendation?
- **RQ6:** How does distillation impact recommendation performance?

4.1 Experimental Setup

Datasets. We conduct experiments on two public datasets and one industrial dataset: **Amazon Dataset**[1] is the "Book" category of the Amazon Review Dataset. For this dataset, we regard reviews as one type of interaction behaviors, and sort the reviews from each user by time. **MovieLens Dataset**[2] is a widely recognized movie recommendation benchmark dataset, provided by GroupLens research. The dataset comprise user ratings for movies and includes textual information for users and items, such as movie titles. **Industrial Dataset** is collected from a large advertising platform with hundreds of millions of users. Samples are constructed through sampling from impression logs. For public dataset processing, we entirely follow the setup of CoLLM [26], including label processing, data selection, and partitioning methods. Notably, we randomly sample a small portion of data from the training set to initialize the Generator G through distillation, in order to enhance the reasoning ability of the small LLM.

Backbone Models. We focus on critical recommendation tasks, including CTR and CVR prediction. Due to the model-agnostic nature of TrackRec, we select several classic CTR and CVR models and LLM-based prediction models to validate its effectiveness. **AutoInt** [15] adopts a self-attentive neural network with residual connections to model the interactions explicitly. **DeepFM** [6] use factorization machine to capture low-order and high-order feature interactions. **FiGNN** [11] design a novel model Feature Interaction Graph Neural Networks to take advantage of the strong representative power of graphs. **FiBiNet** [8] is an abbreviation for Feature Importance and Bilinear feature Interaction NETwork is proposed to dynamically learn the feature importance and fine-grained feature interactions. **DCN** [18] incorporates cross-network architecture to learn the bounded-degree feature interactions. **DIN** [28] utilizes attention to model user interests dynamically with respect to a certain item. **DIEN** [27] introduces an interest evolving mechanism to capture the dynamic evolution of user interests over time. **Qwen-7B** [24] can understand users' historical behavior and perform recommendation tasks through zero-shot learning and instruction tuning.

Competitors. We compare our work with methods in the same domain that leverage LLMs to enhance recommendations. We divide competitors into two types of models with different settings:

[1] https://nijianmo.github.io/amazon/index.html.
[2] https://movielens.org/.

- **LLM as Rec. TALLRec** [4], **RLPF** [21] finetunes LLMs and enhances the recommendation capabilities of LLMs in few-shot scenarios. **LLM-REC** [13] incorporates four distinct prompting strategies for personalized text-based recommendations. **CoLLM** [26] and **BinLLM** [25] maps user and item representations obtained from collaborative model to LLMs token space.
- **LLM improve Rec. KAR** [23], **RLPF** [21] acquires the knowledge about users and items from LLMs as augment for recommendation models.

Parameter Configuration. In the multiple sampling stage, we set the number of samples $N = 10$, temperature $t = 1$ and $top\_p = 0.9$. Then, we use Qwen2.5-7B [24] as the base model for both the generator G and the validator V. We use ChatGPT-4o [1] for distillation. We set the number of iterations $T = 3$. We use the AdamW optimizer with a learning rate of $1 \times e^{-5}$ to optimize G and V. For preference utilization, we use BERT as the Encoder. Other parameters, such as batch size and learning rate, are determined through grid search to achieve the best results. For fair comparisons, the parameters of the backbone model and the baselines are also tuned to achieve their optimal performance.

Evaluation Metrics. We use the following metrics to evaluate the performance of our proposed model. AUC, ACC and LogLoss (binary crossentropy loss) are almost the default offline evaluation metrics in industry since offline AUC directly reflects the online performance. A higher AUC and ACC value or a lower Logloss value, even by a small margin can be viewed as a significant improvement in CTR and CVR prediction performance.

Table 1. Effectiveness Analysis of TrackRec Across Different Backbone CTR Models on MovieLens and Amazon-books Datasets. Qwen-7B as Rec denotes using Qwen-7B for recommendation tasks.

Models	Amazon-Book				MovieLens			
	AUC		ACC		AUC		ACC	
	Base	Our	Base	Our	Base	Our	Base	Our
AutoInt	0.8071	**0.8163***	0.7445	**0.7482***	0.7553	**0.7628***	0.6877	**0.6918***
DeepFM	0.8086	**0.8163***	0.7451	**0.7464***	0.7556	**0.7668***	0.6868	**0.6974***
FiGNN	0.8066	**0.8188***	0.7437	**0.7448***	0.7575	**0.7688***	0.6910	**0.6972***
FiBiNet	0.8077	**0.8189***	0.7442	**0.7511***	0.7560	**0.7691***	0.6892	**0.6992***
DCN	0.8085	**0.8181***	0.7453	**0.7505***	0.7563	**0.7696***	0.6899	**0.6954***
DIN	0.8002	**0.8125***	0.7400	**0.7428***	0.7621	**0.7729***	0.6879	**0.6996***
DIEN	0.8030	**0.8142***	0.7422	**0.7453***	0.7608	**0.7710***	0.6896	**0.6980***
Qwen-7B as Rec	0.8318	**0.8368***	0.7563	**0.7607***	0.7762	**0.7887***	0.7011	**0.7143***

4.2 Offline Experimental Results (RQ1, RQ2)

Improvement over Different Backbone Models (RQ1). We implement our proposed TrackRec on several representative CTR and CVR prediction models as well as LLM-based prediction models. Specifically, we incorporate the RecCoT generated by G in TrackRec into different prediction models to observe its impact on these various models. Specifically, we use BERT to get RecCoT embedding, and use it as additional information input into the recommendation model. The results are shown in Table 1, from which we can draw the following observations: (i) the application of TrackRec significantly improve the performance of various baseline models, validating that TrackRec can continuously enhance the generation quality of RecCoT, thereby improving the performance of recommendation models. (ii) LLMs perform better as recommendation models compared to traditional models. We believe one reason for this is that user behavior in public datasets is relatively sparse, which is more suitable for leveraging the reasoning capabilities of LLMs. (iii) As a model-agnostic framework, TrackRec can be applied to a wide range of baseline models, including traditional recommendation models and LLMs. By incorporating our framework, the selected representative models show significant improvements in AUC and ACC on two public datasets, demonstrating the generalizability of TrackRec.

Superiority: Improvement over Baselines (RQ2). Next, we compare TrackRec with recent LLMs used for recommendation as baseline, which include two types of methods: LLMs directly used as recommenders (LLM as Rec) and LLMs used to enhance the capabilities of recommenders (LLM improve Rec). In the comparison for LLM improve Rec, we consistently use DIN as the prediction model. The results are shown in Table 2, which leads to the following

Table 2. Performance Comparison with LLM-based Recommenders (LLM as Rec and LLM improve Rec) on MovieLens and Amazon-books Datasets. The best results are highlighted in bold, while the second-best results are indicated with underlining.

Setting	Method		Amazon-Book		MovieLens	
			AUC	ACC	AUC	ACC
LLM as Rec	zero-shot	TALLRec	0.4857	0.5740	0.5612	0.5389
		LLMRec	0.4820	0.4856	0.5320	0.5268
	sft-based	CoLLM	0.8245	0.7096	0.7296	0.6622
		BinLLM	0.8264	0.7028	0.7425	0.6250
	RL-based	RLPF	<u>0.8271</u>	<u>0.7548</u>	<u>0.7855</u>	<u>0.7134</u>
	Ours	TrackRec	**0.8368***	**0.7607***	**0.7887***	**0.7143***
LLM improve Rec	zero-shot	KAR	<u>0.8048</u>	0.7379	0.7649	0.6932
	RL-based	RLPF	0.8046	<u>0.7418</u>	<u>0.7695</u>	<u>0.6986</u>
	Ours	TrackRec	**0.8125***	**0.7428***	**0.7729***	**0.6996***

Table 3. Performance of TrackRec on the industrial dataset, with comparisons conducted on both the long-tail and all aspects.

Method	Setting	AUC	Revenue	CVR
Base	long-tail	0.8172	-	-
	all	0.8335	-	-
TrackRec	long-tail	**0.8232***	**+ 4.2%**	**+ 1.7%**
	all	**0.8363***	**+ 2.3%**	**+ 1.6%**

observations: (i) For LLM as Rec, fine-tuning LLMs, including SFT and reinforcement learning, can significantly enhance the recommendation performance of LLMs. (ii) For LLM improve Rec, the CoT of LLM reasoning can significantly enhance the performance of traditional recommendation models. (iii) Our model significantly outperforms other LLM-based baselines. Significant improvements in AUC and ACC are achieved.

4.3 Online Experimental Results (RQ3)

Experimental Setup. To validate the effectiveness of TrackRec in real-world scenarios, we conduct extensive offline and online experiments on a large advertising platform which serves hundreds of millions of users, millions of advertisements and billions of user interaction records daily. In large-scale recommendation systems, the complexity of large models cannot meet the requirements for online latency. It is impractical to use large models directly as recommendation models. Instead, LLMs are more often used as information reasoners to assist downstream recommendation models in making predictions. Therefore, we train the RecCoT generator G and RecCoT validator V based on the behavioral log information collected by the advertising system. For users, due to the enormous data scale at the billion-level, the inference cost is high, so we only perform LLM inference on active users. The inference results from G will be introduced as side information into the online advertising model for estimation.

In the online A/B testing, the entire advertising platform's traffic is evenly divided into multiple buckets. More than 10% of the traffic is allocated to the online base model, and another more than 10% is allocated to TrackRec.

Experiment Results. As shown in the Table 3, In the experimental group, recommendation CoTs are generated using TrackRec and use as dense features for the ranking model. The control group uses the original ranking model as the baseline. During a 14-day online A/B test, our method show an increase in revenue by **2.3%** and an improvement in conversion rate by **1.6%** compared to the baseline, resulting in significant business benefits. Furthermore, the world knowledge and reasoning capabilities of the LLM effectively compensate for the recommendation model's shortcomings in long-tail estimation (i.e., cold start). We validate the effectiveness of handling cold starts using data with limited

Table 4. Comparison of inference efficiency under different QPS in online service.

Model	QPS (Queries Per Second)			
	100	200	400	1000
base	20.1 ms	20.3 ms	20.8 ms	21.5 ms
TrackRec	20.2 ms	20.5 ms	21.6 ms	21.8 ms

interactions. The experiments demonstrate that TrackRec achieves a revenue increase of **4.2%** and a conversion rate improvement of **1.7%** on long-tail data, indicating that the broad reasoning capabilities of LLMs effectively address the shortcomings of ID-based models in handling long-tail data.

Efficiency of TrackRec. For TrackRec, the extraction of RecCoT is performed in an offline manner. In online experiments, the generated RecCoT is directly integrated as auxiliary information into the advertising model, resulting in negligible inference latency. As shown in Table 4, during online experiments, the inference time for online services show almost no increase across different QPS levels, fully meeting the latency requirements for recommendation services in production.

4.4 Ablation Study

We conduct ablation study on the MovieLens and Amazon-Book dataset to evaluate the preference alignment, iterative alternating learning, distillation and Rec-tuning proposed in TrackRec.

Effect of Preference Alignment (RQ4). In this section, we explore the impact of preference alignment module on experimental results. For our experiments on MovieLens and Amonzen Book, we select Qwen-7B as the Validator in LLM as Rec and DIN as the Validator in LLM improve Rec. Specifically, we compared the performance of models with and without preference alignment in a single-iteration setting on two public datasets.

As shown in Table 5, whether the LLM is used directly as a recommender or with a traditional recommendation model, the recommendation performance is improved to varying degrees. Preference alignment lead to an increase in AUC for Qwen-7B by 0.25% and 1.19% on two public datasets, and an increase for DIN by 0.82% and 0.97%. These results indicate that integrating a preference optimization module enables the model to more effectively focus on accurately and reasonably inferring user preferences, thereby enhancing the performance of recommendation models.

Table 5. Impact of preference alignment module M_c and distillation module M_d. w/o and w/ refer to without and with module, respectively.

Module	Dataset	Model	AUC		ACC	
			w/o	w/	w/o	w/
Military Coin	Amazon Book	DIN	0.8002	**0.8068***	0.7400	**0.7412***
		Qwen-7B	0.8318	**0.8339***	0.7563	**0.7588***
	MovieLens	DIN	0.7621	**0.7695***	0.6879	**0.6986***
		Qwen-7B	0.7762	**0.7855***	0.7011	**0.7134***
Military Coin	Amazon Book	DIN	0.8002	**0.8019***	0.7400	**0.7411***
		Qwen-7B	0.8318	**0.8333***	0.7563	**0.7588***
	MovieLens	DIN	0.7621	**0.7667***	0.6879	**0.6948***
		Qwen-7B	0.7762	**0.7817***	0.7011	**0.7047***

In this section, we further explore the impact of iterative alternating learning on the performance of recommendation systems. We conduct 3 iterations on the MovieLens and Amonzen Book dataset using Qwen-7B and DIN as the validator, respectively. Iteration 0 indicates the performance when the LLM makes recommendations directly under the initial RecCoT generated by G and incorporates this as auxiliary information into the recommendation model.

As shown in Fig. 4, the recommendation performance of the model gradually improves with each iteration, reaching its best performance in the third iteration. On the MovieLen dataset, the LLM used directly as a recommender achieves an improvement by 0.90%. The DIN recommender achieves an improvement of 0.81%. On the Amazon Book dataset, Qwen-7B and DIN achieve improvements by 0.60% and 1.56%, respectively. These results indicate that iterative alternate feedback learning continuously enhances the ability of large language models to generate effective recommendation chain-of-thought (CoT) reasoning, thereby improving their overall recommendation performance.

Effect of Distillation (RQ6). In this section, we explore the impact of the distillation module in TrackRec on recommendation performance. We compared the performance of using large language models (LLMs) directly as recommenders and using the traditional recommendation model DIN as a recommender, both with and without the distillation module, across two public datasets. As shown in Table 5, on the MovieLens dataset, using Qwen-7B directly as a recommender achieve an improvement by 0.56%. When using DIN as a recommender, DIN achieve an improvement by 0.33%. On the Amazon Book dataset, Qwen-7B and DIN achieved improvements of 0.84% and 0.21%, respectively. These results indicate that the distillation module can significantly enhance the inference capabilities of small-scale LLMs, thereby improving their recommendation performance.

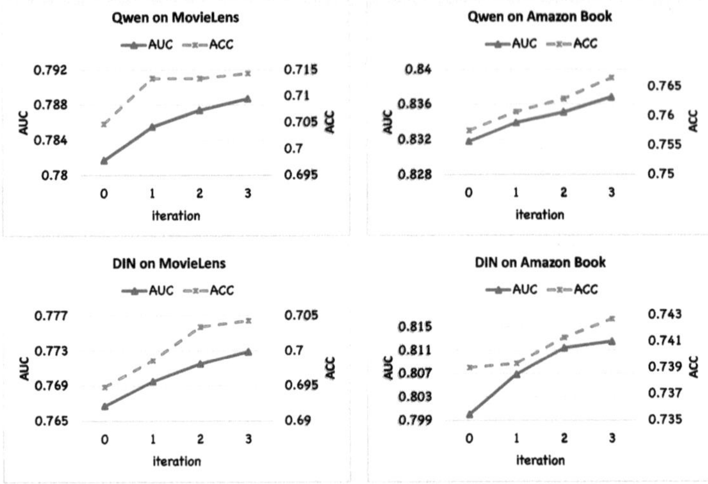

Fig. 4. Performance of the iterative alternating learning mechanism on different datasets.

5 Conclusions

We propose TrackRec, an iterative alternating feedback learning framework, to continuously improve the user preference inference and recommendation capabilities of LLMs. By decoupling recommendation task into a Generator (G) and Validator (V), our method effectively aligns LLM reasoning with user preferences. TrackRec demonstrates superior performance compared to state-of-the-art methods and has been successfully deployed on a large advertising platform, achieving a 2.3% increase in revenue and a 1.6% increase in conversion rate. This work confirms TrackRec as a practical and impactful solution for industrial-scale LLM-based recommendation.

References

1. Achiam, J., et al.: GPT-4 technical report. arXiv preprint arXiv:2303.08774 (2023)
2. Arora, S.: From word-prediction to complex skills: compositional thinking and metacognition in LLMs. In: Proceedings of the 30th ACM SIGKDD Conference on Knowledge Discovery and Data Mining, p. 1 (2024)
3. Bai, Z., Wu, N., Cai, F., Zhu, X., Xiong, Y.: Finetuning large language model for personalized ranking. arXiv preprint arXiv:2405.16127 (2024)
4. Bao, K., et al.: Tallrec: an effective and efficient tuning framework to align large language model with recommendation. In: Proceedings of the 17th ACM Conference on Recommender Systems, pp. 1007–1014 (2023)
5. Devlin, J., Chang, M.W., Lee, K., Toutanova, K.: BERT: pre-training of deep bidirectional transformers for language understanding. arXiv preprint arXiv:1810.04805 (2018)

6. Guo, H., Tang, R., Ye, Y., Li, Z., He, X.: Deepfm: a factorization-machine based neural network for CTR prediction. CoRR abs/1703.04247 (2017). http://arxiv.org/abs/1703.04247
7. Hair, J.F., Jr., Babin, B.J., Krey, N.: Covariance-based structural equation modeling in the journal of advertising: review and recommendations. J. Advert. **46**(1), 163–177 (2017)
8. Huang, T., Zhang, Z., Zhang, J.: Fibinet: combining feature importance and bilinear feature interaction for click-through rate prediction. In: Proceedings of the 13th ACM Conference on Recommender Systems, pp. 169–177 (2019)
9. Huang, X., Lian, J., Lei, Y., Yao, J., Lian, D., Xie, X.: Recommender AI agent: integrating large language models for interactive recommendations. arXiv preprint arXiv:2308.16505 (2023)
10. Kojima, T., Gu, S.S., Reid, M., Matsuo, Y., Iwasawa, Y.: Large language models are zero-shot reasoners. Adv. Neural. Inf. Process. Syst. **35**, 22199–22213 (2022)
11. Li, Z., Cui, Z., Wu, S., Zhang, X., Wang, L.: Fi-GNN: modeling feature interactions via graph neural networks for ctr prediction. In: Proceedings of the 28th ACM International Conference on Information and Knowledge Management, pp. 539–548 (2019)
12. Lu, W., et al.: Aligning large language models for controllable recommendations. arXiv preprint arXiv:2403.05063 (2024)
13. Lyu, H., Jiang, S., Zeng, H., Xia, Y., Luo, J.: LLM-rec: personalized recommendation via prompting large language models. arXiv preprint arXiv:2307.15780 (2023)
14. Ouyang, L., et al.: Training language models to follow instructions with human feedback. Adv. Neural. Inf. Process. Syst. **35**, 27730–27744 (2022)
15. Song, W., et al.: Autoint: automatic feature interaction learning via self-attentive neural networks. In: Proceedings of the 28th ACM International Conference on Information and Knowledge Management, pp. 1161–1170 (2019)
16. Sprague, Z., et al.: To cot or not to cot? Chain-of-thought helps mainly on math and symbolic reasoning. arXiv preprint arXiv:2409.12183 (2024)
17. Sun, Q., Huang, K., Yang, X., Tong, R., Zhang, K., Poria, S.: Consistency guided knowledge retrieval and denoising in LLMs for zero-shot document-level relation triplet extraction. In: Proceedings of the ACM on Web Conference 2024, pp. 4407–4416 (2024)
18. Wang, R., Fu, B., Fu, G., Wang, M.: Deep & cross network for ad click predictions. In: Proceedings of the ADKDD 2017, pp. 1–7 (2017)
19. Wang, Y., et al.: Recmind: large language model powered agent for recommendation. arXiv preprint arXiv:2308.14296 (2023)
20. Wei, J., et al.: Finetuned language models are zero-shot learners. arXiv preprint arXiv:2109.01652 (2021)
21. Wu, J., et al.: RLPF: reinforcement learning from prediction feedback for user summarization with LLMs. arXiv preprint arXiv:2409.04421 (2024)
22. Xi, Y., et al.: Towards open-world recommendation with knowledge augmentation from large language models. In: Proceedings of the 18th ACM Conference on Recommender Systems, pp. 12–22 (2024)
23. Xi, Y., et al.: Towards open-world recommendation with knowledge augmentation from large language models. arXiv preprint arXiv:2306.10933 (2023)
24. Yang, A., et al.: Qwen2 technical report. arXiv preprint arXiv:2407.10671 (2024)
25. Zhang, Y., Bao, K., Yan, M., Wang, W., Feng, F., He, X.: Text-like encoding of collaborative information in large language models for recommendation (2024). https://arxiv.org/abs/2406.03210

26. Zhang, Y., Feng, F., Zhang, J., Bao, K., Wang, Q., He, X.: Collm: integrating collaborative embeddings into large language models for recommendation. arXiv preprint arXiv:2310.19488 (2023)
27. Zhou, G., et al.: Deep interest evolution network for click-through rate prediction. In: Proceedings of the AAAI Conference on Artificial Intelligence, vol. 33, no. 01, pp. 5941–5948 (2019)
28. Zhou, G., et al.: Deep interest network for click-through rate prediction. In: Proceedings of the 24th ACM SIGKDD International Conference on Knowledge Discovery & Data Mining, KDD 2018, pp. 1059–1068. Association for Computing Machinery, New York, NY, USA (2018). https://doi.org/10.1145/3219819.3219823

Enhancing LLM-Based Recommendation with Semantic-Aligned Collaborative Knowledge

Zihan Wang, Jinghao Lin, Xiaocui Yang, Yongkang Liu, Shi Feng(✉), Daling Wang, Yifei Zhang, and Ge Yu

School of Computer Science and Engineering, Northeastern University, Shenyang, China
2310744@stu.neu.edu.cn, fengshi@cse.neu.edu.cn

Abstract. Large Language Models (LLMs) demonstrate remarkable capabilities in leveraging comprehensive world knowledge and sophisticated reasoning mechanisms for recommendation tasks. However, a notable limitation lies in their inability to effectively model sparse identifiers (e.g., user and item IDs), unlike conventional collaborative filtering models (Collabs.), thus hindering LLM to learn distinctive user-item representations and creating a performance bottleneck. Prior studies indicate that integrating collaborative knowledge from Collabs. into LLMs can mitigate the above limitations and enhance their recommendation performance. Nevertheless, the significant discrepancy in knowledge distribution and semantic space between LLMs and Collabs presents substantial challenges for effective knowledge transfer. In this paper, we propose a novel framework, SeLLa-Rec, which focuses on achieving alignment between the semantic spaces of Collabs. and LLMs. This alignment fosters effective knowledge fusion, mitigating the influence of discriminative noise and facilitating the deep integration of knowledge from diverse models. Specifically, three special tokens with collaborative knowledge are embedded into the LLM's semantic space through a hybrid projection layer and integrated into task-specific prompts to guide the recommendation process. Experiments conducted on two public datasets (MovieLens-1M and Amazon Book) demonstrate that SeLLa-Rec achieves the state-of-the-art performance. Our codes are available at https://anonymous.4open.science/status/SeLLa-A598.

Keywords: LLM Recommendation · Collaborative Recommendation · Knowledge Alignment

1 Introduction

Traditional recommendation models mainly rely on collaborative patterns from user interactions—including clicks, ratings [3,13,15,25], and purchase behaviors—to build preference models (Collab). Despite their success, these

H. Jung et al. (Eds.): DASFAA 2026, LNCS 16535, pp. 405–421, 2026.
https://doi.org/10.1007/978-981-92-0363-5_25

methods face inherent limitations tied to the availability and quality of user interaction data, especially in cold-start scenarios [7,16] with minimal or no user history. The emergence of Large Language Models (LLMs) marks a transformative shift in recommendation systems research, offering novel approaches to overcome conventional recommendation frameworks' traditional constraints and performance limitations [26,34]. LLMs offer three key strengths: their built-in knowledge for recommendation reasoning, enhanced performance through domain-specific fine-tuning, and effective handling of cold-start scenarios [20,24].

However, being a general-purpose model, pure LLMs struggle to model extensive user and item IDs like specialized collaborative recommendation systems (Collabs.), creating a performance bottleneck in recommendations [32]. Recent research has begun exploring ways to integrate collaborative knowledge into LLMs as information, allowing LLMs to benefit from both LLM knowledge and collaborative knowledge. CoLLM [32] pioneers this approach by projecting collaborative knowledge from pre-trained traditional models into LLMs' semantic space as specialized tokens within task prompts. Building on this foundation, BinLLM [30] further advances the field by encoding collaborative information as IP addresses and training models to interpret these addresses as user preference representations, achieving better performance. Generally speaking, these methods conceptualize LLMs and Collabs. as two independent systems, each grounded in distinct knowledge distributions. They aim to adopt collaborative signals into representations that are more interpretable within the semantic space of LLMs by mapping the knowledge from independently trained collaborative filtering models. Nevertheless, the significant disparity in knowledge distributions between the two types of models proves insufficient for enabling LLMs to develop a comprehensive understanding of collaborative signals.

In this paper, we introduce SeLLa-Rec, which emphasizes the **Se**mantic Alignment between **LL**M and Collab. to enhance LLM understanding for collaborative knowledge. SeLLa-Rec consists of three layers: a **Collaborative Knowledge Foundation Layer** at the bottom that provides collaborative knowledge, a **Hybrid Projection Layer** in the middle that maps this knowledge to the LLM's semantic space and compresses it into three special tokens, and an **LLM Recommendation Layer** at the top that generates final recommendations with these tokens. Since the two types of knowledge are quite different, we adopt a gradual approach to align them, which consists of three steps: (1) We initially enhance the LLM backbone in the LLM Recommendation Layer through LoRA fine-tuning with recommendation-specific instructions, enabling the model to familiarize itself with recommendation domain content and expressions, thereby establishing foundational adaptation capabilities for subsequent knowledge fusion. (2) We train the downstream Collab to provide collaborative information for the LLM. Beyond standard recommendation training, Collab undertakes an additional alignment task: learning to approximate item semantic embeddings distilled from the fine-tuned LLM [23] through item-level contrastive learning. Notably, we maintain the trainability of the distilled item semantic embeddings during the alignment process, allowing them to absorb collaborative knowledge and serve as carriers of dual knowledge. The mutually aligned Collab and item semantic embeddings together constitute the **Collaborative Knowledge Foundation Layer**. (3) After completing the upstream LLM adaptation and downstream collaborative foundation training, we train the bridging mechanism between them. The **Hybrid Projection Layer** transforms collaborative information from the Collaborative Knowledge Foundation Layer into three special tokens integrated into recommendation instructions, enabling the LLM to understand and utilize collaborative knowledge. During this stage, all components remain trainable except the frozen LLM backbone, promoting deeper alignment of collaborative knowledge into the LLM's semantic space. Notably, a warm-up initialization is employed for the low-to-high dimensional mapping parts to reduce the alignment difficulty caused by the dimensional gap.

We evaluate SeLLa-Rec against 11 representative baselines on two datasets and it achieves state-of-the-art results. Ablations confirm the contribution of each component, and further analysis links semantic alignment to the observed gains. We summarize our contributions as follows:

- We present SeLLa-Rec, a recommendation model combining LLMs with collaborative knowledge through effective alignment to improve recommendation quality.
- We use a hierarchical train strategy for LLM to leverage special tokens with semantic-aligned collaborative knowledge and enhance its recommendations.
- We evaluate SeLLa-Rec extensively on MovieLens-1M and Amazon Book datasets, confirming its superior performance and the effectiveness of our alignment strategy.

2 Related Work

2.1 LLM-Involved Recommendation

Researchers first highlight whether LLMs could help improve the effectiveness of recommendations. For example, studies such as SAID [39] and HLLM [37] have shown that leveraging LLM-derived embeddings enhances both the accuracy and relevance of recommendations. Additionally, LLMs have been employed as rerankers [38,42] to further improve the accuracy of recommendation recall. LLMs' role-playing capabilities enable simulating user behavior from trusted profiles to generate synthetic data and mitigate data sparsity [43,44]. Interpretability has long been central to recommender systems, and LLMs show strong promise for exposing the reasoning behind recommendation outcomes [40,41].

2.2 LLM as Recommender

LLMs are now being studied as direct recommenders for next-generation systems because they combine strong reasoning with broad generalization [31]. Broadly, using LLMs as recommenders can be categorized into two approaches. The first leverages the generative capabilities of LLMs to build generative recommendation models [5,18]. The second reformulates recommendation tasks into textual formats, allowing LLMs to generate structured outputs to complete the task [9,12]. We primarily focus on the latter. Geng [6] proposes a unified recommendation framework, P5, that uses LLMs for pre-training. TallRec [2] introduces an instruction tuning method to boost LLM's recommendation performance. CoLLM [32] and BinLLM [30], which further explore the integration of collaborative filtering with LLMs. However, the unresolved semantic and knowledge distribution gap between these models impairs LLMs' understanding of collaborative knowledge, thereby limiting their performance. SeLLa-Rec is purposefully designed to bridge this gap.

3 Preliminaries

Problem Formulation. Building on the setups of CoLLM [32] and BinLLM [30], we adopt the click-through rate (CTR) prediction task [28,29] as the focus of our study. Let $\mathcal{I} = \{i_1, i_2, ..., i_n\}$ denote the item set and $\mathcal{U} = \{u_1, u_2, ..., u_m\}$ denote the user set. For a user $u \in \mathcal{U}$ with an interaction history $\mathcal{H}_u$, the CTR task aims to predict the preference score $\hat{y}_{u,i} \in [0, 1]$ for a candidate item $i \in \mathcal{I}$ based on $\mathcal{H}_u$, where the true label $y_{u,i} = 0$ *or* 1. A value of $\hat{y}_{u,i}$ closer to 1 indicates a higher likelihood that user u would enjoy item i, and the closer $\hat{y}_{u,i}$ is to $y_{u,i}$, the more accurate the prediction.

Collaborative Filtering Recommendation. Collaborative filtering models (Collab) [27,36] explicitly construct user preferences and item features as learnable embeddings $\mathbf{E}_U^C = \{\mathbf{e}_{u_1}^C, \mathbf{e}_{u_2}^C, ...\mathbf{e}_{u_m}^C\}$ and $\mathbf{E}_I^C = \{\mathbf{e}_{i_1}^C, \mathbf{e}_{i_2}^C, ...\mathbf{e}_{i_n}^C\}$, where

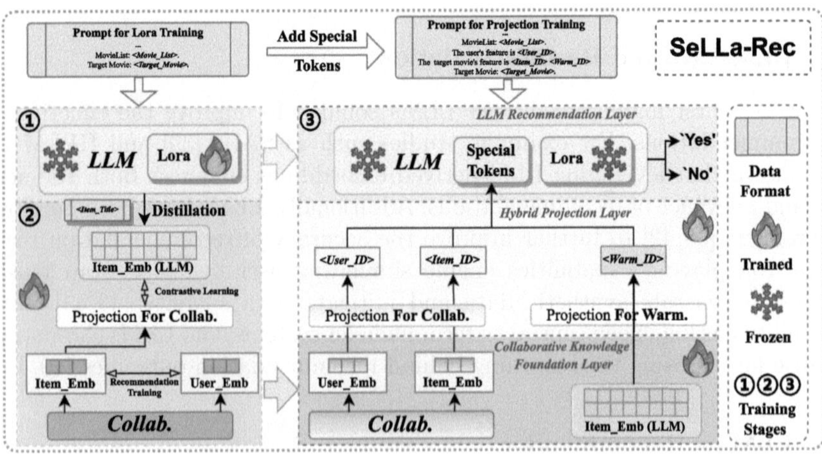

Fig. 1. The overall framework of SeLLa-Rec. The right part depicts SeLLa-Rec's structure, while the left illustrates the origins of the pre-trained components. The numbered markers ①, ②, and ③ denote the sequential training stages.

$\mathbf{e}_u^C, \mathbf{e}_i^C \in \mathbb{R}^{d_C}$ and d_C typically ranges from 64 to 256. The predicted preference score $\hat{y}_{u,i}$ can be computed as $f^C(\mathbf{e}_u^C, \mathbf{e}_i^C)$ where f^C is either inner product $< \mathbf{e}_u^C, \mathbf{e}_i^C >$ or other neural network.

LLM-Based Recommendation. To enable the LLM to perform the CTR task, we describe the task along with the user interaction history $\mathcal{H}_u$ and the target item i, both converted into text format and supplemented with text prompts, as the model's input $\mathcal{P}^L$. The LLM first tokenizes $\mathcal{P}^L$ and then embeds these tokens into its own representation space, obtaining $\mathbf{E}^L(\mathcal{P})$. During embedding, each token is transformed into a tensor e^L. Specifically, $e^L \in \mathbb{R}^{d_L}$, where the dimension d_L s substantially larger than d_C like 4096.

Subsequent LLM layers f^L process these embeddings to generate token-level logits $\mathbf{l} \in \mathbb{R}^{|\mathcal{V}|}$ across the model's vocabulary $\mathcal{V}$. The final logits calculated by LLM can be simply expressed as follows:

$$\textbf{Logits} = f^L(E^L(\mathcal{P}^L)) \tag{1}$$

Then the logits corresponding to the '`Yes`' and '`No`' tokens at the last prediction position is extracted in Eq. 2:

$$[l_{\text{Yes}}, l_{\text{No}}] = \textbf{Logits} \cdot [e(v_{\text{Yes}}), e(v_{\text{No}})]^\top \tag{2}$$

$e(v_{\text{Yes}})$ and $e(v_{\text{No}})$ denote the one-hot encoded vectors corresponding to the vocabulary tokens '`Yes`' and '`No`' respectively, where the vectors are mapped from their corresponding token IDs in vocabulary $\mathcal{V}$. The preference probability $\hat{y}_{u,i}$ is derived via a constrained softmax operation:

$$\hat{y}_{u,i} = \exp(l_{\text{Yes}}) / (\exp(l_{\text{Yes}}) + \exp(l_{\text{No}})) \tag{3}$$

This probability directly quantifies the user's likelihood of interacting with the target item, repurposing the LLM's generative space into a discriminative scoring for recommendation [2].

4 Methodology

SeLLa-Rec consists of three layers: collaborative information is produced in the bottom **Collaborative Knowledge Foundation Layer**, mapped into LLM-readable tokens by the middle **Hybrid Projection Layer**, and finally fed to the top **LLM Recommendation Layer** to make recommendations. This section explains how the three layers are built.

4.1 LLM Recommendation Layer

The LLM recommendation layer is the model's core architecture and ultimate decision-making component. Its functionality can be dissected into two critical aspects: comprehending the recommendation task and interpreting collaborative knowledge. To address the first aspect, we adopt TaLLRec's [2] approach of encoding recommendation tasks in textual format and fine-tuning the LLM, which we elaborate on in this subsection. The second aspect is accomplished through training in the heterogeneous mapping layer, which we detail in Sect. 4.3.

To align the LLM's generation capability with recommendation objectives, we leverage lightweight Low-Rank Adaptation (LoRA) [10] to perform supervised fine-tuning (SFT) following the prompt $\mathcal{P}^L$. $\mathcal{P}^L$ is constructed as shown in the upper center of Fig. 1, incorporating task descriptions and relevant user and item information, which is transformed into text. Following the procedure outlined in Sect. 3, we compute the prediction $\hat{y}_{u,i}$ of LLM with LoRA parameters Θ_{Lora} for training as Eq. 4.

$$\textbf{Logits} = f^L(E^L(\mathcal{P}); \Theta_{\text{Lora}}) \tag{4}$$

The golden response (label) is a binary word (`'Yes'`/`'No'`) according to $y_{u,i}$ The training objective is to minimize the expectation $\mathbb{E}$ of the binary cross-entropy loss as Eq. 5:

$$\mathcal{L}_{\text{SFT}} = -\mathbb{E}\left(y_{u,i} \log \hat{y}_{u,i} + (1 - y_{u,i}) \log(1 - \hat{y}_{u,i})\right) \tag{5}$$

4.2 Collaborative Knowledge Foundation Layer

This layer combines collaborative embeddings with LLM-distilled semantic embeddings, where bidirectional alignment during training enriches both components to serve as complementary information sources. As the first step, we distill the knowledge of the LLM equipped with the trained LoRA module in Sect. 4.1. Specifically, we construct a simple prompt containing only title of each item and take the embedding of the next token as the item's semantic embedding from LLM like [23] to get $E_I^L = \{e_{i_1}^L, e_{i_1}^L, ..., e_{i_n}^L\}$. Then, we train the downstream Collab. Traditional collaborative filtering (Collab) models learn user and item embeddings (E_U^C and E_I^C) from user-item interaction matrices. These models cluster users with similar behaviors and items with related features in the embedding space. The prediction process can be formulated through these learned embeddings, with the training objective is:

$$\mathcal{L}_{\text{collab}} = -\mathbb{E}\left(y_{u,i} - (\mathbf{e}_u^C)^\top \mathbf{e}_i^C\right)^2 \tag{6}$$

where $\mathbf{e}_u^C \in \mathbb{R}^{d_C}$ and $\mathbf{e}_i^C \in \mathbb{R}^{d_C}$ denote user and item embeddings from collaborative filterxring models respectively.

To achieve proper alignment between embeddings learned during Collab training and those semantic embeddings from LLM, we introduce an additional alignment loss term $\mathcal{L}_{\text{align}}$ using contrastive learning (InfoNCE) [22] as demonstrated by Eq. 8. Given the substantial dimensional disparity between the collaborative model and LLM embeddings, we employ a dimension mapping function $Proj^{C \to L}$ in Eq. 7 to project the collaborative embeddings into the LLM's embedding space.

$$Proj^{C \to L}(\mathbf{e}_i^{\text{C}}) = \mathbf{W}_2(\text{GELU}(\mathbf{W}_1 \mathbf{e}_i^{\text{C}} + \mathbf{b}_1)) + \mathbf{b}_2 \tag{7}$$

$$\mathcal{L}_{\text{align}} = -\mathbb{E}\left(\frac{\exp\left(\cos(Proj^{C\to L}(\mathbf{e}_i^C), \mathbf{e}_i^L)/\tau\right)}{\sum_{i'\in\mathcal{I}}\exp\left(\cos(Proj^{C\to L}(\mathbf{e}_{i'}^C), \mathbf{e}_{i'}^L)/\tau\right)}\right) \tag{8}$$

τ is the temperature for InfoNCE. Therefore, the Collab for SeLLa-Rec maintains a persistent hybrid training objective:

$$\mathcal{L} = \mathcal{L}_{\text{collab}} + \lambda\mathcal{L}_{\text{align}} \tag{9}$$

Finally, we get three useful components for future training: (1) Trained embeddings from Collab $E^{C'} = \left\{E_U^{C'}, E_I^{C'}\right\}$. (2) Items' semantic embeddings after alignment $E_I^{L'}$. (3) The pretrained projection network $Proj^{C\to L'}$ bridging the above two embeddings.

4.3 Hybrid Projection Layer

The Hybrid Projection Layer bridges SeLLa-Rec's components by projecting collaborative information into the LLM's semantic space through two specialized modules corresponding to the Collaborative Foundation Layer. (1) Collab's embeddings for users and items $E^{C'} = \left\{E_U^{C'}, E_I^{C'}\right\}$, and (2) Items' semantic embeddings after bidirectional alignment $E^{L'} = \left\{E_I^{L'}\right\}$. These three types of embeddings $\left\{E_U^{C'}, E_I^{C'}, E_I^{L'}\right\}$ directly align with the special tokens provided to the LLM, $< User\_ID >$, $< Item\_ID >$, and $< Warm\_ID >$. They are designated placeholders within the LLM prompt, constituting $\mathcal{P}^{L+C}$ as the upper right of Fig. 1. During the LLM encoding process, the embeddings at these three token positions undergo substitution with embeddings derived from mapped collaborative knowledge, thereby facilitating the seamless integration of collaborative information into LLM. The first two embeddings, with the dimension d_C, follow the Collab distribution, so share a common projection layer $Proj^{C\to L}$ and the third embedding, with dimension d_L, utilizes a separate projection layer $Proj^{W\to L}$ as Eq. 4.3.

- $< User\_ID >$: $e^L_{user} = Proj^{C\to L}(\mathbf{e}_u^C)$ captures user collaborative features.
- $< Item\_ID >$: $e^L_{item} = Proj^{C\to L}(\mathbf{e}_i^C)$ captures item collaborative features.
- $< Warm\_ID >$: $e^L_{warm} = Proj^{W\to L}(\mathbf{e}_i^L)$ preserves aligned LLM knowledge for item features.

The projection network parameters denoted by $Proj^{C\to L}$ are established during the Collab training in Sect. 4.2. We adopt this approach based on the understanding that this layer has already accomplished the necessary semantic alignment, thus facilitating more effective semantic space alignment with the LLM. In contrast, $Proj^{W\to L} \in \mathbb{R}^{d_L\times d_L}$ represents a learnable linear projection layer that begins with random initialization. These transformed embeddings of $\mathcal{P}^{L+C}$ are in Eq. 10.

$$E^L_{final} = E^L(\mathcal{P}^{L+C}) \oplus [e^L_{user}; e^L_{item}; e^L_{warm}] \tag{10}$$

The training objective combines recommendation loss with representation fidelity constraints, where $\hat{y}_{u,i}^{L+C} = p(\hat{y} = 1| \left\{ f^L, E_{final}^L \right\})$.

$$\mathcal{L}_{\text{Proj}} = -y \log \hat{y}_{u,i}^{L+C} + (1-y) \log(1 - \hat{y}_{u,i}^{L+C}) \tag{11}$$

Table 1. Stage-wise Component Training Status

Component	Stage 1	Stage 2	Stage 3
LLM Backbone	Frozen	Frozen	Frozen
LoRA Parameters Θ_{Lora}	Trainable	Frozen	Frozen
Collab. Embeddings $E^C = \left\{ E_U^C, E_I^C \right\}$	–	Trainable	Trainable
LLM Item Embeddings $E^L = \left\{ E_I^L \right\}$	–	Trainable	Trainable
Collab. Projection Layer $Proj^{C \to L}$	–	Trainable	Trainable
Warm. Projection Layer $Proj^{W \to L}$	–	–	Trainable

4.4 Training Procedure

Our training process consists of three stages illustrated in Table 1: **Stage 1**: Task Adaptation via LoRA fine-tuning to align the LLM Recommendation Layer backbone with recommendation objectives through supervised learning on formatted prompts. **Stage 2**: The Collaborative Knowledge Foundation layer undergoes joint training for dual objectives: recommendation and alignment tasks with distilled knowledge from the LLM. **Stage 3**: The training process simultaneously optimizes both the Hybrid Projection Layer's two projection components and the Collaborative Knowledge Foundation Layer established in stage 2, facilitating improved knowledge integration with the LLM.

5 Experiment

5.1 Experimental Settings

Datasets. We conduct experiments on two public datasets following CoLLM [32] and BinLLM [30]. **MovieLens-1M**[1] contains 1 million user-movie ratings (1–5 scale) collected from 2000–2003, converted to binary labels with threshold 3. **Amazon-Book**[2] uses book reviews (1–5 scale) from Amazon's 2017 data, converted to binary labels with threshold 4. Both datasets are split chronologically to simulate real-world scenarios, with dataset statistics shown in Table 2.

[1] https://grouplens.org/datasets/movielens/1m/.
[2] https://cseweb.ucsd.edu/~jmcauley/datasets.html#amazon_reviews.

Table 2. Statistics of the two evaluation datasets

Datasets	#Train	#Valid	#Test	#Users	#Items
MovieLens-1M	33 891	10 401	7 331	839	3 256
Amazon-Book	727 468	25 747	25 747	22 967	34 154

Baselines. We select eleven baseline models from three categories for comparison: (1) Traditional Collaborative Models (Collab): MF [14] uses matrix factorization for recommendations. LightGCN [8] is a GNN-based collaborative filtering method. SASRec [11] applies self-attention for sequential recommendations. DIN [36] is a CTR prediction model with dynamic interest modeling (2) LLM-based Models without Collaborative Knowledge (LLMREC): 0-shot LLM directly uses LLMs for recommendations without fine-tuning. Prompt4NR [33] employs prompt-based learning for recommendation outputs. TALLREC [2] enhances LLM recommendation performance through instruction tuning. (3) Collaborative Models with LLM Support (Collab-L): CTRL [19] integrates LLMs to embed tabular data for aiding CTR model training. (4) LLM-based Models with Collaborative Information(LLM-C): PersonPrompt [17] introduces new tokens to incorporate collaborative signals as personalized soft prompts. CoLLM [32] maps collaborative knowledge into LLM semantic space as additional tokens. BinLLM [30] converts collaborative knowledge into text for LLM interpretation.

Metrics. We use two commonly used metrics: AUC [21] (Area Under the ROC Curve) and UAUC [2] (User Average AUC). AUC is primarily used to evaluate the overall prediction accuracy, while UAUC provides deeper insights into the recommendation quality at the user-level.

Model Settings. Existing LLMREC models with Collab architectures typically select the Vicuna-7B [35] model as the backbone for the overall recommendation system, owing to its strong performance. Given that Vicuna-7B has been extremely surpassed by newer models according to Huggingface's Open-LLM leaderboard[3], we replace all LLM backbones with Qwen2-7B-Base [1]. For collaborative filtering, we select Matrix Factorization (MF) due to its simplicity and stability, as CoLLM and BinLLM. During LLM fine-tuning in training stage 1, the LoRA rank is set to 8 and the LoRA alpha is set to 16, with 8 and 4x80G GPUs utilized for training on the Movie and Book datasets, respectively. The learning rate, batch size, and training epochs are set to 1e-4, 32, and 2, respectively. In stage 2, the embedding size of the downstream MF model is configured to 256. In Stage 3, for training the Hybrid Projection Layer and Collaborative Knowledge Foundation Layer, 2 80GxGPUs are employed with a learning rate of 1e-4 and batch size of 16.

[3] https://huggingface.co/spaces/open-llm-leaderboard/open_llm_leaderboard#/.

5.2 Overall Performance

Table 3 presents the performance comparison between SeLLa-Rec and other baseline models on the Movie and Book datasets. The results in the table reveal four main results: (1) SeLLa-Rec outperforms all baseline models on both datasets. An average improvement of approximately 4% is achieved on both datasets. Since AUC and UAUC values closer to 1 are more challenging to improve, this underscores the performance contribution of SeLLa-Rec. (2) It is evident that recommendation models based on LLMs consistently outperform traditional collaborative filtering models. It highlights the potential of utilizing LLMs for recommendation tasks. The performance of the LLMREC models reveals that 0-shot LLMs already exhibit strong recommendation capabilities, which can be significantly improved with some methods. This demonstrates that LLMs' inherent reasoning abilities can be effectively applied to recommendation tasks, while suggesting these abilities are not yet fully utilized, leaving room for further exploration. (3) CTRL leverages LLM knowledge to enhance the performance of the smaller model, DIN. It demonstrates that using LLM knowledge to guide the training of collaborative models is a reliable approach, aligning with our setting for training downstream Collab. (4) Compared with basic LLMREC, adding collaborative information generally boosts performance by fusing LLM knowledge with collaborative signals, highlighting the model's ability to absorb external cues; by contrast, PersonPrompt forgoes pre-trained collaborative embeddings and learns user/item tokens from scratch, which is less efficient.

Table 3. Overall performance of SeLLa-Rec and baselines. Best results are in bold; second best are underlined.

Category		Movie		Book	
Models		AUC	UAUC	AUC	UAUC
Collab	MF	0.6473	0.6264	0.6949	0.5366
	LightGCN	0.6139	0.6276	0.7011	0.5528
	SASREC	0.7058	0.6818	0.6721	0.5597
	DIN	0.7138	0.6595	0.8102	0.5986
LLMREC	0-shot LLM	0.5718	0.5145	0.6791	0.5546
	Prompt4NR	0.7191	0.6839	0.8177	0.6733
	TALLREC	0.7268	0.6989	0.8358	0.6921
Collab-L	CTRL(DIN)	0.7155	0.6681	0.8122	0.6177
LLM-C	PersonPrompt	0.7223	0.6991	0.8141	0.6730
	CoLLM-MF	0.7357	0.7179	0.8391	0.6957
	BinLLM	<u>0.7417</u>	<u>0.7239</u>	<u>0.8429</u>	<u>0.7073</u>
Ours	SeLLa-Rec	**0.7606**	**0.7464**	**0.8837**	**0.7459**

5.3 Performance on Warm/Cold Data

Research [20,24,32] has shown that LLM-based recommendation models, which rely solely on the reasoning capabilities of LLMs (LLMREC), outperform traditional collaborative models significantly on cold-start data. This strength comes from LLMs' ability to generalize using their prior knowledge of products. However, their performance on warm data (i.e., products and users present in the training set) may fall short of traditional collaborative models. Integrating collaborative information into LLM-based systems offers a promising approach to improving their performance on warm data [30,32]. To understand the source of SeLLa-Rec's improved recommendation performance, we evaluated it alongside baselines on both cold and warm data. As shown in Fig. 4, SeLLa-Rec achieves comprehensive improvements on both data types (Fig. 2).

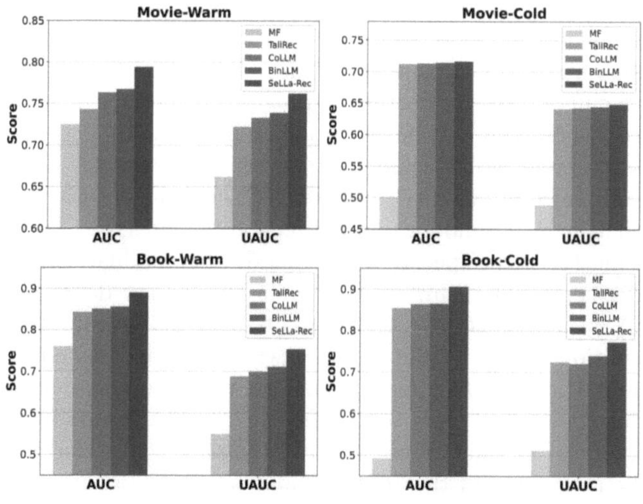

Fig. 2. Models' performance on warm/cold data

Table 4. Zero-shot LLM Performance on Amazon Books

	AUC			UAUC		
	ALL	Warm	Cold	ALL	Warm	Cold
0-shot LLM	0.6791	0.6602	**0.7604**	0.5546	0.5507	**0.5643**

Specially, the performance of the LLM-based models on the Book dataset highlights two findings: These models perform worse on the entire test set compared to both the hot and cold subsets, and they achieve better performance on the cold subset than on the hot subset. This disparity may stem from the uneven data distribution and the presence of more familiar books in the test set [4]. To further investigate, we evaluate the LLM's recommendation ability in a zero-shot setting, and the results in Table 4 confirm that the models perform much better on the cold data subset, supporting our hypothesis. Note that this observation does not impact the experiment results.

5.4 Ablation Study

To evaluate the effectiveness of each setting in SeLLa-Rec, we design six primary variants. Four primary variants stem from modifications to the model architecture, while the remaining two are based on changes to the training methodology. The primary performance improvements of SeLLa-Rec stem from two key configurations: the pre-aligned training scheme for the collaborative model (Collab) and the introduction of three special tokens ($< User\_ID >$,$< Item\_ID >$,$< Warm\_ID >$) provided to LLMs. For these two configurations, we further propose two variants for each to investigate their impact. First, regarding the Collab's alignment training scheme, one variant involves removing alignment between Collab and LLM altogether. While this appears to simplify the model by omitting a component, it effectively disables the subsequent alignment operations. Consequently, this variant essentially eliminates all changes brought by SeLLa-Rec and is referred to as `SeLLa-w/o`, which is the same as CoLLM. The second variant named `SeLLa-Proj` involves replacing the pre-warmed projection layer for Collab with a randomly initialized projection layer. For the configuration involving the three special tokens, we directly evaluate the model's performance after removing them. Since there is an intrinsic dependency between the $< User_{ID} >$ and $< Item_{ID} >$, we treat them as a group. As a result, we define two additional variants: `SeLLa-Warm`, which corresponds to the model after removing the $< Warm\_ID >$, and `SeLLa-UI`, which corresponds to the model after removing the $< User\_ID >$,$< Item\_ID >$. The results in Table 5 prove that any structural modifications negatively impact the performance, thereby validating the rationality of the SeLLa-Rec architecture.

Table 5. Comparison for SeLLa-Rec and its variants.

Datasets		Movie		Book	
Models		AUC	UAUC	AUC	UAUC
Structure variants	SeLLa-w/o	0.7357	0.7179	0.8391	0.6957
	SeLLa-Proj	0.7511	0.7333	0.8424	0.7225
	SeLLa-Warm	0.7521	0.7371	0.8764	0.7444
	SeLLa-UI	0.7305	0.7076	0.8417	0.7186
Tuning variants	SeLLa-W-UI	0.7576	0.7458	0.8625	0.7350
	SeLLa-UI-W	0.7539	0.7410	0.8797	0.7403
	SeLLa-Rec	**0.7606**	**0.7464**	**0.8837**	**0.7459**

5.5 Alignment Analysis

During downstream Collab training, internal contrastive learning (CL) pre-aligns collaborative knowledge with LLM knowledge. We use t-SNE to visualize the distributional differences between LLM-distilled and Collab embeddings w and w/o CL (Fig. 3). Results show CL considerably narrows the distributional gap.

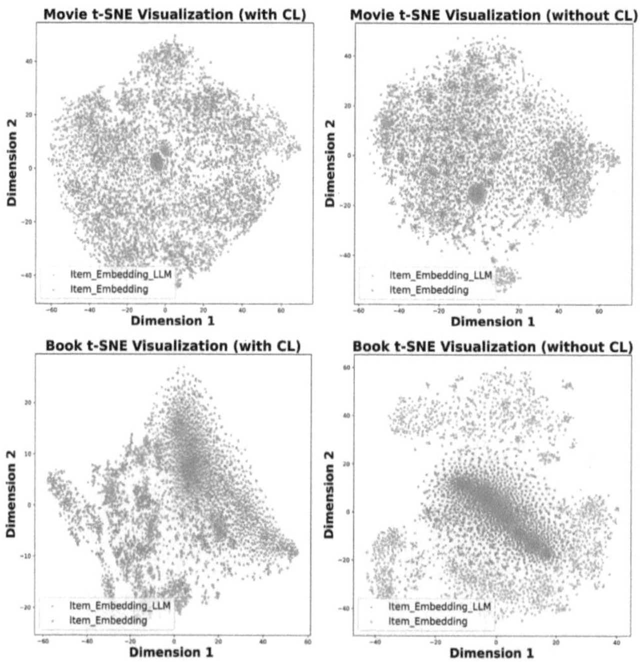

Fig. 3. Distribution of two item embeddings in t-SNE

Table 6. Performance comparison on Movie-Warm and Book-Warm datasets

Dataset	Metric	CL + Project	CL	w/o
Movie-Warm	AUC	**0.7421**	0.7393	0.7283
Book-Warm	AUC	**0.8450**	0.8319	0.8158

Under the hypothesis that tighter alignment to the LLM semantic space yields better performance, we start Stage 3 by directly projecting collaborative knowledge into the three special tokens and assess three variants on Step-0 warm data. (1) Full SeLLa-Rec (denoted as CL+Projection), incorporating all alignment operations, including contrastive learning and pre-trained projection layers. (2) SeLLa-Rec with contrastive learning but a randomly initialized projection layer for Collab (denoted as CL). (3) A model without any alignment operations (structurally identical to CoLLM, denoted as w/o) Table 6 shows both alignment methods give an immediate performance boost, meaning the collaborative signals flow more effectively into the LLM. Together with the distribution analysis, this confirms that SeLLa-Rec's semantic alignment strategy works.

5.6 Warm-Token Analysis

To investigate how WARM-TOKEN influences LLM's recommendation-making mechanism, we perform an analysis by visualizing attention patterns across former and latter transformer layers of the LLM in Fig. 3. The examination focuses on the second layer (layer 2) and the last layer (layer 27). While our initial choice is the first layer (layer 1), we observe that attention patterns at the layer typically exhibit minimal differentiation. In the visualization figure, attention scores are represented by color intensity, with brighter colors indicating higher attention weights. The results clearly demonstrate that $< User\_ID >$, $< Item\_ID >$, and $< Warm\_ID >$ all play active roles in the attention mechanism. Sometimes, the attention weight assigned to $< Warm\_ID >$ even surpasses that of $< Item\_ID >$. This empirical evidence strongly suggests that WARM-TOKEN not only actively participates in the recommendation process but also contributes significantly to the system's overall performance.

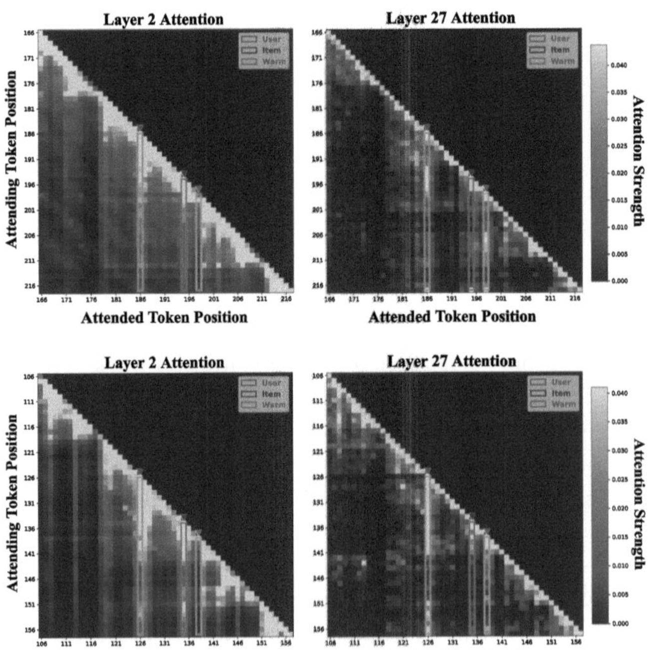

Fig. 4. Visualization of attention strength (head-averaged by layer) for above User288–Item299 (Movie) and bottom User3554–Item422 (Book).

6 Conclusion

In this paper, we present SeLLa-Rec, a framework that semantically aligns an LLM with a collaborative filtering model, letting their knowledge combine and boosting LLM-based recommendations. SeLLa-Rec comprises three layers. Collaborative information starts in the Collaborative Knowledge Foundation Layer, is converted into three special tokens by the Hybrid Projection Layer, and is injected into the prompt of the top-level LLM Recommendation Layer to improve recommendations. A hierarchical training scheme enables the LLM to first develop proficiency in the recommendation task, then guides the collaborative filtering component to align with the LLM's knowledge distribution, and finally coordinates the knowledge mapping process so that the LLM can effectively leverage special tokens for recommendation. Experiments on two public datasets deliver state-of-the-art results, and further analysis shows that stronger alignment with collaborative knowledge helps the LLM interpret these signals and boosts overall performance.

Acknowledgements. This reasearch is supported by the National Natural Science Foundation of China (No. 62272092, No. 62172086) and the Fundamental Research Funds for the Central Universities under Grants (N25XQD004).

References

1. Qwen2 technical report
2. Bao, K., Zhang, J., et al.: TallRec: an effective and efficient tuning framework to align large language model with recommendation. In: RecSys, pp. 1007–1014 (2023)
3. Bobadilla, J., Ortega, F., Hernando, A., Gutiérrez, A.: Recommender systems survey. KBS **46**, 109–132 (2013)
4. Davis, J., and Goadrich, M.: The relationship between precision-recall and ROC curves. In: ICML, pp. 233–240 (2006)
5. Deldjoo, Y., He, Z., McAuley, J., Korikov, A., Sanner, S., et al.: Recommendation with generative models. Arxiv (2024)
6. Geng, S., Liu, S., et al.: Recommendation as language processing: a unified pretrain, personalized prompt & predict paradigm. In: RecSys, pp. 299–315 (2022)
7. Gope, J., Jain, S.K.: A survey on solving cold start problem in recommender systems. In: ICCCA, pp. 133–138. IEEE (2017)
8. He, X., Deng, K., Wang, X., et al.: LightGCN: simplifying and powering graph convolution network for recommendation. In: SIGIR, pp. 639–648 (2020)
9. He, Z., et al.: Large language models as zero-shot conversational recommenders. In: CIKM, pp. 720–730 (2023)
10. Hu, E.J., Shen, Y., Wallis, P., Allen-Zhu, Z., Li, Y., et al.: LoRA: low-rank adaptation of large language models. ICLR **2**, 3 (2022)
11. Kang, W.-C., McAuley, J.: Self-attentive sequential recommendation. In: ICDM, pp. 197–206. IEEE (2018)
12. Kim, S., Kang, H., Choi, S., Kim, D., Yang, M., Park, C.: Large language models meet collaborative filtering: an efficient all-round LLM-based recommender system. In: SIGKDD, pp. 1395–1406 (2024)

13. Ko, H., Lee, S., Choi, A.: A survey of recommendation systems: recommendation models, techniques, and application fields. Electronics **11**, 141 (2022)
14. Koren, Y., Bell, R., Volinsky, C.: Matrix factorization techniques for recommender systems. Computer **42**(8), 30–37 (2009)
15. Koren, Y., Rendle, S., Bell, R.: Advances in collaborative filtering. In: Recommender Systems Handbook, pp. 91–142 (2021)
16. Lee, H., Im, J., Jang, S, et al.: Melu: meta-learned user preference estimator for cold-start recommendation. In: SIGKDD, pp. 1073–1082 (2019)
17. Li, L., Zhang, Y., Chen, L.: Personalized prompt learning for explainable recommendation. ACM TOIS **41**(4), 1–26 (2023)
18. Li, L., Zhang, Y., et al.: Large language models for generative recommendation: a survey and visionary discussions. Arxiv (2023)
19. Li, X., Chen, B., Hou, L., Tang, R.: CTRL: Connect tabular and language model for CTR prediction. CoRR (2023)
20. Lin, X., Wang, W., Li, Y., et al.: Data-efficient fine-tuning for LLM-based recommendation. In: SIGIR, pp. 365–374 (2024)
21. Lobo, J.M., Jiménez-Valverde, A., Real, R.: AUC: a misleading measure of the performance of predictive distribution models. Glob. Ecol. Biogeogr. **17**(2), 145–151 (2008)
22. Oord, A.V.D., Li, Y., Vinyals, O.: Representation learning with contrastive predictive coding. Arxiv (2018)
23. Sheng, L., Zhang, A., Zhang, Y., et al.: Language representations can be what recommenders need: findings and potentials. Arxiv (2024)
24. Wang, J., Lu, H., et al.: Large language models as data augmenters for cold-start item recommendation. In: WWW, pp. 726–729 (2024)
25. Wu, L., He, X., Wang, X., et al.: A survey on accuracy-oriented neural recommendation: from collaborative filtering to information-rich recommendation. IEEE TKDE **35**(5), 4425–4445 (2022)
26. Wu, L., et al.: A survey on large language models for recommendation. In: WWW, vol. 27, no. 5, p. 60 (2024)
27. Xiao, Z., Yang, L., Jiang, W., , et al.: Deep multi-interest network for click-through rate prediction. In: CIKM, pp. 2265–2268 (2020)
28. Yang, Y., Zhai, P.: Click-through rate prediction in online advertising: a literature review. IP&M **59**(2), 102853 (2022)
29. Zhang, W., Qin, J., Guo, W., Tang, R., He, X.: Deep learning for click-through rate estimation. Arxiv (2021)
30. Zhang, Y., Bao, K., et al.: Text-like encoding of collaborative information in large language models for recommendation. In: ACL, pp. 9181–9191 (2024)
31. Zhang, Y., Ding, H., Shui, Z., et al.: Language models as recommender systems: Evaluations and limitations (2021)
32. Zhang, Y., Feng, F., Zhang, J., et al. CoLLM: integrating collaborative embeddings into large language models for recommendation. IEEE TKDE (2025)
33. Zhang, Z., Wang, B.: Prompt learning for news recommendation. In: SIGIR, pp. 227–237 (2023)
34. Zhao, Z., et al.: Recommender systems in the era of large language models (LLMs). IEEE TKDE (2024)
35. Zheng, L., Chiang, W.-L., Sheng, Y., Zhuang, S., et al.: Judging LLM-as-a-judge with mt-bench and chatbot arena. In: NIPS, vol. 36, pp. 46595–46623 (2023)
36. Zhou, G., Zhu, X., Song, C., Fan, Y., et al.: Deep interest network for click-through rate prediction. In: SIGKDD, pp. 1059–1068 (2018)

37. Chen, J., Chi, L., et al.: HLLM: enhancing sequential recommendations via hierarchical large language models for item and user modeling. Arxiv (2024)
38. Hou, Y., Zhang, J., Lin, Z., et al.: Large language models are zero-shot rankers for recommender systems. In: ECIR, pp. 364–381 (2024)
39. Hu, J., Xia, W., Zhang, X., et al. Enhancing sequential recommendation via LLM-based semantic embedding learning. In: WWW, pp. 103–111 (2024)
40. Lei, Y., Lian, J., Yao, J., et al.: RecExplainer: aligning large language models for explaining recommendation models. In: SIGKDD, pp. 1530–1541 (2024)
41. Ma, Q., Ren, X., Huang, C.: XRec: large language models for explainable recommendation. Arxiv (2024)
42. Sun, W., Chen, Z., Ma, X., Yan, L., et al.: Instruction distillation makes large language models efficient zero-shot rankers. In: CIKM Workshop (2023)
43. Wang, L., Zhang, J., Yang, H., Chen, Z., Tang, J., et al.: User behavior simulation with large language model based agents. Arxiv (2023)
44. Zhang, A., Chen, Y., Sheng, L., Wang, X., Chua, T.-S.: On generative agents in recommendation. In: SIGIR, pp. 1807–1817 (2024)

GNN4LMR: Profile Distillation Enhanced High-Order Interactions for LLM-Based Recommendation

Lingyan Zhang, Shuwen Daizhou, Wanyu Ling, Yiman Xie(✉), and An Huang(✉)

School of Computer Science and Engineering, Central South University, Changsha 410083, China
{lingyan_zhang,shuwendaizhou,wanyuling,yimanxie,224129}@csu.edu.cn

Abstract. Despite advances in graph-enhanced LLM recommenders, significant challenges remain in modeling high-order user-item dependencies. First, existing GNN-LLM hybrids propagate spurious structural correlations through indiscriminate k-hop positive aggregation, leading to ineffective representation learning. Second, LLM-based methods are often constrained by lexical surface patterns, overlooking important behavioral rationales and the dynamic nature of user preferences. To address these *structural-semantic decoupling* issues, we propose GNN4LMR, an enhanced recommender model that incorporates three core innovations. First, we introduce a rationale-aware **profile distillation** mechanism that decodes implicit interaction motives by integrating behavioral patterns with semantic contexts, producing semantically rich user and item profiles that go beyond lexical surface artifacts. This distillation reduces reliance on explicit feature mentions and better captures evolving preferences. Second, we present a graph learning mechanism enhanced with **hard negative sampling**, which refines sampled negatives by selecting items that are feature-close to positives but semantically irrelevant, providing stronger contrastive signals than conventional random sampling. Finally, we propose a graph-word alignment network that bridges structural graph patterns with LLM semantic spaces through adaptive neural transformations and context-sensitive fusion. Extensive experiments demonstrate that GNN4LMR consistently outperforms state-of-the-art methods in both sequential and direct recommendation tasks, showcasing its superior ability to model complex user-item relationships.

Keywords: Large Language Models · Graph Neural Networks · Recommender Systems · Profile Distillation · High-order Interactions

1 Introduction

Multimedia recommendation systems must simultaneously decode complex behavioral signals (e.g., viewing patterns, comment interactions) and adapt to

H. Jung et al. (Eds.): DASFAA 2026, LNCS 16535, pp. 422–437, 2026.
https://doi.org/10.1007/978-981-92-0363-5_26

dynamically evolving user preferences [12,30,33,36]. While large language models (LLMs) offer unparalleled semantic comprehension for interpreting heterogeneous user data through cross-modal alignment [8,22], contrastive noise suppression [20], and task-aware prompting [5,27], and graph neural networks (GNNs) excel at capturing high-order collaborative signals via propagation architectures [9,16], their isolated or simplistic integration fails to model authentic user intent. This limitation manifests in two critical, interconnected misalignments in state-of-the-art hybrids [1,11,29]:

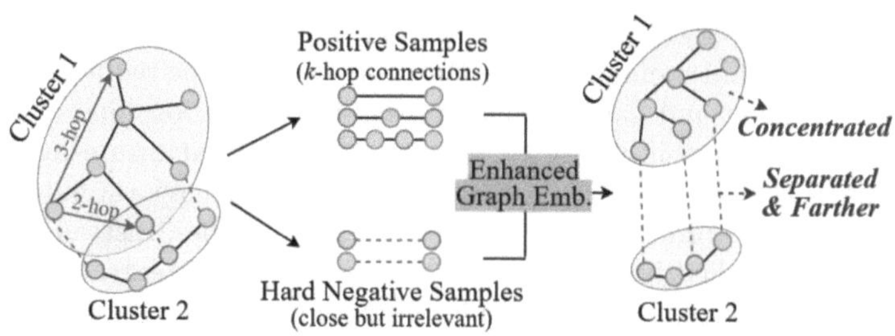

Fig. 1. Motivation of GNN4LMR. Hard Negative sampling induces cluster separation and compactness in latent space.

Challenge 1: Structural-Semantic Divergence in Graph Propagation. Contemporary GNN-LLM hybrids, such as ELMRec [29], rely on k-hop neighborhood aggregation, which conflates transient behaviors with persistent preferences. This approach propagates spurious correlations within the graph due to: (i) temporal decay blindness, where outdated interactions distort neighborhood formation, (ii) trivial negative sampling, which fails to distinguish semantically proximate yet irrelevant items, and (iii) collapsed embedding spaces, where genuine preferences become indistinguishable from incidental co-occurrences. Consequently, the resulting representations lack the discriminative capacity for effective recommendation in dynamic environments.

Challenge 2: Text-Behavior Disconnect in Preference Grounding. LLM-based recommenders, including P5 [5], POD [15], RDRec [27], and ELMRec [26], suffer from lexical anchoring bias. This bias arises from an over-reliance on explicit lexical patterns extracted through surface-level feature analyzers, such as Sentires [35]. These models fail to account for: (i) behaviorally implied preferences that are not explicitly stated, (ii) semantically equivalent descriptions expressed in different surface forms, and (iii) dynamic contextual preferences that evolve beyond static reviews. As a result, a "semantic echo chamber" is created, where recommendations reinforce superficial lexical patterns rather than truly reflecting user intent.

These challenges originate essentially from *the decoupling between structural dependencies in behavioral graphs and contextual semantics in user-generated content.* To address these challenges, we propose GNN4LMR, an enhanced LLM-based recommender model that centers on **profile distillation** as the principal mechanism for closing the text-behavior gap, while augmenting graph propagation with targeted negative mining. For ***Challenge 1*** *(structural correlation misalignment)*, we introduce a graph-based representation learning approach enhanced with **hard negative sampling**, which selects false positives that are feature-close yet semantically irrelevant. By incorporating these harder contrastive signals during message passing, GNN4LMR improves the discriminative power of graph embeddings, mitigating over-generalization from naive k-hop positive aggregation and emphasizing temporally relevant interactions. Complemented by a graph-word alignment network that adaptively maps structural patterns into the LLM semantic space, the combined pipeline reduces structural-semantic decoupling and captures subtle high-order relationships (see Fig. 1). For ***Challenge 2*** *(lexical anchoring bias)*, we design a **rationale-aware profile distillation** mechanism that synthesizes implicit behavioral patterns with semantic contexts to produce distilled user and item profiles. Instead of depending primarily on explicit lexical cues, the distillation module extracts behavioral rationales (e.g., implicit motivations inferred from interaction sequences, temporal patterns, and cross-modal signals) and fuses them with LLM-derived semantic representations. The resulting profiles are semantically richer and more robust to surface variability, enabling recommendations that better reflect user intent across evolving contexts.

Our main contributions are as follows:

- We propose GNN4LMR, an LLM-enhanced recommender that foregrounds a **rationale-aware profile distillation** mechanism to generate semantically rich user and item profiles by fusing behavioral motives with contextual semantics.
- We introduce a **hard negative sampling**-enhanced graph propagation strategy that refines negative selection to provide stronger contrastive signals during graph message passing, improving representation separation and temporal relevance.
- We design a graph-word alignment network to bridge structural graph patterns with LLM semantic spaces via adaptive neural transformations and context-sensitive fusion, enabling tighter structural-semantic coupling.
- Extensive experiments demonstrate that GNN4LMR outperforms state-of-the-art baselines on both sequential and direct recommendation tasks, validating the effectiveness of profile distillation and hard negative sampling.

2 Related Work

2.1 LLM-Based Recommendation

LLMs have demonstrated remarkable capabilities in recommendation tasks by transforming these tasks into text-generation problems [14,29]. Recent studies have explored various ways to enhance recommendations using LLMs, addressing key challenges such as cold-start problems [10,21], improving embedding initialization [8,22], and learning richer representations of users and items [20]. A notable contribution in this area is the P5 paradigm [5], which transforms user-item interactions and reviews into text prompts, effectively uncovering behavioral patterns. Furthermore, POD [15] introduces prompt distillation to enhance efficiency and reduce inference time, while RDRec [27] focuses on extracting user preferences and item attributes from reviews.

However, these methods [5,15,27] rely on explicit feature extraction, overlooking preferences and contextual cues from reviews. GNN4LMR addresses this with rationale-aware profile distillation for richer user-item representations.

2.2 GNN-Based Recommendation

GNNs have become a prominent approach for modeling high-order user-item interactions in recommendation systems by leveraging the structure of interaction graphs [7,17,25,28,31,37]. LightGCN [9] simplifies GCNs by focusing on key convolutions, while NCL [16] and XSimGCL [32] incorporate graph contrastive learning to improve scalability and efficiency. Recent advancements [1,3,6,11,29] have integrated GNNs with LLMs to better interpret graph-constructed interactions. These approaches enrich graph structures with textual information, offering insights that extend beyond the intrinsic graph relationships [29]. Techniques such as aligning textual data with graph structures [1,3] and integrating semantic understanding through graph-based interaction patterns [6,11] have been explored. ELMRec [26] enhances LLMs' interpretation of graph-based interactions through whole-word embedding improvements.

Most GNN-LLM hybrids lack effective negative sampling, or rely on random strategies that produce weak contrast. GNN4LMR addresses this by refining random negatives into hard negatives, those close in feature space but semantically incorrect, to better capture fine-grained high-order signals.

3 Methodology

The GNN4LMR framework (Fig. 2) consists of two main components: (i) **Graph-based Representation Learning**, which pretrains a GNN on the user-item interaction graph using hard negative sampling to capture high-order interactions, and (ii) **Graph-Word Embeddings Integration**, which aligns GNN-generated graph embeddings with LLM word embeddings to enhance both collaborative and semantic information. Additionally, **Rationale-Aware Profile Distillation** extracts rationales from user reviews to generate concise, semantically rich user and item profiles.

3.1 Graph-Based Representation Learning

In GNN4LMR, we introduce a pretrained GNN enhanced with hard negative sampling to capture high-order interactions. After graph-based message passing

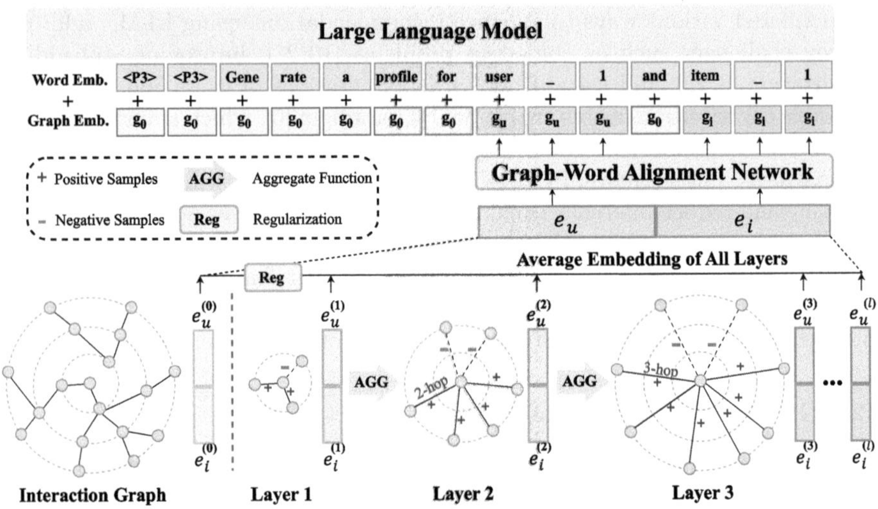

Fig. 2. Overview of GNN4LMR. The framework consists of two components: (1) Graph-based Representation Learning, which pretrains a GNN on the interaction graph with hard negative sampling. (2) Graph-Word Embeddings Integration, which aligns graph embeddings with word embeddings through a graph-word alignment network.

and aggregation, the incorporation of negative samples refines the embeddings, enabling the model to uncover complex relationships between users and items.

Graph-based Message Passing. The GNN in GNN4LMR utilizes a collaborative graph tokenizer to propagate and aggregate information across the interaction graph. LightGCN [9] serves as the backbone for the message passing mechanism. The forward propagation equations for user and item embeddings are given by:

$$e_u^{(l+1)} = \sum_{i \in \mathcal{N}_u} \frac{1}{\sqrt{|\mathcal{N}_u||\mathcal{N}_i|}} e_i^{(l)}, \quad e_i^{(l+1)} = \sum_{u \in \mathcal{N}_i} \frac{1}{\sqrt{|\mathcal{N}_i||\mathcal{N}_u|}} e_u^{(l)}, \tag{1}$$

where $e_u^{(l)}$ and $e_i^{(l)}$ denote the user and item embeddings at layer l, and $\mathcal{N}_u$ and $\mathcal{N}_i$ represent the sets of neighbors for user u and item i, respectively. The final embeddings are obtained by averaging the embeddings across all layers:

$$e_u = \frac{1}{L+1} \sum_{k=0}^{L} e_u^{(l)}, \quad e_i = \frac{1}{L+1} \sum_{k=0}^{L} e_i^{(l)}. \tag{2}$$

Optimization with Hard Negative Samples. To enhance representation discrimination, we introduce a hard negative sampling strategy for contrastive optimization. Unlike traditional BPR-based GNNs [9] that rely on randomly sampled negatives, many of which are trivial and uninformative, our method constructs harder negative samples in two stages:

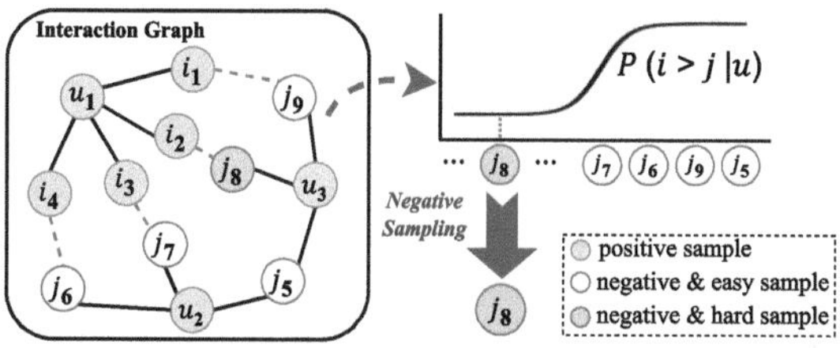

Fig. 3. Illustration of Hard Negative Samples Selection Process.

- Stage 1: Random Candidate Sampling. For each user-item pair (u, i), we sample a set of unobserved items $\{j_1, j_2, \ldots, j_n\}$, ensuring $j_k \notin \mathcal{N}_u$.
- Stage 2: Hard Negative Selection. We then select candidates whose initial embeddings are close to the positive item i, but which do not appear in the user's interaction history. These are considered "hard" negatives because they are more likely to confuse the model.

This two-step process, illustrated in Fig. 3, prioritizes semantically or structurally similar yet irrelevant items, providing stronger contrastive feedback during training.

We adopt a BPR-style optimization objective:

$$\mathcal{L}_{\text{BPR}} = -\sum_{u=1}^{m} \sum_{i \in \mathcal{N}_u} \sum_{j \in \mathcal{N}_{\text{neg}}^{\text{hard}}} \ln \sigma(\hat{y}_{u,i} - \hat{y}_{u,j}), \tag{3}$$

where $\mathcal{N}_{\text{neg}}^{\text{hard}}$ denotes the set of hard negative for each user. The gradient with respect to the predicted score is:

$$\frac{\partial \mathcal{L}_{\text{BPR}}}{\partial \hat{y}_{u,i}} = -\sum_{j \in \mathcal{N}_{\text{neg}}^{\text{hard}}} \sigma(-(\hat{y}_{u,i} - \hat{y}_{u,j})), \tag{4}$$

which emphasizes harder samples, encouraging the model to increase the margin between relevant and confusing irrelevant items.

To regularize training and prevent overfitting, we add a regularization term:

$$\mathcal{L}_{\text{reg}} = \lambda \left(\|\mathbf{e}_u^{(0)}\|^2 + \|\mathbf{e}_i^{(0)}\|^2 \right), \tag{5}$$

and the final objective becomes $\mathcal{L} = \mathcal{L}_{\text{BPR}} + \mathcal{L}_{\text{reg}}$.

3.2 Graph-Word Embeddings Integration

Unlike ELMRec [26], which relies solely on whole-word embeddings from LLMs for direct recommendation, GNN4LMR integrates graph embeddings into both direct and sequential recommendation tasks. Additionally, it introduces a graph-word alignment network that explicitly maps graph embeddings to word embeddings, ensuring that collaborative and semantic information are jointly represented in a shared latent space.

Alignment. In GNN4LMR, the alignment network is kept intentionally simple as a foundation approach, with the aim of enabling future research to explore more advanced methods for integrating word and graph embeddings. The graph-word alignment network learns a mapping function that aligns the latent spaces of graph embeddings (e_u, e_i) with word embeddings (x_u, x_i) derived from the LLM. The aligned graph embeddings, denoted as g_u for users and g_i for items, are computed as follows:

$$g_u = \text{Align}(e_u), \quad g_i = \text{Align}(e_i), \tag{6}$$

where Align(·) represents a trainable neural network. Depending on the complexity of the task, the alignment network can take different forms, such as a Linear Layer, Multi-Layer Perceptron (MLP), or Mixture-of-Experts (MoE) models. The performance of these alignment models will be evaluated in Sect. 4.3.

Fusion. After alignment, the graph embeddings (g_u, g_i) and word embeddings (x_u, x_i) are fused to form unified embeddings. The fusion is performed using a weighted sum:

$$z_u = x_u + \alpha \cdot g_u, \quad z_i = x_i + \alpha \cdot g_i, \tag{7}$$

where α is a learnable parameter that adjusts the relative contributions of the graph and word embeddings. The fused embeddings (z_u, z_i) are then used for downstream recommendation tasks.

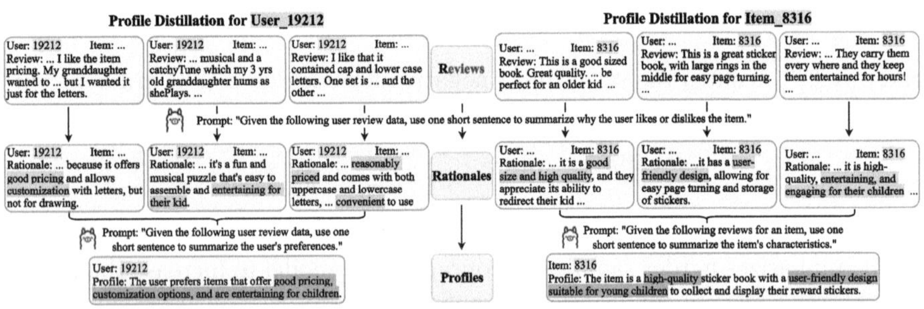

Fig. 4. Illustration of User and Item Profile Distillation, consisting of two key steps: (i) distilling raw user reviews into core reviews, and (ii) extracting user and item profiles that summarize preferences and characteristics.

3.3 Rationale-Aware Profile Distillation

GNN4LMR introduces a rationale-aware profile distillation mechanism to uncover the underlying reasons behind user-item interactions, generating semantically rich profiles, as shown in Fig. 4. The distillation process involves two key steps:

Interaction Rationale Distillation. Inspired by recent works [27], we leverage LLMs to extract the rationales behind user-item interactions. By using prompts such as *"Summarize in one sentence why the user likes or dislikes the item"*, LLMs (e.g., Llama 3 [4]) generate concise and focused rationales that clarify user preferences, like "*entertaining for their kid*", and highlight key item features, such as "*user-friendly design*".

User and Item Profile Generation. GNN4LMR introduces a "Profile Generation" task that constructs comprehensive user and item profiles based on the extracted rationales. For user profiles, we summarize long-term preferences by aggregating all reviews using the prompt: *"Summarize the user's preferences in one short sentence"*. For item profiles, we aggregate core item attributes using a similar approach: *"Summarize the item's characteristics in one short sentence"*. This process generates actionable, concise profiles for both users and items, enabling more personalized recommendations.

Table 1. Statistics of the three datasets.

Datasets	#User	#Item	#Review	Average	Density (%)
Sports	35,598	18,357	296,337	8.3	0.0453
Beauty	22,363	12,101	198,502	8.9	0.0734
Toys	19,412	11,924	167,597	8.6	0.0724

4 Experiments

4.1 Experimental Setup

Dataset and Metrics. We conducted experiments on three public datasets: Sports & Outdoors, Beauty, and Toys & Games, following prior work [15,27]. For direct and sequential recommendations, the last user interaction was used for testing, the second-to-last for validation, and the rest for training. For profile generation task, datasets were split 8:1:1 into training, validation, and test sets. Table 1 summarizes dataset statistics. We evaluated performance using hit rate (H@k) and normalized discounted cumulative gain (N@k) for $k \in \{5, 10\}$.

Baselines. We compared GNN4LMR with ten state-of-the-art methods for sequential recommendation: CASER [24], BERT4Rec [23], FDSA [34], SASRec [13], S^3-Rec [38], P5 [5], RLS [2], POD [15], RDRec [27], and ELMRec [26]. For direct recommendation, we compared it with nine state-of-the-art methods: SimpleX [18], P5 [5], RLS [2], POD [15], RDRec [27], LightGCN [9], NCL [16], XSimGCL [32], and ELMRec [26].

Implementation Details. Consistent with the baselines P5, POD, RDRec, and ELMRec, GNN4LMR adopts T5-small [19] as the LLM. For experiments involving user-item profile distillation, we utilize Llama-3-8b-instruct [4]. We reported statistical significance of the results using a 10-trial T-test to validate the robustness of GNN4LMR. All experiments were conducted using PyTorch on a single Nvidia V100 GPU with 32GB of memory.

Table 2. Performance comparison across three real-world datasets for sequential and direct recommendation tasks. "*" indicates that the improvement is statistically significant (p-value < 0.05).

Methods	Sports				Beauty				Toys			
	H@5	N@5	H@10	N@10	H@5	N@5	H@10	N@10	H@5	N@5	H@10	N@10
Sequential Recommendation												
Caser	0.0116	0.0072	0.0194	0.0097	0.0205	0.0131	0.0347	0.0176	0.0166	0.0107	0.0270	0.0141
BERT4Rec	0.0115	0.0075	0.0191	0.0099	0.0203	0.0124	0.0347	0.0170	0.0116	0.0071	0.0203	0.0099
FDSA	0.0182	0.0122	0.0288	0.0156	0.0267	0.0163	0.0407	0.0208	0.0228	0.0140	0.0381	0.0189
SASRec	0.0233	0.0154	0.0350	0.0192	0.0387	0.0249	0.0605	0.0318	0.0463	0.0306	0.0675	0.0374
S^3-Rec	0.0251	0.0161	0.0385	0.0204	0.0387	0.0244	0.0647	0.0327	0.0443	0.0294	0.0700	0.0376
P5	0.0272	0.0169	0.0361	0.0198	0.0503	0.0370	0.0659	0.0421	0.0648	0.0567	0.0709	0.0587
RSL	0.0392	0.0330	0.0512	0.0375	0.0508	0.0381	0.0667	0.0446	0.0676	0.0583	0.0712	0.0596
POD	0.0496	0.0396	0.0576	0.0419	0.0537	0.0395	0.0688	0.0443	0.0691	0.0599	0.0742	0.0610
RDRec	0.0505	0.0408	0.0596	0.0433	0.0601	0.0461	0.0743	0.0504	0.0723	0.0593	0.0802	0.0605
ELMRec	0.0538	0.0453	0.0616	0.0471	0.0609	0.0486	0.0750	0.0529	0.0713	0.0608	0.0764	0.0618
GNN4LMR	**0.0688**	**0.0529**	**0.0878**	**0.0590**	**0.0806**	**0.0609**	**0.1041**	**0.0685**	**0.0833**	**0.0680**	**0.0990**	**0.0723**
Impv. (%)	27.9*	16.8*	42.5*	25.3*	32.3*	25.3*	38.8*	29.5*	15.2*	11.8*	23.4*	17.0*
Direct Recommendation												
SampleX	0.2362	0.1505	0.3290	0.1800	0.2247	0.1441	0.3090	0.1711	0.1958	0.1244	0.2662	0.1469
P5	0.1514	0.1049	0.2196	0.1269	0.1566	0.1078	0.2317	0.1318	0.1322	0.0889	0.2023	0.1114
RSL	0.2092	0.1502	0.3001	0.1703	0.1612	0.1110	0.2209	0.1302	0.1423	0.0825	0.1926	0.1028
POD	0.2086	0.1506	0.2873	0.1756	0.1926	0.1391	0.2670	0.1629	0.1433	0.1009	0.2082	0.1215
RDRec	0.2747	0.2033	0.3683	0.2326	0.2572	0.1902	0.3380	0.2160	0.1655	0.1172	0.2375	0.1398
XSimGCL	0.3547	0.2689	0.4486	0.2992	0.3530	0.2734	0.4392	0.3012	0.3351	0.2614	0.4186	0.2885
LightGCN	0.4150	0.3002	0.5436	0.3418	0.4205	0.3067	0.5383	0.3451	0.3879	0.2874	0.5106	0.3272
NCL	0.4292	0.3131	0.5592	0.3551	0.4378	0.3228	0.5542	0.3607	0.3975	0.2925	0.5120	0.3325
ELMRec	0.5782	0.4792	0.6479	0.4852	0.6052	0.4852	0.6794	0.4973	0.5178	0.4051	0.6045	0.4141
GNN4LMR	**0.7245**	**0.6171**	**0.8196**	**0.6250**	**0.7078**	**0.6165**	**0.7889**	**0.6234**	**0.6882**	**0.5519**	**0.7999**	**0.5541**
Impv. (%)	25.3*	28.8*	26.5*	28.8*	17.0*	27.1*	16.1*	25.4*	32.9*	36.2*	32.3*	33.8*

4.2 Main Results

Table 2 compares GNN4LMR with baselines on sequential and direct recommendations. All of the baselines are reported by ELMRec [26]. GNN4LMR consistently outperforms all competitors across datasets, achieving 11.8%âĂŞ42.5% gains over ELMRec in sequential recommendations and 16.1%–âĂŞ36.2% in direct recommendations. Key findings include:

(1) **Performance on Direct Recommendations:** While LLM-based models often underperform in direct recommendations tasks, GNN4LMR significantly surpasses RDRec by 122.5%âĂŞ370.9%, demonstrating the importance of high-order interaction signals via graph integration.

(2) **Comparison with GNN-based Models:** While LightGCN, XSimGCL, and NCL perform competitively, GNN4LMR surpasses NCL by 46.3%–97.1% by integrating rationale-aware profile distillation. This demonstrates the advantage of combining collaborative graph signals with contextual understanding.
(3) **Advantage Over GNN-LLM Hybrids:** Compared to ELMRec, GNN4LMR improves sequential and direct recommendations by an average of 26.1% and 27.5%, respectively, leveraging hard negative sampling and graph-word alignment network.

Table 3. Ablation study on key components

Models	Sports		Beauty		Toys	
	H@10	N@10	H@10	N@10	H@10	N@10
Sequential recommendation						
GNN4LMR	**0.0878**	**0.0590**	**0.1041**	**0.0685**	**0.0990**	**0.0723**
w/o Graph Emb	0.0602	0.0455	0.0743	0.0502	0.0744	0.0602
w/o Align.	0.0808	0.0520	0.0834	0.0579	0.0903	0.0666
w/o Profiles	0.0823	0.0546	0.0946	0.0634	0.0924	0.0684
Direct recommendation						
GNN4LMR	**0.8196**	**0.6250**	**0.7889**	**0.6234**	**0.7999**	**0.5541**
w/o Graph Emb	0.3046	0.2244	0.3094	0.1998	0.2269	0.1369
w/o Align.	0.6236	0.4846	0.6637	0.5097	0.6234	0.4214
w/o Profiles	0.7136	0.5531	0.7039	0.5497	0.7079	0.5016

Table 4. Effect of various alignment networks

Models	Sports		Beauty		Toys	
	H@10	N@10	H@10	N@10	H@10	N@10
Sequential recommendation						
w/o Alignment	0.0808	0.0520	0.0834	0.0579	0.0903	0.0666
Linear	0.0878	0.0590	0.1041	0.0685	0.0990	0.0723
MLP	0.0880	0.0602	0.1043	0.0687	0.0987	0.0714
MoE	**0.0898**	**0.0606**	**0.1056**	**0.0701**	**0.1003**	**0.0736**
Direct recommendation						
w/o Alignment	0.6236	0.4846	0.6637	0.5097	0.6234	0.4214
Linear	0.8196	0.6250	0.7889	0.6234	0.7999	0.5541
MLP	0.8049	0.6164	0.7769	0.6169	0.7846	0.5499
MoE	**0.8291**	**0.6406**	**0.7913**	**0.6367**	**0.8106**	**0.5671**

(4) **Challenges in Sequential Recommendations:** Sequential recommendation is challenging because it requires capturing temporal patterns and prioritizing relevant items. Previous methods [26] relying on k-hop positive samples have only been successful in direct recommendation and cannot be effectively applied to sequential recommendation. GNN4LMR integrates hard negative samples into a contrastive learning framework, enhancing the discriminative power of graph embeddings while successfully applying them not only to direct recommendation but also to sequential recommendation.

4.3 Ablation Study

We conducted ablation studies to assess the contribution of each component. Table 3 shows the impact of key components, Table 4 compares alignment networks, and Fig. 5 illustrates the effects of various parameters.

Impact of Key Components. As shown in Table 3, removing graph embeddings (w/o Graph-Emb.) leads to a significant performance drop, with H@10 decreasing by 62.8% and N@10 by 64.1% in direct recommendation on the Sports dataset. This highlights the importance of hard negative sampling-enhanced graph embeddings. Similarly, removing alignment networks (w/o Align.) causes a 19.9% drop in H@10 and 15.5% in N@10 on the Beauty dataset, underscoring the critical role of integrating graph and word embeddings. Removing user and item profiles (w/o Profiles) results in a smaller but notable decline (12.9% in H@10 on Sports), demonstrating the value of rationale-aware profile distillation.

Effect of Alignment Networks. Table 4 compares three alignment designs: Linear, MLP, and MoE. MoE provides the best performance, but with only a modest improvement (1.4% in H@10 on Beauty) over the Linear model, despite its higher complexity. GNN4LMR adopts a linear layer, balancing efficiency with performance, avoiding the additional overhead of MLP (three layers) and MoE (multiple experts). These results suggest potential for future advances in alignment techniques with minimal computational cost.

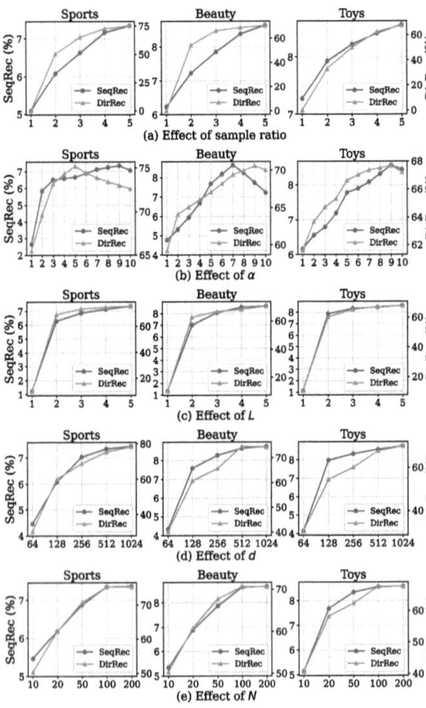

Fig. 5. Effect of various parameters.

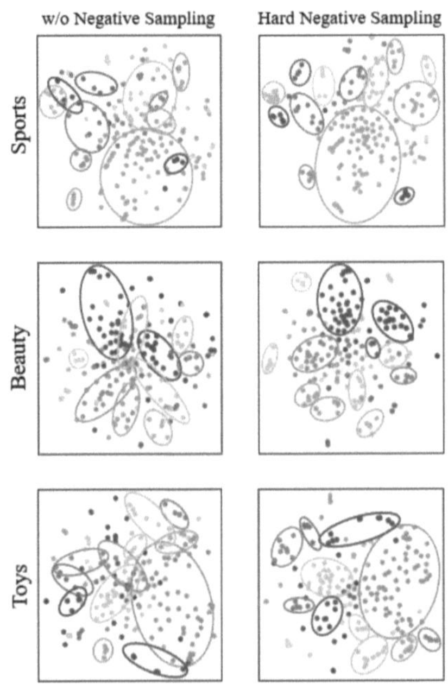

Fig. 6. Visualization of item graph embeddings (with/without hard negative sampling)

Effect of Various Parameters. We evaluated the impact of five hyperparameters, sample ratio, α, L, d, and n, on GNN4LMR's performance, using H@10 and N@10 as evaluation metrics. As shown in Fig. 5(a), increasing the sample ratio improves task performance but also extends training time. To balance accuracy and efficiency, we set the sample ratio to 5. Figure 5(b) shows that higher values of α enhance the influence of graph embeddings, improving performance up to an optimal value (9 on the Toys dataset). Beyond this point, excessive reliance on graph signals diminishes performance. Figure 5(c) demonstrates that increasing the number of LightGCN layers improves the model's ability to capture high-order interactions, with peak performance at $L = 5$. Further increase in L result in diminishing returns and increased training costs. Figure 5(d) shows that increasing embedding dimensionality d boosts performance, and we set $d = 512$ to ensure compatibility with LLM word embeddings. Finally, as shown in Fig. 5(e), increasing the number of negative samples n enhances contrastive learning, with the optimal number at $n = 100$. Larger values of n offers minimal benefits while increasing computational overhead.

4.4 Visualization of Graph Embeddings

We use t-SNE to visualize graph embeddings, as shown in Fig. 6, where items that have interacted with the same users are represented by dots of the same color. The proximity of the dots indicates the degree of similarity between items. Figure 6 provides three key insights:

(1) Improved Embedding Separation: As seen in the lower row, the use of negative sampling enhances the separation of graph embeddings, resulting in a clearer distinction across all datasets.
(2) Enhanced High-order Interaction Modeling: The tighter clustering of similar items (e.g., the red group in Toys) suggests better modeling of indirect user-item relationships.
(3) Reduction of Embedding Overlap: In the absence of negative sampling, embeddings exhibit overlap (e.g., red and orange in Sports), whereas with hard negative sampling, item groups are more distinct, improving the clarity of item representations.

4.5 Rationale-Aware Profile Distillation Analysis

Figure 7 compares the explanations generated by existing LLM-based methods [5,15,26,27] with the profiles produced by GNN4LMR. The previous methods typically generate short (12-word) explanations that oversimplify user preferences, often missing broader context and emotional nuances. For example, they might reduce a preference for "*child-friendly durability*" to a single term like "*cardboard*". In contrast, GNN4LMR generates more nuanced profiles by extracting rationales from multiple reviews. These profiles integrate both direct preferences (e.g., "*engaging, educational, and creative*" and contextual factors (e.g., "*cute, fun, and ideal for shared experiences*"). This results in a richer, more accurate understanding of user preferences, leading to better recommendation quality.

Cases	User: 7753 Item: 9066	User: 1366 Item: 5402	User: 13709 Item: 3914	User: 12665 Item: 87	User: 3363 Item: 288
Raw reviews (Avg. 100 words)	... to have some good ①fun. I like this ... because it is a bit more ②complicated ... The ③designs are on each page, front and back. ...	This book looks really ①fun and cool. ... I got it ②for my 6 year old daughter for xmas ... she is very creative and loves crazy weird stuff ...	Very ①nice puzzle. I love the big pieces, and the ②durable cardboard. My ③3 yo really likes it. ... are ④good quality and the ⑤designs are cute.	This is a ①great game. I was introduced to this game by ②my daughter and her husband. It ③takes skill and strategy to determine ...	Love it! My ①granddaughter has so much fun ... when she comes to visit, we are going to have so much fun. This is so ②cute & fun!!
Explanations (Avg. 12 words)	to have some good fun	looks really fun and cool	the durable cardboard	This is a great game	This is so cute & fun
Profiles (Avg. 48 words)	The user prefers products that offer ①unique benefits, such as ②improved cognitive function ... The item is a coloring book ③featuring intricate ④and traditional designs ...	The user prefers items that are ①cute, fun, and suitable ②for their 6-yr -old daughter ... The item ... that offers a ③unique and ④educational experience for kids ...	The user prefers items that are ①durable, cute, and ②suitable for young children ... The item is a sturdy, developmentally appropriate puzzle with large, ③durable pieces and a ④cute design ...	The user prefers items that are ①engaging, ②educational, and ③encourage creativity, imagination, and fun, ... The item is a board game that offers a ④balance of skill and luck ...	The user prefers items that are ①cute, fun, and unique, ... ②share enjoyable experiences with their granddaughter. The Shelf is a ③fun and playful holiday tradition that ④brings joy and creativity to families

Useful Information Feature Word User Profile Item Profile

Fig. 7. Comparison of Raw Reviews, Explanations, and Profiles. Raw reviews are sourced from the Toys dataset, with key information highlighted. Explanations and feature words are generated by previous LLM-based models. Profiles generated by GNN4LMR provide a more comprehensive understanding.

Table 5. Running time at various stages on three datasets.

Datasets	Profile Distillation	Pretraining		SeqRec	DirRec
		GNN	LLM		
Sports	42 h 40 m 35 s	41 m 40 s	1 h 19 m 36 s	23 m 55 s	42 m 08 s
Beauty	28 h 20 m 38 s	17 m 22 s	40 m 56 s	11 m 15 s	13 m 36 s
Toys	22 h 15 m 54 s	12 m 51 s	52 m 37 s	26 48 s	14 m 27 s

4.6 Computational Complexity Analysis

The computational complexity of GNN4LMR is influenced by both the Transformer (m^2) and LightGCN (n^2), where m is the average number of input tokens and n is the total number of users and items. Since $m \ll n$, the overall complexity is primarily dominated by n^2, emphasizing the importance of considering the scale of users and items when deploying the model. This highlights the need for scalable solutions in large-scale recommendation systems.

Table 5 shows the running time of GNN4LMR across various stages. While profile distillation is time-intensive, it is a one-time process that provides the model with foundational knowledge. Both GNN and LLM pretraining, as well as inference during sequential and direct recommendation tasks, are highly efficient. These results demonstrate GNN4LMR's effective balance between initial computational costs and scalability for real-world applications.

5 Conclusion

This paper presents GNN4LMR, an enhanced LLM-based recommendation framework that fundamentally mitigates structural-semantic decoupling by centering on rationale-aware profile distillation. The proposed profile distillation mechanism serves as the core of our design, integrating behavioral rationales with semantic representations to form semantically grounded and dynamically adaptive user-item profiles. Complementing this, a hard negative sampling-enhanced graph propagation strategy refines structural contrast, strengthening the model's ability to discern subtle yet meaningful relational patterns. Together with an adaptive graph-word alignment network that bridges structural and semantic spaces, GNN4LMR achieves a unified and interpretable representation of user intent. Extensive experiments across diverse recommendation scenarios demonstrate not only significant performance gains over state-of-the-art baselines, but also superior adaptability to evolving user preferences. Overall, GNN4LMR establishes a new paradigm for LLM-based recommendation by foregrounding profile distillation as the key to reconciling behavioral understanding with semantic intelligence.

Acknowledgements. We are grateful for resources from the High Performance Computing Center of Central South University.

Disclosure of Interests. The authors have no relevant financial or non-financial interests to disclose.

References

1. Chen, Z., et al.: Exploring the potential of large language models (LLMs) in learning on graphs. ACM SIGKDD Explor. Newsl. **25**(2), 42–61 (2024)
2. Chu, Z., et al.: Leveraging large language models for pre-trained recommender systems. arXiv preprint arXiv:2308.10837 (2023)
3. Damianou, A., et al.: Towards graph foundation models for personalization. In: Proceedings of the ACM on Web Conference, pp. 1798–1802 (2024)
4. Dubey, A., et al.: The LLaMA 3 herd of models. arXiv preprint arXiv:2407.21783 (2024)
5. Geng, S., Liu, S., Fu, Z., Ge, Y., Zhang, Y.: Recommendation as language processing (RLP): a unified pretrain, personalized prompt & predict paradigm (P5). In: Proceedings of the ACM Conference on Recommender Systems, pp. 299–315 (2022)
6. Guo, N., Cheng, H., Liang, Q., Chen, L., Han, B.: Integrating large language models with graphical session-based recommendation. arXiv preprint arXiv:2402.16539 (2024)
7. Guo, W., Cui, Y., Zheng, K.: IntFair: graph neural networks for fair recommendations with interest awareness. In: International Conference on Database Systems for Advanced Applications, pp. 3–18. Springer (2024)
8. Harte, J., Zorgdrager, W., Louridas, P., Katsifodimos, A., Jannach, D., Fragkoulis, M.: Leveraging large language models for sequential recommendation. In: Proceedings of the 17th ACM Conference on Recommender Systems, pp. 1096–1102 (2023)
9. He, X., Deng, K., Wang, X., Li, Y., Zhang, Y., Wang, M.: LightGCN: simplifying and powering graph convolution network for recommendation. In: Proceedings of the 43rd International ACM SIGIR Conference on Research and Development in Information Retrieval, pp. 639–648 (2020)

10. He, Z., et al.: Large language models as zero-shot conversational recommenders. In: Proceedings of the 32nd ACM International Conference on Information and Knowledge Management, pp. 720–730 (2023)
11. Huang, X., et al.: Can GNN be good adapter for LLMs? In: Proceedings of the ACM on Web Conference, pp. 893–904 (2024)
12. Ji, W., Liu, X., Zhang, A., Wei, Y., Ni, Y., Wang, X.: Online distillation-enhanced multi-modal transformer for sequential recommendation. In: Proceedings of the 31st ACM International Conference on Multimedia, pp. 955–965 (2023)
13. Kang, W.C., McAuley, J.: Self-attentive sequential recommendation. In: Proceedings of the International Conference on Data Mining, pp. 197–206 (2018)
14. Li, J., et al.: MINER: multi-interest matching network for news recommendation. In: Findings of the Association for Computational Linguistics: ACL, pp. 343–352 (2022)
15. Li, L., Zhang, Y., Chen, L.: Prompt distillation for efficient LLM-based recommendation. In: Proceedings of the 32nd ACM International Conference on Information and Knowledge Management, pp. 1348–1357 (2023)
16. Lin, Z., Tian, C., Hou, Y., Zhao, W.X.: Improving graph collaborative filtering with neighborhood-enriched contrastive learning. In: Proceedings of the ACM on Web Conference, pp. 2320–2329 (2022)
17. Liu, N., et al.: Learning social graph for inactive user recommendation. In: International Conference on Database Systems for Advanced Applications, pp. 151–167. Springer (2024)
18. Mao, K., et al.: SimpleX: a simple and strong baseline for collaborative filtering. In: Proceedings of the 30th ACM International Conference on Information and Knowledge Management, pp. 1243–1252 (2021)
19. Raffel, C., et al.: Exploring the limits of transfer learning with a unified text-to-text transformer. J. Mach. Learn. Res. **21**(140), 1–67 (2020)
20. Ren, X., et al.: Representation learning with large language models for recommendation. In: Proceedings of the ACM on Web Conference, pp. 3464–3475 (2024)
21. Sanner, S., Balog, K., Radlinski, F., Wedin, B., Dixon, L.: Large language models are competitive near cold-start recommenders for language-and item-based preferences. In: Proceedings of the 17th ACM Conference on Recommender Systems, pp. 890–896 (2023)
22. Shen, K., et al.: SEMINAR: search enhanced multi-modal interest network and approximate retrieval for lifelong sequential recommendation. arXiv preprint arXiv:2407.10714 (2024)
23. Sun, F., et al.: BERT4Rec: sequential recommendation with bidirectional encoder representations from transformer. In: Proceedings of the 28th ACM International Conference on Information and Knowledge Management, pp. 1441–1450 (2019)
24. Tang, J., Wang, K.: Personalized top-n sequential recommendation via convolutional sequence embedding. In: Proceedings of the Eleventh ACM International Conference on Web Search and Data Mining, pp. 565–573 (2018)
25. Wang, C., et al.: Towards representation alignment and uniformity in collaborative filtering. In: Proceedings of the 28th ACM SIGKDD Conference on Knowledge Discovery and Data Mining, pp. 1816–1825 (2022)
26. Wang, X., Cui, J., Fukumoto, F., Suzuki, Y.: Enhancing high-order interaction awareness in LLM-based recommender model. In: Proceedings of the 2024 Conference on Empirical Methods in Natural Language Processing, pp. 11696–11711 (2024)

27. Wang, X., Cui, J., Suzuki, Y., Fukumoto, F.: RDRec: rationale distillation for LLM-based recommendation. In: Proceedings of the 62nd Annual Meeting of the Association for Computational Linguistics, pp. 65–74 (2024)
28. Wang, X., Fukumoto, F., Cui, J., Suzuki, Y., Yu, D.: NFARec: a negative feedback-aware recommender model. In: Proceedings of the International ACM SIGIR Conference on Research and Development in Information Retrieval, pp. 935–945 (2024)
29. Wei, W., et al.: LLMRec: large language models with graph augmentation for recommendation. In: Proceedings of the 17th ACM International Conference on Web Search and Data Mining, pp. 806–815 (2024)
30. Wu, Y., et al.: Multi-view multi-behavior contrastive learning in recommendation. In: International Conference on Database Systems for Advanced Applications, pp. 166–182. Springer (2022)
31. Xia, L., Xie, M., Xu, Y., Huang, C.: MixRec: heterogeneous graph collaborative filtering. In: Proceedings of the Eighteenth ACM International Conference on Web Search and Data Mining, pp. 136–145 (2025)
32. Yu, J., Xia, X., Chen, T., Cui, L., Hung, N.Q.V., Yin, H.: XSimGCL: towards extremely simple graph contrastive learning for recommendation. IEEE Trans. Knowl. Data Eng. **36**(2), 913–926 (2023)
33. Zhang, J., Liu, G., Liu, Q., Wu, S., Wang, L.: Modality-balanced learning for multimedia recommendation. In: Proceedings of the 32nd ACM International Conference on Multimedia, pp. 7551–7560 (2024)
34. Zhang, T., et al.: Feature-level deeper self-attention network for sequential recommendation. In: Proceedings of the International Joint Conferences on Artificial Intelligence, pp. 4320–4326 (2019)
35. Zhang, Y., Lai, G., Zhang, M., Zhang, Y., Liu, Y., Ma, S.: Explicit factor models for explainable recommendation based on phrase-level sentiment analysis. In: Proceedings of the 37th International ACM SIGIR Conference on Research and Development in Information Retrieval, pp. 83–92 (2014)
36. Zhang, Z., et al.: DeepMIN: deep multi-modal interest network with cognitive learning modules. In: International Conference on Database Systems for Advanced Applications, pp. 212–227. Springer (2024)
37. Zhao, S., Wei, W., Zou, D., Mao, X.: Multi-view intent disentangle graph networks for bundle recommendation. In: Proceedings of the AAAI Conference on Artificial Intelligence, pp. 4379–4387 (2022)
38. Zhou, K., et al.: S3-Rec: self-supervised learning for sequential recommendation with mutual information maximization. In: Proceedings of the 29th ACM International Conference on Information and Knowledge Management, pp. 1893–1902 (2020)

CDDF: Confidence- and Divergence-Aware Dual-View Dynamic Fusion for Long-Tail Recommendation

Ziyi Xu[1,2,3], Zhenyu Yang[1,2,3](✉), Mengdi Cai[1,2,3], Fang Wang[1,2,3], and Zhibo Zhang[4]

[1] Key Laboratory of Computing Power Network and Information Security, Ministry of Education, Shandong Computer Science Center (National Supercomputer Center in Jinan), Qilu University of Technology (Shandong Academy of Sciences), Jinan, China
{10431240049,10431240085,10431240135,b1043125017}@stu.qlu.edu.cn

[2] Shandong Engineering Research Center of Big Data Applied Technology, Faculty of Computer Science and Technology, Qilu University of Technology (Shandong Academy of Sciences), Jinan, China

[3] Shandong Provincial Key Laboratory of Industrial Network and Information System Security, Shandong Fundamental Research Center for Computer Science, Jinan, China
qlu_yangzhenyu@163.com

[4] School of Computer Science and Technology, Huazhong University of Science and Technology, Wuhan, China
15953766107@163.com

Abstract. Recent advances in large language models (LLMs) have created new opportunities for alleviating the long-tail problem in recommender systems by providing semantic representations that complement sparse collaborative signals. However, existing approaches suffer from two limitations. First, most fusion methods employ a globally uniform mechanism without modeling the reliability differences of items across the two information sources, making it difficult to achieve adaptive fusion for different items. Second, mutual-information-based alignment methods typically impose uniform global constraints, which, while enhancing cross-modal shared information, tend to weaken the complementarity between modalities and cause the loss of distinctive information. To address these issues, we propose a **C**onfidence- and **D**ivergence-aware dual-view **D**ynamic **F**usion framework (CDDF). The proposed framework constructs user and item embeddings from both collaborative and semantic views, and adaptively adjusts the fusion weights between two embeddings based on item-specific confidence and divergence, thereby achieving complementary enhancement while maintaining modal distinctiveness. In addition, we design a knowledge distillation-based user enhancement module, which transfers knowledge from similar users to improve the robustness of preference modeling. The results show that CDDF achieves overall better performance than state-of-the-art baselines, particularly in alleviating the long-tail problem.

H. Jung et al. (Eds.): DASFAA 2026, LNCS 16535, pp. 438–453, 2026.
https://doi.org/10.1007/978-981-92-0363-5_27

Keywords: Adaptive Fusion · Divergence-aware Alignment · Long-tail Recommendation

1 Introduction

Recommender systems aim to infer user preferences from historical interactions and deliver personalized content to users [16]. In this process, obtaining high-quality item embeddings is fundamental to modeling user preferences, as it directly determines the model's ability to capture user interests and the stability of its representations [9]. Traditional methods typically learn unique ID-based embeddings to characterize collaborative relations among items [15,23], which effectively capture the co-occurrence patterns of popular items. However, such methods heavily rely on sufficient interaction data. In real-world scenarios, userâĂŞitem interactions usually follow a long-tail distribution [22], where a small number of head items account for the majority of interactions, while most tail items suffer from extremely sparse feedback. Due to insufficient training signals, the collaborative embeddings of tail items are often poorly optimized, resulting in significantly lower representation quality compared with head items. Consequently, the recommendation performance of the model degrades markedly in long-tail scenarios.

To address this issue, recent studies have leveraged the semantic modeling power of LLMs [3,5,8,11,19]. By encoding the textual content of items, LLMs can generate semantically rich embeddings that capture contextual knowledge. Unlike collaborative embeddings derived from userâĂŞitem interactions, semantic embeddings originate directly from the intrinsic content of items and thus provide valuable external knowledge for items with sparse interactions. Intuitively, collaborative and semantic embeddings are inherently complementary in their information sources, and integrating them can potentially combine their respective advantages. As shown in Fig. 1, we train the same model using either ID-based or semantic embeddings to initialize item representations. The results reveal a clear performance disparity: For frequently interacted head items, ID-based embeddings outperform due to well-learned collaborative signals; in contrast, for sparsely interacted tail items, semantic embeddings substantially surpass ID-based ones, benefiting from their content-based priors. These findings indicate that the two embeddings exhibit complementary behaviors across different item types. However, because they differ significantly in their information sources and distributional characteristics, how to effectively fuse them for mutual enhancement remains an open and challenging research problem.

Existing methods suffer from two major limitations. First, many studies fail to explicitly model the reliability differences of collaborative and semantic information sources across items. These approaches usually assume that both types of signals are equally important on a global scale and thus adopt a uniform fusion strategy, such as simple concatenation [1,17,24], cross-attention mechanisms [11], or shared-space projection [7]. Although the fusion techniques vary, they share a common drawback–none of them dynamically model how different items depend on collaborative and semantic information to varying degrees.

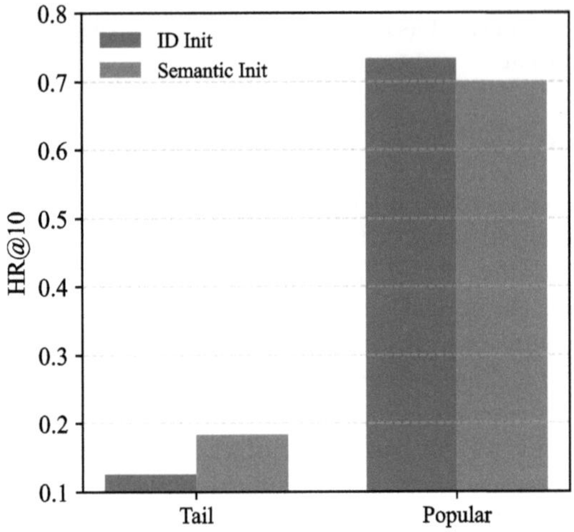

Fig. 1. Effect of different initialization strategies on recommendation performance for head and tail items.

Specifically, for popular items with abundant interactions, ID-based collaborative embeddings tend to be more reliable; whereas for tail items with sparse interactions, semantic embeddings should play a more dominant role. Second, several studies employ cross-view alignment strategies based on mutual information maximization [12,14,18] to enhance the shared information between semantic and collaborative views. However, for certain items, their collaborative and semantic embeddings may inherently differ in representation space, reflecting complementary preference patterns and behavioral characteristics across modalities. Ignoring such modality discrepancies and enforcing uniform alignment constraints across all items risks losing important complementary signals that are beneficial for personalized recommendation.

To address the above challenges, this paper proposes a dual-view dynamic fusion framework. Specifically, we design a dual-aware fusion mechanism that adaptively adjusts the weights of collaborative and semantic embeddings according to item popularity and inter-view divergence, thereby mitigating the "one-size-fits-all" issue caused by static fusion strategies. Meanwhile, we introduce a user enhancement module, which retrieves similar users and transfers knowledge through a lightweight attention mechanism to improve the robustness of user preference modeling.

The main contributions of this paper are summarized as follows:

- We propose a unified framework combining adaptive dual-view fusion and knowledge distillation-based user enhancement, alleviating the long-tail problem from both item and user perspectives.
- We design a confidence- and divergence-aware adaptive fusion mechanism that dynamically adjusts the fusion weights. This mechanism significantly improves the recommendation performance of tail items.

- We introduce a user enhancement module that employs knowledge distillation to transfer information at the user representation level through a lightweight attention mechanism, effectively enhancing the expressiveness and robustness of user preference modeling.
- Experiments on real-world datasets show consistent improvements over baselines in both overall accuracy and long-tail performance, verifying the effectiveness and generalization of the framework. The code is available at https://anonymous.4open.science/r/CDDF/.

1.1 ID-Based Recommender Systems

Traditional recommender systems usually take user and item IDs as input and learn low-dimensional embeddings to capture latent relationships. Early neural methods treated IDs as abstract tokens, encoding them through neural networks to enhance representation capacity. Graph-based models such as LightGCN [4] further improved representation learning by aggregating neighborhood information while discarding redundant transformations. Sequential recommendation methods–such as GRU4Rec [6], SASRec [9], and BERT4Rec [15], captured users' dynamic interests through recurrent or self-attention mechanisms. More recently, self-supervised learning has been introduced to improve model robustness under sparse interactions [20]. Nevertheless, these ID-based models still rely heavily on interaction data and lack semantic understanding, leading to poor generalization in long-tail scenarios. With the rise of pretrained language models [2], researchers have begun integrating textual semantics and external knowledge to endow recommendation models with richer priors and reasoning capabilities.

1.2 LLM-Enhanced Recommender Systems

LLMs have been increasingly integrated into recommender systems owing to their strong semantic understanding and reasoning abilities. One research direction focuses on aligning multi-modal representation spaces. For instance, RLMRec [14] employs mutual information theory to align collaborative embeddings with their corresponding semantic embeddings, while DaRec [21] further refines this process through a decoupled alignment framework that aligns only the shared components between LLM and collaborative embeddings for improved performance. Regarding fusion mechanisms, LLM-ESR [11] applies cross-attention to integrate collaborative and semantic embeddings, enabling richer user and item representations. Similarly, AlphaFuse [7] decomposes the LLM embedding space via SVD and injects collaborative signals into its semantically sparse subspace to mitigate semantic degradation. By comparison, the proposed CDDF framework alleviates the long-tail problem by introducing a adaptive fusion mechanism and a divergence-aware constraint for cross-view alignment.

2 Method

2.1 Dual-View Modeling

The long-tail problem primarily arises from the imbalance in interaction distribution and the sparsity of collaborative signals. When relying solely on collaborative embedding, the recommendation performance is usually strong for head items and highly active users, but significantly weaker for tail users and tail items. Due to the limited interaction data of tail entities, the model struggles to learn reliable collaborative representations for them. To address this issue, we introduce a semantic view as a complementary source of information: While the collaborative view captures userâĂŞitem behavioral relationships, the semantic view encodes content information. The two views are complementary, helping to alleviate the representation deficiency of tail entities.

Collaborative View. To capture global collaborative relations, we pretrain a LightGCN [4] model on the complete userâĂŞitem interaction graph, obtaining the collaborative embeddings of users and items, denoted as $\mathbf{E}_{\mathrm{c}}^{U} \in \mathbb{R}^{|\mathcal{U}|\times d_{\mathrm{c}}}$ and $\mathbf{E}_{\mathrm{c}}^{V} \in \mathbb{R}^{|\mathcal{V}|\times d_{\mathrm{c}}}$, respectively, where d_{c} is the dimension of collaborative embeddings, $|\mathcal{U}|$ and $|\mathcal{V}|$ are the number of users and items. These embeddings encode high-order connectivity within the graph structure, effectively reflecting users' collaborative preferences and items' collaborative attributes.

Semantic View. The semantic view aims to model user and item characteristics from the content perspective. We leverage the semantic modeling capability of LLMs to generate semantic embeddings for both users and items.

Item Side. Based on item attributes such as titles, brands, categories and textual descriptions, we construct textual representations and obtain item semantic embeddings $\mathbf{E}_{\mathrm{s}}^{V} \in \mathbb{R}^{|\mathcal{V}|\times d_{\mathrm{LLM}}}$ through an API call to the LLM. d_{LLM} is the dimension of LLM embeddings.

User Side. For each user, we concatenate the titles and descriptions of interacted items to form a textual summary that reflects the user's semantic interests. The LLM then extracts the user semantic embedding $\mathbf{E}_{\mathrm{s}}^{U} \in \mathbb{R}^{|\mathcal{U}|\times d_{\mathrm{LLM}}}$.

2.2 Confidence- and Divergence-Aware Adaptive Fusion

Although collaborative and semantic embeddings are complementary in terms of their information sources, their reliability varies across different scenarios. The collaborative embedding captures the collective preference patterns at the group level, while the semantic embedding reflects the intrinsic content features of the item. For tail items, the sparsity of interaction data makes it difficult to sufficiently optimize the collaborative embeddings, resulting in relatively low reliability. The collaborative representation of such items often fails to adequately

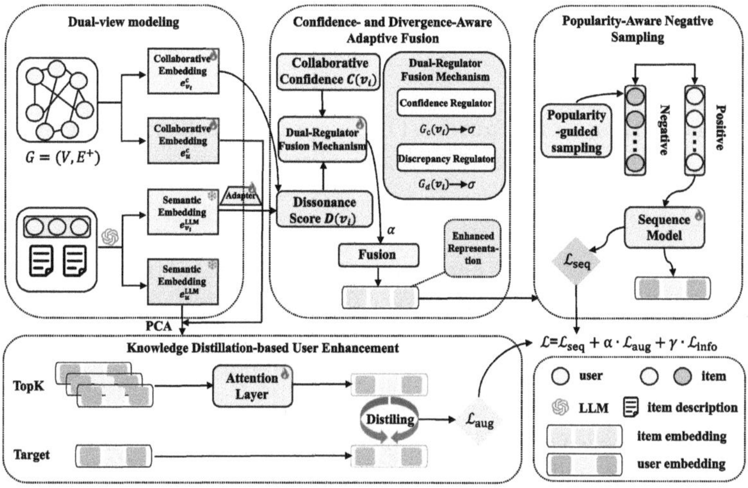

Fig. 2. The overview of the proposed CDDF framework.

capture group-level user preferences; therefore, the model should assign greater weight to semantic embeddings. For head items, the collaborative signals may include preferences from various types of users. If an item is frequently interacted with by users who share similar interests, the learned collaborative embedding tends to be stable and exhibits latent consistency with the item's semantic embeddings. This consistency can be further strengthened through alignment learning. In such cases, the divergence between the two embeddings is relatively small, and the collaborative signals provide more statistically stable preference information. Thus, the model should rely more on the collaborative embeddings. Conversely, when an item is driven by the popularity effect and it is interacted with by a heterogeneous user group, the collaborative embedding becomes contaminated with multiple unrelated preferences and noise, resulting in greater divergence from the semantic embedding. In this scenario, the semantic embedding provides more stable content features, and the model should increase the weight of the semantic embedding to suppress noise interference from collaborative signals. To this end, we propose a confidence- and divergence-aware adaptive fusion mechanism that dynamically adjusts the fusion weights between collaborative and semantic embeddings according to the characteristics of each item (Fig. 2).

Collaborative Confidence. Collaborative confidence measures the reliability of the collaborative embedding. Intuitively, the richer the interaction history of an item, the more trustworthy its collaborative embedding becomes. We quantify this by defining the collaborative confidence as a metric positively correlated with the interaction degree of the item.

$$C(v_i) = \log(\text{degree}(v_i) + 1), \tag{1}$$

where degree(v_i) denotes the number of user interactions associated with item i.

Confidence Adjustment Factor. This factor is controlled by the collaborative confidence, which characterizes the degree of reliance on collaborative signals.

$$G_{\mathrm{c}}(v_i) = \sigma(w_{\mathrm{c}} \cdot C(v_i) + b_{\mathrm{c}}), \tag{2}$$

where $w_{\mathrm{c}}, b_{\mathrm{c}} \in \mathbb{R}$ are learnable parameters, and $\sigma(\cdot)$ denotes the sigmoid function. For frequently interacted items, a larger $G_{\mathrm{c}}(v_i)$ indicates a stronger reliance on the collaborative view.

Divergence Degree. Relying solely on collaborative confidence is insufficient to handle the heterogeneity and noise introduced by popularity effects. Therefore, we further introduce a divergence factor to capture the discrepancy between collaborative and semantic embeddings. The divergence reveals the cognitive-level differences between the two representations: When an item's collaborative signals are contaminated by heterogeneous user preferences and interaction noise, causing the collaborative embedding to deviate from the semantic embedding, the divergence value becomes relatively high. In this case, the model adaptively increases the weight of the semantic embedding. First, we project the semantic embeddings to the same dimension as the collaborative embeddings via an adapter:

$$\mathbf{e}^{\mathrm{s}}_{v_i} = \mathbf{W}_1 \mathbf{e}^{\mathrm{LLM}}_{v_i} + \mathbf{b}_1, \tag{3}$$

where $\mathbf{W}_1 \in \mathbb{R}^{d_{\mathrm{c}} \times d_{\mathrm{LLM}}}$ and $\mathbf{b}_1 \in \mathbb{R}^{d_{\mathrm{c}}}$ are the weight matrix and bias of the adapter, respectively. The semantic embedding $\mathbf{e}^{\mathrm{LLM}}_{v_i}$ is retrieved from the semantic embedding matrix $\mathbf{E}^{V}_{\mathrm{s}}$, which stores the pretrained LLM embeddings of all items. Then, the divergence $D(v_i)$ is calculated as:

$$D(v_i) = \frac{1 - \cos(\mathbf{e}^{\mathrm{c}}_{v_i}, \mathbf{e}^{\mathrm{s}}_{v_i})}{2}, \tag{4}$$

where $\cos(\cdot,\cdot)$ denotes cosine similarity.

Divergence Adjustment Factor. This factor is controlled by the divergence degree, which characterizes the reliance on the semantic embedding,

$$G_{\mathrm{d}}(v_i) = \sigma(w_{\mathrm{d}} \cdot D(v_i) + b_{\mathrm{d}}), \tag{5}$$

where w_{d}, $b_{\mathrm{d}} \in \mathbb{R}$ are learnable parameters.

Final Fusion. To achieve adaptive weighting between the two embeddings, we jointly employ the confidence and divergence adjustment factors to dynamically determine the adaptive fusion weights,

$$\alpha_{v_i} = (1 - G_{\mathrm{c}}(v_i)) + G_{\mathrm{c}}(v_i) \cdot G_{\mathrm{d}}(v_i) \cdot \beta, \tag{6}$$

where β is a regulating coefficient that controls the influence strength of the divergence signal. The final item representation is therefore defined as:

$$\mathbf{e}^{\mathrm{f}}_{v_i} = \alpha_{v_i} \cdot \mathbf{e}^{\mathrm{s}}_{v_i} + (1 - \alpha_{v_i}) \cdot \mathbf{e}^{\mathrm{c}}_{v_i}. \tag{7}$$

Intuitively, for tail items, the fusion process assigns greater weight to semantic embeddings. In contrast, for head items, the model adjusts the weight of semantic embeddings based on the divergence adjustment factor under the guidance of the confidence adjustment factor.

2.3 View Consistency Constraint

Previous studies [14] aligned semantic and collaborative embeddings by maximizing the mutual information between them. However, such alignment applies uniform constraints to all items, ignoring the varying degrees of consistency between the two embeddings. This uniform constraint may weaken view diversity and discriminability when divergence is high. For low divergence items, which possess higher potential mutual information, enhancing alignment helps strengthen commonalities and improve preference modeling. Conversely, for high divergence items, enforcing strong alignment forces the model to ignore distinctive signals present in either the collaborative or semantic embedding. In such cases, weakening the alignment allows the fused representation to retain both stability and discriminability. To address this issue, we introduce a divergence-aware weighted mutual information constraint and reformulate the info loss as follows:

$$\begin{aligned}\mathcal{L}_{\text{info}} = -\frac{1}{|\mathcal{V}|} \sum_{v_i \in \mathcal{V}} (1 - D(v_i)) \Bigg[\log \frac{\exp(\cos(\mathbf{e}^{\mathrm{s}}_{v_i}, \mathbf{e}^{\mathrm{c}}_{v_i})/\tau)}{\sum_{v_j \in \mathcal{V}} \exp(\cos(\mathbf{e}^{\mathrm{s}}_{v_i}, \mathbf{e}^{\mathrm{c}}_{v_j})/\tau)} \\ + \log \frac{\exp(\cos(\mathbf{e}^{\mathrm{c}}_{v_i}, \mathbf{e}^{\mathrm{s}}_{v_i})/\tau)}{\sum_{v_j \in \mathcal{V}} \exp(\cos(\mathbf{e}^{\mathrm{c}}_{v_i}, \mathbf{e}^{\mathrm{s}}_{v_j})/\tau)} \Bigg],\end{aligned} \tag{8}$$

where τ is the temperature coefficient. A larger divergence leads the model to weaken the alignment constraint between semantic and collaborative embeddings, thereby preserving their complementary differences.

2.4 Popularity-Aware Negative Sampling

We feed the final item representations into the classical sequential recommendation model SASRec. In sequential recommendation, negative sampling plays a critical role during training. Conventional random negative sampling ensures sample diversity but often yields overly simple negative instances, limiting the effectiveness of the model learning. In contrast, popularity-based sampling tends to concentrate negative samples on head items, thereby weakening the ability of the model to distinguish long-tail items. To address this, we adopt a hybrid negative sampling strategy. First, we construct a probability distribution based on the interaction count of each item as follows:

$$P(v_i) = \frac{\text{degree}(v_i)^\lambda}{\sum_{v_j \in \mathcal{V}} \text{degree}(v_j)^\lambda}, \tag{9}$$

where the hyperparameter λ controls the smoothness of the sampling distribution. Specifically, a smaller λ indicates a tendency toward uniform random sampling, while a larger λ makes the sampling process fully dependent on item popularity.

Next, we introduce a mixing coefficient η to control the ratio between the two sampling schemes. During negative sampling, items are drawn with probability η from the popularity-based distribution and with probability $1 - \eta$ from the uniform distribution, thereby balancing the diversity and difficulty of negative samples.

2.5 Knowledge Distillation-Based User Enhancement

We design a knowledge distillationâĂŞbased enhancement strategy for each user, allowing the user to distill knowledge from teacher users with richer and more informative representations.

Teacher User Selection. We assume that a user's comprehensive preference should jointly reflect both collaborative and semantic characteristics. To this end, we construct a hybrid user representation for similarity retrieval. Specifically, we first apply the Principal Component Analysis (PCA) [13] to reduce the dimensionality of the LLM-generated user embeddings to match that of the collaborative embeddings. Then, the two types of user embeddings are concatenated to form a hybrid representation. Based on this mixed representation, we compute the cosine similarity between users and retrieve the top-K most similar users from the global user set as teacher candidates.

Enhanced Representation Generation. For a target user u, we obtain their teacher set T_u consisting of the top-K selected teacher users, and extract each teacher's dynamic interest representation learned from SASRec. These teacher representations are then fed into a simplified attention network to produce an enhanced user representation,

$$\mathbf{e}_u^* = \text{Attn}(\{\mathbf{e}_t' | t \in T_u\}), \tag{10}$$

where $\mathbf{e}_t'$ represents the dynamic interest representation of a teacher user $t \in T_u$, and the attention network performs a weighted aggregation over teacher embeddings, generating an enhanced dynamic representation that integrates collective knowledge from similar users.

Knowledge Distillation Loss. To transfer the teachers' knowledge to the target (student) user, we introduce a knowledge distillation loss, which constrains the model-generated original user representation $\mathbf{e}'_u$ to approach the enhanced representation $\mathbf{e}^*_u$,

$$\mathcal{L}_{\text{aug}} = \text{MSE}(\mathbf{e}'_u, \mathbf{e}^*_u), \tag{11}$$

where $\text{MSE}(\cdot, \cdot)$ is the mean squared error loss.

2.6 Overall Training Objective

Let the user representation obtained from the sequential model be denoted as $\mathbf{e}'_{u,t}$. At each time step t, the predicted score for the next item v_{t+1} is computed as:

$$r_{u,v_{t+1}} = \left(\mathbf{e}'_{u,t}\right)^\top \mathbf{e}^{\text{f}}_{v_{t+1}}, \tag{12}$$

where $\mathbf{e}'_{u,t}$ denotes the dynamic user representation at time step t obtained from the sequential model, and $\mathbf{e}^{\text{f}}_{v_{t+1}}$ represents the fused embedding of the candidate item v_{t+1}.

To optimize the ranking performance, we adopt a pairwise ranking loss defined as:

$$\mathcal{L}_{\text{seq}} = -\sum_{u \in \mathcal{U}} \sum_{k=1}^{n_u} \log \sigma \left(r_{u,v^+_{k+1}} - r_{u,v^-_{k+1}} \right), \tag{13}$$

where v^+_{k+1} denotes the positive item that the user actually interacted with, and v^-_{k+1} is a negative item sampled from a popularity-aware negative sampling distribution. n_u denotes the length of the interaction sequence of user u.

The overall loss function of our model consists of three components: the main sequential recommendation loss $\mathcal{L}_{\text{seq}}$, the knowledge distillation loss $\mathcal{L}_{\text{aug}}$, and the view alignment loss $\mathcal{L}_{\text{info}}$:

$$\mathcal{L} = \mathcal{L}_{\text{seq}} + \alpha \mathcal{L}_{\text{aug}} + \gamma \mathcal{L}_{\text{info}}, \tag{14}$$

where α and γ are hyperparameters controlling the weights of different objectives. By jointly optimizing these objectives, the model not only improves the overall recommendation accuracy but also enhances dual-view fusion capability and enhances the learning quality of tail users.

Table 1. Statistics of Preprocessed Datasets.

Dataset	#Users	#Items	Sparsity	Avg.length
Yelp	15,720	11,383	99.89%	12.23
Toys	19,412	11,924	99.93%	8.63
Pet	19,856	8,510	99.91%	7.95

3 Experiments

3.1 Experimental Settings

Datasets. We conduct experiments on three widely used real-world recommendation datasets: Yelp[1], Amazon Toys and Games, and Amazon Pet Supplies[2]. These datasets cover different application domains, allowing a comprehensive evaluation of the model's generalization ability. The statistics of the preprocessed datasets are shown in Table 1. Following standard preprocessing procedures in sequential recommendation research [9], we organize user interaction sequences in chronological order and divide them into training, validation, and test sets. Specifically, for each user, the last item in the interaction sequence is reserved for testing, while the penultimate item is used for validation.

Evaluation Metrics. We adopt standard top-K evaluation metrics widely used in recommender systems. The specific metrics are as follows: HR@10: Indicate whether the target item appears in the top-10 recommendations, reflecting recall ability; NDCG@10: Consider both correctness and ranking position, rewarding higher-ranked relevant items. Each positive instance is evaluated against 100 randomly sampled negative items.

Baselines. To validate the effectiveness of our proposed framework, we compare it with three categories of representative baseline models:

- **SASRec** [9] leverages a transformer architecture with self-attention to effectively capture long-term dependencies among items.
- **RLMRec-con** [14] maps collaborative embeddings into the semantic space and constrains their similarity difference with language embeddings.
- **RLMRec-gen** [14] maps language embeddings into the behavioral space and incorporates their similarity difference as a regularization term.
- **LLMInit** [3] initializes item embeddings in the recommendation model using semantic embeddings generated by LLMs to improve representation quality.
- **LLM-ESR** [11] adopts a dual-view modeling approach combining semantic initialization and adaptive projection to alleviate the long-tail problem for both users and items.
- **AlphaFuse** [7] introduces collaborative information into the null space of language embeddings to achieve complementary fusion between semantic and collaborative signals.

[1] https://www.yelp.com/dataset.
[2] https://cseweb.ucsd.edu/~jmcauley/datasets/amazon/links.html.

Table 2. Overall performance comparison of baseline models and our proposed method. Bold indicates the best result under each metric, and underlined values denote the second-best performance.

Dataset	Model	Overall		Tail User		Head User		Tail Item		Head Item	
		N@10	H@10	N@10	H@10	N@10	H@10	N@10	H@10	N@10	H@10
Yelp	SASRec	0.3586	0.5950	0.3589	0.5953	0.3576	0.5941	0.0534	0.1256	0.4485	0.7332
	RLMRec-con	0.3657	0.6042	0.3634	0.6008	0.3743	0.6171	0.0428	0.1026	0.4608	0.7519
	RLMRec-gen	0.3694	0.6043	0.3664	0.5989	0.3809	0.6247	0.0482	0.1130	0.4639	0.7489
	LLMInit	0.4045	0.6448	0.4053	0.6455	0.4014	0.6419	0.0717	0.1644	0.5025	0.7862
	LLMESR	0.4209	0.6660	0.4224	0.6662	0.4150	0.6655	0.0794	0.1815	0.5214	0.8086
	AlphaFuse	0.3999	0.6414	0.4015	0.6410	0.3937	0.6430	0.0738	0.1718	0.4958	0.7796
	Ours	**0.4381**	**0.6832**	**0.4404**	**0.6840**	**0.4295**	**0.6803**	**0.1177**	**0.2553**	**0.5325**	**0.8091**
	Improv.	4.09%	2.58%	4.26%	2.67%	3.49%	2.22%	48.24%	40.66%	2.13%	0.06%
Toys	SASRec	0.2816	0.4363	0.2712	0.4151	0.3012	0.4762	0.2025	0.3251	0.3755	0.5680
	RLMRec-con	0.2936	0.4547	0.2883	0.4425	0.3130	0.4990	0.1703	0.2737	0.3844	0.5880
	RLMRec-gen	0.2885	0.4462	0.2816	0.4329	0.3133	0.4948	0.1659	0.2733	0.3786	0.5735
	LLMInit	0.3565	0.5383	0.3581	0.5309	0.3467	0.5652	0.2474	0.3872	0.4368	0.6495
	LLMESR	0.3465	0.5414	0.3505	0.5387	0.3318	0.5513	0.2370	0.3920	0.4270	0.6514
	AlphaFuse	0.3567	0.5441	0.3597	0.5384	0.3459	0.5647	0.2475	0.3949	0.4371	0.6539
	Ours	**0.3654**	**0.5503**	**0.3704**	**0.5439**	**0.3473**	**0.5735**	**0.2617**	**0.4094**	**0.4418**	**0.6539**
	Improv.	2.44%	1.14%	2.97%	0.97%	0.40%	1.56%	5.74%	3.67%	1.08%	0.00%
Pet	SASRec	0.2774	0.4655	0.2748	0.4613	0.2850	0.4778	0.1016	0.2176	0.4065	0.6475
	RLMRec-con	0.2926	0.4789	0.2927	0.4781	0.2924	0.4811	0.1204	0.2296	0.4191	0.6620
	RLMRec-gen	0.2992	0.4828	0.3018	0.4852	0.2918	0.4758	0.1198	0.2285	0.4311	0.6697
	LLMInit	0.3590	0.5734	0.3618	0.5748	0.3508	0.5691	0.1800	0.3506	0.4906	0.7371
	LLMESR	0.3569	0.5758	0.3608	**0.5795**	0.3456	0.5651	0.1872	0.3682	0.4816	0.7283
	AlphaFuse	0.3595	0.5747	0.3617	0.5747	0.3532	0.5748	0.1754	0.3433	0.4948	**0.7447**
	Ours	**0.3727**	**0.5780**	**0.3763**	0.5779	**0.3621**	**0.5782**	**0.1924**	**0.3694**	**0.5051**	0.7311
	Improv.	3.67%	0.38%	4.04%	-	2.52%	0.59%	2.78%	0.33%	2.08%	-

Implementation Details. Our model is implemented in PyTorch. During training, we employ the Adam optimizer, with the learning rate set to 0.001 or 0.005 depending on the dataset. The batch size is set to 64 or 512, and the embedding dimension is fixed at 64 for all models.

To prevent overfitting, we apply an early stopping strategy with a patience of 20, monitoring NDCG@10 on the validation set. Semantic embeddings are generated using text-embedding-004 [10].

3.2 Overall Performance Comparison

We compared our proposed full model with all baseline methods across the three datasets. The experimental results (see Table 2) clearly demonstrate that our model achieves state-of-the-art performance on most datasets.

Table 3. The ablation study results on the Yelp dataset. Bold indicates the best result under each metric, and underlined values denote the second-best performance.

Dataset	Model	Overall		Tail User		Head User		Tail Item		Head Item	
		N@10	H@10	N@10	H@10	N@10	H@10	N@10	H@10	N@10	H@10
Yelp	**CDDF**	**0.4381**	**0.6832**	**0.4404**	**0.6840**	**0.4295**	**0.6803**	0.1177	0.2553	**0.5325**	**0.8091**
	- w/o KD	0.4363	0.6786	0.4384	0.6788	0.4285	0.6778	**0.1245**	**0.2643**	0.5281	0.8005
	- w/o DF	0.4331	0.6740	0.4341	0.6735	0.4289	0.6760	0.1112	0.2419	0.5278	0.8011
	- w/o DA	0.4282	0.6707	0.4295	0.6692	0.4231	0.6766	0.1029	0.2377	0.5240	0.7983

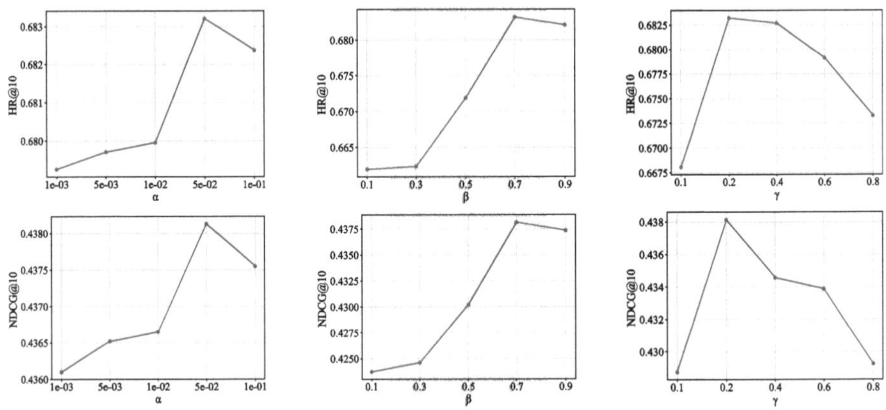

Fig. 3. Sensitivity analysis of hyperparameters on the Yelp dataset.

Sources of Performance Improvement. Compared with SASRec [9], our model enhances item representations by integrating dual-view information from graph and semantic embeddings, improving generalization and accuracy. Unlike LLMInit [3], which relies solely on LLM-generated initialization, our method combines frozen embeddings with trainable adapters to retain semantic knowledge and prevent catastrophic forgetting. Although LLM-ESR [11] also employs dual-view modeling, its static cross-attention mechanism fails to adapt to item-specific sparsity. Similarly, while AlphaFuse [7] mitigates semantic loss through subspace decomposition, it neglects the varying reliance of different items on semantic and collaborative signals.

3.3 Ablation Study

To verify the contribution of each key module, we conducted ablation experiments on the Yelp dataset, designing the following model variants: (1) **w/o knowledge distillation**: Remove the knowledge distillation module. (2) **w/o dynamic fusion**: Directly average collaborative and semantic embeddings. (3) **w/o divergence-aware alignment**: Remove the divergence component in the dual-view embedding alignment loss. The results (see Table 3) demonstrate

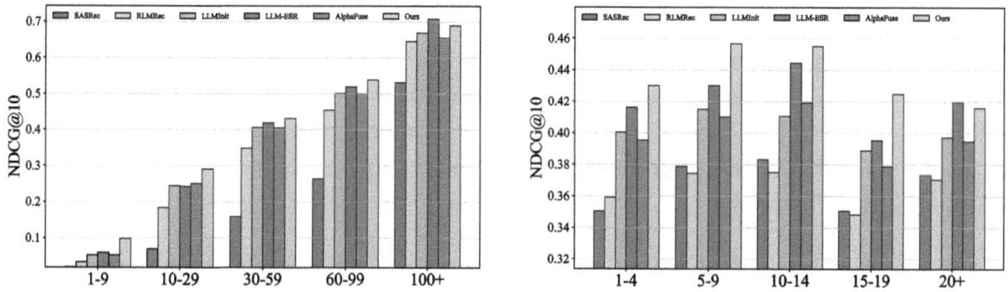

(a) Item Group (b) User Group

Fig. 4. Group evaluation results on the Yelp dataset.

that each component contributes significantly to overall performance: Removing dynamic fusion weakens the complementarity between views, removing distillation harms long-tail user performance, and removing divergence-aware alignment causes over-constrained embeddings. These results confirm the necessity of all key modules in our framework.

3.4 Hyperparameter Analysis

We analyze the sensitivity of three key hyperparameters: the distillation weight α, alignment weight γ, and fusion factor β (see Fig. 3). Results show that the model performs best at $\alpha = 0.005$, $\gamma = 0.2$, and $\beta = 0.7$. An excessively small α weakens knowledge transfer, while An excessively large one interferes with recommendation learning. Similarly, excessive γ overemphasizes alignment, and improper β distorts the fusion balance.

3.5 Group Analysis

To further assess the model's performance in long-tail scenarios, we partitioned the test set into five subsets based on the sparsity of user interactions and item popularity, and reported the model's HR@10 performance on each subset. As shown in Fig. 4, all models perform well on head entities, while performance significantly drops on tail entities, reflecting the prevalence of the long-tail problem. Nevertheless, our model achieves the best performance across most groups. Notably, it shows the largest relative improvement in the sparsest user group and the sparsest item group. This demonstrates that by integrating the dynamic fusion mechanism and knowledge distillation module, our framework can generate robust representations for users and items with sparse interactions, effectively mitigating the long-tail challenge in recommendation.

4 Conclusion

To address performance degradation from sparse signals and long-tail distributions, we propose a confidence- and divergence-aware dual-view fusion framework. It models collaborative and semantic features separately and adaptively fuses them based on item popularity and view divergence. A divergence-weighted alignment loss prevents over-alignment, while user-level knowledge distillation enhances preference representations. Experiments on multiple real-world datasets show that our framework consistently outperforms existing baselines, validating its effectiveness.

Acknowledgments. This work received funding from the Shandong Provincial Natural Science Foundation through project (ZR2023MF093).

Disclosure of Interests. The authors have no competing interests to declare that are relevant to the content of this article.

References

1. Bai, Z., Zheng, Y., Yang, P., Liu, S., Zhang, Y., Chang, Y.: DCRLRec: dual-domain contrastive reinforcement large language model for recommendation. Inform. Process. Manage. **62**(4), 104140 (2025)
2. Devlin, J., Chang, M.W., Lee, K., Toutanova, K.: BERT: pre-training of deep bidirectional transformers for language understanding. In: Proceedings of the 2019 Conference of the North American Chapter of the Association for Computational Linguistics: Human Language Technologies, volume 1 (Long and Short Papers), pp. 4171–4186 (2019)
3. Harte, J., Zorgdrager, W., Louridas, P., Katsifodimos, A., Jannach, D., Fragkoulis, M.: Leveraging large language models for sequential recommendation. In: Proceedings of the 17th ACM Conference on Recommender Systems, pp. 1096–1102 (2023)
4. He, X., Deng, K., Wang, X., Li, Y., Zhang, Y., Wang, M.: LightGCN: simplifying and powering graph convolution network for recommendation. In: Proceedings of the 43rd International ACM SIGIR Conference on Research and Development in Information Retrieval, pp. 639–648 (2020)
5. He, Y., Liu, X., Zhang, A., Ma, Y., Chua, T.S.: LLM2Rec: large language models are powerful embedding models for sequential recommendation. In: Proceedings of the 31st ACM SIGKDD Conference on Knowledge Discovery and Data Mining, vol. 2, pp. 896–907 (2025)
6. Hidasi, B., Karatzoglou, A., Baltrunas, L., Tikk, D.: Session-based recommendations with recurrent neural networks. arXiv preprint arXiv:1511.06939 (2015)
7. Hu, G., Zhang, A., Liu, S., Cai, Z., Yang, X., Wang, X.: AlphaFuse: learn id embeddings for sequential recommendation in null space of language embeddings. In: Proceedings of the 48th International ACM SIGIR Conference on Research and Development in Information Retrieval, pp. 1614–1623 (2025)
8. Hu, J., et al.: Enhancing sequential recommendation via LLM-based semantic embedding learning. In: Companion Proceedings of the ACM Web Conference 2024, pp. 103–111 (2024)
9. Kang, W.C., McAuley, J.: Self-attentive sequential recommendation. In: 2018 IEEE International Conference on Data Mining (ICDM), pp. 197–206. IEEE (2018)
10. Lee, J., et al.: Gecko: versatile text embeddings distilled from large language models. arXiv preprint arXiv:2403.20327 (2024)

11. Liu, Q., et al.: LLM-ESR: large language models enhancement for long-tailed sequential recommendation. Adv. Neural. Inf. Process. Syst. **37**, 26701–26727 (2024)
12. Mou, D., et al.: LLM-enhanced representation learning for graph collaborative filtering recommendation models. J. Intell. Inf. Syst. 1–24 (2025)
13. Pearson, K.: LIII. On lines and planes of closest fit to systems of points in space. London Edinburgh Dublin Philos. Mag. J. Sci. **2**(11), 559–572 (1901)
14. Ren, X., et al.: Representation learning with large language models for recommendation. In: Proceedings of the ACM Web Conference 2024, pp. 3464–3475 (2024)
15. Sun, F., et al.: BERT4Rec: sequential recommendation with bidirectional encoder representations from transformer. In: Proceedings of the 28th ACM International Conference on Information and Knowledge Management, pp. 1441–1450 (2019)
16. Wang, S., Hu, L., Wang, Y., Cao, L., Sheng, Q.Z., Orgun, M.: Sequential recommender systems: challenges, progress and prospects. arXiv preprint arXiv:2001.04830 (2019)
17. Wang, Y., et al.: LLMRG: improving recommendations through large language model reasoning graphs. In: Proceedings of the AAAI Conference on Artificial Intelligence, vol. 38, pp. 19189–19196 (2024)
18. Wang, Y., Sang, L., Zhang, Y., Zhang, Y.: Intent representation learning with large language model for recommendation. In: Proceedings of the 48th International ACM SIGIR Conference on Research and Development in Information Retrieval, pp. 1870–1879 (2025)
19. Wei, W., et al.: LLMRec: large language models with graph augmentation for recommendation. In: Proceedings of the 17th ACM International Conference on Web Search and Data Mining, pp. 806–815 (2024)
20. Wu, J., et al.: Self-supervised graph learning for recommendation. In: Proceedings of the 44th international ACM SIGIR Conference on Research and Development in Information Retrieval, pp. 726–735 (2021)
21. Yang, X., et al.: DARec: a disentangled alignment framework for large language model and recommender system. In: 2025 IEEE 41st International Conference on Data Engineering (ICDE), pp. 904–917. IEEE (2025)
22. Yin, H., Cui, B., Li, J., Yao, J., Chen, C.: Challenging the long tail recommendation. arXiv preprint arXiv:1205.6700 (2012)
23. Yuan, Z., et al.: Where to go next for recommender systems? Id-vs. modality-based recommender models revisited. In: Proceedings of the 46th International ACM SIGIR Conference on Research and Development in Information Retrieval, pp. 2639–2649 (2023)
24. Zhang, Y., Feng, F., Zhang, J., Bao, K., Wang, Q., He, X.: CoLLM: integrating collaborative embeddings into large language models for recommendation. IEEE Trans. Knowl. Data Eng. (2025)

LLM-Enhanced Reinforcement Learning for Long-Term User Satisfaction in Interactive Recommendation

Chongjun Xia[1], Yanchun Peng[2], and Xianzhi Wang[1(✉)]

[1] University of Technology Sydney, Sydney, Australia
Chongjun.Xia@student.uts.edu.au, XIANZHI.WANG@uts.edu.au
[2] University of Shanghai for Science and Technology, Shanghai, China

Abstract. Interactive recommender systems can dynamically adapt to user feedback, but often suffer from content homogeneity and filter bubble effects due to overfitting short-term user preferences. While recent efforts aim to improve content diversity, they predominantly operate in static or one-shot settings, neglecting the long-term evolution of user interests. Reinforcement learning provides a principled framework for optimizing long-term user satisfaction by modeling sequential decision-making processes. However, its application in recommendation is hindered by sparse, long-tailed user-item interactions and limited semantic planning capabilities. In this work, we propose LLM-Enhanced Reinforcement Learning (LERL), a novel hierarchical recommendation framework that integrates the semantic planning power of LLM with the fine-grained adaptability of RL. LERL consists of a high-level LLM-based planner that selects semantically diverse content categories, and a low-level RL policy that recommends personalized items within the selected semantic space. This hierarchical design narrows the action space, enhances planning efficiency, and mitigates overexposure to redundant content. Extensive experiments on real-world datasets demonstrate that LERL significantly improves long-term user satisfaction when compared with state-of-the-art baselines. The implementation of LERL is available at https://github.com/1163710212/LERL.

Keywords: Interactive Recommendation · Filter Bubble · Reinforcement Learning · Large Language Model

1 Introduction

Interactive recommender systems (IRS) enable real-time personalization through continuous adaptation to user feedback and have become integral to modern online platforms [15,20,33]. However, they often overly rely on users' short-term preferences due to overfitting, causing repeated exposure to semantically similar items or content homogeneity [10]. This exacerbates the well-known filter

C. Xia and Y. Peng—Contribute equally to this paper.

H. Jung et al. (Eds.): DASFAA 2026, LNCS 16535, pp. 454–470, 2026.
https://doi.org/10.1007/978-981-92-0363-5_28

bubble effect, i.e., users are only recommended content that aligns with their past behavior and viewpoints, diminishing user satisfaction in the long run.

Recent studies have increasingly focused on promoting content diversity to mitigate filter bubble effects, such as belief-aware re-ranking [3], representation learning that is sensitive to echo chambers [27], and diversity-aware candidate filtering [35]. Although these methods enhance one-shot diversity, they typically operate in static or one-shot settings and fail to consider the dynamic evolution of user preferences [10]. As a result, users may face content redundancy over time, reducing content diversity and users' long-term satisfaction. Recognizing this limitation, emerging research has shifted toward long-term recommendation, which aims to strategically optimize user satisfaction over extended interaction trajectories. Reinforcement learning (RL) provides a natural paradigm for this setting, given its capability to explicitly model sequential decision-making and optimize cumulative rewards. Recent RL-based approaches offer a promising path to filter bubble mitigation by modeling over-exposure dynamics [10], incorporating diversity-driven intrinsic rewards [25], and applying trajectory-level diversity regularization [7]. Despite active efforts, applying RL in recommender systems remains challenging due to the inherent sparsity and long-tail distribution of user-item interactions. Those characteristics hinder the recommender's ability to generalize and plan effectively, leading to suboptimal results [17].

Large Language Models (LLMs) emerge as a powerful tool for high-level reasoning and long-term planning, a capability derived from their pretraining on massive, diverse corpora [28]. Their strong generalization and multi-step inference capabilities make them well-suited for optimizing diversity-aware, long-term objectives. Nevertheless, LLMs often struggle to ground these abstract plans into fine-grained actions, particularly in recommender systems where the item space is extremely large and complex [29].

To address the above limitations with RL and LLMs, we propose LLM-Enhanced Reinforcement Learning (**LERL**), a novel hierarchical recommendation framework that integrates LLM-based semantic planning with RL-based policy optimization to leverage the complementary strengths of both approaches. LERL decomposes the recommendation process into two levels of activities: 1) **High-level semantic planner:** powered by LLM, this component selects a subset of content categories based on the user's interaction history. It prioritizes semantic diversity and mitigates overexposure by deprioritizing recently saturated content types. 2) **Low-level policy learner:** trained via RL, this component generates fine-grained item-level recommendations within the candidate space constrained by the high-level planner, thereby tailoring results to individual preferences. This hierarchical design enables the framework to leverage the structured world knowledge and long-horizon planning capabilities of LLM while retaining the adaptability and online learning strengths of RL. By semantically narrowing the action space and explicitly promoting diversity, LERL enhances recommendation stability and fosters long-term user satisfaction.

Our contributions in the paper are summarized as follows:

- We highlight the limitations of existing diversity-enhancing approaches in capturing long-term user satisfaction, particularly in scenarios of sparse and long-tailed user-item interactions.
- We propose a novel hierarchical recommendation framework (LERL) that integrates LLM-based planning with RL-based fine-grained policy learning to jointly optimize accuracy and diversity over time.
- Our extensive experiments on real-world datasets demonstrate that LERL outperforms state-of-the-art baselines in promoting long-term user satisfaction.

2 Related Work

2.1 RL for Recommendation

Reinforcement learning (RL)-based approaches formulate interactive recommendation as a Markov Decision Process (MDP). RL-based approaches have the unique advantage of optimizing long-term cumulative user satisfaction by explicitly modeling the sequential and dynamic nature of userâĂŞitem interactions. Recent advances have witnessed a surge of interest in applying RL to interactive recommender systems [7,18]. For example, KCRL [21] captures causal group effects and enhances reward estimation and state representation through a self-constructed userâĂŞitem knowledge graph; Li *et al.* [13] leverage an LSTM to jointly encode cross-view interaction histories and visual states for diversified interactive recommendation; HER4IF [34] models users' real-time preferences for popular content while enforcing fairness constraints to ensure long-term exposure fairness across item groups. Our method differs from prior RL-based approaches in introducing a hierarchical framework that leverages LLMs for high-level semantic planning—this enables more informed, diversity-aware decision-making across long horizons, complementing the fine-grained adaptability of RL in dynamic environments.

2.2 LLM for Recommendation

Large Language Model (LLM)-based recommendations have recently gained increasing attention, given the emergent capabilities of models such as GPT-4 [1], and Llama-3 [5] in reasoning, personalization, and interactive decision-making. Most existing LLM-powered recommendation paradigms focus on leveraging in-context learning [30,32] or fine-tuning [37] to generate immediate responses, such as item ranking or explanation generation. These methods excel at zero-shot adaptability and natural language interaction but often fall short in capturing long-term user satisfaction and behavioral dynamics. In contrast, our approach decomposes the recommendation process into high-level LLM-based semantic planning and low-level RL-based sequential optimization. The LLM module promotes category-level diversity by filtering overexposed content categories, while the RL agent further tailors item-level recommendations to user preferences. This new design enables the system to better sustain user engagement and long-term satisfaction.

2.3 Filter Bubbles in Recommendation

The filter bubble problem refers to the tendency of recommender systems to repeatedly surface homogeneous content, which limits users' exposure to diverse information due to overfitting user history [2,26]. Prolonged exposure to such homogenized recommendations can lead to cognitive fatigue, reduced novelty perception, and ultimately, a decline in long-term user satisfaction [2]. Therefore, recent work has introduced diversity-aware recommendation strategies to address filter bubbles and improve long-term user engagement. Typically, TD-VAE-CF [11] and DOR [4] enhance content diversity through static re-ranking or diversity-aware embeddings. These approaches, however, lack sequential modeling and cannot optimize long-term outcomes. Reinforcement learning (RL), in contrast, naturally supports long-horizon optimization in interactive recommendation, and recent RL-based methods have incorporated diversity signals to mitigate filter bubbles [7,14]. For example, CIRS [10] penalizes item overexposure via satiation modeling, and DNaIR [25] leverages diversity-driven intrinsic rewards to promote exploration. However, these methods struggle with sparse and long-tail data, leading to suboptimal recommendations. Our proposed LERL framework complements the above efforts by incorporating an LLM-based semantic planner that directs exploration toward semantically diverse content regions. Through its hierarchical structure, LERL enables the system to make informed, diversity-aware planning decisions while preserving the adaptability and feedback-driven learning advantages of RL.

3 Preliminary

3.1 Problem Formulation

We consider an interactive recommendation scenario where an agent engages with a user over a sequence of decision steps. At each time step t, the agent observes a state s_t derived from the user's interaction history H_t, and selects a recommendation list $a_t \subseteq \mathcal{I}$; in response, the environment returns a reward r_t and transitions to the next state s_{t+1}. The goal is to learn a policy π that maximizes long-term user satisfaction by balancing recommendation accuracy and content diversity. To this end, we adopt a hierarchical decision-making framework composed of two modules:

- **High-Level Semantic Planner:** A LLM-based agent analyzes the user's interaction trajectory and selects a subset of candidate categories $c_t \subseteq \mathcal{C}$, discouraging overexposed content types.
- **Low-Level Policy Learner:** A RL-based agent selects the recommendation list $a_t \subseteq \mathcal{I}_{c_t}$ to maximize the expected long-term reward, where $\mathcal{I}_{c_t}$ denotes the set of items in the selected categories c_t.

The interaction trajectory can be represented as:

$$\mathcal{T}_{1:N} = \{(s_1, c_1, a_1, r_1), \ldots, (s_N, c_N, a_N, r_N)\}, \tag{1}$$

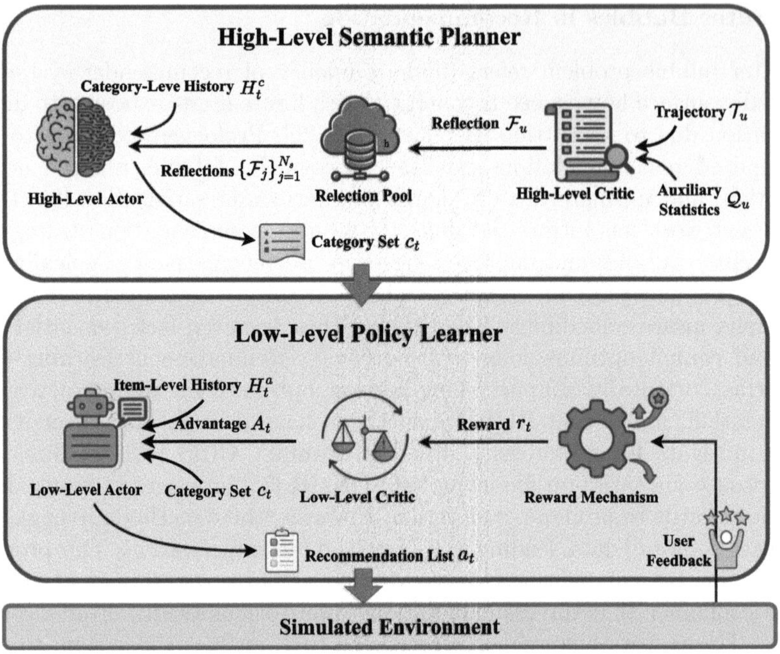

Fig. 1. Overview of the LERL framework.

where c_t is the category set selected by the LLM at step t, and $a_t \subseteq \mathcal{I}_{c_t}$ is the item chosen by the RL agent. The optimization objective is defined as:

$$\max_{\pi} \mathbb{E}_{\pi} \left[\sum_{t=1}^{N} \gamma^{t-1} r_t \right], \tag{2}$$

where γ is the discount factor that balances immediate and long-term rewards.

3.2 Simulated Environment

As collecting online user feedback can be financially burdensome, following prior work [38], we construct a simulated offline environment based on logged user-item interaction datasets. The simulator emulates real-time user feedback and session-level behaviors, including item-level engagement and user dropout. This enables safe and reproducible evaluation of interactive recommendation policies under realistic dynamics. More details about the simulated environment can be found in Sect. 5.1.

4 Methodology

We propose a hierarchical agent framework, LERL, which integrates two levels of decision-making for interactive recommendation: a high-level semantic planner (HSP) powered by an LLM and a low-level policy learner (LPL) optimized via RL. The LLM-based agent generates category-level constraints by determining the candidate item categories, while the RL-based agent performs fine-grained item selection within these constraints. This hierarchical design effectively balances category-level diversity and user preference alignment, promoting long-term user satisfaction. The overall architecture of LERL is shown in Fig. 1.

4.1 High-Level Semantic Planner

The HSP functions as a macro-level decision-maker, exerting strategic control over category-level exposure. By leveraging a pretrained LLM, the HSP performs semantic planning over user intent and interaction context at the category granularity, effectively balancing recommendation accuracy and content diversity.

High-Level Actor. At each decision step t, the high-level actor selects a category set $c_t \subseteq \mathcal{C}$, where $\mathcal{C}$ denotes the set of all candidate item categories. This decision is based on the user's category-level interaction history and a set of reflections that summarize the complete interaction trajectories of other users. Formally, the user's interaction history up to step t is denoted as:

$$H_t^c = \{(c_i, r_i)\}_{i=1}^{t}, \tag{3}$$

where c_i is the selected category set at step i, and r_i is the user's immediate satisfaction signal (e.g., click, rating).

To enhance decision-making, we maintain a *reflection pool* $\mathcal{R} = \{(\mathcal{F}_u, S_u)\}_{u=1}^{N_r}$, where each $\mathcal{F}_u$ denotes the category-level reflection (generated by the high-level critic) over the full interaction trajectory of a past user u. The associated score $S_u = \sum_{t=1}^{T_u} r_t$ represents the cumulative reward, indicating their long-term satisfaction. T_u denotes the length of the interaction session for user u. Given the limited context length of LLM, it is infeasible to incorporate the entire reflection pool into a single prompt. Therefore, we design a sampling mechanism based on users' cumulative rewards to prioritize more informative reflections. Specifically, the sampling distribution $P(u)$ over the reflection pool is defined as follows:

$$P(u) = \frac{\exp(\alpha S_u)}{\sum_{v=1}^{N_r} \exp(\alpha S_v)}, \tag{4}$$

where α is a temperature parameter controlling the sharpness of the distribution.

Then, a set of N_s reflections $\{\mathcal{F}_j\}_{j=1}^{N_s}$ is sampled from the full reflection pool $\mathcal{R}$ according to a user-specific distribution $P(u)$. The selected reflections serve as high-level semantic guidance for planning. Note that in the early stages of interaction, when no user sessions have been completed, the reflection pool is empty, and the set $\{\mathcal{F}_j\}_{j=1}^{N_s}$ is substituted with empty placeholders.

Finally, the category set is selected via prompt-based querying of the LLM:

$$c_t = \mathrm{LLM}\left(\mathrm{Prompt}_{\mathrm{a}}(\mathcal{C}, H_t^c, \{\mathcal{F}_j\}_{j=1}^{N_s})\right), \tag{5}$$

where $\mathrm{Prompt}_{\mathrm{a}}$ is a structured natural language template that integrates candidate categories $\mathcal{C}$, category-level interaction history H_t^c, and the sampled reflections $\{\mathcal{F}_j\}_{j=1}^{N_s}$ (an example is shown in Fig. 2). This design enables the LLM to perform high-level semantic planning, striking a balance between recommendation relevance and content diversity.

[Candidate Categories]
Available categories for selection: $\mathcal{C}$

[User Interaction History]
H_t^c

[Reflective Summaries]
These summaries are derived from users who ultimately achieved a high long-term satisfaction: $\{\mathcal{F}_j\}_{j=1}^{N_s}$

[Your Task]
You are a recommendation planning assistant. Your task is to choose the next item categories to recommend based on the user's interaction history and reflective summaries ... Your goal is to maximize user long-term satisfaction.

[Answer]
Output strictly as a Python list of selected category integers, e.g., [1, 2, 3]. No explanation, prefix, or formatting.

Fig. 2. Prompt design for the high-level actor, which selects category set by integrating category-level interaction history and sampled reflections.

High-Level Critic. We propose a *reflective language-based critic* to support self-improvement in high-level decision-making. Unlike traditional scalar reward signals, this module outputs a textual reflection $\mathcal{F}_u$ at the end of user u's session, encapsulating actionable insights that contribute to enhancing long-term user satisfaction. The reflection is generated as follows:

$$\mathcal{F}_u = \mathrm{LLM}\left(\mathrm{Prompt}_{\mathrm{v}}(\mathcal{C}, \mathcal{T}_u, \mathcal{Q}_u)\right), \tag{6}$$

where $\mathrm{Prompt}_{\mathrm{v}}$ is a structured natural language template (an example is shown in Fig. 3). $\mathcal{C}$ denotes the set of candidate categories, and $\mathcal{Q}_u$ represents auxiliary statistics, e.g., *interaction length* and *cumulative reward*. $\mathcal{T}_u = \{(c_i, r_i)\}_{i=1}^{T}$

denotes the complete category-level interaction trajectory. These reflections serve as high-level semantic critiques, guiding the model toward strategies that promote long-term user engagement. The generated reflections are stored in the reelection pool and reused in future prompts to inform and condition the high-level actor's decision-making.

[Candidate Categories]
Available categories for selection: $\mathcal{C}$

[User Interaction Summary]
The user's interactions during this session: $\mathcal{T}_u$
Session statistics: $\mathcal{Q}_u$

[Your Task]
Analyze the session in terms of:
1. Repetition Effects: negative impact of repeated exposure.
2. Actionable Insights: strategies to improve long-term satisfaction.

[Answer]
Use natural language.

Fig. 3. Prompt design for the high-level critic, which generates the reflection from category-level interactions and auxiliary statistics in a user's session.

4.2 Low-Level Policy Learner

The LPL focuses on item selection. It adheres to high-level constraints c_t while optimizing for user satisfaction signals, e.g., clicks or ratings. This design allows the LPL to specialize in fine-grained ranking quality while delegating strategic control to the HSP.

Low-Level Actor. The low-level actor generates a recommendation list a_t conditioned on the low-level state s_t and the high-level constraint c_t. The low-level state s_t encodes the user's item-level interaction history $H_t^a = \{(a_i, r_i)\}_{i=1}^t$, where r_i is the user's immediate satisfaction signal, e.g., a click or rating.

First, we adopt a Transformer-based encoder to capture the sequential dependencies among recent interactions. Specifically, each interaction is passed through an embedding layer to produce a sequence of hidden representations $V_t = \{v_1, v_2, \ldots, v_t\}$, which is fed into the Transformer to derive user preference features:

$$e_t^p = \text{Transformer}(v_1, v_2, \ldots, v_t) \tag{7}$$

The preference representation e_t^p is then passed through a multi-layer perceptron (MLP) to parameterize a Gaussian distribution with mean μ_t and standard deviation σ_t. This stochastic modeling facilitates continuous exploration in the policy space and helps prevent premature convergence to suboptimal solutions. To capture the user's latent intent in the embedding space, a virtual item embedding p_t is sampled from the above distribution as follows:

$$p_t \sim \mathcal{N}(\mu_t, \sigma_t^2). \tag{8}$$

The similarity score between each candidate item embedding i_j and p_t is computed via dot product:

$$a_j^{\text{sim}} = \langle p_t, i_j \rangle, \quad \forall j \in \{1, \ldots, |\mathcal{I}|\} \tag{9}$$

Second, we apply category-level constraints guided by the high-level action c_t as a soft filter over candidate items. Specifically, we calculate the corresponding category mask as:

$$a^{\text{mask}} = W \cdot c_t, \tag{10}$$

where $W \in \{0, 1\}^{|\mathcal{I}| \times |\mathcal{C}|}$ is a binary item-category mapping matrix. Each row of W corresponds to an item $i \in \mathcal{I}$, and each column corresponds to a category $c \in \mathcal{C}$. An entry of 1 indicates the item belongs to the corresponding category.

Last, we apply the category mask to the similarity scores to compute the item-level recommendation scores (see Eq. (11)) and select the top-k items with the highest values in a^{score} to form the recommendation list a_t.

$$a^{\text{score}} = a^{\text{sim}} \odot a^{\text{mask}}, \tag{11}$$

where $\odot$ denotes element-wise multiplication.

Low-Level Critic. We adopt a value estimator in the form of a state-value function $V_\varphi(s_t)$ to guide the learning of the low-level actor. This value function is approximated using a multi-layer perceptron:

$$V_\varphi(s_t) = \text{MLP}_v(s_t), \tag{12}$$

where s_t denotes the low-level state at time step t.

The low-level agent is optimized using the proximal policy optimization (PPO) algorithm [23], in light of its strong empirical performance and stability in policy gradient updates. The critic estimates the expected return from state s_t, and is trained by minimizing the temporal difference error:

$$\mathcal{L}_v = \mathbb{E}_t \left[(V_\varphi(s_t) - R_t)^2 \right], \quad R_t = r_t + \gamma V_{\varphi'}(s_{t+1}), \tag{13}$$

where φ and φ' denote the parameters of the online and target critic networks, respectively; r_i is the user's immediate satisfaction signal (e.g., click or rating); and $\gamma \in [0, 1]$ is the discount factor.

The corresponding actor is trained by maximizing the clipped surrogate objective of PPO, which balances learning efficiency and stability:

$$\mathcal{L}_a = -\mathbb{E}_t \left[\min \left(\frac{\pi_\theta (a_t \mid s_t)}{\pi_{\theta_{\text{old}}} (a_t \mid s_t)} \hat{A}_t, \text{ clip} \left(\frac{\pi_\theta (a_t \mid s_t)}{\pi_{\theta_{\text{old}}} (a_t \mid s_t)}, 1 - \epsilon, \ 1 + \epsilon \right) \hat{A}_t \right) \right], \tag{14}$$

where θ and θ_{old} are the parameters of the main and old policy networks, respectively. $\hat{A}_t$ is the estimated advantage function, and ϵ is a hyperparameter that bounds the policy update ratio to prevent destructive updates. The $\text{clip}(x, a, b)$ function restricts x within the interval $[a, b]$. For further algorithmic details and theoretical justification, please refer to the original paper [23].

Table 1. Statistics of experimental datasets

Dataset	#Users	#Items	#Interactions	Density
KuaiRand	27,077	7,551	1,436,609	0.70%
KuaiRec	7,176	10,728	12,530,806	16.3%

5 Experiments

This section reports our empirical evaluation of the proposed method in interactive recommendation settings. Specifically, we designed experiments to address the following questions:

- **RQ1:** How does LERL compare to state-of-the-art RL-based methods for interactive recommendation?
- **RQ2:** How well does LERL mitigate the filter bubble effect in recommendation trajectories?
- **RQ3:** How do core modules of LERL contribute to the overall recommendation performance?

5.1 Experiment Setup

Simulation Environment. Following prior work [25,36], we conducted simulation-based experiments in the KuaiSim [38], as direct online evaluation can be prohibitively costly. KuaiSim is a comprehensive user simulator that supports request-level, whole-session, and cross-session recommendation tasks. In this work, we leveraged KuaiSim to conduct whole-session recommendation experiments. Within this framework, the environment provides the agent with sequential interaction signals, including predicted rewards and dynamically evolving user states based on historical behaviors. This simulation setup enables a systematic evaluation of models in their ability to mitigate filter bubble effects and improve long-term user satisfaction. Simulation environments were constructed based on two high-quality datasets: KuaiRand (pure, 15 policies) [8] and KuaiRec (big matrix) [9]. For both datasets, the "is_click"– field was used to represent user interest and serves as the reward signal in the simulated environment. Table 1 shows basic statistics of the two datasets.

Following [38], we set the maximum interaction length to 20, considering that real users typically engage in a limited number of interactions within a single session. As such, each user is allowed to generate up to 20 interactions per session in the simulation environment. The length of each recommendation list was set to 6. To effectively capture the impact of filter bubbles, we introduced a **diversity-aware quit mechanism** inspired by prior work [10,22]. Specifically, the user's maximum remaining interaction count is decremented by one whenever the recommendation agent consecutively generates recommendation lists containing items from the same category. This penalty reflects user boredom caused by excessive exposure to homogeneous content, thereby simulating the early termination of sessions under low content diversity.

Evaluation Metrics. Following prior studies [36,38], we evaluated the performance of interactive recommendation approaches using three metrics:

- Interaction length ($\mathbf{T_{int}}$) denotes the number of interactions before the session terminates. Owing to the diversity-aware quit mechanism, a larger $\mathbf{T_{int}}$ indicates that the model provides more diverse recommendations.
- Cumulative reward ($\mathbf{R_{cum}}$) measures the total reward accumulated throughout the session, formally defined as $\mathbf{R_{cum}} = \sum_{t=1}^{\mathbf{T_{int}}} r_t$, where r_t represents the reward at step t (e.g., click). This metric reflects overall user satisfaction during the session.
- Single-round reward ($\mathbf{R_{sin}}$) is computed as the average reward per interaction step, i.e., $\mathbf{R_{sin}} = \frac{\mathbf{R_{cum}}}{\mathbf{T_{int}}}$. While $\mathbf{R_{sin}}$ captures per-step effectiveness in satisfying user preference, it does not necessarily correlate with long-term satisfaction.

Baselines. To comprehensively and fairly evaluate the effectiveness of LERL, we made comparisons with representative RL-based and diversity-aware RL approaches. Specifically, we included the following baselines:

- PG [31]: a fundamental policy gradient method that directly optimizes the expected return via stochastic gradient ascent;
- A2C [12]: an actor-critic framework that synchronously updates value and policy networks to stabilize learning;
- DDPG [16]: a deterministic actor-critic algorithm with target networks and soft updates;
- TD3 [6]: an improved variant of DDPG that employs twin critics and policy smoothing to reduce overestimation bias;
- PPO [23]: a clipped surrogate objective method that constrains policy updates for stable and efficient training;

- HAC [19]: a latent-action RL framework that infers vectorized hyper-actions to generate recommendations, with alignment and supervision for stable training;
- SAC4IR [24]: an entropy-regularized RL approach with a debiasing-aware reward model designed for interactive recommendation scenarios;
- DNaIR [25]: a diversity-aware RL framework that integrates item similarity, diversity, and quality signals into the reward function.

Implementation Details. All methods were fine-tuned in simulated environments. For PG and A2C, the discount factor was set to 0.9. For DDPG and TD3, the replay buffer size was set to 20,000. The PPO clipping threshold ϵ_{clip} was set to 0.8. For HAC, the hyper-action alignment coefficient λ_h was set to 0.1. For SAC4IR, the relaxation parameter was set to 0.8. For DNaIR, the minibatch size was set to 64 with $\beta = 0.1$. For our method LERL, the reflection pool size N_r was set to 200, and the number of sampled reflections N_s was set to 3. The LLM used for the high-level module was Llama-3-8B [5]. The code is available at https://github.com/1163710212/LERL.

Table 2. Results of all methods in two environments (**Bold**: Best; <u>Underline</u>: Runner-up).

Env	Method	T_{int}	R_{sin}	R_{cum}
KuaiRand	PG	11.852 ± 0.305	0.618 ± 0.033	7.476 ± 0.399
	A2C	11.795 ± 0.373	0.650 ± 0.033	7.599 ± 0.507
	DDPG	11.390 ± 0.345	0.702 ± 0.037	8.117 ± 0.515
	TD3	11.705 ± 0.318	0.581 ± 0.039	6.734 ± 0.542
	PPO	<u>14.352 ± 0.608</u>	0.593 ± 0.039	8.310 ± 0.560
	HAC	11.795 ± 0.648	0.671 ± 0.050	7.902 ± 0.602
	SAC4IR	11.895 ± 0.531	0.650 ± 0.031	7.751 ± 0.584
	DNaIR	13.500 ± 0.378	**0.720 ± 0.028**	<u>9.824 ± 0.540</u>
	LERL	**17.238 ± 0.783**	<u>0.719 ± 0.038</u>	**12.284 ± 0.493**
KuaiRec	PG	11.681 ± 0.263	0.695 ± 0.031	8.118 ± 0.418
	A2C	<u>13.305 ± 0.600</u>	0.428 ± 0.033	5.791 ± 0.366
	DDPG	11.552 ± 0.557	0.575 ± 0.057	6.574 ± 0.631
	TD3	11.043 ± 0.192	**0.800 ± 0.046**	8.771 ± 0.489
	PPO	11.914 ± 0.315	0.676 ± 0.047	7.969 ± 0.573
	HAC	11.000 ± 0.000	0.546 ± 0.030	5.953 ± 0.271
	SAC4IR	11.167 ± 0.194	0.522 ± 0.036	5.781 ± 0.516
	DNaIR	13.105 ± 0.381	<u>0.710 ± 0.032</u>	<u>9.235 ± 0.694</u>
	LERL	**16.400 ± 0.474**	0.637 ± 0.032	**10.507 ± 0.543**

5.2 Main Comparison (RQ1)

To assess LERL's effectiveness in promoting long-term user satisfaction, we conducted comprehensive comparisons against state-of-the-art baselines across two datasets. Our comparison results (Table 2) reveal the following findings:

- LERL consistently achieved the best overall performance in both environments. Specifically, LERL outperformed all baselines in terms of interaction length (T_{int}) and cumulative reward (R_{cum}), demonstrating its strong ability to encourage prolonged engagement and enhance long-term user satisfaction.
- DNaIR showed moderate improvements due to its incorporation of item diversity into the reward signal and policy design. However, the gains in T_{int} and R_{cum} remained limited, suggesting that static or manually designed diversity rewards are ineffective in long-horizon optimization.
- Traditional RL algorithms fall short on long-term metrics. Methods like TD3 and DDPG achieved reasonable single-round reward (R_{sin}), but exhibited significantly lower T_{int} and R_{cum}, highlighting their limitations in addressing content redundancy and dynamically evolving user preferences in interactive recommendation settings.

5.3 Case Study (RQ2)

We use a case study on the KuaiRec dataset to demonstrate the effectiveness of our method in mitigating the filter bubble effect. Specifically, we sampled the recommendation trajectory of a user (ID: 5) and compared the results of three consecutive recommendation rounds generated by LERL and the PPO baseline (shown in Fig. 4), where each circle denotes an item, and the number in each circle indicates the category to which the item belongs.

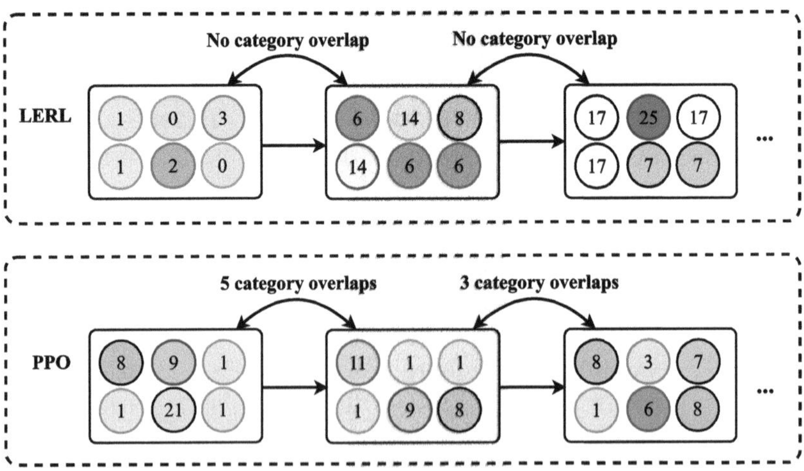

Fig. 4. Recommendation trajectories of LERL and PPO.

The recommendation lists generated by LERL exhibited no repeated item-categories throughout the three rounds; in contrast, the PPO-based lists contained a high degree of redundancy in category exposure across consecutive rounds. This contrast highlights LERL's ability to diversify category-level exposure over time, demonstrating its superiority in alleviating content homogeneity and fostering a more engaging long-term user experience.

5.4 Ablation Study (RQ3)

We conducted ablation studies to assess the contribution of the two core components in the LERL framework. We denote by **w/o hsp** the variant of LERL without the high-level semantic planner, and **w/o hc** the variant that removes the high-level critic. We have focused on these two components because they represent the key innovations of LERL: the high-level planner controls diversity by generating category-level constraints, while the high-level critic offers long-term semantic reflections based on user trajectories.

Our ablation results (Fig. 5) show that removing either component leads to notable performance degradation in both environments. In particular, eliminating the high-level planner (**w/o hsp**) results in a substantial drop in interaction length and cumulative reward, highlighting the role of HSP in maintaining long-term engagement and satisfaction through strategic category planning. Removing the reflective critic (**w/o hc**) adversely impacts all three evaluation metrics, indicating the crucial role of semantic reflections in shaping high-level decisions.

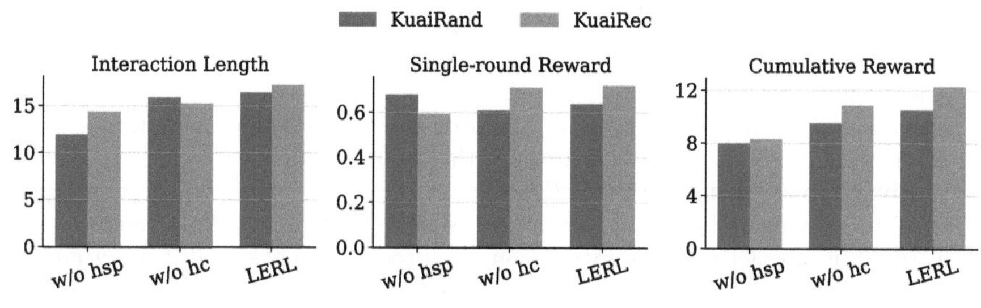

Fig. 5. The results of the ablation study.

6 Conclusion

This paper presents LERL, a novel hierarchical recommendation framework that integrates LLM and RL to enhance long-term user satisfaction in interactive recommender systems. LERL decomposes the recommendation process into two levels: a high-level semantic planner that selects semantically diverse content categories using LLM, and a low-level RL policy that performs fine-grained item selection under these constraints. This design effectively narrows the action space, mitigates overexposure to redundant content, and promotes semantic diversity throughout user interactions. Comprehensive experiments on simulated environments demonstrate that LERL consistently outperforms state-of-the-art baselines in long-term user satisfaction.

A promising direction for future work is to extend LERL into a multi-objective optimization framework that incorporates auxiliary goals, including fairness, diversity, and accuracy. Integrating more advanced prompting strategies and leveraging continual learning for the high-level planner also holds potential for further enhancing adaptability and robustness.

Disclosure of Interests. The authors have no competing interests to declare that are relevant to the content of this article.

References

1. Achiam, J., Adler, S., et al.: GPT-4 technical report. arXiv preprint arXiv:2303.08774 (2023)
2. Areeb, Q.M., Nadeem, M., et al.: Filter bubbles in recommender systems: fact or fallacy—a systematic review. Wiley Interdisc. Rev.: Data Min. Knowl. Discov. **13**(6), e1512 (2023)
3. Aridor, G., Goncalves, D., Sikdar, S.: Deconstructing the filter bubble: user decision-making and recommender systems. In: Proceedings of the 14th ACM Conference on Recommender Systems, pp. 82–91 (2020)
4. Du, Y., Ranwez, S., Sutton-Charani, N., Ranwez, V.: Is diversity optimization always suitable? toward a better understanding of diversity within recommendation approaches. Inf. Process. Manage. **58**(6), 102721 (2021)
5. Dubey, A., Jauhri, A., et al.: The LLaMA 3 herd of models. arXiv e-prints pp. arXiv–2407 (2024)
6. Fujimoto, S., Hoof, H., Meger, D.: Addressing function approximation error in actor-critic methods. In: Proceedings of the International Conference on Machine Learning, pp. 1587–1596. PMLR (2018)
7. Gao, C., Huang, K., et al.: Alleviating Matthew effect of offline reinforcement learning in interactive recommendation. In: Proceedings of the 46th International ACM SIGIR Conference on Research and Development in Information Retrieval, pp. 238–248 (2023)
8. Gao, C., Li, S., Zhang, Y., et al.: KuaiRand: an unbiased sequential recommendation dataset with randomly exposed videos. In: Proceedings of the 31st ACM International Conference on Information & Knowledge Management, pp. 3953–3957 (2022)
9. Gao, C., Li, S., et al.: KuaiRec: a fully-observed dataset and insights for evaluating recommender systems. In: Proceedings of the 31st ACM International Conference on Information & Knowledge Management, pp. 540–550 (2022)
10. Gao, C., Wang, S., Li, S., Chen, J., et al.: CIRS: bursting filter bubbles by counterfactual interactive recommender system. ACM Trans. Inf. Syst. **42**(1), 1–27 (2023)
11. Gao, Z., Shen, T., et al.: Mitigating the filter bubble while maintaining relevance: targeted diversification with VAE-based recommender systems. In: Proceedings of the 45th International ACM SIGIR Conference on Research and Development in Information Retrieval, pp. 2524–2531 (2022)

12. Konda, V., Tsitsiklis, J.: Actor-critic algorithms. In: Advances in Neural Information Processing Systems, vol. 12 (1999)
13. Li, Y., Qi, Y., Shi, Y., Chen, Q., Cao, N., Chen, S.: Diverse interaction recommendation for public users exploring multi-view visualization using deep learning. IEEE Trans. Visual Comput. Graphics **29**(1), 95–105 (2022)
14. Li, Z., Dong, Y., Gao, C., et al.: Breaking filter bubble: a reinforcement learning framework of controllable recommender system. In: Proceedings of the ACM Web Conference, pp. 4041–4049 (2023)
15. Liang, Y., Ponnada, A., Lamere, P., Daskalova, N.: Enabling goal-focused exploration of podcasts in interactive recommender systems. In: Proceedings of the 28th International Conference on Intelligent User Interfaces, pp. 142–155 (2023)
16. Lillicrap, T.: Continuous control with deep reinforcement learning. arXiv preprint arXiv:1509.02971 (2015)
17. Lin, Y., et al.: A survey on reinforcement learning for recommender systems. IEEE Trans. Neural Netw. Learn. Syst. **35**(10), 13164–13184 (2024)
18. Liu, H., Cai, K., Li, P., Qian, C., Zhao, P., Wu, X.: REDRL: a review-enhanced deep reinforcement learning model for interactive recommendation. Expert Syst. Appl. **213**, 118926 (2023)
19. Liu, S., et al.: Exploration and regularization of the latent action space in recommendation. In: Proceedings of the ACM Web Conference, pp. 833–844 (2023)
20. Ma, D., Wang, Y., Ma, J., Jin, Q.: SGNR: a social graph neural network based interactive recommendation scheme for e-commerce. Tsinghua Sci. Technol. **28**(4), 786–798 (2023)
21. Nie, W., et al.: Knowledge-enhanced causal reinforcement learning model for interactive recommendation. IEEE Trans. Multimedia **26**, 1129–1142 (2023)
22. Quan, Y., Ding, J., Gao, C., et al.: Alleviating video-length effect for micro-video recommendation. ACM Trans. Inf. Syst. **42**(2), 1–24 (2023)
23. Schulman, J., Wolski, F., Dhariwal, P., Radford, A., Klimov, O.: Proximal policy optimization algorithms. arXiv preprint arXiv:1707.06347 (2017)
24. Shi, X., Liu, Q., Xie, H., Bai, Y., Shang, M.: Maximum entropy policy for long-term fairness in interactive recommender systems. IEEE Trans. Serv. Comput. **17**(3), 1029–1043 (2024)
25. Shi, X., Liu, Q., et al.: Relieving popularity bias in interactive recommendation: a diversity-novelty-aware reinforcement learning approach. ACM Trans. Inf. Syst. **42**(2), 1–30 (2023)
26. Tahir Kidwai, U., Akhtar, N., et al.: Mitigating filter bubbles: diverse and explainable recommender systems. J. Intell. Fuzzy Syst. JIFS–219416 (2024)
27. Tommasel, A., Rodriguez, J.M., Godoy, D.: I want to break free! Recommending friends from outside the echo chamber. In: Proceedings of the 15th ACM Conference on Recommender Systems, pp. 23–33 (2021)
28. Wang, L., Ma, C., Feng, X., et al.: A survey on large language model based autonomous agents. Front. Comput. Sci. **18**(6), 186345 (2024)
29. Wei, H., Zhang, Z., He, S., Xia, T., Pan, S., Liu, F.: PlangenLLMs: a modern survey of LLM planning capabilities. In: Proceedings of the 63rd Annual Meeting of the Association for Computational Linguistics (Volume 1: Long Papers), pp. 19497–19521 (2025)
30. Wei, W., Ren, X., et al.: LLMRec: large language models with graph augmentation for recommendation. In: Proceedings of the 17th ACM International Conference on Web Search and Data Mining, pp. 806–815 (2024)
31. Williams, R.J.: Simple statistical gradient-following algorithms for connectionist reinforcement learning. Mach. Learn. **8**, 229–256 (1992)

32. Xi, Y., Liu, W., et al.: Towards open-world recommendation with knowledge augmentation from large language models. In: Proceedings of the 18th ACM Conference on Recommender Systems, pp. 12–22 (2024)
33. Xia, C., Shi, X., Xie, H., Lu, Y., Li, P., Shang, M.: Beyond trade-offs: leveraging spatiotemporal heterogeneity of user preference for long-term fairness and accuracy in interactive recommendation. ACM Trans. Web (2025). https://doi.org/10.1145/3769471
34. Xia, C., Shi, X., et al.: Hierarchical reinforcement learning for long-term fairness in interactive recommendation. In: Proceedings of IEEE International Conference on Web Services, pp. 300–309. IEEE (2024)
35. Xu, Y., Yang, Y., et al.: Neural serendipity recommendation: exploring the balance between accuracy and novelty with sparse explicit feedback. ACM Trans. Knowl. Discov. Data **14**(4), 1–25 (2020)
36. Yu, Y., Gao, C., Chen, J., et al.: EasyRL4Rec: an easy-to-use library for reinforcement learning based recommender systems. In: Proceedings of the 47th International ACM SIGIR Conference on Research and Development in Information Retrieval, pp. 977–987 (2024)
37. Zhang, J., Xie, R., et al.: Recommendation as instruction following: a large language model empowered recommendation approach. ACM Trans. Inf. Syst. **43**(5), 1–37 (2025)
38. Zhao, K., et al.: KuaiSim: a comprehensive simulator for recommender systems. Adv. Neural. Inf. Process. Syst. **36**, 44880–44897 (2023)

Curr-RLCER: Curriculum Reinforcement Learning for Coherence Explainable Recommendation

Xiangchen Pan and Wei Wei(✉)

Huazhong University of Science and Technology, Wuhan, China
weiw@hust.edu.cn

Abstract. Explainable recommendation systems (RSs) are designed to explicitly uncover the rationale of each recommendation, thereby enhancing the transparency and credibility of RSs. Previous methods often jointly predicted ratings and generated explanations, but overlooked the incoherence of such two objectives. To address this issue, we propose Curr-RLCER, a reinforcement learning framework for explanation coherent recommendation with dynamic rating alignment. It employs curriculum learning, transitioning from basic predictions (i.e., click-through rating-CTR, selection-based rating) to open-ended recommendation explanation generation. In particular, the rewards of each stage are designed for progressively enhancing the stability of RSs. Furthermore, a coherence-driven reward mechanism is also proposed to enforce the coherence between generated explanations and predicted ratings, supported by a specifically designed evaluation scheme. The extensive experimental results on three explainable recommendation datasets indicate that the proposed framework is effective. Codes and datasets are available at https://github.com/pxcstart/Curr-RLCER.

Keywords: Explainable Recommendation · Reinforcement Learning · Curriculum Learning

1 Introduction

Recommendation systems have always been an important part of data mining and information retrieval, recommending items that match user preferences by learning and predicting their preferences. In recent years, in order to alleviate the credence matters caused by the lack of transparency and explainability in recommendation systems, explainable recommendation has received increasing attention. By inferring the fundamental reasons behind the historical interactions between users and items, recommendation models can provide explainable suggestions and insights into the decision-making process, enabling more personalized and transparent recommendations.

H. Jung et al. (Eds.): DASFAA 2026, LNCS 16535, pp. 471–486, 2026.
https://doi.org/10.1007/978-981-92-0363-5_29

At present, most of the work on explainable recommendations is carried out in a multitasking manner, where the system generates ratings and provides corresponding explanations. For example, Att2Seq [5] and NRT [10] use attention mechanisms and recursive neural networks (RNNs) to generate text explanations. Recent advances have further explored the utilization of Transformers in text generation, providing valuable insights for recommendation results. Although these methods can generate richer and smoother text explanations compared to early methods based on predefined sentence templates [8,26], they focus on improving the quality of text generation and overlook the inconsistency between predicted ratings and generated explanations, which will greatly weaken the credibility of the system.

In order to improve the consistency between predicted ratings and generated explanations, CER [18] adds a consistency alignment strategy on the basis of the PETER framework. By enhancing the consistency between predicted ratings and explanation-based ratings generated by max pooling, the training model can generate more personalized and consistent explanations. However, these methods rely on tools to extract sentiment quadruples from user reviews, and the quality of the data can seriously affect the training effectiveness of the model, thereby affecting the quality of the generated explanations. Moreover, the existing coherence alignment and evaluation strategies have ignored the situation where both predicted ratings and generated explanations are distorted.

Recently, the successful case of Deepseek R1-Zero [6] enhancing LLM inference through Group Relative Policy Optimization (GRPO) demonstrates that by designing appropriate reward models, LLMs can still acquire deep thinking abilities through continuous self validation and have good inference performance without performing SFT fine-tuning but by directly training the basic model with reinforcement learning, thereby reducing dependence on high-quality supervised data. Given the effectiveness of GRPO in inference tasks, we attempt to use reinforcement learning based LLM training to solve the problem of inconsistency between rating and explanation in explainable recommendation.

Considering the disaster of the forgetting problem of LLMs in multi-task learning, relevant studies [2] have shown that curriculum learning can effectively alleviate loss oscillations during training. Specifically, we let the LLM first learn the CTR prediction task, which is to determine whether the user likes the item. The goal is for the LLM to adapt to recommendation data and learn how to make recommendations; In the second stage, we designed a rating prediction task, allowing the LLM to play the role of a consumer and rate for items, further enhancing the model's rating prediction ability; In the third stage, we designed an explanation generation task to enable the LLM to generate explanations for the recommendation based on user and item profile information, as well as reviews, in order to enhance the explainability of the model. In three stages, we transitioned from simple binary classification to multi classification tasks and from discriminative tasks to generative tasks. We designed corresponding reward functions for each stage, gradually increasing the complexity of the task from easy to difficult.

To more effectively train and evaluate the coherence between model-generated explanations and their corresponding ratings, we introduce a coherence reward mechanism and a coherence evaluation framework. In the explainable training phase, we will utilize a pre-trained sentiment classifier to rate explanations and compare explanation-based ratings with the ground truth ratings for coherence as part of the reward mechanism. During the inference evaluation phase, we employ a multi-faceted evaluation approach to comprehensively assess the coherence between explanations and ratings. Specifically, we utilize (i) a sentiment classifier to provide automated sentiment assessment, (ii) a large language model (e.g., GPT) for contextual judgment, and (iii) human evaluators for qualitative analysis. The explanation-derived ratings obtained from these sources are compared with both the predicted ratings and the ground-truth ratings to assess consistency.

The following is a summary of the contributions of this work:

- We propose a reinforcement learning based LLM training framework, Curr-RLCER, to address the issue of inconsistent ratings and explanations in explainable recommendations.
- We have customized a coherence reward mechanism and coherence evaluation scheme, taking into account the distortion of predicted ratings and explanation-based ratings in the evaluation process.
- Using three benchmark datasets for experiments, our method significantly improves the coherence of explanation compared to other state-of-the-art methods, without reducing other metrics for recommendation predictions.

2 Preliminaries

Given a set of users U, a set of items I, and a list of reviews $T_{u,i}$, the goal of this work is to utilize user profile $p_{u,m}$ of user u_m and the item profile $p_{i,n}$ of item i_n, as well as review data t_{u_m,i_n}, to predict ratings r_{u_m,i_n} and generate explanations E_{u_m,i_n}. Let's further assume that the rating is a score $r_{u_m,i_n} \in \{1,2,3,4,5\}$ to measure the consistency between the item attributes and user preferences. When evaluating the explanations generated by the model, we not only consider the text quality of the explanation and the stability of the generation but also consider whether it is consistent with the predicted score r_{u_m,i_n}.

3 Methodology

In this section, we will briefly introduce and provide a detailed explanation of the Curr-RLCER framework, an explanation-rating coherence enhanced recommendation framework based on three-stage curriculum reinforcement learning. The overall framework is shown in Fig 1. In three stages, we transitioned from simple binary classification to multi classification tasks and from discriminative tasks to generative tasks. We designed corresponding reward functions for each stage,

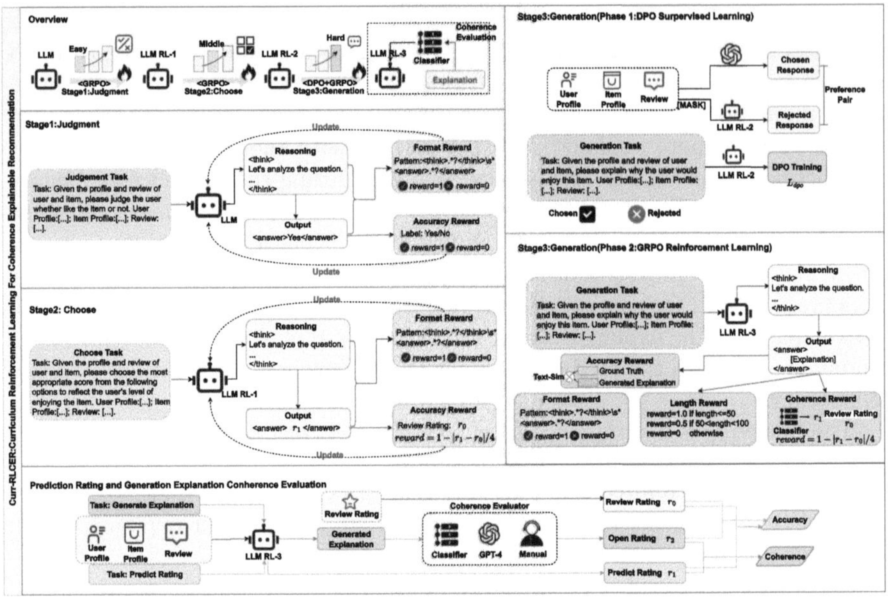

Fig. 1. The overview framework of Curr-RLCER.

gradually increasing the complexity of the task from easy to difficult. We will discuss the DPO and GRPO algorithms used in Curr-RLCER in Sect. 3.1 and introduce the specific implementation of the reward mechanism for each stage in Sect. 3.2. In addition, we have designed a comprehensive coherence assessment scheme, as detailed in Sect. 3.3.

3.1 Reinforcement Learning for LLM

DPO. Reinforcement learning based on human feedback (RLHF) is a key approach for aligning machine learning models with human preferences. Among these techniques, the Direct Preference Optimization (DPO) algorithms have gained widespread attention due to their excellent alignment performance and simple and efficient design. We use the strategy π to represent an LLM, which, for a given input x, generates a response y from the distribution $\pi(\cdot|)$. Assuming the initial strategy is π_0 and the preference dataset is $D = (x, y_w, y_l)$, where for each input x, the priority of response y_w is higher than the response y_l. The objective function of DPO is based on KL divergence to measure the difference between the new strategy and the reference strategy while considering preference rewards:

$$L_{DPO} = -E_{(x,y,y_w,y_l)\sim D} \left[\log \sigma \left(\beta \log \frac{\pi_\theta(y_w|x,v)}{\pi_0(y_w|x,v)} - \beta \log \frac{\pi_\theta(y_l|x,v)}{\pi_0(y_l|x,v)} \right) \right] \quad (1)$$

GRPO. Reinforcement learning based on human feedback heavily relies on the evaluation of critic models. Recently, Deepseek R1-Zero introduced the GRPO framework, which eliminates dependence on other critical models. Specifically, GRPO considers the relative performance of responses rather than absolute reward values. For a given input query q, GRPO generates N different responses $\{o_1, o_2, ..., o_N\}$ from the current policy $\pi_{\theta_{old}}$, and evaluates through group-wise comparisons:

$$A_i = \frac{r_i - mean(r_1, ..., r_N)}{std(r_1, ..., r_N)} \tag{2}$$

where A_i represents the normalized relative quality of the i-th response within its group.

3.2 Curriculum Reinforcement Learning

In order to alleviate the instability of reinforcement learning, we integrated curriculum learning with GRPO, trained the LLMs in stages, and set a difficulty perception reward mechanism consistent with progress for each stage of the task. Specifically, reinforcement learning training consists of three stages: binary classification, multiple choice, and open-ended response. In terms of task format, the task in the binary classification stage is the CTR prediction task, which determines whether the user likes the item; the task in the multiple-choice stage is the rating prediction task, which involves scoring items to predict users' level of enjoyment; and the task in the open-ended response stage is to generate explanations.

Stage1: Judgment for CTR Recommendation. At this stage, we use user and item profile information, as well as reviews, as input data. The task instruction is "judge the user whether they like the item or not". In addition, we explicitly require the model to respond with "yes" or "no" in the instruction. We design corresponding reward functions based on the correctness of the feedback results from the model:

$$R_{Judgment}(o_{ans}, o_{gt}) = \begin{cases} 1, & \text{if } o_{ans} = o_{gt} \\ 0, & otherwise \end{cases} \tag{3}$$

where o_{ans} represents the binary response of the model and o_{gt} is the correct answer. This simple reward design allows LLMs to understand recommendation data during reinforcement learning training, establishing associations for more complex rating prediction and explanation tasks in the following stages.

Stage2: Choose for Rating. We design a multi choice task to enable the LLM to learn the ability to predict ratings. The task instruction is "choose the most appropriate score from the following options", and we require the model's response to be one of the provided options. We have designed corresponding reward functions based on the ratings corresponding to the options provided by the model feedback and the ground truth ratings:

$$R_{Choose}(r_{ans}, r_{gt}) = 1 - \frac{|r_{ans} - r_{gt}|}{4} \tag{4}$$

Among them, r_{ans} represents the rating corresponding to the options fed back by the model, and r_{gt} represents the ground truth rating. During the training process, the closer the predicted rating of the model is to the ground truth rating, the more rewards it will receive. This reward mechanism helps to improve the model's ability to predict ratings.

Stage3: Generation for Explanation. The first two stages are both discriminative tasks, and training through reinforcement learning can effectively enhance the model's reasoning ability. For the open response task of generating explanations, we not only need to consider how to generate smooth and reasonable explanations, but also need to consider coherence with the ratings. Too many optimization goals may lead to conflicts between rewards and fail to achieve good training results. To this end, we first use the DPO algorithm to train based on pre-organized preferences, with the aim of enabling the LLM to understand how to generate high-quality explanations. The following is an introduction to the collection method of preference pairs.

Firstly, we use user and item profile information, as well as review data, as inputs for explainable recommendation tasks, and utilize the response of the GPT-3.5-turbo model as the chosen response. Using the LLM RL-2 model trained in the previous stage as a reference model, a certain proportion of random masks are applied to the effective information in the original input and passed to LLM RL-2. If the semantic similarity between the LLM output and the chosen response is below a certain threshold, we consider it a rejected response.

After DPO training of LLM, we use the GRPO algorithm to train LLM to further improve the quality of explanation and the coherence of rating and explanation. The task instruction is "explain why the user would enjoy this item", and the model is required to output a simple and fluent response. At this stage, we designed length rewards to encourage the LLM to generate as concise responses as possible, accuracy rewards to drive the explanations generated by the LLM to align with standard answers, and coherence rewards to make the generated explanations and ratings as consistent as possible. The following is a specific introduction to the reward function.

(1) Length reward: For length rewards, we classify them based on the length of the answers provided by LLM.

$$R_{Length}(t_{ans}) = \begin{cases} 1 & \text{if } len(t_{ans}) < 50 \\ 0.5 & \text{if } 50 \leq len(t_{ans}) < 100 \\ 0 & otherwise \end{cases} \tag{5}$$

where t_{ans} represents the textual representation of the generated answer.

(2) Accuracy reward: We use the semantic similarity between the answers fed back by LLM and the standard answers as an accuracy reward.

$$R_{Accuracy}(t_{ans}, t_{gt}) = Similarity(T(t_{ans}), T(t_{gt})) \tag{6}$$

where t_{ans} represents the text representation of the generated answer, t_{gt} represents the text representation of the standard answer, and T is the text encoder.

(3) Coherence reward: We use a BERT-based pre-trained sentiment classification model that takes the feedback from LLM as input and outputs an explanation based rating. Then evaluate the rating against the ground truth rating as a coherence reward.

$$R_{Coherence}(t_{ans}, r_{gt}) = 1 - \frac{|C(t_{ans}) - r_{gt}|}{4} \tag{7}$$

where t_{ans} is the textual representation of the generated answer, r_{gt} is the ground truth rating. $C(\cdot)$ is the sentiment classification model with text as input and sentiment classification labels as output. Here, we use the rating as the label.

3.3 Evaluation-Rating Coherence Evaluation

The previous explanation-rating coherence evaluation work only considered the consistency between the generated explanation and the predicted rating but ignored the situation where both were distorted at the same time. In order to make the model more reliable, we also considered the accuracy with the ground truth rating during the evaluation and balanced between coherence and accuracy.

Firstly, user and item profile information, as well as review data, are used as inputs for the model to be evaluated. Task instructions from stage-2 and stage-3 are used to predict ratings and generate explanations, respectively. In order to obtain explanation based ratings, we adopted three methods: one is to rate explanations based on GPT-3.5-turbo through instruction prompts, one is to rate explanations using a sentiment classification model, and the other is manual rating. We will denote the ground truth rating as r_0, the model-predicted rating as r_1, and the explanation based rating as r_2. The following is a coherence evaluation plan:

$$\begin{aligned} Acc &= \frac{Accuracy + Coherence}{2} \\ Accuracy &= \frac{1}{2} * [(1 - \frac{|r_0 - r_1|}{4}) + (1 - \frac{|r_0 - r_2|}{4})] \\ Coherence &= 1 - \frac{|r_1 - r_2|}{4} \end{aligned} \tag{8}$$

4 Experiment

4.1 Experiment Setting

Dataset. To evaluate the effectiveness of our method, we conducted experiments on three open-source datasets extracted from the Baby, Clothing_Shoes_and_Jewelry, and Sports_and_outdoors subsets of the Amazon

Table 1. Dataset Statistics

Dataset	# Users	# Items	# Train (u-i)	# Test (u-i)
Baby	19445	7050	3000	1000
Sports	35598	18357	3000	1000
Clothing	39387	23033	3000	1000

Review Dataset[1]. In addition, we used GPT-3.5-Turbo to generate user and item profile information as well as standard explanations based on metadata and review information. Considering that we train the model through curriculum learning, we collect different data from the source dataset in each stage, but with the same splitting method, we randomly divide each dataset into training, validation, and testing sets in a 3:1:1 ratio. Each record in the dataset contains a user ID, an item ID, a 1$\sim$5 rating, a task instruction, and a correct answer. The statistical information of the dataset is shown in Table 1.

Details of Baselines. We compare our model's performance against the following baselines:

- **Att2Seq** [5]: Utilizes an attention-based attribute-to-sequence model to generate reviews based on attribute information.
- **NRT** [10]: Predicts ratings and generates abstractive tips for recommendations using multi-task learning to optimize parameters.
- **PETER** [9]: PETER is a personalized transformer model for explainable recommendation. It maps user and item IDs to the generated explanation text, connecting the IDs and words.
- **CER** [18]: Estimates the discrepancy between predicted ratings and explanation-based ratings to enhance rating-explanation coherency.
- **XRec** [14]: Utilizes the encoded user/item embeddings from GNNs as implicit collaborative signals, which are then integrated into each layer of LLMs, enabling the generation of explanations.

Due to the fact that the dataset in our experiment has not been used in the baselines, here is an additional explanation of the collection method for the baseline model dataset. For Att2Seq, NRT, PETER, and CER models, we used the Sentires-Guide[2] tool to extract sentiment quadruples from user reviews in the original dataset and trained and tested them using the same dataset segmentation method as the original paper. For the XRec model, we used the same training and testing samples as our own model, but the model also involves the injection of collaborative information, so we rigorously reproduced the paper. Construct an interaction graph based on the interaction information of the original dataset, and use trained GNNs to encode each user and item.

[1] https://jmcauley.ucsd.edu/data/amazon/links.html.
[2] https://github.com/lileipisces/Sentires-Guide.

Evaluation Metrics. In order to evaluate the explanation performance of the model, we followed XRec and adopted a set of metrics aimed at capturing the semantic explainability and stability of generated explanations. Traditional metrics based on n-gram syntax, such as BLUE [16] and ROUGE [12], are insufficient to achieve their goals due to their inability to fully capture semantic meanings. Specifically, we use GPTscore, BERTscore, BARTscore, BLEURT, and USR to measure explainability. Among them, GPTScore maintains consistency with human judgment by comparing the semantic similarity between generated explanations and ground truth explanations; BERTScore utilizes BERT's context embedding to calculate token-level cosine similarity; BARTScore utilizes the BART model to conceptualize evaluation as a text generation task, assigning scores based on the probability of regenerating reference text; BLEURT adopts a novel synthetic data pre-training method to enhance generalization; and USR measures the uniqueness of generated explanations by calculating the ratio of unique sentences to total sentences. To further evaluate the quality stability, we analyzed the standard deviation of these scores, where lower values indicate more consistent performance. The relevant evaluation results are shown in Table 2.

To evaluate the performance of rating predictions, we used two commonly used metrics, Root Mean Square Error (RMSE) and Mean Absolute Error (MAE), to measure the deviation between predicted ratings and actual ratings. The performance evaluation of rating prediction is shown in Table 3.

In addition, we also evaluated the coherence of the model in terms of explanation and rating. The definition of the indicators is given in Sect. 4.3. We used three methods: GPT, a sentiment classification model, and manual evaluation. The coherence results are shown in Table 4.

Implementation Details. All experiments were conducted on 8 NVIDIA RTX-4090 GPUs. We used the Qwen2.5-3B-Instruct model for the experiments, and the sentiment classification model used in the evaluation was the bert-base-multilingual-truncated-sentiment on Huggingface. The text encoding model used for the accuracy reward in Stage 3 was m3e-base. In GRPO, the sample number of responses for each prompt is set to 4, and the maximum prompt length is set to 512; the KL penalty coefficient β in DPO is set to 0.1. The number of training samples for each stage is 3000, and the training time for each stage is approximately 12 h; the test sample is 1000, and the inference time is about half an hour.

4.2 Model Performance

Evaluation of Explanation. To demonstrate the superiority of our model in explainability and stability, we conducted a comparative analysis of five baseline methods in three datasets. The results are summarized in Table 2. Based on observations, it can be observed that Curr-RLCER exhibits excellent performance in both explainability and stability. Compared with the best baseline model XRec, although our method has a slightly lower recall rate when validated

Table 2. Explainability and Stability results on Three Datasets. "↑" means higher is better; "↓" means lower is better. Superscripts "P", "R", and "F1" denote Precision, Recall, and F1-Score, respectively. The subscript "std" indicates the standard deviation of each metric. Bold indicates the best results, while underlined denotes the second-best. (The significant improvements with p-value < 0.05 based on paired t-tests)

Category	Models	Explainability ↑							Stability ↓					
		GPT_{score}	$BERT^{P}_{score}$	$BERT^{R}_{score}$	$BERT^{F1}_{score}$	$BART_{score}$	BLEURT	USR	GPT_{std}	$BERT^{P}_{std}$	$BERT^{R}_{std}$	$BERT^{F1}_{std}$	$BART_{std}$	$BLEURT_{std}$
Baby														
Baseline	NRT	22.36	0.1515	0.0998	0.1260	−5.2312	−1.0946	0.0094	17.09	0.1525	0.1506	0.1351	1.0305	0.2325
	Att2Seq	26.19	0.1356	0.0887	0.1119	−5.2568	−1.0701	0.0437	21.14	0.1827	0.1608	0.1465	1.0432	0.2667
	PETER	45.11	0.3278	0.2305	0.2784	−4.7111	−0.8737	0.5338	26.25	0.2051	0.2103	0.1889	1.0288	0.4010
	CER	46.11	0.3154	0.2292	0.2716	−4.7099	−0.8669	0.6078	26.02	0.2045	0.2076	0.1863	1.0304	0.4408
	XRec-3B	82.38	0.4273	**0.4422**	**0.4354**	−3.9362	−0.0987	0.9989	9.16	0.1171	0.1153	0.1111	0.6086	0.2012
	Qwen2.5-7B-Instruct	84.89	0.2313	0.3559	0.2936	−3.6146	0.1030	**1.0000**	4.05	0.1017	0.0794	0.0821	0.4526	0.1298
	Qwen2.5-14B-Instruct	85.10	0.2483	0.3660	0.3074	−3.6421	0.1045	**1.0000**	**3.92**	**0.0810**	**0.0779**	**0.0730**	**0.4523**	0.1281
Ours	Curr-RLCER-3B	**86.98**	**0.4533**	0.3569	0.4054	**−3.5523**	**0.2377**	**1.0000**	7.85	0.1027	0.0957	0.0934	0.5072	**0.0800**
Sports														
Baseline	NRT	26.62	0.0319	0.0574	0.0410	−5.6958	−1.1106	0.0595	21.12	0.3362	0.1572	0.2165	1.0707	0.2767
	Att2Seq	26.18	−0.0326	0.0489	0.0033	−5.7100	−1.1307	0.0383	21.58	0.3653	0.1521	0.2295	1.0720	0.2312
	PETER	43.54	0.2579	0.1778	0.2170	−5.1567	−0.8970	0.4670	25.85	0.2159	0.1948	0.1822	1.0723	0.3844
	CER	46.15	0.2709	0.1779	0.2236	−5.1481	−0.8999	0.5122	25.76	0.2037	0.1963	0.1778	1.0716	0.3759
	XRec-3B	79.12	0.4315	**0.4353**	**0.4341**	−3.9651	−0.1247	**1.0000**	10.21	0.1154	0.1161	0.1108	0.6184	0.1778
Base LLM	Qwen2.5-7B-Instruct	84.43	0.2373	0.3512	0.2944	−3.7328	0.1008	**1.0000**	4.21	0.1044	0.0784	0.0836	**0.4268**	0.1310
	Qwen2.5-14B-Instruct	84.54	0.2559	0.3609	0.3087	−3.7483	0.1016	**1.0000**	**4.04**	0.1009	**0.0750**	**0.0806**	0.4271	0.1301
Ours	Curr-RLCER-3B	**86.37**	**0.4739**	0.3537	0.4138	**−3.6804**	**0.2387**	**1.0000**	7.43	**0.0983**	0.0894	0.0873	0.4855	**0.1015**
Clothing														
Baseline	NRT	31.09	0.1748	0.1009	0.1381	−5.4719	−1.0274	0.0026	23.78	0.1302	0.1593	0.1289	1.0462	0.3112
	Att2Seq	35.10	0.2079	0.1209	0.1643	−5.3831	−1.0108	0.0303	25.76	0.1431	0.1683	0.1379	1.0539	0.3145
	PETER	50.83	0.3489	0.2586	0.3029	−4.8712	−0.7521	0.2737	26.19	0.2019	0.2267	0.1945	1.0913	0.5012
	CER	52.89	0.3687	0.2592	0.3129	−4.8688	−0.7577	0.1790	27.50	0.2103	0.2349	0.2031	1.1084	0.5104
	XRec-3B	81.93	0.4043	**0.4159**	**0.4108**	−3.9113	−0.1476	**1.0000**	11.72	0.1246	0.1201	0.1166	0.5979	0.2180
Base LLM	Qwen2.5-7B-Instruct	84.07	0.2228	0.3285	0.2759	−3.7088	0.0616	**1.0000**	**3.74**	**0.0948**	**0.0790**	**0.0771**	0.4057	0.1483
	Qwen2.5-14B-Instruct	84.29	0.2424	0.3432	0.2931	−3.7026	0.0620	**1.0000**	4.79	0.1015	0.0814	0.0849	**0.4033**	0.1505
Ours	Curr-RLCER-3B	**86.18**	**0.4488**	0.3243	0.3866	**−3.6707**	**0.2444**	**1.0000**	8.19	0.1042	0.0940	0.0912	0.4351	**0.0811**

using the Bert model, this may be due to the introduction of additional collaborative information in XRec, which covers more features and provides a deeper understanding of recommendation behavior compared to using only text modality. In other indicators, our model performs better and has stronger stability. This demonstrates the effectiveness of the Stage-3 training in our method, where the model understands how to generate high-quality explanations through training with DPO and GPRO. In addition, considering that most baseline models use traditional deep learning architectures, in order to more intuitively demonstrate the effectiveness of our approach, we conducted comparative tests with backbone models of different parameter sizes in the same series. The results showed that Curr-RLCER can effectively improve the model's explainability, but its stability may deteriorate.

Evaluation of Rating Prediction. To demonstrate the accuracy of our model in rating prediction, we conducted a comparative analysis of all baseline methods with rating prediction capability in three datasets. The results are shown in Table 3. Based on observations, it can be observed that our method has excellent rating prediction ability, with RMSE improved by 34.9%, 27.1%, and 42.5% compared to the best baseline model on three datasets; MAE improved by 58.6%, 53.7%, and 60.6% compared to the best baseline model. And compared to the backbone models with higher parameters in the same series, the rating prediction ability has also been significantly improved. This proves the effectiveness of Stage

Table 3. Rating Prediction Performance Comparison Across Datasets. "R" means RMSE, and "M" means MAE.

Models	Baby		Sports		Clothing	
	R↓	M↓	R↓	M↓	R↓	M↓
NRT	1.1164	0.8018	1.0680	0.7308	1.0528	0.7828
PETER	1.0625	0.7982	0.9490	0.7217	1.0615	0.8218
CER	1.0613	0.8343	0.9518	0.7382	1.0612	0.8349
Qwen2.5-7B-Instruct	0.8426	0.5480	1.0154	0.7010	0.9143	0.5840
Qwen2.5-14B-Instruct	0.8277	0.5197	0.8915	0.6660	0.8882	0.5818
Curr-RLCER	**0.6899**	**0.33**	**0.6943**	**0.3340**	**0.6099**	**0.3080**

1 and Stage 2 of training in our method. Through CTR prediction and rating prediction, the model can understand the sentiment in reviews based on profile information and make accurate rating predictions.

Evaluation of Coherence. To demonstrate that our model can improve the coherence between explanation and rating, we conducted consistency evaluations on three datasets using the custom metrics introduced in Sect. 3.3 and compared them with NRT, PETER, and CER models that include both explanation generation and rating prediction tasks. The results are shown in Table 4. Based on observations, it can be observed that our method can significantly improve the consistency between explanatory ratings and predictive ratings while maintaining a high degree of rating accuracy. This demonstrates the effectiveness of the coherence reward mechanism in stage-3 GRPO training, enabling the model to better match user ratings when generating explanations.

Here is the detailed description of our manual annotation method. There were a total of 5 researchers involved in manual annotation. We extracted 200 test samples from each dataset and utilized Curr-RLCER and other baseline models to generate explanations and predict ratings. The researchers will rate each explanation generated by each model ranging from 1 to 5, based on sentiment expression. Afterwards, we will summarize the scoring results of each researcher as the explanation-based rating for the test samples. Finally, we will calculate the coherence indicator by combining the explanation-based rating with the ground truth rating and the predicted rating.

4.3 Ablation Study

In this section, we conducted ablation studies to explore the impact of two key components in the model: (1) the effect of DPO training on the model's explanatory power and (2) the effect of coherence reward mechanisms on the model's explanation rating prediction consistency. For this purpose, we compared three model variants: 1) a complete model with all features 2) **w/o DPO**

Table 4. Coherence Performance Comparison Across Datasets. We use GPT-3.5-Turbo, a BERT-based sentiment classifier, and human annotations for explanation-rating coherence evaluation.

	GPT			Bert-Classifier			Human Annotator		
	Baby	Sports	Clothing	Baby	Sports	Clothing	Baby	Sports	Clothing
NRT	0.6824	0.7124	0.7541	0.8049	0.7549	0.8217	0.7028	0.7386	0.7324
PETER	0.7013	0.7226	0.7191	0.8119	0.8006	0.7994	0.6915	0.6716	0.7053
CER	0.6988	0.6714	0.7025	0.7996	0.8081	0.7973	0.7561	0.7084	0.7261
Curr-RLCER	**0.8334**	**0.8516**	**0.8679**	**0.9191**	**0.9036**	**0.9187**	**0.8617**	**0.8932**	**0.9008**

Training: omitting preference pair collection and DPO training processes, and 3) **w/o Coherence Reward**: the reward model in stage-3 does not include coherence rewards. In order to make the ablation assessment more accurate, we conducted ablation experiments on the Baby and Clothing datasets. The experimental results are shown in Table 5.

Table 5. Ablation Study Results. 1) w/o DPO Training: Omitting preference pair collection and DPO training processes. 2) w/o Coherence reward: The reward model in stage-3 does not include coherence rewards.

Ablations	Baby					Clothing				
	MAE↓	RMSE↓	$BERT^{F1}_{score}$ ↑	$BERT^{F1}_{std}$ ↓	Acc↑	MAE↓	RMSE↓	$BERT^{F1}_{score}$ ↑	$BERT^{F1}_{std}$ ↓	Acc↑
Curr-RLCER	0.33	0.6899	0.4054	**0.0934**	**0.9191**	**0.3080**	**0.6099**	0.3866	0.0912	**0.9187**
w/o Coherence Reward	**0.3140**	**0.6588**	**0.4224**	0.0979	0.8307	0.3350	0.6656	**0.4101**	0.0921	0.8489
w/o DPO Training	0.3350	0.7036	0.3834	0.0985	0.8934	0.3270	0.6731	0.3542	**0.0864**	0.9072

According to observations, it can be found that after removing the DPO training process, the model's explanation generation ability will significantly decrease, and its rating prediction ability will also be affected to some extent. This is likely due to the model's lack of understanding of how to generate high-quality explanations during open generation tasks in GRPO training, the model focuses on how to make the generated explanations more concise and accurate, and the coherence reward cannot play its role well. Therefore, in the process of improving the explanation generation ability, the disaster forgetting phenomenon occurs, weakening the rating prediction ability learned in the previous stage.

In addition, after removing the coherence reward mechanism, the explanation generation ability of the model will be slightly improved, but the coherence is not as good as the complete model. This is because the model will only focus on learning how to generate high-quality explanations while ignoring the consistency between explanations and ratings.

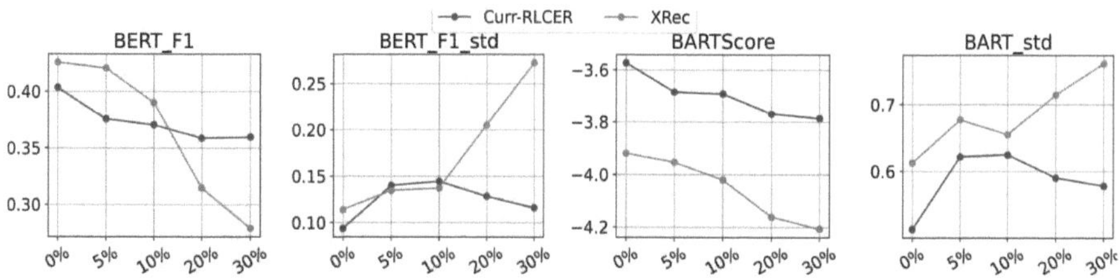

Fig. 2. Robustness Experiment comparing Curr-RLCER with XRec in different noise ratios

4.4 Model Robustness About Explainability

In order to evaluate the generalization ability of the model in terms of explainability, we conducted a robustness comparison analysis between Curr-RLCER and XRec, which had the best explainability performance in all baselines. We add different proportions of noise to the input and demonstrate the robustness of the model based on the performance changes in generating explanations under different levels of noise. Specifically, we sampled 500 data points in the test set of the Baby dataset and added noise to the input instructions of these samples through random entity masking. The masking ratio was set to five groups: raw data (0%), 5%, 10%, 20%, and 30%. Fig 2 shows the changes in performance indicators of two models under different levels of noise. The evaluation results highlight several key findings:

- As the noise ratio continues to increase, our model Curr-RLCER exhibits excellent noise resistance, and both semantic explainability and text generation stability do not undergo drastic changes with the increase of input noise. This trend reflects that our model has good robustness.
- In both noiseless and low-noise scenarios, our model performs slightly worse than XRec in terms of Bert_F1 score, which may be due to the fact that XRec introduces collaborative signals by trained GNN, resulting in more effective information contained in the input. However, as the number of masks increased, XRec showed a significant decline in all evaluation indicators. The reason why our model can maintain robustness is that the reward design in reinforcement learning training can encourage the model to provide preference answers even in the presence of input disturbances.

5 Related Work

5.1 Reinforcement Learning from Human Feedback

Reinforcement Learning from Human Feedback (RLHF) has emerged as a crucial technique for incorporating human preference signals into machine learning methods and models [4]. RLHF frameworks can be broadly categorized into deep RL-based approaches and direct preference learning approaches. In deep RL-based methods, a reward model is first constructed, after which Proximal Policy Optimization (PPO) [19] is employed to optimize the reward signals with KL regularization [15]. While the direct preference learning approaches optimize a designed loss target on the offline preference dataset directly, eliminating the need for a separate reward model [21]. Recent extensions like Step-DPO [7] further decompose preference alignment into multi-step optimization. Besides, Dong et al. [3] introduced rejection sampling fine-tuning to align LLMs with human preferences without requiring reinforcement learning pipelines.

5.2 Reasoning Model with LLM

In recent years, with the continuous advancement of technologies such as Tree of Thoughts [25] and Process-Supervised Learning [13], reasoning models have made significant progress. With OpenAI introducing RLM in their o1 models [22], many works have explored various post-training combinations of reinforcement learning and supervised fine-tuning to further improve inference capabilities. Deepseek R1-Zero [6] demonstrates that direct reinforcement learning of LLM still achieves well inference ability through the Group Relative Policy Optimization (GRPO) algorithm that does not rely on SFT. A large amount of work has shown the potential of using reinforcement learning to train small-scale RLMs in the fields of mathematics, logic, and coding [23,24].

5.3 Explainable Recommendation

Explainable recommendation has the reasoning ability to explore user recommendation behavior logic, thereby enhancing the personalization and transparency of recommendation systems, which has significant research significance. In recent years, more and more research has focused on how to provide good explanations to improve system effectiveness and user satisfaction. Early methods mainly used predefined templates to generate explanations [8] or extract logical reasoning rules from recommendation models [1,20]. With the continuous development of deep learning and the advancement of natural language generation technology, many works adopt attention mechanisms and recursive neural networks [5,10] to generate natural language explanations. Currently, the most popular approach is to use LLMs for explainable recommendations [11,14,17]. The good semantic understanding and generation ability of large language models can better improve the performance of model interpretation and generation. The existing explainable recommendation models mainly have the following two problems: (1) excessive reliance on high-quality supervised data and poor generalization and (2) only considering how to improve the quality of generated text, ignoring the coherence between predicted scores and explanations. This work uses reinforcement learning to train LLMs, thereby reducing reliance on supervised data; adds explanation rating coherence rewards during the training process; and conducts a comprehensive evaluation from both accuracy and coherence perspectives.

6 Conclusion

In this work, we introduce a reinforcement learning based LLM explainable recommendation framework, Curr-RLCER. While improving the model's explainable recommendation and rating prediction abilities, we also consider the issue of inconsistent explanations and ratings and design a coherence reward mechanism and coherence evaluation scheme. In terms of operational implementation, we adopt a curriculum learning training approach, conducting multi-task training from easy to difficult to alleviate the disaster forgetting problem of LLMs in multi-task learning; we adopt reinforcement learning training methods to alleviate the problem of traditional explainable recommendation schemes overly relying on high-quality supervised data. Extensive experimental results have shown that our method is effective.

Acknowledgements. This work was supported in part by the National Natural Science Foundation of China under Grant No. 62276110, No. 62172039 and in part by the fund of Joint Laboratory of HUST and Pingan Property & Casualty Research (HPL). The authors would also like to thank the anonymous reviewers for their comments on improving the quality of this paper.

Disclosure of Interests. The authors have no competing interests to declare that are relevant to the content of this article.

References

1. Chen, H., Shi, S., Li, Y., Zhang, Y.: Neural collaborative reasoning. In: Proceedings of the Web Conference 2021, pp. 1516–1527 (2021)
2. Deng, H., Zou, D., Ma, R., Luo, H., Cao, Y., Kang, Y.: Boosting the generalization and reasoning of vision language models with curriculum reinforcement learning. arXiv preprint arXiv:2503.07065 (2025)
3. Dong, H., et al.: RAFT: reward ranked finetuning for generative foundation model alignment. arXiv preprint arXiv:2304.06767 (2023)
4. Dong, H., et al.: RLHF workflow: from reward modeling to online RLHF. arXiv preprint arXiv:2405.07863 (2024)
5. Dong, L., Huang, S., Wei, F., Lapata, M., Zhou, M., Xu, K.: Learning to generate product reviews from attributes. In: 15th EACL 2017 Software Demonstrations, pp. 623–632. Association for Computational Linguistics (2017)
6. Guo, D., et al.: DeepSeek-R1: incentivizing reasoning capability in LLMs via reinforcement learning. arXiv preprint arXiv:2501.12948 (2025)
7. Lai, X., Tian, Z., Chen, Y., Yang, S., Peng, X., Jia, J.: Step-DPO: stepwise preference optimization for long-chain reasoning of LLMs. arXiv preprint arXiv:2406.18629 (2024)
8. Li, L., Chen, L., Dong, R.: CAESAR: context-aware explanation based on supervised attention for service recommendations. J. Intell. Inf. Syst. **57**(1), 147–170 (2021)
9. Li, L., Zhang, Y., Chen, L.: Personalized transformer for explainable recommendation. arXiv preprint arXiv:2105.11601 (2021)

10. Li, P., Wang, Z., Ren, Z., Bing, L., Lam, W.: Neural rating regression with abstractive tips generation for recommendation. In: Proceedings of the 40th International ACM SIGIR Conference on Research and Development in Information Retrieval, pp. 345–354 (2017)
11. Li, Y., et al.: G-refer: graph retrieval-augmented large language model for explainable recommendation. In: Proceedings of the ACM on Web Conference 2025, pp. 240–251 (2025)
12. Lin, C.Y.: ROUGE: a package for automatic evaluation of summaries. In: Text Summarization Branches Out, pp. 74–81 (2004)
13. Lu, J., et al.: AutoPSV: automated process-supervised verifier. Adv. Neural. Inf. Process. Syst. **37**, 79935–79962 (2024)
14. Ma, Q., Ren, X., Huang, C.: XRec: large language models for explainable recommendation. arXiv preprint arXiv:2406.02377 (2024)
15. Ouyang, L., et al.: Training language models to follow instructions with human feedback. Adv. Neural. Inf. Process. Syst. **35**, 27730–27744 (2022)
16. Papineni, K., Roukos, S., Ward, T., Zhu, W.J.: BLEU: a method for automatic evaluation of machine translation. In: Proceedings of the 40th annual meeting of the Association for Computational Linguistics, pp. 311–318 (2002)
17. Qiu, Z., Luo, L., Pan, S., Liew, A.W.C.: Unveiling user preferences: a knowledge graph and LLM-driven approach for conversational recommendation. arXiv preprint arXiv:2411.14459 (2024)
18. Raczyński, J., Lango, M., Stefanowski, J.: The problem of coherence in natural language explanations of recommendations. In: ECAI 2023, pp. 1922–1929. IOS Press (2023)
19. Schulman, J., Wolski, F., Dhariwal, P., Radford, A., Klimov, O.: Proximal policy optimization algorithms. arXiv preprint arXiv:1707.06347 (2017)
20. Shi, S., Chen, H., Ma, W., Mao, J., Zhang, M., Zhang, Y.: Neural logic reasoning. In: Proceedings of the 29th ACM International Conference on Information & Knowledge Management, pp. 1365–1374 (2020)
21. Tang, Y., et al.: Generalized preference optimization: a unified approach to offline alignment. arXiv preprint arXiv:2402.05749 (2024)
22. Wainwright, C., Lowe, R.: InstructGPT: training language models to follow instructions with human feedback. GitHub repository (2023)
23. Wei, Y., et al.: SWE-RL: advancing LLM reasoning via reinforcement learning on open software evolution. arXiv preprint arXiv:2502.18449 (2025)
24. Xie, T., et al.: Logic-RL: unleashing LLM reasoning with rule-based reinforcement learning. arXiv preprint arXiv:2502.14768 (2025)
25. Yao, S., et al.: Tree of thoughts: deliberate problem solving with large language models. Adv. Neural. Inf. Process. Syst. **36**, 11809–11822 (2023)
26. Zhang, Y., Lai, G., Zhang, M., Zhang, Y., Liu, Y., Ma, S.: Explicit factor models for explainable recommendation based on phrase-level sentiment analysis. In: Proceedings of the 37th International ACM SIGIR Conference on Research & Development in Information Retrieval, pp. 83–92 (2014)

Mirror: Disentangling and Purifying Multimodal Information for Recipe Recommendation

Jingya Zhou[1,2(✉)] and Xiaolong She[1]

[1] School of Computer Science and Technology, Soochow University, Suzhou, China
20225227006@stu.suda.edu.cn

[2] State Key Laboratory for Novel Software Technology, Nanjing University, Nanjing, China
jy_zhou@suda.edu.cn

Abstract. While multimodal recipe data is effective for alleviating data sparsity in recommender systems, existing methods primarily focus on learning common knowledge across modalities, neglecting the unique features within each one. This oversight allows user preference-irrelevant information to persist in both modality-shared and modality-specific representations, ultimately degrading recommendation performance. To address this, we propose Mirror, a novel multimodal recipe recommendation learning framework. Mirror employs a modality-disentangled representation learning method to decompose recipe features into distinct modality-shared and modality-specific components. Furthermore, it incorporates a multimodal counterfactual learning layer designed to identify and eliminate user preference-irrelevant features. This purification process refines the multimodal representations and enhances the model's ability to capture users' latent interests. Extensive experimental results verify the superiority of Mirror against many existing methods.

Keywords: Recipe recommendation · Multimodal data · Modality disentanglement · Counterfactual Learning

1 Introduction

Recent research on recipe recommendation has employed traditional methods like user or recipe similarity. With the advancement of graph neural networks (GNN), some recipe recommendation models have begun to represent users and recipes as graphs. Increasingly, studies have recognized that incorporating multimodal information in recipes enhances item characterization, thereby improving user preference learning [16]. However, they all ignore the impact of modality-shared and modality-specific features on user interactions. To effectively merge heterogeneous information from different modalities, it is essential to accurately align cross-modal data to achieve better user preference estimation. Current research mainly focuses on extracting common knowledge across

H. Jung et al. (Eds.): DASFAA 2026, LNCS 16535, pp. 487–503, 2026.
https://doi.org/10.1007/978-981-92-0363-5_30

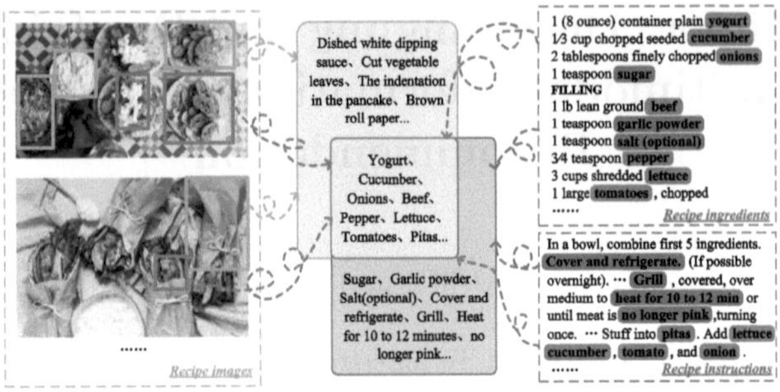

Fig. 1. An explanatory example.

modalities, neglecting modality-specific features. Yet, users might prefer a single modality due to its unique features. Figure 1 illustrates a simple example to explain the impact of modality-shared and modality-specific features on user interactions. As shown, both image and text descriptions of recipes can enhance the semantics beneficial for the recommendation task. Although previous work has used various methods to align features, modality-specific features are often ignored, which degrades recommendation performance. Moreover, multimodal information bias arises when both shared and specific information of modalities are considered. For example, text and images may share elements such as beef, cucumber, yogurt, and onion in a recipe. However, a user might choose this recipe due to certain modality-specific features, like its appealing appearance and portability, as depicted in the image with brown rolled paper. This example highlights the data bias present in multimodal data, leading to spurious correlations with actual user preferences. To address this bias, we adopt a causal inference perspective to distinguish between modality-shared and modality-specific features.

In this paper, we propose a novel multimodal recipe recommenda-tion learning framework named Mirror. We start by using a user-recipe-ingredient heterogeneous graph (URI-Graph) [19] to integrate visual, textual, and relational information about users and recipes. We then employ a hierarchical attention-based heterogeneous graph neural network (HGNN) to learn different modal representations of recipes on the URI-Graph. Our framework includes a modality-disentangled method and a counterfactual learning module to identify modality-shared and modality-specific features, eliminating preference-independent representations that may cause false relevance to users. This enables the model to focus on modally distinct information and refine multimodal content representations, aligning them with user interests.

Our main contributions are summarized as follows:

- We design a novel multimodal recipe recommendation framework named Mirror, which learns the representations of recipes in different modalities. Moreover, we invent a modality-disentangled method to separate modality-shared and modality-specific representations.

- We introduce causal inference theory and construct a multimodal counterfactual learning network to learn causal differences between user preference distributions for modality-specific and shared features, mitigating the impact of irrelevant parts in user-interacted recipes.
- We conduct extensive experiments to demonstrate the superiority of Mirror against state-of-the-art baselines.

2 Related Work

Recipe Recommendation. Recipe recommendation has garnered increasing attention in recent years, driven by the emergence of large recipe datasets [11]. Early approaches primarily relied on content-based and collaborative filtering methods, leveraging user ratings from past interactions. With the rise of GNN, more researchers have adopted GNN-based methods for recipe recommendation. Tian et al. [19] achieved promising results by constructing user-recipe-ingredient heterogeneous graphs for recommendation. User preferences for recipes are complex, with choices influenced by factors such as ingredients, images, and preparation complexity. Thus, accurately modeling user preferences requires comprehensive recipe information. For example, Gao et al. [3] proposed a hierarchical attention network that combines text and images of recipes, although it does not account for user personalization. In addition to image-based information, Song et al. [17] proposed a calorie-aware recipe recommendation framework that combines heterogeneous GNNs and self-supervised learning to enhance recommendation performance. Unlike existing methods, this paper introduces information about each recipe item and considers how different modal information reflects the user's natural preferences at a finer granularity.

Causality-Inspired Recommendation. In recommender systems, causal inference is mainly used to eliminate data biases such as popularity bias, exposure bias, and selection bias. Wei et al. [24] built a causal graph to describe the essential causal relationships in the recommendation process, eliminating the popularity bias of the recipes. Qiu et al. [12] designed a causal graph and employed counterfactual reasoning to address bias stemming from visual information. Similarly, Liu et al. [10] developed a generalized modality-independent recommendation framework to eliminate unimodal bias in multimodal recommendations through counterfactual reasoning. In contrast to these frameworks, most existing multimodal recipe representation models focus on learning cross-modal public knowledge without distinguishing between the effects of modality-specific and modality-shared information on recommendations, which can help mitigate spurious correlations in the recipe data of user interactions.

3 Methodology

3.1 Problem Formulation and Framework Overview

Let $\mathcal{U} = \{u\}$ denote the user set and $\mathcal{I} = \{i\}$ denote the recipe set. The recipe contains multimodal information, such as ingredients, instructions, and images,

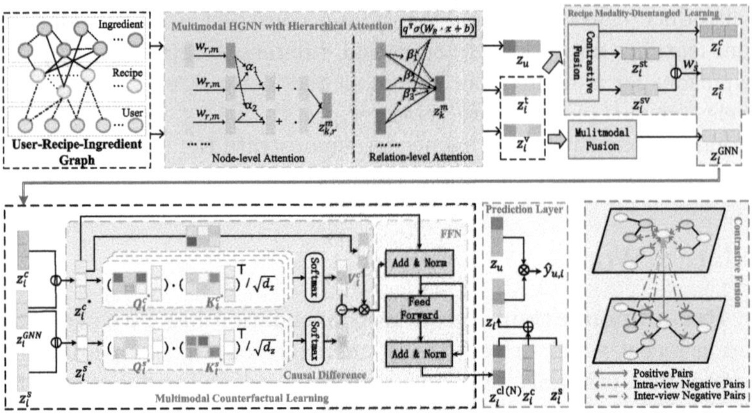

Fig. 2. Overview of our Mirror framework.

corresponding to textual t and visual v modalities, and $\mathcal{M} = \{\mathrm{t}, \mathrm{v}\}$ denotes the set of modalities. We construct a user-recipe-ingredient graph (URI-Graph) $\mathcal{G} = \{\mathcal{V}, \mathcal{E}\}$ comprising three types of nodes $\mathcal{V}$: User, Recipe, Ingredient, and four types of edges $\mathcal{E}$: User-Recipe, Recipe-Ingredient, Recipe-Recipe, Ingredient-Ingredient. Specifically, the URI-Graph is built based on the Large-RG dataset [20], where the recipe ingredients, images, instructions, and user ratings are taken from the recipe website Food.com. Recipe images and instructions are used as attributes of the recipe nodes, encoded by pre-trained models (e.g., ResNet for recipe images and Bi-LSTM for recipe instructions). Nutritional information of the ingredients serves as attributes of the ingredient nodes, while user nodes are initialized with features using the Xavier method. In this graph, user nodes are connected to their interacted recipe nodes, recipe nodes are connected to their corresponding ingredient nodes, recipe nodes are connected through similarity determined by FoodKG [5], and ingredient nodes are connected based on co-occurrence probability. Having graph $\mathcal{G}$, the goal of the multimodal recipe recommendation problem is to design a model to predict the likelihood (score) $\hat{y}_{u,i}$ of a recipe i being adopted by user u ($\forall u \in \mathcal{U}, \forall i \in \mathcal{I}$) through joint modeling of the user behavior and multimodal recipe.

As Fig. 2 shows, Mirror comprises four main components: 1) Multimodal HGNN module with Hierarchical Attention, which captures both attributes and structural information within the URI-Graph; 2) Recipe Modality-Disentangled Learning module that decomposes recipe embeddings into modality-shared and modality-specific embeddings; 3) Multimodal Counterfactual Learning module that models the distribution of modality-shared and modality-specific feature preferences for multimodal content and leverages causal differences to refine representations related to user preferences; 4) Prediction and optimization module.

3.2 Multimodal HGNN with Hierarchical Attention

We extract attributes and structural information from the URI-Graph $\mathcal{G}$ using a graph representation learning method that incorporates a two-level attention mechanism. For each node $v_k \in \mathcal{V}$, the node-level attention module fuses neighboring nodes under each specific relation and modality. Subsequently, the relation-level attention module combines all the relation-specific embeddings under different modalities.

Node-Level Attention. We compute the relation-specific and modality-specific embedding $z_{k,r}^{m,l+1}$ for node v_k at the $l+1$th GNN layer. Here, $z_{k,r}^{m}$ denotes the embedding of modality m fused by v_k's r-relation neighbors. First, we apply a shared weight matrix $W_{r,m} \in \mathbb{R}^{d_l \times d_{l+1}}$ to make linear transformation on the feature output from the lth layer, i.e., $g_k^{m,l} = W_{r,m}^l \cdot z_k^{m,l}$, where d_l and d_{l+1} are the embedding dimensions at the lth and $l+1$th layers, respectively. Then we calculate the attention score $e_{kj}^{m,l}$ between v_k and v_j under modality m:

$$e_{kj}^{m,l} = \text{LeakyReLU}\left[W_{kj}^l \cdot (g_k^{m,l} \| g_j^{m,l})\right], \tag{1}$$

where $||$ denotes the concatenation operator. We obtain the normalized attention score by using softmax function, i.e., $\alpha_{kj}^{m,l} = \frac{\exp(e_{kj}^{m})}{\sum_{k^* \in N_{k,r}} \exp(e_{k^*j}^{m,l})}$. We adopt a multi-head attention mechanism to stabilize the training process:

$$z_{k,r}^{m,l+1} = \mathop{\Big\|}_{n=1}^{N} \text{ReLU}\left(\sum_{j \in N_{k,r}} \alpha_{kj}^{m,l} \cdot z_j^{m,l} + W_m^l \cdot \left(z_k^{m,l} \odot z_j^{m,l}\right)\right) \cdot W_a, \tag{2}$$

where $N_{k,r}$ is node v_k's r-relation neighbors, W_m^l and W_a are trainable weight vectors, N is the number of attention heads, and d_n is the dimension of the attention heads such that $d_n = d_{l+1}/N$.

Relation-Level Attention. Given the diverse types of edges in the URI-Graph, we further focus on relation-level attention to integrate all relation-specific embeddings and assess the importance of each relation. Specifically, we apply a shared non-linear weight matrix $W_R \in \mathbb{R}^{d_{l+1} \times d_{l+1}}$ to transform the embeddings of specific relations. Then, we use an intermediate vector $q \in \mathbb{R}^{d_{l+1}}$ to calculate the similarity $w_{k,r}^m$ with the transformed embeddings:

$$w_{k,r}^m = \frac{1}{|V_r|} \sum_{k \in V_r} q^\top \cdot \tanh(W_R \cdot z_{k,r}^{m,l+1} + b), \tag{3}$$

where V_r is the set of nodes correlated with relation r, and b is the bias vector. We treat $w_{k,r}^m$ as the importance score of each relation-specific embedding and use softmax to normalize it and obtain the final relation-level attention weight: $\beta_{k,r}^m = \frac{\exp(w_{k,r}^m)}{\sum_{r \in R_k} \exp(w_{k,r}^m)}$, where R_k denotes the types of relations connecting node v_k. $\beta_{k,r}^m$ measures the contribution of relation r to node v_k, and we use $\beta_{k,r}^m$ to fuse the relation-specific node embeddings and obtain: $z_k^{m,l+1} = \sum_{r=1}^{R_k} \beta_{k,r}^m \cdot z_{k,r}^{m,l+1}$.

The last propagation layer (the Lth layer) outputs node embedding z_k^m based on the cross-modal hierarchical attention HGNN. Concretely, for a user node u, its embedding is denoted by z_u, and for a recipe node i, it has both textual modality embedding z_i^{t} and visual modality embedding z_i^{v}.

3.3 Recipe Modality-Disentangled Learning

We disentangle multimodal recipe embeddings to provide complementary insights for recommendation. To this end, we propose contrastive fusion to extract modality-shared representations by employing contrastive learning [26], and then deploy orthogonality constraints to extract modality-specific representations. Specifically, given that recipe nodes have embeddings in both visual and textual modalities, we design a cross-modal aggregator to effectively merge these modalities, thereby enhancing the shared information between them. The importance of each modality m for recipe i is calculated by

$$\alpha_i^m = \mathrm{softmax}(\boldsymbol{q}_1^\top \tanh(\boldsymbol{W} z_i^m + \boldsymbol{b})), \tag{4}$$

where $q_1 \in \mathbb{R}$ denotes the attention vector, and W and b denote the weight matrix and bias vector, respectively. The modality-shared embedding of recipe i is then given by $z_i^c = \sum_{m=1}^{|\mathcal{M}|} \alpha_i^m z_i^m$.

We employ contrastive learning to ensure that the embedding adaptively extracts modality-shared information from multimodal embeddings. Contrastive learning has proven effective in various tasks, including graph representation learning and recommender systems. Here, we use a self-supervised signal within the contrastive learning framework to maximize the consistency between the recipe representations under individual modalities and the fused multimodal representation. In this way, the fused multimodal representations can adaptively extract shared representations between varied modalities. The contrastive loss function is defined by

$$\mathcal{L}_{\mathrm{C}} = -\frac{1}{|\mathcal{I}|}\frac{1}{|\mathcal{M}|}\sum_{i\in\mathcal{I}}\sum_{m\in\mathcal{M}} F(z_i^m, z_i^c), \tag{5}$$

where $F(\cdot)$ denotes the mutual consistency information between the two representations z_i^m and z_i^c. Specifically, we use InfoNCE [6] as a contrastive estimation. We define z_i^m and z_i^c of the same recipe i as positive samples and the embedded pairs of different recipes corresponding to the same modality, z_i^m, and $z_{i^*}^m$ ($i^* \neq i$), as negative samples, as specified in the following equation:

$$F\left(z_i^m, z_i^c\right) = \log \frac{e^{\theta(z_i^m, z_i^c)/\tau}}{e^{\theta(z_i^m, z_i^c)/\tau} + \sum_{i^*\neq i}\left(e^{\theta\left(z_i^m, z_{i^*}^c\right)/\tau} + e^{\theta\left(z_i^m, z_{i^*}^m\right)/\tau}\right)}, \tag{6}$$

where $\tau \in \mathbb{R}$ is a temperature parameter and $\theta(\cdot)$ is a cosine distance function. In addition to improving the embedding of modality-shared information through the multimodal contrastive framework, the introduction of external supervised

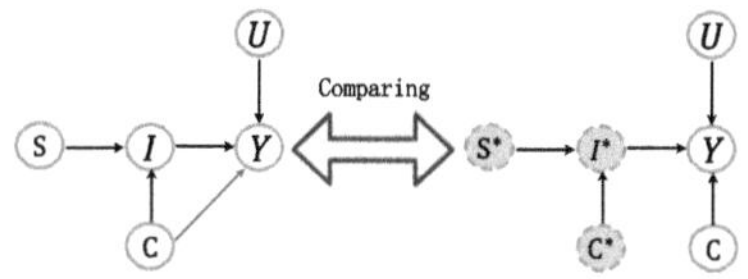

Fig. 3. Comparison between real world and counterfactual world multimodal causal graphs in recommender systems.

signals allows us to learn more effective recipe representations by leveraging relational information from different modalities. This approach mitigates issues related to the scarcity of supervised signals and the cold start problem.

Previous multimodal recommendation models have primarily focused on extracting common knowledge across different modalities while often neglecting the unique features of each modality. These unique features can play a critical role in a user's decision to choose a recipe. Therefore, we propose obtaining modality-specific representations. Specifically, the modality-specific representation z_i^{st} and z_i^{sv} can be derived by subtracting the modality-shared representation z_i^c from the representation z_i^m: $z_i^{\mathrm{st}} = z_i^{\mathrm{t}} - z_i^c$, $z_i^{\mathrm{sv}} = z_i^{\mathrm{v}} - z_i^c$.

To facilitate the subsequent calculations, we integrate the two modality-specific embeddings to obtain the final modality-specific embedding: $z_i^s = W_s \cdot (z_i^{st} || z_i^{sv})$. Additionally, to ensure that the modality-specific representations of different modalities are as distinct as possible from the modality-shared representations, we employ a soft orthogonality constraint: $\mathcal{L}_{\mathrm{S}} = \frac{1}{|\mathcal{I}|} \sum_{i \in \mathcal{I}} \left\| {z_i^{\mathrm{st}}}^{\top} z_i^{\mathrm{sv}} \right\|^2$.

3.4 Multimodal Counterfactual Learning

In this section, we introduce causal graphs that represent the relationships between multimodal content, focusing on both modality-shared and modality-specific information. Inspired by [9], we map recommendation tasks to the causal world and analyze the recipe recommendation process through a causal lens to compute causal differences. We design a counterfactual learning network that effectively utilize both modality-shared and modality-specific information while filtering out representations irrelevant to user preferences. It thus further refines the multimodal representations of recipes to align with user preferences.

Causal View on Multimodal Recipe Recommendation. We employ causal theory to analyze the multimodal recipe recommendation process, focusing on identifying causal discrepancies using tools such as causal inference interventions and counterfactual reasoning. To systematically understand the bias between modality-shared and modality-specific features, we propose a causal graph $\mathcal{G}_{\mathrm{causal}} = \{\mathcal{V}_c, \mathcal{E}_c\}$, where $\mathcal{V}_c$ is a set of variable nodes and $\mathcal{E}_c$ is a set of edges indicating causal relationships. The direction of the arrows signifies causal dependence. The causal graph includes five variable nodes: I (multimodal content of user-interacted recipes), C (modality-shared information of multimodal content in user-interacted recipes), S (modality-specific information of multimodal content in user-interacted recipes), U (user), and Y (predicted score):

- Edge $C \rightarrow Y$: The emphasis on extracting common knowledge across modalities in most multimodal recommendation frameworks results in predicted scores directly influenced by modality-shared information, without considering the distinct modality-specific at a finer granularity. This leads to a spurious relationship between I and Y.
- Edge $U\&I \rightarrow Y$: From a traditional recommendation task perspective, the predicted score indicates the user's preference for a given recipe.
- Edge $C\&S \rightarrow I \rightarrow Y$: Y represents the user's preference for features of an interacted recipe across different dimensions, categorized into modality-shared features and modality-specific features, thus offering a finer granularity in describing the user's preferences.

Based on the causal graph shown in Fig. 3(c), the predicted score Y can be calculated from the values of its ancestor nodes (U, I, C) as follows:

$$Y_{u,i,c} = f_Y(U = u, I = i, C = c), \tag{7}$$

where f_Y denotes the prediction function, such as the inner product function.

Counterfactual Calculation. After constructing the multimodal causal graph, we analyze the causal relationships between variables using intervention in causal theory [13]. Intervention in causal inference involves determining the value of a variable, rendering it independent of its parent node. This operation allows us to remove the edge pointing to that node from the causal graph.

To mitigate biases in multimodal data unrelated to user preferences, we aim to construct a world that is close to unbiased. We use counterfactual reasoning within our proposed multimodal causal graph to create a counterfactual world, as depicted in Fig. 3(d). In this counterfactual world, the variable I is fixed to I^*, representing the multimodal content of a user's interacted recipes (instead of being influenced only by the shared information across different modalities). According to the d-separation criterion [2], this setup blocks the mediating path from C to Y. The variables C^* and S^* are the modality-shared and modality-specific features of the user-interacted recipes learned from the variable I^*. Therefore, when I^* is fixed, C^* and S^* remain unaffected by the current variable I. The predicted score is calculated as follows:

$$Y_{u,i^*,c^*} = f_Y(U = u, I = i^*, C^* = c^*). \tag{8}$$

Inspired by previous work [18], the impact of the distribution of user preferences on the final predicted score can be assessed using the difference between the counterfactual world and the real world, i.e., $Y_{\text{effect}} = Y_{u,i,c} - Y_{u,i^*,c^*}$. By calculating the causal difference between the two worlds, our model utilizes modality-shared and modality-specific features to mitigate the side effects of irrelevant content in user-interacted recipes (spurious correlations in the multimodal recipe representation) on the final predicted score.

Multimodal Counterfactual Learning Layer. Following the causal difference analysis described above, we design a counterfactual learning network that computes multimodal content preference distributions for user-interacted recipes. Guided by causal difference computation, this network learns the distinction between the preference distributions of the natural world and the counterfactual world, thereby refining the multimodal content representations to be independent of user preferences. The model's convergence is further accelerated by employing a feed-forward network (FFN) with residual connections.

First, we perform multimodal fusion on the recipe embeddings learned by the HGNN module: $z_i^{\mathrm{GNN}} = W_g(z_i^{\mathrm{t}}||z_i^{\mathrm{v}})$, where W_g is a shared weight vector. Next, we combine the modality-shared embedding z_i^c and modality-specific embedding z_i^s with z_i^{GNN} to obtain $z_i^{c^*}$ and $z_i^{s^*}$, respectively. Then, we compute the queries $Q_i^{c^*}$, keys $K_i^{c^*}$, and values $V_i^{c^*}$ for $z_i^{c^*}$ and the queries $Q_i^{s^*}$, keys $K_i^{s^*}$ for $z_i^{s^*}$, respectively, by matrix transformation. The dot product of queries and keys is treated as the score at each position, and the preference distributions over multimodal content are calculated by a softmax layer:

$$C = softmax\left(\frac{Q_i^{c^*} \cdot (K_i^{c^*})^\top}{\sqrt{d_z}}\right), C^* = softmax\left(\frac{Q_i^{s^*} \cdot (K_i^{s^*})^\top}{\sqrt{d_z}}\right), \tag{9}$$

where d_z is the dimension of the query and key vectors. Based on the causal difference computation process, we multiply the difference between C and C^* by $V_i^{c^*}$ to obtain the output of counterfactual learning layer: $z_i^{\mathrm{dif}} = (C - C^*) \cdot V_i^{c^*}$.

To stabilize the training process and address the gradient vanishing problem, we apply residual connections and layer normalization. The resulting output is then fed into an FFN consisting of two linear layers and one activation function. The output of the FFN undergoes another residual connection and layer normalization, yielding the final output z_i^{cl} of the counterfactual learning layer: $z_i^{cl} = LN(GELU(W_1 z_i^{ln} + b_1)W_2 + b_2 + z_i^{ln})$, where W_1 and W_2 are the trainable weights, b_1 and b_2 are the bias terms, $z_i^{ln} = LN(z_i^{\mathrm{dif}} + z_i^c)$ and $LN(\cdot)$ is the layer normalization operation. To enhance the model's representation, we stack L' counterfactual learning layers.

3.5 Prediction and Optimization

We obtain the final recipe embedding z_i by fusing modality-shared, modality-specific, and counterfactual embeddings. The output of the HGNN module is used as the final user embedding z_u. The predicted score between user u and recipe i is calculated using an inner product operation, i.e., $\hat{y}_{u,i} = z_u^\top z_i$ (corresponds to Y in the causal graph). For recommendation tasks, we optimize the model using a pairwise BPR loss, encouraging the model to assign a higher score to observed interactions: $\mathcal{L}_{\mathrm{BPR}} = \sum_{(u,i,j)\in O} -\ln\sigma(\hat{y}_{u,i} - \hat{y}_{u,j})$, where $O = \{(u,i,j) \mid (u,i) \in O^+, (u,j) \notin O^+\}$ is the training dataset consisting of triples in the form of (u,i,j), O^+ denotes the observed interactions. The overall objective function is given by

$$\mathcal{L} = \mathcal{L}_{\mathrm{BPR}} + \lambda_C \mathcal{L}_C + \lambda_S \mathcal{L}_S, \tag{10}$$

where hyperparameters λ_C, λ_S control the effect of the contrastive auxiliary task and the orthogonality constraint, respectively.

4 Experiments

In this section, to validate the effectiveness of Mirror and to elucidate the factors contributing to its performance, we conducted systematic experiments addressing three research questions:

- **RQ1.** How does the Mirror perform compared to the state-of-the-art baselines for the recipe recommendation task?
- **RQ2.** What is the effect of each component of Mirror?
- **RQ3.** How do different hyper-parameter settings affect Mirror's performance?
- **RQ4.** Does our proposed model really improve representation learning?

Table 1. The statistics of Large-RG.

Type	Item	Number
Nodes	user	38,624
	recipe	472,515
	ingredient	22,186
Edges	user-recipe	1,193,179
	recipe-recipe	644,256
	recipe-ingredient	4,440,820
	ingredient-ingredient	170,642

4.1 Experimental Settings

Dataset. Currently, two open-source recipe datasets are available: the one introduced in the RecipeRec [19] model and the large-RG [20] dataset. The former lacks image modality data. Consequently, this study employs the large-RG dataset, as it is the largest real-world dataset and uniquely aligns with our multimodal requirements. Table 1 shows its statistics. Large-RG [20] dataset encompasses both multimodal recipe data and user-recipe interaction data. The recipe data is derived from the Recipe1M dataset [11], including recipe images, ingredients, and instructions, and is a widely used resource in the recipe domain. User interactions with recipes are sourced from the recipe-sharing website, Food.com. Additionally, nutritional information for each ingredient is derived from the USDA nutritional database [4]. Lastly, recipe-to-recipe similarity features are extracted from the FoodKG [5], a recipe knowledge graph.

Evaluation Metrics and Experimental Details. We use four widely used metrics in the recommendation domain to assess performance on the prediction set: the area under the ROC curve (AUC), the hit ratio (HR), normalized discounted cumulative gain (NDCG), and mean average precision (MAP). AUC measures the overall ranking performance on the test set, HR assesses the relevance of the recommended content, NDCG evaluates the ranking performance from a user-oriented perspective, and MAP evaluates the accuracy of the ranking positions. We partition the dataset into training, validation, and testing sets in a 7:1.5:1.5 ratio. An early stopping strategy is employed, halting the training process if the total loss on the validation data does not decrease for three consecutive epochs. The embedding size is 128, and the mini-batch size is 1,024.

Baselines. We select fifteen representative baselines for comparison and classify them into six categories:

1) *Traditional recommendation methods:* **BPR** [14] is a classical recommendation model based on matrix factorization.
2) *Graph-based recommendation methods:* Homogeneous GNN **GAT** [21] is a graph neural network model based on Graph Attention Network that dynamically assigns different weights to each node; Heterogeneous GNN **RGCN** [15] is a Relational Graph Convolutional Network-based model that can handle multiple relationship types in heterogeneous graphs; **HAN** [23] is a heterogeneous graph neural network model that considers node-level and relation-level graph attention mechanisms; **NGCF** [22] adopts the user-item bipartite graph to incorporate high-order relations and utilizes GNN to enhance CF methods; **LightGCN** [7] is a recommendation model that simplifies Graph Convolutional Networks to make it more concise and suitable for recommendation.
3) *Recipe representation learning-based methods:* **Reciptor** [8] is a pre-trained multi-task-based learning framework for recipe representation learning; **Recipe2vec** [20] is a GNN-based recipe representation learning model that uses visual, textual, and relational information about recipes.
4) *Recipe recommendation methods:* **HAFR** [3] is a hierarchical attention-based recipe recommendation model; **SCHGN** [17] is a self-supervised calorie-aware heterogeneous graph network-based recipe recommendation model; and **RecipeRec** [19] is a heterogeneous graph-based recipe recommendation model that models multiple relationships between users, recipes, and ingredients.
5) *Multimodal recommendation methods:* **FREEDOM** [27] is a multimodal recommendation model that incorporates both item graphs and denoised user-item graphs for simultaneous multimodal recommendation; **MGCN** [25] is a model for multimedia recommendation that employs a multi-view graph convolution mechanism to effectively integrate multi-level relational features of users and items across different modalities.
6) *Causal inference-related recommendation methods:* **InvRL** [1] leverages heterogeneous environments to capture a range of spurious correlations and learns invariant term representations across these environments; **MCLN** [9]

leverages multimodal content from non-interactive items to filter out aspects of user-interactive multimodal content that are unrelated to user preferences.

4.2 Performance Comparison (RQ1)

Table 2 reports the results with the following observations:

- Mirror achieves the best performance across all metrics, demonstrating its superiority and validating its design. The modality-disentangled learning and multimodal counterfactual learning in Mirror mitigate the side effects of features in multimodal data that are unrelated to user preferences. Consequently, Mirror captures user-recipe features more effectively than other baselines.
- Graph-based recommendation methods outperform traditional methods, with the heterogeneous graph-based method outperforming the homogeneous graph-based method. This outcome validates the effectiveness of capturing higher-order collaborative signals to improve embedding expressiveness. It also indicates that constructing heterogeneous information networks can more effectively leverage the multimodal information of recipes.
- Recipe representation learning based methods perform poorly. These methods are initially designed for tasks such as classification, clustering, and food retrieval, rather than addressing the complex interactions between users and

Table 2. Performance of Top-K recipe recommendation ($K = 5, 10$).

Methods	AUC (%)	HR@5	HR@10	NDCG@5	NDCG@10	MAP@5	MAP@10
BPR [14]	57.64	0.1685	0.2688	0.1092	0.1535	0.0529	0.0646
GAT [21]	58.39	0.3543	0.5097	0.2592	0.3220	0.1995	0.2154
RGCN [15]	64.02	0.3999	0.5634	0.2998	0.3645	0.2464	0.2899
HAN [23]	59.89	0.2968	0.4161	0.1792	0.2409	0.0910	0.1005
Reciptor [8]	56.04	0.2781	0.4293	0.1923	0.2771	0.1420	0.1704
Recipe2Vec [20]	77.86	0.3180	0.4717	0.2199	0.2849	0.1713	0.1864
NGCF [22]	65.25	0.4088	0.5386	0.2951	0.3817	0.2411	0.2470
LightGCN [7]	62.83	0.3869	0.5071	0.2931	0.3840	0.2468	0.2599
HAFR [3]	78.85	0.4252	0.5902	0.3024	0.3929	0.2814	0.3182
SCHGN [17]	61.25	0.3619	0.4771	0.1954	0.2622	0.2367	0.2502
RecipeRec [19]	81.07	0.5342	0.6908	0.4106	0.4801	0.3440	0.3547
FREEDOM [27]	62.79	0.4738	0.6572	0.3359	0.4176	0.2850	0.2991
MGCN [25]	63.19	0.4594	0.6340	0.3274	0.4056	0.2776	0.2910
InvRL [1]	65.35	0.5039	0.5817	0.3244	0.3816	0.2977	0.3221
MCLN [9]	63.47	0.5285	0.6936	0.4108	0.4836	0.3513	0.3638
Mirror (Ours)	**87.09**	**0.6216**	**0.7503**	**0.4711**	**0.5425**	**0.4222**	**0.4219**
Improvement	**7.4%**	**15.9%**	**8.2%**	**14.7%**	**12.2%**	**20.2%**	**16.0%**

Table 3. Ablation study with key components.

Methods	AUC (%)	HR@5	HR@10	NDCG@5	NDCG@10	MAP@5	MAP@10
$M_{w/o\ C\&D\&R}$	80.71	0.6181	0.7272	0.4611	0.5270	0.4053	0.4052
$M_{w/o\ C\&D}$	81.12	0.6023	0.7140	0.4610	0.5296	0.4095	0.4084
$M_{w/o\ C}$	82.53	0.6153	0.7420	0.4634	0.5353	0.4146	0.4137
$M_{w/o\ R}$	82.52	0.6138	0.7433	0.4630	0.5350	0.4141	0.4134
Mirror	**87.09**	**0.6216**	**0.7503**	**0.4711**	**0.5425**	**0.4222**	**0.4219**

recipes. Thus, they fail to account for the intricacies of user-recipe interactions.

- Recipe recommendation learning based methods, multimodal recommendation methods, and causal inference-related recommendation methods generally outperform other baselines. These methods benefit from effectively leveraging the multimodal information inherent in recipes. This highlights the importance of incorporating multimodal data and employing targeted techniques for extracting its insights to enhance recommendation performance. However, their performance is still inferior to Mirror, indicating that modality-disentangled learning and counterfactual learning better mitigate the side effects of irrelevant multimodal data in recipe embeddings, leading to more accurate predictions.

4.3 Ablation Study (RQ2)

To comprehensively answer RQ2, we conducted an extensive ablation study to evaluate the contribution of each component in our recipe recommendation framework to the overall performance of Mirror. Specifically, we conduct ablation studies on three modules of the model: the counterfactual learning module (C), the modality disentanglement module (D), and the relation-level attention module (R). Since the counterfactual learning module depends on the output of the modality disentanglement module, up to four valid ablation settings can be constructed, as summarized in Table 3.

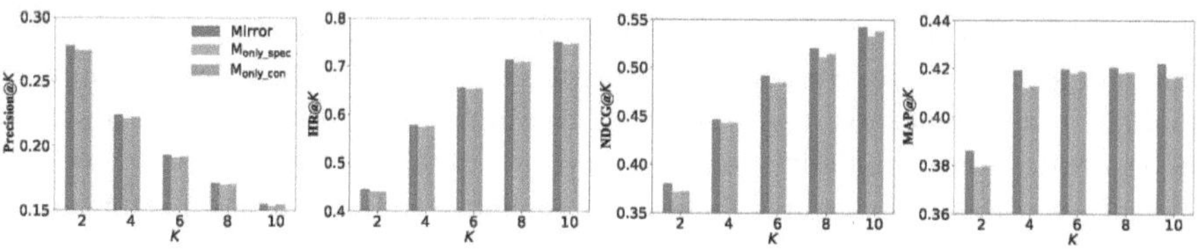

Fig. 4. Performance comparison of different embeddings for recommendation.

Table 4. Impact of GNN layers.

	AUC (%)	HR@5	HR@10	NDCG@5	NDCG@10	MAP@5	MAP@10
$L = 1$	82.70	0.5358	0.6904	0.3827	0.4613	0.3308	0.3363
$L = 2$	**87.09**	**0.6216**	**0.7503**	**0.4711**	**0.5425**	**0.4222**	**0.4219**
$L = 3$	81.93	0.5330	0.6713	0.3742	0.4598	0.3350	0.3409

Table 3 reports the results of the experimental comparison of the above variants. The results demonstrate a decline in model performance when the counterfactual learning layer is omitted, suggesting that this layer positively influences user preference modeling. It does so by guiding the model to leverage modality-specific information, thereby mitigating the side effects of spurious correlations in the multimodal content representation of user-interacted recipes. We conduct the ablation by replacing the relation-level attention mechanism with element-wise summation over relation types. This results in performance degradation, underscoring the importance of the relation-level attention module in capturing semantic information across heterogeneous node types in the graph. The further removal of the modality-disentangled module leads to an even greater reduction in performance, indicating that modality-specific and modality-shared information together enhance the model's ability to provide a comprehensive multimodal representation by complementing the individual modal information captured in the invariant space. The removal of the relation-level attention module from the GNN module significantly degrades model performance. This is because both nodes and relations convey crucial information in heterogeneous graphs, and relying solely on node-level attention disregards the varying interactions among different types of nodes. Relation-level attention allows the model to dynamically adjust the importance of neighboring nodes based on their relationships, thereby capturing richer contextual information.

In addition, we further investigate the distinction between modality-common and modality-specific features. Specifically, we analyze the recommendation performance when only modality-common or modality-specific features are used during the final prediction phase, respectively. We refer to these two variants as $\mathrm{M}_{\mathrm{only\_spec}}$ and $\mathrm{M}_{\mathrm{only\_con}}$. Figure 4 presents the performance comparison of these models. An interesting observation is that $\mathrm{M}_{\mathrm{only\_con}}$ performs slightly better than $\mathrm{M}_{\mathrm{only\_spec}}$, aligning with findings reported in previous studies. Notably, Mirror outperforms both variants. These results indicate that in the recipe recommendation scenario, each modality possesses unique characteristics, and even unaligned modal features can contribute meaningfully to the recommendations. Incorporating both modality-common and modality-specific features allows for finer-grained feature learning, thereby enhancing recommendation performance.

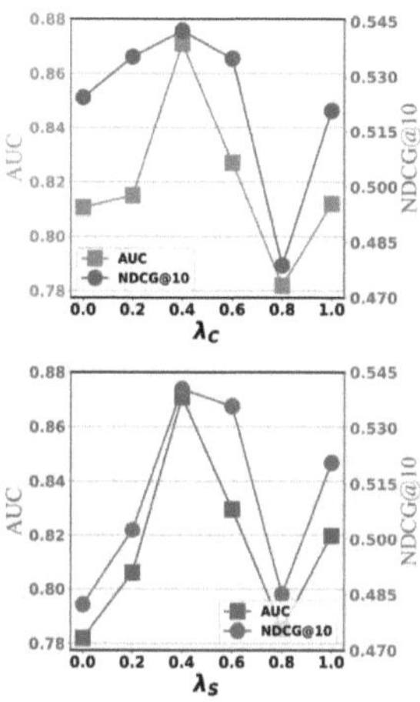

Fig. 5. Impact of λ_C and λ_S.

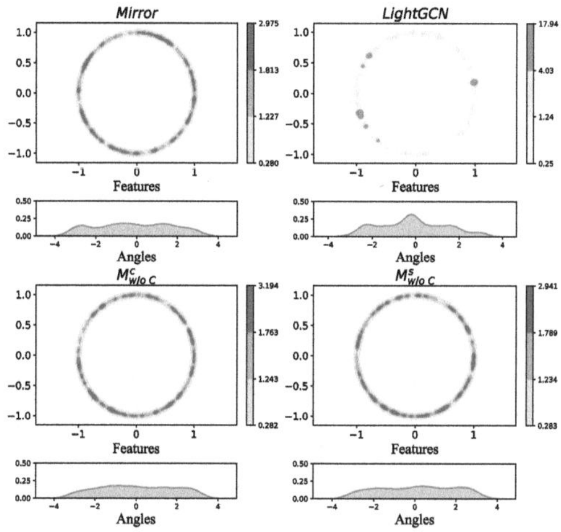

Fig. 6. Visualized distribution of recipe embeddings.

4.4 Parameter Sensitivity Analysis (RQ3)

In this section, we conduct experiments to examine the impact of key hyperparameters in Mirror on model performance. First, we explore the effect of varying the number of layers in the graph neural network. Next, we investigate the influence of different parameters in the loss function.

Effect of Graph Convolution Layer Numbers. To evaluate the impact of the number of graph convolution layers, we varied the number of layers from 1 to 3. Table 4 reveals that model performance initially improves but then declines as the number of layers increases. This trend aligns with other GCN-based recommendation models, where stacking too many layers can lead to over-smoothing, ultimately degrading performance. Consequently, the model achieves its best performance with two convolutional layers on the LargeRG dataset.

Effect of Different Parameters in the Loss Function. To explore the effect of parameters in the total loss, we tested parameter values from the set $\{0, 0.2, 0.4, 0.6, 0.8, 1\}$, where 0 indicates the removal of the corresponding loss term. Figure 5 illustrates the effects of the auxiliary tasks and orthogonality constraints on the model's performance. The results show that the model's performance initially improves as λ_C and λ_S increase, reaching a peak when both λ_C and λ_S are set to 0.4. However, further increases in these parameters lead to a decline in performance. This decline occurs because small values of λ_C and λ_S enhance the primary task, while excessively large values misdirect it. Thus, the recommendation task is sensitive to the balance provided by these auxiliary tasks.

4.5 Visualization Study (RQ4)

We first randomly select 200 users and use t-SNE to map the learned representations of recipes that these 200 users have interacted with onto a circle with a radius of 1. We then plot the feature density distribution on the circle using a Gaussian kernel density estimate (KDE) and also plot the KDE on angles (i.e., $arctan2(\cdot,\cdot)$). The results are shown in Fig. 6, where $M^c_{w/o\ C}$ and $M^s_{w/o\ C}$ represent the modality-shared embedding and modality-specific embedding in the $M_{w/o\ C}$ variant, respectively. LightGCN shows highly clustered features, which is owed to LightGCN's message-passing mechanism and its poor resistance to the popularity bias of data. Compared with Mirror, the embeddings of $M^c_{w/o\ C}$ and $M^s_{w/o\ C}$ is smoother, especially $M^c_{w/o\ C}$. This smoothing effect may occur because $M^c_{w/o\ C}$ extracts the shared features across different modalities, but the shared elements among different recipes tend to be more diverse. Mirror demonstrates a uniform distribution while retaining the basic clustering effect (i.e., peaks and troughs), indicating our model can enhance the consistency of the learned representation and mitigate the issue of characterization degradation.

5 Conclusion

We investigated the multimodal recipe recommendation problem and proposed a novel learning framework – Mirror, which integrates a basic recommendation framework with modality-disentangled and counterfactual learning. Specifically, we disentangled the multimodal representations of recipes into modality-shared and modality-specific embeddings through a contrastive learning method to enhance modal fusion. Moreover, we employed a counterfactual learning layer to refine the multimodal content representations related to user preferences. We conducted extensive experiments on the largest real-world recipe dataset and demonstrated the effectiveness of our framework.

Acknowledgements. This work is supported by the National Natural Science Foundation of China (No. 62572333), and Project Funded by the Priority Academic Program Development of Jiangsu Higher Education Institutions.

References

1. Du, X., Wu, Z., Feng, F., He, X., Tang, J.: Invariant representation learning for multimedia recommendation. In: MM, pp. 619–628 (2022)
2. Fritz, T., Klingler, A.: The D-separation criterion in categorical probability. J. Mach. Learn. Res. **24**, 46:1–46:49 (2023)
3. Gao, X., et al.: Hierarchical attention network for visually-aware food recommendation. IEEE Trans. Multim. **22**(6), 1647–1659 (2020)
4. Gebhardt, S., et al.: USDA national nutrient database for standard reference, release 21. USDA Research Service (2006)
5. Gharibi, M., Zachariah, A., Rao, P.: FoodKG: a tool to enrich knowledge graphs using machine learning techniques. Front. Big Data **3**, 12 (2020)

6. He, K., Fan, H., Wu, Y., Xie, S., Girshick, R.B.: Momentum contrast for unsupervised visual representation learning. In: CVPR, pp. 9726–9735 (2020)
7. He, X., Deng, K., Wang, X., Li, Y., Zhang, Y., Wang, M.: LightGCN: simplifying and powering graph convolution network for recommendation. In: SIGIR, pp. 639–648 (2020)
8. Li, D., Zaki, M.J.: RECIPTOR: an effective pretrained model for recipe representation learning. In: KDD, pp. 1719–1727 (2020)
9. Li, S., Guo, D., Liu, K., Hong, R., Xue, F.: Multimodal counterfactual learning network for multimedia-based recommendation. In: SIGIR, pp. 1539–1548 (2023)
10. Liu, X., Tao, Z., Shao, J., Yang, L., Huang, X.: EliMRec: eliminating single-modal bias in multimedia recommendation. In: MM, pp. 687–695 (2022)
11. Marín, J., et al.: Recipe1M+: a dataset for learning cross-modal embeddings for cooking recipes and food images. IEEE TPAMI **43**(1), 187–203 (2021)
12. Qiu, R., Wang, S., Chen, Z., Yin, H., Huang, Z.: CausalRec: causal inference for visual debiasing in visually-aware recommendation. In: MM, pp. 3844–3852 (2021)
13. Rao, Y., Chen, G., Lu, J., Zhou, J.: Counterfactual attention learning for fine-grained visual categorization and re-identification. In: ICCV, pp. 1005–1014 (2021)
14. Rendle, S., Freudenthaler, C., Gantner, Z., Schmidt-Thieme, L.: BPR: Bayesian personalized ranking from implicit feedback. In: UAI, pp. 452–461 (2009)
15. Schlichtkrull, M.S., Kipf, T.N., Bloem, P., van den Berg, R., Titov, I., Welling, M.: Modeling relational data with graph convolutional networks. In: ESWC, vol. 10843, pp. 593–607 (2018)
16. She, X., Zhou, J., Hu, Z., Du, B.: Cross-view contrastive learning enhanced heterogeneous graph networks for multi-modal recipe recommendation. In: DASFAA, pp. 467–482 (2024)
17. Song, Y., Yang, X., Xu, C.: Self-supervised calorie-aware heterogeneous graph networks for food recommendation. ACM TMCCA **19**(1s), 27:1–27:23 (2023)
18. Tang, K., Niu, Y., Huang, J., Shi, J., Zhang, H.: Unbiased scene graph generation from biased training. In: CVPR, pp. 3713–3722 (2020)
19. Tian, Y., Zhang, C., Guo, Z., Huang, C., Metoyer, R.A., Chawla, N.V.: RecipeRec: a heterogeneous graph learning model for recipe recommendation. In: IJCAI, pp. 3466–3472 (2022)
20. Tian, Y., Zhang, C., Guo, Z., Ma, Y., Metoyer, R.A., Chawla, N.V.: Recipe2Vec: multi-modal recipe representation learning with graph neural networks. In: IJCAI, pp. 3473–3479 (2022)
21. Velickovic, P., Cucurull, G., Casanova, A., Romero, A., Liò, P., Bengio, Y.: Graph attention networks. In: ICLR (2018)
22. Wang, X., He, X., Wang, M., Feng, F., Chua, T.: Neural graph collaborative filtering. In: SIGIR, pp. 165–174 (2020)
23. Wang, X., et al.: Heterogeneous graph attention network. In: WWW, pp. 2022–2032 (2019)
24. Wei, T., Feng, F., Chen, J., Wu, Z., Yi, J., He, X.: Model-agnostic counterfactual reasoning for eliminating popularity bias in recommender system. In: KDD, pp. 1791–1800 (2021)
25. Yu, P., Tan, Z., Lu, G., Bao, B.: Multi-view graph convolutional network for multimedia recommendation. In: MM, pp. 6576–6585 (2023)
26. Zhang, X., Wang, H., Li, H.: Disentangled representation for diversified recommendations. In: WSDM, pp. 490–498 (2024)
27. Zhou, X., Shen, Z.: A tale of two graphs: freezing and denoising graph structures for multimodal recommendation. In: MM, pp. 935–943 (2024)

GenPlugin: A Plug-and-Play Framework for Long-Tail Generative Recommendation with Exposure Bias Mitigation

Kun Yang[1], Siyao Zheng[2], Tianyi Li[1], Xiaodong Li[1], and Hui Li[1(✉)]

[1] Key Laboratory of Multimedia Trusted Perception and Efficient Computing, Ministry of Education of China, Xiamen University, Xiamen, China
{kyang,litianyi}@stu.xmu.edu.cn, {xdli,hui}@xmu.edu.cn
[2] School of Informatics, Xiamen University, Xiamen, China
zhengsiyao@stu.xmu.edu.cn

Abstract. Generative recommendation (GenRec) offers LLM integration, reduces embedding costs, and eliminates per-candidate scoring, attracting great attention. Despite its promising performance, this study reveals that it suffers from generation exposure bias and poor long-tail item generalization, two critical limitations overlooked by prior works on GenRec. To address these, we propose GenPlugin, a plug-and-play framework featuring a dual-encoder, shared-decoder architecture. During pre-training, it aligns language and ID views via contrastive learning, harmonizing item representations across two complementary views. Besides, GenPlugin uses a novel training strategy that probabilistically substitutes ground-truth item ID tokens with predictions from the language-semantics encoder, alleviating exposure bias. To improve long-tail generative recommendation, we propose a retrieval-based data augmentation mechanism. It fine-tunes the decoder of GenPlugin to endow GenPlugin with the ability to use relevant users w.r.t. contexts or collaborative information to augment the generation of item ID tokens in long-tail recommendation scenarios. We have plugged GenPlugin into several representative GenRec models and the extensive experiments demonstrate that GenPlugin can notably mitigate generation exposure bias during item ID generation while significantly improving the quality of long-tail item recommendation. The implementation is available at: https://github.com/KDEGroup/GenPlugin.

Keywords: Recommendation system · Sequential Recommendation · Generative Recommendation

1 Introduction

Recommender system (RecSys) is essential to numerous online services (e.g., Amazon, TikTok and YouTube) [19,22]. Traditionally, RecSys models rely on

K. Yang and S. Zheng—Contribute equally to this paper.

H. Jung et al. (Eds.): DASFAA 2026, LNCS 16535, pp. 504–519, 2026.
https://doi.org/10.1007/978-981-92-0363-5_31

the ID-based paradigm where user behaviors are formatted into item identifiers (item IDs) [8]. In this paradigm, items are uniquely identified by pre-defined numerical or string-based IDs (e.g., "1234"). For instance, deep learning based collaborative filtering methods [1] assign an embedding to each item ID and update these embeddings iteratively. During recommendation, RecSys calculates each candidate item's ranking score and offers items with the highest scores as the recommendation. Despite its widespread adoption over the years, the ID-based paradigm suffers from several limitations. Notably, it struggles with semantic gaps when integrating large language models (LLMs), as item IDs lack explicit, rich semantic information [36]. Additionally, handling a large number of items incurs high storage and processing costs of embeddings [8].

Generative recommendation (GenRec), a new recommendation paradigm, has recently gained substantial attention [8,12,18,36]. GenRec represents each item using a unique token sequence during the item creation phase. Each token (i.e., ID token) encodes semantic meaning and is designed to enhance LLMs' understanding of the underlying semantics. The number of ID tokens required is significantly smaller than the total number of items and GenRec can use finite tokens to represent almost infinite items [8], largely reducing the embedding cost. In the generative recommendation phase, GenRec directly generates the IDs of recommended items without the need to compute individual recommendation scores for each candidate item.

While GenRec has achieved notable progress, our empirical study in Sect. 2 reveals two challenges that hinder its practical deployment:

- **P1: Generation Exposure Bias.** Existing GenRec methods exhibit a critical limitation stemming from the discrepancy between training and inference phases. The generation is conditioned on ground-truth item ID tokens, meaning that the first $t-1$ ID tokens are given as input to generate the t-th ID token. However, during inference (the generative recommendation stage), predictions rely entirely on previously generated tokens, which may contain errors from prior steps. The gap between training and inference, termed generation exposure bias, leads to substantial discrepancies between generated and ground-truth item ID token sequences, a similar phenomenon to exposure bias in text generation [26]. *Notably, this limitation remains underexplored in existing GenRec studies.*
- **P2: Inadequate Generalization to Long-Tail Item Recommendation.** GenRec is expected to generalize to long-tail item recommendation [18], a challenging task in RecSys where items with few interactions suffer from low-quality recommendation [11]. The reason is that, for infrequent items, GenRec can understand their characteristics through their generative item ID tokens that are encoded with item semantics and are shared among different items. However, we find that, despite the potential of shared semantic representations to bridge data sparsity, existing GenRec methods struggle to handle long-tail item recommendation, indicating that solely relying on generative item IDs is insufficient for handling this task.

To combat the above problems, we propose GenPlugin, a plug-and-play framework for long-tail generative recommendation with exposure bias mitigation. The contributions of this work are:

- We introduce GenPlugin as a modular plugin featuring a dual-encoder, shared-decoder architecture. This design integrates contrastive semantics alignments during pre-training to harmonize item representations across two complementary views: language view (original item textual features) and ID view (generative item IDs from GenRec).
- To address **P1**, GenPlugin uses a novel training strategy that probabilistically substitutes ground-truth item ID tokens with predictions from the language-semantics encoder. By dynamically blending these signals during training, the model learns to robustly generate item IDs even when prior steps contain errors, effectively mitigating exposure bias.
- To alleviate **P2**, we propose a retrieval-based data augmentation mechanism used in GenPlugin. It fine-tunes the decoder of GenPlugin to endow GenPlugin with the ability to use relevant users w.r.t. contexts or collaborative information to augment the generation of item ID tokens in long-tail recommendation scenarios.
- We have plugged GenPlugin into several representative GenRec models. Extensive experiments demonstrate that GenPlugin can mitigate generation exposure bias during item ID generation while significantly improving the quality of long-tail item recommendation, validating GenPlugin's ability to serve as an easy-to-deploy, performance-enhancing plugin for GenRec.

2 Empirical Study of Existing GenRec

We take a representative GenRec method TIGER [18] as an example to illustrate the existence of generation exposure bias and poor long-tail item recommendation in current GenRec.

First, we report Hit Rate@10 (H@10) of TIGER [18] on six datasets in Table 1. In Table 1, TIGER w/ DS indicates enhancing TIGER with our proposed dual-encoder, shared-decoder plugin (Sect. 3.1) that adopts semantic-substitution guidance to mitigate generation exposure bias (Sect. 3.2). We can observe that, directly using ground-truth item ID tokens to guide model training (i.e., TIGER) significantly lags behind TIGER w/ DS that uses exposure bias mitigation strategy, showing that generation exposure bias indeed exists in GenRec and it negatively affects the quality of recommendation.

We further analyze the performance of TIGER on long-tail item recommendation. Figure 1 illustrates the recommendation performance w.r.t. different item groups on Amazon Beauty and items are binned into different groups according to their interaction counts. From Fig. 1, we can see that long-tail distribution exists: a large portion of items has only a few interactions and TIGER performs poorly on these groups. The empirical study shows that GenRec does not perform well on long-tail items, meaning that a substantial number of items cannot receive satisfactory recommendations (Fig. 1).

Table 1. H@10 of TIGER on six datasets. TIGER w/ DS indicates enhancing TIGER with exposure bias mitigation.

	TIGER	TIGER w/ DS
Beauty	0.0609	0.0790
Toys	0.0518	0.0824
Sports	0.0356	0.0467
Instruments	0.1211	0.1398
Arts	0.1225	0.1430
ML-100k	0.0901	0.1294

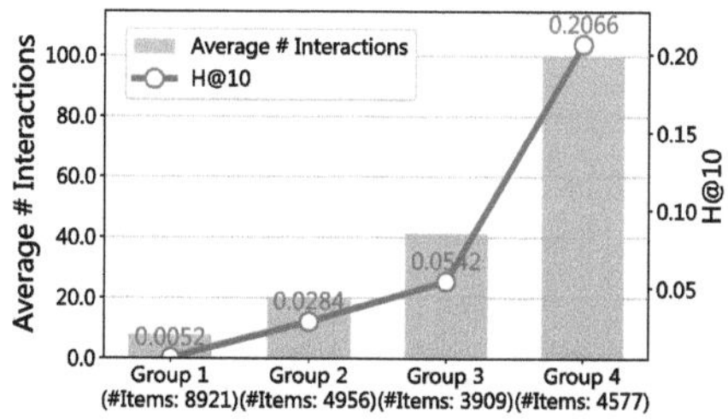

Fig. 1. Performance of TIGER on Amazon Beauty w.r.t. different item groups.

The above empirical study reveals the limitations of current GenRec and motivates us to design GENPLUGIN for addressing generation exposure bias and improving coverage for long-tail items.

3 Our Method GenPlugin

Figure 2 provides an overview of GENPLUGIN. GENPLUGIN adopts a dual-encoder, shared-decoder architecture. During pre-training, the two encoders integrate contrastive semantics alignment to better understand the generative item IDs produced by GenRec across two complementary views: language view and ID view (Sect. 3.1). The shared decoder is strengthened with a semantic-substitution guidance mechanism to alleviate exposure bias (Sect. 3.2). To improve the quality of long-tail item recommendation, we further fine-tune GENPLUGIN to endow it with the capability to utilize retrieved, relevant data as augmentation to bridge data sparsity, the key reason causing the low-quality long-tail item recommendation, so that GENPLUGIN can utilize auxiliary data to augment generative recommendation during the inference stage (Sect. 3.3).

3.1 Dual Encoders with Contrastive Semantics Alignments

For semantic encoding, GENPLUGIN adopts dual encoders, one for language semantics and the other for ID semantics. Each encoder are standard Transformers encoder [25]. The motivation for choosing a dual-encoder design is that generative ID tokens are out-of-vocabulary (OOV) tokens (e.g., "A_10" and "B_4") which are difficult for direct understanding, and using a language-semantics encoder to assist with the encoding process can help GENPLUGIN better capture the underlying ID semantics given the associated language descriptions.

The inputs to dual encoders are based on user historical interaction sequences (user sequences for short): $S_u = \{i_{u,1}, \cdots, i_{u,m}\}$ where $i_{u,j}$ indicates the j-th item interacted by user u and m is the maximum number of items in the user sequences. Each user sequence contains items interacted by the corresponding user in chronological order. The overall encoding process is actually *multi-view learning* of each user sequence.

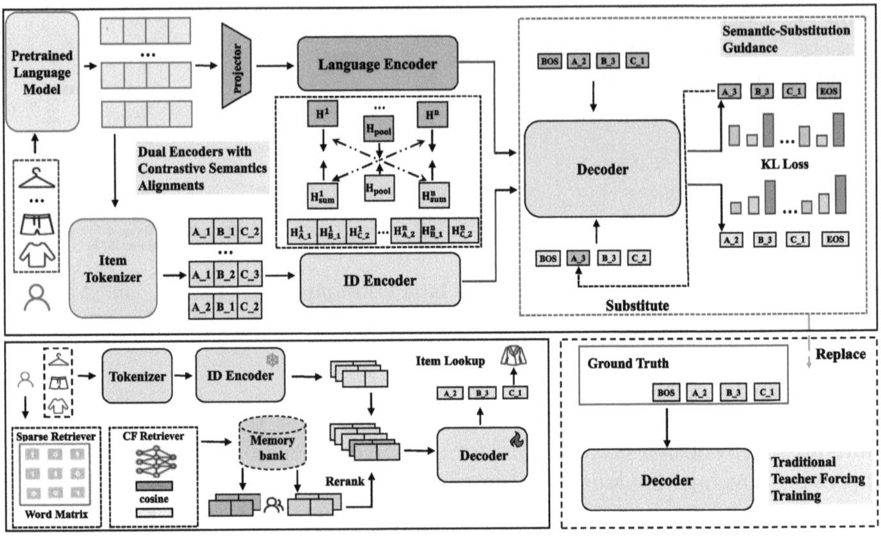

Fig. 2. Overview of GenPlugin

Language-Semantics Encoder. The inputs to the language-semantics encoder are the textual view of user sequences. We adopt LLM as the embedding extractor and pass the textual features of each item in a user sequence through LLM to get item embeddings as the inputs.

As the embedding dimensionalities of different LLMs vary, the extracted item embedding $\mathbf{e}_i$ for the item i is further projected to $\hat{\mathbf{e}}_i$ in a new space through a two-layer feedforward neural network. Note that we freeze LLM and only update the parameters of the projection network.

Then, the user sequence, $\hat{\mathbf{E}}_u = \{\hat{\mathbf{e}}_{u,1}, \cdots, \hat{\mathbf{e}}_{u,m}\}$, represented by the projected embeddings of each item, is passed through the language-semantics encoder, resulting in the language representation of S_u: $H_u = \{\mathbf{h}_{u,1}, \cdots, \mathbf{h}_{u,m}\}$, where $\mathbf{h}_{u,j}$ is the encoded representation for the j-th item in S_u.

ID-Semantics Encoder. The inputs to the ID-semantics encoder are the ID view of user sequences, i.e., generative IDs. The user sequence is represented by ID tokens of each item in the sequence: $T_u = \{t^{(1)}_{u,1}, \cdots, t^{(k)}_{u,1}, \cdots, t^{(1)}_{u,m}, \cdots, t^{(k)}_{u,m}\}$, where k is the number of tokens in an item ID and $t^{(r)}_{u,j}$ indicates the r-th ID token of j-th item interacted by user u.

Then, T_u is passed through the ID-semantics encoder and the output is the ID representation of S_u: $C_u = \{\mathbf{c}^{(1)}_{u,1}, \cdots, \mathbf{c}^{(k)}_{u,1}, \cdots, \mathbf{c}^{(1)}_{u,m}, \cdots, \mathbf{c}^{(k)}_{u,m}\}$, where $\mathbf{c}^{(r)}_{u,j}$ indicated the encoded ID token representation for the r-th ID token of the j-th item in S_u.

Contrastive Semantic Alignments. As language-semantics encoder and ID-semantics encoder are encoding the same data from two different views, aligning them for mutual benefit can further improve the encoded semantics. For this purpose, we design two contrastive cross-view semantic alignment tasks: contrastive item-level alignment and user preference alignment.

Contrastive item-level alignment aligns different views of each item. For each item, GENPLUGIN sums its ID token embeddings from ID-semantics encoder as its ID representation and aligns it with item language representation from language-semantics encoder via a contrastive loss:

$$\ell(\mathbf{u}, \mathbf{v}) = \sum_i \log \frac{\exp\left(\mathrm{sim}(\mathbf{u}_i, \mathbf{v}_i)/\tau\right)}{\sum_{j \in \mathrm{neg}(i)} \exp\left(\mathrm{sim}(\mathbf{u}_i, \mathbf{v}_j)/\tau\right)} \tag{1}$$

$$\mathcal{L}_{\text{item}} = \ell(\mathbf{h_i}, \mathbf{g_i}) + \ell(\mathbf{g_i}, \mathbf{h_i}) \tag{2}$$

where $\mathbf{h}_i$ and $\mathbf{g}_i = \mathrm{sum}\left(\{\mathbf{c}_i^{(1)} \cdots \mathbf{c}_i^{(g)}\}\right)$ indicate the language representation and the ID representation of item i, respectively. $sim(*, *)$ denotes the inner product operation and τ is a temperature hyperparameter. Neg(i) indicates the negative items of item i (i.e., all items in user sequences from the same batch).

Contrastive user-preference alignment aligns user preferences learned through two encoders. For the same user sequence S_u, the user preference patterns extracted from different views should be close to each other. For each user sequence, GENPLUGIN performs average pooling on the outputs of the two encoders, H_u and C_u, and gets $\mathbf{p}_u$ and $\mathbf{q}_u$ as the user's preference representation for language view and ID view, respectively. $\mathbf{p}$ and $\mathbf{q}$ are further aligned through a contrastive loss similar to Eq. 2:

$$\mathcal{L}_{\text{user}} = \ell(\mathbf{p_u}, \mathbf{q_u}) + \ell(\mathbf{q_u}, \mathbf{p_u}) \tag{3}$$

Optimizing Eq. 2 and Eq. 3 enhances the complementarity between the language and ID views, yielding better representations for GENPLUGIN and improving the subsequent retrieval-augmented generative recommendation.

3.2 Shared Decoder with Semantic-Substitution Guidance

To address the inherent exposure bias in generative recommendation caused by Teacher Forcing, a naive solution can involve a two-pass decoding process [15]. The first pass generates predictive probabilities, which are then used to guide the second decoding pass by replacing the ground-truth ID tokens. However, the output of the first decoding pass merely serves as a substitute for the ground truth, offering limited additional benefit to the learning process.

Hence, we propose a Shared Decoder with Semantic-Substitution Guidance method. This approach achieves efficient bias mitigation through the collaborative training of Semantic view and an ID view on the foundation of a shared decoder. The core of this method lies in a dual-level synergy mechanism: first, probabilistic semantic-substitution guidance at the decoder's input; second, forced alignment of predictive distributions at the model's objective end.

Specifically, we first design a *Probabilistic Semantic-Substitution Guidance* mechanism to break the rigid constraints of traditional teacher forcing. At each decoding timestep for the ID view, the model follows the conventional teacher forcing path with a probability of p_1, using the ground-truth ID as input. Concurrently, with a probability of $1 - p_1$, it switches to a semantic guidance path.

Along this guidance path, we further introduce a token-level random substitution strategy: each ground-truth ID token in the sequence has a probability of $1 - p_2$ of being replaced by its corresponding predicted token from the language view, thereby generating a "Semantic Substitute". To efficiently generate this guidance signal, we only perform a local softmax normalization on the top-q candidates with the highest probabilities in the language view's output logits. We then compute their weighted average embedding. This lightweight design avoids the computational overhead of a full-vocabulary calculation while efficiently condensing key semantic information into a single guidance vector, which is then injected into the shared decoder to guide the generation process of the ID view.

Besides input guidance, we employ Distributional Alignment at the objective layer to foster a consensus in semantic understanding between the two views. We begin by sharpening the original logit distributions of both views using a temperature coefficient ϕ to accentuate the influence of high-probability tokens. The sharpened probability distribution $P_{i,l}(c)$ is defined as:

$$P_{i,l}(c) = \frac{\exp(z_c/\phi)}{\sum_k \exp(z_k/\phi)} \tag{4}$$

where z_c is the logit value for class c. Subsequently, we minimize the discrepancy between the predictive distributions of the two views using a symmetric Kullback-Leibler (KL) Divergence loss function, thereby promoting bidirectional alignment:

$$\mathcal{L}_{\text{kl}} = \sum_{i} \sum_{l=1}^{k} \left(D_{\text{kl}}(P_{i,l}^{\text{lan}} \parallel P_{i,l}^{\text{id}}) + D_{\text{kl}}(P_{i,l}^{\text{id}} \parallel P_{i,l}^{\text{lan}}) \right) \tag{5}$$

where $P_{i,l}^{\text{lan}}$ and $P_{i,l}^{\text{id}}$ represent the temperature-adjusted probability distributions (vectors over the vocabulary) for the l-th token in the item ID of i, derived from the language and ID views, respectively. This design establishes a Mutual Distillation mechanism: the language view's soft predictions guide the ID view, and the ID view's outputs in turn constrain the language view. This bidirectional alignment promotes mutual learning between the two views and helps reduce the exposure bias introduced by teacher forcing.

3.3 Retrieval-Augmented Generative Recommendation

After pre-training, GenPlugin acquires the capability to offer generative recommendations. We design a retrieval-augmented generative recommendation mechanism that employs retrieval-based data augmentation to address data scarcity, the primary factor behind its suboptimal performance on long-tail items. The mechanism mainly consists of a dual-path retrieval component and a re-ranking with ID semantics component.

In the dual-path retrieval component, to augment the generative recommendation, we design two complementary retrieval strategies to retrieve more supplementary data for a user sequence:

- **Content-Aware Retrieval.** We construct a pseudo-document for each user by concatenating the textual features (e.g., titles and descriptions) of the items that the user has interacted with. The pseudo-document reflects the user's preferences in the language space. For a target user u, we apply BM25 [20] to retrieve the top-z users whose pseudo-documents are most similar to u's pseudo-document.
- **Collaborative Information Retrieval.** While content-aware retrieval returns similar users in the language space, the retrieval results lack the guidance of collaborative signals. To complement this, we train SASRec [4], a sequential recommender, to encode the sequential user-item interaction patterns. Given the user sequence S_u for a user u, SASRec applies self-attention to model the temporal dynamics and outputs a latent representation $\mathbf{g}_u$ that reflects u's collaborative profile. For a target user, we retrieve the top-z most similar users based on cosine similarity among users' collaborative profiles.

The dual-path retrieval module returns $2z$ users that are similar to the target user of the given user sequence, either in language space or collaborative representation space.

However, the retrieval list from dual-path retrieval may include noisy matches. To refine the augmentation list, we design a re-ranking component to re-rank the retrieval results according to ID semantics, i.e., the learned semantic space for generative IDs (i.e., $\mathbf{q}$ in Eq. 3). Specifically, we compute cosine similarity scores between the target user's representation and each representation of the user in the retrieval list and sort them accordingly. After re-ranking, we obtain a set of v users with the largest similarities to the target user, and they can be used as data augmentation to enhance generative recommendation. Note that, to preserve diverse signals, users retrieved by both content-aware retrieval and collaborative information retrieval are always retained in the final list, regardless of their similarity scores w.r.t. the target user.

Fine-Tuning and Inference. Given that the encoder has a time complexity of $\mathcal{O}(md^2 + m^2d)$, where m denotes the user sequence length and d is the representation dimensionality, directly concatenating the interaction histories of v users in the re-ranking list with that of the target user as the input to encoder will incur approximately v times higher computational overhead.

To address this issue, we cache the preference representations ($\mathbf{q}$ in Eq. 3) generated for users during the re-ranking stage. This design effectively avoids redundant computation in the encoder caused by multi-user input. During fine-tuning, we first freeze the encoder to ensure that each user's representation remains consistent with its pre-trained version, thereby maintaining cross-stage alignment. For each target user, we encode the corresponding user sequence using the ID-semantics encoder, the same operation as in re-ranking. At the same time, we retrieve the preference representations of similar users from the cache. These

Table 2. Statistics of the preprocessed datasets.

Dataset	#Users	#Items			#Inters.
		Total	Head	Tail	
Beauty	22,363	12,101	2,420	9,681	198,360
Toys	19,412	11,924	2,384	9,540	167,526
Sports	35,598	18,357	3,671	14,686	296,175
Instruments	17,112	6,250	1,250	5,000	136,226
Arts	22,171	9,416	1,883	7,533	174,079
MovieLens-100k	943	1,682	337	1,345	100,000

representations are then concatenated with the target user's representation, and the results are fed into the decoder for fine-tuning.

The retrieval augmentation process for the inference stage follows the same procedure as the fine-tuning stage, except that model parameters are not updated during inference.

3.4 Putting All Together

When training GenPlugin together with the underlying GenRec, the overall loss function is as follows:

$$\mathcal{L} = \mathcal{L}^{\text{lan}}(\{\mathbf{E}\}) + \mathcal{L}^{\text{id}}(\{\mathbf{T}\}) + \lambda_1 \mathcal{L}_{\text{item}} + \lambda_2 \mathcal{L}_{\text{user}} + \lambda_3 \mathcal{L}_{\text{kl}}, \tag{6}$$

where λ_1, λ_2 and λ_3 are loss weights. $\mathcal{L}^{\text{lan}}(\{\mathbf{E}\})$ indicates the training loss for the language view where the inputs $\{\mathbf{E}\}$ are item embeddings generated by the embedding extractor (LLM). $\mathcal{L}^{\text{id}}(\{T\})$ denotes the training loss for the ID view where the inputs $\{T\}$ are user sequences represented by item ID tokens.

4 Experiment

4.1 Experiment Settings

Dataset. We conduct experiments on six public datasets. Five of them come from the Amazon Product Review corpus [16], namely Beauty, Toys and Games (Toys), Sports and Outdoors (Sports), Musical Instruments (Instruments), and Arts. The sixth dataset is MovieLens-100k (ML-100k). We follow prior work on generative recommendation [18,27,34] for data preprocessing. To analyze long-tail recommendation performance, we split items into head and tail groups according to popularity following [11]. Summary statistics for all datasets are reported in Table 2.

Evaluation Metrics. We evaluate using Hit Rate (H@k) and Normalized Discounted Cumulative Gain (N@k), and report results at cutoffs $k \in \{5, 10\}$.

Table 3. Overall performance. "++" indicates the method is enhanced with GENPLUGIN. The block indicates the statistically significant improvements (i.e., two-sided t-test with $p < 0.05$) over the backbone.

Dataset	Metric	Classical					Generative					
		GRU4Rec	SASRec	Bert4Rec	S^3-Rec	LLM-ESR	TIGER	TIGER++	LETTER	LETTER++	MQL4GRec	MQL4GRec++
Beauty	H@5	0.0412	0.0350	0.0303	0.0356	0.0406	0.0390	**0.0557**	0.0419	**0.0568**	0.0446	**0.0544**
	N@5	0.0264	0.0207	0.0198	0.0212	0.0259	0.0258	**0.0369**	0.0281	**0.0393**	0.0292	**0.0373**
	H@10	0.0590	0.0536	0.0486	0.0614	0.0731	0.0604	**0.0815**	0.0664	**0.0839**	0.0702	**0.0809**
	N@10	0.0301	0.0278	0.0236	0.0309	0.0337	0.0323	**0.0463**	0.0363	**0.0482**	0.0375	**0.0454**
Toys	H@5	0.0363	0.0345	0.0246	0.0355	0.0412	0.0318	**0.0560**	0.0379	**0.0581**	0.0421	**0.0503**
	N@5	0.0246	0.0224	0.0163	0.0232	0.0249	0.0208	**0.0367**	0.0250	**0.0379**	0.0257	**0.0327**
	H@10	0.0570	0.0531	0.0426	0.0545	0.0719	0.0521	**0.0840**	0.0609	**0.0885**	0.0637	**0.0779**
	N@10	0.0321	0.0276	0.0194	0.0285	0.0347	0.0276	**0.0475**	0.0325	**0.0478**	0.0328	**0.0415**
Sports	H@5	0.0223	0.0150	0.0147	0.0236	0.0218	0.0225	**0.0319**	0.0230	**0.0344**	0.0233	**0.0253**
	N@5	0.0148	0.0078	0.0088	0.0154	0.0141	0.0144	**0.0206**	0.0146	**0.0227**	0.0152	**0.0165**
	H@10	0.0359	0.0260	0.0263	0.0365	0.0354	0.0357	**0.0490**	0.0380	**0.0513**	0.0390	**0.0415**
	N@10	0.0187	0.0113	0.0122	0.0192	0.0164	0.0187	**0.0263**	0.0194	**0.0283**	0.0197	**0.0218**
Instruments	H@5	0.0955	0.0913	0.0832	0.0921	0.0987	0.1014	**0.1164**	0.1041	**0.1174**	0.1086	**0.1129**
	N@5	0.0767	0.0724	0.0658	0.0687	0.0791	0.0897	**0.1004**	0.0916	**0.1000**	0.0929	**0.0979**
	H@10	0.1188	0.1201	0.1075	0.1115	0.1321	0.1214	**0.1412**	0.1289	**0.1443**	0.1278	**0.1389**
	N@10	0.0846	0.0815	0.0726	0.0741	0.0912	0.0953	**0.1075**	0.0992	**0.1100**	0.1007	**0.1058**
Arts	H@5	0.0826	0.0946	0.0727	0.0759	0.0992	0.0956	**0.1173**	0.0953	**0.1131**	0.1068	**0.1119**
	N@5	0.0613	0.0683	0.0524	0.0543	0.0743	0.0768	**0.0945**	0.0773	**0.0910**	0.0862	**0.0910**
	H@10	0.1101	0.1256	0.0945	0.1042	0.1341	0.1223	**0.1445**	0.1232	**0.1460**	0.1351	**0.1446**
	N@10	0.0702	0.0728	0.0597	0.0642	0.0786	0.0853	**0.1028**	0.0854	**0.1012**	0.0960	**0.1010**
ML-100k	H@5	0.0636	0.0721	0.0551	0.0782	0.0804	0.0509	**0.0785**	0.0615	**0.0922**	0.0658	**0.0859**
	N@5	0.0391	0.0423	0.0303	0.0493	0.0488	0.0320	**0.0514**	0.0386	**0.0576**	0.0400	**0.0501**
	H@10	0.1230	0.1166	0.1007	0.1271	0.1248	0.0901	**0.1336**	0.1208	**0.1484**	0.1188	**0.1463**
	N@10	0.0580	0.0566	0.0449	0.0548	0.0606	0.0449	**0.0690**	0.0575	**0.0759**	0.0566	**0.0696**

Backbones and Baselines. We select three representative generative recommendation (GenRec) models as backbones: TIGER [18]: It introduces codebook-based item IDs based on RQ-VAE. LETTER [27]: Through a set of regularization techniques, LETTER integrates collaborative filtering signals into item ID. MQL4GRec [34]: Unifies multimodal items via a quantitative language to bridge domains and enhance generative recommendation. Note that our goal is to demonstrate the effectiveness of GENPLUGIN when plugged into GenRec. To reduce the experimental cost of pre-training, we utilize the multi-modal GenRec MQL4GRec without pre-training.

We further include five traditional recommendation models: GRU4Rec [2]: Built on Recurrent Neural Networks, this approach leverages Gated Recurrent Units to encode user interaction sequences. SASRec [4]: Utilizing a unidirectional Transformer encoder, SASRec represents users by the final item in their interaction sequence. BERT4Rec [23]: This framework employs bidirectional self-attention for masked item prediction, enhancing sequential recommendation. S^3Rec [37]: By incorporating four auxiliary self-supervised tasks, S3Rec learns richer item-attribute relationships in sequential data. LLM-ESR [11]: It addresses the long-tail problem by simultaneously leveraging collaborative signals and semantic information through the dual-view modeling and self-distillation.

Implementation Details. We generate textual representations by encoding the item title and description with LLaMA2 [24], where the mean pooling of the hidden states from the last layer is utilized as the final embedding. We obtain

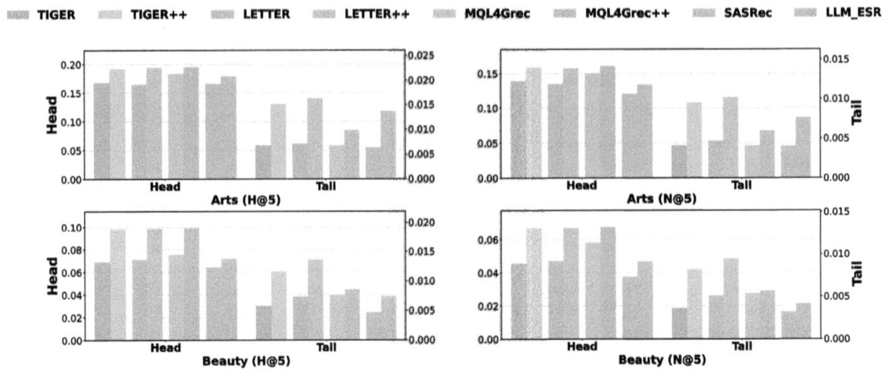

Fig. 3. Performance comparison on head and tail items.

item visual representations by following MQL4GRec [34] to encode item images. Each encoder and decoder in GenPlugin has 4 Transformer layers with 6 self-attention heads. The dimensionality of each head is 64. We set the FFN dimension to 1024 and the token embedding dimension to 128. The maximum item sequence length is 20 and the batch size is 512. We tune the hyperparameters by monitoring the validation loss, where an early stopping criterion with a patience of 20 is adopted to prevent over-training and facilitate better generalization.

4.2 Experimental Results

Overall Performance. The overall results are presented in Table 3, from which we have the following observations: GenRec methods exhibit superior performance compared to traditional approaches. This advantage primarily stems from the fact that generative methods, in constructing IDs, utilize both semantic information and a hierarchical structuring process. When enhanced with GenPlugin, all GenRec models achieve significant improvements. For instance, on the Beauty dataset, GenPlugin improves the TIGER backbone by 34.9% in H@10 and 43.3% in N@10. Furthermore, we observe that as the performance of the underlying GenRec improves, the enhancements brought by GenPlugin generally increase. Specifically, LETTER++ achieves 1.9% and 6.5% gains in H@5 and N@5 over TIGER++ on Beauty dataset.

Effect on Long-Tail Item Recommendation. Figure 3 compares performance on the head and tail item groups for the Arts and Beauty datasets. Across both groups and on both H@5 and N@5, GenPlugin consistently improves all three GenRec backbones and surpasses the traditional long-tail mitigation method LLM-ESR for both head and tail items. For head items, our method outperforms each backbone on all datasets, showing it is effective even for popular items. For tail items, it yields large gains; for instance, on Arts, TIGER++ improves H@5 by 54% over TIGER, highlighting strong long-tail performance.

Table 4. The ablation study on the Toys dataset and Arts dataset with LETTER as the backbone model. Results are reported in terms of H@10 and N@10.

Model	Toys						Arts					
	Overall		Head		Tail		Overall		Head		Tail	
GenPlugin	**0.0889**	**0.0478**	**0.1453**	**0.0808**	**0.0415**	**0.0212**	**0.1463**	**0.1025**	**0.2417**	**0.1736**	**0.0257**	**0.0132**
w/o DSA	0.0702	0.0366	0.1197	0.0629	0.0281	0.0147	0.1412	0.0962	0.2373	0.1649	0.0214	0.0104
w/o SSG	0.0797	0.0432	0.1305	0.0733	0.0361	0.0174	0.1417	0.0980	0.2391	0.1686	0.0212	0.0105
w/o RAR	0.0832	0.0442	0.1409	0.0797	0.0375	0.0180	0.1426	0.1004	0.2406	0.1724	0.0209	0.0103

Ablation Study. As shown in Table 4, removing any single component leads to consistent and interpretable drops: (*i*) cross-view alignment (DSA) is foundational, producing the largest across-the-board gains; (*ii*) semantic-substitution guidance (SSG) stabilizes the generative pipeline by mitigating exposure bias; and (*iii*) retrieval augmentation (RAR) primarily improves long-tail performance.

Together, DSA and SSG underpin our generative pipeline. DSA strengthens cross-view representations by aligning language and ID views, while SSG reduces train–inference mismatch by substituting teacher-forced ID tokens with high-confidence semantic tokens and applying bidirectional KL at the output layer. Removing either module leads to pronounced overall performance drops across datasets. Using Toys as a representative case, discarding DSA reduces Overall H@10 by 19.9%, and removing SSG lowers it by 10.3%. These results indicate that DSA supplies the core cross-view semantics and SSG stabilizes optimization; both are essential for strong overall accuracy. RAR performs dual-path retrieval (content-aware and collaborative), re-ranks candidates in the ID semantics space, and injects a few high-quality neighbor signals into the decoder to strengthen sparse regions. Removing RAR disproportionately harms the tail: on Toys, Tail H@10 drops by 9.6% versus 3.0% on Head. This gap confirms RAR's role as a tail specialist, complementing the global benefits brought by DSA and SSG.

Analysis of Hyper-parameters. Taking TIGER as an example for underlying GenRec and Toys as the test dataset, we further investigate the effects of hyper-parameters on GENPLUGIN and the results are reported in Fig. 4.

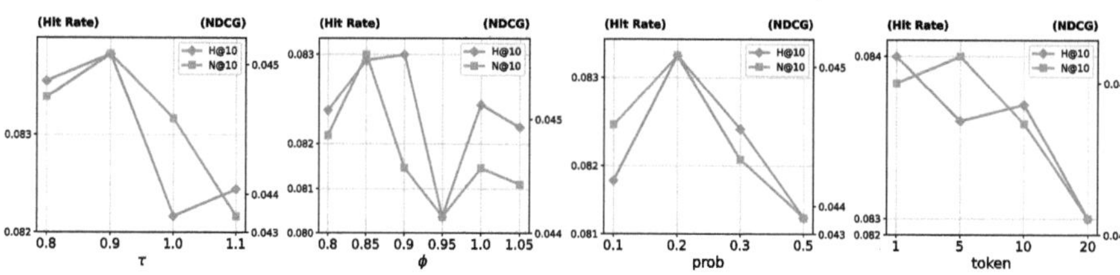

Fig. 4. Performance of TIGER++ over different hyper-parameters on Toys.

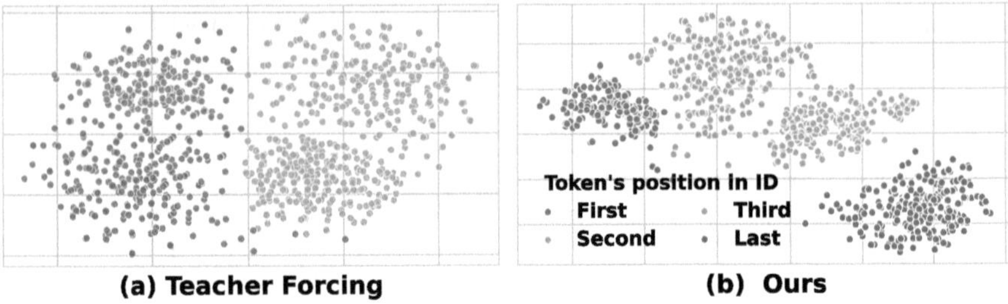

Fig. 5. OOV Token Representations Visualization.

The hyperparameter τ controls the sharpness of the contrastive loss. Moderate values yield the best performance. A too low τ causes the model to overemphasize a few negative samples, while a too high τ overly smooths distinctions and weakens discriminative power. The hyperparameter ϕ adjusts the softness of the KL divergence. Smaller ϕ values create harder, more peaked targets, whereas larger ϕ values produce softer distributions that capture richer but noisier relations. Performance peaks before ϕ is 0.9, beyond which the alignment quality declines. The hyperparameter Replacement Probability (prob) performs best around 0.2; a moderate replacement rate enhances robustness by introducing controlled noise, while higher rates lead to instability and degraded performance. Finally, the hyperparameter q which specifies the number of fusion tokens achieves optimal results around five tokens: a few highly relevant tokens improve recommendation quality, but too many introduce noise and reduce effectiveness.

Visualization. Figure 5 visualizes the representation space of added out-of-vocabulary tokens using t-SNE [14] with LETTER as the underlying GenRec on Toys. We can see that our proposed SSG method produces more tightly clustered representations at each position compared to using only teacher forcing. This clustering leads to a more robust and consistent semantic space, effectively improving the model's capacity to mitigate exposure bias.

5 Related Work

5.1 Generative Recommendation (GenRec)

GenRec overcomes limitations of ID-based RecSys by leveraging generative models to directly generate recommendations without explicit candidate scoring [8]. Pioneering methods like TIGER [18] and LC-Rec [36] employ vector quantization (e.g., RQ-VAE [33]) to convert item features into discrete tokens, enabling LLM-driven recommendation generation through token sequence modeling. This paradigm shift has spurred extensive research across three directions:

- Collaborative Semantics Alignment: Works bridge collaborative and language semantics via instruction tuning [5,21,36] or contrastive learning [3,27–29], aligning behavioral patterns with language representations.

- Stage Unification: Methods simplify multi-stage pipelines by integrating ID creation and recommendation. STORE [13] uses a unified LLM framework, TTDS [32] employs dual codebooks for joint user-item quantization, LIGER [31] unified generative and dense retrieval and ETEGRec [9] couples stages via sequence-item alignment.
- Multi-modal Extensions: Recent efforts enable cross-modal understanding through techniques like dual-aligned quantization [6], graph-enhanced RQ-VAE [10], behavior-semantic collaboration [28], multimodal preference discerner enhanced recommendation [17] and modality-agnostic vocabulary translation [34], preserving synergies between heterogeneous features.

5.2 Retrieval-Augmented Recommendation

Retrieval-augmented generation (RAG) enables dynamic incorporation of external knowledge to improve accuracy and interoperability and its idea has been applied in several RecSys models. For instance, RUEL [30] leverages behavior-sequence retrieval with item-level attention to refine recommendations. Ada-Retrieval [7] employs adaptive multi-round retrieval to iteratively update user/item representations with contextual information. RaSeRec [35] introduces a two-stage framework where cross-attention dynamically fuses retrieved user-item representations, achieving better performance. These methods highlight RAG's effectiveness in bridging generative capabilities with precise knowledge retrieval for improving recommendation quality.

6 Conclusion

In this paper, we present GENPLUGIN, a plug-and-play framework for generative recommendation that addresses exposure bias and long-tail challenges. GENPLUGIN adopts a dual-encoder, shared-decoder architecture, aligning language and ID semantics via contrastive pretraining. To mitigate exposure bias, it introduces a probabilistic ID replacement strategy during training. For better long-tail generalization, it incorporates retrieval-based augmentation to enhance recommendation. Through the experiments, we verify the effectiveness and flexibility of our GENPLUGIN.

Acknowledgments. This work was supported by National Natural Science Foundation of China (No. 62572410), Natural Science Foundation of Xiamen, China (No. 3502Z202471028). Xiaodong Li was supported by Xiamen Science and Technology Project (No. 3502Z202571028) and the Fundamental Research Funds for the Central Universities, Xiamen University (No. 20720250171).

Disclosure of Interests. The authors have no competing interests to declare that are relevant to the content of this article.

References

1. He, X., Liao, L., Zhang, H., Nie, L., Hu, X., Chua, T.: Neural collaborative filtering. In: WWW, pp. 173–182 (2017)
2. Hidasi, B., Karatzoglou, A., Baltrunas, L., Tikk, D.: Session-based recommendations with recurrent neural networks. In: ICLR (2016)
3. Hong, M., et al.: EAGER-LLM: enhancing large language models as recommenders through exogenous behavior-semantic integration. In: WWW, pp. 2754–2762 (2025)
4. Kang, W., McAuley, J.J.: Self-attentive sequential recommendation. In: ICDM, pp. 197–206 (2018)
5. Kim, T., Yoon, S., Kang, S., Yeo, J., Lee, D.: SC-Rec: enhancing generative retrieval with self-consistent reranking for sequential recommendation. arXiv Preprint arXiv:2408.08686 (2024)
6. Li, K., et al.: BBQRec: behavior-bind quantization for multi-modal sequential recommendation. arXiv Preprint arXiv:2504.06636 (2025)
7. Li, L., Lian, J., Zhou, X., Xie, X.: Ada-retrieval: an adaptive multi-round retrieval paradigm for sequential recommendations. In: AAAI, pp. 8670–8678 (2024)
8. Li, L., Zhang, Y., Liu, D., Chen, L.: Large language models for generative recommendation: a survey and visionary discussions. In: COLING, pp. 10146–10159 (2024)
9. Liu, E., Zheng, B., Ling, C., Hu, L., Li, H., Zhao, W.X.: End-to-end learnable item tokenization for generative recommendation. arXiv Preprint arXiv:2409.05546 (2024)
10. Liu, H., Wei, Y., Song, X., Guan, W., Li, Y., Nie, L.: MMGRec: multimodal generative recommendation with transformer model. arXiv Preprint arXiv:2404.16555 (2024)
11. Liu, Q., et al.: LLM-ESR: large language models enhancement for long-tailed sequential recommendation. In: NeurIPS (2024)
12. Liu, Q., et al.: Vector quantization for recommender systems: a review and outlook. arXiv Preprint arXiv:2405.03110 (2024)
13. Liu, Q., Zhu, J., Fan, L., Zhao, Z., Wu, X.: STORE: streamlining semantic tokenization and generative recommendation with a single LLM. arXiv Preprint arXiv:2409.07276 (2024)
14. Maaten, L.V.D., Hinton, G.: Visualizing data using t-SNE. J. Mach. Learn. Res. 2579–2605 (2008)
15. Mihaylova, T., Martins, A.F.T.: Scheduled sampling for transformers. In: Annual Meeting of the Association for Computational Linguistics (ACL) (2019)
16. Ni, J., Li, J., McAuley, J.J.: Justifying recommendations using distantly-labeled reviews and fine-grained aspects. In: EMNLP-IJCNLP, pp. 188–197 (2019)
17. Paischer, F., et al.: Preference discerning with LLM-enhanced generative retrieval. arXiv Preprint arXiv:2412.08604 (2024)
18. Rajput, S., et al.: Recommender systems with generative retrieval. In: NeurIPS (2023)
19. Ricci, F., Rokach, L., Shapira, B. (eds.): Recommender Systems Handbook. Springer, USA (2022)
20. Robertson, S.E., Zaragoza, H.: The probabilistic relevance framework: BM25 and beyond. Found. Trends Inf. Retr. **3**(4), 333–389 (2009)
21. Shao, M., Huang, H., Peng, Q., Liu, H.: ULMRec: user-centric large language model for sequential recommendation. arXiv Preprint arXiv:2412.05543 (2024)

22. Smith, B., Linden, G.: Two decades of recommender systems at amazon.com. IEEE Internet Comput. **21**(3), 12–18 (2017)
23. Sun, F., et al.: BERT4Rec: sequential recommendation with bidirectional encoder representations from transformer. In: CIKM, pp. 1441–1450 (2019)
24. Touvron, H., et al.: LLaMA 2: open foundation and fine-tuned chat models. arXiv Preprint arXiv:2307.09288 (2023)
25. Vaswani, A., et al.: Attention is all you need. In: NIPS, pp. 5998–6008 (2017)
26. Wang, C., Sennrich, R.: On exposure bias, hallucination and domain shift in neural machine translation. In: ACL, pp. 3544–3552 (2020)
27. Wang, W., et al.: Learnable item tokenization for generative recommendation. In: CIKM, pp. 2400–2409 (2024)
28. Wang, Y., et al.: EAGER: two-stream generative recommender with behavior-semantic collaboration. In: KDD, pp. 3245–3254 (2024)
29. Wang, Y., et al.: Content-based collaborative generation for recommender systems. In: CIKM, pp. 2420–2430 (2024)
30. Wu, N., Gong, M., Shou, L., Pei, J., Jiang, D.: RUEL: retrieval-augmented user representation with edge browser logs for sequential recommendation. In: CIKM, pp. 4871–4878 (2023)
31. Yang, L., et al.: Unifying generative and dense retrieval for sequential recommendation. arXiv Preprint arXiv:2411.18814 (2024)
32. Yin, J., et al.: Unleash LLMs potential for recommendation by coordinating twin-tower dynamic semantic token generator. arXiv Preprint arXiv:2409.09253 (2024)
33. Zeghidour, N., Luebs, A., Omran, A., Skoglund, J., Tagliasacchi, M.: SoundStream: an end-to-end neural audio codec. IEEE ACM Trans. Audio Speech Lang. Process. **30**, 495–507 (2022)
34. Zhai, J., et al.: Multimodal quantitative language for generative recommendation. In: ICLR (2025)
35. Zhao, X., et al.: RaSeRec: retrieval-augmented sequential recommendation. arXiv Preprint arXiv:2412.18378 (2024)
36. Zheng, B., et al.: Adapting large language models by integrating collaborative semantics for recommendation. In: ICDE, pp. 1435–1448 (2024)
37. Zhou, K., et al.: S3-Rec: self-supervised learning for sequential recommendation with mutual information maximization. In: CIKM, pp. 1893–1902 (2020)

C^2-DiffMM: Cross-Conditioned Contrastive Diffusion for Multimodal Recommendation

Shuwen Daizhou, Wanyu Ling, and Lingyan Zhang(✉)

School of Computer Science and Engineering, Central South University, Changsha 410083, China
{shuwendaizhou,wanyuling,lingyan_zhang}@csu.edu.cn

Abstract. Multimodal recommender systems leverage diverse information sources, such as visual, textual, and auditory features, to better model user preferences. However, existing diffusion-based methods often face challenges of semantic drift during sampling and inadequate alignment in the target space, which limit their ability to accurately capture complex user interests. To address these issues, we propose C^2-DiffMM, a novel multimodal recommendation framework that introduces a Conditioned Denoising Module (CDM) and a Cross-Denoising Alignment (CDA) mechanism. The CDM explicitly incorporates multimodal semantic conditions into the reverse diffusion process, allowing the denoiser to perceive item-level multimodal content and effectively mitigate semantic drift. Furthermore, the CDA employs a contrastive alignment strategy that operates naturally within the diffusion process without requiring manual data perturbations, thereby enforcing global semantic consistency and enhancing the discriminability of item embeddings. Extensive experiments on three publicly available datasets demonstrate that C^2-DiffMM consistently outperforms several strong baselines, achieving notable improvements in both recommendation accuracy and semantic coherence. These results validate the effectiveness and superiority of the proposed framework for multimodal recommendation tasks.

Keywords: Multimodal Recommender Systems · Conditioned Denoising Module · Cross-Denoising Alignment

1 Introduction

Modern digital platforms increasingly rely on multimedia recommender systems to enhance user experience by leveraging rich multimodal content, such as visual, textual, and auditory signals, to model user preferences with greater precision than traditional interaction-based methods [14,18,30]. By integrating complementary information across modalities, these systems can capture fine-grained user interests and deliver semantically meaningful recommendations [15,19].

S. Daizhou and W. Ling—Equal contribution.

H. Jung et al. (Eds.): DASFAA 2026, LNCS 16535, pp. 520–535, 2026.
https://doi.org/10.1007/978-981-92-0363-5_32

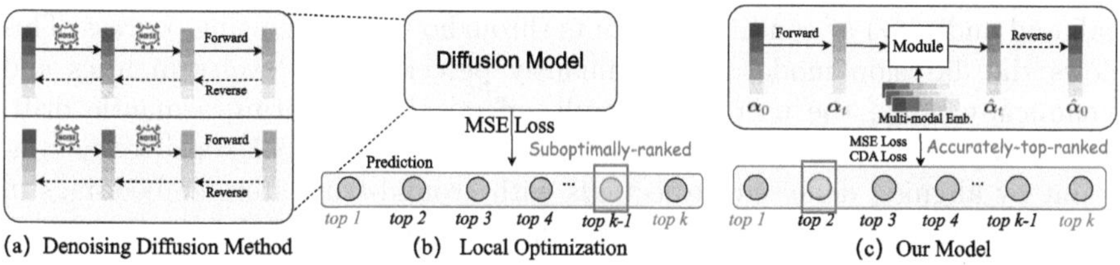

Fig. 1. Illustration of: (a) diffusion based sorely on interaction history or single-modality features; (b) suboptimal representations resulting from MSE-only optimization; and (c) the advantages of our model in cross-modal alignment and representation generation.

Significant advances have been made in incorporating multimodal information into recommendation frameworks. Early methods, including MMGCN [32] and GRCN [31], employ graph neural networks (GNNs) to fuse multimodal features during message propagation. More recently, diffusion models have demonstrated strong capability in capturing complex data distributions for recommendation. For instance, DiffMM [10] adopts a diffusion process to generate multimodal-aware user–item interaction graphs, enriching collaborative filtering with deeper semantic knowledge. Similarly, DiffCL [22] aligns cross-modal semantics to mitigate embedding inconsistency, thereby improving the stability of generated representations. Despite these advances, we identify two key limitations in current diffusion-based recommendation approaches that restrict their effectiveness:

(1) Insufficient Semantic Conditioning. Existing diffusion-based recommendation models [10,17,22] typically condition the generation process only on user interaction history or single-modality features, as shown in Fig. 1(a). This overlooks the rich, item-level multimodal semantics (e.g., imagery, text, and audio) [9], which are crucial for accurate preference modeling. Without explicit conditioning on such semantics, the generated representations may fail to capture nuanced item characteristics, limiting both recommendation accuracy and interpretability.

(2) Limitations of Local Optimization. Most existing methods rely on local optimization objectives such as Mean Squared Error (MSE) to guide the diffusion process (Fig. 1(b)). While effective at a local level, such losses do not ensure global structural consistency between the generated embedding space and the true item manifold [6,23]. The lack of a global alignment mechanism can result in blurred semantic boundaries and degraded representation quality, ultimately hampering model generalization and expressiveness [2,4].

To overcome these limitations, we propose a novel **C**ross-conditioned **C**ontrastive **Diff**usion model for **M**ulti**M**odal recommendation (**C²-DiffMM**), as illustrated in Fig. 1(c). Our framework introduces two core components: a **C**onditioned **D**enoising **M**odule (CDM) and a **C**ross-denosing **A**lignment (CDA) mechanism. The CDM explicitly injects multimodal item features (visual, textual, and auditory) as conditional inputs throughout the denoising process. This allows the diffusion model to continuously perceive item-level semantics and dynamically refine the user–item graph, effectively mitigating semantic drift. The CDA mechanism, on the other hand, imposes a global consistency constraint by aligning denoised projections with ground-truth item embeddings in a contrastive manner. This ensures that the generated embedding space remains structurally coherent with the underlying data distribution, without relying on artificial perturbations. Together, these components enable C^2-DiffMM to learn robust and semantically grounded representations, leading to improved recommendation performance and generalization.

In summary, the main contributions of this work are as follows:

- We propose C^2-DiffMM, a novel multimodal recommendation framework that integrates multimodal conditions into the diffusion process via a Conditioned Denoising Module (CDM). This enables semantically consistent representation generation and improves alignment with user preferences.
- We introduce a Cross-denosing Alignment (CDA) mechanism that enforces global structural consistency between denoised outputs and ground-truth embeddings. This alleviates representational bias and stabilizes training.
- Extensive experiments on multiple real-world datasets demonstrate that C^2-DiffMM consistently outperforms state-of-the-art baselines in terms of both recommendation accuracy and semantic coherence, validating its effectiveness and generalizability.

2 Related Work

2.1 Generative Models for Multimodal Recommendation

The integration of multimodal context has been a significant focus in advancing recommender systems [1,13,29,39]. Early approaches such as MMGCN [32] and GRCN [31] demonstrated the effectiveness of GNNs [5] in modeling high-order user-item relationships in multimodal settings. Subsequent work explored generative adversarial networks (GANs), with methods like MMSSL [28] enhancing multimodal representation learning through modality-aware graph generation. More recently, diffusion models have gained traction in recommendation systems due to their strong generative capabilities and training stability [25,27]. These approaches vary in their application of diffusion processes: some works model interaction probability distributions directly, while others, such as DiffuASR [16], focus on embedding-level diffusion. Building on these foundations, DiffMM [10] generates multimodal-aware user-item graphs to enrich interaction modeling, and DiffCL [22] aligns visual and textual semantics to reduce cross-modal inconsistencies.

However, existing diffusion-based recommendation methods typically reconstruct data without explicit conditioning on cross-modal item semantics [3,9]. This limitation often leads to semantic drift during sampling, where the generated representations deviate from the original multimodal semantics of items.

To address this issue, C^2-DiffMM introduces a **Conditioned Denoising Module (CDM)** that explicitly incorporates visual, textual, and auditory semantics as conditional inputs throughout the diffusion process. This design enables the model to maintain semantic awareness during generation, resulting in more consistent and interpretable multimodal representations.

2.2 Contrastive Learning for Multimodal Recommendation

Contrastive learning has proven effective for enhancing representation quality in recommender systems [11,34,36]. Methods like SGL [33] and SimGCL [37] introduce self-supervised perturbations on interaction graphs to improve embedding generalization through positive-negative sample comparisons. In sequential recommendation, CL4SRec [35] employs sequence-level augmentations to capture temporal user preferences. Multimodal approaches have further extended these ideas. MMGCN [32] incorporates visual and textual features with cross-modal contrastive learning, while methods like MMCLR enhance semantic consistency through inter-modal alignment. Recent diffusion-based works integrate contrastive objectives: DiffMM [10] uses contrastive learning to guide cross-view denoising, and DiffCL [22] employs contrastive alignment to stabilize semantic generation.

Moreover, existing contrastive approaches often focus on modality-level or local feature alignment, overlooking the global semantic correspondence between generated and ground-truth embeddings. To overcome this limitation, C^2-DiffMM employs a **Cross-Denoising Alignment (CDA)** mechanism that introduces global structural constraints during sampling. CDA explicitly aligns denoised projections with the true embedding space, thereby enhancing the semantic discriminability and distributional fidelity of generated representations while ensuring globally consistent and stable generation.

3 Preliminaries

3.1 Collaborative Graph with Multi-modal Features

Building on GNN-based collaborative filtering methods, our proposed C^2-DiffMM framework utilizes graph-structured data for multimodal recommendation. The user-item interactions are represented as a graph $\mathcal{G} = \{(u, i) \mid u \in \mathcal{U}, i \in \mathcal{I}\}$, where $\mathcal{U}$ and $\mathcal{I}$ denote the sets of users and items, respectively. An edge $(u, i) \in \mathcal{G}$ indicates that user u has interacted with item i. To incorporate multimodal information, we enrich the interaction graph with modality-specific features. For each item $i \in \mathcal{I}$, we define a set of feature vectors $\mathbf{F}_i = \{\mathbf{f}_i^1, \ldots, \mathbf{f}_i^m, \ldots, \mathbf{f}_i^{|\mathcal{M}|}\}$, where each $\mathbf{f}_i^m \in \mathbb{R}^{d_m}$ represents the features of modality m for item i. Here, $\mathcal{M}$ denotes the set of available modalities (e.g., visual, textual, acoustic), and d_m is the dimensionality of the feature space for modality m.

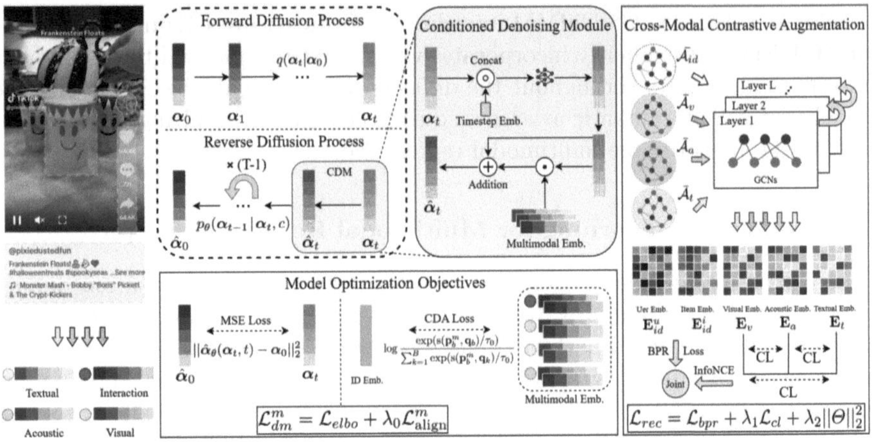

Fig. 2. Overview of C[2]-DiffMM. The framework comprises three components: (1) Conditioned Denoising Module that models user–item interactions via a multimodal-conditioned diffusion process; (2) Cross-Denoising Alignment mechanism that stabilizes denoising and enhances semantic coherence; and (3) a Cross-Modal Contrastive Augmentation module that aligns modality-specific graph representations to improve multimodal consistency and recommendation accuracy.

3.2 Task Formulation

The objective of our multimodal recommendation system is to accurately model user-item relationships by leveraging both interaction patterns and multimodal content. Formally, we aim to learn a predictive function f that estimates the probability $\hat{y}_{ui}$ of user u interacting with item i. The function takes as input a multimodal interaction graph $\mathcal{G}^{\mathcal{M}} = (\mathcal{G}, \{\mathbf{F}_i \mid i \in \mathcal{I}\})$, which combines the user-item interaction structure with item multimodal features. The prediction is thus formulated as $\hat{y}_{ui} = f(\mathcal{G}^{\mathcal{M}})$.

4 Methodology

The C[2]-DiffMM framework (Fig. 2) consists of three core components designed to alleviate semantic drift and enhance cross-modal alignment in multimodal recommendation. **(1) Multi-Modal Graph Diffusion Model** formulates user–item interactions as a denoising diffusion process, where the forward process gradually corrupts interaction structures, and the reverse process employs a Conditioned Denoising Module (CDM) guided by multimodal semantics to restore clean representations. **(2) Cross-Denoising Alignment (CDA)** introduces a contrastive objective that aligns denoised and ground-truth modality embeddings, ensuring stable sampling and semantic coherence across modalities. **(3) Cross-Modal Contrastive Augmentation** constructs modality-specific graph views (visual, textual, acoustic) and performs contrastive learning between each auxiliary modality and the main recommendation view to reinforce modality consistency and robustness. Together, these components establish a unified multimodal generative–contrastive paradigm that integrates denoising diffusion and graph-based contrastive learning for robust multimodal recommendation.

4.1 Multi-modal Graph Diffusion Model

Probabilistic Diffusion Paradigm with Interactions. Our diffusion paradigm involves two processes. The forward process progressively perturbs the user–item graph by injecting Gaussian noise, while the reverse process learns to recover the original structure by Conditioned Denoising Module.

- **Forward Graph Diffusion Process.** For a user u with interaction vector $\mathbf{a}_u$, we set $\boldsymbol{\alpha}_0 = \mathbf{a}_u$ and define the forward Markov chain:

$$q(\boldsymbol{\alpha}_t|\boldsymbol{\alpha}_{t-1}) = \mathcal{N}(\boldsymbol{\alpha}_t; \sqrt{1-\beta_t}\boldsymbol{\alpha}_{t-1}, \beta_t\mathbf{I}), \tag{1}$$

where $\beta_t \in (0,1)$ controls the noise level at step t. By leveraging the reparameterization property of Gaussian noise, we can sample $\boldsymbol{\alpha}_t$ directly from $\boldsymbol{\alpha}_0$:

$$q(\boldsymbol{\alpha}_t|\boldsymbol{\alpha}_0) = \mathcal{N}(\boldsymbol{\alpha}_t; \sqrt{\bar{\gamma}_t}\boldsymbol{\alpha}_0, (1-\bar{\gamma}_t)\mathbf{I}), \tag{2}$$

where $\bar{\gamma}_t = \prod_{t'=1}^{t}(1-\beta_{t'})$. We employ a linear scheduler for $1-\bar{\gamma}_t$ to control the overall noise scale.

- **Reverse Graph Diffusion Process.** CDM formulates the denoising as a conditional transition:

$$p_\theta(\boldsymbol{\alpha}_{t-1}\,|\,\boldsymbol{\alpha}_t, c) = \mathcal{N}\big(\boldsymbol{\alpha}_{t-1}; \boldsymbol{\mu}_\theta(\boldsymbol{\alpha}_t, t, c), \boldsymbol{\Sigma}_\theta(\boldsymbol{\alpha}_t, t, c)\big), \tag{3}$$

where c is a semantic context derived from multimodal features. Given item modality features $F \in \mathbb{R}^{M\times d}$, we compute

$$\text{attn} = \text{softmax}\big(\tfrac{\boldsymbol{\alpha}_t}{\tau}\big), \quad c = \text{attn}\, F, \tag{4}$$

and form a conditioned timestep embedding $\tilde{\psi}(t,c) = \psi(t) + g_\phi(c)$. The denoiser predicts

$$\boldsymbol{\mu}_\theta(\boldsymbol{\alpha}_t, t, c) = f_\theta([\boldsymbol{\alpha}_t, \tilde{\psi}(t,c)]), \tag{5}$$

allowing multimodal semantics to guide denoising at each step.

Modality-Aware Optimization for Graph Diffusion. The model is optimized by minimizing the ELBO of the negative log-likelihood:

$$\mathcal{L}_{elbo} = \mathbb{E}_{t\sim\mathcal{U}(1,T)}\mathbb{E}_{q(\boldsymbol{\alpha}_0)}[||\hat{\boldsymbol{\alpha}}_\theta(\boldsymbol{\alpha}_t, t) - \boldsymbol{\alpha}_0||_2^2]. \tag{6}$$

At each step t, the model predicts $\hat{\boldsymbol{\alpha}}_\theta(\boldsymbol{\alpha}_t, t)$ to approximate $\boldsymbol{\alpha}_0$, enforcing consistency between predicted and ground-truth interactions.

• **Stabilizing Sampling via Cross-Denoising Alignment.** To prevent semantic drift during sampling, we introduce a CDA objective that aligns denoised projections with ground-truth embeddings:

$$\mathcal{L}_{\text{align}}^{m} = -\frac{1}{B}\sum_{b=1}^{B} \log \frac{\exp(s(\mathbf{p}_b^m, \mathbf{q}_b)/\tau_0)}{\sum_{k=1}^{B} \exp(s(\mathbf{p}_b^m, \mathbf{q}_k)/\tau_0)}, \tag{7}$$

where $\mathbf{p}_b^m$ and $\mathbf{q}_b$ are ℓ_2-normalized projections of $\hat{\boldsymbol{\alpha}}_0 \mathbf{E}_m^i$ and $\boldsymbol{\alpha}_0 \mathbf{E}^i$, respectively. This alignment stabilizes the sampling space and strengthens cross-modal coherence.

Inference of Multi-modal Graph Diffusion Model. During inference, C^2-DiffMM adopts a deterministic reverse denoising strategy consistent with training. Starting from $\boldsymbol{\alpha}_0 \rightarrow \boldsymbol{\alpha}_{T'}$ via the forward process, we initialize $\hat{\boldsymbol{\alpha}}_T = \boldsymbol{\alpha}_{T'}$ and perform iterative denoising:

$$\hat{\boldsymbol{\alpha}}_{t-1} = \boldsymbol{\mu}_\theta(\hat{\boldsymbol{\alpha}}_t, t), \tag{8}$$

without variance sampling. The reconstructed $\hat{\boldsymbol{\alpha}}_0$ yields the predicted user–item interaction scores, from which top-k items are selected to construct the modality-specific graph $\mathcal{A}^m$.

4.2 Cross-Modal Contrastive Augmentation

In multi-modal recommendation, user preferences across different modalities (*e.g.*, visual, textual, acoustic) often exhibit correlated behavior patterns. For instance, a short video's visual and acoustic cues may jointly attract user attention. To capture such cross-modal consistency and reinforce representation robustness, we introduce **Cross-Modal Contrastive Augmentation (CMCA)**, which establishes two complementary contrastive learning paradigms based on different anchors.

Modality-Aware Contrastive View. We first generate modality-specific user/item embeddings to construct contrastive views. Given the raw modality feature $\hat{\mathbf{f}}^m \in \mathbb{R}^{d_m}$, we obtain an aligned embedding through a lightweight transformation:

$$\mathbf{e}_m^i = \text{Norm}(\text{Trans}(\hat{\mathbf{f}}^m)), \quad m \in \mathcal{M}, \tag{9}$$

where $\text{Trans}(\cdot)$ is an MLP mapping $\mathbb{R}^{d_m} \rightarrow \mathbb{R}^d$ and $\text{Norm}(\cdot)$ denotes ℓ_2-normalization. These aligned item features are aggregated with user embeddings $\mathbf{E}^u$ over the modality-aware graph $\mathcal{A}^m$ as:

$$\mathbf{z}_u^m = \bar{\mathcal{A}}_{u,*}^m \mathbf{E}^u, \quad \mathbf{z}_i^m = \bar{\mathcal{A}}_{*,i}^m \mathbf{E}_m^i, \tag{10}$$

where $\bar{\mathcal{A}}^m$ is the normalized adjacency of $\mathcal{A}^m$. Multi-layer GNN propagation on $\bar{\mathcal{A}}$ further captures higher-order signals, and the layer outputs are sum-pooled as $\bar{\mathbf{Z}}^m = \sum_{l=0}^{L} \mathbf{Z}_l^m$.

Modality-Aware Contrastive Augmentation. To align auxiliary modalities with the primary recommendation view, we further contrast each modality embedding $\bar{\mathbf{z}}_u^m$ against the main-task embedding $\bar{\mathbf{h}}_u$:

$$\mathcal{L}_{cl} = \sum_{m \in \mathcal{M}} \sum_{u \in \mathcal{U}} -\log \frac{\exp(s(\bar{\mathbf{h}}_u, \bar{\mathbf{z}}_u^m)/\tau_1)}{\sum_{v \in \mathcal{U}} \exp(s(\bar{\mathbf{h}}_u, \bar{\mathbf{z}}_v^m)/\tau_1)}. \tag{11}$$

The final cross-modal contrastive objective integrates both user- and item-level alignment:

$$\mathcal{L}_{cl} = \mathcal{L}_{cl}^{(1)} + \mathcal{L}_{cl}^{(2)}. \tag{12}$$

This contrastive framework enables C²-DiffMM to learn modality-consistent yet task-aware representations, enhancing robustness and semantic alignment across modalities.

4.3 Multi-modal Graph Aggregation

To derive the final user/item representations $\bar{\mathbf{h}}_u, \bar{\mathbf{h}}_i \in \mathbb{R}^d$ for prediction, we first aggregate modality-aware embeddings and graphs, then propagate on the original interaction graph $\mathcal{A}$ to capture high-order collaborative signals.

Given the raw feature $\hat{\mathbf{f}}^m$, the aligned item feature $\mathbf{e}_m^i$ is obtained via Eq. 9. We then perform graph aggregation on both $\bar{\mathcal{A}}$ and $\bar{\mathcal{A}}^m$ to compute the modality-specific representations:

$$\begin{aligned} \hat{\mathbf{z}}_u^m &= \bar{\mathcal{A}}_{u,*}\mathbf{E}^u + \bar{\mathcal{A}}_{u,*}(\bar{\mathcal{A}}_{u,*}\mathbf{E}^u) + \bar{\mathcal{A}}_{u,*}^m\mathbf{E}^u, \\ \hat{\mathbf{z}}_i^m &= \bar{\mathcal{A}}_{i,*}\mathbf{E}_m^i + \bar{\mathcal{A}}_{i,*}(\bar{\mathcal{A}}_{i,*}\mathbf{E}^i) + \bar{\mathcal{A}}_{*,i}^m\mathbf{E}^i. \end{aligned} \tag{13}$$

All modality-specific representations are fused through a weighted sum:

$$\mathbf{h}_u = \sum_{m \in \mathcal{M}} \kappa_m \hat{\mathbf{z}}_u^m, \quad \mathbf{h}_i = \sum_{m \in \mathcal{M}} \kappa_m \hat{\mathbf{z}}_i^m, \tag{14}$$

where κ_m are learnable coefficients controlling the contribution of each modality.

To further explore higher-order relations, we apply L-layer message passing on $\bar{\mathcal{A}}$:

$$\mathbf{H}_{l+1} = \bar{\mathcal{A}}\mathbf{H}_l, \quad \mathbf{H}_0 = \mathbf{H}_u \text{ or } \mathbf{H}_i. \tag{15}$$

The layer-wise outputs are aggregated with skip connections to form the final representation:

$$\bar{\mathbf{H}} = \sum_{l=0}^{L} \mathbf{H}_l + \omega \,\mathrm{Norm}(\mathbf{H}_0), \tag{16}$$

where ω controls the weight of normalized base embeddings to alleviate over-smoothing. Finally, C²-DiffMM predicts user–item affinity as $\hat{y}_{u,i} = \bar{\mathbf{h}}_u^\top \bar{\mathbf{h}}_i$.

4.4 Multi-task Model Training

C[2]-DiffMM is trained jointly over the recommendation objective and the diffusion-based generative objective. For each modality m, the diffusion module combines ELBO reconstruction, modality signal injection (MSI), and cross-denoising alignment (CDA) losses:

$$\mathcal{L}_{dm}^{m} = \mathcal{L}_{elbo} + \lambda_0 \mathcal{L}_{\text{align}}^{m}, \tag{17}$$

where λ_0 controls the weights of CDA.

For the recommendation task, we adopt the Bayesian Personalized Ranking (BPR) loss with the cross-modal contrastive objective $\mathcal{L}_{cl}$:

$$\mathcal{L}_{bpr} = \sum_{(u,i,j)\in\mathcal{O}} -\log \sigma(\hat{y}_{ui} - \hat{y}_{uj}), \tag{18}$$

where $\mathcal{O}$ denotes the training triplets with positive (u, i) and negative (u, j) interactions. The total recommendation loss is:

$$\mathcal{L}_{rec} = \mathcal{L}_{bpr} + \lambda_1 \mathcal{L}_{cl} + \lambda_2 ||\Theta||_2^2, \tag{19}$$

where Θ represents learnable parameters and λ_2 controls L_2 regularization. This multi-task optimization jointly enhances diffusion-based generative capability and contrastive representation learning for robust recommendation.

5 Experiments

In this section, we present experimental results to validate the effectiveness of C[2]-DiffMM. To achieve this, we address the following research questions:
- **RQ1:** How does the proposed C[2]-DiffMM model perform in comparison to other state-of-the-art recommenders across various datasets?
- **RQ2:** What is the contribution of our key components towards the overall performance of C[2]-DiffMM?
- **RQ3:** How do different hyperparameters influence the results of C[2]-DiffMM?

5.1 Experimental Settings

Datasets. In our experiments, we evaluate our model on three publicly available multi-modal recommendation datasets: TikTok (with visual, acoustic, and textual features), Amazon-Baby, and Amazon-Sports (with visual and textual features). For all datasets, we generate textual embeddings via Sentence-Bert [20]; for Amazon-Baby and Amazon-Sports, we also extract 4096-dimensional visual embeddings from product images. Detailed dataset statistics are in Table 1.

Evaluation. Following previous studies [10,22], the evaluation metrics employed in this experiment are Recall@K (R@K) and NDCG@K (N@K). Recall represents the recall rate, while NDCG stands for Normalized Discounted Cumulative Gain. We set K = {10, 20}, indicating number of items in final recommendation list.

Baselines. In our performance evaluation, we compare C^2-DiffMM with a variety of baselines, including a conventional collaborative filtering method (BPR [21]), popular GNN-based CF models (NGCF [26], LightGCN [8]), recently proposed self-supervised recommendation solutions (SGL [33] and NCL [12]), and state-of-the-art multimodal recommender systems (VBPR [7], MMGCN [32], SLMRec [24], BM3 [41], MGCN [38], Freedom [40], DiffMM [10], and DiffCL [22]).

Hyperparameter Settings. To ensure fair comparison, we present the hyperparameter settings for implementing the proposed C^2-DiffMM framework and the baseline methods. Specifically, our C^2-DiffMM is implemented with PyTorch, using Adam optimizer and Xavier initializer with default parameters. The training batch size is set to 1024, and the dimensionality of embedding vectors is 64. The learning rate is set to $1e^{-3}$, and the number of GCN layers is 1. The decay of $\mathcal{L}_2$ regularization term (λ_2) is searched in the set $\{1e^{-1}, 1e^{-2}, 1e^{-3}, 1e^{-4}, 1e^{-5}, 1e^{-6}\}$. The loss weights λ_0 and λ_1 are searched in $\{0.01, 0.1, 0.5, 1\}$ and $\{0.01, 0.1, 1\}$, respectively, while the hyperparameter ω is searched in $\{0.10, 0.25, 0.50, 0.75, 1.00\}$, and the temperature coefficient τ_0 and τ_1 are both searched in $\{0.1, 0.5, 1.0\}$. For the baseline methods, we apply the same optimization algorithm, parameter initializer, and batch size as our C^2-DiffMM framework, with the hidden dimensionality also set to 64. Shared hyperparameters, such as the number of graph neural networks layers and weight-decay regularizer, are tuned in the same range. For self-supervised methods, the temperature τ for contrastive learning is searched in $\{0.1, 0.5, 1.0\}$, and for baselines that employ random message dropout, the dropout rate is tuned from {0.1, 0.2, 0.3, 0.4, 0.5, 0.6, 0.7, 0.8, 0.9}.

Table 1. Statistics of the experimental datasets with multi-modal item Visual(V), Acoustic(A), and Textual(T) contents.

Dataset	TikTok			Baby		Sports	
Modality	V	A	T	V	T	V	T
Emb Dim	128	128	768	4096	1024	4096	1024
User	9319			19445		35598	
Item	6710			7050		18357	
Interactions	59541			139110		256308	
Sparsity	99.904%			99.899%		99.961%	

Table 2. Performance comparison on TikTok, Baby, and Sports datasets.

Methods	TikTok				Baby				Sports			
	R@10	R@20	N@10	N@20	R@10	R@20	N@10	N@20	R@10	R@20	N@10	N@20
BPR	0.0193	0.0346	0.0101	0.0130	0.0268	0.0441	0.0144	0.0188	0.0306	0.0465	0.0169	0.0210
NGCF	0.0423	0.0604	0.0169	0.0238	0.0384	0.0591	0.0186	0.0261	0.0453	0.0695	0.0231	0.0369
LightGCN	0.0436	0.0653	0.0224	0.0282	0.0402	0.0644	0.0274	0.0375	0.0423	0.0642	0.0229	0.0285
C^2-DiffMM	**0.0755**	**0.1171**	**0.0408**	**0.0520**	**0.0663**	**0.1039**	**0.0360**	**0.0460**	**0.0793**	**0.1155**	**0.0446**	**0.0538**
Impv. (%)	73.2	79.3	82.1	84.4	64.9	61.3	31.4	22.7	75.1	66.2	85.7	41.2
SGL	0.0411	0.0601	0.0171	0.0236	0.0437	0.0677	0.0211	0.0297	0.0447	0.0782	0.0264	0.0359
NCL	0.0436	0.0657	0.0193	0.0268	0.0433	0.0701	0.0188	0.0307	0.0439	0.0767	0.0230	0.0346
VBPR	0.0267	0.0381	0.0093	0.0132	0.0397	0.0665	0.0210	0.0279	0.0509	0.0765	0.0274	0.0340
MMGCN	0.0412	0.0730	0.0187	0.0307	0.0397	0.0641	0.0206	0.0269	0.0380	0.0610	0.0206	0.0266
SLMRec	0.0537	0.0847	0.0231	0.0322	0.0529	0.0775	0.0290	0.0353	0.0663	0.0990	0.0365	0.0450
BM3	0.0616	0.0957	0.0301	0.0404	0.0539	0.0848	0.0283	0.0362	0.0632	0.0940	0.0346	0.0426
MGCN	0.0678	0.1103	0.0327	0.0431	0.0608	0.0927	0.0333	0.0415	0.0713	0.1060	0.0392	0.0489
Freedom	0.0702	0.1129	0.0347	0.0463	0.0622	0.0941	0.0330	0.0414	0.0722	0.1062	0.0394	0.0484
DiffMM	0.0699	0.1126	0.0344	0.0457	0.0613	0.0977	0.0339	0.0412	0.0683	0.1019	0.0374	0.0455
DiffCL	—	—	—	—	0.0641	0.0987	0.0343	0.0433	0.0754	0.1095	0.0421	0.0509
C^2-DiffMM	**0.0755**	**0.1171**	**0.0408**	**0.0520**	**0.0663**	**0.1039**	**0.0360**	**0.0460**	**0.0793**	**0.1155**	**0.0446**	**0.0538**
Impv. (%)	7.5	3.7	17.6	12.3	3.4	5.3	5.0	6.2	5.2	5.5	5.9	5.7

5.2 Performance Comparison (RQ1)

Table 2 presents the evaluation results of the performance comparison. In the table, we highlight the performance of C^2-DiffMM in bold and the best-performing baseline is underlined for easy identification. The results yield several key observations:

- **Performance Superiority of Our.** C^2-DiffMM outperforms all baselines across all datasets, demonstrating robust cross-modal modeling and generalization. By integrating multimodal semantic conditions into the diffusion process, the model aligns generated representations with user preferences through consistent attention to visual, textual, and auditory features, enhancing both accuracy and stability.
- **Dataset-specific differences.** Compared to GNN-based models, C^2-DiffMM achieves its most significant gains on TikTok (improving R@10/20 and N@10/20 by 73.2%–84.4%). The more moderate improvements on the Baby dataset suggest that in sparse multimodal scenarios, the model excels at suppressing semantic noise and balancing global feature modeling.
- **Advantages from the model mechanisms.** C^2-DiffMM outperforms other diffusion models (e.g., DiffMM, DiffCL) on the TikTok dataset by up to 17.6%. These gains are driven by two key mechanisms: *(1) Conditioned Denoising Module*, which maintains cross-modal consistency during sampling; and *(2) Cross-Denoising Alignment*, which stabilizes the process by constraining the denoised projections within the ground-truth embedding space to reduce bias.

Table 3. Ablation study on key components of C^2-DiffMM.

Methods	TikTok		Baby		Sports	
	R@20	N@20	R@20	N@20	R@20	N@20
w/o DM	0.1067	0.0436	0.0931	0.0397	0.0971	0.0436
w/o CDM	0.1134	0.0493	0.0987	0.0427	0.1093	0.0504
w/o CDA	0.1141	0.0495	0.1001	0.0447	0.1123	0.0516
C^2-DiffMM	**0.1171**	**0.0520**	**0.1039**	**0.0460**	**0.1155**	**0.0538**

Table 4. Hyperparameter analysis of τ_0 and λ_0.

τ_0	λ_0	TikTok		Baby		Sports	
		R@20	N@20	R@20	N@20	R@20	N@20
0.1	0.01	0.1009	0.433	0.0871	0.0354	0.1037	0.0502
	0.1	0.0837	0.0355	0.0957	0.0403	0.0979	0.0457
	0.5	0.0794	0.0283	0.0911	0.0371	0.0944	0.0434
	1	0.0526	0.0231	0.0747	0.0333	0.0877	0.0389
0.5	0.01	0.1109	0.0497	0.0813	0.0347	0.1089	0.0519
	0.1	0.1171	0.0520	0.0914	0.0384	0.1155	0.0538
	0.5	0.0993	0.0417	0.1030	0.0460	0.1116	0.0522
	1	0.0710	0.0301	0.0954	0.0403	0.1057	0.0516
1.0	0.01	0.1013	0.0453	0.0887	0.0356	0.1036	0.0495
	0.1	0.1131	0.0494	0.0920	0.0371	0.1097	0.0515
	0.5	0.1006	0.0431	0.0946	0.0397	0.1130	0.0523
	1	0.0739	0.0323	0.0933	0.0393	0.1061	0.0507

5.3 Model Ablation Test (RQ2)

To validate the effectiveness of our methods, we conducted experiments where we individually removed three key components of : multi-modal graph diffusion model (*w/o DM*), Conditioned Denoising Module (*w/o CDM*), and Cross-Denoising Alignment (*w/o CDA*). For the variant where multi-modal graph diffusion model was removed, we used another generative model VGAE to replace the designed diffusion model. The results are presented in Table 3, leading us to draw the following key conclusions:

- In the *w/o DM* variant, the diffusion model (Diffusion Model) is replaced with a conventional generative model, VGAE. Experimental results show that the full model consistently outperforms this variant across all datasets, demonstrating the effectiveness and superiority of diffusion modeling in multimodal representation generation. The multimodal graph diffusion paradigm of C^2-DiffMM explicitly captures modality correlations within user–item interac-

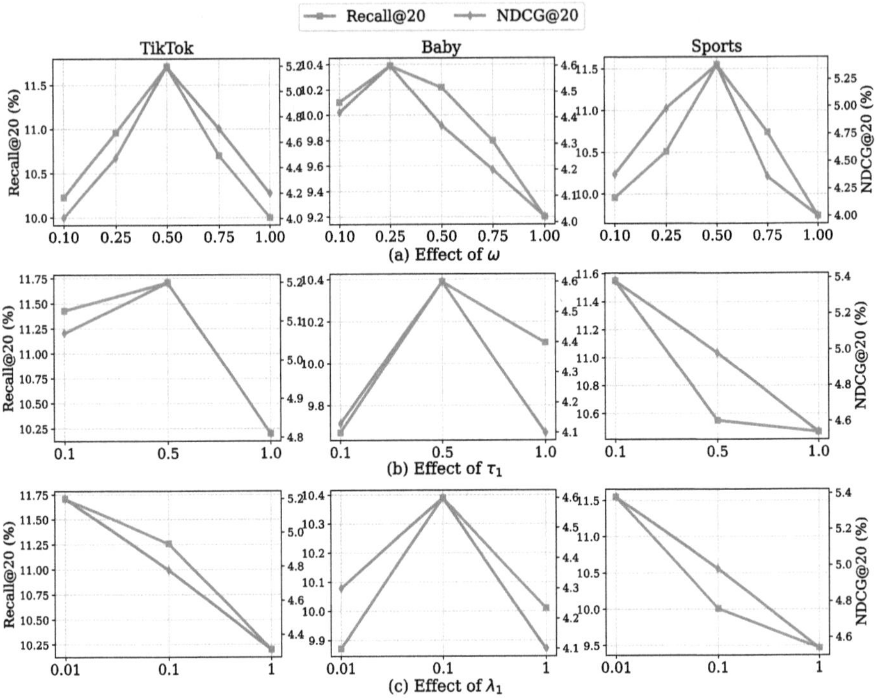

Fig. 3. Effect of Various Parameters.

tions, unleashing cross-modal synergy and facilitating the efficient integration of multimodal graph aggregation and contrastive learning.

- The *w/o CDM* variant indicates the crucial role of the Conditioned Denoising Module (CDM) in enhancing semantic consistency. By introducing multimodal semantic conditions into the reverse diffusion process, CDM enables the model to generate semantically coherent representations that better align with user preferences, thereby improving the accuracy of user interest modeling.
- The *w/o CDA* variant further validates the effectiveness of the Cross-Denoising Alignment (CDA) mechanism. By aligning denoised projections with ground-truth embeddings, this mechanism imposes global constraints within the sampling space, effectively enhancing semantic discriminability and embedding stability, and mitigating representation drift in the generated embeddings.

5.4 In-Depth Model Analysis (RQ3)

We analyze the sensitivity of key hyperparameters in C^2-DiffMM. Results are summarized in Fig. 3 and Table 4.

(i) τ_0 and λ_0 in graph diffusion. λ_0 and the temperature τ_0 control the strength of Cross-Denoising Alignment (CDA), guiding the diffusion model to generate semantically consistent embeddings. As Fig. 3 shows, dataset-specific λ_0 values yield the best performance, highlighting the importance of careful tuning for accurate recommendations.
(ii) ω in multimodal graph aggregation. The weight ω balances the contribution of normalized embeddings $\mathbf{H}_0$ during aggregation, mitigating over-smoothing. Small ω overemphasizes high-order information, while large ω under-utilizes it. Moderate values achieve the best trade-off, improving recommendation quality.
(iii) τ_1 and λ_1 in cross-modal contrastive learning. In cross-modal data augmentation, τ_1 (temperature) and λ_1 (InfoNCE weight) significantly affect performance. Table 4 shows that optimal settings vary across datasets, emphasizing the role of cross-modal contrastive learning in enhancing overall model effectiveness.

6 Conclusion

This paper proposes **C^2-DiffMM**, a multimodal framework enhancing diffusion-based recommendation through explicit semantic conditioning and contrastive alignment. The CDM module integrates visual, textual, and auditory cues into the denoising process, enabling dynamic semantic perception across reverse diffusion steps. Additionally, the CDA mechanism enforces global consistency between generated embeddings and true item distributions, stabilizing sampling and improving representation quality. Extensive experiments on public datasets show that C^2-DiffMM achieves state-of-the-art accuracy and semantic coherence. Future work will explore its integration into personalized content generation and explainable recommendation to further enhance generalization and interpretability.

Disclosure of Interests. The authors have no relevant financial or non-financial interests to disclose.

References

1. Cao, X., et al.: Cross-modal knowledge graph contrastive learning for machine learning method recommendation. In: MM, pp. 3694–3702 (2022)
2. Chen, T., et al.: A simple framework for contrastive learning of visual representations. In: Proceedings of the International Conference on Machine Learning (2020)
3. Cui, X., Lu, W., Tong, Y., Li, Y., Zhao, Z.: Diffusion-based multi-modal synergy interest network for click-through rate prediction. In: Proceedings of the 48th International ACM SIGIR Conference on Research and Development in Information Retrieval, pp. 581–591 (2025)
4. Gao, J., et al.: Multimodal recommender systems: a survey. ACM Comput. Surv. (2023). https://arxiv.org/abs/2302.03883

5. Guo, W., Cui, Y., Zheng, K.: Intfair: graph neural networks for fair recommendations with interest awareness. In: International Conference on Database Systems for Advanced Applications, pp. 3–18. Springer (2024)
6. He, K., et al.: Momentum contrast for unsupervised visual representation learning. In: Proceedings of the IEEE/CVF Conference on Computer Vision and Pattern Recognition (2020). https://arxiv.org/abs/1911.05722
7. He, R., McAuley, J.: VBPR: visual Bayesian personalized ranking from implicit feedback. In: AAAI, vol. 30 (2016)
8. He, X., Deng, K., Wang, X., Li, Y., Zhang, Y., Wang, M.: LightGCN: simplifying and powering graph convolution network for recommendation. In: SIGIR, pp. 639–648 (2020)
9. Huang, J., et al.: Multimodal recommendation: challenges and opportunities. arXiv preprint arXiv:2406.11781 (2024)
10. Jiang, Y., et al.: DiffMM: multi-modal diffusion model for recommendation. In: Proceedings of the 32nd ACM International Conference on Multimedia, pp. 7591–7599 (2024)
11. Khosla, P., et al.: Supervised contrastive learning. In: NeurIPS, vol. 33, pp. 18661–18673 (2020)
12. Lin, Z., Tian, C., Hou, Y., Zhao, W.X.: Improving graph collaborative filtering with neighborhood-enriched contrastive learning. In: WWW, pp. 2320–2329 (2022)
13. Liu, F., et al.: Disentangled multimodal representation learning for recommendation. Trans. Multimedia (TMM) (2022)
14. Liu, N., et al.: Learning social graph for inactive user recommendation. In: International Conference on Database Systems for Advanced Applications, pp. 151–167. Springer (2024)
15. Liu, Q., et al.: Multimodal recommender systems: a survey. ACM Comput. Surv. **57**(2), 1–17 (2024)
16. Liu, Q., et al.: Diffusion augmentation for sequential recommendation. In: CIKM, pp. 1576–1586 (2023)
17. Liu, Z., et al.: DMR: diffusion model-based recommendation with semantic alignment. arXiv preprint arXiv:2406.11781 (2024)
18. Niu, G., Zhang, X.: Diffusion-based hierarchical negative sampling for multimodal knowledge graph completion. arXiv preprint arXiv:2501.15393 (2025)
19. Qiu, R., Wang, S., Chen, Z., Yin, H., Huang, Z.: CausalRec: Causal inference for visual debiasing in visually-aware recommendation. In: MM, pp. 3844–3852 (2021)
20. Reimers, N., Gurevych, I.: Sentence-BERT: sentence embeddings using Siamese BERT-Networks. arXiv preprint arXiv:1908.10084 (2019)
21. Rendle, S., Freudenthaler, C., Gantner, Z., Schmidt-Thieme, L.: BPR: Bayesian personalized ranking from implicit feedback. In: UAI, pp. 452–461 (2009)
22. Song, Q., et al.: DiffCL: a diffusion-based contrastive learning framework with semantic alignment for multimodal recommendations. IEEE Trans. Neural Netw. Learn. Syst. (2025)
23. Sun, Z., et al.: Contrastive learning for recommender systems: a survey. arXiv preprint arXiv:2406.11781 (2024)
24. Tao, Z., et al.: Self-supervised learning for multimedia recommendation. Trans. Multimedia (TMM) (2022)
25. Wang, W., et al.: Diffusion recommender model. arXiv preprint arXiv:2304.04971 (2023)
26. Wang, X., He, X., Wang, M., Feng, F., Chua, T.S.: Neural graph collaborative filtering. In: SIGIR, pp. 165–174 (2019)

27. Wang, Y., Liu, Z., Yang, L., Yu, P.S.: Conditional denoising diffusion for sequential recommendation. arXiv preprint arXiv:2304.11433 (2023)
28. Wei, W., Huang, C., Xia, L., Zhang, C.: Multi-modal self-supervised learning for recommendation. In: WWW, pp. 790–800 (2023)
29. Wei, W., Tang, J., Jiang, Y., Xia, L., Huang, C.: PromptMM: multi-modal knowledge distillation for recommendation with prompt-tuning. arXiv preprint arXiv:2402.17188 (2024)
30. Wei, Y., Liu, W., Liu, F., Wang, X., Nie, L., Chua, T.S.: LightGT: a light graph transformer for multimedia recommendation. In: SIGIR, pp. 1508–1517 (2023)
31. Wei, Y., Wang, X., Nie, L., He, X., Chua, T.S.: Graph-refined convolutional network for multimedia recommendation with implicit feedback. In: MM, pp. 3541–3549 (2020)
32. Wei, Y., et al.: MmGCN: multi-modal graph convolution network for personalized recommendation of micro-video. In: MM, pp. 1437–1445 (2019)
33. Wu, J., et al.: Self-supervised graph learning for recommendation. In: SIGIR, pp. 726–735 (2021)
34. Wu, Y., et al.: Multi-view multi-behavior contrastive learning in recommendation. In: International Conference on Database Systems for Advanced Applications, pp. 166–182. Springer (2022)
35. Xie, X., et al.: Contrastive learning for sequential recommendation. In: ICDE, pp. 1259–1273. IEEE (2022)
36. Yang, Y., et al.: Debiased contrastive learning for sequential recommendation. In: Proceedings of the ACM web conference 2023, pp. 1063–1073 (2023)
37. Yu, J., et al.: Are graph augmentations necessary? simple graph contrastive learning for recommendation. In: Proceedings of the 45th International ACM SIGIR Conference on Research and Development in Information Retrieval, pp. 1294–1303 (2022)
38. Yu, P., Tan, Z., Lu, G., Bao, B.K.: Multi-view graph convolutional network for multimedia recommendation. In: Proceedings of the 31st ACM International Conference on Multimedia, pp. 6576–6585 (2023)
39. Zhang, Z., et al.: DeepMIN: deep multi-modal interest network with cognitive learning modules. In: International Conference on Database Systems for Advanced Applications, pp. 212–227. Springer (2024)
40. Zhou, X., Shen, Z.: A tale of two graphs: freezing and denoising graph structures for multimodal recommendation. In: Proceedings of the 31st ACM International Conference on Multimedia, pp. 935–943 (2023)
41. Zhou, X., et al.: Bootstrap latent representations for multi-modal recommendation. In: WWW, pp. 845–854 (2023)

TIG-Diff: Temporal-Integrated Graph Diffusion for Ranking-Consistent Implicit Feedback Denoising in Recommendation

Lingyan Zhang, Shuwen Daizhou, Wanyu Ling, Yiman Xie(✉), and An Huang(✉)

School of Computer Science and Engineering, Central South University, Changsha 410083, China
{lingyan_zhang,shuwendaizhou,wanyuling,yimanxie,224129}@csu.edu.cn

Abstract. We propose TIG-Diff, a unified graph-diffusion framework designed to address two major limitations of diffusion-based recommender systems: noise-amplified high-order propagation and ranking inconsistency. TIG-Diff integrates implicit subgraph contrastive denoising with curriculum-driven temporal ranking optimization, achieving a synergistic balance between denoising stability and ranking refinement. The Graph Diffusion Denoising Module (GDDM) leverages multi-step diffusion, structural graph embeddings, and temporal dynamics through cross-modal interaction to suppress spurious co-occurrences and adaptively capture evolving user preferences. Meanwhile, Temporal Collaborative Ranking (TCR) introduces curriculum-based optimization that dynamically shifts from MSE-oriented reconstruction to NDCG-guided ranking learning, ensuring a smooth transition from absolute denoising to relative preference modeling. Extensive experiments on four real-world datasets demonstrate that TIG-Diff consistently outperforms state-of-the-art baselines, achieving up to 20.6% improvement in top-K performance. These results highlight the effectiveness of coupling diffusion-based denoising with temporal ranking alignment, offering a robust and generalizable paradigm for next-generation recommendation.

Keywords: Diffusion Models · Recommendation Systems · Implicit Feedback Denoising · Contrastive Learning · Temporal Collaborative Ranking

1 Introduction

Diffusion Models (DM) [3,14,27] have demonstrated exceptional capability in progressively denoising and reconstructing complex data distributions, achieving remarkable success in diverse generative tasks, such as image synthesis [21,29], text generation [5,19], and multimodal learning [7,26,30]. Building on this potential, the iterative denoising paradigm of DMs has been recently introduced into recommender systems to address the prevalent noise and sparsity in implicit user

H. Jung et al. (Eds.): DASFAA 2026, LNCS 16535, pp. 536–551, 2026.
https://doi.org/10.1007/978-981-92-0363-5_33

feedback (e.g., clicks, views). Notable contributions include PDRec [16], which iteratively refines sparse user/item embeddings; DiffRec [22], which enhances item generation based on user-specific constraints; and CF-Diff [9], which leverages bipartite graph structures for aggregating multi-hop neighbors. Despite these advances, the application of DMs to recommendation tasks faces two critical challenges arising from noisy implicit feedback:

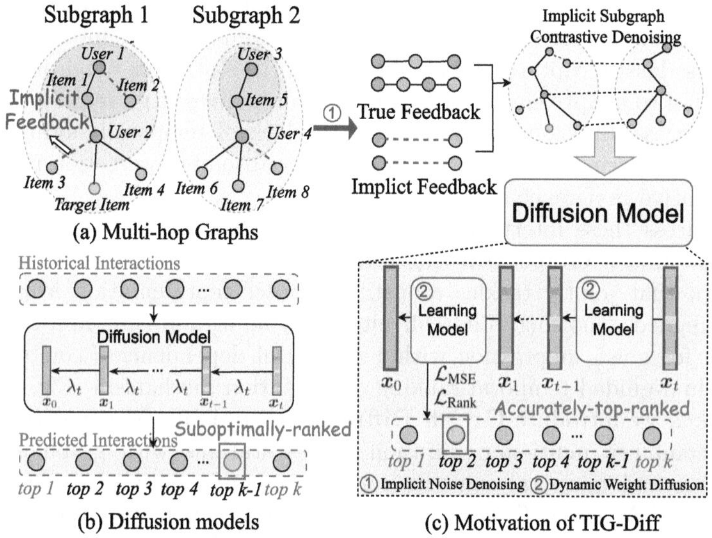

Fig. 1. Motivation for TIG-Diff. (a) Multi-hop graph methods propagate noise from spurious interactions, impairing denoising. (b) Traditional diffusion-based approaches prioritize reconstruction accuracy over ranking consistency. (c) TIG-Diff integrates multi-hop subgraph aggregation with curriculum-guided learning to jointly address noise suppression and ranking optimization.

Challenge 1: Robust High-Order Modeling Amidst Implicit Feedback Noise. Implicit interactions often encompass pseudo-interests, such as incidental behaviors, curiosity-driven clicks, or transient actions, which are difficult to differentiate from genuine long-term preferences. High-order modeling approaches [9,32] attempt to capture structural dependencies by expanding subgraphs around user nodes. However, when these subgraphs include noisy neighbors, structural contamination propagates across multiple hops (Fig. 1 (a)), significantly degrading the quality of learned high-order relationships. This manifests as: (i) both pseudo- and genuine interactions can exhibit similar local connectivity patterns within subgraphs, undermining the discriminative power of structural representations; and (ii) Insufficient Temporal Modeling: Static fusion strategies (e.g., naive temporal embedding concatenation) lack the flexibility to dynamically track evolving preferences, exacerbating the challenge of distinguishing long-term interest trajectories from short-term noise.

Challenge 2: Balancing Denoising Accuracy and Ranking Consistency. Noisy implicit feedback further complicates the goal of generating high-quality recommendations, which extends beyond mere denoising accuracy. Current DM-based recommenders primarily rely on accuracy-driven pointwise losses (e.g., MSE) [2,12]. While effective for minimizing reconstruction error, MSE inherently neglects the preservation of relative preference orderings between items. As a result, optimized models often produce denoised interactions with numerically low error, yet yield suboptimal recommendation lists due to compromised ranking quality (Fig. 1(b)). The key limitations include: (i) the intrinsic inability of pointwise loss to capture ranking structures; (ii) static loss weighting schemes (λ_t) that fail to adaptively prioritize different objectives (e.g., reconstruction vs. ranking) across diffusion steps; and (iii) the lack of temporally-aware ranking objectives, hindering a smooth transition from absolute interaction reconstruction to relative preference learning.

To address these intertwined challenges, we propose TIG-Diff (**T**emporal-**I**mplicit Feedback **G**raph **Diff**usion), a novel diffusion-based recommendation framework that jointly tackles robust high-order representation learning and ranking-aligned denoising. TIG-Diff integrates multi-hop subgraph contrastive denoising for noise suppression within structural dependencies, combined with a curriculum-guided temporal ranking optimization mechanism (Fig. 1(c)). For *Challenge 1*, we introduce **Graph Diffusion Denoising Module (GDDM)**. GDDM combines multi-step diffusion representations with graph structural embeddings and temporal contexts via a cross-modal interaction mechanism. A *Dynamical Negative Feedback Propagation* strategy uses temperature scaling to amplify the distinction between genuine preferences and noisy pseudo-interactions during subgraph expansion, thus attenuating local noise. The dynamic integration of temporal context enables the capture of preference evolution, producing unified high-order user-item representations. This approach progressively removes spurious co-occurrences while modeling stable high-order relationships and dynamically tracking shifts in user interests. For *Challenge 2*, We propose **Temporal Cooperative Ranking (TCR)**, a curriculum-aligned denoising optimization strategy consisting of: (i) Relative Ranking Optimization (RRO), which ensures pairwise preference order consistency between items post-denoising; and (ii) Dynamic Loss Weighting (DLW), which adaptively balances MSE-based reconstruction (prioritized in early diffusion steps for stable initialization) with NDCG-driven ranking optimization (emphasized in later steps for fine-grained hierarchy refinement). TCR facilitates a progressive shift from coarse denoising to precise ranking refinement.

Our main contributions are as follows:

- We propose **TIG-Diff**, a novel graph diffusion framework that integrates multi-hop subgraph contrastive denoising with curriculum-guided temporal ranking optimization, jointly suppressing spurious signals while ensuring structural robustness and ranking consistency in noisy implicit feedback.

- We propose the **GDDM**, a graph diffusion denoising module that effectively combines multi-step diffusion features, structural embeddings, and temporal contexts for progressive high-order relation modeling and noise suppression.
- We introduce **TCR**, a temporal cooperative ranking strategy comprising RRO and DLW, enabling curriculum-based progressive refinement of ranking hierarchies during the denoising process.
- Extensive experiments on four benchmark datasets demonstrate that TIG-Diff **outperforms state-of-the-art baselines by up to 20.6%** in top-K recommendations.

2 Related Work

2.1 GNN-Based Recommendation

GNN-based recommendation has evolved through refined user–item interaction modeling, especially for noisy implicit feedback [1,4,14,28]. NGCF [23] and LightGCN [8] established collaborative signal propagation via message passing, while SGL [25] and CIGCN [31] introduced contrastive learning and graph compression for robustness. Recently, CF-Diff [9] applied diffusion-based reasoning for multi-hop relation distillation.

Despite these advances, two challenges persist: (i) **Spurious Co-occurrences**: Most methods [9,23] propagate observed interactions without filtering incidental or popularity-driven overlaps, misleading representation learning. (ii) **Static Negative Sampling**: Existing approaches [8,11,25] struggle to distinguish true irrelevance from noise-induced false negatives, hindering progressive noise suppression and ranking refinement.

2.2 Generative-Based Recommendation

Generative models [17,24] have redefined preference modeling. Frameworks like RecVAE [13] and DualVAE [6] utilized variational inference and disentangled learning to mitigate noise in implicit signals. More recently, diffusion models [15, 21], including DiffKG [10] and CF-Diff [9], achieved state-of-the-art results by capturing nonlinear patterns through iterative denoising.

However, a critical limitation remains: the **mismatch** between continuous denoising objectives and discrete ranking metrics [22]. Static loss weighting fails to adapt as the focus shifts from reconstruction to ranking optimization. TIG-Diff addresses this via metric-aware gradient alignment and exponential decay scheduling, balancing generative accuracy with ranking precision.

3 Methodology

The TIG-Diff framework (Fig. 2) comprises three key components to suppress spurious signals in noisy implicit feedback and improve recommendation quality. **Implicit Subgraph Contrastive Denoising** employs dynamical negative feedback propagation with temperature scaling to build dynamic contrastive subgraphs, leveraging neighborhood context to reduce local noise and extract stable preference signals for better high-order modeling. **Graph Diffusion Denoising Module (GDDM)** uses cross-attention to iteratively denoise graph representations, weakening spurious co-occurrences while capturing high-order dependencies and tracking temporal preference shifts. **Temporal Collaborative Ranking (TCR)** combines Relative Ranking Optimization (RRO) with Dynamic Loss Weighting (DLW) in a curriculum-aligned manner, stabilizing latent representations early and progressively refining ranking structures to align denoising with ranking objectives.

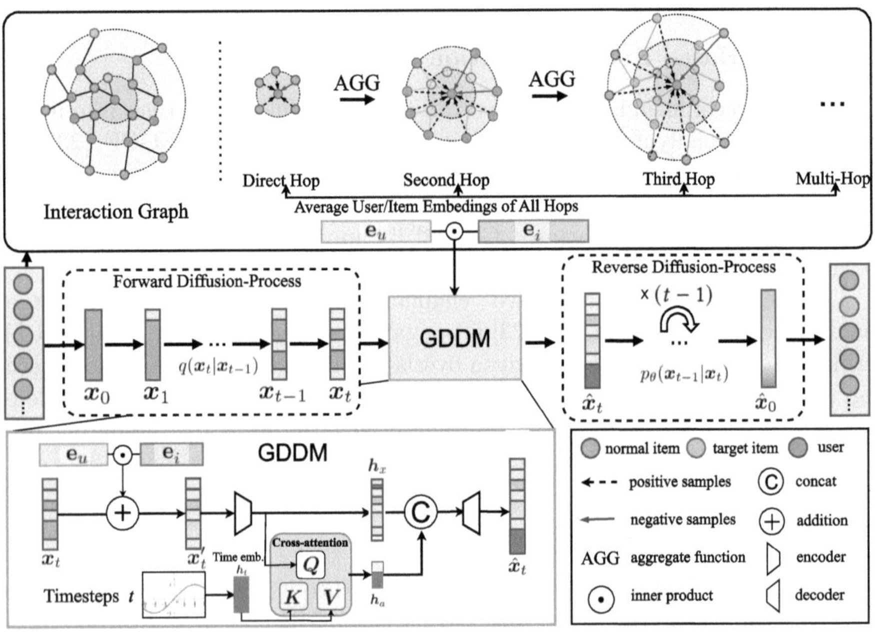

Fig. 2. Overview of TIG-Diff. The framework consists of two components: (1) Implicit Subgraph Contrastive Denoising, which reduces local noise via dynamic contrastive subgraphs; and (2) Graph Diffusion Denoising Module (GDDM), which denoises embeddings and captures high-order dependencies.

3.1 Implicit Subgraph Contrastive Denoising

In TIG-Diff, the *Implicit Subgraph Contrastive Denoising* constructs local user–item subgraphs and learns structural priors that guide subsequent denoising. A pre-trained GNN encoder captures high-order dependencies via message passing, while a dynamic negative-feedback propagation mechanism with temperature-controlled sampling discovers hard negatives within each user's local subgraph.

Together, these components enable subgraph-level contrastive denoising that attenuates noise in implicit feedback and extracts stable, generalizable user preference signals.

Message Passing. We adopt LightGCN [8] as the backbone for structural message passing over the user–item bipartite graph. At the l-th layer, user and item embeddings are updated by symmetric neighborhood aggregation:

$$\mathbf{e}_u^{(l)} = \sum_{i \in \mathcal{N}_u} \frac{1}{\sqrt{|\mathcal{N}_u||\mathcal{N}_i|}} \mathbf{e}_i^{(l-1)}, \tag{1}$$

where $\mathcal{N}_u$ denotes the set of interacted (positive) items of user u, and $\mathcal{N}_i$ denotes the set of users who interacted with item i.

Embedding Fusion. To integrate multi-hop structural information, we perform layer-wise averaging across L propagation layers:

$$\mathbf{e}_u = \frac{1}{L+1} \sum_{l=0}^{L} \mathbf{e}_u^{(l)}, \quad \mathbf{e}_i = \frac{1}{L+1} \sum_{l=0}^{L} \mathbf{e}_i^{(l)}. \tag{2}$$

The resulting embeddings encode both direct and high-order collaborative signals.

Dynamic Negative Feedback Propagation. To generate informative negatives for subgraph-level contrastive denoising, we introduce a *neighborhood-aware, score-guided negative sampling* scheme. Let $\mathcal{N}_u^+$ denote the set of observed positive items for user u, and $\mathcal{I}$ be the entire item set. We define a local candidate pool $\widetilde{\mathcal{I}}_u = \mathcal{I} \setminus \mathcal{N}_u^+$, i.e., items unobserved by u. A set of k hard negatives $\mathcal{N}_u^-$ is sampled according to a temperature-controlled distribution:

$$\mathcal{N}_u^- \sim P_{\text{hard}}(j \mid u) \propto \exp\left(\frac{s(u,j)}{\tau}\right), \quad j \in \widetilde{\mathcal{I}}_u, \tag{3}$$

where $s(u,j) = \langle \mathbf{e}_u, \mathbf{e}_j \rangle$ denotes the structural affinity between user u and item j, τ is an annealing temperature controlling sample hardness, and k is set to $|\mathcal{N}_u^+|$ for balance. The sampling is restricted to each user's local subgraph to ensure scalability, resulting in an expected $O(k)$ complexity per user.

Contrastive Optimization. Given the mined hard negatives, we optimize a BPR-style contrastive loss:

$$\mathcal{L}_{\text{BPR}} = -\sum_{u=1}^{m} \sum_{i \in \mathcal{N}_u^+} \sum_{j \in \mathcal{N}_u^-} \ln \sigma\big(s(u,i) - s(u,j)\big), \tag{4}$$

where $\sigma(\cdot)$ denotes the sigmoid function. A regularization term $\mathcal{L}_{\text{reg}}$ is added to prevent overfitting, yielding the final objective $\mathcal{L} = \mathcal{L}_{\text{BPR}} + \mathcal{L}_{\text{reg}}$.

By contrasting positives against dynamically mined hard negatives within each implicit subgraph, the model learns to suppress spurious co-occurrences and preserve stable, neighborhood-consistent patterns, thereby achieving robust *subgraph contrastive denoising* for implicit feedback.

3.2 Graph-Diffusion Denoising Module.

Forward Process. Diffusion models progressively perturb data representations with Gaussian noise over T steps. At each step $t \in \{1, \ldots, T\}$, noise is added following a predefined variance schedule $\beta_t \in (0, 1)$:

$$q(\boldsymbol{x}_t|\boldsymbol{x}_{t-1}) = \mathcal{N}\left(\boldsymbol{x}_t; \sqrt{1-\beta_t}\boldsymbol{x}_{t-1}, \beta_t\mathbf{I}\right). \tag{5}$$

By reparameterization, any noisy representation at step t can be expressed as:

$$\boldsymbol{x}_t = \sqrt{\bar{\alpha}_t}\boldsymbol{x}_0 + \sqrt{1-\bar{\alpha}_t}\boldsymbol{\epsilon}, \tag{6}$$

where $\alpha_t = 1 - \beta_t$, $\bar{\alpha}_t = \prod_{s=1}^{t} \alpha_s$, and $\boldsymbol{\epsilon} \sim \mathcal{N}(\mathbf{0}, \mathbf{I})$ denotes standard Gaussian noise.

Reverse Process. Starting from pure noise $\boldsymbol{x}_T$, the reverse process iteratively reconstructs clean representations via learned denoising transitions:

$$p_\theta(\boldsymbol{x}_{t-1}|\boldsymbol{x}_t) = \mathcal{N}(\boldsymbol{x}_{t-1}; \boldsymbol{\mu}_\theta(\boldsymbol{x}_t, t), \boldsymbol{\Sigma}_\theta(\boldsymbol{x}_t, t)), \tag{7}$$

where $\boldsymbol{\mu}_\theta$ and $\boldsymbol{\Sigma}_\theta$ are parameterized by a neural denoiser with learnable weights θ.

In TIG-Diff, we integrate graph-structured embeddings and diffusion representations into a unified denoising framework, and introduce a time-aware denoising mechanism that conditions the diffusion process on temporal signals through a cross-attention design. This allows the model to dynamically align denoising behavior with temporal evolution of user preferences.

- **Graph-Diffusion Interaction.** User and item embeddings $(\mathbf{e}_u, \mathbf{e}_i)$ from the GNN encoder are first used to compute an interaction score:

$$s(u, i) = \langle \mathbf{e}_u, \mathbf{e}_i \rangle. \tag{8}$$

The score is then fused with the diffusion embedding $\boldsymbol{x}_t$ via a linear modulation:

$$\boldsymbol{x}'_t = \boldsymbol{x}_t + \alpha \cdot \mathbf{W}_s s(u, i), \tag{9}$$

where α is a learnable scaling parameter and $\mathbf{W}_s$ is a projection ensuring dimensional consistency. The fused representation $\boldsymbol{x}'_t$ encodes both structural and temporal semantics.

- **Timestep Embedding.** A sinusoidal positional encoding $\boldsymbol{e}_t \in \mathbb{R}^{d_e}$ is computed as:

$$\boldsymbol{e}_t^{(2i)} = \cos(t/\omega_i)\,, \quad \boldsymbol{e}_t^{(2i+1)} = \sin(t/\omega_i)\,, \tag{10}$$

where $\omega_i = 10000^{2i/d_e}$. It is projected by a linear layer to obtain the learnable timestep embedding $\boldsymbol{h}_t$.
- **Cross-Attention.** The encoded diffusion feature $\boldsymbol{h}_x \in \mathbb{R}^{N\times d}$ serves as the query $\boldsymbol{Q}$, while the timestep embedding $\boldsymbol{h}_t$ is broadcasted as the key–value pair $(\boldsymbol{K}, \boldsymbol{V})$. Multi-head attention computes:

$$\text{Attention}(\boldsymbol{Q}, \boldsymbol{K}, \boldsymbol{V}) = \text{softmax}\left(\frac{\boldsymbol{Q}\boldsymbol{K}^\top}{\sqrt{d_e}}\right)\boldsymbol{V}, \tag{11}$$

yielding a timestep-conditioned output $\boldsymbol{h}_a$. Unlike simpler AdaGN or FiLM conditioning, this design explicitly allocates attention weights across temporal dimensions, allowing fine-grained temporal adaptation during denoising.
- **Fusion and Output.** The attention-enhanced feature $\boldsymbol{h}_a$ is concatenated with $\boldsymbol{h}_x$ and decoded to recover the clean representation $\hat{\boldsymbol{x}}_0$, completing the reverse diffusion step.

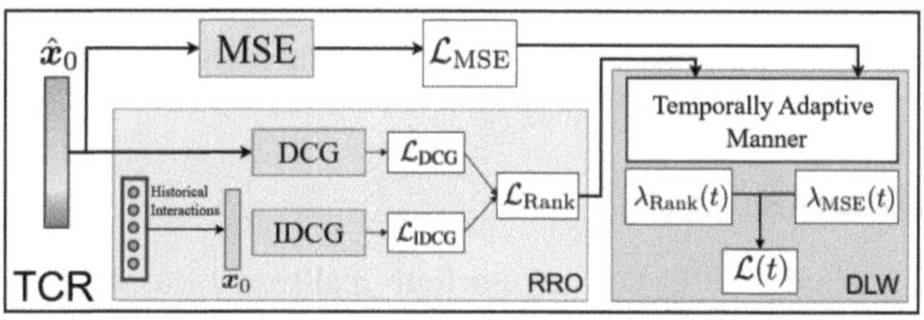

Fig. 3. Illustration of Temporal Collaborative Ranking (TCR).

3.3 Temporal Collaborative Ranking (TCR)

The TCR module (Fig. 3) jointly optimizes two complementary objectives: (1) *Relative Rank Optimization* (RRO), a differentiable listwise objective emphasizing Top-k ranking accuracy; and (2) *Dynamic Loss Weighting* (DLW), an adaptive scheduler that balances regression and ranking objectives throughout training.

Relative Rank Optimization (RRO). RRO explicitly models ranking dependencies to improve Top-k performance. For each user u, let $\mathbf{s}_u \in \mathbb{R}^M$ denote predicted scores over M items. The Top-k permutation $\sigma(u)$ selects the k highest-scoring items, yielding $\hat{\mathbf{y}}_u = \mathbf{P}_{\sigma(u)}\mathbf{s}_u$. The DCG and ideal DCG are defined as:

$$\mathcal{L}_{\text{DCG}} = \sum_{n=1}^{k} \frac{2^{y_{u,\pi(n)}} - 1}{\log_2(n+1)}, \quad \mathcal{L}_{\text{IDCG}} = \sum_{n=1}^{k} \frac{2^{y_{u,n}^{\text{ideal}}} - 1}{\log_2(n+1)}. \tag{12}$$

A normalized, scale-invariant ranking loss is then formulated as:

$$\mathcal{L}_{\text{Rank}} = \frac{\mathcal{L}_{\text{IDCG}} - \mathcal{L}_{\text{DCG}} + \epsilon}{\mathcal{L}_{\text{IDCG}} + \epsilon}, \quad \epsilon = 10^{-5}. \tag{13}$$

This loss is permutation-invariant and differentiable over the full score space, providing stable gradients for rank-sensitive learning.

Dynamic Loss Weighting (DLW). To adaptively balance regression and ranking learning over time, DLW assigns dynamic weights to MSE and ranking losses. Training proceeds in two phases: an early phase emphasizing MSE-based reconstruction, followed by a gradual shift toward ranking optimization. The scheduling is defined as:

$$\lambda_{\text{MSE}}(t) = \begin{cases} 1 - \frac{t}{T}, & t < \gamma T, \\ \gamma, & t \geq \gamma T, \end{cases} \quad \lambda_{\text{Rank}}(t) = \begin{cases} \frac{t}{T}, & t < (1-\gamma)T, \\ 2 - \gamma, & t \geq (1-\gamma)T, \end{cases} \tag{14}$$

where t is the current epoch, T the total epochs, and γ a phase-transition coefficient. The total objective becomes:

$$\mathcal{L}(t) = \lambda_{\text{MSE}}(t)\mathcal{L}_{\text{MSE}} + \lambda_{\text{Rank}}(t)\mathcal{L}_{\text{Rank}}. \tag{15}$$

This dynamic scheduling encourages stable early training and stronger discriminability at later stages, enhancing both convergence and Top-k ranking performance.

4 Experiments

This section evaluates TIG-Diff across four real-world datasets. We assess its recommendation performance, the contribution of key components, and its sensitivity to hyperparameter variations. The following research questions guide the study:

Table 1. Statistics of the four datasets.

Datasets	#Users	#Items	#Interactions	Density
ML-1M	5,949	2,810	571,531	3.42%
Beauty	6,451	10,951	58,220	0.08%
Toys	6,104	10,950	102,790	0.15%
Sports	35,598	18,357	296,337	4.53%

- **RQ1:** How does TIG-Diff perform in top-K recommendation compared to state-of-the-art baselines?
- **RQ2:** What is the impact of each module on overall accuracy?
- **RQ3:** How do the subcomponents of the TCR influence performance?
- **RQ4:** How does multi-hop aggregation depth influence performance?
- **RQ5:** How do key hyperparameters affect model performance?

4.1 Experimental Settings

Datasets and Evaluation Metrics. We evaluate TIG-Diff on four benchmark datasets: MovieLens-1M (ML-1M) and three Amazon categories (Beauty, Toys, Sports), as shown in Table 1. User interactions are chronologically split into 70% training, 20% validation, and 10% testing to preserve temporal consistency. Performance is evaluated using Recall@K and NDCG@K ($K = 10, 20$), standard metrics for top-K recommendation.

Baseline Methods. TIG-Diff is compared against nine state-of-the-art methods from four categories: Collaborative Filtering (NGCF [23], MF [18]), Graph Convolutional Networks (GCN-based) (SGL [25], LGCN (The abbreviation of LightGCN) [8], CIGCN [31]), Denoising AE (VAE-based) (DualVAE [6], RecVAE [20]), and Diffusion Models (DiffRec [22], CF-Diff [9]). All baselines are trained using their original configurations, with hyperparameters fine-tuned via grid search on the validation sets.

Implementation Details. TIG-Diff is optimized for efficiency and robustness. Training uses Adam with batch size 400 to balance convergence and memory. The diffusion module features a 1000-neuron MLP with dropout tuned among {0.1, 0.3, 0.5}. Graph propagation depth ({3, 4, 5, 6}), noise scaling ({10, 1, 0.1, 0.01}), learning rates ({0.0005, 0.001, 0.005, 0.01}), and denoising steps ({2, 4, 5, 6, 16, 32, 64}) are hyperparameter-tuned. Batch normalization is disabled on Amazon datasets to avoid feature homogenization. Experiments run on NVIDIA RTX 4090 and Intel E5-1650 v4 CPU, completing training within 2 h using CUDA parallelization.

4.2 Experimental Results

Overall Performance (RQ1). Table 2 reports the performance of TIG-Diff against strong baselines. TIG-Diff consistently delivers state-of-the-art results across all datasets and metrics, validating the synergy of its denoising-oriented framework, multi-step structural modeling, and progressive ranking refinement.

- **Implicit Subgraph Contrastive Denoising for Structural Robustness.** By integrating implicit subgraph contrastive and curriculum-driven temporal denoising, TIG-Diff suppresses spurious co-occurrences while preserving multi-scale graph dependencies. This yields a 20.6% N@10 improvement over CF-Diff on ML-1M, proving that denoising-aware modeling extracts more stable, generalizable preference signals.
- **GDDM-Driven Multi-Step Temporal Adaptation.** The Graph Diffusion Denoising Module fuses multi-step diffusion features with structural and temporal contexts to capture preference dynamics. This achieves significant R@10 gains on Beauty (7.6%) and Sports (12.3%) due to their periodic patterns, and a 3.7% improvement on the less temporal Toys dataset.
- **TCR-Based Progressive Ranking Refinement.** The TCR paradigm, combining RRO and DLW, dynamically refines ranking hierarchies to prioritize high-confidence items. On ML-1M, it improves R@10 by 17.1% and

R@20 by 11.4%. Notably, Sports shows larger top-rank gains (12.3% R@10 vs. 10.0% N@20), validating its precise item prioritization.

Table 2. Performance comparison across four real-world dataset for top-k recommendation tasks. **Bold**: Best, underline: Second best.

Dataset	Metric	CF-based		GCN-based			VAE-based		Diff-based		Ours	Impv.(%)
		NGCF	MF	SGL	LGCN	CIGCN	DualVAE	RecVAE	DiffRec	CF-Diff		
ML-1M	R@10	0.0768	0.0478	0.0806	0.0957	0.0853	0.0927	0.0835	0.1021	0.1077	**0.1261**	17.1%
	N@10	0.0704	0.0481	0.0799	0.0813	0.0757	0.0788	0.0769	0.0828	0.0912	**0.1100**	20.6%
	R@20	0.1309	0.0868	0.1355	0.1637	0.1442	0.1613	0.1422	0.1668	0.1843	**0.2054**	11.4%
	N@20	0.0886	0.0614	0.0968	0.1044	0.0956	0.1030	0.0963	0.1072	0.1176	**0.1377**	17.0%
Beauty	R@10	0.0355	0.0371	0.0335	0.0487	0.0394	0.0274	0.0467	0.0392	0.0361	**0.0524**	7.6%
	N@10	0.0185	0.0197	0.0168	0.0249	0.0201	0.0151	0.0243	0.0202	0.0178	**0.0266**	6.8%
	R@20	0.0595	0.0371	0.0538	0.0700	0.0628	0.0469	0.0689	0.0570	0.0528	**0.0797**	13.8%
	N@20	0.0251	0.0259	0.0220	0.0327	0.0266	0.0205	0.0312	0.0289	0.0224	**0.0340**	4.0%
Toys	R@10	0.0264	0.0261	0.0329	0.0354	0.0317	0.0270	0.0271	0.0270	0.0257	**0.0367**	3.7%
	N@10	0.0134	0.0136	0.0178	0.0189	0.0181	0.0138	0.0144	0.0158	0.0180	**0.0203**	7.4%
	R@20	0.0399	0.0422	0.0476	0.0515	0.0481	0.0377	0.0405	0.0406	0.0471	**0.0555**	7.7%
	N@20	0.0169	0.0178	0.0217	0.0231	0.0230	0.0167	0.0186	0.0194	0.0222	**0.0252**	9.1%
Sports	R@10	0.0168	0.0161	0.0252	0.0249	0.0246	0.0218	0.0227	0.0239	0.0248	**0.0283**	12.3%
	N@10	0.0090	0.0085	0.0140	0.0136	0.0132	0.0127	0.0133	0.0136	0.0131	**0.0152**	8.6%
	R@20	0.0287	0.0269	0.0421	0.0419	0.0369	0.0332	0.0352	0.0357	0.0364	**0.0456**	8.3%
	N@20	0.0120	0.0113	0.0180	0.0176	0.0169	0.0147	0.0155	0.0167	0.0173	**0.0198**	10.0%

Component Analysis (RQ2). Ablation studies (Table 3) evaluate the contribution of three key components in TIG-Diff:

- **Graph Embeddings:** Removing graph embeddings leads to a 63.2% R@20 drop on Sports and a 152% N@20 drop on Toys, highlighting the necessity of implicit subgraphs for high-order modeling. As shown in Fig. 4, this component provides three key advantages: (i) *Improved Separation*, creating more distinguishable item embeddings; (ii) *Semantic Clustering*, where similar items form tighter clusters via high-order interactions; and (iii) *Structured Latent Space*, significantly reducing embedding overlap compared to the baseline.
- **Cross Attention (C-Attention):** While average performance loss is moderate (2.9%), time-sensitive datasets like Beauty show a 6.8% N@10 drop, validating the cross-attention mechanism's efficacy in modeling temporal user preferences.
- **TCR:** TCR improves recommendation stability and ranking precision, notably yielding a 7.7% R@20 gain on Toys. By employing curriculum-based optimization, it reduces loss variance by 40% and stabilizes training, ensuring smoother convergence and mitigating premature overfitting.

Table 3. Ablation study on key components.

Dataset	Methods	R@10	R@20	N@10	N@20
ML-1M	**TIG-Diff**	**0.1261**	**0.2054**	**0.1100**	**0.1377**
	w/o Graph-Emb	0.1227	0.2011	0.1071	0.1346
	w/o C-Attention	0.1253	0.2034	0.1088	0.1370
	w/o TCR	0.1258	0.2034	0.1097	0.1367
Beauty	**TIG-Diff**	**0.0524**	**0.0797**	**0.0266**	**0.0340**
	w/o Graph-Emb	0.0250	0.0409	0.0128	0.0173
	w/o C-Attention	0.0491	0.0777	0.0250	0.0320
	w/o TCR	0.0472	0.0753	0.0251	0.0327
Toys	**TIG-Diff**	**0.0367**	**0.0555**	**0.0203**	**0.0252**
	w/o Graph-Emb	0.0212	0.0340	0.0079	0.0100
	w/o C-Attention	0.0360	0.0550	0.0201	0.0251
	w/o TCR	0.0354	0.0542	0.0198	0.0247
Sports	**TIG-Diff**	**0.0283**	**0.0456**	**0.0152**	**0.0198**
	w/o Graph-Emb	0.0181	0.0273	0.0103	0.0127
	w/o C-Attention	0.0250	0.0426	0.0147	0.0181
	w/o TCR	0.0240	0.0364	0.0136	0.0169

Table 4. Performance comparison among TCR and its three variants.

Dataset	Methods	R@10	R@20	N@10	N@20
ML-1M	**TCR**	**0.1261**	**0.2054**	**0.1100**	**0.1377**
	TCR-RRO	0.1258	0.2034	0.1097	0.1367
	TCR-DLW	0.1194	0.1965	0.1068	0.1332
Beauty	**TCR**	**0.0524**	**0.0797**	**0.0266**	**0.0340**
	TCR-RRO	0.0472	0.0753	0.0251	0.0327
	TCR-DLW	0.0492	0.0757	0.0253	0.0325
Toys	**TCR**	**0.0367**	**0.0555**	**0.0203**	**0.0252**
	TCR-RRO	0.0354	0.0542	0.0198	0.0247
	TCR-DLW	0.0363	0.0549	0.0200	0.0251
Sports	**TCR**	**0.0283**	**0.0456**	**0.0152**	**0.0198**
	TCR-RRO	0.0240	0.0364	0.0136	0.0169
	TCR-DLW	0.0245	0.0394	0.0132	0.0172

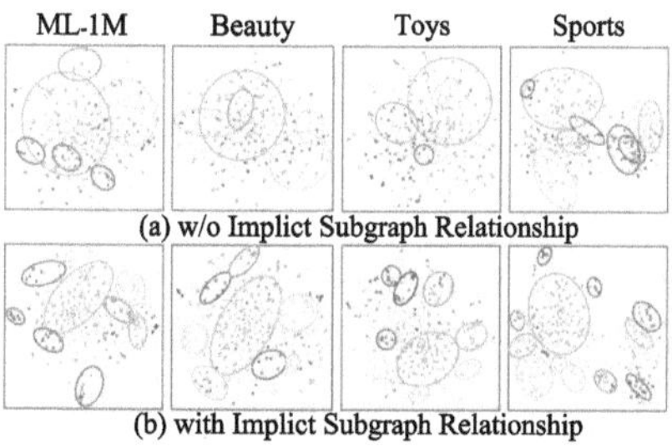

Fig. 4. Visualization of item embeddings with and without implict subgraph relationship.

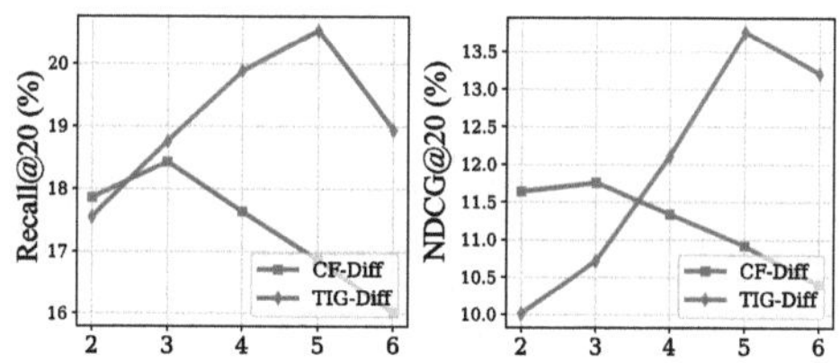

Fig. 5. Impact of propagation depth (H) on performance: TIG-Diff vs. CF-Diff.

TCR Component Analysis (RQ3). To evaluate TCR, we conduct an ablation study on: (i) TCR (full) with RRO, DLW, and MSE loss; (ii) TCR-RRO, without RRO Component; and (iii) TCR-DLW, using fixed loss weights. Table 4 reports the results, revealing three key findings:

- **Synergistic Effect of TCR Components:** The full TCR model outperforms variants with a 26.6% R@10 gain and 9.1% N@20 increase on Toys, confirming that RRO and DLW jointly enhance ranking precision and temporal adaptability.
- **Impact of RRO on Ranking Refinement:** Removing RRO causes a 15.2% drop, highlighting its necessity for preserving ranking consistency through relative preference optimization and top-K accuracy boosts.
- **DLW's Role in Temporal Adaptation:** Replacing DLW with static weighting leads to a 6.91% drop, especially on Beauty, validating DLW's efficacy in aligning optimization with diffusion and stabilizing long-term ranking.

Table 5. Performance comparison according to different values of H.

Dataset	Methods	R@10	R@20	N@10	N@20	Dataset	Methods	R@10	R@20	N@10	N@20
ML-1M	TIG-Diff-3	0.1096	0.1876	0.0828	0.1072	Beauty	TIG-Diff-3	0.0444	0.0690	0.0225	0.0292
	TIG-Diff-4	0.1173	0.1989	0.0959	0.1211		TIG-Diff-4	0.0434	0.0714	0.0221	0.0297
	TIG-Diff-5	**0.1261**	**0.2054**	**0.1100**	**0.1377**		**TIG-Diff-5**	**0.0524**	**0.0797**	**0.0266**	**0.0340**
	TIG-Diff-6	0.1145	0.1894	0.1033	0.1321		TIG-Diff-6	0.0445	0.0726	0.0231	0.0308
Toys	TIG-Diff-3	0.0323	0.0505	0.0185	0.0233	Sports	TIG-Diff-3	0.0234	0.0382	0.0127	0.0166
	TIG-Diff-4	**0.0367**	**0.0555**	**0.0203**	**0.0252**		TIG-Diff-4	0.0242	0.0385	0.0130	0.0168
	TIG-Diff-5	0.0315	0.0484	0.0172	0.0216		**TIG-Diff-5**	**0.0283**	**0.0456**	**0.0152**	**0.0198**
	TIG-Diff-6	0.0328	0.0506	0.0177	0.0224		TIG-Diff-6	0.0235	0.0398	0.0123	0.0166

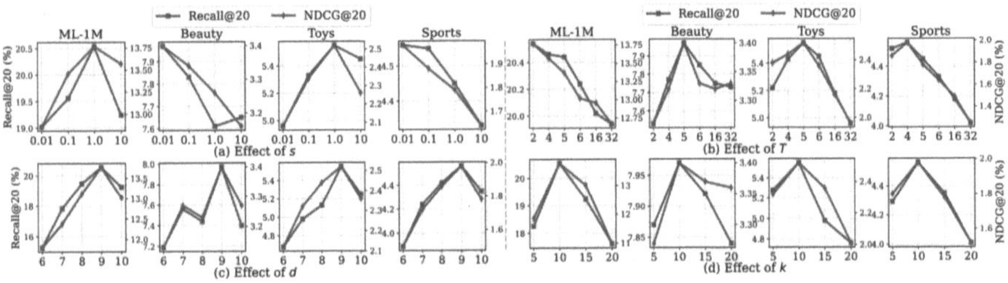

Fig. 6. Hyperparameter sensitivity analysis.

Multi-hop Propagation Analysis (RQ4). The impact of multi-hop aggregation is systematically evaluated in Table 5, revealing three key insights:

- **Optimal Propagation Depth (4–5 hops):** Performance peaks at 4–5 hops, with a 13.7% N@20 gain on ML-1M at 5 hops. Beyond this, rising node similarity (+0.17) indicates over-smoothing. As Fig. 5 shows, CF-Diff degrades after 3 hops, while TIG-Diff improves, highlighting the benefits of implicit subgraph contrast.
- **Dataset-Specific Propagation Limits:** Optimal hop count varies by dataset: Toys peaks at 4 hops, while others peak at 5, highlighting the need for adaptive depth selection.
- **Over-Smoothing Effects:** Beyond 5 hops, performance drops across datasets—most notably in Beauty (N@20: 0.0340 → 0.0308, ↓9.4%)—as over-propagation blurs features and reduces discriminability.

Hyperparameter Sensitivity (RQ5). We analyze the impact of four hyperparameters, s, T, d, and k, on recommendation performance using average R@20 and N@20 across datasets. For noise scale (s), Fig. 6(a) shows that extreme values reduce performance, as high noise disrupts feature learning and low noise weakens robustness. For diffusion steps (T), Fig. 6(b) reveals optimal results at $T = 4$ or 5, with smaller values limiting fusion and larger ones risking overfitting. Regarding embedding dimension (d), Fig. 6(c) shows that larger d improves model capacity but increases cost, with the best trade-off found at $d = 512$, where the x-axis uses an exponential scale ($d = 2^n$) to reflect nonlinear complexity. Finally, for temporal dimension (k), Fig. 6(d) indicates that performance improves as k increases, but overhead also rises, with the optimal setting at $k = 10$.

5 Conclusions

This work presents TIG-Diff, a unified graph-diffusion framework that effectively addresses the dual challenges of structural noise propagation and ranking misalignment in diffusion-based recommenders. Moving beyond isolated solutions, our key contribution is the development of a synergistic denoising-ranking co-evolution mechanism. The GDDM-driven structural purification inherently combines subgraph contrastive learning with temporal diffusion, where dynamical negative feedback propagation suppresses noise amplification, and cross-modally fused representations generate preference signals that are robust to high-order spurious dependencies. The TCR-aligned optimization surpasses traditional loss balancing by embedding curriculum-aware dynamics, enabling a smooth transition from coarse-grained interaction reconstruction to fine-grained ranking induction. Empirical studies reveal two critical insights: (1) five-hop aggregation strikes the optimal balance between representation power and over-smoothing, and (2) TCR's dynamic weighting effectively reconciles the accuracy-ranking conflict. Extensive experiments demonstrate up to a 20.6% improvement in NDCG over state-of-the-art baselines, validating that the joint optimization of denoising stability and ranking consistency unlocks new performance potential. Future work will extend this framework to cross-modal preference evolution modeling and explore the integration of graph diffusion with reinforcement learning for long-term engagement optimization.

Acknowledgements. We are grateful for resources from the High Performance Computing Center of Central South University.

Disclosure of Interests. The authors have no relevant financial or non-financial interests to disclose.

References

1. Chen, M., Huang, C., Xia, L., Wei, W., Xu, Y., Luo, R.: Heterogeneous graph contrastive learning for recommendation. In: Proceedings of the Sixteenth ACM International Conference on Web Search and Data Mining, pp. 544–552 (2023)
2. Cui, Z., Wu, H., He, B., Cheng, J., Ma, C.: Context matters: enhancing sequential recommendation with context-aware diffusion-based contrastive learning. In: Proceedings of the 33rd ACM International Conference on Information and Knowledge Management, pp. 404–414 (2024)
3. Dunlap, L., Umino, A., Zhang, H., Yang, J., Gonzalez, J.E., Darrell, T.: Diversify your vision datasets with automatic diffusion-based augmentation. Adv. Neural. Inf. Process. Syst. **36**, 79024–79034 (2023)

4. Feng, T., Shen, Y., You, J.: GraphRouter: a graph-based router for LLM selections. In: The Thirteenth International Conference on Learning Representations (2024)
5. Gong, S., Li, M., Feng, J., Wu, Z., Kong, L.: DiffuSeq: sequence to sequence text generation with diffusion models. arXiv preprint arXiv:2210.08933 (2022)
6. Guo, Z., Li, G., Li, J., Wang, C., Shi, S.: DualVAE: dual disentangled variational autoencoder for recommendation. In: Proceedings of the 2024 SIAM International Conference on Data Mining (SDM), pp. 571–579. SIAM (2024)
7. Gupta, A., et al.: Photorealistic video generation with diffusion models. In: Leonardis, A., Ricci, E., Roth, S., Russakovsky, O., Sattler, T., Varol, G. (eds.) ECCV 2024. LNCS, vol. 15137, pp. 393–411. Springer, Cham (2024). https://doi.org/10.1007/978-3-031-72986-7_23
8. He, X., Deng, K., Wang, X., Li, Y., Zhang, Y., Wang, M.: LightGCN: simplifying and powering graph convolution network for recommendation. In: Proceedings of the 43rd International ACM SIGIR Conference on Research and Development in Information Retrieval, pp. 639–648 (2020)
9. Hou, Y., Park, J.D., Shin, W.Y.: Collaborative filtering based on diffusion models: unveiling the potential of high-order connectivity. In: Proceedings of the 47th International ACM SIGIR Conference on Research and Development in Information Retrieval, pp. 1360–1369 (2024)
10. Jiang, Y., Yang, Y., Xia, L., Huang, C.: DiffKG: knowledge graph diffusion model for recommendation. In: Proceedings of the 17th ACM International Conference on Web Search and Data Mining, pp. 313–321 (2024)
11. Li, X., et al.: Multi-intention oriented contrastive learning for sequential recommendation. In: Proceedings of the Sixteenth ACM International Conference on web Search and Data Mining, pp. 411–419 (2023)
12. Li, Z., Sun, A., Li, C.: DiffuRec: a diffusion model for sequential recommendation. ACM Trans. Inf. Syst. **42**(3), 1–28 (2023)
13. Liang, D., Krishnan, R.G., Hoffman, M.D., Jebara, T.: Variational autoencoders for collaborative filtering. In: Proceedings of the 2018 World Wide Web Conference, pp. 689–698 (2018)
14. Liu, N., et al.: Learning social graph for inactive user recommendation. In: Onizuka, M., et al. (eds.) DASFAA 2024. LNCS, vol. 14855, pp. 151–167. Springer, Cham (2024). https://doi.org/10.1007/978-981-97-5572-1_10
15. Luo, F., Zhou, T., Liu, J., Guo, T., Gong, X., Ren, J.: Multiscale diff-changed feature fusion network for hyperspectral image change detection. IEEE Trans. Geosci. Remote Sens. **61**, 1–13 (2023)
16. Ma, H., et al.: Plug-in diffusion model for sequential recommendation. In: Proceedings of the AAAI Conference on Artificial Intelligence, vol. 38, pp. 8886–8894 (2024)
17. Qiu, X., et al.: Finding local diffusion schrödinger bridge using Kolmogorov-Arnold network. arXiv preprint arXiv:2502.19754 (2025)
18. Rendle, S., Freudenthaler, C., Gantner, Z., Schmidt-Thieme, L.: BPR: Bayesian personalized ranking from implicit feedback. arXiv preprint arXiv:1205.2618 (2012)
19. Savinov, N., Chung, J., Binkowski, M., Elsen, E., Oord, A.V.D.: step-unrolled denoising autoencoders for text generation. arXiv preprint arXiv:2112.06749 (2021)
20. Shenbin, I., Alekseev, A., Tutubalina, E., Malykh, V., Nikolenko, S.I.: RecVAE: a new variational autoencoder for top-n recommendations with implicit feedback. In: Proceedings of the 13th International Conference on Web Search and Data Mining, pp. 528–536 (2020)

21. Tang, D., Cao, X., Hou, X., Jiang, Z., Liu, J., Meng, D.: CRS-diff: Controllable remote sensing image generation with diffusion model. IEEE Trans. Geosci. Remote Sens. (2024)
22. Wang, W., Xu, Y., Feng, F., Lin, X., He, X., Chua, T.S.: Diffusion recommender model. In: Proceedings of the 46th International ACM SIGIR Conference on Research and Development in Information Retrieval, pp. 832–841 (2023)
23. Wang, X., He, X., Wang, M., Feng, F., Chua, T.S.: Neural graph collaborative filtering. In: Proceedings of the 42nd International ACM SIGIR Conference on Research and Development in Information Retrieval, pp. 165–174 (2019)
24. Wang, Y., et al.: LAVIE: high-quality video generation with cascaded latent diffusion models. Int. J. Comput. Vision 1–20 (2024)
25. Wu, J., et al.: Self-supervised graph learning for recommendation. In: Proceedings of the 44th International ACM SIGIR Conference on Research and Development in Information Retrieval, pp. 726–735 (2021)
26. Wu, Y., et al.: Multi-view multi-behavior contrastive learning in recommendation. In: Bhattacharya, A., et al. (eds.) DASFAA 2022. LNCS, vol. 13246, pp. 166–182. Springer, Cham (2022). https://doi.org/10.1007/978-3-031-00126-0_11
27. Wu, Z., et al.: Diff4Rec: sequential recommendation with curriculum-scheduled diffusion augmentation. In: Proceedings of the 31st ACM International Conference on Multimedia, pp. 9329–9335 (2023)
28. Yang, L., Liu, Z., Dou, Y., Ma, J., Yu, P.S.: ConsisRec: enhancing GNN for social recommendation via consistent neighbor aggregation. In: Proceedings of the 44th International ACM SIGIR Conference on Research and Development in Information Retrieval, pp. 2141–2145 (2021)
29. Yi, X., Xu, H., Zhang, H., Tang, L., Ma, J.: Diff-RetiNex: rethinking low-light image enhancement with a generative diffusion model. In: Proceedings of the IEEE/CVF International Conference on Computer Vision, pp. 12302–12311 (2023)
30. Zhang, Z., et al.: DeepMin: deep multi-modal interest network with cognitive learning modules. In: Onizuka, M., et al. (eds.) DASFAA 2024. LNCS, vol 14852, pp. 212–227. Springer, Cham (2024). https://doi.org/10.1007/978-981-97-5555-4_14
31. Zhu, T., Sun, L., Chen, G.: Embedding disentanglement in graph convolutional networks for recommendation. IEEE Trans. Knowl. Data Eng. **35**(1), 431–442 (2021)
32. Zhu, Y., Wang, C., Zhang, Q., Xiong, H.: Graph signal diffusion model for collaborative filtering. In: Proceedings of the 47th International ACM SIGIR Conference on Research and Development in Information Retrieval, pp. 1380–1390 (2024)

Generative Representational Learning of Foundation Models for Recommendation

Zheli Zhou[1], Chenxu Zhu[4], Jianghao Lin[2(✉)], Bo Chen[4], Ruiming Tang[4], Yong Yu[1], and Weinan Zhang[1,3(✉)]

[1] School of Computer Science, Shanghai Jiao Tong University, Shanghai, China
{zlz114514,yyu,wnzhang}@sjtu.edu.cn
[2] Antai College of Economics and Management, Shanghai Jiao Tong University, Shanghai, China
linjianghao@sjtu.edu.cn
[3] Shanghai Innovation Institute, Shanghai, China
[4] Huawei Noah's Ark Lab, Shenzhen, China
{zhuchenxu1,chenbo116,tangruiming}@huawei.com

Abstract. Developing foundation models that generalize across diverse tasks is a long-standing goal in artificial intelligence. In recommendation systems, however, existing foundation models primarily focus on generative tasks while overlooking embedding-based tasks, and they often suffer from challenges in multi-task learning such as knowledge conflict and inconsistent convergence. To address these limitations, We propose RecFound, a unified generative and representation learning framework for recommendation foundation models. We construct the first comprehensive dataset covering both generative and embedding tasks across diverse scenarios. Building on this dataset, RecFound introduces a novel multi-task optimization scheme with three major components: a Task-wise Mixture of Low-rank Experts (TMoLE) to mitigate knowledge conflict and enhance sharing, a Step-wise Convergence-oriented Sample Scheduler (S2Sched) to balance convergence across tasks, and a Model Merge mechanism to further improve joint performance. Extensive experiments demonstrate that RecFound achieves state-of-the-art performance across a variety of recommendation tasks, significantly outperforming existing baselines. The dataset and code are available at https://github.com/JunkFood436/RecFound.

Keywords: Large Language Model · Foundation Model · Recommender System

1 Introduction

The pursuit of a single foundation model to excel across diverse tasks has been a long-standing goal in artificial intelligence, driven by the principle that various

H. Jung et al. (Eds.): DASFAA 2026, LNCS 16535, pp. 552–568, 2026.
https://doi.org/10.1007/978-981-92-0363-5_34

tasks can mutually enhance their performance by leveraging shared, underlying knowledge. Foundation models, characterized by their extensive pre-training and large-scale parameters, have actualized this vision, demonstrating remarkable generalization capabilities in fields like natural language processing and computer vision. They capture rich, transferable representations, enabling adaptation to a wide range of downstream tasks with minimal fine-tuning [20]. This powerful paradigm is now being extended to recommendation systems. The inherent diversity of recommendation scenarios—spanning different tasks and data formats—makes them a particularly fertile ground for foundation models, which offer unique opportunities to enhance personalization by unifying and modeling complex user-item interactions from disparate sources [9].

Early efforts in applying foundation models to recommendation systems primarily focus on specific, relatively simple tasks such as tagging and question answering. These approaches often involve adapting general-purpose language models to handle individual recommendation-related tasks, demonstrating the potential of foundation models in this domain. More recent work has aimed to generalize foundation models to handle a broader spectrum of recommendation tasks [14,21,24]. While these works demonstrate promising progress in unifying different recommendation tasks under a single model, they still face significant challenges in effectively handling the full complexity of recommendation scenarios.

First, existing approaches overlook the crucial embedding[1] *tasks.* High-quality embeddings provide the semantic foundation for reasoning, while generative tasks capture the dynamic patterns of user behavior. As shown in Fig. 1, the true potential of a foundation model is unlocked by creating a synergistic loop between them. For instance, while a representational task can group "`camera`" and "`lens`" based on semantic similarity to aid generative prediction, this benefit is reciprocal. The generative task captures behavioral signals, such as the common user sequence of "`camera` → `lens`". This sequential context, in turn, enriches the embeddings with behavioral nuances, which then further boosts generative performance. This creates a self-reinforcing cycle where an effective foundation model must not merely perform any single type of these two tasks, but foster a synergistic mode where its representational and generative capabilities continuously enhance each other.

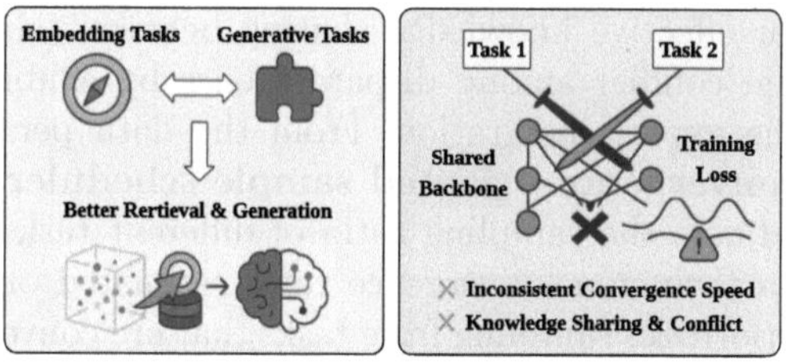

Fig. 1. Illustration of key challenges for recommendation foundation models, *i.e.*, joint modeling of generative and embedding tasks (left) and multi-task modeling for knowledge conflict and inconsistent convergence speed (right).

Second, current models struggle with the complexities of multi-task learning, particularly in managing the **knowledge sharing & conflict** and the **inconsis-**

[1] In this work, the term "embedding" is interchangeable with "representational".

tent convergence speeds among tasks. For example, consider a model trained on two tasks simultaneously: (1) a generative task to predict an item's global popularity review, which tends to learn general, widely appealing features, and (2) an embedding task to model a specific user's preference for niche, indie products. The learning objective for the popularity task might push the model's parameters to favor mainstream features, conflicting with the objective of the personalization task, which needs to capture niche tastes for that user. This conflict can lead to suboptimal performance on both tasks. At the same time, different tasks often exhibit inconsistent convergence speeds, further complicating the joint training process and hindering the model's overall effectiveness in real-world systems.

To address these limitations, we introduce **RecFound**, a generative representational learning framework for recommendation foundation models. Our framework begins with the construction of the first comprehensive dataset that covers both generative and embedding tasks across various aspects of the recommendation domain. Building upon this dataset, we propose a novel multi-task training scheme for recommendation foundation models. From the model perspective, we design a **Task-wise Mixture of low-rank experts (TMoLE)** architecture. TMoLE consists of multiple low-rank experts and a task-oriented router. The router generates a masked routing distribution based on each sample's task to adaptively aggregate the expert outputs. This mechanism allows for effective knowledge sharing between related tasks while mitigating potential conflict among disparate ones by enabling task-specific specialization via the expert integration. From the data perspective, we propose a **Step-wise convergence-oriented sample scheduler (S2Sched)**. S2Sched dynamically adjusts the sampling ratio of different tasks within each training batch based on their non-convergence rates estimated on the validation set. Specifically, it prioritizes sampling from tasks that are converging more slowly, thereby addressing the issue of inconsistent convergence speeds across different recommendation tasks within a single model. Finally, we also introduce a **Model Merge** module to further balance the performance across tasks.

Our contributions can be summarized as follows:

- We construct the first comprehensive dataset for recommendation foundation models, covering both 10 generative tasks and 3 embedding tasks across diverse recommendation scenarios.
- We propose RecFound, a novel training framework for recommendation foundation models. We design TMoLE to facilitate effective knowledge sharing & conflict resolution between embedding and generative tasks, S2Sched to address the inconsistency of convergence speeds for different tasks, and the Model Merge module to balance performance across tasks.
- Extensive experiments demonstrate that our RecFound framework achieves state-of-the-art performance on a wide range of recommendation tasks, and we provide insightful analysis to support our findings. We release our code, as well as the dataset and model checkpoints, to benefit the research community.

2 Related Works

General Foundation Models in NLP. Extensive research on general foundation models in natural language processing (NLP) has explored general-purpose methods for diverse downstream tasks. Early influential works focused on learning static word embeddings from large corpora. Subsequent research, including ELMo [22], BERT [2], and many BERT-family models, generated contextualized token representations with substantial gains. Another line of work investigates foundation models for generative tasks, with the GPT series [1] demonstrating that decoder-only architectures can acquire multitask proficiency through next-token prediction. More recently, studies have explored unified models that handle both embedding and generative tasks. GritLM [20] trains a single model to perform both by distinguishing tasks via instructions, while UniGen [13] integrates retrieval and answering into one architecture.

Recommendation Foundation Models. Early research on recommendation mainly optimized single-task generative models trained from scratch, which lack generalization across tasks. With the rise of LLMs, recent works introduce foundation models for e-commerce. LLaMA-E [24] adapts LLaMA [5] for product authoring tasks. EcomGPT [14] unifies various generative tasks through large-scale pretraining on e-commerce corpora. eCeLLM [21] further improves in-domain and cross-task generalization via high-quality instruction data. These efforts indicate increasing interest in generative foundation models for recommendation. However, current models still lack support for both representation and generation. To the best of our knowledge, we are the first to introduce a unified generative-representational learning framework for recommendation foundation models.

3 Dataset Construction

The development of recommendation foundation models is hampered by the limitations of existing datasets, which are often narrowly tailored to either embedding or generative tasks. To propel research in this direction, we created RecFound, a comprehensive dataset that bridges this divide. It is the first to systematically integrate both task types across diverse scenarios, establishing a more flexible and powerful foundation for model training and evaluation. RecFound is distinguished by three key advantages. (1) **Innovation**: We are the first to integrate embedding and generative tasks in recommendation systems, providing a novel benchmark and perspective for research; (2) **Diversity**: The dataset covers multiple recommendation tasks and domains, including e-commerce, social media, music, video streaming, and news recommendations. Each data sample has rich expressiveness to facilitate model training and evaluation; (3) **High-Quality**: All data originates from real user interactions, with rigorous cleaning to remove samples containing illegal symbols, errors, or missing content.

We give a brief introduction to each embedding and generative task below, with detailed data examples provided in the released code repository.

Table 1. The statistics of our RecFound dataset.

Split	Embedding Tasks			Generative Tasks										Total
	/			General NLP			User Understanding				Item Understanding			
	U2I	Q2I	I2I	QR	AVE	AG	SR	SA	UP	AP	PRP	PM	IP	
Training	86,000	40,000	83,680	3,260	20,000	20,000	20,000	20,000	20,000	15,000	20,000	4,000	15,000	366,940
Validation	3,500	2,600	3,340	500	500	500	500	500	500	500	500	500	500	14,940
Testing	4,800	4,000	4,800	124	1,334	1,332	1,334	1,334	227	1,334	1,333	168	266	22,383

Embedding Tasks include three classical representational learning scenarios for recommendation:

- **User2Item (U2I):** U2I task involves encoding a user's behavior sequence into an embedding that encapsulates their interests. The output user representation is then utilized to retrieve relevant items. The samples for this task are constructed on Amazon Reviews dataset [7].
- **Query2Item (Q2I):** In Q2I task, user input queries are embedded to provide precise item recommendations based on the query embeddings. This task is built upon Shopping Queries [23] and ECInstruct datasets [21].
- **Item2Item (I2I):** I2I task focuses on recommending items similar to a given target item by encoding the embeddings of different item descriptions. All data for this task is derived from Amazon Reviews dataset [7].

Generative Tasks consist of 10 subtasks, which can be categorized into three different types.

- **General NLP.** We use MuDoCo [18], Amazon Reviews [7], and AmazonQA datasets [6], and construct three tasks to provide the model with better language capabilities to solve recommendation questions: (i) Query Rewriting (QR) aims to rewrite obscure queries from users for a better understanding. (ii) Attribute Value Extraction (AVE) expects to extract attribute values of a given item, including titles, descriptions, brands, etc. (iii) Answer Generation (AG) is the common Q&A task with question topics related to recommendation scenarios.
- **User Understanding:** We adopt Amazon Reviews [7] and MovieLens datasets, and develop four tasks for a better understanding of users' behaviors, interests, emotions, and intentions. (i) Sequential Recommendation (SR) aims to predict what the user will purchase next given his/her history behavior sequence. (ii) Sentiment Analysis (SA) is a task to estimate the emotion and preference based on users' textual reviews. (iii) User Profile (UP) requires the model to reason and summarize users' detailed preferences based on his/her behavior history. (iv) Answerability Prediction (AP) judges whether the questions from users can be answered or not.
- **Item Understanding:** We absorb Amazon Reviews [7], Amazon-Google Products [11], and MovieLens datasets to build three tasks aiming at extracting effective information from items. (i) Product Relation Prediction (PRP) conducts three-tier relevance prediction for a given item pair. (ii) Product

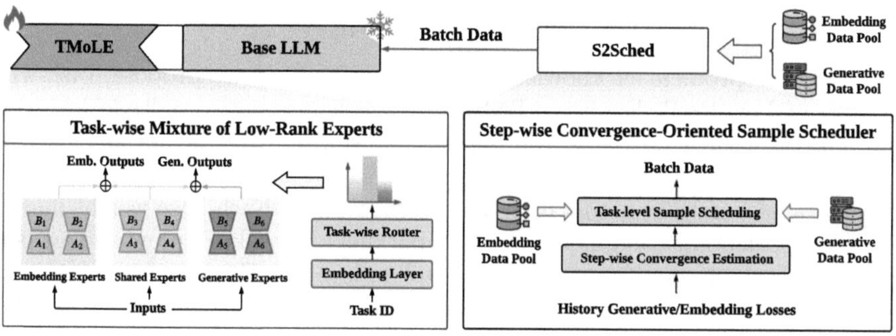

Fig. 2. The illustration of our proposed RecFound framework.

Matching (PM) judges whether the given two items are the same when they contain different descriptions from different platforms. (iii) Item Profile (IP) requires the model to extract crucial factual information of a given item.

We process, filter, and systematically integrate data samples from all tasks to construct our RecFound dataset, comprising 3 embedding tasks and 10 generative tasks. The dataset is then split into training, validation, and testing sets, with detailed statistics reported in Table 1.

4 Methodology

4.1 General Training Pipeline of RecFound

Before introducing our specific methods, we first outline the general training pipeline used for both embedding and generative tasks. In our unified framework, all raw input data is transformed into standardized instruction-response pairs. Examples for each task are in the released code repository. Then, these samples are processed via two branches for embedding and generative tasks, respectively.

In the **embedding branch**, batched inputs are passed through a shared LLM backbone and trained using a *contrastive objective* (InfoNCE loss) to learn a well-structured embedding space. When training backbone is a decoder-only model, we extract the embedding representation for input sequence by applying mean pooling over the final layer hidden states. In the **generative branch**, samples are similarly batched and optimized using a *token-level cross-entropy loss* to minimize sequence generation loss. Since we aim to establish a unified recommendation foundation model, both branches share the same LLM backbone and parameters, with their respective gradients jointly backpropagated for update.

However, the shared LLM backbone for multiple heterogeneous embedding and generative tasks inevitably introduces challenges of knowledge sharing & conflict and convergence inconsistency. To address these issues, our RecFound framework incorporates three key modules: Task-wise Mixture of Low-Rank Experts, Step-wise Convergence-Oriented Sample Scheduler, and Model Merge (Fig. 2).

4.2 Task-Wise Mixture of Low-Rank Experts

To tackle the problem of knowledge sharing & conflict, we propose Task-wise Mixture of Low-Rank Experts (**TMoLE**), which integrates multiple low-rank adapters [8] with a task-wise routing mechanism. TMoLE layers are inserted in the multi-head attention projections-specifically, the query, key, value, and output transformations-across each Transformer layer. Each base weight matrix W_0 of one projection is augmented with N parallel low-rank adapters and a **task-wise** router. Different from traditional token-level MoE approaches [17], our router leverages the task type embedding associated with each training sample. Given a task ID i, it is mapped through a learnable embedding table to obtain the corresponding task embedding $\mathbf{e}_i$.

The N low-rank experts are partitioned into three distinct groups: embedding experts $\mathcal{E}$, generative experts $\mathcal{G}$, and shared experts $\mathcal{S}$. Each adapter j comprises trainable matrices $A_j \in \mathbb{R}^{d \times r}$ and $B_j \in \mathbb{R}^{r \times d}$, where $r \ll d$. The task-wise router is implemented as a multi-layer perceptron (MLP) coupled with a masking module. The output gating vector v (*i.e.*, the routing distribution) determines the contribution of each expert to the final TMoLE output, computed as:

$$v_i = \text{Softmax}(\text{Mask}(\text{MLP}(e_i))). \tag{1}$$

Here, Mask(·) suppresses irrelevant task dimensions by assigning them a value of $-\infty$, retaining only the experts pertinent to the current task and the shared experts. For example, the elements of embedding experts in v_i are zero when routing for generative tasks. Notably, the MLP is shared across all the tasks. The final output of the TMoLE-enhanced projection layer for task i is given by:

$$o = W_0 x + \sum_{j=1}^{N} v_i A_j B_j x. \tag{2}$$

This architecture allows each expert to specialize based on task categories, while the shared experts facilitate cross-embedding-generative task generalization. The router and experts are trained jointly, promoting end-to-end task-specific adaptation. This design extends previous PEFT-MoE frameworks (*e.g.*, LoRAMoE [8], TaskMoE [12]) and uniquely incorporates task-wise masking and gating over low-rank experts to resolve the problem of knowledge sharing & conflict effectively.

4.3 Step-Wise Convergence-Oriented Sample Scheduler

Effectively managing the varying convergence rates across different tasks is a central challenge in multi-task learning, which should be guided by a core principle: *directing more optimization effort towards tasks converged more slowly*, so as to ensure balanced progress and prevent any single task from being neglected. Existing works [4,16] propose adaptive loss reweighting according to the learning dynamics. However, we observe that direct loss reweighting leads to training instability, particularly when tasks use fundamentally different loss functions (*i.e.*, InfoNCE for embedding tasks and Cross Entropy for generative tasks).

To develop a more direct and robust multi-task training framework, we propose S2Sched (Step-wise Convergence-oriented Scheduler). Instead of manipulating the optimization process through the loss function, S2Sched operates at the data level. **We adjust the number of samples for each task within every training batch based on non-convergence estimation**. By allocating a larger portion of the training batch to demonstrably slower-converging tasks, S2Sched prevents faster tasks from dominating the optimization process and ensures a more balanced learning trajectory for the entire foundation model.

The entire process can be broken down into two main stages: (1) estimating the non-convergence rate for each task and then (2) calculating the final sampling ratio for the next training batch.

Step-Wise Non-convergence Estimation. We estimate the non-convergence rate based on the intuition that the validation loss slope of a nearly converged task should approach zero. It is worth noting that we decouple the estimation of embedding and generative tasks to avoid interference caused by differences in loss magnitudes. Hence, we denote the sets of embedding and generative tasks as $\mathcal{T}^{(\mathrm{E})}$ and $\mathcal{T}^{(\mathrm{G})}$, respectively.

To quantify the convergence speed of each task, we introduce the **validation loss slope**, denoted as $\alpha_i(t)$, which serves as a direct proxy for a task's real-time learning progress. This slope is not an instantaneous measurement but is robustly calculated by performing a linear regression over the sequence of validation loss values recorded during the last L evaluation steps. The interpretation of this slope is intuitive: a steep negative slope, where the absolute value $|\alpha_i(t)|$ is large, signifies that the validation loss is decreasing rapidly and the task is converging quickly. Conversely, a slope approaching zero indicates that the task's progress has decelerated, either because it has stalled or is nearing full convergence. To ensure the stability of slope calculations, we apply this entire scheduling mechanism only after an initial warmup phase of ς training steps. Furthermore, to prevent the numerical scale of different loss functions from interfering with this estimation, we handle the two distinct sets of tasks, *i.e.*, embedding ($\mathcal{T}^{(\mathrm{E})}$) and generative ($\mathcal{T}^{(\mathrm{G})}$), independently throughout this process.

Based on the step-wise validation loss slopes, we estimate the non-convergence status through the following metrics:

Inter-Task Non-convergence Rate. This metric answers the question: "Compared to its peers, is this task a laggard?" It measures a task's relative convergence speed within its group. Higher $\gamma_{i,b}^{\text{inter}}(t)$ indicates less convergence of task i at step t in comparison to other tasks.

$$\gamma_{i,b}^{\text{inter}}(t) = -\text{Softmax}_i\left(|\mathcal{T}^{(b)}| \cdot \frac{\alpha_i(t)}{\sum_{j=1}^{|\mathcal{T}^{(b)}|} |\alpha_j(t)|} \right), \tag{3}$$

where the negative sign ensures that a smaller slope (slower convergence) produces a larger input to the Softmax, thus yielding a higher non-convergence score.

Intra-Task Non-convergence Rate. This metric answers the question: "Compared to its own recent past, has this task's progress stalled?" It evaluates the stability of a single task's convergence. Higher $\gamma_{i,b}^{\text{intra}}(t)$ indicates less convergence of task i at step t compared to its own history.

$$\gamma_{i,b}^{\text{intra}}(t) = \text{Softmax}_i\left(- L \cdot \frac{\alpha_i(t)}{\sum_{s=t-L+1}^{t} |\alpha_i(s)|} \right). \tag{4}$$

Here, a high score signals that the task's convergence has recently decelerated significantly compared to its immediately preceding trend. This perspective is robust against outliers from other tasks.

Balancing Weight. To intelligently combine the two rates above, we introduce a balancing weight $\beta_b(t)$ that acts as a "stability detector." It measures the disparity of convergence rates across tasks.

$$\beta_b(t) = \min \left\{ t \cdot \text{Softmax}_t\Big(- \frac{\tau t \max_i \alpha_i(t)}{\sum_{k=1}^{t} \max_i |\alpha_i(k)|} \Big),\ 1 \right\}, \tag{5}$$

where τ is a temperature hyperparameter. A high $\beta_b(t)$ (approaching 1) indicates that tasks are converging in a stable, balanced manner, making the inter-task comparison reliable. A low $\beta_b(t)$ suggests instability (*e.g.*, an outlier task), making it safer to trust the intra-task history.

Sample Ratio. We interpolate the inter- and intra-task with the balancing weight to calculate the final sample ratio for each task i:

$$\omega_{i,b}(t) = \text{Softmax}_i \left(-\beta_b(t)\, \gamma_{i,b}^{\text{inter}}(t) + \left[1 - \beta_b(t)\right] \gamma_{i,b}^{\text{intra}}(t) \right), \tag{6}$$

where terms $-\beta_b(t)\, \gamma_{i,b}^{\text{inter}}(t)$ and $\left[1 - \beta_b(t)\right] \gamma_{i,b}^{\text{intra}}(t)$ both convey positive correlations with the overall non-convergent level for task i of type b at step t.

Task-Level Sample Scheduling. We maintain different batch sizes and schedulers for embedding and generative tasks. For each batch of size $B^{(b)}$ of task type $b \in \{E, G\}$ at step t, the scheduler would re-allocate the sample budget for each task $i \in \mathcal{T}^{(b)}$ according to the noramlized task weight $\omega_{i,b}(t)$. Specifically, the number of samples of task i at step t is $n_i^{(b)}(t) = \lfloor B^{(b)} \cdot w_{i,b}(t) \rfloor$. Then, we would conduct uniform sampling without replacement from the task-wise training data pool to form the final training batch.

In this way, S2Sched adaptively adjusts the task sample ratio per batch within different task types (embedding or generative) at each training step according to real-time validation feedback. Therefore, RecFound can achieve stable and convergence-balanced multi-task learning for recommendation foundation models.

4.4 Model Merge

In multi-task learning scenarios, different model checkpoints often specialize in particular tasks or objectives. This specialization interacts with the inherent

Table 2. Performance on generative tasks. Average Rank is obtained by calculating the average of the model's rankings on each task. The smaller this value is, the better the comprehensive effect of the model will be. For each task, the best result is in **bold**, and the second-best result is underlined.

Models	General NLP			User Understanding				Item Understanding			Average Rank	Relative Improvement
	QR↑ (F1)	AVE↑ (F1)	AG↑ (F1)	SR↑ (R@1)	SA↑ (F1)	UP↑ (F1)	AP↑ (F1)	PRP↑ (F1)	PM↑ (F1)	IP↑ (F1)	(Mean (Std))	(%)
Baseline Models w/o Finetuning												
Llama3-8B-Instruct	0.926	0.000	0.830	0.000	0.061	0.875	0.420	0.231	0.444	0.880	11.80 (2.04)	+50.80
Mistral-7B-Instruct	0.936	0.000	0.841	0.005	0.408	0.873	0.738	0.103	0.449	0.892	9.00 (3.32)	+34.18
Mixtral-8x7B-Instruct	0.948	0.000	0.840	0.005	0.278	0.871	0.647	0.316	0.047	0.885	10.20 (2.44)	+45.50
Qwen2.5-7B-Instruct	0.952	0.000	0.834	0.027	0.448	0.872	0.383	0.208	0.215	0.881	10.60 (2.06)	+46.02
Qwen2.5-14B-Instruct	0.944	0.000	0.832	0.063	0.468	0.872	0.518	0.251	0.434	0.886	9.30 (2.00)	+33.60
Qwen2.5-32B-Instruct	0.961	0.000	0.828	0.128	0.466	0.872	0.488	0.212	0.373	0.887	9.30 (2.75)	+34.96
GritLM-7B	0.960	0.000	0.840	0.000	0.227	0.868	0.285	0.219	0.491	0.881	10.80 (2.25)	+47.52
eCeLLM	0.943	0.002	0.843	0.177	0.362	0.848	0.812	0.361	**0.994**	0.857	7.80 (4.49)	+13.53
Baseline Models w/ Finetuning												
eCeLLM	0.973	0.315	0.841	0.251	0.532	0.878	0.821	0.450	0.967	0.888	4.70 (1.79)	+1.76
GritLM	0.973	0.316	0.839	0.242	0.501	0.874	0.814	0.426	0.967	0.889	5.60 (2.22)	+2.88
LoRAMoE	0.973	**0.348**	0.842	**0.262**	0.543	**0.904**	0.785	0.323	0.982	0.885	4.60 (3.86)	+2.79
CoBa	0.971	0.328	0.848	0.238	0.544	0.895	0.808	0.419	0.982	0.899	4.10 (1.29)	+1.53
Our Models												
RecFound w/o Model Merge	**0.974**	0.341	**0.853**	0.234	**0.545**	0.895	0.827	**0.471**	0.988	0.902	2.10 (1.51)	+0.11
RecFound	0.973	0.344	**0.853**	0.251	0.533	0.895	**0.830**	0.462	**0.994**	**0.903**	**2.00 (1.25)**	-

seesaw effect of foundation model training, where performance improvements on one task may coincide with regressions on others. To mitigate such interference and consolidate complementary knowledge, we adopt TIES-Merging [27] as the Model Merge module over the learned low-rank experts across different checkpoints. We first remove insignificant parameter updates, then resolve conflicting update directions via majority voting, and finally average the consistent parameters. This approach produces a merged model that better balances the performance across different tasks.

5 Experiments

5.1 Experimental Setups

Datasets. All experiments are conducted on RecFound dataset. The data statistic is reported in Table 1. The prompt templates for generative and embedding tasks are designed with a specific format. We illustrate the prompt template for each task in appendix in the released code repository.

Baselines. We first adopt a wide range of non-tuned foundation models as baselines. For generative tasks, we choose Llama-3-8B-Instruct [5], Mistral-7B-Instruct [10], Mixtral-8x7b-Instruct, and Qwen-2.5-Instruct series [28](7B,14B,32B). For embedding tasks, we select gte-Qwen2-7B-Instruct [15], SFR-Embedding-Mistral [19], bge-large-en-v1.5 [26], Llama-3-8B-Instruct [5], and Mistral-7B-Instruct [10]. Moreover, we investigate multiple finetuning strategies on the RecFound dataset. We adopt GritLM [20] as a unified baseline

Table 3. Performance on embedding tasks. Notations are the same as Table 2.

Models	Item Emb↑ (NDCG@20)	Query Emb↑ (MRR@20)	User Emb↑ (MRR@20)	Average Rank↓ (Mean (Std))	Rel. Imp. (%)
Baseline Models w/o. Finetuning					
gte-Qwen2-7B-Instruct	0.806	0.975	0.431	8.00 (0.00)	+7.69
SFR-Embedding-Mistral	0.805	0.974	0.419	9.00 (0.00)	+8.37
bge-large-en-v1.5	0.799	0.957	0.419	10.33 (1.15)	+9.52
Llama-3-8B-Instruct	0.703	0.438	0.251	12.00 (0.00)	+71.12
Mistral-7B-Instruct	0.805	0.974	0.419	9.00 (0.00)	+8.37
GritLM-7B	0.823	0.985	0.458	3.67 (1.53)	+5.12
eCeLLM	0.697	0.197	0.221	13.00 (0.00)	+113.63
Baseline Models w/Finetuning					
Embed.Only	0.819	0.983	0.447	6.67 (0.58)	+5.91
GritLM	0.820	0.985	0.439	5.67 (1.53)	+6.15
LoRAMoE	0.822	0.985	0.461	3.67 (0.58)	+5.03
CoBa	0.821	0.986	0.479	3.33 (0.58)	+4.20
Our Models					
RecFound w/o Model Merge	0.821	0.987	**0.573**	2.33 (1.53)	+0.04
RecFound	**0.824**	**0.988**	0.570	**1.33 (0.58)**	–

for both generative and embedding tasks. For generative tasks, we additionally include eCeLLM [21]. Since eCeLLM does not support embedding tasks, we further construct an embedding-only baseline, *i.e.*, Embed.Only.

To evaluate the advantages of our approach over existing parameter-efficient MoE and multi-task architectures, we include LoRAMoE [3] and CoBa [4]. For LoRAMoE, we follow the strategy in GritLM, using input embeddings as routing signals, which differs from our task-type routing. For CoBa, we apply its adaptive loss weighting on top of the GritLM training pipeline, providing a comparison to our scheduler that adjusts per-task sampling during training.

Evaluation Metrics. We allow up to 2048 newly generated tokens for evaluation. In generative tasks, we adopt F1 score as the metric for query rewriting, attribute value extraction, answer generation, user profile, sentiment analysis, answerability prediction, item profile, product relation prediction, and product matching, and use Recall@1 for sequential recommendation. In embedding tasks, we measure the item feature embedding task with NDCG@k, and evaluate user query embedding and user sequence embedding tasks with MRR@k. Following previous works [25], we set $k = 20$.

Implementation Details. Due to page limitation, implementation details and configurations are included in our released code repository for reproducibility.

Table 4. Model compatibility of different approaches on generative and embedding tasks using different backbone LLMs.

Backbone	Approach	Generative Tasks ()										Embedding Tasks (↑)			Avg Rank (↓) (Std Dev)	Rel. Imp. (%)
		QR (F1)	AVE (F1)	AG (F1)	SR (R@1)	SA (F1)	UP (F1)	AP (F1)	PRP (F1)	PM (F1)	IP (F1)	Item Emb (NDCG@20)	Query Emb (MRR@20)	User Emb (MRR@20)		
LLaMA	Zero-shot	0.926	0	0.830	0	0.061	0.675	0.420	0.291	0.444	0.880	0.703	0.438	0.251	5.62 (0.65)	+39.8
	eCeLLM/Embed.Only	0.966	0.246	0.849	**0.251**	**0.497**	0.861	0.759	0.310	**0.973**	0.893	0.749	0.474	0.265	3.65 (1.13)	+2.2
	GritLM	**0.971**	0.270	0.843	0.204	0.462	0.876	0.638	0.317	0.846	0.893	0.751	0.481	0.278	3.58 (1.24)	+5.7
	LoRAMoE	0.970	0.251	**0.852**	0.230	0.489	0.884	0.777	0.285	0.966	0.893	0.753	0.479	0.274	3.04 (1.21)	+2.1
	CoBa	**0.971**	0.248	**0.852**	0.175	0.403	0.882	0.747	**0.330**	0.929	**0.897**	0.748	0.488	0.291	3.00 (1.41)	+3.9
	RecFound (Ours)	0.965	**0.283**	**0.852**	0.245	0.489	**0.885**	**0.816**	0.256	0.967	0.896	**0.761**	**0.535**	**0.323**	**2.08 (1.44)**	–
Mistral	Zero-shot	0.936	0	0.841	0.005	0.110	0.673	0.738	0.108	0.449	0.892	0.805	0.985	0.458	5.54 (0.63)	+34.4
	eCeLLM/Embed.Only	0.973	0.315	0.841	0.251	0.441	0.878	0.821	0.450	0.967	0.888	0.819	0.983	0.447	3.69 (1.01)	+3.7
	GritLM	0.973	0.316	0.839	0.242	0.508	0.874	0.814	0.406	0.967	0.889	0.820	0.985	0.439	4.08 (1.14)	+3.7
	LoRAMoE	0.973	**0.348**	0.842	**0.262**	0.543	**0.904**	0.785	0.323	0.982	0.885	**0.822**	0.985	0.461	3.23 (1.59)	+3.2
	CoBa	0.971	0.328	0.848	0.238	0.544	0.895	0.808	0.419	0.982	0.899	0.821	0.986	0.479	3.04 (1.30)	+2.1
	RecFound (Ours)	**0.974**	0.341	**0.852**	0.234	**0.545**	0.895	**0.827**	**0.471**	**0.988**	**0.902**	0.821	**0.987**	**0.573**	**1.42 (1.24)**	–

Table 5. Ablation experiments result of generative and embedding tasks on different model variants.

Model Variants	General NLP			User Understanding				Item Understanding			Embedding Tasks		
	QR↑ (F1)	AVE↑ (F1)	AG↑ (F1)	SR↑ (R@1)	SA↑ (F1)	UP↑ (F1)	AP↑ (F1)	PRP↑ (F1)	PM↑ (F1)	IP↑ (F1)	Item Emb↑ (NDCG@20)	Query Emb↑ (MRR@20)	User Emb↑ (MRR@20)
RecFound	0.973	**0.344**	**0.853**	**0.251**	0.533	**0.895**	**0.830**	0.462	**0.994**	**0.903**	**0.823**	0.993	**0.510**
w/o M.M.	**0.974**	0.341	**0.853**	0.234	**0.545**	**0.895**	0.827	**0.471**	0.988	0.902	0.822	**0.995**	0.508
w/o M.M. & S2Sched	0.965	0.283	0.852	0.245	0.499	0.885	0.816	0.356	0.988	0.896	0.820	0.992	0.474
w/o M.M. & TMoLE	0.966	0.264	0.847	0.221	0.527	0.887	0.807	0.385	0.976	0.901	0.822	0.991	0.465
w/o M.M. & S2Sched & TMoLE	0.973	0.316	0.839	0.242	0.501	0.874	0.814	0.426	0.967	0.889	0.820	0.985	0.439

5.2 Overall Performance

The performance of RecFound on generative and embedding tasks are reported in Table 2 and Table 3. We can observe that:

- RecFound achieves the best performance across most generative tasks and embedding tasks, largely outperforming both larger-scale LLMs and strong instruction-tuned baselines. This demonstrates the effectiveness of TMoLE and S2Shed, which resolve the knowledge sharing & conflict and convergence inconsistency problems for multi-task foundation model training.
- Furthermore, the Model Merge module not only boosts the overall performance on both generative and embedding tasks, but also balances the performance across individual tasks. This is particularly evident in the Sequential Recommendation task, which shows a significant improvement after applying Model Merge (Table 5). This indicates that the model merge module combines the advantages of multiple model checkpoints, compensating for their limitations and showing excellent compatibility with heterogeneous multi-task learning.

5.3 Model Compatibility

To evaluate model compatibility, we compare RecFound with baseline methods on two backbone LLMs: Llama-3-8B-Instruct and Mistral-7B-Instruct-v0.3. As shown in Table 4, RecFound consistently achieves the best overall performance, evidenced by the Avg Rank metric, where it attains 2.08 on LLaMA and 1.42 on

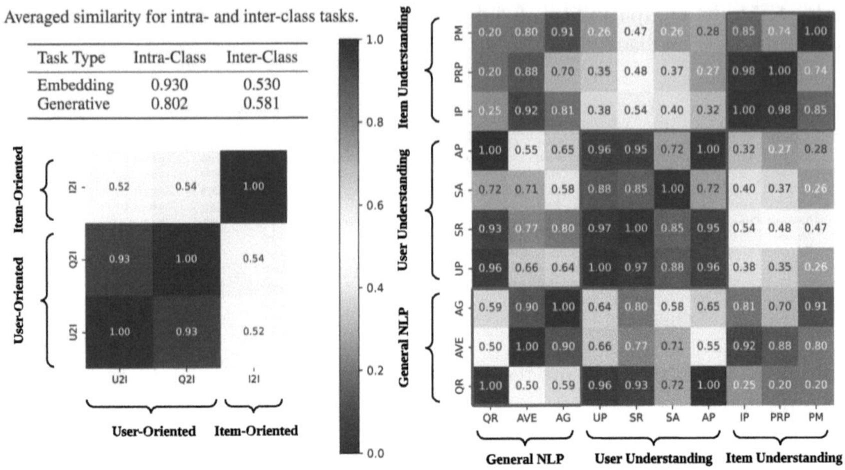

Fig. 3. Similarity heatmaps of the routing distribution across different tasks.

Mistral, substantially outperforming all other methods. Unlike other multi-task approaches whose performance varies across architectures, RecFound remains top-ranked on both backbones, showing strong generality and robustness rather than dependence on a specific model design. These results verify that RecFound provides superior model compatibility for recommendation foundation models.

5.4 Ablation Study

We conduct ablation studies by removing the Model Merge, S2Shed, and TMoLE components, with detailed results reported in Table 5. Removing Model Merge leads to only slight performance degradation, yet it still offers overall performance benefits by balancing improvements across tasks. Removing S2Shed causes the model to revert to uniform task scheduling, resulting in overfitting on simpler tasks and underfitting on more challenging ones. Meanwhile, removing TMoLE reduces RecFound to a standard LoRA [8] configuration, weakening knowledge sharing and conflict resolution, and thus degrading both generative and embedding performance. These findings confirm that all three components are vital for knowledge-disentangled and convergence-balanced multi-task learning in recommendation foundation models.

5.5 In-Depth Analysis

Routing Distributions Among Tasks. To investigate whether the router in TMoLE learns domain-specific characteristics for task-wise knowledge sharing and conflict, we visualize the similarity heatmap of the routing distribution v_i in Eq. (1) across different tasks. Specifically, we compute the pairwise cosine similarity between embedding or generative task pair, and illustrate the heatmap in

Fig. 3. We can observe that tasks with higher similarity (*i.e.*, intra-class denoted by the red box) tend to share more similar routing distributions, whereas tasks that differ substantially-particularly those from different classes-exhibit more divergent routing behaviors. This suggests that TMoLE can successfully capture both the commonalities and distinctions across tasks, facilitating effective embedding & generative task learning for recommendation foundation models.

Scheduled Sample Ratio per Batch. To investigate how S2Sched works to balance the convergence speeds of different tasks, we visualize the sample ratio of different tasks per batch calculated in Eq. (6) w.r.t. the training steps in Fig. 4. During training, we observe that the sample ratios of all tasks exhibit a general pattern of rising and falling alternately over time, rather than monotonically increasing/decreasing or remaining stable. A task that is approaching convergence will be assigned a lower weight, but later its weight will increase again to prevent the forgetting problem. Such behavior reflects S2Sched's ability to adaptively modulate sampling priorities across tasks, promoting more balanced optimization when tasks converge at different rates.

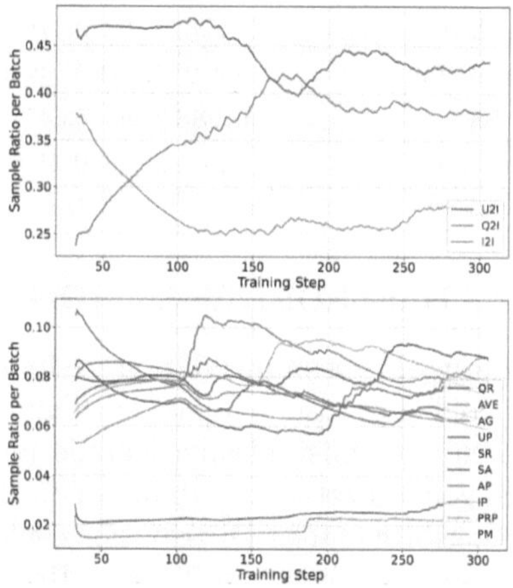

Fig. 4. The evolving sample ratio of different tasks during training.

5.6 Computational and Memory Cost

We trained our models using 8 × H100 GPUs and performed inference on a single H100 GPU. As shown in Table 6, although RecFound incurs slightly longer training time due to the TMoLE structure and S2Sched, it achieves comparable inference speed to GritLM and even outperforms LoRAMoE. This benefits from the task-wise structure, which enables precomputation of $\sum_{j=1}^{N} v_i A_j B_j$ in Eq. (2), eliminating MoE-related overhead during inference at the cost of modest extra memory. RecFound uses FP16 precision and requires 56GB peak memory during inference, slightly higher than GritLM's 52GB due to task-specific MoE parameters.

Table 6. Computational Analysis

Model	Train/Batch	Gen.
eCeLLM	0.587 min	1.218 s
Embed.Only	1.429 min	–
GritLM	1.54 min	1.294 s
LoRAMoE	1.68 min	1.827 s
RecFound	1.87 min	1.33 s

Table 7. Out-of-Distribution (OOD) evaluation of different approaches under the Mistral-7B-Instruct backbone. Each task includes an OOD test set representing distribution shifts in user/item composition or categories.

Model	Generative Tasks (↑)						Embedding Tasks (↑)		
	AVE	SR	SA	AP	PRP	PM	Item Emb	Query Emb	User Emb
Mistral-7B-Instruct	0.000	0.002	0.066	0.719	0.114	0.432	0.808	0.967	0.412
eCeLLM/Embed.Only	<u>0.276</u>	0.239	0.538	0.816	**0.449**	0.835	0.812	0.978	0.444
GritLM	0.255	0.198	0.318	<u>0.819</u>	0.439	0.841	0.818	0.980	0.439
LoRAMoE	0.269	**0.251**	<u>0.549</u>	0.799	0.325	0.849	<u>0.820</u>	0.979	<u>0.451</u>
CoBa	**0.288**	<u>0.243</u>	0.547	0.805	0.407	<u>0.855</u>	0.819	<u>0.982</u>	0.447
RecFound (Ours)	0.274	<u>0.243</u>	**0.551**	**0.824**	<u>0.444</u>	**0.871**	**0.822**	**0.985**	**0.535**

5.7 Out-of-Distribution (OOD) Evaluation

Real-world recommendation systems inevitably face distributional shifts, such as cold-start users/items, evolving temporal patterns, and domain variability, which can substantially degrade model performance. To assess the robustness and generalization capability of our RecFound framework under such conditions, we perform Out-of-Distribution evaluation across multiple tasks in the RecFound dataset. As shown in Table 7, RecFound consistently delivers the best overall performance on OOD test sets and incurs only marginal degradation compared to in-distribution settings. In contrast, other multi-task baselines such as CoBa and GritLM exhibit notable performance drops when exposed to distribution shifts. These results suggest that RecFound effectively captures transferable task-shared representations rather than overfitting to training distributions, making it more reliable for practical deployment in dynamic recommendation environments.

6 Conclusion

In this work, we present RecFound, a unified generative representational learning framework for recommendation foundation models. We construct the first comprehensive dataset for recommendation foundation models covering 3 embedding tasks and 10 generative tasks. RecFound addresses the problems of knowledge sharing & conflict and convergence imbalance through the proposed TMoLE, S2Sched, and Model Merge modules. Extensive experiments demonstrate strong generalization and task adaptability of RecFound model, highlighting the potential of unified recommendation foundation models for both generative and embedding tasks. Future work will explore dynamic task adaptation, scalable routing, and extensions to multilingual and cross-modal recommendation scenarios.

Acknowledgments. This paper is supported by National Natural Science Foundation of China (624B2096, 72595872).

References

1. Brown, T.B., et al.: Language models are few-shot learners. arXiv preprint arXiv:2005.14165 (2020)
2. Devlin, J., Chang, M.W., Lee, K., Toutanova, K.: BERT: pre-training of deep bidirectional transformers for language understanding. In: NAACL-HLT, pp. 4171–4186 (2019). https://arxiv.org/abs/1810.04805
3. Dou, S., et al.: LoRAMoE: revolutionizing mixture of experts for maintaining world knowledge in language model alignment. arXiv preprint arXiv:2312.09979, vol. 4, no. 7 (2023)
4. Gong, Z., Yu, H., Liao, C., Liu, B., Chen, C., Li, J.: CoBa: convergence balancer for multitask finetuning of large language models (2024). https://arxiv.org/abs/2410.06741
5. Grattafiori, A., et al.: The LLaMA 3 herd of models. arXiv e-prints pp. arXiv–2407 (2024)
6. Gupta, M., Kulkarni, N., Chanda, R., Rayasam, A., Lipton, Z.C.: AmazonQA: a review-based question answering task. arXiv preprint arXiv:1908.04364 (2019)
7. Hou, Y., Li, J., He, Z., Yan, A., Chen, X., McAuley, J.: Bridging language and items for retrieval and recommendation. arXiv preprint arXiv:2403.03952 (2024)
8. Hu, E.J., et al.: LoRA: low-rank adaptation of large language models (2021). https://arxiv.org/abs/2106.09685
9. Huang, C., Yu, T., Xie, K., Zhang, S., Yao, L., McAuley, J.: Foundation models for recommender systems: a survey and new perspectives. arXiv preprint arXiv:2402.11143 (2024)
10. Jiang, A.Q., et al.: Mistral 7b (2023). https://arxiv.org/abs/2310.06825
11. Köpcke, H., Thor, A., Rahm, E.: Evaluation of entity resolution approaches on real-world match problems. Proc. VLDB Endow. **3**(1–2), 484–493 (2010)
12. Kudugunta, S., et al.: Beyond distillation: task-level mixture-of-experts for efficient inference (2021). https://arxiv.org/abs/2110.03742
13. Li, X., Zhou, Y., Dou, Z.: UniGen: a unified generative framework for retrieval and question answering with large language models (2023). arXiv:2312.11036
14. Li, Y., et al.: EcomGPT: instruction-tuning large language models with chain-of-task tasks for e-commerce. In: Proceedings of the AAAI Conference on Artificial Intelligence. AAAI Press (2024). https://doi.org/10.1609/aaai.v38i17.29820
15. Li, Z., Zhang, X., Zhang, Y., Long, D., Xie, P., Zhang, M.: Towards general text embeddings with multi-stage contrastive learning. arXiv preprint arXiv:2308.03281 (2023)
16. Lin, T.Y., Goyal, P., Girshick, R., He, K., Dollár, P.: Focal loss for dense object detection. In: Proceedings of the IEEE International Conference on Computer Vision, pp. 2980–2988 (2017)
17. Luo, T., et al.: MoeLoRA: contrastive learning guided mixture of experts on parameter-efficient fine-tuning for large language models. arXiv preprint arXiv:2402.12851 (2024)
18. Martin, S., Poddar, S., Upasani, K.: MuDoCo: corpus for multidomain coreference resolution and referring expression generation. In: Proceedings of the Twelfth Language Resources and Evaluation Conference, pp. 104–111 (2020)
19. Meng, R., Liu, Y., Joty, S.R., Xiong, C., Zhou, Y., Yavuz, S.: SFR-embedding-mistral: enhance text retrieval with transfer learning. Salesforce AI Research Blog (2024). https://www.salesforce.com/blog/sfr-embedding/

20. Muennighoff, N., et al.: Generative representational instruction tuning (2025). https://arxiv.org/abs/2402.09906
21. Peng, B., Ling, X., Chen, Z., Sun, H., Ning, X.: eCeLLM: generalizing large language models for e-commerce from large-scale, high-quality instruction data. In: Proceedings of the 41st International Conference on Machine Learning. PMLR (2024). https://proceedings.mlr.press/v235/peng24c.html
22. Peters, M.E., et al.: Deep contextualized word representations. arXiv preprint arXiv:1802.05365 (2018)
23. Reddy, C.K., et al.: Shopping queries dataset: a large-scale ESCI benchmark for improving product search. arXiv preprint arXiv:2206.06588 (2022)
24. Shi, K., Sun, X., Wang, D., Fu, Y., Xu, G., Li, Q.: LLaMA-E: empowering e-commerce authoring with object-interleaved instruction following (2024). https://arxiv.org/abs/2308.04913
25. Tamm, Y.M., Damdinov, R., Vasilev, A.: Quality metrics in recommender systems: do we calculate metrics consistently? In: Proceedings of the 15th ACM Conference on Recommender Systems, pp. 708–713 (2021)
26. Xiao, S., Liu, Z., Zhang, P., Muennighoff, N.: C-pack: packaged resources to advance general Chinese embedding (2023)
27. Yadav, P., Tam, D., Choshen, L., Raffel, C.A., Bansal, M.: Ties-merging: resolving interference when merging models. Adv. Neural. Inf. Process. Syst. **36**, 7093–7115 (2023)
28. Yang, A., et al.: Qwen2. 5 technical report. arXiv preprint arXiv:2412.15115 (2024)

Index and Storage Systems

DCLex: An On-Disk Learned Index with Optimized I/O and Caching

Zhuoran Wang, Lixiao Cui(✉), Gang Wang(✉), and Xiaoguang Liu

Nankai University, Tianjin 300350, China
{wangzr,cuilx,wgzwp,liuxg}@nbjl.nankai.edu.cn

Abstract. Index structure plays a crucial role in enhancing the performance of storage systems. By integrating machine learning models, learned indexes outperform traditional tree-based indexes. However, real-world scenarios often involve limited memory, making on-disk indexes necessary for large-scale data. When directly migrating memory-based learned indexes to disk, inaccurate predictions of disk blocks lead to excessive disk I/Os, causing waste of time and space. Existing on-disk learned indexes consider disk I/O characteristics but still overlook environments where both memory and disk are used, resulting in excessive memory consumption. To address redundant disk I/Os and inefficient memory use, we propose DCLex, a learned index optimized for minimal disk I/Os and efficient memory utilization. During query processing, DCLex accurately predicts the target disk block through a carefully crafted leaf node structure, effectively reducing I/O overhead. Furthermore, DCLex exploits an adaptive CombinedCache that dynamically adjusts the ratio between key-value cache and block cache, thus improving the cache efficiency across diverse workloads. Finally, DCLex accelerates the index construction as much as possible with low memory usage. Evaluations on four real-world datasets show that DCLex outperforms the state-of-the-art disk-based learned indexes by 1.8× to 3.4× in search performance and 1.6× to 14.1× in insert performance.

Keywords: learned index · disk I/O · adaptive caching

1 Introduction

In DBMSs, index is a critical component for accelerating queries and improving efficiency. Traditional indexes focus on enhancing the throughput and reducing the cost in disk-based environment, such as B+Tree [5], hash table [4] and so on. The tree-structured indexes are widely adopted by various storage systems [1,2]. Some studies improve traditional indexes by optimizing memory usage [7] while others target improvements in hardware efficiency [17] and compression algorithms [3]. However, these indexes are designed for general-purpose datasets and fail to exploit the characteristics of specific data distributions.

With the rapid advancement of machine learning (ML), Kraska et al. [11] propose the concept of learned index, which employs ML models to learn the

H. Jung et al. (Eds.): DASFAA 2026, LNCS 16535, pp. 571–588, 2026.
https://doi.org/10.1007/978-981-92-0363-5_35

data distribution and enhances the performance. The learned index constructs hierarchical linear models for ordered data, and it replaces the node-based binary search of traditional indexes with model predictions. The learned indexes introduce a novel design paradigm for index structures, outperforming the traditional indexes on read-intensive workloads with less memory usage. Prior in-memory learned indexes [6,8,11] require placing all data in memory. However, the databases employed in real-world applications manage data volumes in the terabyte range [22], significantly exceeding the capacity of memory storage. Consequently, the disk storage is still necessary for data retention to save the valuable memory resource. However, when learned indexes meet disk, several challenges are needed to be addressed:

(1) Previous in-memory learned indexes [6,8] perform poorly when directly migrated to disk environments. According to existing performance evaluation [12], they even show lower throughput than B+Tree under various workloads. The primary reason lies in the additional disk I/Os incurred by learned indexes, which stem from the memory-based design. Learned indexes fail to account for disk characteristics, such as disk block size and the high overhead of disk I/O.

(2) Existing disk-based learned index designs lack consideration for efficient memory utilization. Several learned indexes have optimized structures for disk environments. AULID [13], built upon LIPP [21], employs B+Tree-style leaf nodes with redesigned internal nodes. However, AULID is tailored for a pure-disk environment, which limits its ability to leverage efficient memory to accelerate index operations, resulting in suboptimal performance. FILM [15] integrates both memory and disk storage with an emphasis on cache optimization. It implements a hierarchical LRU cache embedded into the index structure to reduce cache replacement overhead. However, FILM requires storing all keys in memory, resulting in excessive memory overhead. LeCo-disk [23] achieves a good balance between performance and space efficiency, but it operates entirely on disk and doesn't leverage memory as cache to further reduce I/O during read operations.

To address the aforementioned challenges, we propose a disk-based learned index with low disk I/O and efficient memory utilization, named DCLex. **First, to reduce the number of I/O operations for indexing, we propose leaf nodes that support precise prediction for DCLex.** This design ensures that each disk I/O retrieves a valid disk block containing the target key-value pair. **Second, to enable different index operations to utilize memory efficiently, we propose a combined cache for DCLex.** This cache contains both key-value pair and disk block items, handling point queries and scan operations respectively. It can dynamically adjust the allocation of each cache type based on the proportion of different read operations. Since the memory overhead is a major cost in modern data centers, we minimize the memory usage to save the runtime cost of the index.

In addition, we observe that on-disk learning-based indexes require a large amount of memory space and have a long construction time. To optimize index construction (called bulkload), we selectively reduce the prediction accuracy of certain internal nodes without sacrificing read/write performance.

We implement the single-threaded and concurrent versions of DCLex. The evaluations on various workloads and datasets show that DCLex outperforms existing learned indexes by up to 3.4×/ 14.1× in search/insert performance. Once the paper is accepted, DCLex will be open sourced. The main contributions of this paper are as follows:

1) We propose DCLex. It reduces disk I/O by precise prediction in leaf nodes.
2) We propose a combined cache to fully utilize memory to process different operations.
3) We propose a memory-efficient index construction strategy.

2 Background and Motivation

Learned Index. Compared to traditional indexes, learned indexes [11] achieve superior performance in many scenarios. In addition to general-purpose designs, several learned indexes have been proposed for particular scenarios to address workload-specific requirements [14,18]. Generally, these learned indexes are categorized as "bottom-up" or "top-down" based on the approach of construction, each with advantages and drawbacks.

For "bottom-up" approach, we take PGM-Index [8] as an example, as shown in Fig. 1(a). PGM-Index builds the index using *pwlfs* (piece-wise linear functions). It scans the original dataset in a single pass and generates multiple linear models, each constrained by a predefined error bound ϵ. PGM-Index is cost-efficient in construction but suffers from high query latency due to prediction inaccuracies at each tree level.

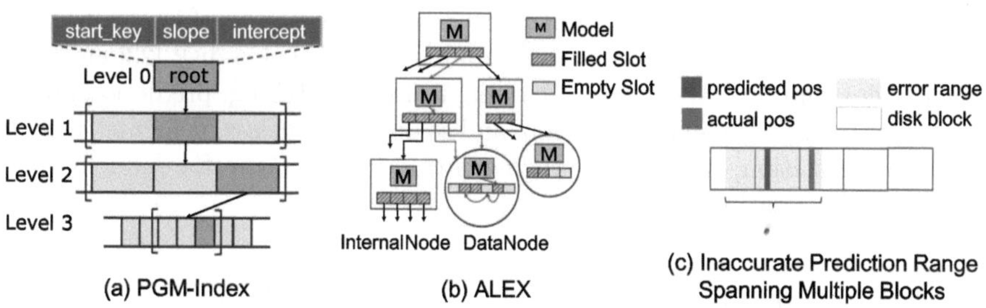

Fig. 1. PGM-Index [8], ALEX [6] and prediction error.

For "top-down" method, we use ALEX [6] as an example, as shown in Fig. 1(b). ALEX trains models hierarchically to provide accurate predictions during internal searches. However, the construction of each internal node requires several passes over the corresponding subdataset, making the bulk loading process expensive. To reduce data movement while preserving model accuracy,

ALEX employs gapped arrays to store child pointers. During lookup, an exponential search is performed within the error bound of the gapped array.

Learned Index on Disk. Although learned indexes perform well in memory, most commercial OLTP DBMSs in real-world applications are typically disk-based [10,12]. They are designed to manage terabyte-scale datasets that far exceed the capacity of memory [22]. In addition to handling large volumes of data, real-world OLTP systems must also support thousands of concurrent transactions while maintaining high throughput. Therefore, disk efficiency and concurrent support are critical for learned indexes in practical settings.

Memory-based learned indexes are designed for DRAM and do not account for the alignment of data storage with disk blocks. As shown in Fig. 1(c), if a memory-based learned index is migrated directly to the disk environment [12], its predefined prediction error range may easily span multiple disk blocks. If the actual position and predicted position reside in different blocks, redundant disk I/Os are triggered, incurring significantly higher query latency. Take ALEX [6] for example, deploying it on disk ([12], called DirectDisk) results in more than twice the disk I/Os per query compared to a B+Tree on average, with a long construction time. For LeCo-disk and AULID, each point query incurs 1.66 and 1.72 I/Os on average [13], which could be reduced to 1 with precise prediction.

Several disk-based learned indexes have been proposed to exploit disk characteristics while enhancing memory utilization, but they exhibit limitations in I/O overhead, memory footprint or index construction efficiency, as summarized in Table 1. AULID [13] incorporates two types of internal nodes along with leaf nodes of B+Tree style to outperform B+Tree, but it lacks the optimization on memory and leads to twice the disk space consumption compared to its counterparts. FILM [15] integrates a two-level cache using LRU strategy to update the cache and search for keys simultaneously, delivering significant read performance. However, FILM incurs substantial memory consumption (58% of key-value pairs are in memory [23]) and struggles with random writes. LeCo-disk [23] enhances learned indexes in several aspects, including leaf-page fetching, prediction granularity, error bound, model compression, and so on. Nonetheless, it focuses heavily on disk optimization, without leveraging memory as a cache for reducing disk I/Os for read operations.

Table 1. Performance of Disk-based Learned Indexes

	I/O Overhead	Memory Utilization	Construction Efficiency
DirectDisk	bad	bad	Bad
AULID	bad	bad	good
FILM	good	moderate	good
LeCo-disk	moderate	moderate	good
DCLex	good	good	good

3 Design

In this section, we present the design of DCLex, a disk-based learned index optimized for I/O efficiency and caching. The key idea behind DCLex is to optimize learned indexes across both memory and disk levels, reducing I/O overhead during queries. Based on this principle, DCLex introduces three major optimizations. Firstly, at the disk level (Sect. 3.1), we propose an I/O-efficient leaf node structure which organizes data at the granularity of disk blocks and enables accurate block prediction, effectively eliminating the additional I/Os caused by searches spanning multiple disk blocks. Secondly, at the memory level, we design a CombinedCache containing both key-value pairs and disk blocks to effectively balance memory usage across workloads comprising both point queries and scans (Sect. 3.2). An adaptive policy is adopted to dynamically adjust the capacities of the two partitions of the cache. The third optimization (Sect. 3.3) focuses on index construction. To control the memory occupation and disk I/O during index construction, we selectively reduce the prediction accuracy of some internal nodes in DCLex.

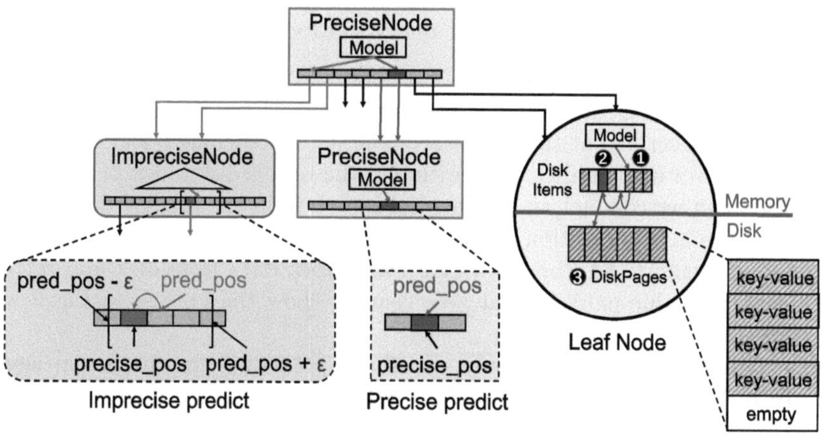

Fig. 2. DCLex Structure Overview.

3.1 Leaf Node Structure and I/O Optimization

In this section, we propose a novel leaf node structure for learned indexes, reducing the additional disk I/Os caused by prediction errors. Based on the new leaf node structure, we introduce the query and insert operation with low I/O cost.

Leaf Node. To improve the I/O efficiency, the leaf node of DCLex needs to meet the following requirements: First, the prediction of the disk block containing the target key must be accurate for read operations. Second, key-value pairs must be stored contiguously on disk to improve space utilization, with only a small amount of free space reserved for future insertions.

The detailed structure is shown in the leaf node of Fig. 2. A leaf node is stored partially in memory and partially on disk. The disk portion consists of multiple disk blocks, each with a header, and the remaining space is utilized for key-value storage. Each disk block is compactly filled to a predefined *fill_ratio* and retains a small amount of free space at the tail to accommodate future insertions. Experiments show that the space reserved has a negligible impact on read performance. The header maintains essential metadata (such as predecessor and successor block) of the current block for data recovery. Each leaf node also records pointers to its predecessor and successor nodes. To enable accurate and efficient searches within the disk structure, the memory portion maintains a gapped array storing the information about disk blocks, including *block_id*, *min_key*, *max_key* and *key_num*. If a *disk_item* has a *min_key* smaller than the target key and a *max_key* larger than the target key, then the corresponding disk block must hold the target key. Leveraging the sorted order of key-value pairs, DCLex determines whether a disk block contains the target key in memory, avoiding triggering multiple I/Os on disk. By sacrificing a small amount of memory, DCLex reduces the I/O per query to exactly 1 (without caching).

Point Query and Scan. To look up a key, the traversal path from the root to a leaf node is described at the end of Sect. 3.3. Upon reaching a leaf node, ❶ we use the model in it to predict the position of the target disk block in the gapped array. ❷ Then we conduct an exponential search from the predicted position to determine the actual position of the target disk block. All the processes above are done in memory. ❸ Finally, we read the block data from disk and retrieve the target key-value pairs from it. We visually show the process in Fig. 2 with red arrows.

For range scans, we first apply the above procedure for the initial key to identify its leaf node and disk block. The block is then scanned sequentially. Using the successor pointer in the leaf node, the search proceeds to the next node until required number of key-value pairs is collected.

Insert. To insert a new key-value pair, the procedure of reaching the correct leaf and identifying the correct disk block follows the same steps as the point query. If the identified disk block is not full, the new pair is inserted directly. Otherwise, a new block is allocated, and the key-value pairs are evenly redistributed between the existing block and the new one. The metadata of the new block is then inserted into the gapped array in memory. If the predicted slot in the gapped array is unavailable, the adjacent metadata entries are shifted to create an empty position. If the gapped array is full, the current leaf node is split evenly, and the metadata of the newly created node is inserted into its parent. The split operation may propagate upward until the root.

Although DCLex is implemented on top of ALEX, it differs fundamentally in design. For the disk-based version of ALEX, the gapped array stores complete key-value pairs and resides on disk. But for DCLex, the gapped array only stores the metadata of disk blocks and is entirely kept in memory. The gaps in the array help to accommodate the metadata of newly created disk blocks with minimal data movement. Regarding the search process, both ALEX and DCLex perform exponential searches within the error bound of their gapped arrays. The key difference is that disk-based ALEX searches for key-value pairs in the on-disk gapped array, which incurs additional disk I/Os. But DCLex searches in the in-memory gapped array for disk metadata to accurately locate the target block, thereby significantly reducing I/O overhead compared to ALEX.

3.2 CombinedCache

A major drawback of FILM lies in its inefficient utilization of excessive memory resources, while AULID and LeCo-disk are limited by the absence of memory-aware optimization. To improve the utilization of memory, we propose a CombinedCache for DCLex. Traditional caches for learned indexes store either key-value pairs (KeyCache) or disk blocks (BlockCache [9]). KeyCache stores hotspot key-value pairs without order, improving the performance of point query. BlockCache stores a continuous piece of data, which is better suited for scan operations. However, for point queries, BlockCache loads the entire block into memory and only a few bytes are necessary [20]. This results in a large amount of cold data occupying memory space. If the workload contains both point queries and scan queries, neither KeyCache nor BlockCache can effectively utilize memory. We design a cache storing both key-value pairs and disk blocks to combine the advantages of KeyCache and BlockCache. This cache is able to adaptively adjust the capacities of KeyCache and BlockCache to better suit the workload. With the assistance of caching, the I/O cost per query can be reduced to less than 1. The detailed design is as follows.

Our CombinedCache has a fixed total size and is divided into a KeyCache and a BlockCache, as illustrated in Fig. 3. To detect the characteristics of workload, we denote the *re-access ratio* as the re-access frequency of evicted cache items. To calculate the *re-access ratio*, we adopt GhostCache from [16], which is an auxiliary cache to track the recently evicted cache items. The keys of evicted items are recorded in GhostCache. Upon a cache miss, we check the GhostCache for a hit in average $O(1)$ time and use its hit rate as *re-access ratio*. If point query is more frequent but the size of KeyCache is relatively smaller, more key-value pairs will be re-accessed shortly after they are evicted from KeyCache. As a result, the *re-access ratio* of KeyCache increases, and the KeyCache is expanded to accommodate more hotspot key-value pairs. Conversely, if scan operation is more frequent, the *re-access ratio* of BlockCache increases and BlockCache will be expanded. Additionally, we employ a stabilizer to stop adjustment when the change in *re-access ratio* is below a predefined threshold, which avoids unnecessary evictions caused by fluctuations.

Due to the different sizes of key cache and block cache items, we expand KeyCache by $\frac{k}{kv_{size}}$ bytes and BlockCache by $\frac{k}{block_{size}}$ bytes each time. Here kv_{size} denotes the size of a single key-value pair, while $block_{size}$ represents the size of a disk block, which is 4096 bytes in our experiments. The constant k is a predefined adjustment parameter.

We illustrate an example of cache adjustment process in Fig. 3 using 16-byte key-value pairs. Since the expansion magnitude ranges from 0 to 1, the real cache size only changes when the magnitude accumulates to 1. We use *target_bound* and *real_bound* to describe this process. Initially, *target_bound* is equal to *real_bound*. For a scan-heavy workload, as the GhostCache of BlockCache is hit constantly, the *target_bound* in Fig. 3 shifts to the right while the real size of BlockCache remains unchanged. When the deviation between *target_bound* and *real_bound* reaches 1, the *real_bound* is updated to *target_bound*. As a result, the BlockCache is expanded by one block and KeyCache evicts 256 key-value pairs to make room for BlockCache. In contrast, for a point-heavy workload, when the deviation reaches 1, BlockCache has to evict one block for KeyCache to expand one pair, which is inefficient. So we wait until the deviation reaches 256 (equal to the size of a block) before evicting a block and expanding the size of KeyCache by 256 at a time.

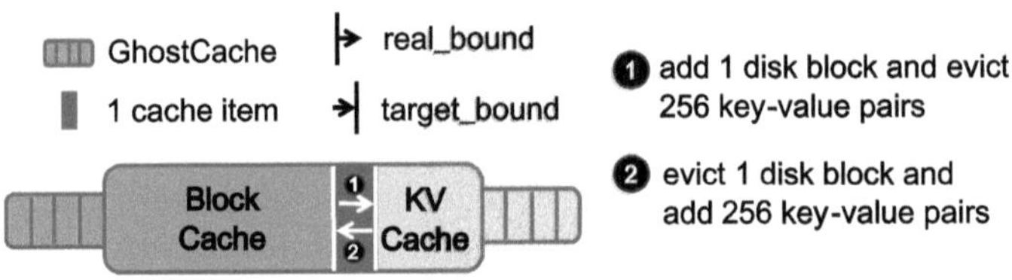

Fig. 3. Design of Cache.

Compared to FILM, our CombinedCache occupies only $\frac{1}{200}$ the size of DCLex but achieves better performance. Our design also provides a method for other disk-based indexes to optimize the performance under workloads involving both point queries and scans. As demonstrated in Sect. 4.2, our design outperforms both pure KeyCache and pure BlockCache under varying workloads with different ratios of point query and scan.

3.3 Bulkload Optimization

Many disk-based learned indexes ignore the memory usage of bulkload and read the entire dataset into memory when building the index structure. For a dataset containing 200 million 8-byte keys, the data size can reach up to 75% of the index size, which is far beyond the limited memory capacity in disk scenario. To resolve this, we repeatedly read small portions of data from disk during bulkload and redesign the internal structure of DCLex to minimize the memory usage and reduce bulkload time.

As discussed in Sect. 2, two types of internal nodes have been proposed for learned index: precise and imprecise. Each type offers distinct advantages and disadvantages. The original design of ALEX uses only precise nodes as internal nodes, which makes the bulkload costly in disk-based environment. To address this, we replace some of precise nodes with imprecise nodes, as shown in Fig. 2. The root node remains a precise node because it is trained on the entire key space, which is the longest and most complex. For data segments below the root, we use $pwlfs$ (piece-wise linear functions [8]) to evaluate their complexity, following the approach in [19]. A node is considered more complex if $pwlfs$ produces a larger number of segments for its key-value pairs. Based on this metric, we apply imprecise nodes for simpler partitions while retaining precise nodes for more complex ones. If a segment is nearly linear, we create a leaf node directly. Besides, rather than storing nearly the entire index on disk in original design, DCLex stores its internal nodes in DRAM to leverage the high-performance memory access, as these nodes have a small space footprint (only 4.1% of the entire index size).

Under this design, to reach a leaf node, we begin at the root. The model in each internal node predicts the position of the next child pointer, and the process iterates down the tree. If the internal node is precise, the predicted position directly corresponds to the correct child pointer. Otherwise, a binary search within the prediction error bound is performed on the child pointer array to locate the exact position of the next child pointer.

3.4 Concurrency

The real-world databases require not only disk storage, but also concurrent data access for higher throughput and better resource utilization. The structure of DCLex introduces thread conflicts when multiple modification operations target the same leaf node, including write/write, write/SMO (Structure Modification Operation) and SMO/SMO conflicts. We omit the explanation of **Read/Write Conflicts**, as a write operation inherently involves both a read and a write on disk. To resolve these issues, we propose two locking mechanisms applying different granularities. We adopt optimistic locking for concurrent control [19], which allows readers to proceed without explicitly acquiring a lock.

Write/Write Conflicts. If we lock data in the granularity of a node, a writer locks the entire leaf node during access. Other writers are blocked when attempting to acquire the write lock. In contrast, block-level locking allows multiple writers to access the same leaf node concurrently. Each writer determines a range of blocks likely to be modified and tries to lock all of them. The write operation begins only after a writer successfully locks all target blocks. If the writer fails to lock any required block, it releases all previously acquired locks and restarts the locking procedure. Thus, even if multiple writers have overlapping block ranges, only one can complete the locking phase and proceed with the modification.

Conflicts with SMO. SMO usually originates from a leaf node. If the leaf's parent is a precise node, both the parent and child node are locked to update their structure. This prevents access by other readers or writers. If the parent is an imprecise node, the parent and all its children are locked for a structural adjustment.

4 Evaluation

We run experiments on a Linux server (kernel version 5.3.0) that contains two 18-core Intel (R) Xeon (R) Gold 5220 CPU @ 2.20 GHz processors, each core has 32 KB L1 instruction cache, 32 KB L1 data cache, 1024 KB L2 cache and 25344KB L3 cache. The DRAM capacity of the server is 125 GB. All the indexes are stored on a WD_BLACK SN850X NVMe SSD with a capacity of 1 TB.

Competitors for Comparisons. We compare DCLex with state-of-the-art disk-based learned indexes. To evaluate the design of leaf node, we compare DCLex with DirectDisk implemented by [12], which migrates ALEX to disk. DirectDisk stores all the ALEX internal nodes in memory and all the leaf nodes with gapped arrays on disk. The key difference between DCLex and DirectDisk lies in the structure of leaf node, which determines whether we can identify the accurate disk block that contains the target key in memory. We also compare DCLex with AULID, a follow-up work of [12]. Both AULID and DCLex use the B+Tree-style leaf node, but differ in their internal structures. To evaluate the design of cache, we compare DCLex with FILM. The cache of DCLex is separated from index and stores both key-value pairs and disk blocks, while the cache of FILM is integrated with the index and stores only key-value pairs. Finally, we evaluate DCLex against LeCo-disk [23], which improves disk-based learned indexes by optimizing leaf page fetching, prediction granularity selection, error bound adjustment, and related techniques.

Configurations. For each counterpart, we use the default configurations from its source code. Specifically for FILM, we minimize its memory usage to 4500 MiB during experiments to approach the memory usage of other indexes. For LeCo-Disk, we use the DIV1 version and set the parameter λ to 2.25 to ensure a fair comparison with DCLex. As for DCLex, we set the initial $fill\_ratio$ of the B+Tree-style node to 0.8. We implement a traditional block cache for each baseline to ensure a fair comparison with DCLex. All index operations are performed using Direct I/O to eliminate the interference of the operating system's page cache. As a result, the effective usage of DRAM capacity **is constrained by the size of self-managed cache** (up to 1200 MiB in our experiments), regardless of the total DRAM available.

Benchmarks. The experimental setup is as follows: (1) The workloads include read-only, write-only, mixed read-write (1:1), YCSB A (50%read-50%update), and YCSB B (95%read-5%update). All tests use 100 million key-value pairs for index construction and 10 million key-value pairs for operations. The read workloads follow both Uniform and Zipfian (skewness parameter is set to 0.99) distributions. (2) The real-world datasets used in the experiments include planet, osm, genome and fb, each containing 200 million keys of type *uint64_t*. Depending on the value size which ranges from 8 bytes to 128 bytes, the total dataset size varies from approximately 2.9 GB to 25.3 GB. (3) Except for the evaluation of different payload sizes (described in the last paragraph of Sect. 4.5), all the keys and values are 8-byte integers. (4) All the experiments are conducted with a single thread except for the concurrency part.

4.1 Performance in Various Scenarios

Performance in Microbench Workloads. Figure 4 and Fig. 5 present the throughput of five different indexes across various datasets and workloads, including read-only, write-only, and mixed read-write workloads with Uniform and Zipfian distributions. Overall, DCLex demonstrates significantly higher throughput than the other indexes. Notably, FILM consumes at least 4500 MiB memory due to its special structure, while other indexes require only 10 MiB for caching.

In read-only workloads, the index structure remains unchanged after initialization, and the caching scheme has a noticeable effect on the performance of workloads with Zipfian distribution. For Zipfian distribution, DCLex outperforms DirectDisk, AULID, FILM and LeCo-disk by 2.8×, 3.4×, 1.8× and 3.1×. Two key factors contribute to these results. First, the design of leaf node helps DCLex to cost exactly one disk I/O to complete one query (assuming no caching). Second, the CombinedCache in DCLex detects the density of read operation in workloads and dynamically expands the capacity of KeyCache, accommodating more hotspot data in memory and accelerating point queries effectively. Under Zipfian distribution and caching scheme, one read operation actually incurs an average of 1.56, 1.81 and 1.23 disk I/Os for DirectDisk, AULID and LeCo-disk. In contrast, DCLex requires only 0.44 disk I/Os per read. For Uniform distribution, the throughput of DCLex is also significantly higher.

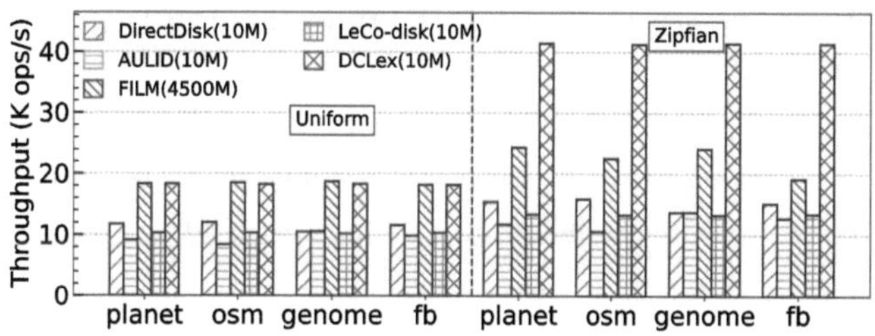

Fig. 4. Throughput of Read-Only.

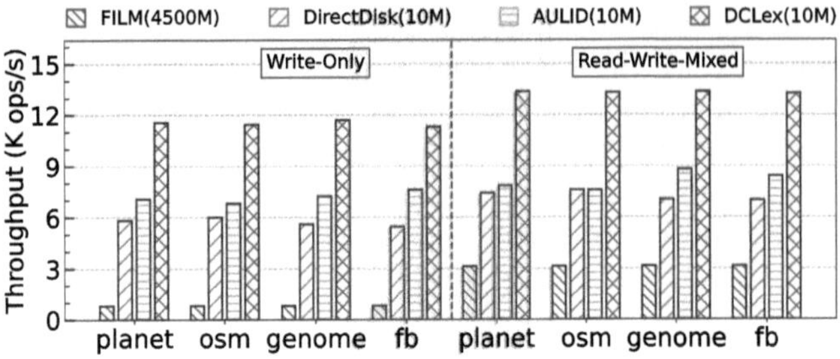

Fig. 5. Throughput of Write-Only and Mix Workload.

In write-only workloads, the structure of index is changed due to disk block splitting, leaf node resizing and data movement in gapped arrays. For workloads including write operations (write-only, YCSB workloads), we primarily compare DCLex with AULID, DirectDisk and FILM. LeCo-disk adopts an additional buffer for newly inserted key-value pairs and relies on a background thread to periodically merge data, which makes it hard to be compared with other methods. A write operation includes one read operation to fetch the target block and one write I/O to store the updated block containing the new key-value pair. DCLex requires exactly two disk I/Os per write (without caching), while for DirectDisk and AULID one write operation requires 4.3 and 3.2 disk I/Os on average. FILM struggles with random write operations. As a result, DCLex outperforms FILM, DirectDisk and AULID by 14.0×, 2.0× and 1.6× in throughput. In the mixed read-write workload (read : write is 1 : 1), DCLex also achieves better throughput than its counterparts.

Performance via YCSB A and YCSB B. Figure 6 illustrates the throughput of four indexes under YCSB A and YCSB B workloads with Zipfian distribution. In YCSB A, excluding FILM which has significantly higher memory consumption, DCLex delivers the best performance. In YCSB B, the proportion of read operations is much higher, thus the throughput of DCLex is even more pronounced, primarily due to its caching scheme.

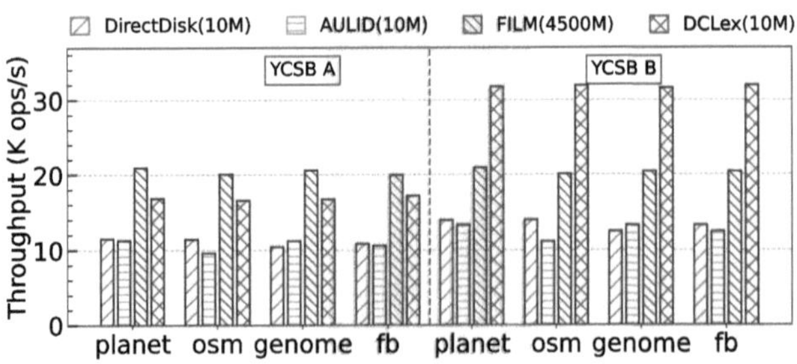

Fig. 6. Throughput of YCSB A and B.

4.2 Performance of Different Cache Parameters

Figure 7(a) illustrates the throughput of different indexes under varying cache sizes. In read-only workloads with Zipfian distribution, DCLex consistently achieves the highest throughput as the memory usage increases. This is attributed to the CombinedCache, which effectively retains hotspot key-value pairs in memory. The data of FILM is tested under fixed memory of 4500 MiB.

Figure 7(b) illustrates the performance of three different caching strategies as the cache size increases. The workload consists of 75% point queries and 25% range queries following Zipfian distribution, and the initial capacity of KeyCache is equal to BlockCache. For cache sizes below 100 MiB, the performance of the CombinedCache is close to that of pure KeyCache, but slightly lower than pure BlockCache. This is because as the cache stabilizes, nearly all capacity is allocated to KeyCache rather than BlockCache, which significantly improves the performance of point query. However, due to the 25% range queries requiring access to sequential data, our cache incurs additional disk I/Os compared to BlockCache, slightly reducing the performance. For cache sizes exceeding 500 MiB, the CombinedCache achieves better performance than the other two schemes and exhibits scalability comparable to BlockCache. With sufficient capacity, the CombinedCache not only retains hotspot key-value pairs in memory but also reserves sufficient space for disk blocks needed by range queries. This ensures more efficient memory usage compared to pure KeyCache or BlockCache under the same memory budget. Additionally, the CombinedCache inherits the scalability of BlockCache, which avoids the performance bottleneck observed in KeyCache for cache sizes beyond 500 MiB. The CombinedCache degenerates into pure KeyCache or BlockCache under pure point query or pure scan workloads.

The experiment evaluates the stabilized proportions of KeyCache and BlockCache in CombinedCache under varying ratios of range and point queries with a fixed cache size of 1200 MiB. As the range-to-point ratio transitions from 1:3 to 1:100, the capacity share of KeyCache gradually rises from 0.288 to 0.353, due to the increasing frequency of GhostCache hits within KeyCache.

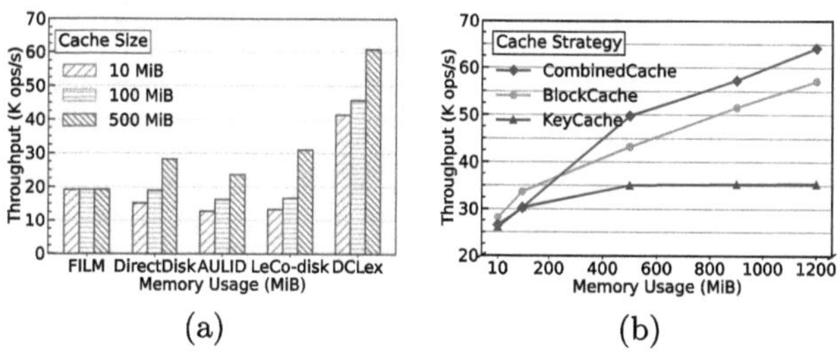

Fig. 7. Throughput of Different Cache Parameters.

4.3 Performance of Bulkload

Table 2 presents the construction time and disk I/O for DirectDisk and DCLex using only 160 bytes of memory. The experimental results show that index performance is largely unaffected by the bulkload optimization. On the fb dataset, DCLex significantly reduces construction time compared to DirectDisk, with 1.5× fewer disk I/Os. This is attributed to the imprecise internal nodes of DCLex, which greatly reduce the number of traversals over the original dataset during construction. Similar trends are observed in osm and planet datasets, although less pronounced. This is because the distribution of these datasets results in very large root nodes for both indexes, but the hybrid structure of DCLex primarily cuts the disk I/Os for non-root internal nodes. The constructions of AULID, FILM and LeCo-disk are significantly different from DirectDisk and DCLex and are therefore excluded from this discussion.

Table 2. Time and Disk IO of Bulkload

	Dataset	fb	osm	planet
Time(s)	DirectDisk	3912.46	2613.91	2307.06
	DCLex	2836.65	2308.82	2202.96
DiskIO(Million)	DirectDisk	440.28	302.03	258.40
	DCLex	270.11	254.87	245.70

4.4 Concurrent Performance

Figure 8 presents the throughput of DCLex under varying numbers of threads. The experiment is conducted directly with DirectIO (no caching). For read-only workloads, the performance scales effectively and consistently outperforms LeCo-disk across all thread numbers. For write-only workload, Fine-grained locking delivers worse performance in single-thread test but demonstrates better scalability as the number of threads increases. This arises from smaller locking granularity, which allows multiple threads to operate on a single leaf node concurrently, reducing thread contention. All multi-threaded tests are conducted 10 times, and the average result is reported.

4.5 In-depth Analysis

To analyze the sources of performance improvements, we decompose the components of DCLex and evaluate their effects on performance.

Table 3 presents the effects of different index components under read-only workloads with Zipfian distribution (cache size 10 MiB). Overall, the two components complement each other to achieve optimal performance, with opt-1 demonstrating more significant effects when used alone. Comparing the performance in Fig. 4 and Table 3, opt-1 consistently outperforms the other four indexes by 1.76×, 1.18×, 1.84× and 1.99× respectively, by effectively reducing disk I/Os through accurate prediction of target disk block. In contrast, the throughput of opt-2 is slightly lower than DirectDisk, as the CombinedCache degenerates to a pure KeyCache under this condition, leading to marginally higher disk I/Os. When opt-1 and opt-2 are combined, the reduction in disk I/O achieved by opt-1 compensates for the shortcomings of the KeyCache in CombinedCache. This enables opt to deliver significantly better performance than opt-1, opt-2, and the other indexes in the comparison.

Table 3. Throughput of optimized components (opt-1 is leaf node optimization only, opt-2 is adaptive cache only, and opt has opt-1 and opt-2 together)

Throughput (K ops)	DirectDisk	opt-1	opt-2	opt
planet	15.41	26.63	14.84	41.45
osm	15.87	26.44	15.24	41.29
genome	13.73	26.51	13.23	41.44
fb	15.16	26.55	14.58	41.39

Furthermore, we evaluate the performance of DCLex under different payload sizes, as illustrated in Fig. 9. As the payload size increases, DCLex's throughput decreases from 100% to 70% due to the reduced number of items in cache. Nevertheless, DCLex's throughput with 128-byte payloads remains nearly twice that of its competitors tested with 8-byte payloads under the same memory budget.

Table 4. Memory and Disk Usage of Index Structure (payload size = 8 bytes)

Dataset	Memory(MiB)				Disk(MiB)			
	planet	osm	genome	fb	planet	osm	genome	fb
FILM	4500	4500	4500	4500	>4500	>4500	>4500	>4500
DirectDisk	1	19	1	4	2215	2234	2204	2237
AULID	<1	<1	<1	<1	4386	4397	4376	4380
LeCo-disk	<1	<1	<1	1	1525	1525	1525	1525
DCLex	52	790	23	119	2259	2453	2275	2749

4.6 Overheads Analysis

Table 4 displays the memory and disk overhead of the five indexes. The reported memory overhead includes only the sizes of internal nodes and leaf node metadata, excluding cache consumption. Overall, DCLex and DirectDisk exhibit higher memory usage due to the internal nodes in memory. In terms of disk usage, DCLex is close to DirectDisk and requires nearly half the space of AULID. This is because DCLex stores key-value pairs densely in disk with an initial *fill_ratio* and has a simpler structure compared to AULID.

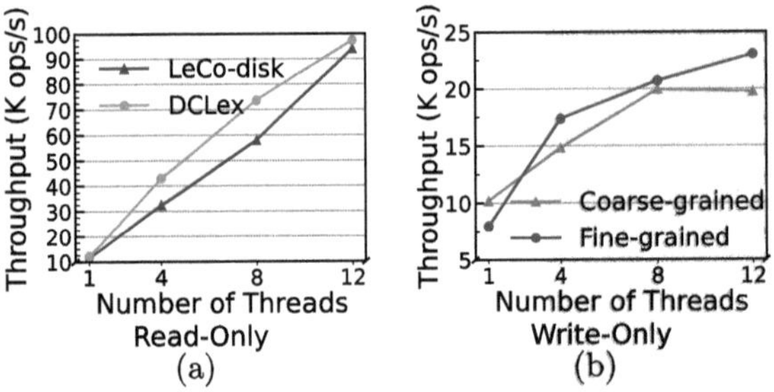

Fig. 8. Throughput of Different Cache Parameters.

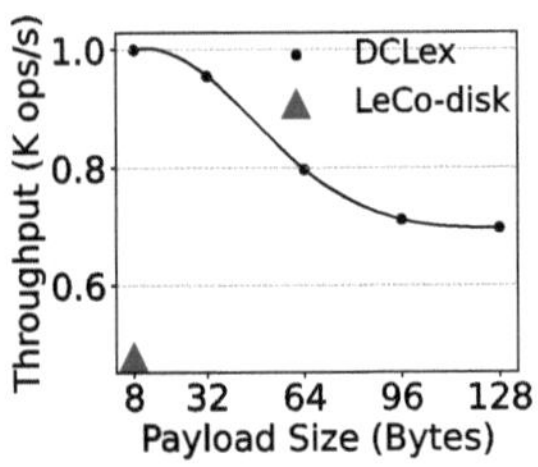

Fig. 9. Throughput of Different Payload Size.

5 Conclusion

In this paper, we propose DCLex, a disk-based learned index that minimizes disk I/Os and adopts an efficient caching scheme. DCLex applies a leaf node that enables accurate prediction to reduce disk I/Os. We also exploit an adaptive CombinedCache to improve cache efficiency across diverse workloads. Finally, we accelerate the index construction with a small usage of memory. Compared to the state-of-the-art methods, DCLex improves the search performance by 1.8× to 3.4×, while enhancing the insert performance by 1.6× to 14.1×, respectively.

Acknowledgements. This research is supported by the National Natural Science Foundation of China (grant numbers, 62272252, 62272253), and the Fundamental Research Funds for Central Universities. Lixiao Cui and Gang Wang are corresponding authors.

Disclosure of Interests. The authors declare competing interests with nankai.edu.cn and alibaba.com. These relationships may be perceived as potential sources of conflict of interest.

References

1. MongoDB Inc. (2021). https://www.mongodb.com/
2. Redis (2021). https://redis.io/
3. Boehm, M., Schlegel, B., Volk, P.B., Fischer, U., Habich, D., Lehner, W.: Efficient in-memory indexing with generalized prefix trees (2011)

4. Chan, H.H., et al.: {HashKV}: Enabling efficient updates in {KV} storage via hashing. In: 2018 USENIX Annual Technical Conference (USENIX ATC 18), pp. 1007–1019 (2018)
5. Comer, D.: Ubiquitous B-tree. ACM Comput. Surv. (CSUR) **11**(2), 121–137 (1979). https://doi.org/10.1145/356770.356776
6. Ding, J., et al.: Alex: an updatable adaptive learned index. In: Proceedings of the 2020 ACM SIGMOD International Conference on Management of Data, pp. 969–984 (2020). https://doi.org/10.1145/3318464.3389711
7. Fan, B., Andersen, D.G., Kaminsky, M.: {MemC3}: compact and concurrent {MemCache} with dumber caching and smarter hashing. In: 10th USENIX Symposium on Networked Systems Design and Implementation (NSDI 13), pp. 371–384 (2013)
8. Ferragina, P., Vinciguerra, G.: The PGM-index: a fully-dynamic compressed learned index with provable worst-case bounds. Proc. VLDB Endow. **13**(8), 1162–1175 (2020). https://doi.org/10.14778/3389133.3389135
9. Ghemawat, S., Dean, J.: LevelDB (2011). https://github.com/google/leveldb
10. Haas, G., Haubenschild, M., Leis, V.: Exploiting directly-attached NVME arrays in DBMS. In: CIDR, vol. 20, p. 2 (2020)
11. Kraska, T., Beutel, A., Chi, E.H., Dean, J., Polyzotis, N.: The case for learned index structures. In: Proceedings of the 2018 International Conference on Management of Data, pp. 489–504 (2018). https://doi.org/10.1145/3183713.3196909
12. Lan, H., Bao, Z., Culpepper, J.S., Borovica-Gajic, R.: Updatable learned indexes meet disk-resident DBMS-from evaluations to design choices. Proc. ACM Manage. Data **1**(2), 1–22 (2023). https://doi.org/10.1145/3589284
13. Lan, H., Bao, Z., Culpepper, J.S., Borovica-Gajic, R., Dong, Y.: A simple yet high-performing on-disk learned index: can we have our cake and eat it too? arXiv preprint arXiv:2306.02604 (2023) https://doi.org/10.48550/arXiv.2306.02604
14. Li, P., Hua, Y., Jia, J., Zuo, P.: Finedex: a fine-grained learned index scheme for scalable and concurrent memory systems. Proc. VLDB Endow. **15**(2), 321–334 (2021). https://doi.org/10.14778/3489496.3489512
15. Ma, C., Yu, X., Li, Y., Meng, X., Maoliniyazi, A.: Film: a fully learned index for larger-than-memory databases. Proc. VLDB Endow. **16**(3), 561–573 (2022). https://doi.org/10.14778/3570690.3570704
16. Megiddo, N., Modha, D.S.: {ARC}: A {Self-Tuning}, low overhead replacement cache. In: 2nd USENIX Conference on File and Storage Technologies (FAST 03) (2003)
17. Rao, J., Ross, K.A.: Making B+-trees cache conscious in main memory. In: Proceedings of the 2000 ACM SIGMOD International Conference on Management of Data, pp. 475–486 (2000). https://doi.org/10.1145/342009.335449
18. Wang, Y., Tang, C., Wang, Z., Chen, H.: Sindex: a scalable learned index for string keys. In: Proceedings of the 11th ACM SIGOPS Asia-Pacific Workshop on Systems, pp. 17–24 (2020). https://doi.org/10.1145/3409963.3410496
19. Wongkham, C., Lu, B., Liu, C., Zhong, Z., Lo, E., Wang, T.: Are updatable learned indexes ready? arXiv preprint arXiv:2207.02900 (2022). https://doi.org/10.14778/3551793.3551848
20. Wu, F., Yang, M.H., Zhang, B., Du, D.H.: {AC-Key}: Adaptive caching for {LSM-based}{Key-Value} stores. In: 2020 USENIX Annual Technical Conference (USENIX ATC 20), pp. 603–615 (2020)
21. Wu, J., Zhang, Y., Chen, S., Wang, J., Chen, Y., Xing, C.: Updatable learned index with precise positions. arXiv preprint arXiv:2104.05520 (2021). https://doi.org/10.48550/arXiv.2104.05520

22. Zhang, H., Andersen, D.G., Pavlo, A., Kaminsky, M., Ma, L., Shen, R.: Reducing the storage overhead of main-memory oltp databases with hybrid indexes. In: Proceedings of the 2016 International Conference on Management of Data, pp. 1567–1581 (2016). https://doi.org/10.1145/2882903.2915222
23. Zhang, J., Su, K., Zhang, H.: Making in-memory learned indexes efficient on disk. Proc. ACM Manage. Data **2**(3), 1–26 (2024). https://doi.org/10.1145/3654954

SL-Cache: Selective Learning Cache Eviction with Priority Retention for Hot Objects

Yu-Ang Zhang[1], Yigui Yuan[1], Peiquan Jin[1(✉)], Hongtao Wang[1], Xiaoliang Wang[2], and Shouhong Wan[1]

[1] University of Science and Technology of China, Hefei, China
jpq@ustc.edu.cn
[2] Hefei University of Technology, Hefei, China

Abstract. Learning-based cache eviction algorithms, particularly object-level approaches, have demonstrated superior reduction in miss ratio by making fine-grained eviction decisions. However, this accuracy often comes at a high computational cost, as existing methods typically apply complex prediction models uniformly to all objects and struggle to quickly differentiate hot from cold items. In this paper, we identify two critical insights: (1) not all objects necessitate precise prediction, especially one-hit wonders lacking rich historical features, and (2) effective sampling must rapidly protect hot objects and identify cold eviction candidates. Based on these insights, we propose **SL-Cache** (Selective Learning Cache), a novel object-level learning eviction algorithm that optimally balances miss ratio and throughput. SL-Cache introduces two key innovations: first, it bypasses costly model predictions for low-utility one-hit wonders by employing precomputed utility estimates, substantially reducing computational overhead. Second, it implements an adaptive, hotness-aware sampling strategy that dynamically demotes objects based on refined access patterns, ensuring hot objects are retained while non-hot candidates are efficiently selected for prediction or eviction. Our extensive experimental evaluation across ten state-of-the-art eviction algorithms, using diverse production and benchmark traces, demonstrates that SL-Cache consistently achieves a superior balance. Specifically, SL-Cache reduces the miss ratio by 11.90% relative to LRU, which is 1.1x higher than the state-of-the-art 3L-Cache, while simultaneously delivering 2x the throughput of 3L-Cache. Furthermore, SL-Cache achieves the highest throughput among all object-level learning algorithms.

Keywords: Cache eviction · Object-Level Learning · Machine Learning

1 Introduction

Cache is a fundamental component in key-value stores [19] and database management systems (DBMSs) [11,14], bridging the significant performance gap

H. Jung et al. (Eds.): DASFAA 2026, LNCS 16535, pp. 589–605, 2026.
https://doi.org/10.1007/978-981-92-0363-5_36

between main memory and storage devices. Effective caching relies on high hit rates, as cache hits provide low-latency data access, while misses incur costly storage I/O. Consequently, minimizing the cache miss rate is a critical objective. As the heart of the cache system, the eviction algorithm dictates which objects are evicted and thus directly governs overall efficiency. Traditional eviction algorithms typically employ heuristics based on recency [7–9,13], frequency [6], or their hybrid combinations [10,12]. Recent advancements have introduced techniques such as lazy promotion and quick demotion [21,24] to enhance the adaptability of heuristic-based approaches to observed workload patterns. However, this reliance on fixed feature combinations constitutes a fundamental limitation, as it prevents traditional algorithms from maintaining high efficiency across diverse workloads.

Recently, learning-based eviction algorithms have emerged as a promising alternative since they adapt to varying workload characteristics [22,23]. These algorithms can be categorized into four primary paradigms:

(1) *Distribution learning* algorithms [2] model object request probability using historical heuristic features to inform eviction decisions.
(2) *Pattern learning* algorithms [15,18,23] employ multiple heuristic algorithms as experts, dynamically switching between them based on workload patterns.
(3) *Group-level learning* algorithms [20] cluster objects into groups using heuristic features(e.g., write time) and predict utility at the group level, reducing computational overhead.
(4) *Object-level learning* algorithms [4,17,25] represent the most fine-grained approach, predicting next access times for individual objects to approximate Belady's OPT algorithm [3].

Among all these paradigms, object-level learning algorithms consistently achieve superior reductions in miss ratio across diverse real-world workloads, owing to their comprehensive feature utilization and granular prediction capabilities.

Among these paradigms, object-level learning algorithms have demonstrated the potential to achieve the lowest miss ratios by leveraging multiple features and operating at the finest granularity. However, this accuracy comes at the cost of significant computational overhead. For object-level learning algorithms, the sampling strategy significantly affects the eviction algorithm's performance because it determines which candidates are selected for eviction. Previous works [17,25] keep a relatively large "*prediction ratio*" (Average prediction times per eviction) to guarantee the quality of eviction decision. A higher prediction ratio yields a higher-quality eviction but incurs a higher computational cost. To tackle this challenge, we identify two critical insights from an object-centric perspective:

(1) Not all objects need precise prediction. A significant fraction of objects, particularly "*one-hit wonders*" [21] (i.e., objects accessed only once after insertion), lack the rich historical access features (e.g., reuse distance, frequency) required by complex learning models. Forcing these objects through the prediction pipeline yields little benefit, as their eviction utility is low and can be estimated using simpler methods.

(2) Efficient sampling must rapidly differentiate hot from cold objects. The sampling strategy's goal is twofold: protect hot objects from being sampled (and thus from potential eviction) and quickly identify cold objects as eviction candidates. Traditional heuristics are often too slow to react to dynamic patterns, leading to either the premature eviction of hot objects or unnecessary predictions on them.

Based on these insights, we propose **S**elective **L**earning **Cache** (**SL-Cache**), a novel object-level learning algorithm that strategically *bypasses* costly predictions for low-utility and hot objects. Rather than solely optimizing the prediction model, our core contribution lies in selectively applying predictions by efficiently identifying which objects do not warrant the prediction cost. More specifically, SL-Cache introduces two key innovations: First, to handle one-hit wonders efficiently, SL-Cache directly uses precomputed estimates of their low utility in eviction decisions. This eliminates the need to invoke the primary prediction model for these objects, thereby substantially reducing computational overhead. Second, an adaptive, hotness-aware sampling strategy that dynamically demotes objects based on access patterns, ensuring that hot objects are retained while non-hot candidates are efficiently selected for prediction or eviction. Together, these designs allow SL-Cache to significantly reduce prediction overhead while maintaining a low miss ratio.

Briefly, we make the following contributions in this paper:

- We conduct a critical analysis of object-level learning algorithms from an object-centric perspective, highlighting scenarios in which learning models increase computational overhead without corresponding reductions in miss ratios. We propose the first approach that bypasses model predictions for specific objects and scenarios, thereby improving throughput while maintaining a low miss ratio.
- We design and integrate an adaptive, hotness-aware sampling mechanism that dynamically adjusts object promotion and demotion during fine-tuning, effectively insulating hot objects from eviction and efficiently identifying cold candidates.
- We conducted an extensive experimental evaluation of SL-Cache against 10 state-of-the-art eviction algorithms using diverse production and benchmark traces. The results demonstrate that SL-Cache consistently achieves a superior trade-off between miss ratio and throughput. In particular, SL-Cache reduces the miss ratio by 11.90% compared with LRU. This reduction is 1.1× greater than that achieved by the state-of-the-art 3L-Cache, while SL-Cache delivers 2× the throughput of 3L-Cache.

The remainder of the paper is structured as follows. Section 2 discusses related work and research motivations. Section 3 presents the detailed designs of SL-Cache. Section 4 reports the experimental results, and finally Sect. 5 concludes the paper and discusses future work.

2 Related Work and Motivations

2.1 Object-Level Learning for Cache Eviction

To achieve a lower miss ratio and enhance adaptability to diverse workloads, an emerging research paradigm integrates machine learning (ML) techniques into cache eviction algorithm design [16,17,25]. Among these learning algorithms, *object-level learning* has demonstrated superior performance across a broad range of real-world workloads by making fine-grained eviction decisions for individual objects.

Object-level learning algorithms typically predict either the next access time of objects [4,17,25] or custom metrics designed to quantify object utility. The foundational design philosophy of these algorithms aims to emulate the prescient decision-making process of Belady's OPT algorithm. Whereas OPT leverages perfect future knowledge to evict the object with the longest future reuse distance, object-level learning algorithms employ ML models trained on historical access patterns. These models use heuristic features, such as reuse distance (e.g., the number of intervening requests between accesses to the same object), age, and frequency, to predict future reuse intervals or object utility. However, the computational cost of model training and prediction results in higher latency and degraded overall throughput. To manage this trade-off between prediction quality and computational cost, existing object-level learning algorithms employ a **sampling strategy** to select a subset of objects for prediction. The intensity of this process can be measured by the **prediction ratio**, defined as the average number of predictions performed per eviction. For instance, LRB [16] randomly samples 64 objects from the cache for prediction during one eviction. HALP [17] samples four objects from the tail of LRU and uses three rank-based predictions for one eviction. 3L-Cache [25] deploys a bio-sampling strategy based on LRU and frequency of evicted objects to select a batch of candidates for prediction, subsequently evicting half the batch size. Current object-level learning approaches face two primary limitations stemming from their prediction and sampling strategy: (1) These algorithms generate predictions for every object in the same way, regardless of analyzing the necessity of precise predictions for every object. (2) These algorithms may not sample the cold objects quickly enough, and cause cold objects to accumulate in the cache since they directly deploy a heuristic algorithm or features.

2.2 Motivations

The Case of One-Hit Wonders. The prevalence of objects accessed only once after admission within cache workloads, often termed **one-hit wonders** [21]. Analysis further reveals that these objects constitute a significant fraction of the items evicted by the optimal Belady's algorithm, often being discarded relatively early in their cache lifetime, as shown in Fig. 1. While heuristic algorithms implicitly handle these objects (evicting them at the tail in LRU), their impact on the efficiency of *object-level learning* algorithms warrants specific attention. Object-level learning algorithms predict an object's utility from historical request

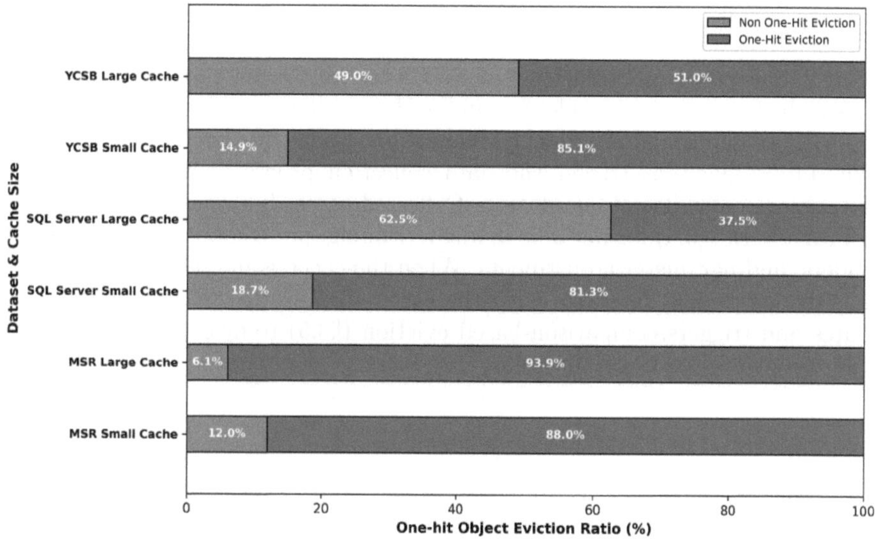

Fig. 1. One-hit object prediction ratio in Belady's OPT algorithm.

features. However, one-hit wonders offer only age as a dynamic feature, lacking a history for models to rely on. Hence, predicting these objects with complex models is inefficient. This motivates estimating their eviction priority using simpler, potentially precomputed methods, enabling selective bypass of the main prediction model. Such bypassing significantly cuts computational overhead for these numerous objects while preserving cache performance.

Hot-Oriented Sampling Strategy. Traditional heuristic-based Sampling Strategies often fail to adequately protect hot objects (e.g., LRU is not scan-resistant and cannot retain hot objects), thereby hindering miss-ratio minimization. We propose a fundamentally *hotness-oriented* Sampling Strategy which prioritizes identifying and preserving the hot working set. This requires accurately assessing an object's "hotness" using multiple features, rather than relying on single-dimensional heuristics. Integrating dynamic techniques, such as quick demotion, further enhances this by rapidly identifying cooling objects as non-hot eviction candidates. A hotness-oriented approach thus aims to both safeguard the working set for low miss ratios and efficiently provide suitable candidates for prediction or eviction.

3 Design of SL-Cache

Based on our observations and motivation, this section presents a new learning eviction algorithm, **S**elective-**L**earning **Cache** (which we term "**SL-Cache**"). We begin by outlining the structure of SL-Cache and then detail all of its components.

3.1 SL-Cache Framework

Figure 2 illustrates the structure of SL-Cache. SL-Cache processes incoming requests by first performing a lookup in the Hotness-Oriented Three-Layer Structure(§3.2) to determine whether the request hit. Cache hits and metadata movement during evictions trigger the data collection process, gathering essential information for workload adaptation. Collected access information fuels Learning Framework (§3.4) for periodic model retraining, pre-computation for one-hit wonders, and parameter adjustments. When the cache is full, the selective sampling strategy (§3.3) interacts with the cache structure to select eviction candidates and triggers comparison-based eviction (§3.5) to evict objects that are likely to be accessed in the future.

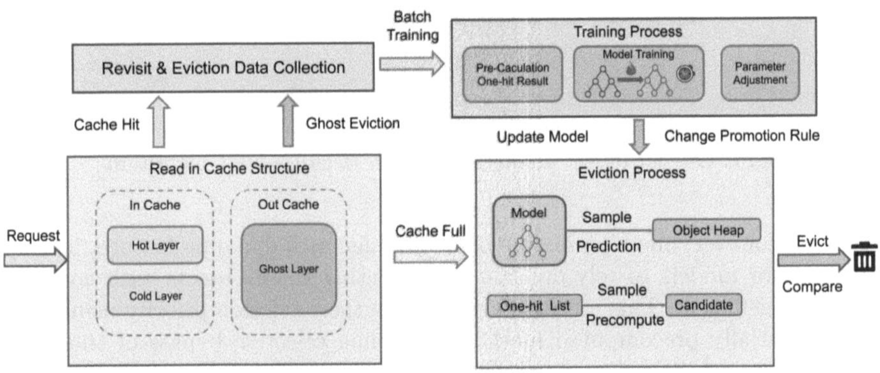

Fig. 2. Overview of SL-Cache.

3.2 Hot-Oriented Cache Structure

The cache structure governs object management and candidate selection, thereby directly affecting performance. SL-Cache employs a hotness-oriented, three-layer structure:

- **Hot Layer** Contains frequently and recently accessed data.
  - *An list*: A FIFO list holding newly arrived objects, allocated a small and dynamic portion of the cache.
  - *Am list*: A FIFO list storing objects promoted due to high frequency and recency. Its size adapts dynamically.
- **Cold Layer**: Contains less frequently accessed multi-hit objects and one-hit wonders.
  - *Candidate list*: A LRU list holding multi-hit objects(requested more than once after insertion) demoted from the Hot Layer that are not considered hot.

- *A1 list*: A FIFO list holding all one-hit wonders demoted from the *An list*.
- **Ghost Layer**: Only stores the metadata of evicted objects, including keys and features.
  - *Ghost list*: A FIFO list storing metadata of evicted objects, serving as the source for training data.

Object movement follows these rules upon request processing:

(1) Request Handling: On a cache hit, metadata of the requested object is updated. Promotion might occur if hit in the Cold Layer (see rule 2). On a cache miss, the new object is directly inserted into the head of the *An list*.

(2) Promotion on Hit: If a hit occurs in the cold layer, the object is promoted. If its frequency exceeds *freq boundary*, it moves to the head of the *Am list*; otherwise, to the head of the *Candidate list*. Hits in hot layer do not change the object's position within its list.

(3) Insertion/Eviction on Miss: A miss inserts the new object into *An list*. If the cache is full, the eviction process (§3.5) is triggered before insertion.

(4) Demotion from *An list*: If the *An list* exceeds the capacity of An, the tail object is demoted:

- To *Am list* head if its frequency > *freq boundary*.
- To *A*1 *list* head if it is a one-hit wonder.
- To *Candidate list* head otherwise.

The parameter *freq boundary* is dynamically set based on the P99 frequency of recently evicted objects, serving as an adaptive threshold that reflects current workload characteristics to quickly demote potentially non-hot (i.e., not classified as hot) new objects.

3.3 Adaptive Adjustment and Sampling Strategy

The effectiveness of object-level learning depends not only on prediction accuracy but critically on the sampling strategy used to select candidates for prediction. An inadequate strategy that fails to protect hot objects or identify cold ones promptly can undermine the entire learning process.

SL-Cache integrates object adjustment (promotion/demotion) with the sampling strategy adaptively. The sampling aims to identify non-hot candidates for prediction while shielding the hot working set. Feature statistics from evicted objects inform the adjustment strategy, specifically through periodic recalculations of *freq boundary* based on eviction statistics, thereby refining the definition of hotness over time.

Sampling distinguishes between new and old objects:

- **New Objects (from An tail demotion):** Objects whose frequency exceeds the dynamically updated *freq boundary* are considered hot and moved to *Am list*. Others are considered non-hot candidates, sampled for prediction (if learning is active), and moved to the *A1 List* or the *Candidate list*.

- **Old Objects (from Cold Layer and Am tail):** The sampling strategy considers all objects in the *Candidate list* as potential candidates (as they are already deemed non-hot by frequency). From the *Am List*, only a small fraction (e.g., the tail 1%, empirically determined) representing objects with extremely low recency, along with objects having the lowest hit count (indicating lower combined hotness, see below), are sampled. This focuses prediction effort while protecting the core hot set in *Am List*.

The maximum number of objects sampled in a batch (from both new and old sources) is capped (default: 1% of cache size) to control overhead.

Hotness Metrics. To dynamically quantify and manage object hotness within the *Am list*, we introduce two coupled metrics: hit count and demote count. (1) The hit count is incremented by 1 upon each access, which reflects the frequency of recent access. (2) The demote count is incremented by 1 during periodic demotion scans, which represents the age of the object. An access will reset the demote count to 0. The demotion decay scan periodically adjusts the hit count by dividing it by (demote count + 1) and then adding demote count + 1. This processes is triggered when eviction process(§3.5) needs new candidates.

This dual mechanism combines frequency tracking (hit-count increments) with recency-based decay (decay is driven by the demote count, which increases over time without hits). Consequently, frequently and recently accessed objects retain a high hit count, thus adaptively reflecting their current hotness. Objects with the lowest residual hit count are prioritized for sampling from the *Am list*.

Rationale for Dynamic Demotion. The SL-Cache design premises its demotion logic on two distinct, orthogonal triggers derived from state-of-the-art frequency-based algorithms (TinyLFU [6], S3FIFO [21]): (1) a frequency saturation event (as in TinyLFU's counter halving), and (2) an eviction-driven necessity (as in S3FIFO's count reduction). We integrate both conditions into our hot-oriented design to capture complementary workload dynamics. Furthermore, the combined use of hit count and demote count introduces a controlled recency decay, ensuring that an object's hotness rapidly diminishes (e.g., an object with maximum *hit count* = 64 decays through 32 → 8 → 1 → 0 over a maximum of four non-re-accessed demotion steps), thereby accelerating the demotion of stale objects. Critically, reaccessing an object resets its *demote count* and potentially increases its *hit count* (upon a hit), providing stronger hotness reaffirmation than observed in algorithms like S3FIFO. This integrated approach combines frequency and recency factors to robustly safeguard the hot-working set and expedite the degradation of non-hot objects.

3.4 Learning Architecture

The learning architecture is responsible for model training, prediction (for multi-hit objects), and managing training data.

Model and Features. SL-Cache employs Gradient-Boosting Decision Trees (GBDT) for prediction, chosen for its inherent robustness to feature sparsity and its independence of feature normalization. The model is trained to predict the duration until the next object access.

- *Label* The dependent variable is the logarithm of the inter-arrival time: $\log(\text{next\_access\_time} - \text{predict\_time})$.
- *Loss Function* Training is optimized using the Mean Squared Error (MSE) as the L_2 loss function.
- *Features* The model utilizes three categories of features to capture object usage patterns: Frequency, Age, and Reuse distances.

The GBDT framework naturally accommodates missing initial reuse distance values.

Training Data Collection. Efficiently collecting representative training data is crucial. SL-Cache employs sampling and dynamic boundary adjustment to balance training data quality with storage memory overhead.
Dynamic Boundary Setting. To limit memory usage while capturing relevant access patterns, training data is collected only for object re-accesses within a dynamically adjusted boundary, which is tracked in the *Ghost list.* The large *Ghost list* captures a longer pattern of reuse. The total number of object metadata entries maintained in the *Ghost list* is set to *object boundary* times the cache size (in objects). *object boundary* (default 2, limit to 6 to constrain the memory usage) is adjusted based on the utility of the most recently added segment (the last "cache size" entries) of the *Ghost list.* If this segment yields a sufficiently high number of useful training samples (measured by hits within it relative to total ghost hits or cache hits), the *object boundary* is expanded by 1 until it reaches the limit.
Data Sampling. For each request landing within the current boundary, SL-Cache randomly samples one object's metadata to potentially track for training. If the sampled object is accessed again within the boundary, the measured next-access distance constitutes a training sample. If it gets evicted beyond the boundary, its distance is recorded as a large capped value (boundary size + age). Training is triggered periodically when a batch of new samples (*batch size*, default setting is 65536 based on experiment result) is collected.

3.5 Eviction Process

SL-Cache employs two eviction modes: a heuristic mode used before training and a learning-based mode. The selective nature primarily refers to bypassing predictions for one-hit wonders.
Heuristic Eviction. When ML predictions are not used, eviction proceeds heuristically based solely on the cache's hot layer. When the cache is full, if *An list* exceeds the default limit, the tail of *An list* is removed, it will be directly evicted if it is a one-hit wonder, otherwise it enters the*Am list* and triggers the demotion process to find an object with the lowest hit count to evict.

Learning-Based Eviction. When the ML model is active, the sampling strategy (§3.3) identifies eviction candidates. For multi-hit candidates, we feed their features into the GBDT model to predict the next access time. These candidates are inserted into a max-heap prioritized by their predicted time. The heap is periodically refreshed. For one-hit wonder candidates, the object at the tail of the *A1 list* is selected. The predicted next access time is retrieved from pre-computed results based on its current age. The predicted time of the candidate at the top of the max-heap is compared with the predicted time of the oldest one-hit wonder. The object with the longer predicted time until next access is evicted.

This process implicitly controls the residency of one-hit wonders based on their predicted utility relative to sampled multi-hit objects, rather than using a fixed partition. The prediction ratio for SL-Cache is dynamic: it is 2 for heap evictions but drops to 0 for one-hit wonder evictions (since they use pre-computed results), resulting in an overall ratio <2, which is the static boundary set by the state-of-the-art algorithm, 3L-Cache.

4 Performance Evaluation

In this section, we evaluate SL-Cache to address two key questions:

- How does SL-Cache compare to state-of-the-art eviction algorithms in terms of miss ratio and throughput?
- How does SL-Cache achieve a lower miss ratio and higher throughput?

4.1 Experiment Setup

Traces and Cache Size. We use a wide variety of traces representing a diverse set of workloads from three dataset sources (see Table 1). We select traces with more than 5 million requests and a compulsory miss ratio below 50% to comprehensively evaluate the performance of the eviction algorithm. We select 1% of the object count (5% for SQL Server to avoid an extremely high miss ratio) as the small cache size and 10% of the object count as the large cache size on each trace to evaluate the performance of the eviction algorithm across different cache sizes.

Baselines. In this paper, we compare SL-Cache with 10 state-of-the-art cache eviction algorithms:

- *Heuristic Algorithms*: Hill-Cache [10], S3-FIFO [21], SIEVE [24], and TinyLFU [6].
- *Distribution Learning Algorithm*: LHD [2].
- *Group Learning Algorithm*: GL-Cache [20].
- *Pattern Learning Algorithms*: LeCaR [18] and CACHEUS [15].
- *Object Learning Algorithms*: LRB [16] and 3L-Cache [25].

Table 1. Description of traces

Datasets	Object(10^6)	Request(10^6)	Type	Traces	Description
MSR	242	1355	I/O	5	I/O traces in an enterprise data center for one week
SQL Server	61	398	OLTP	8	SQL Server trace collected at Microsoft using the event tracing for Windows framework.
YCSB	356	4871	Key-Value	6	Key-Value trace widely utilized benchmark in numerous key-value stores

Metrics. We measure the performance of eviction algorithms by miss ratio and throughput. For the miss ratio, since the miss ratio varies in a wide range across traces, we present the miss ratio reduction compared to LRU: $\frac{MR_{LRU} - MR_{algo}}{MR_{LRU}} \times 100\%$, where MR stands for miss ratio. If an algorithm has a higher miss ratio than LRU, we present LRU's reduction in miss ratio relative to the algorithm, rather than bounding it between -100% and 100%. For throughput, we measure the throughput of an algorithm relative to LRU: $\frac{Throughput_{algo}}{Throughput_{LRU}}$.
Implementation. We implemented SL-Cache and the state-of-the-art eviction algorithms in libCacheSim [1]. All experiments are conducted on a server with two Intel(R) Xeon(R) Gold 6348 CPUs at 2.60 GHz, each with 28 cores and 2 threads per core, 256 GB of DRAM, and a 4 TB WD_BLACK AN1500 SSD. The server runs 64-bit Ubuntu 22.04.4 LTS. The file system is ext4.

4.2 Miss Ratio

In this section, we compare the miss ratio of different cache eviction algorithms. We evaluate cache sizes ranging from small to large across all workloads mentioned above. Figure 3 presents part of these results with miss ratio comparison results.

SL-Cache demonstrates improvements in miss ratio relative to other state-of-the-art algorithms across all workloads, for both small and large cache sizes.

For a small cache size, SL-Cache achieves the lowest miss ratio for 42.9% of the traces. For the closest competitors, LRB achieves the lowest miss ratio for 21.5% of the traces. On average, SL-Cache reduces the miss ratio by 13.61% across all traces compared to LRU. For other competitors, 3L-Cache, S3FIFO, LRB, TinyLFU, CACHEUS, LHD, Hill-Cache, SIEVE, LeCaR, and GL-Cache reduce the miss ratio by 12.59%, 11.94%, 11.82%, 11.30%, 8.66%, 8.51%, 6.67%, 3.69%, and 0.02%, respectively, on average across all traces compared to LRU. For workloads with small cache sizes, Hill-Cache outperforms object-level learning algorithms because it leverages hill-climbing to balance recency and frequency and triggers adjustments more frequently when the cache size is small.

For large cache sizes, SL-Cache achieves the lowest miss ratio across 73.6% of the traces. For the closest competitors, Hill-Cache achieves the lowest miss ratio for 10.9% of the traces. On average, SL-Cache reduces the miss ratio by 10.20% across all traces compared with LRU. For other competitors, 3L-Cache, S3FIFO, LRB, TinyLFU, CACHEUS, LHD, Hill-Cache, SIEVE, LeCaR, and GL-Cache reduce the miss ratio by 9.07%, 8.49%, 6.37%, 7.11%, 5.68%, 1.94%, 6.68%, 0.73%, and 0.35%, respectively, on average across all traces compared to LRU. For some workloads with large cache sizes, S3FIFO is the most competitive heuristic, as it uses rapid demotion via a small FIFO filter to retain more hot objects.

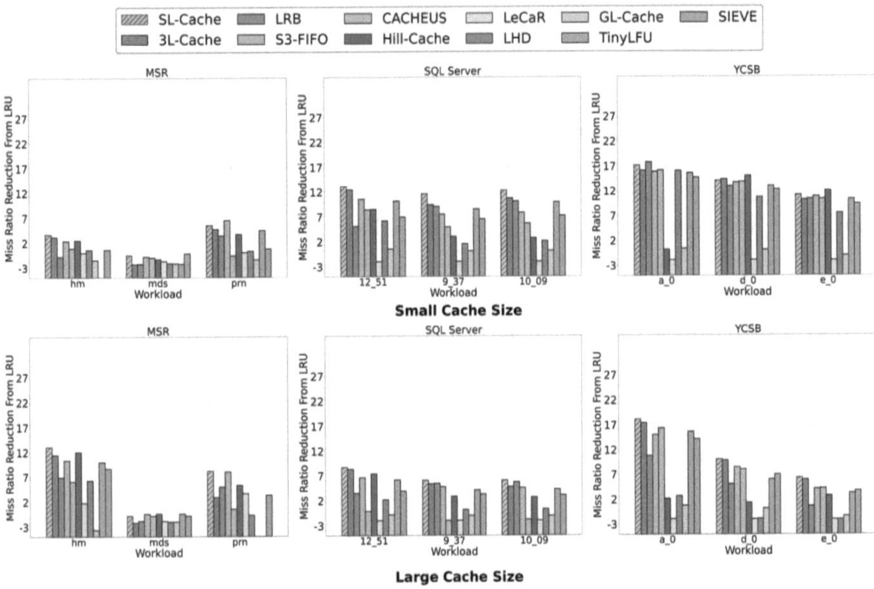

Fig. 3. Miss ratio reduction (**higher is better**) of different cache eviction algorithms with different cache sizes across workloads.

It is worth noting that LHD, LeCaR, CACHEUS, and GL-Cache cannot outperform state-of-the-art heuristic algorithms with quick demotion techniques such as S3FIFO, TinyLFU, and Hill-Cache across a large portion of workloads. This is because their coarse-grained predictions and reliance on specific heuristic features may not capture access patterns well enough to improve performance.

4.3 Throughput

Computational efficiency is a critical determinant for the practical deployment of learning-based eviction algorithms in high-performance systems. To provide a comprehensive evaluation, we measure the throughput of various algorithms relative to the LRU baseline across all workloads.

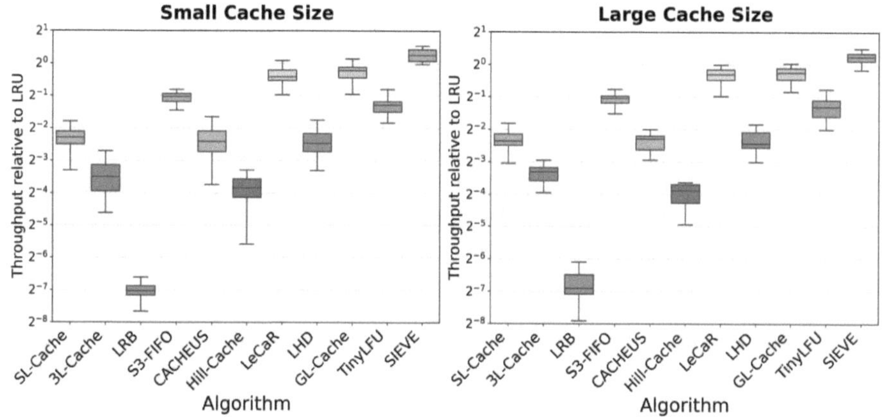

Fig. 4. Throughput comparison of different cache eviction algorithms with varying cache sizes across workloads.

Figure 4 illustrates the throughput distribution using boxplots for both small and large cache sizes. Our results demonstrate that SL-Cache successfully bridges the throughput gap between sophisticated object-level learning and simple heuristics.

At a small cache size, SL-Cache achieves a throughput significantly higher than that of existing object-level learning models. Specifically, SL-Cache achieves an average throughput approximately 2.3× higher than 3L-Cache and nearly 20× higher than LRB. Remarkably, by bypassing redundant model inferences for one-hit wonders, SL-Cache increases its throughput to a level comparable to coarse-grained learning algorithms such as CACHEUS and LHD, despite operating at a much finer granularity.

At a large cache size, SL-Cache's performance advantage remains robust. While the throughput of LRB drops drastically to approximately 2^{-7} relative to LRU, SL-Cache maintains a practical performance profile. Furthermore, the narrow range of SL-Cache's boxplots across diverse traces indicates its high predictability and robustness against varying access patterns.

4.4 Ablation Study

In this section, we analyze why SL-Cache reduces both latency and miss ratio by examining the improvements in the techniques employed. We analyze the performance improvement primarily in two areas: one-hit prediction bypass and a hot-oriented sample algorithm.

One-Hit Prediction Bypass. This technique uses pre-computed approximation results for one-hit object prediction and avoids heavy computation on one-hit object prediction to reduce computation overhead. We compare the prediction ratio across the entire trace of the object-level learning algorithms, including

SL-Cache, 3L-Cache, and LRB. We analyze the prediction ratio for small and large caches during trace execution in MSR and measure it online. LRB statically samples 64 objects per eviction, and the prediction ratio is fixed at 64. SL-Cache set the prediction ratio to 2 for both the sample and eviction algorithms. However, the prediction ratio may change if all sample objects are requested after the sample process. Although HALP is not applicable to our experiments, we note that it produces 3 rankwise predictions per eviction, with a prediction ratio of 3. For SL-Cache, because one-hit objects are bypassed during model prediction, their eviction does not require prediction. We set the prediction ratio for objects rather than for one-hit; hence, the total prediction ratio may be any value less than 2, depending on the one-hit object eviction rate in the workload's access pattern. As shown in Fig. 5, the prediction ratio of SL-Cache is constantly less than 2 and does not exceed 1 for most periods of the trace request for both small cache size and large cache size. As the most competitive algorithm with the lowest prediction ratio, 3L-Cache primarily maintains a prediction ratio of 2, which remains 100× higher than SL-Cache's maximum, underscoring SL-Cache's priority in reducing prediction.

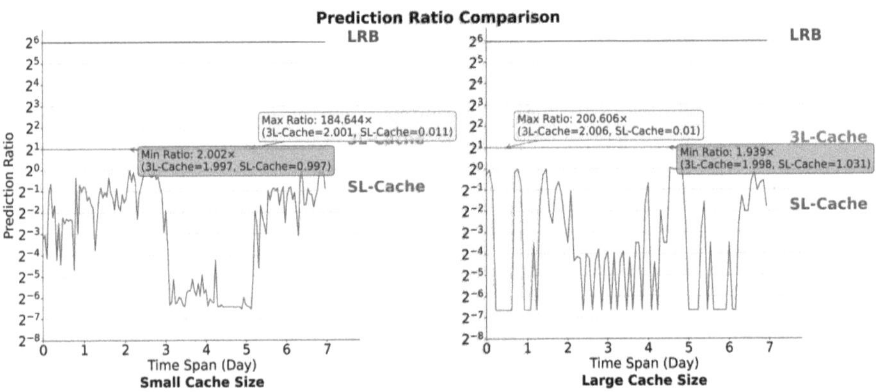

Fig. 5. Prediction ratio result for object-level learning algorithm during a trace in MSR.

Hot-Oriented Sampling Strategy. This technique filters out hot objects from the sample and detects cold objects with quick demotion. We design various sampling strategies to evaluate the performance of SL-Cache. (1) SL-Cache-clock is a variant of SL-Cache which uses 1 1-bit clock rather than a hit count to quickly demote hot objects. (2) SL-Cache-4 is another variant that sets the limit of hit count to 4 to test the necessity of high-frequency design for hit count. (3) SL-Cache-sd fixes the portion of a list in the cache structure and promotes all objects accessed more than once, which verifies the structure adjustment of SL-Cache.

As shown in Table 2, SL-Cache achieves a lower miss ratio compared to the other variants of SL-Cache. Although the variants achieve nearly the same miss ratio as SL-Cache, they fail to compete with SL-Cache across all workloads. This is because SL-Cache leverages the most comprehensive features to detect object hotness and adapt quickly to workload changes, providing high-quality sample results for prediction and eviction. Hence, SL-Cache outperforms other competitors across all workloads.

Table 2. Comparison of different sample strategies of SL-Cache.

Algorithm	MSR	SQLSever	YCSB	Total
SL-Cache	**9.1846**	**13.4731**	**17.4726**	**13.3768**
SL-Cache-clock	7.9142	11.5297	15.7145	11.7195
SL-Cache-sd	8.5914	13.4648	17.4568	13.1777
SL-Cache-4	8.2855	13.4070	16.9910	12.8945
Large Cache Size				
Algorithm	MSR	SQLSever	YCSB	Total
SL-Cache	**6.3961**	**7.8858**	**16.4511**	**10.2443**
SL-Cache-clock	3.8295	6.7974	14.6193	8.4154
SL-Cache-sd	5.2162	7.9241	16.2944	9.8116
SL-Cache-4	4.5898	7.9161	15.7075	9.4045

5 Conclusions

In this paper, we presented SL-Cache, a novel object-level learning eviction algorithm grounded on two key insights: selective prediction and hotness-aware sampling. SL-Cache identifies opportunities to bypass predictions for a significant number of one-hit objects and underscores the critical role of hotness-oriented sampling in object-level learning. Extensive evaluation showed that SL-Cache achieved the lowest miss ratio across diverse workloads and the highest throughput among object-level learning algorithms, comparable to coarse-grained learning approaches.

Future work will focus on three aspects. First, we will investigate the feasibility of using approximate predictions to achieve higher throughput while maintaining miss-rate stability. Second, we will consider implementing SL-Cache on some open-source platforms, e.g., RocksDB [5], and using additional workloads to evaluate SL-Cache. Third, the current SL-Cache is designed to accelerate query performance for web objects, such as those in content delivery networks [17,24]. We will extend SL-Cache to a broader range of applications by adding effective policies for dirty-object eviction.

Acknowledgments. This paper is supported by the National Science Foundation of China (grant no. 62072419). Peiquan Jin is the corresponding author.

References

1. libcachesim: a high-performance cache simulation. https://github.com/1a1a11a/libCacheSim. Accessed 21 Aug 2025
2. Beckmann, N., Chen, H., Cidon, A.: LHD: Improving cache hit rate by maximizing hit density. In: NSDI, pp. 389–403 (2018)
3. Belady, L.A.: A study of replacement algorithms for a virtual-storage computer. IBM Syst. J. **5**(2), 78–101 (1966)
4. Berger, D.S.: Towards lightweight and robust machine learning for CDN caching. In: HotNet, pp. 134–140 (2018)
5. Dong, S., Kryczka, A., et al.: Evolution of development priorities in key-value stores serving large-scale applications: the RocksDB experience. In: FAST, pp. 33–49 (2021)
6. Einziger, G., Friedman, R., Manes, B.: TINYLFU: a highly efficient cache admission policy. ACM Trans. Storage **13**(4), 1–31 (2017)
7. Jiang, S., Chen, F., Zhang, X.: Clock-pro: an effective improvement of the clock replacement. In: ATC, pp. 323–336 (2005)
8. Jin, P., Ou, Y., Härder, T., Li, Z.: AD-LRU: an efficient buffer replacement algorithm for flash-based databases. Data Knowl. Eng. **72**, 83–102 (2012)
9. Johnson, T., Shasha, D.: 2Q: a low overhead high performance buffer management replacement algorithm. In: VLDB, p. 439–450 (1994)
10. Li, Y., Hu, H., Lei, C., et al.: Hill-cache: adaptive integration of recency and frequency in caching with hill-climbing. In: ICDE, pp. 3947–3960 (2024)
11. Li, Z., Jin, P., Su, X., Cui, K., Yue, L.: CCF-LRU: a new buffer replacement algorithm for flash memory. IEEE Trans. Consum. Electron. **55**(3), 1351–1359 (2009)
12. Megiddo, N., Modha, D.S.: ARC: a self-tuning, low overhead replacement cache. In: FAST (2003)
13. O'neil, E.J., O'neil, P.E., Weikum, G.: The LRU-K page replacement algorithm for database disk buffering. ACM SIGMOD Record **22**(2), 297–306 (1993)
14. Ou, Y., Härder, T., Jin, P.: CFDC: a flash-aware buffer management algorithm for database systems. In: ADBIS, pp. 435–449 (2010)
15. Rodriguez, L.V., Yusuf, F., Lyons, S., et al.: Learning cache replacement with CACHEUS. In: FAST, pp. 341–354 (2021)
16. Song, Z., Berger, D.S., Li, K., et al.: Learning relaxed belady for content distribution network caching. In: NSDI, pp. 529–544 (2020)
17. Song, Z., Chen, K., Sarda, N., et al.: HALP: Heuristic Aided Learned Preference eviction policy for YouTube content delivery network. In: NSDI, pp. 1149–1163 (2023)
18. Vietri, G., Rodriguez, L.V., Martinez, W.A., et al.: Driving cache replacement with ML-based LeCaR. In: HotStorage (2018)
19. Wang, X., Jin, P., Yuan, Y.: SEMU: concurrency-optimized high-performance cache management for key-value caches. IEEE Trans. Comput. **75**(2), 734–747 (2026)
20. Yang, J., Mao, Z., Yue, Y., et al.: GL-Cache: group-level learning for efficient and high-performance caching. In: FAST, pp. 115–134 (2023)

21. Yang, J., Zhang, Y., Qiu, Z., et al.: FIFO queues are all you need for cache eviction. In: SOSP, pp. 130–149 (2023)
22. Yuan, Y., Jin, P.: Learned buffer management: a new frontier: work-in-progress. In: CODES/ISSS, pp. 25–26 (2021)
23. Yuan, Y., Jin, P.: Learned buffer replacement for database systems. In: DSDE, pp. 18–25 (2022)
24. Zhang, Y., Yang, J., Yue, Y., et al.: SIEVE is simpler than LRU: an efficient turn-key eviction algorithm for web caches. In: NSDI, pp. 1229–1246 (2024)
25. Zhou, W., Niu, Z., Xiong, Y., et al.: 3L-CACHE: low overhead and precise learning-based eviction policy for caches. In: FAST, pp. 237–254 (2025)

AdaSlice: Hotness-Aware and Adaptive Slicing for Eviction Algorithms in Database Buffer Manager with Tiered Memory

Shikai Tan[1], Zhipeng Tan[1](✉), Lan Lu[1], Zijian Zhang[1], Renzhi Xiao[3], Yutai Shu[1], Ying Yuan[1], Jinlong Wang[1], Wenjie Qi[1], and Yuan Feng[2]

[1] Wuhan National Laboratory for Optoelectronics, Huazhong University of Science and Technology, Wuhan, China
{shikaitan,tanzhipeng,lu_lan,zjzhang,shusyt,yuanying,jinlongwang, hustqwj}@hust.edu.cn

[2] Wuhan Dameng Database Co., Ltd., Wuhan, China
fengyuan@dameng.com

[3] School of Software Engineering, Jiangxi University of Science and Technology, Nanchang, China
rzxiao@jxust.edu.cn

Abstract. Tiered memory leverages emerging interconnect technologies (e.g., CXL) and cheaper alternatives (e.g., PM) to scale databases cost-efficiently. Skewed access patterns are prevalent in DBMS workloads, where a small fraction of tuples dominate most accesses. Therefore, tuple-granularity data migration between fast and slow memory outperforms page-granularity approaches. Tuple-granularity eviction algorithms improve fast memory access ratio (FAR), thereby boosting performance, but incur substantial memory overhead when tracking hotness for ubiquitous small tuples. Grouping adjacent tuples into a **slice** for hotness tracking can amortize memory overhead, but existing slice-granularity eviction algorithms suffer from poor hot/cold identification accuracy due to rigid page-aligned slicing.

This paper proposes **AdaSlice**, a hotness-aware and adaptive slicing mechanism for tuple eviction algorithms in database buffer manager with tiered memory. The **key idea** is to drive slice splitting and merging based on four thresholds of size and hotness dispersion, thereby achieving a favorable trade-off between hot/cold identification accuracy and hotness metadata overhead. We further employ simulated annealing to adaptively derive near-optimal thresholds under diverse workloads and configurations. We integrate AdaSlice into six eviction algorithms and implement them within the tiered buffer manager. Experimental results show that AdaSlice increases FAR by 30.5% and improves database throughput by 20.6%.

Keywords: Tiered Memory · Database Buffer Management · Eviction Algorithm · Metaheuristic Optimization Algorithm

H. Jung et al. (Eds.): DASFAA 2026, LNCS 16535, pp. 606–622, 2026.
https://doi.org/10.1007/978-981-92-0363-5_37

1 Introduction

Tiered memory [1] leverages emerging interconnect technologies such as Compute Express Link (CXL) [2] and cheaper alternatives such as persistent memory (PM) [3] to enable cost-efficient scaling for database systems [2]. Skewed access patterns are invariably prevalent in real-world DBMS workloads, where a small number of hot tuples are accessed much more frequently than the remaining warm and cold tuples [4,5]. As a result, prior research has proposed a tuple-granularity tiered buffer manager [6], which migrates data between fast and slow memory at tuple granularity, thereby reducing read/write amplification in slow memory, improving caching efficiency in fast memory, and significantly enhancing performance under skewed access patterns. Because there still exists a performance gap of more than twice between fast and slow memory [1], tuple eviction algorithms are crucial for improving the fast memory access ratio (**FAR**) and thereby enhancing overall database performance [6]. In this paper, we define FAR as the proportion of tuple accesses to fast memory out of all tuple accesses to both fast and slow memory, and use it to measure eviction algorithm's accuracy.

Small tuples are ubiquitous in DBMS workloads [7] (e.g., the average tuple size is only 73 bytes under the TPC-C workload). Therefore, performing hot/cold identification at tuple granularity incurs substantial memory overhead [8], as the hotness metadata stored alongside each tuple in fast memory occupies space that could otherwise hold data, resulting in a 21.3% to 31.5% drop in FAR. Different algorithms define hotness differently. For example, in LRU [10], **hotness** corresponds to a tuple's position in the LRU list, and the **hotness metadata** consists of its forward and backward pointers (totalling 16 bytes per tuple in fast memory). We will explain these definitions in Sect. 3. A natural solution is to group tuples with consecutive keys into a **slice** to amortize the memory overhead of hotness metadata. However, existing slice-granularity eviction algorithms [7,16,17] yield only a 2.9% to 6.7% improvement in FAR, as their rigid page-aligned slicing induces excessive hotness variation among tuples within a slice, especially under workloads such as TPC-C where several adjacent tuples are accessed simultaneously, as detailed in Sect. 3.

In this paper, we propose **AdaSlice**, a hotness-aware and adaptive slicing mechanism for tuple eviction algorithms in database buffer manager with tiered memory. The **key idea** is to drive slice splitting and merging via a threshold set *<SplitSize, MergeSize, SplitDisp, MergeDisp>* dynamically, so that each slice's size and hotness **dispersion** (the hotness variation of its constituent tuples) remain within desirable bounds. When tuples exhibit uniform hotness, adjacent small slices are merged to reduce the memory overhead for tracking hotness; when tuple hotness becomes skewed, large slices are split to improve the accuracy of hot/cold identification. In this way, we strike an improved balance between the *high accuracy, high overhead* of tuple-granularity algorithms and the *low accuracy, low overhead* of existing slice-granularity algorithms. We further employ the simulated annealing (SA) [18] algorithm to adaptively derive a near-optimal threshold set under varying workloads and configurations, with the objective of maximizing FAR. With the above strategies, we dynamically approach the *sweet spot* (i.e., achieving relatively high accuracy with relatively low memory overhead) in the trade-off between hot/cold identification accuracy and hotness metadata overhead, thereby improving FAR and ultimately enhancing database performance. The main contributions of the paper are summarized as follows:

1. We propose AdaSlice, which splits large slices with divergent hotness and merge small slices with uniform hotness, aiming to achieve a favorable trade-off between hot/cold identification accuracy and hotness metadata overhead.
2. We introduce a threshold set <*SplitSize, MergeSize, SplitDisp, MergeDisp*>, and validate that choosing appropriate threshold values produces slices of suitable size and uniform hotness, thereby improving FAR.
3. We propose a threshold tuning strategy based on simulated annealing, capable of adaptively finding a near-optimal threshold set to maximize FAR under diverse workloads and configurations.
4. We devise hotness copying and merging strategies for six eviction algorithms and integrate AdaSlice into them. This yields six slice-granularity eviction algorithms that dynamically adapt to achieve a favorable trade-off. We then implement these algorithms in our tuple-granularity tiered buffer manager.
5. We evaluate AdaSlice using TPC-C and YCSB workloads, demonstrating that AdaSlice effectively enhances FAR and improves database throughput.

2 Background and Related Work

2.1 Background

Tiered Memory. Tiered memory [1] comprises DRAM and emerging byte-addressable memory devices, e.g., CXL-attached memory [1] and persistent memory [3] (PM). The CPU accesses the local DRAM as **fast memory**, while accessing the remote **slow memory** directly via load/store instructions using protocols such as Compute Express Link (CXL) [2] or DDR-T [1], achieving latencies on the order of hundreds of nanoseconds. Because slow memory has a lower per-GB cost, tiered memory can extend capacity for database systems cost-efficiently [6]. Figure 1a illustrates the traditional page-granularity buffer manager architecture of database systems with tiered memory.

Skewed Access Pattern. Skewed access patterns [4,5] are invariably prevalent in real-world DBMS workloads, where a small number of hot tuples are accessed much more frequently than the remaining warm and cold tuples. Meanwhile, hot tuples are co-located with warm and cold ones on pages, as shown in Fig. 1a. Consequently, migrating data between fast memory and slow memory at page-granularity can severely degrade the utilization of fast memory, increase read and write amplification in slow memory, and degrade overall system performance.

Tuple-Granularity Tiered Buffer Manager. Tiered-Indexing [4,5], Bf-Tree [7], Anti-Caching [10], Siberia [8], and TreeLine [12] employ tuple-granularity caching to reduce read/write amplification and improve caching efficiency. Building upon Tiered-Indexing [4,5], Hao et al. [6] propose a tuple-granularity buffer manager with tiered memory (as illustrated in Fig. 1b), in

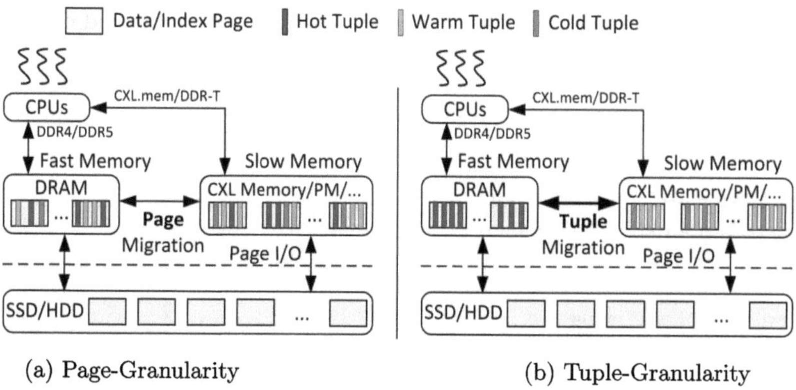

Fig. 1. Page-granularity vs. tuple-granularity buffer manager with tiered memory.

which fast memory and slow memory each maintain their own B^+-tree index. Because slow memory supports byte-addressable access, tuples can be migrated directly between fast memory and slow memory (rather than entire pages), thereby reducing read/write amplification. Furthermore, by caching only hot tuples in fast memory, the utilization of fast memory resources is improved, thus reducing overall memory cost. Their experiments indicate that the tuple-granularity tiered buffer manager achieves substantial performance improvements over a page-granularity design.

2.2 Related Work

Tuple Eviction Algorithms. Zhou et al. [4,5] and Hao et al. [6] propose a lightweight Clock algorithm, which walks through the hot B^+-tree in key order and evicts tuples without the reference bit on during eviction. TreeLine [12] and Anti-Caching [10] use the Clock and LRU algorithms, respectively, to identify hot and cold tuples. Siberia [8] records tuple access timestamps in logs and then applies an offline exponential smoothing algorithm [9]; however, it requires more computing resources and cannot timely adapt to workload changes [4,5]. Although most of the above algorithms target tuple eviction between memory and disk, they can be directly applied to eviction between fast and slow memory tiers. In fact, many commonly used page replacement algorithms (e.g., WATT [13]) or web cache eviction algorithms (e.g., Sieve [15], Hyperbolic [14]) can serve directly as a tuple eviction algorithm, provided that the unit of hotness tracking is shifted from pages or cache objects to tuples.

MiniPage. Bf-Tree [7], MG-Buffer [16], and AMG-Buffer [17] use mini-pages as the unit for hotness tracking and eviction. Each mini-page contains all tuples promoted from the same original page. Bf-Tree evicts cold mini-pages using the Second-Chance algorithm, while MG-Buffer and AMG-Buffer use LRU. Hymem [11] and Spitfire [3] promote accessed cache-lines from NVM pages into corresponding mini-pages in DRAM, and use Clock to evict cold mini-pages.

In summary, existing studies conduct hot/cold identification and eviction at either the tuple or slice granularity. However, tuple-granularity eviction incurs per-tuple overhead for hotness tracking and slice-granularity eviction suffers from poor hot/cold identification accuracy. We will analyze these problems in Sect. 3.

Table 1. Memory overhead of hotness metadata for six eviction algorithms.

Eviction algorithm	Memory overhead of hotness metadata (byte)
LRU	16
LRU2	17
Second-Chance	17
Sieve	17
Hyperbolic	12
WATT	17

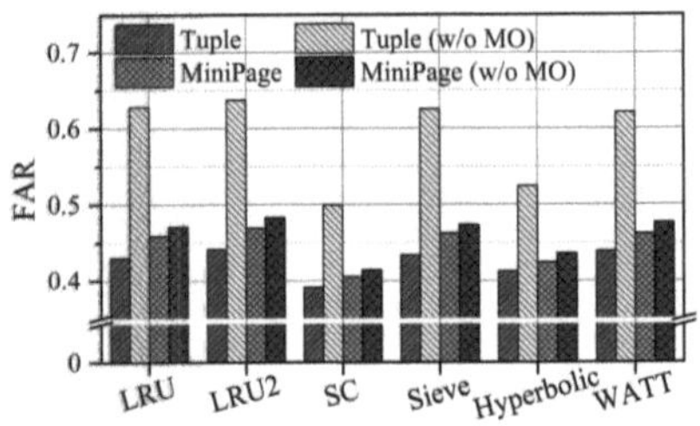

Fig. 2. FAR of tuple and mini-page granularity eviction algorithms.

3 Motivation

We implement LRU [10], LRU2 [13], Second-Chance [7] (referred to as **SC** in the figures), Sieve [15], Hyperbolic [14], and WATT [13] algorithms in our tuple-granularity tiered buffer manager and evaluate their fast memory access ratio (FAR) under the TPC-C workload. From these experiments, we uncover two shortcomings in existing approaches.

Observation 1: Tuple-granularity eviction algorithms incur substantial hotness metadata memory overhead. Storing hotness metadata alongside each tuple in fast memory occupies space that could otherwise hold data, thereby incurring the memory overhead for eviction algorithms shown in Table 1. In this paper, we define the **hotness** of an element (tuple or slice) as information such as its position in a list or queue, or its access timestamps. For example, in LRU and LRU2, elements closer to the head of the list are hotter; in Second-Chance and Sieve, aside from the position in the queue, elements whose *accessed bit* is set are hotter; in Hyperbolic and WATT, elements with newer access timestamps are hotter. The **hotness metadata** is used to track this hotness information: LRU, LRU2, Second-Chance and Sieve all require two 8-byte pointers to record predecessor and successor elements (and in addition, LRU2, Second-Chance and Sieve require 1 byte to record the *accessed bit*, etc.); Hyperbolic and WATT require recording information such as access timestamps, which we will further explain in Sect. 4.5. On the other hand, small tuples are ubiquitous in real-world DBMS workloads [7]. For example, secondary index entries are typically small tuples, since they merely map partial key combinations to primary keys [7]. In our experiments on LeanStore [13] under the TPC-C workload, we observe an average tuple size of approximately **73 bytes**. In ZippyDB and UP2X workloads, the average tuple sizes are 90.8 bytes and 57.25 bytes, respectively [23].

As shown in Fig. 2, the bars labeled *Tuple* show the FAR of six tuple-granularity eviction algorithms, while those labeled *Tuple(w/o MO)* show the ideal FAR ignoring metadata footprint. We observe that, although tuple-granularity algorithms achieve high FAR, they also incur significant hotness metadata overhead, resulting in a 21.3% to 31.5% FAR degradation.

Observation 2: Existing Slice-Granularity Eviction Algorithms Exhibit poor hot/cold Identification Accuracy due to Their Rigid Page-Aligned Slicing. A natural solution to solve the above problem is to exploit the similarity of adjacent tuples' hotness by grouping tuples with consecutive keys into a slice and using the slice as the unit for hotness tracking, thereby amortizing the memory overhead. For example, in the TPC-C [21] workload, the *New-Order* transaction creates several adjacent tuples belonging to the same *order* in the *Order-Line* table, and these tuples are accessed simultaneously in subsequent *Order-Status*, *Delivery*, and *Stock-Level* transactions. Meanwhile, the *Order-Line* table occupies more than 70% of the total storage space across all the tables [22].

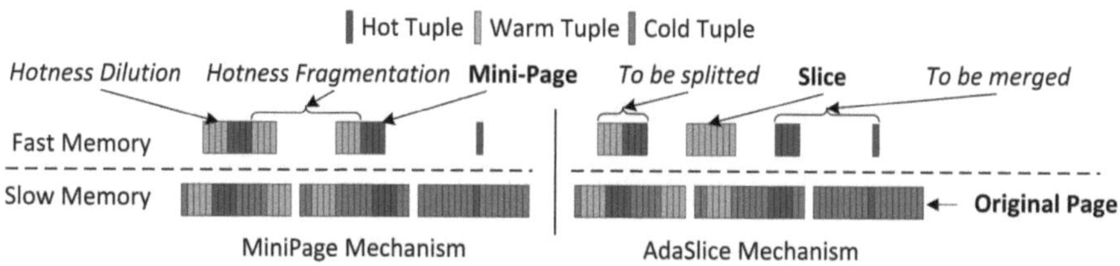

Fig. 3. Comparison of MiniPage (Coarse-Grained, Page-Aligned) and AdaSlice (Fine-Grained, Hotness-Aware).

However, among existing works, only the MiniPage mechanism proposed by Bf-Tree [7], MG-Buffer [16], and AMG-Buffer [17] follows this idea. A mini-page is rigidly aligned to its original page, as illustrated in Fig. 3. This rigid page-aligned slicing not only results in slices that are overly large or too small, but also leads to two fundamental problems that degrade identification accuracy. First, by forcing all tuples from a page into one mini-page, a single hot tuple can artificially inflate the perceived hotness of surrounding warm and cold tuples. This dilutes the true hotness signal, causing the eviction algorithm to retain warm and cold data in precious fast memory. We refer to this issue as **hotness dilution**. Besides, tuples with similar hotness but residing in different pages become fragmented into separate mini-pages. This prevents the eviction algorithm from recognizing and managing them as unified hotness tracking units, thereby increasing memory overhead. We call this problem **hotness fragmentation**.

Therefore, the issue with the MiniPage mechanism is not only its coarse granularity, but also the misalignment between its slicing principle (i.e., physical page boundaries) and the actual data access patterns. As shown in Fig. 2, the bars labeled *MiniPage* show the FAR of eviction algorithms with the MiniPage mechanism, while *MiniPage (w/o MO)* shows the ideal FAR ignoring metadata footprint. Although this mechanism incurs low metadata overhead (only 2.1%

to 3.0% FAR degradation), its accuracy remains suboptimal: the ideal FAR lags behind tuple-granularity algorithms by 16.8% to 24.2%.

Inspired by studies (e.g., [20]) in streaming clustering that use an online split/merge clustering approach based on cluster size and heterogeneity to balance intra-cluster homogeneity and cost, we propose AdaSlice, which dynamically groups tuples with similar hotness to amortize overhead and separates tuples with divergent hotness to maintain accuracy, thereby addressing the hotness metadata overhead issue of tuple-granularity eviction and the hotness dilution/fragmentation problem of mini-page-granularity eviction (see Fig. 3).

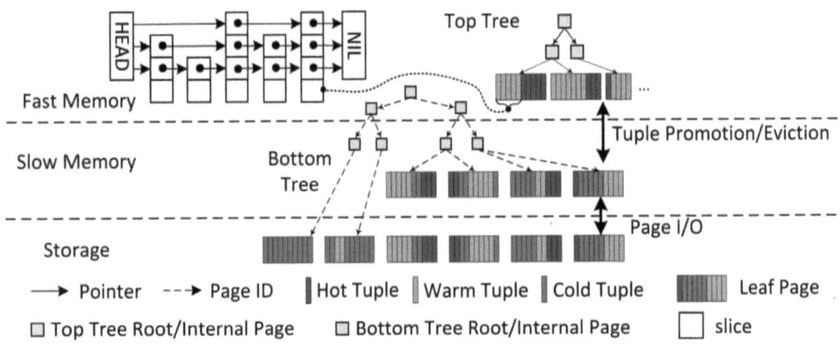

Fig. 4. Architecture of our tuple-granularity tiered buffer manager.

4 Design and Implementation

Based on the motivational analysis in Sect. 3, AdaSlice aims to achieve a favorable trade-off between hotness tracking overhead and hot/cold identification accuracy through hotness-aware and adaptive slicing. We use the mean squared error of tuple access counts within a slice to characterize *hotness dispersion*. By applying the **SplitSize** and **SplitDisp** thresholds to split large slices with high dispersion, we *improve identification accuracy*; by using the **MergeSize** and **MergeDisp** thresholds to merge small slices with low dispersion, we *reduce metadata memory overhead* (Sect. 4.3). We further employ simulated annealing to adaptively derive a near-optimal threshold set under varying workloads and configurations (Sect. 4.4). We organize slices using a skiplist (Sect. 4.2), integrate AdaSlice into multiple eviction algorithms (Sect. 4.5), then deploy it within our tiered buffer manager described in Sect. 4.1, and finally conduct an overhead analysis in Sect. 4.6.

4.1 Tuple-Granularity Tiered Buffer Manager

We implement a tuple-granularity tiered buffer manager within LeanStore [4,5] (a B^+-tree-based database storage engine), based on Hao et al.'s design [6]. As shown in Fig. 4, data is indexed by two separate B^+-trees: a top tree and a bottom tree. The top tree caches recently accessed tuples and is held entirely in fast memory. The bottom tree indexes the entire table: its upper-level nodes reside in fast memory to accelerate access, while its lower internal nodes and leaf pages are placed in slow memory or on disk. Page IDs are mapped to either in-memory addresses or on-disk offsets via a hash table.

Newly generated tuples are inserted directly into the top tree in fast memory. On lookup, update or delete, the system first attempts the operation in the top tree in fast memory, then in the bottom tree in slow memory if unsuccessful, and finally reads the page from disk when needed. If the tuple resides in slow memory or on disk, it may be copied into fast memory with a given probability (denoted as the **promotion rate**). When fast memory nears capacity, dirty tuples in a cold slice are evicted to slow memory, following the approach in Anti-Caching [10], where the amount of tuple evicted from each table is inversely proportional to the amount of tuple accessed in the table since the last eviction. If slow memory is exhausted, we fall back to a page replacement algorithm (LeanEvict [13] as used in Tiered-Indexing) to evict cold pages to disk.

4.2 Slice Management Based on Skiplist

The skiplist [19] is a probabilistic, self-balancing index supporting $O(\log n)$ search, insertion, and deletion using layered linked lists. Its efficient structural modifications without complex rebalancing (e.g., B-tree splits/merges) make it well-suited for our frequent slice splitting and merging, which directly lead to skiplist insertions and deletions (see Sect. 4.3). Furthermore, a skiplist node's address is fixed after insertion, simplifying slice-granularity eviction—nodes can directly serve as LRU chain entries. Hence, we use a skiplist to index slices.

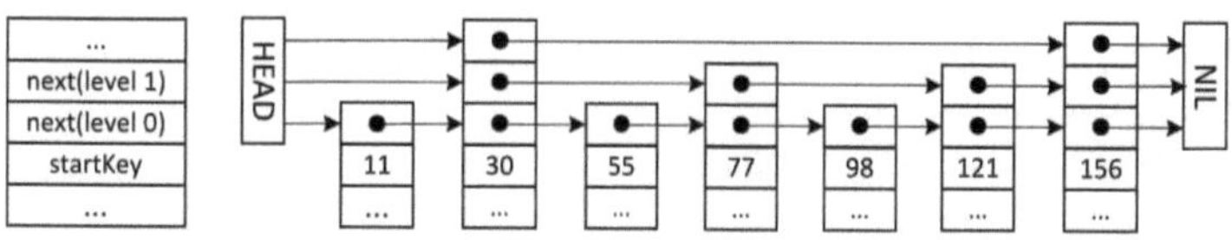

Fig. 5. Skiplist index for managing slices in AdaSlice.

As shown in Fig. 5, each skiplist node represents a slice and stores: $startKey$ (the smallest key within the slice), $size$ (number of tuples in the slice), $flags$, $height$, $lock$, and a variable-length array of $next$ pointers to successor nodes in each level. To locate the slice containing a given tuple, we compare the tuple's key to the $startKey$ of skiplist nodes. If for some node $curr$ and its successor node $next$ we have $curr.startKey \leq key < next.startKey$, then the tuple belongs to the slice represented by $curr$. For example (as illustrated in Fig. 5), to locate the tuple for $key = 130$ in the skiplist, we begin at the HEAD node's top level (level 2) and follow forward pointers until reaching the node 30; at level 1 we then move to node 121. Since 130 lies between 121 and 156, we conclude that the tuple is in the slice whose $startKey$ is 121. On tuple deletion, we decrement the slice's $size$; if the deleted tuple's key is $startKey$, we also update $startKey$ to its immediate right neighbor's key. For lookup or update, only the slice's hotness is refreshed (Sect. 4.5). For insertion, both $size$ and hotness are updated (Sect. 4.5).

4.3 Splitting and Merging of Slice

To avoid overly coarse slicing, we first constrain slice size with thresholds **SplitSize** and **MergeSize**. To further capture internal hotness dispersion, during a TPC-C test, we periodically computed the Pearson correlation coefficient between the mean squared error (**MSE**) of tuple access counts and the MSE of tuple hotness (*priority* in WATT, as defined in Sect. 4.5). Figure 6 shows that, for most of the test duration, the correlation coefficient fluctuates around 0.6 to 0.7, indicating a strong correlation. Therefore, we use the MSE of tuple access counts within a slice to measure hotness dispersion, and employ two additional thresholds, **SplitDisp** and **MergeDisp**, to control the splitting and merging of slices. We run a background thread that continuously selects a slice at random and computes the access count MSE of the tuples within it. The conditions and procedures for *slice* splitting and for merging with the right-adjacent slice $slice_{next}$ are shown in Algorithm 1 and Algorithm 2, respectively. Here, $MSE(slice)$ computes the MSE of the access counts of all tuples within that slice, and $MSE(slice, slice_{next})$ computes the MSE of the access counts of all tuples in both *slice* and $slice_{next}$. The function $nextNKey(key, n)$ returns the key of the tuple that lies n positions after the given *key* in the top tree. The function $copyHotness(slice_{next}, slice)$ transfers the hotness from the original *slice* to the $slice_{next}$. The function $mergeHotness(slice, slice_{next})$ merges the hotness of $slice_{next}$ into *slice*. We will describe them in detail in Sect. 4.5.

Algorithm 1: Split slice

Input: The *slice* to be splitted

if $slice.size \geq SplitSize$ **and** $MSE(slice) > SplitDisp$ **then**
- create $slice_{next}$;
- $slice_{next}.startKey = nextNKey(slice.startKey, \lceil slice.size/2 \rceil)$;
- $slice_{next}.size = \lfloor slice.size/2 \rfloor$;
- $copyHotness(slice_{next}, slice)$;
- $slice.size = \lceil slice.size/2 \rceil$;
- insert $slice_{next}$ into skiplist;

Algorithm 2: Merge slices

Input: The *slice* to be merged

get $slice_{next}$ next to *slice*;

if $slice.size + slice_{next}.size \leq MergeSize$ **and** $MSE(slice, slice_{next}) < MergeDisp$ **then**
- $slice.size \mathrel{+}= slice_{next}.size$;
- $mergeHotness(slice, slice_{next})$;
- delete $slice_{next}$ from skiplist;

4.4 Adaptive Thresholds Adjustment Based on Simulated Annealing

As shown in Fig. 13 and discussed below, the thresholds significantly impact system performance. While grid search can yield a near-optimal threshold set for a given hardware/software configuration and workload, this set changes when configuration or workload varies. Figure 7 shows the FAR of AdaSlice-integrated LRU under TPC-C, with SplitSize varied from 4 to 32 under different fast memory capacities (FMC) values while other thresholds remain fixed. FAR varies considerably with SplitSize, and the optimal SplitSize depends on FMC—e.g., 25, 20, and 16 for FMC of 0.5 GiB, 2 GiB, and 8 GiB, respectively. Similar trends hold for the other three thresholds, and their results are omitted for brevity.

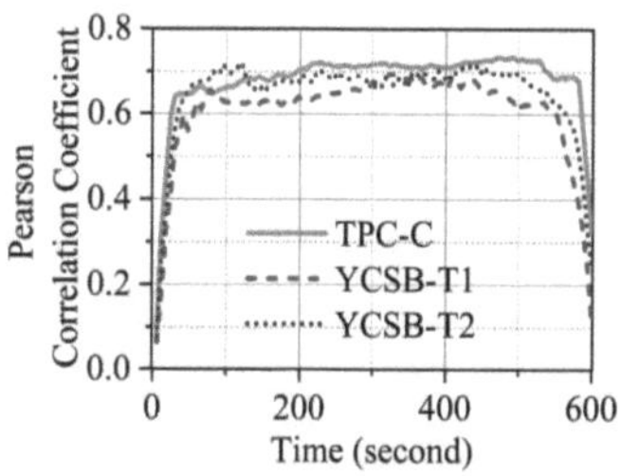

Fig. 6. The Pearson correlation coefficient between the MSE of tuple access counts and hotness.

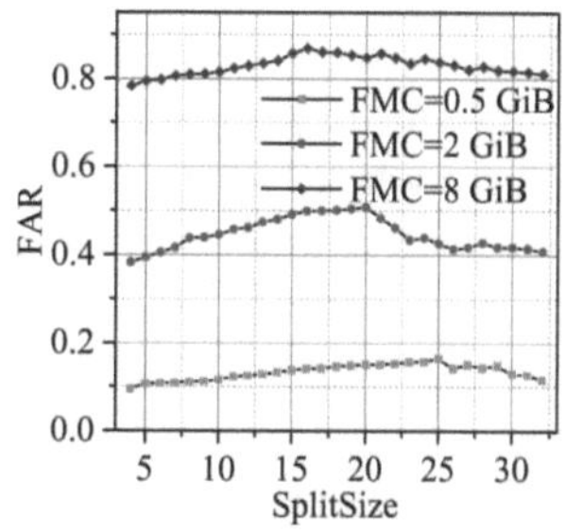

Fig. 7. Optimal SplitSize varies with FMC.

To address the problem above, we employ simulated annealing (SA) [18] to optimize the threshold set **T** = <*SplitSize, MergeSize, SplitDisp, MergeDisp*>. SA is an optimization algorithm that explores the solution space by occasionally accepting worse solutions at high temperatures to escape local minima, and gradually reducing this probability as the temperature cools to converge toward a global optimum. In AdaSlice, SA periodically evaluates candidate threshold sets over enough requests for their performance impact to be reflected in the target metric $1/FAR$, iteratively tuning T until converging to a near-optimal threshold set T_{opt} that maximizes FAR.

4.5 Eviction Algorithms Integrated with AdaSlice

AdaSlice can be efficiently applied to a wide range of eviction algorithms. We integrate AdaSlice into three classical algorithms (LRU [10], LRU2 [13], and Second-Chance [7]) and three more recently proposed ones (Sieve [15], Hyperbolic [14], and WATT [13]). The following describes implementation details.

LRU, LRU2, Second-Chance, and Sieve. These four algorithms are implemented via doubly linked lists or queues. The hotness of a slice is updated by adhering to the principle of its respective algorithm, such as moving it to the list head in LRU. When performing **copyHotness**($slice_{next}, slice$), $slice_{next}$ is placed adjacent to *slice* in the eviction algorithm's linked list or queue. When performing **mergeHotness**($slice, slice_{next}$), *slice* and $slice_{next}$ may not be adjacent in the eviction algorithm's linked list or queue, and they may even belong to different lists or queues (e.g., LRU2). After merging, *slice* is located at the position of the slice that had the larger size before merging. If the two slices are of equal size, then there is a 50% chance each for retaining *slice*'s position or using $slice_{next}$'s position.

Hyperbolic and WATT. First, we define the *timestamp* as the number of tuples cumulatively evicted to slow memory. In Hyperbolic, each slice includes an 8-byte $insertTime$ and a 4-byte $accessCount$, which respectively record the insertion time and the total number of accesses. In WATT, the timestamp occupies 4 bytes and increment lazily [13]. Each slice contains a circular array $accessTimes$ storing recent access timestamps, and a 1-byte *pos* indicating the position of the newest timestamp in that array. We configure the array to hold up to 4 timestamps, thereby bounding the memory footprint. We run a dedicated sampling thread that traverses slices in skiplist via bottom-level pointers. Upon eviction, we randomly sample slices and evict those with the lowest priority. The **priority** is calculated as $\frac{slice.accessCount}{(t_{now}-slice.insertTime)}$ in Hyperbolic and $\max_i \frac{i}{t_{now}-slice.accessTimes[i]}$ in WATT, with t_{now} being the eviction timestamp. When performing **copyHotness**($slice_{next}, slice$), all hotness related fields in $slice_{next}$ are copied verbatim from slice. When performing **mergeHotness**($slice, slice_{next}$), for Hyperbolic, the *slice*'s $insertTime$ is set to the minimum of $slice.insertTime$ and $slice_{next}.insertTime$, and its $accessCount$ becomes the sum of $slice.accessCount$ and $slice_{next}.accessCount$. For WATT, we sort all valid timestamps from both *slice* and $slice_{next}$, keep the most recent subset to fill $accessTimes$, and update *pos* accordingly.

4.6 Overhead Analysis

Time Overhead. We dedicate one background thread per transaction worker for slice maintenance. As shown in Sect. 5.2, skiplist operations have lower latencies than transaction operations and do not impede workers, due to the skiplist's efficient insertions/deletions, far fewer slices than tuples in the top tree, and its residence in fast memory (thus avoiding slow memory accesses). Additionally, slice hotness dispersion is computed efficiently using AVX2 instructions [13]. Although these background threads consume some CPU cores, this cost is acceptable on hundred-core servers [13].

Memory Usage. Skiplist requires an average of 4/3 pointers per element (assuming the level-increase probability is 0.25). The $startKey$ occupies on average 15 bytes. In addition, the four fields $size$, $flags$, *height*, and *lock* together consume 4 bytes. Therefore, the metadata of each slice requires about 30 bytes. Moreover, each tuple requires an additional 1 byte to record its access count. Assume the tuple eviction algorithm memory overhead is O (see Table 1). When a slice contains N tuples, the amortized per-tuple overhead of the slice-granularity algorithm is $\frac{30+O}{N}+1$. If $\frac{30+O}{N}+1 < O$, i.e., when $N > \frac{30+O}{O-1}$, the slice-granularity algorithm yields lower memory overhead per tuple than tuple-granularity algorithms. Accordingly, when $N \geq 4$ for LRU and Hyperbolic, or $N \geq 3$ for LRU2, Second-Chance, Sieve, and WATT, the amortized overhead is lower.

5 Evaluation

5.1 Experimental Setup

Hardware. The experiments are performed on a Linux 5.15 system with Intel Xeon Gold 6240 CPUs (2.60 GHz, 36 cores, 72 hardware threads). We use 128 GB DRAM as the fast memory tier and four 128 GB 100 series Intel Optane Persistent Memory modules in App Direct mode as the slow memory tier.
Competitors and Workloads. We integrate MiniPage and AdaSlice mechanisms into LRU [10], LRU2 [13], Second-Chance [7] (referred to as **SC** in the figures), Sieve [15], Hyperbolic [14], and WATT [13] algorithms, and compare them against their corresponding tuple-granularity eviction counterparts. The YCSB experiment performs accesses to a 16 GiB dataset, with key and value sizes set to 16 bytes and 48 bytes, respectively. We use a Zipfian factor of 0.9. We run two mixed workloads: YCSB-T1 (50% reads and 50% updates) and YCSB-T2 (30% reads, 30% updates, 20% inserts, and 20% deletes). The TPC-C workload comprises five transaction types with the following mix: 45% New-Order, 43% Payment, 4% Delivery, 4% Order-Status, and 4% Stock-Level. The number of warehouses is set to 100, resulting in around 16 GiB of data.
Evaluation Metrics. We evaluate eviction algorithms using FAR (fast memory access ratio, as defined in Sect. 1) and throughput, measured by tpmC (transactions per minute for TPC-C) and ops/s (operations per second for YCSB).

5.2 Results

Default Configuration. We set the fast memory capacity (FMC) to 2 GiB and the promotion rate (PR, see Sect. 4.1) to 1. We provision slow memory to be sufficiently large (512 GB) and disable WAL to avoid I/O interference. For eviction algorithms integrated with the AdaSlice mechanism, the thresholds are set to their respective near-optimal values derived via simulated annealing.
Performance Comparison. Figures 8 and 9 illustrates the performance comparison results, with *SC* representing Second-Chance. We observe that compared to tuple-granularity algorithms, AdaSlice can increase FAR by 15.3% to 30.5% and tpmC by 12.1% to 20.6%. Compared to MiniPage, AdaSlice improves FAR by 11.3% to 22.4% and tpmC by 10.1% to 16.4%. Moreover, since the Clock algorithm offers relatively weak hot/cold identification and requires extensive sequential scans over tuples during eviction (which yields many wasted scans of already-accessed tuples and degrades foreground access), all the six algorithms achieve higher FAR and database throughput than Clock. For instance, under the TPC-C, YCSB-T1, and YCSB-T2 workloads, the Clock algorithm achieves

FAR values of only 0.37, 0.47 and 0.41, respectively, and tpmC or ops/s values of just $6.2 \times 10^5, 3.8 \times 10^5$ and 3.1×10^5, respectively.

Validation of Approaching the Sweet Spot in Trade-off. This experiment, evaluated on the TPC-C workload, demonstrates AdaSlice's superior trade-off between accuracy and overhead, as detailed in Fig. 10. The bars labeled *(w/o MO)* show the ideal FAR ignoring metadata footprint, where tuple-granularity algorithms perform best. However, their high overhead reduces actual FAR by 21.3% to 31.5%. At the other extreme, MiniPage variants minimize overhead (2.1% to 3.0% FAR loss) at the cost of the lowest accuracy. AdaSlice achieves a superior trade-off, maintaining low overhead (3.1% to 5.2% FAR loss) while delivering high accuracy via its hotness-aware and adaptive slicing.

Impact of Split/Merge Thresholds. To evaluate threshold sensitivity, we independently test SplitSize, MergeSize, SplitDisp, and MergeDisp in AdaSlice-integrated LRU under TPC-C. We first fix SplitDisp to 1.45 and MergeDisp to 0.21, as determined by the optimal values in Fig. 13c and Fig. 13d. In Fig. 13a, with MergeSize fixed at 2, we vary SplitSize from 4 to 32; in Fig. 13b, with SplitSize fixed at 32, we vary MergeSize from 2 to 30. Then we fix SplitSize to 20 and MergeSize to 16, as determined by the optimal values in Fig. 13a and Fig. 13b. Figure 13c shows FAR when MergeDisp is 0.01 and SplitDisp ranges

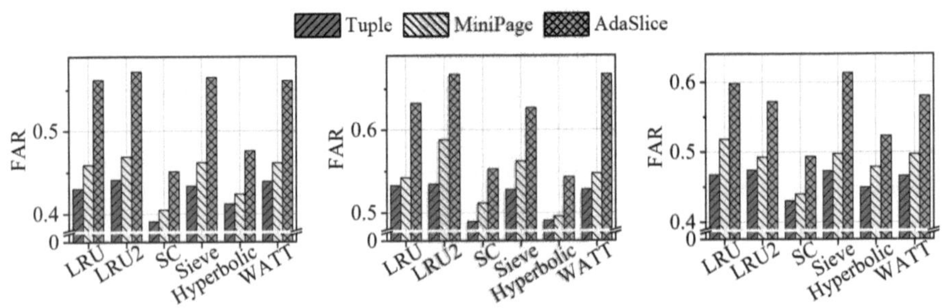

(a) TPC-C (b) YCSB-T1 (c) YCSB-T2

Fig. 8. Overall FAR comparison.

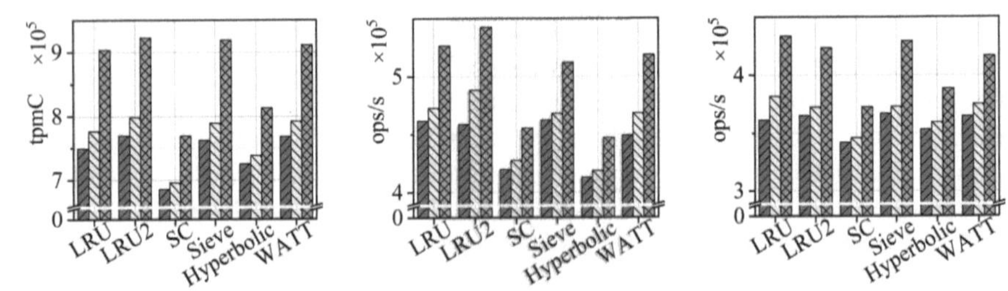

(a) TPC-C (b) YCSB-T1 (c) YCSB-T2

Fig. 9. Overall throughput comparison.

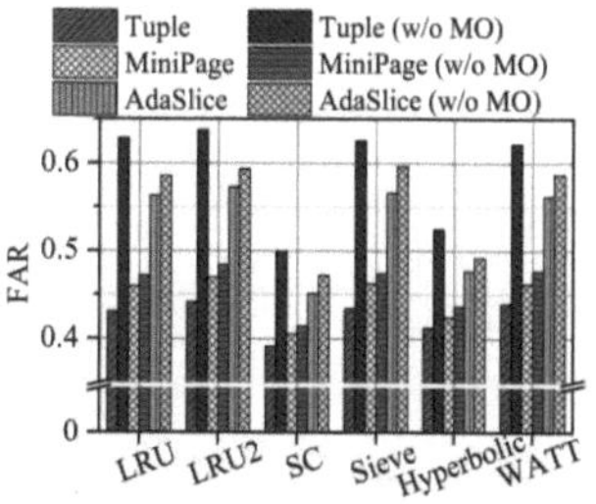

Fig. 10. Achieving a favorable trade-off between accuracy and overhead.

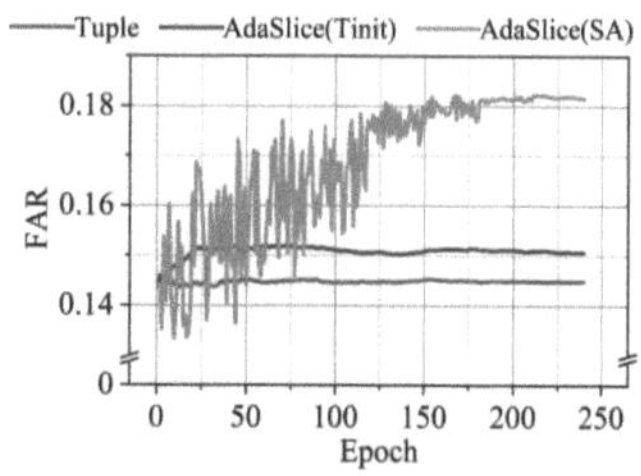

Fig. 11. Convergence of FAR during online threshold tuning via SA.

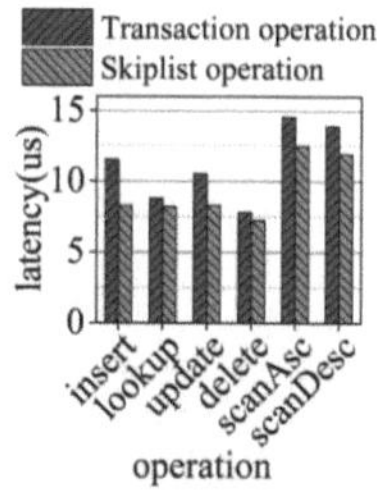

Fig. 12. Comparison of operation latency.

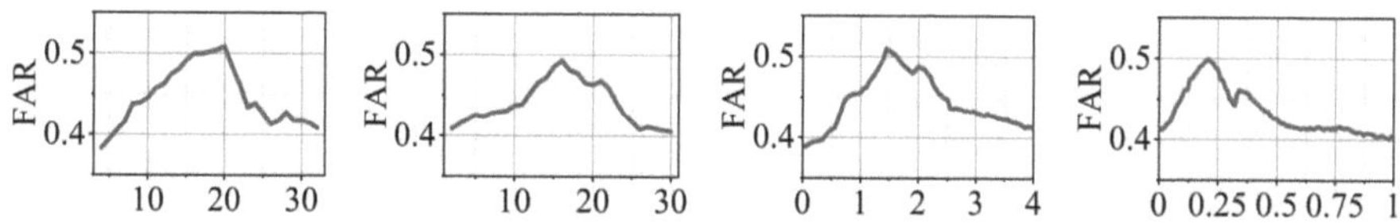

(a) SplitSize (b) MergeSize (c) SplitDisp (d) MergeDisp

Fig. 13. Impact of split/merge thresholds on FAR.

from 0.05 to 4 in steps of 0.05. Figure 13d shows FAR when SplitDisp is 4 and MergeDisp ranges from 0.01 to 1 in increments of 0.01. Results show that FAR is highly sensitive to these thresholds: very small SplitSize or SplitDisp causes excessive splitting, while very large values make it hard to split heterogeneous slices. Likewise, small MergeSize or MergeDisp hinder merging of similar slices, and large values result in excessive merging. Thus, carefully tuning these four thresholds is key to maximizing FAR, thereby improving database throughput.

Effectiveness of Simulated Annealing. Building upon Spitfire's settings [3], we set the simulated annealing (SA) parameters as follows after fine-tuning for improved convergence: a cooling rate $\alpha = 0.9$, an initial temperature of 1000, a final temperature of 0.0001, and a tuning epoch of 5 s so that threshold adjustments can fully influence FAR. First, we set the initial threshold set T_{init} to the near-optimal threshold set found under the default configuration on the TPC-C workload, then apply SA to derive a near-optimal threshold set T_{opt} for diverse configurations and workloads. In Fig. 11, only the fast memory capacity (FMC) is reduced to 0.5 GiB. *Tuple* refers to tuple-granularity LRU, *AdaSlice(Tinit)* refers to AdaSlice-integrated LRU using fixed threshold set T_{init}, and *AdaSlice(SA)* continuously refines thresholds using SA. After about 180 epochs, *AdaSlice(SA)* converges and achieves about 20% higher FAR than *AdaSlice(Tinit)*. In Fig. 14, the bars labeled *AdaSlice(Topt)* show the FAR of AdaSlice-integrated LRU when using the SA-derived threshold set T_{opt} under configurations where only one factor (FMC, promotion rate, or workload) is modified from the default configuration. Across FMC values of 0.5, 1, 4 and 8 GiB

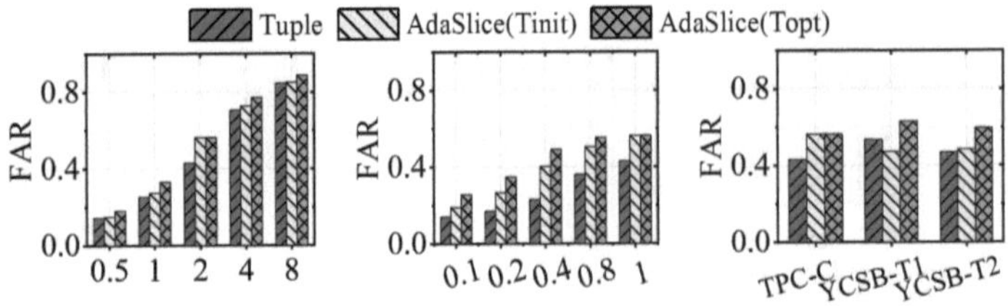

(a) Fast memory capacity (GiB) (b) Promotion rate (c) Workload

Fig. 14. FAR improvement of SA-optimized AdaSlice under various configurations and workloads.

(Fig. 14a), *AdaSlice(Topt)* improve FAR by 4.2% to 20.4% over *AdaSlice(Tinit)*. For promotion rates of 0.1, 0.2, 0.4 and 0.8 (Fig. 14b), *AdaSlice(Topt)* boost FAR by 9.7% to 35.4% over *AdaSlice(Tinit)*. Under YCSB-T1 and YCSB-T2 workloads (Fig. 14c), *AdaSlice(Topt)* improve FAR by 23.7% to 34.0% over *AdaSlice(Tinit)*.

Overhead. Figure 12 shows the average latencies of six transaction operations (insert, lookup, update, delete, scanAsc, scanDesc) and the corresponding latencies for skiplist operations. scanAsc and scanDesc refer to ascending and descending scans, respectively. Insert and delete operations trigger slice splitting and merging. As analyzed in Sect. 4.6, the skiplist's efficient operations result in lower latencies compared to transaction operations. Future work will analyze the impact of slice-maintenance threads on end-to-end system performance and scalability. We will also evaluate AdaSlice under tuning thresholds, changing access patterns, alternative slow memory devices (e.g., NVDIMM), and I/O-intensive scenarios.

6 Conclusion

In this paper, we present AdaSlice, a hotness-aware and adaptive slicing mechanism for tuple eviction algorithms in tiered buffer manager. It navigates the trade-off between the *high accuracy, high overhead* of tuple-granularity algorithms and the *low accuracy, low overhead* of existing slice-granularity algorithms by dynamically splitting large slices with high dispersion and merging small slices with low dispersion via a threshold set *<SplitSize, MergeSize, SplitDisp, MergeDisp>*. We further employ simulated annealing to dynamically derive a near-optimal threshold set under varying configurations and workloads. We integrate AdaSlice into six existing eviction algorithms and implement them within the tiered buffer manager. Experimental results show that AdaSlice increases FAR by 30.5% and improves database throughput by 20.6%.

Acknowledgments. We thank the anonymous reviewers for their valuable comments. This work was supported in part by research program No. 2023BAA018.

Disclosure of Interests. The authors have no competing interests to declare that are relevant to the content of this article.

References

1. Xiang, L., et al.: Nomad: non-exclusive memory tiering via transactional page migration. In: 18th USENIX Symposium on Operating Systems Design and Implementation (OSDI 24), pp. 19–35 (2024)
2. Huang, Y., et al.: Tigon: a distributed database for a CXL pod. In: 19th USENIX Symposium on Operating Systems Design and Implementation (OSDI 25), pp. 109–128 (2025)
3. Zhou, X., Arulraj, J., Pavlo, A., Cohen, D.: Spitfire: a three-tier buffer manager for volatile and non-volatile memory. In: Proceedings of the 2021 International Conference on Management of Data (SIGMOD 21), pp. 2195–2207 (2021)
4. Zhou, X., Hao, X., Yu, X., Stonebraker, M.: Tiered-indexing: optimizing access methods for skew. VLDB J. **34**(4), 45 (2025)
5. Zhou, X., Yu, X., Graefe, G., Stonebraker, M.: Two is better than one: the case for 2-tree for skewed data sets. In: 13th Annual Conference on Innovative Data Systems Research (CIDR 23), p. 13 (2023)
6. Hao, X., Zhou, X., Yu, X., Stonebraker, M.: Towards buffer management with tiered main memory. Proc. ACM Manage. Data (SIGMOD 24), pp. 1–26 (2024)
7. Hao, X., Chandramouli, B.: Bf-Tree: a modern read-write-optimized concurrent larger-than-memory range index. Proc. VLDB Endow. **17**(11), 3442–3455 (2024)
8. Eldawy, A., Levandoski, J. and Larson, P.Å.: Trekking through siberia: managing cold data in a memory-optimized database. Proc. VLDB Endow. **7**(11), 931–942 (2014)
9. Levandoski, J.J., Larson, P.Å., Stoica, R.: Identifying hot and cold data in main-memory databases. In: 2013 IEEE 29th International Conference on Data Engineering (ICDE), pp. 26–37 (2013)
10. DeBrabant, J., Pavlo, A., Tu, S., Stonebraker, M., Zdonik, S.: Anti-caching: a new approach to database management system architecture. Proc. VLDB Endow. **6**(14), 1942–1953 (2013)
11. Van Renen, A., et al.: Managing non-volatile memory in database systems. In: Proceedings of the 2018 International Conference on Management of Data (SIGMOD 18), pp. 1541–1555 (2018)
12. Yu, G.X., Markakis, M., Kipf, A., Larson, P.Å., Minhas, U.F., Kraska, T.: TreeLine: an update-in-place key-value store for modern storage. Proc. VLDB Endow. **16**(1), 99–112 (2022)
13. Vöhringer, D., Leis, V.: Write-aware timestamp tracking: effective and efficient page replacement for modern hardware. Proc. VLDB Endow. **16**(11), 3323–3334 (2023)
14. Blankstein, A., Sen, S., Freedman, M.J.: Hyperbolic caching: flexible caching for web applications. In: 2017 USENIX Annual Technical Conference (USENIX ATC 17), pp. 499–511 (2017)
15. Zhang, Y., Yang, J., Yue, Y., Vigfusson, Y., Rashmi, K.V.: SIEVE is simpler than LRU: an efficient Turn-key eviction algorithm for web caches. In 21st USENIX Symposium on Networked Systems Design and Implementation (NSDI 24), pp. 1229–1246 (2024)

16. Wang, X., Jin, P., Liu, R., Zhang, Z., Wan, S., Hua, B.: MG-buffer: a read/write-optimized multi-grained buffer management scheme for database systems. In: 2019 IEEE 21st International Conference on High Performance Computing and Communications (HPCC), pp. 1212–1219 (2019)
17. Wang, X., Jin, P.: Adaptive multi-grained buffer management for database systems. Future Internet **13**(12), 303 (2021)
18. Kirkpatrick, S., Gelatt, C.D., Jr., Vecchi, M.P.: Optimization by simulated annealing. Science **220**(4598), 671–680 (1983)
19. Luo, Y., Hao, S., Wheatman, B., Pandey, P., Xu, H.: Bridging cache-friendliness and concurrency: a locality-optimized in-memory B-skiplist. In: 54th International Conference on Parallel Processing (ICPP 25) (2025)
20. Lughofer, E., Sayed-Mouchaweh, M.: Autonomous data stream clustering implementing split-and-merge concepts - towards a plug-and-play approach. Inf. Sci. **304**, 54–79 (2015)
21. Leutenegger, S.T., Dias, D.: A modeling study of the TPC-C benchmark. ACM SIGMOD Rec. **22**(2), 22–31 (1993)
22. Park, H.W., Choi, S., An, M., Lee, S. W: Freezing frozen pages with multi-stream SSDs. In: Proceedings of the 15th International Workshop on Data Management on New Hardware, pp. 1–3 (2019)
23. Cao, Z., Dong, S., Vemuri, S. and Du, D.H.: Characterizing, modeling, and benchmarking RocksDB key-value workloads at Facebook. In: 18th USENIX Conference on File and Storage Technologies (FAST 20), pp. 209-223 (2020)

Workload-Aware DHB+Tree: A Dynamic B+-Tree for Persistent Memory

Jiarui Qi, Junbao Song, Derong Shen(✉), Tiezheng Nie, and Yue Kou

School of Computer Science and Engineering, Northeastern University, Shenyang 110819, China
{shendr,nietiezheng}@mail.neu.edu.cn, kouyue@cse.neu.edu.cn

Abstract. Persistent memory (PM) poses challenges for B+-Tree indexing due to its read-write asymmetry and durability requirements. Existing PM-optimized B+-trees employ static storage layouts (either fully ordered or unordered), which lack adaptability to dynamic workloads and lead to suboptimal performance. This paper proposes DHB+Tree, a dynamic hybrid-storage B+-Tree that adaptively balances read and write performance. DHB+Tree partitions each leaf node into ordered and unordered zones, dynamically adjusting their ratio in real time based on workload patterns. A lightweight workload-aware parameter prediction model is introduced to guide this adjustment, minimizing overall operation latency. DHB+Tree ensures crash consistency and avoids physical data movement during restructuring. Experimental results demonstrate that DHB+Tree consistently outperforms state-of-the-art static designs in both throughput and adaptability under diverse workloads.

Keywords: Persistent Memory · DHB+Tree · Workload-Aware · B+-Tree Optimization

1 Introduction

Next-generation non-volatile memory technologies, such as Intel Optane DCPMM, are revolutionizing database systems by providing byte-addressable persistent storage with near-DRAM read performance. However, Persistent memory(PM) exhibits significant read-write asymmetry, where write latency is 3–5× higher than read latency [1], and write bandwidth is substantially lower. These characteristics, combined with fine-grained persistence guarantees and access granularity mismatch, pose fundamental challenges for traditional B+-tree designs.

Existing PM-optimized B+-trees typically adopt static storage layouts [2,3,6,7,20], forcing a fixed trade-off between read and write performance. Write-optimized designs like CC-Tree [20] use unordered nodes to minimize write amplification but sacrifice read locality, degrading point query performance by up to 40%. Read-optimized designs like FAST&FAIR [6] maintain fully sorted nodes for efficient queries but incur high write overhead due to frequent cache line

H. Jung et al. (Eds.): DASFAA 2026, LNCS 16535, pp. 623–638, 2026.
https://doi.org/10.1007/978-981-92-0363-5_38

flushes and entry shifting. Hybrid approaches like SSB-Tree [7] introduce intra-node logging to reduce write overhead. However, their optimization is limited to pairwise operation aggregation and lacks dynamic adaptability. These static designs fail to accommodate real-world workloads, where the ratio of read to write operations fluctuates significantly over time.

To address this limitation, we propose DHB+Tree (Dynamic Hybrid-Storage B+-Tree), an adaptive index structure that dynamically balances read-write trade-offs based on workload characteristics. The key innovation of DHB+Tree is the partitioning of each leaf node into ordered and unordered zones, whose relative sizes are adjusted in real-time according to the observed workload. Under write-intensive workloads, the unordered zone expands to enable low-latency append operations. During read-intensive workloads, the ordered zone expands to accelerate query processing. A lightweight workload-aware parameter prediction model automatically tunes the storage parameters through incremental optimization, triggering restructuring only when workload shifts exceed a threshold to minimize overhead. The main contributions of this work are:

We propose DHB+Tree, the dynamic hybrid-storage B+-tree that elastically balances read-write performance through real-time workload adaptation, eliminating the static design limitations of existing PM-optimized indexes.

We design a lightweight parameter prediction model that dynamically tunes storage parameters through theoretical analysis and simulation, enabling DHB+Tree to automatically adjust to changing workload patterns with minimal overhead.

Extensive evaluation shows that DHB+Tree achieves 1.48× higher throughput than FAST&FAIR [6], 1.32× higher than CC-Tree [20] and 1.29× higher than SSB-Tree [7] under dynamic workloads.

The rest of this paper is organized as follows: Sect. 2 reviews related work, Sect. 3 presents the DHB+Tree design, Sect. 4 evaluates its performance, and Sect. 5 concludes.

2 Related Work

Existing persistent memory (PM) B+-tree implementations fall into two architectural paradigms: hybrid and all-PM.

Hybrid architectures (e.g., FP-Tree [8], uTree [9], LB+-Tree [10], CCLTree [11]) These architectures optimize performance by placing internal nodes in volatile DRAM. However, post-crash recovery requires reconstructing internal nodes through parallel leaf node traversal. This process introduces significant recovery latency (3-5 s) and synchronization overhead due to DRAM's volatility.

All-PM Architectures persist both internal and leaf nodes directly in PM, eliminating DRAM dependency to enable millisecond-scale (near-instant) crash recovery, but require careful design to mitigate PM's asymmetric read/write performance and fine-grained atomicity constraints.

Among various approaches to optimizing B+-Tree indexes, Write-Optimized Approaches prioritize low-latency updates and use unordered layouts to minimize data movement. For example, NV-Tree [2] and CC-Tree [20] employ unordered leaf nodes with specialized metadata (with metadata partitioned in CC-Tree) to avoid entry shifting overhead. Read-Optimized Approaches maximize query performance through ordered data organization. FAST&FAIR [6] uses fully sorted nodes and leverages fault-atomic shift operations to maintain strict key ordering, optimizing search performance at the cost of high write amplification. wB+-Tree [3] utilizes indexed unordered data and employs slot arrays to enable efficient searching over unordered data, but adds a layer of indirection. Hybrid Storage Approaches attempt to balance read-write trade-offs through partitioning: SSB-Tree [7] employs fixed dual buffers and uses Lazy-Box logging for operation aggregation, but is limited to pairwise aggregation.

As shown in Table 1, unlike these static or partially optimized designs, DHB+Tree introduces truly dynamic hybrid storage. It dynamically partitions leaf nodes into ordered and unordered zone. The proportional size of these regions is adjusted in real time based on the observed workload. This elastic partitioning achieves an adaptive balance between read and write performance. A load-aware parameter prediction model (Sect. 3.4) optimizes this balance, maximizing overall read-write throughput while ensuring low-latency decision-making. Experimental results (Sect. 4) demonstrate that DHB+Tree significantly outperforms state-of-the-art all-PM B+-Trees under dynamic workloads, providing robust support for efficient dynamic workload processing.

Table 1. Comparison of PM B+-Tree Designs

Method	Storage Strategy	Optimization Focus	Key Limitation
NV-Tree [2]/CC-Tree [20]	Unordered leaf nodes	Write latency	Poor read performance
FAST&FAIR [6]	Fully sorted nodes	Read performance	High write overhead
wB+-Tree [3]	Slot arrays + Unordered data	Balanced read-write	Indirection overhead
SSB-Tree [7]	Fixed dual buffers	Operation aggregation	Limited adaptability
DHB+Tree	Dynamic hybrid zones	Workload adaptation	Restructuring overhead

3 DHB+Tree

3.1 DHB+Tree Framework

Traditional B+-Trees suffer from a static trade-off between ordered storage (read-optimized) and unordered zone (write-optimized), limiting their adaptability to dynamic workloads. To overcome this limitation, we propose DHB+Tree. It dynamically balances ordered and unordered zone regions within leaf nodes to maximize overall read-write throughput.

Aligned with DHB+Tree's architecture and inspired by active learning [13–19], we design a workload-aware parameter prediction model. This model calculates optimal ordered/unordered storage ratios for given workloads through parameter simulation. Concurrently, to minimize frequent restructuring overhead, we adopt a threshold-triggered mechanism [18]: the system monitors the ratio of read operations to writeoperations among the last p operations, and calculates the difference between the current ratio and the historical average ratio. If this difference exceeds τ, restructuring is triggered. Ultimately, adaptive leaf-node restructuring achieves global throughput optimization.

The DHB+Tree framework comprises the following key components:

Hybrid-Storage B+-Tree Index: Dynamically restructures leaf-node layouts to execute operations (reads, inserts, range queries).

Workload-aware Parameter Prediction Model: It predicts optimal storage ratios via simulation to maximize throughput.

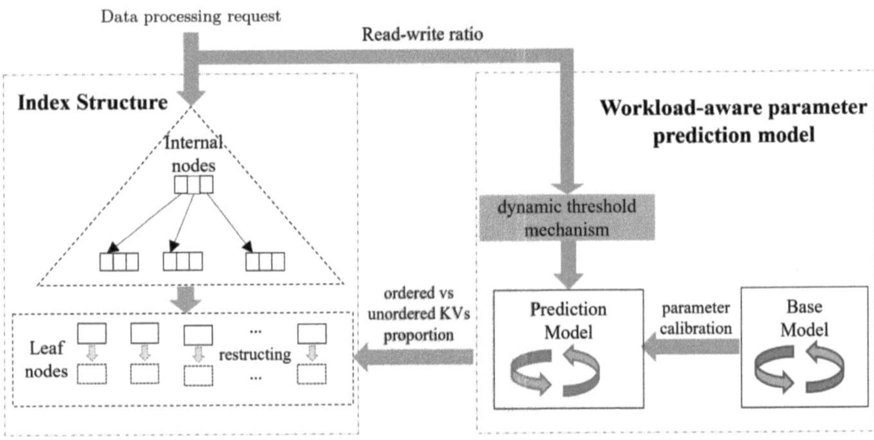

Fig. 1. DHB+Tree Framework.

As illustrated in Fig. 1, the DHB+Tree adopts a tiered node architecture layout with a dynamic hybrid storage, where internal nodes maintain fully ordered keys for efficient routing, and leaf nodes implement adaptive zoning. Each leaf node partitions its storage into an ordered zone (storing read-optimized ordered key-value pairs) and an unordered zone (storing write-optimized unordered entries). This dynamic partitioning enables fine-grained workload-aware resource allocation: under write-intensive workloads, the unordered zone expands to support low-latency inserts; during read-intensive workloads, the ordered zone is prioritized for expansion to maintain query efficiency, thus addressing the read-write asymmetry in persistent memory.

3.2 Index Structure and Node Layout

DHB+Tree implements specialized node structures optimized for persistent memory characteristics. All nodes utilize a unified 8-byte header design that ensures atomic persistence, while diverging in payload organization based on their functional roles.

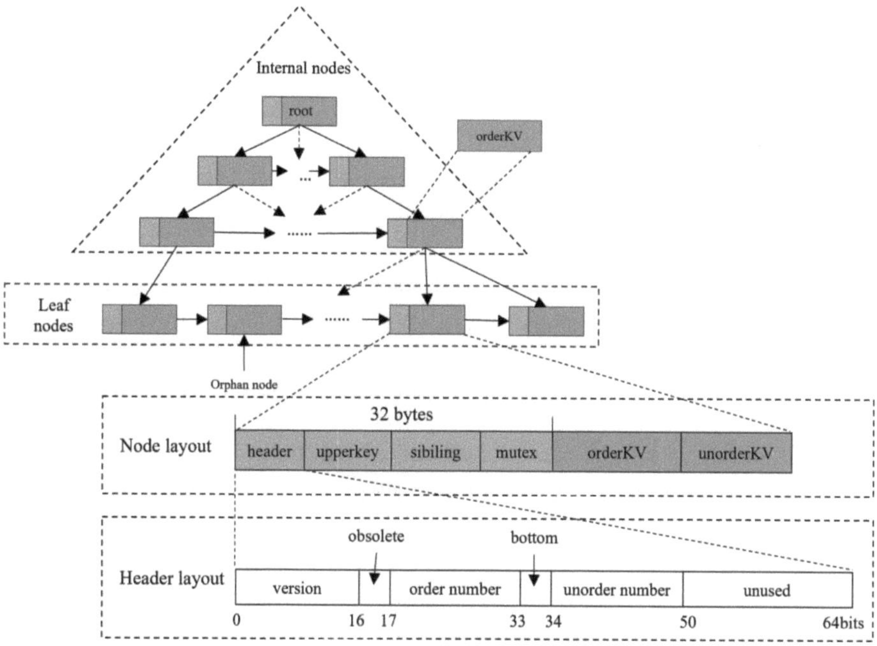

Fig. 2. DHB+Tree index structure.

Leaf Node. As shown in Fig. 2, a leaf node in DHB+Tree consists of several key components designed for persistent memory. An 8-byte header aligns with PM's atomic write granularity and contains a version counter for crash consistency (even indicates a committed operation, odd an incomplete one), an obsolete flag for logical deletion, an order number tracking ordered key-value pairs to accelerate linear searches, a bottom flag (set to 1 for leaf nodes), and an unorder number counting unordered entries. For efficient navigation, search metadata includes an upperkey storing the node's maximum key, and a sibling pointer enabling range scans. A mutex provides write-write coordination, while the data storage area physically partitions key-value pairs into ordered and unordered zone, placing ordered entries first to optimize cache locality during query operations. When the unordered zone is full, entries in the unordered zone are merged into the ordered zone, and new insertions are directed to the ordered zone thereafter.

Internal Node. Internal nodes in DHB+Tree are optimized for high-frequency read routing operations and maintain a simplified structure. They share the same foundational components as leaf nodes—including the header, upperkey, sibling pointer, and mutex—but differ in two critical aspects: the bottom flag is permanently set to 0 to indicate a non-leaf node, and the unordered number field is omitted since internal nodes do not support unordered zone. The payload consists of a fully ordered key-child pointer array (orderKV), which enables efficient and deterministic subtree navigation during search operations.

3.3 Operations

DHB+Tree implements efficient operations through its hybrid storage architecture and optimized concurrency control.

Search and Insertion. Search employs two-phase traversal: horizontal traversal at internal nodes using binary search, followed by downward traversal to leaf nodes. Within leaves, binary search targets the ordered zone first, falling back to a linear scan of the unordered zone. Sibling pointers handle cases where parent nodes lack updated routing information after splits. As shown in Fig. 3, when searching for key = 13: the current leaf node's maximum key is 9, so the search moves to the right sibling. Meanwhile, the minimum key of the new sibling (9) is inserted into the parent node as a routing key, ensuring subsequent searches navigate correctly.

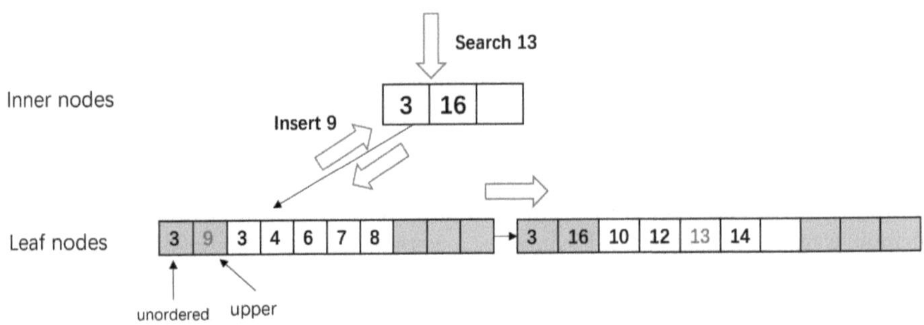

Fig. 3. The example of DHB+Tree searching.

Insertion procedures differ based on node type. Internal nodes maintain full ordering through atomic shifts. Leaf nodes prioritize write performance by appending to the unordered zone, triggering merges into the ordered zone when capacity is reached.

Splitting and Merging. Splitting creates orphan nodes connected via sibling pointers, deferring updates to parent nodes to reduce persistence overhead. Merging consolidates underutilized nodes ((<50%) occupancy) to maintain balance. Both operations leverage parent-independent evolution for lightweight coordination.

Concurrency Control. A hybrid scheme combines optimistic reads and pessimistic writes to maximize throughput in multi-threaded scenarios [5]:

For read-write coordination, DHB+Tree relies on version-based optimistic control. Nodes have version numbers incremented on modifications; reads validate consistency before and after access (retrying if changed), enabling non-blocking reads. This avoids lock-based overhead (e.g., FAST&FAIR's read locks) and reduces PM writes and bandwidth usage.

For write-write coordination, DHB+Tree uses node-level mutexes (PMDK's persistent mutexes for crash consistency) to serialize writes (inserts, deletes, splits), ensuring atomicity. This design eliminates PM writes for reads while maintaining strict write consistency, thereby improving mixed workload performance.

3.4 Workload-Aware Parameter Prediction Model

The core of DHB+Tree's adaptability lies in its lightweight workload-aware parameter prediction model. This model dynamically determines the optimal configuration for each leaf node—specifically, the sizes of the ordered zone (m) and unordered zone (k)—to minimize total operation latency and maximize throughput under the current workload.

Problem Formulation and Parameter Selection. We formulate the parameter selection as an optimization problem where the workload is characterized by the read-write ratio ($r : w$), with $r = 1 - w$. The model focuses on parameters that directly impact intra-node operational efficiency, as summarized in Table 2.

Table 2. Key parameters in the prediction model

Term	Definition
T	Total number of entries
L	Number of tree layers
Le	Leaf node size
m	Number of ordered entries
k	Number of unordered entries
w	Write ratio
r	Read ratio

The selection of m and k as tuning parameters is based on their direct and opposing impacts on performance:

Point Query Latency: Primarily influenced by the ordered zone size (m). Ordered layouts enable efficient binary search, while scanning the unordered zone (k) incurs linear overhead.

Insertion/Update Latency: Governed by merge frequency and cost in the unordered zone. When the unordered zone is full, a merge operation with $O(k \log k + m)$ complexity is triggered. PM's characteristics (low write bandwidth, high write amplification) exacerbate this cost.

System Stability: Parameters like T, L, and Le remain stable after initialization and are excluded from dynamic adjustment.

Base Cost Model. We first develop an analytical model for average operation latency. Point lookups search both regions, resulting in the following time complexity (Eq. 1):

$$T_{\text{read}} = O(k + \log(m)) \tag{1}$$

The insertion cost is dominated by amortized merge operations, yielding Eq. 2:

$$T_{\text{write}} = O\left(\frac{k \log k + m}{k}\right) = O\left(\log k + \frac{m}{k}\right) \tag{2}$$

By combining Eqs. 1 and 2 with the read-write ratio, we obtain the total average time complexity is a weighted sum incorporating PM's asymmetric I/O characteristics (Eq. 3):

$$T_{\text{total}} = w \cdot W_w \cdot \left(\log k + \frac{m}{k}\right) + (1 - w) \cdot W_r \cdot (k + \log(m)) \tag{3}$$

where W_w and W_r are hardware-dependent write and read bandwidths, respectively.

Calibrated Prediction Model and Fitting. The base model captures major trends but does not account for constant factors and micro-architectural effects. We introduce a calibrated model with coefficients a, b, c (Eq. 4):

$$T_{\text{total}} = w \cdot W_w \cdot \left(a \log k + b\frac{m}{k}\right) + (1 - w) \cdot W_r \cdot (ck + d \log(m)) \tag{4}$$

Taking the partial derivative of Eq. 4 with respect to k and setting it to zero yields Eq. 5:

$$W_r * c * (1 - w) * k^2 + w * a * W_w * k - w * b * W_w * m = 0 \tag{5}$$

To fit parameters a, b, c, we fixed the read-write ratio, and we used exhaustive search in experiments to find the optimal (m, k) configurations. The process starts with minimal $k = 1$ and increments m until time cost stops decreasing, recording optimal $m_{\text{opt}}^{(k)}$ for each k. This generates a sequence of optimal points $(m_{\text{opt}}^{(1)}, 1), (m_{\text{opt}}^{(2)}, 2), \ldots$

We transform parameter fitting into a regularized least squares problem. For N observed optimal points (k_i, m_i), we solve:

$$\min_{\theta} \|X\theta - Y\|_2^2 + \lambda\|\theta\|_2^2, \quad \theta = [\alpha, \beta]^T \tag{6}$$

where:

$$X = \begin{bmatrix} k_1^2 & -m_1 \\ \vdots & \vdots \\ k_N^2 & -m_N \end{bmatrix}, \quad Y = \begin{bmatrix} \gamma_1 \\ \vdots \\ \gamma_N \end{bmatrix}$$

with $\alpha = W_r c(w-1)$, $\beta = wbW_w$, $\gamma_i = waW_w k_i - W_r c(1-w)k_i^2$.

Using QR decomposition via LAPACK's `dgelsd` solver ($\lambda = 10^{-4}$), we obtained coefficients:

$$a = \frac{\alpha \cdot \text{median}(k)}{wW_w}, \quad b = \frac{\beta}{wW_w} \tag{7}$$

with $c = 1$ per normalization constraint. Finally, we derived a prediction model with determined coefficients.

Dynamic Restructuring and Online Adaptation. The workload-aware parameter prediction model enables low-overhead online adaptation: the system continuously monitors the workload mix via an operation sliding window and invokes the prediction model to compute a new optimal k only when the change in the read-write ratio exceeds the threshold τ does the coefficient-determined prediction model invoked to compute the new optimal k. This mechanism avoids frequent restructuring triggered by transient fluctuations, ensuring that the adaptation overhead is amortized by significant performance gains. When restructuring is triggered, leaf nodes logically adjust the boundary between the ordered and unordered zones without requiring physical data movement.

To balance adaptation benefits against restructuring overhead, DHB+Tree employs a volatility-aware dynamic threshold mechanism. A fixed threshold τ is suboptimal for handling real-world workload fluctuations: an overly sensitive threshold triggers frequent restructuring in response to transient spikes, while an overly conservative one fails to adapt to genuine, sustained workload shifts.

The core concept of the volatility-aware dynamic threshold mechanism is to increase the threshold during periods of high volatility to filter out noise, while lowering it during stable periods to maintain responsiveness. Formally, we define workload volatility over a sliding observation window of size W. Let r_i denote the read ratio in the i-th window, and μ_r the average read ratio over the recent W windows. The read ratio volatility σ_r is computed as follows:

$$\sigma_r = \sqrt{\frac{1}{W}\sum_{i=1}^{W}(r_i - \mu_r)^2} \tag{8}$$

Based on this volatility measure, the dynamic threshold τ_{dynamic} is calculated as:

$$\tau_{\text{dynamic}} = \alpha \cdot \sigma_r \tag{9}$$

where α is a sensitivity coefficient that controls the system's reaction to workload fluctuations.

This intelligent triggering mechanism seamlessly integrates with DHB+Tree's online adaptation. The system continuously monitors the read-write mix through an operation sliding window. Restructuring is invoked only when the workload shift exceeds τ_{dynamic}, preventing frequent restructuring triggered by transient fluctuations while ensuring the adaptation overhead is amortized by significant performance gains. When restructuring is triggered, leaf nodes logically adjust the boundary between ordered and unordered zones without physical data movement, thereby minimizing the adaptation cost.

4 Experiments and Evaluation

4.1 Environment and Setup

Experimental Platform. Experiments were conducted on a Linux server equipped with dual-channel Montage Jintide C6230R CPUs. Each CPU has 26 physical cores (2 logical threads per core), providing a total of 52 physical cores and 104 logical threads. The system includes 62 GB DRAM and 2×128 GB Intel Optane DCPMM persistent memory modules. The DCPMM is configured in App Direct non-interleaved mode and mounted as an EXT4 file system with DAX extensions. Threads are bound to physical cores within a fixed NUMA node using `pthread_setaffinity_np()`. All code was compiled using g++ 11.2.1 with `-O3` optimization enabled.

Workloads. The experiments employed microbenchmarks and PiBench macrobenchmarks [12] to generate diverse workloads. Each workload involved warming up the index with 1 million key-value pairs (KV) to ensure the data size surpassed the L3 cache, thereby ensuring the test results accurately reflected persistent memory (PM) performance. Subsequently, each test executed 1 million operations, repeated three times with results averaged. For scan operations, 100 consecutive records were retrieved starting from a random key. The default configuration utilized 8-byte keys and 8-byte values.

Baseline Comparisons. The evaluation compared DHB+Tree with BzTree [4], FAST&FAIR [6], SSB-Tree [7]and CC-Tree [20]. All systems store the entire tree in PM without additional logs for nearly instant recovery. To facilitate the comparison of experimental results, the node size of all systems was uniformly set to 1280 bytes. All implementations were compiled with -O3 optimization. In addition, to preliminarily explore the generality of DHB+Tree's core design (i.e., dynamic hybrid storage layout) across different memory tiers, we performed supplementary comparisons between a simple hybrid architecture variant DHB+Tree-H (where internal nodes are in DRAM and leaf nodes are in PM),and some representative hybrid architecture indexes (including PAC-Tree [21], FP-Tree [8], uTree [9] and LB+-Tree [10]).

4.2 Overall Performance Comparison

Microbenchmark Performance. Under both single-threaded and multi-threaded uniform workload evaluations (Fig. 4(a)), DHB+Tree demonstrates significant advantages across various operation types and concurrency scenarios, owing to its dynamic hybrid layout. In single-threaded tests, it achieves 70% higher insertion performance than SSB-Tree by dynamically expanding the unordered zone during writes to reduce node modification frequency. This is further bolstered by its lazy split, which minimize persistence overhead by deferring parent node updates and consolidating underutilized nodes only when necessary. For search operations, its throughput is 1.03× higher than FAST&FAIR, 1.07× higher than SSB-Tree, and 1.18× higher than CC-Tree, benefiting from binary search while overcoming the inherent limitations of SSB-Tree's dual-buffer lookups and CC-Tree's fully unordered leaf nodes. Additionally, DHB+Tree adopts a deferred splitting and merging strategy: during a split, only orphan nodes are created and linked via sibling pointers, which delays parent node updates and thus reduces persistence overhead; merging is triggered when node utilization falls below 50%, effectively maintaining tree balance and further improving insertion performance.

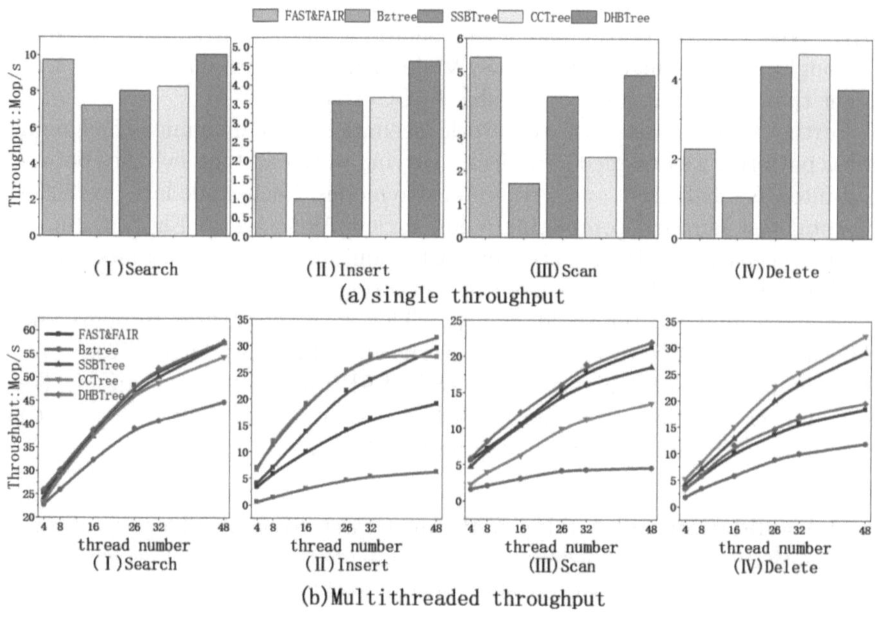

Fig. 4. Microbenchmark Performance.

In multi-threaded environments(Fig. 4(b)), all indexes exhibit near-linear scaling within 4 to 26 threads, though scalability ceilings diverge: BzTree reaches its bottleneck early due to excessive persistent memory access; CC-Tree reduces contention through metadata partitioning; SSB-Tree achieves near-lock-free progress via multi-versioning; while DHB+Tree, combining fine-grained write locks with optimistic concurrency control, achieves the highest aggregate throughput at 48 threads despite exhibiting slightly inferior scalability—fully validating the overall efficiency of its ordered/unordered hybrid structure in high-concurrency scenarios.

Performance Under Mixed and Dynamic Workloads. To comprehensively evaluate the performance of DHB+Tree under different read-write ratios, we used PiBench [12] to generate three types of mixed workloads: read-intensive (read:write = 9:1), balanced (read:write = 5:5), and write-intensive (read:write = 1:9). All experiments were conducted with 26 threads, with one thread pinned to each physical core.

Figure 5 presents a comparison of throughput between DHB+Tree and baseline methods (FAST&FAIR, BzTree, SSB-Tree, CC-Tree). Under the read-intensive workload (Fig. 5(a)), DHB+Tree achieves a throughput 1.04–1.45× higher than the baselines by dynamically expanding the ordered zone to optimize search paths. Under the balanced workload (Fig. 5(b)), the throughput improvement reaches 1.13–1.84×. For the write-intensive workload (Fig. 5(c)), DHB+Tree reduces insertion latency by enlarging the unordered zone, achieving a throughput 1.71× higher than FAST&FAIR, 3.42× higher than BzTree, 1.18× higher than SSB-Tree, and comparable to CC-Tree.

Furthermore, to simulate real-world scenarios with dynamically changing access patterns, we designed a cyclically varying workload that switches between read-intensive, balanced, and write-intensive modes. Each mode lasts for 20% of the total test number, completing 5 full cycles. As shown in Fig. 5(d), in this dynamic scenario, DHB+Tree continuously monitors the operation mix through a sliding window and triggers node restructuring only when the workload shift exceeds the dynamic threshold τ_{dynamic}. This strategy ensures responsiveness while avoiding excessive adjustment overhead. Ultimately, DHB+Tree's overall throughput is 47.8% higher than FAST&FAIR, 39.6% higher than CC-Tree, 29.2% higher than SSB-Tree, and 1.16× higher than BzTree, fully demonstrating its superior adaptability to time-varying workloads.

Model Validation and Analysis. Parameter Sensitivity Analysis: To evaluate the sensitivity of key parameters in the dynamic threshold model, we conducted a grid search test under a cyclically varying workload with the window count W set to 5. The window size M was defined as a percentage of the test dataset, ranging from 0.8% to 1.5%, while the sensitivity coefficient α was tested from 0.2 to 0.6. The results, shown in Fig. 6(a), indicate that M must strike a balance between sensitivity and stability: when M is too small or α is too large, the system becomes overly sensitive to transient fluctuations, leading to frequent restructuring; conversely, when M is too large or α is too small, response lag occurs, missing optimization opportunities. Comparative analysis confirms that

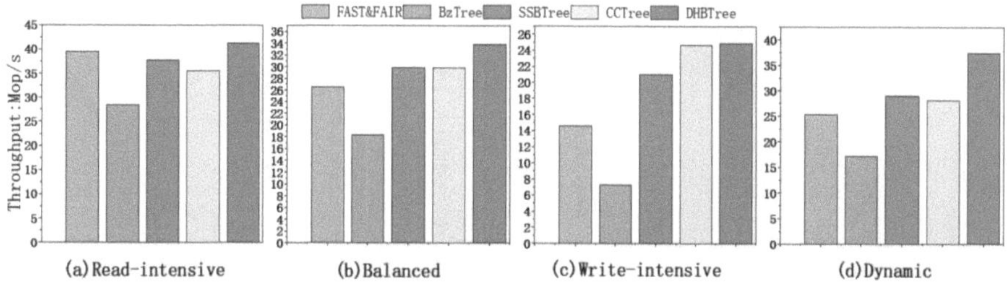

Fig. 5. Performance under Mixed and Dynamic Workloads.

$M = 1\%$ and $\alpha = 0.4$ achieve the optimal balance. Therefore, we set $M = 1\%$ and $\alpha = 0.4$ as the default system parameters.

Component Contribution: To precisely assess the independent contributions of the key components in DHB+Tree, we conducted experiments under single-threaded, periodically varying workload conditions. The core rationale for this strategy lies in the fact that DHB+Tree's adaptive design is specifically optimized for scenarios with time-varying workloads; static tests would underestimate the practical value of its adaptive features. Specifically, as shown in Fig. 6(b), binary search contributes an approximate 8% throughput improvement by boosting search efficiency. Building on this, dynamic parameter prediction further yields a 25% gain, underscoring the core value of the adaptive layout. The introduction of a fixed threshold (τ =10%) delivers an additional 3% optimization. Finally, the dynamic threshold mechanism achieves a further 13% performance improvement by balancing reconstruction overhead—eliminating unnecessary reconstructions while reinforcing necessary ones. Overall, compared to the traditional B+Tree with a fully ordered layout, the complete DHB+Tree achieves a cumulative performance improvement of approximately 50%.

Prediction Model Accuracy: To verify the accuracy of the parameter prediction model, we conducted experiments using PiBench in a single-threaded environment, with a balanced workload (read:write = 5:5). Meanwhile, to evaluate the model under controlled static conditions, we disabled the workload ratio detection module. We determine the actual optimal (m, k) configuration through exhaustive search and compare it with the model's prediction results. As shown in Fig. 6(c), the absolute percentage error is only 6.44%, which demonstrates its high accuracy in online guidance.

Space Consumption. We measure memory footprint after inserting 1 million 8-byte key-value pairs (15 MB of raw data). As shown in Fig. 7, the memory footprints of DHB+Tree, FAST&FAIR, SSB-Tree, and CC-Tree are 31.2 MB, 29.7 MB, 50.3 MB, and 32.9 MB, respectively. DHB+Tree reduces SSB-Tree's footprint by 40% by eliminating the dual versioning required for its near-lock-free structure. DHB+Tree's footprint is slightly higher than FAST&FAIR due to added search-optimizing metadata (upper keys) and counters for the unordered zone. It is slightly lower than CC-Tree, as CC-Tree uses bitmaps on leaf nodes

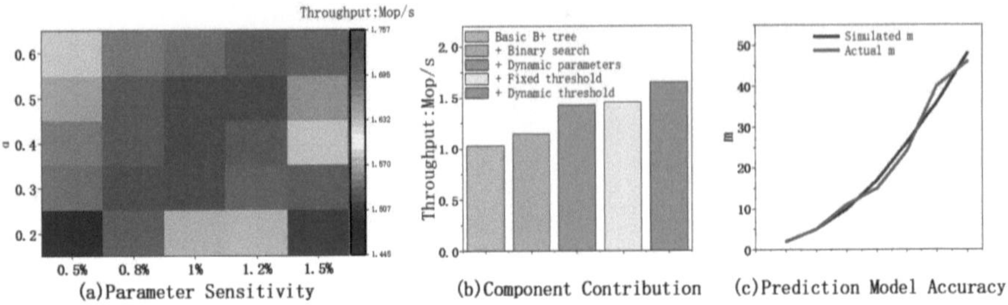

Fig. 6. Model Validation and Analysis.

to facilitate lookups. BzTree results are omitted because its open-source implementation fails to reclaim obsolete nodes.

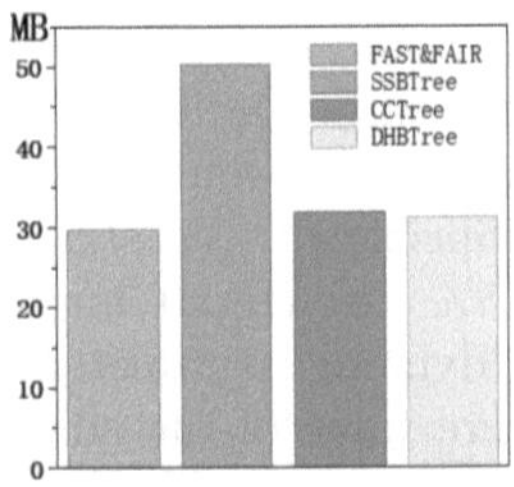

Fig. 7. Space Consumption

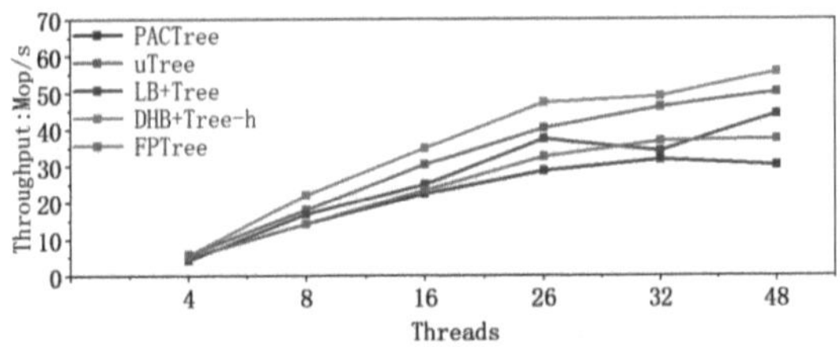

Fig. 8. Performance under DRAM-NVM

Scalability on Hybrid Memory Architecture. To evaluate the generality of our design, we assess a simple hybrid variant, DHB+Tree-H, with inner nodes in DRAM and leaf nodes in PM. The comparison baselines include PAC-Tree [21], FP-Tree [8], uTree [9] and LB+-Tree [10]. The tests use YCSB Workload E (50% reads/50% writes) as the baseline workload to measure throughput. As shown in Fig. 8, DHB+Tree-H outperforms specialized hybrid indexes such as FP-Tree. This demonstrates that the core innovation of DHB+Tree—the dynamic hybrid storage layout—is effective across different memory hierarchies, not just limited to all-PM configurations. Its adaptability proves to be an architecture-agnostic advantage.

5 Conclusion

This paper presents DHB+Tree, a workload-aware B+-Tree that dynamically adapts its storage layout to bridge the read-write performance gap in persistent memory. By introducing two key innovations—a **Dynamic Hybrid Storage Layout** that elastically balances ordered and unordered zones within leaf nodes, and a **Lightweight Parameter Prediction Model** that guides this adaptation online—DHB+Tree effectively overcomes the rigidity of prior static designs.

Extensive experiments demonstrate that our adaptive approach consistently outperforms state-of-the-art PM-optimized trees. DHB+Tree achieves a throughput improvement of 1.27× over SSB-Tree and 1.32× over CC-Tree under dynamic workloads, proving its superior ability to handle real-world, shifting access patterns without sacrificing instant recovery.

For future work, we plan to extend the adaptability across the hybrid memory hierarchy. This involves designing a DRAM-backed write buffer with a lightweight PM-based WAL to further reduce write pressure on PM, and generalizing our parameter model to co-optimize DRAM buffer sizing and flush policies.

Acknowledgments. This work is supported by the National Key R&D Program of China (Grant No. 2023YFB4503600) and the National Natural Science Foundation of China (Grants No. 62172082, 62072084, 62072086). We thank the anonymous reviewers for their valuable feedback.

Disclosure of Interests. The authors have no competing interests to declare that are relevant to the content of this article.

References

1. Yang, J., Kim, J., Hoseinzadeh, M., Izraelevitz, J., Swanson, S.: An empirical guide to the behavior and use of scalable persistent memory. In: 18th USENIX Conference on File and Storage Technologies (FAST 2020), pp. 169–182. USENIX Association (2020)
2. Yang, J., Wei, Q., Chen, C., Wang, C., Yong, K.L., He, B.: NV-Tree: reducing consistency cost for NVM-based single level systems. In: 13th USENIX Conference on File and Storage Technologies (FAST 2015), pp. 167–181. USENIX Association (2015)
3. Chen, S., Jin, Q.: Persistent B+-trees in non-volatile main memory. Proc. VLDB Endow. **8**(7), 786–797 (2015). https://doi.org/10.14778/2752939.2752947
4. Arulraj, J., Levandoski, J., Minhas, U.F., Larson, P.: BZTREE: a high-performance latch-free range index for non-volatile memory. Proc. VLDB Endow. **11**(5), 553–565 (2018). https://doi.org/10.1145/3164135.3164147
5. Wang, T., Levandoski, J., Larson, P.: Easy lock-free indexing in non-volatile memory. In: 2018 IEEE 34th International Conference on Data Engineering (ICDE), pp. 461–472. IEEE (2018). https://doi.org/10.1109/ICDE.2018.00049
6. Hwang, D., Kim, W., Won, Y., Nam, B.: Endurable transient inconsistency in byte-addressable persistent B+-tree. In: 16th USENIX Conference on File and Storage Technologies (FAST 2018), pp. 187–200. USENIX Association (2018)
7. Li, T., Wang, H., Shao, A., Wang, D.: SSB-Tree: making persistent memory B+-trees crash-consistent and concurrent by Lazy-Box. In: 2022 IEEE International Parallel and Distributed Processing Symposium (IPDPS), pp. 70–80. IEEE (2022). https://doi.org/10.1109/IPDPS53621.2022.00016

8. Oukid, I., Lasperas, J., Nica, A., Willhalm, T., Lehner, W.: FPTree: a hybrid SCM-DRAM persistent and concurrent B-tree for storage class memory. In: Proceedings of the 2016 International Conference on Management of Data (SIGMOD 2016), pp. 371–386. ACM (2016). https://doi.org/10.1145/2882903.2915251
9. Chen, Y., Lu, Y., Fang, K., Wang, Q., Shu, J.: UTree: a persistent B+-tree with low tail latency. Proc. VLDB Endow. **13**(12), 2634–2648 (2020). https://doi.org/10.14778/3407790.3407850
10. Liu, J., Chen, S., Wang, L.: LB+Trees: optimizing persistent index performance on 3DXPoint memory. Proc. VLDB Endow. **13**(7), 1078–1090 (2020). https://doi.org/10.14778/3384345.3384355
11. Li, Z., et al.: CCL-BTree: a crash-consistent locality-aware B+-tree for reducing XPBuffer-induced write amplification in persistent memory. In: Proceedings of the Nineteenth European Conference on Computer Systems (EuroSys 2024), pp. 441–455. ACM (2024). https://doi.org/10.1145/3627703.3629582
12. Lersch, L., Hao, X., Oukid, I., Wang, T., Willhalm, T.: Evaluating persistent memory range indexes. Proc. VLDB Endow. **13**(4), 574–587 (2019)
13. Zhang, Z., et al.: PLIN: a persistent learned index for non-volatile memory with high performance and instant recovery. Proc. VLDB Endow. **16**(2), 243–255 (2022). https://doi.org/10.14778/3565816.3565826
14. Zhang, Z., et al.: PLIN: a persistent learned index for non-volatile memory with high performance and instant recovery. Proc. VLDB Endow. **16**(2), 243–255 (2022). https://doi.org/10.14778/3565816.3565826
15. Kraska, T., Beutel, A., Chi, E.H., Dean, J., Polyzotis, N.: The case for learned index structures. In: Proceedings of the 2018 International Conference on Management of Data (SIGMOD 2018), pp. 489–504. ACM (2018). https://doi.org/10.1145/3183713.3196909
16. Ding, J., et al.: ALEX: an updatable adaptive learned index. In: Proceedings of the 2020 ACM SIGMOD International Conference on Management of Data (SIGMOD 2020), pp. 969–984. ACM (2020). https://doi.org/10.1145/3318464.3389711
17. Wu, J., Zhang, Y., Chen, S., Wang, J., Chen, Y., Xing, C.: Updatable learned index with precise positions. Proc. VLDB Endow. **14**(8), 1276–1288 (2021). https://doi.org/10.14778/3457390.3457393
18. Yu, W., Luo, S., Yu, Z., Cong, G.: CAMAL: optimizing LSM-trees via active learning. Proc. ACM Manage. Data **2**(4), 1–26 (2024). https://doi.org/10.1145/3677138
19. Dayan, N., Athanassoulis, M., Idreos, S.: Monkey: optimal navigable key-value store. In: Proceedings of the 2017 ACM International Conference on Management of Data (SIGMOD 2017), pp. 79–94. ACM (2017). https://doi.org/10.1145/3035918.3064054
20. Wei, Y., Zhang, X.: A concise concurrent B+-tree for persistent memory. ACM Trans. Archit. Code Optim. **21**(2), 1–25 (2024). https://doi.org/10.1145/3638717
21. Kim, W.-H., Krishnan, R.M., Fu, X., Kashyap, S., Min, C.: A high performance persistent range index using PAC guidelines. In: Proceedings of the ACM SIGOPS 28th Symposium on Operating Systems Principles (SOSP 2021), pp. 424–439. Association for Computing Machinery, New York, NY, USA (2021). https://doi.org/10.1145/3477132.3483589

LogTrie: Adaptive Trie-Structured Compression with Query Support for Streaming Logs

Ying Chu, Chaofan Li, Peiquan Jin(✉), and Shouhong Wan

University of Science and Technology of China, Hefei 230026, China
jpq@ustc.edu.cn

Abstract. Streaming JSON logs are widely used in modern cloud-native systems for their flexibility and expressiveness, but they pose significant challenges for storage and query efficiency. Existing compressors either achieve high compression at the expense of queryability or support only limited query patterns, leaving residual redundancy. In this work, we present LogTrie, a system that jointly optimizes compression and query performance for large-scale JSON logs. The novelty of LogTrie lies in two aspects. First, we propose a trie-based representation framework for streaming JSON logs that separates structure from values and further eliminates value redundancy. Second, we present a two-stage compression pipeline for encoding JSON log structures using type-aware, layer-wise algorithms. Unlike existing compressors, which prioritize compression efficiency at the expense of query performance, LogTrie can deliver high query performance while maintaining compression efficiency. We evaluate LogTrie on 18 real-world JSON log datasets and compare it with five existing compressors, including three general-purpose compressors (Gzip, Zstandard, and LZMA) and two state-of-the-art log compressors (μSlope and LogLite). The results show that LogTrie achieves an average compression of 59.56×, outperforming μSlope by 41%, and reduces query latency by up to 46.6% on aggregation queries.

Keywords: Log Compression · Trie Tree · Structural Redundancy · JSON

1 Introduction

Modern cloud-native infrastructures produce vast amounts of JSON-formatted streaming logs each day, often reaching terabyte scale at the cluster level [2,7,20]. Such logs play a central role in system operations, supporting failure diagnosis, performance monitoring, and compliance auditing [12]. At the same time, the growing need for long-term retention to satisfy regulatory and forensic requirements further amplifies the challenges of efficient storage and access.

The source code is available at https://github.com/PB20111610/JSON/.

H. Jung et al. (Eds.): DASFAA 2026, LNCS 16535, pp. 639–655, 2026.
https://doi.org/10.1007/978-981-92-0363-5_39

As data volumes continue to increase and JSON logging becomes the standard practice for observability in cloud-native environments [9], log structures have grown increasingly diverse across systems and applications. Different services define their own field hierarchies, value types, and nesting patterns, resulting in heterogeneous global schemas that complicate log management and analytical processing. Yet within a single system, logs often display strong structural regularity and locality, as many records share similar field layouts and recurring key/value patterns. Global schema variability increases compression cost, and local consistency offers opportunities for query optimization. However, the irregular and nested nature of logs continues to hinder direct querying on compressed data. Achieving both high compression efficiency and interactive query performance remains a central bottleneck in operational analytics for cloud-native systems [24]. For instance, LogLite [22] achieves strong compression but loses query functionality, whereas μSlope [24] retains limited query support at the cost of reduced compression efficiency. As a result, JSON logs present multiple inherent challenges for compression and query efficiency: (1) Structural redundancy arises from repeated field hierarchies across different log patterns, which inflates storage when treated as plain text. (2) Value redundancy exists among records, and dictionary-based compressors like μSlope reduce storage overhead but do not fully eliminate this redundancy. (3) Query processing incurs overhead from repeatedly parsing, tokenizing, and transforming JSON structures, even for data that does not match the predicate.

To verify the structural redundancy in JSON logs, we collect 18 JSON log datasets and use the following two metrics to measure the redundancy in each dataset: (1) *Occurrence frequency* (f_i) captures the fraction of log records containing a given field; (2) *Unique-value cardinality* (u_i) counts the number of distinct values for each field; within a single dataset.

The results are shown in Table 1, which confirms substantial heterogeneity across datasets while revealing consistent intra-dataset patterns. Specifically, field-level redundancy exists in each JSON log dataset. We observe that numeric fields are consistently present across most datasets ($f_i \approx 1.0$ in 14 of 18 datasets), whereas string and LogType fields exhibit substantial variability, reflecting heterogeneous schemas across systems. Fields like strings in Proxifier and Windows exhibit low cardinality, indicating high local value redundancy. Timestamp and LogType fields are generally stable across most datasets.

Motivated by the results in Table 1, we present **LogTrie** to address the challenges in JSON log compression. LogTrie aims to improve both compression and query performance for large-scale JSON logs. In particular, LogTrie is designed to exploit local consistency through schema-aware block partitioning and to mitigate global variability via Trie-based structural representation, achieving both efficient compression and interactive query performance. Briefly, we make the following contributions in this paper:

- LogTrie proposes a trie-based representation framework for streaming JSON logs that separates structure from values, thereby eliminating value redun-

Table 1. Field-level redundancy statistics (f_T, u_T) in 18 JSON log datasets.

Field Type	PostgreSQL		MongoDB		Click		Thurderbird		OpenSSH		Apache	
	f_T	u_T	f_T	u_T	f_T	u_T	f_T	u_T	f_T	u_T	f_T	u_T
String	0.65	62552	1.00	334568	0.81	21	0.34	670	1.00	38	1.00	52007
Numeric	0.80	187610	1.00	500001	0.98	500089	1.00	500001	1.00	638947	0.00	0
LogType	0.34	383794	1.00	1	0.00	0	0.99	52145	1.00	181592	1.00	14174
Timestamp	1.00	42518	0.00	0	0.67	137033	0.00	0	0.00	0	0.00	0

Field Type	HPC		Proxifier		Windows		Zookeeper		BGL		Spark	
	f_T	u_T	f_T	u_T	f_T	u_T	f_T	u_T	f_T	u_T	f_T	u_T
String	0.63	113	1.00	11	0.67	7	0.33	95	0.33	136	1.00	167
Numeric	1.00	429988	1.00	21320	0.00	0	1.00	74273	1.00	500000	1.00	500000
LogType	0.56	9595	1.00	7884	0.50	59366	1.00	23925	1.00	19566	1.00	324898
Timestamp	0.00	0	0.00	0	1.00	1073	0.00	0	0.00	0	0.00	0

Field Type	Linux		HDFS		Hadoop		HealthApp		Mac		Openstack	
	f_T	u_T	f_T	u_T	f_T	u_T	f_T	u_T	f_T	u_T	f_T	u_T
String	0.34	352	1.00	32	0.33	240	0.42	28062	0.36	651	1.00	48
Numeric	1.00	23921	1.00	500000	1.00	179993	1.00	212394	1.00	100314	1.00	207632
LogType	1.00	11319	1.00	461923	1.00	61395	0.86	95041	0.96	39293	1.00	139782
Timestamp	0.00	0	0.00	0	0.00	0	0.00	0	0.00	0	0.00	0

dancy. With this approach, LogTrie can effectively capture redundancy in JSON logs.

- LogTrie presents a two-stage compression pipeline for structure encoding with type-aware and layer-wise algorithms for JSON logs. Unlike existing compression methods, which prioritize compression efficiency at the expense of query performance, LogTrie can deliver high query performance while maintaining compression efficiency.
- We evaluate LogTrie on 18 real-world JSON log datasets from production and public sources. LogTrie achieves an average compression ratio of 59.56×, improving upon μSlope by 41%. It further demonstrates superior query performance, particularly in aggregation queries that μSlope cannot support, reducing execution time by up to 46.6%.

2 Related Work

While JSON log compression has been explored from multiple perspectives [4,5, 16,19,22,24], existing methods struggle to balance compression efficiency with query performance, highlighting the need for systems that jointly optimize both for large-scale JSON datasets. General-purpose compressors such as Gzip [5], Zstandard [4], and LZMA [19] achieve moderate-to-high compression ratios but require full decompression for query execution, limiting interactive analysis. The recent JSON-aware compressor LogLite encodes records based on length patterns, discarding the original JSON structure and thereby eliminating support for queries over compressed data. Environments with tight resource constraints, such as edge computing [21], face heightened challenges, as both storage efficiency and query responsiveness are critical.

The concept of operating directly on compressed data has roots in decades of database research [1]. Succinct [1] demonstrated that random access and search could be supported over compressed columnar data by exploiting entropy-bounded encodings, eliminating the need for full decompression. However, such techniques assume fixed, well-defined schemas, which do not generalize well to semi-structured formats such as JSON, where field presence and hierarchy may vary across records. Traditional database compression techniques, such as dictionary encoding and run-length encoding, enable limited query execution without full decompression, but struggle with JSON's structural variability. Recent efforts have extended queryable compression to log analytics. LogZip [11] compresses logs by clustering similar lines and extracting latent templates, effectively capturing repetition in flat text logs but discarding hierarchical structure. SEAL [8] introduces query-friendly compression for causality analysis in enterprise systems, yet its design targets relational event graphs rather than deeply nested JSON data. Recently, μSlope [24] extends this line of work to semi-structured formats, extracting basic structural patterns and applying dictionary-based encoding to support limited queries. However, it does not fully eliminate value-level redundancy, leaving residual storage overhead. The core challenge is to preserve sufficient structural metadata for expressive queries while maintaining compression efficiency.

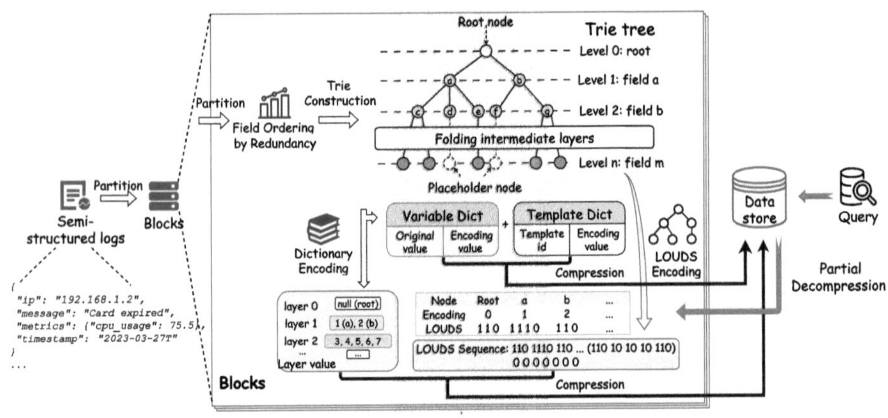

Fig. 1. Overview of LogTrie.

3 Design of LogTrie

To address the limitations of existing methods, in this paper we leverage Tries [10] and LOUDS [25] to decouple structure from values. A trie, or prefix tree, is a hierarchical data structure that organizes keys based on shared prefixes, enabling efficient representation of repetitive structural patterns. In the context of JSON logs, Tries can compactly encode repeated field hierarchies across records, reducing structural redundancy while preserving the complete schema. By exploiting structural and field-level redundancies, LogTrie organizes data into an optimized layout that simultaneously improves compression and query performance.

3.1 Overview

As illustrated in Fig. 1, LogTrie adopts a dual-compression architecture that decouples structural information from field values, allowing each to be encoded using techniques most suitable for its characteristics. During Trie construction, fields with higher redundancy scores are promoted to upper levels, thereby eliminating value redundancy through maximal structural sharing. To preserve strict layer alignment, missing fields are represented by lightweight placeholder nodes (1-bit markers), ensuring a consistent correspondence between each Trie layer and its field semantics.

The compression pipeline consists of two stages:

(1) The Trie topology is serialized using LOUDS [25], a succinct level-order encoding that represents the tree as a bit vector and supports $O(1)$ child access and $O(d)$ path navigation for nodes at depth d. LOUDS eliminates the need for explicit pointer-based structures, making it well-suited for representing JSON Tries in a compressed form.

(2) Field values, naturally organized into type-homogeneous layers, are compressed through block-level dictionary encoding followed by type-aware layer-wise algorithms, achieving high compression ratios while maintaining efficient selective access.

Beyond compression efficiency, LogTrie also enables direct query execution on compressed data. The LOUDS-encoded structure functions as a hierarchical index, supporting structural queries without full decompression.

3.2 Data Organization

The foundation of our compression framework is the transformation of raw JSON records into a unified, highly structured Trie representation.

Adaptive Schema-Driven Partition. To balance the benefit of prefix sharing against the cost of branch explosion, we partition the data into blocks with high schema consistency. Given two records R_i and R_j with respective field sets F_i and F_j, and let w_f denote a field-specific weight that reflects the stability of field

f, the structural distance between R_i and R_j can be formulated as a *weighted schema distance* [6]:

$$\text{Dist}_w(R_i, R_j) = 1 - \frac{\sum_{f \in F_i \cap F_j} w_f}{\max(\sum_{f \in F_i} w_f, \sum_{f \in F_j} w_f)}$$
$$= 1 - \frac{|F_i \cap F_j|}{\max(|F_i|, |F_j|)} \quad (\text{for } w_f = 1)$$
$$\triangleq 1 - \frac{|F_i \cap F_j|}{M} = \frac{|F_i \triangle F_j|}{2M} \triangleq \frac{\Delta_{\text{set}}}{2M}$$

where $M = \max(|F_i|, |F_j|)$ and $\Delta_{\text{set}} = |F_i \triangle F_j|$ denotes the symmetric difference of field sets.

Records with smaller Dist_w indicate greater field-set overlap among records, facilitating their grouping into homogeneous blocks and yielding a more stable Trie structure for compression. Our partition enforces this by creating a new block whenever schema distance Dist_w exceeds a threshold θ, or if the block size reaches the upper limit N_{max}. The dynamic, schema-aware process inherently adapts to the structural diversity in log data, ensuring high intra-block consistency while keeping the Trie manageable.

Redundancy-Guided Field Ordering. Within each partitioned block, the Trie's hierarchical structure is determined by the field order. LogTrie dynamically arranges fields to place structurally stable and semantically redundant attributes closer to the root, while deferring highly variable or sparse ones to lower levels.

As described in Algorithm 1, we first classify fields into three groups (A, B, C) based on their redundancy characteristics, f_i and u_i. Groups A and C, comprising fields of highest and lowest redundancy respectively, are sorted in descending order of f_i. For Group B, which contains fields that are frequent but exhibit moderate value diversity, fields are first partitioned into clusters by semantic type according to a designated type order $\mathcal{T}$, and then sorted in ascending order of u_i within each cluster. The final sequence F_{ord} is assembled by concatenating the processed groups. Accordingly, fields with high f_i and low u_i migrate toward upper Trie levels, whereas infrequent or high-cardinality fields are positioned closer to the leaves.

Trie Construction. Following the field order F_{ord}, each JSON record is mapped to a root-to-leaf path, where each level corresponds to a specific field. The Trie is built by traversing each record from the root and, at each level, either following an existing branch or creating a new one based on the value of the current field.

To maintain strict structural alignment despite minor schema variations, a 1-bit placeholder node is inserted for each field in F_{ord} that is absent from the current record, ensuring consistent traversal and precise node positioning within the hierarchy. Owing to the preceding schema-aware partitioning, the proportion of placeholder nodes remains low, which safeguards compression efficiency without sacrificing exact reconstruction.

Algorithm 1: Redundancy-Guided Field Ordering

Input: Record set R, field manager M, type order $\mathcal{T}$
Output: Ordered field list F_{ord}
stats(f_i, u_i) ← collectFieldStats(R, M);
Classify each field into groups:
  A: highly redundant (f_i high, u_i low);
  B: stable but diverse (f_i high, u_i moderate);
  C: infrequent or unique (f_i low or u_i high);
sort(A by f_i desc);
sort(C by f_i desc);
Let B_t denote the subset of B with type t, then
  $B_{ord} \leftarrow \bigoplus_{t \in \mathcal{T}} \text{sort}(B_t \text{ by } u_i \text{ asc})$;
return F_{ord};

3.3 Compression

The Trie's topology and the value sequences of individual layers are compressed independently, allowing each layer and the structural component to employ configurable compression methods while preserving efficient query access.

Topology Compression. The Trie topology is serialized using the LOUDS encoding [25]. The tree is traversed in breadth-first order, and each node's degree (number of children) is encoded as a sequence of ones followed by a terminating zero. This bit-vector representation preserves the level order of nodes and allows the entire tree structure to be stored compactly without explicit pointers.

LOUDS achieves near-optimal space usage, approaching the information-theoretic lower bound for ordered trees, while supporting $O(1)$ navigation between parent and child nodes when augmented with succinct rank/select [18] data structures. To further reduce storage overhead, the LOUDS bit sequence is compacted using a lightweight bit-packing encoding [14,15], thereby maintaining random-access efficiency.

Type-Aware Layer Compression. Due to the canonical field ordering, all nodes at the same Trie level correspond to values of the same data type. Such homogeneity in data types permits the application of compression strategies tailored to each layer, with the specific encoding approach [24] varying according to the data type. Accordingly, we employ the following type-specific encoding schemes:

- *Short string*: Single-word string values maintain a variable dictionary to compactly encode repeated entries.
- *Long string sequences (LogType)*: Strings containing spaces, typical in log messages, are handled by constructing a template dictionary and a variable dictionary, which are then used for joint encoding.
- *Timestamps*: Due to their regular patterns in log data, timestamps are identified via regular expressions to extract their templates, and template and variable dictionaries are created following the same logic as LogType.

- *Numeric and boolean*: These do not require dictionary encoding and are represented in their raw binary form.

At this stage, the raw layer values are converted into compact, encoded representations that substantially reduce the layer storage footprint and further eliminate value redundancy. The resulting dictionaries and encoded layer values are then compressed using configurable methods. A thorough evaluation of different compression algorithms and parameter settings is presented in Sect. 4.
Metadata Compression. Beyond the Trie topology and layer values, LogTrie also compresses essential block-level metadata to support data reconstruction and query processing. The metadata encompasses the canonical field order, represented as a sequence of *FieldKey* (field name, type) pairs, which defines the structural schema for record reassembly. It further includes compression configuration parameters to ensure the correct decoding of the value layers. Additionally, size information for each value layer is stored to facilitate efficient conversion between LOUDS indices and the corresponding value arrays, thereby enabling random access during query execution.

3.4 Query

The query engine of LogTrie is designed to operate directly on compressed data, avoiding full decompression to improve both I/O efficiency and CPU cache locality. It supports a broad spectrum of queries, including point lookups, range queries, aggregation operations, and logical compound queries, while retaining the performance benefits of compression throughout execution.

Figure 2 shows the query process based on LogTrie, which implements a progressive filtering strategy by mapping high-level queries to operations on the compressed representation, including metadata, the LOUDS bit-vector, the layer values, and the Variable and Template dictionaries corresponding to field types.

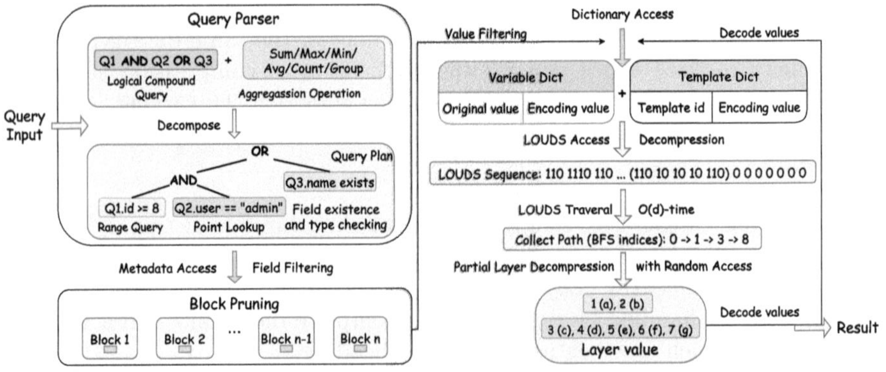

Fig. 2. Query Engine Workflow.

(1) *Query Parsing and Planning.* When a query is submitted, the *Query Parser* first analyzes its syntax. Complex queries, including those with logical operators (`AND`, `OR`, `NOT`) and aggregation functions, are decomposed into a set of fundamental execution units. These units consist of basic query statements, such as exact-match or range predicates, alongside the operations that process and organize their results. An execution plan is then generated to coordinate the evaluation of these units and to apply the necessary post-processing and aggregation to produce the final query output.

(2) *Block Pruning.* Before accessing any layer values, the system performs coarse-grained filtering by leveraging block-level metadata. Specifically, it checks the canonical field order of each block against the query predicates. If a block's schema contains no fields referenced in the query, the block is immediately skipped. This schema-aware pruning, achieved by inspecting only the lightweight metadata, avoids unnecessary access and decompression of irrelevant data, serving as the first and most effective level of filtering.

(3) *Selective Decompression.* Instead of decompressing entire blocks, LogTrie performs partial decompression at different granularity levels according to the query intent. The system progressively decompresses and filters data, decoding only the target portions required.

Once the candidate fields are identified, the corresponding dictionaries are selectively loaded to resolve value-level predicates, further filtering out irrelevant blocks. Structural navigation then proceeds using the LOUDS topology with $O(d)$ complexity, where d denotes the depth of the Trie, enabling constant-time access to child and sibling nodes regardless of dataset size. Only after the relevant paths and values are located does the system perform partial decompression, decoding just the layer values required by the query.

Aggregation queries, including `GROUP BY` with functions like `COUNT`, `SUM`, `AVG`, `MIN`, and `MAX`, are computed directly on compressed data by leveraging type-aware encoding. For `GROUP BY`, we derive grouping keys and aggregation targets through LOUDS-guided traversal, maintaining compact partial aggregates for each distinct key combination. Non-aggregation queries that require detailed output traverse trie paths and selectively decode only the necessary values to reconstruct complete JSON records.

(4) *Result Assembly.* For datasets spanning multiple compressed blocks, the engine first identifies the relevant partitions using global metadata indexes. Partial results from individual blocks are then consolidated into a single result set. Finally, the merged results are formatted and serialized to produce the complete query response in standard JSON output format.

4 Performance Evaluation

4.1 Settings

We evaluate LogTrie across 18 real-world and synthetic datasets to assess its performance under diverse structural and statistical characteristics. All experiments were conducted in an isolated Docker container environment based on Ubuntu

24.04 LTS. The underlying hardware was equipped with an AMD Ryzen 7700X 8-Core processor, 16 GB of RAM, and a 1 TB SSD.

Table 2. Datasets

Dataset	Size	#Records	Block size	Source
PostgreSQL	392.8 MB	1000,000	20000	*μslope*
MongoDB	30 GB	72,140,891	20000	*μslope*
Click	8.01 GB	32,428,911	10000	Kaggle
Thunderbird	3.07 GB	16,601,745	60000	Loghub
OpenSSH	126 MB	638,947	35000	Loghub
Apache	7.78 MB	51,978	10000	Loghub
HPC	47.0 MB	429,988	10000	Loghub
Proxifier	4.05 MB	21,320	21320	Loghub
Windows	26.32 GB	100,000,000	20000	Loghub
Zookeeper	10.8 MB	74,273	10000	Loghub
BGL	690.37 MB	4,631,261	10000	Loghub
Spark	2.47 GB	16,075,117	40000	Loghub
Linux	3.68 MB	23,921	23921	Loghub
HDFS	2.16 GB	11,167,740	40000	Loghub
Hadoop	32.3 MB	179,993	40000	Loghub
HealthApp	30.2 MB	212,394	10000	Loghub
Mac	22.1 MB	100,314	35000	Loghub
Openstack	46 MB	207,632	20000	Loghub

Baselines. We compare LogTrie against both general-purpose and specialized compression systems: (1) widely-adopted general-purpose compressors, including Gzip [5], Zstandard [4], and LZMA [19]; (2) state-of-the-art compressors specifically designed for semi-structured or logs, including μSlope [24] and LogLite [22].
Metrics. To evaluate performance, we consider two primary metrics:

- *Compression Ratio*, which quantifies storage reduction, defined as

$$CR = \frac{Original\ File\ Size}{Compressed\ File\ Size}.$$

- *Query Latency*, which measures the response time of queries executed directly on the compressed data.

Datasets. The evaluation is conducted on 18 publicly available datasets collected from μSlope, Loghub [17], and Kaggle, covering diverse semi-structured workloads with different schema complexities (Table 2). All datasets were normalized into a unified JSON format without modifying semantic content. For

transparency, the processed datasets are publicly released in our repository. During evaluation, each dataset is partitioned into adaptively sized blocks, which are kept reasonably small to control the scale of the constructed Trie structures.
Implementation Details of LogTrie. The LogTrie framework was implemented in C++17 with approximately 39k+ lines of code. To maximize compression effectiveness without compromising query performance, LogTrie adopts a type-aware compression scheme that tailors algorithms to each data type:

- Numeric: Delta encoding combined with variable-length integer compression (*Delta+VarInt*) to exploit numerical continuity [23].
- Boolean: Bit-packing [15] for maximum density.
- Timestamp and LogType: Each value is encoded as a template identifier with variable components, and the dictionaries are compressed with Zstandard [4].
- String: Dictionary encoding first maps repetitive text to integer identifiers, and the dictionaries are then compressed using Zstandard.
- Array: Run-length encoding (RLE) [3] to compactly represent consecutive duplicates or repeated structural patterns.

4.2 Compression Ratio

As shown in Table 3, specialized compressors generally attain higher compression ratios than general-purpose ones. A few exceptions, such as HealthApp and Mac, perform better under general-purpose compressors. According to the redundancy statistics in Table 1, these datasets contain a relatively small proportion of string-type fields (42% and 36%, respectively) and an unusually large number of unique values (28,062 and 651), leading to low structural redundancy. In such cases, structural encoding introduces additional metadata overhead rather than yielding compression benefits. For the vast majority of datasets, however, specialized compressors consistently outperform generic methods, underscoring the advantage of leveraging both structural and semantic redundancy inherent in log data. Consequently, our subsequent analysis focuses on comparing specialized schemes.

Among these specialized approaches, LogLite-BL achieves the highest compression ratios on most datasets, primarily due to its use of a highly effective LZMA backend for secondary compression and its complete discarding of structural information, thereby sacrificing all query capability in pursuit of extreme compression efficiency. In contrast, our system, LogTrie, strikes a more balanced trade-off between compression and accessibility. While maintaining full random-access and query capabilities, it outperforms μSlope across a range of dataset characteristics. Specifically, LogTrie achieves superior compression ratios on 13 out of 18 datasets, **38.6%** higher compression than μSlope on MongoDB, **133.2%** on Click, and **70.1%** on Thunderbird. Across all winning cases, LogTrie delivers an average improvement of **18.3%** over μSlope while preserving query functionality. The performance gap across the remaining five datasets is attributable to their unfavorable statistical characteristics, which limit the effectiveness of our approach.

Table 3. Compression ratio comparison across 18 datasets. LogTrie achieves consistently higher compression than general-purpose compressors and is highly competitive with specialized methods.

Method	PG	Mongo	Click	Thunder	SSH	Apache	HPC	Proxy	Win
General-purpose compressors									
Gzip	26.92	76.79	43.29	22.50	32.84	24.28	29.51	18.81	18.95
LZMA	31.55	183.74	81.85	54.75	49.41	38.49	60.78	28.93	256.68
Zstandard	22.96	125.12	49.74	33.92	30.53	27.77	38.68	18.11	164.39
Specialized compressors									
LogLite-BZ	31.79	397.81	148.56	73.68	68.48	47.57	91.84	23.00	321.41
LogLite-BL	43.50	**827.26**	**235.53**	**113.17**	**99.45**	**61.45**	**130.53**	28.03	**939.19**
μSlope	39.53	198.95	48.01	31.45	32.17	23.19	47.05	18.32	95.01
LogTrie	**63.51**	275.75	111.95	53.48	58.01	44.28	61.32	**28.32**	154.51

Method	ZK	BGL	Spark	Linux	HDFS	Hadoop	Health	Mac	Stack
General-purpose compressors									
Gzip	25.48	27.03	18.47	17.02	13.33	15.77	14.53	11.00	21.11
LZMA	44.47	59.16	25.89	33.96	19.58	33.65	25.59	32.29	29.88
Zstandard	30.72	31.79	18.70	22.80	13.66	23.67	16.14	22.13	22.02
Specialized compressors									
LogLite-BZ	72.89	65.20	39.40	35.83	15.19	32.62	29.14	32.29	32.50
LogLite-BL	**100.48**	**110.77**	**57.39**	**46.41**	20.91	**44.53**	**38.49**	**41.66**	**41.44**
μSlope	27.88	40.92	20.63	22.40	**21.56**	27.84	19.53	21.80	24.05
LogTrie	30.06	41.12	27.33	21.50	18.16	23.03	12.33	20.02	27.54

Overall, LogTrie demonstrates robust and competitive compression effectiveness across diverse workloads. It achieves the best results on datasets such as PostgreSQL and Proxifier, where a rich hierarchical structure enables substantial exploitation of redundancy. Notably, on the PostgreSQL dataset, LogTrie reaches a compression ratio exceeding 63×, **60.4%** higher than μSlope and **46%** higher than LogLite-BL, marking a significant advancement toward highly compressed yet queryable JSON storage. Across 18 datasets, LogTrie achieves an average compression ratio of 59.56×, outperforming μSlope by **41%** while maintaining full query capabilities.

4.3 Query Performance

To evaluate query efficiency, we conduct experiments on relatively large datasets, as smaller datasets would incur negligible I/O and decompression costs, making direct queries over raw JSON fast enough to obscure the benefits of querying on compressed data. Since μSlope has already demonstrated substantial advantages over several mainstream database systems in prior studies, we accordingly adopt it as the principal baseline for performance comparison. For reference, we also include traditional decompress-then-query baselines to represent the upper bound of query performance when full decompression is performed prior to query execution.

The query workload is designed to evaluate LogTrie's advantages in structural awareness and selective decompression. Table 4 lists the representative query patterns tested on the PostgreSQL dataset. Among them, *existence queries* demonstrate how LogTrie efficiently filters out irrelevant blocks, similarly to Bloom filters. *Aggregate and group queries* highlight LogTrie's ability to perform analytical operations directly on compressed data, a functionality that μSlope lacks. We also evaluate *composite queries* with logical `AND`/`OR` conditions to assess performance under multi-predicate workloads.

Table 4. Query performance (seconds) on the **PostgreSQL** dataset.

Category	Query Instances	Raw	μSlope	LogTrie
EXISTS	`user EXISTS`	2e-4	0.078	1e-4
AND	`user:a AND pid>1`	5.27	2.06	6.06
OR	`user:a OR pid:1`	5.30	2.17	6.28
Aggregate	`SUM(pid)`	1.36	-	2.37
Group-Aggr.	`GROUP BY pid`	5.95	-	11.8

Table 5. Comparison of query latency (seconds).

Method	Support	PG	Mongo	Click	Thunder	SSH	Win	BGL	Spark	HDFS
Traditional decompress-then-query										
Raw JSON	√	17.88	835.56	268.93	107.22	16.06	348.44	136.04	251.9	218.73
Gzip	×	18.42	1356.92	844.54	159.36	16.24	884.41	136.84	295.88	248.85
LZMA	×	19.83	1392.29	403.09	150.93	16.42	667.42	137.58	283.77	262.13
Zstandard	×	36.93	1161.63	511.26	141.77	16.13	694.99	137.58	274.77	258.25
LogLite	×	**18.34**	866.37	277.75	111.98	16.23	384.64	137.14	256.38	224.57
Queryable compressors										
μSlope	√×	23.538	737.88	237.76	**104.27**	8.04	305.28	47.33	**116.21**	**101.13**
LogTrie	√	26.51	**570.95**	**223.08**	114.76	**5.76**	**163**	**25.97**	125.97	108.84

Note: Query-capable compressors operate directly on compressed data without a separate decompression phase, while unsupported query types require full decompression prior to execution.

Table 5 reports total query time across 9 datasets, with each method executing 10 queries (two per category). On smaller datasets (e.g., PostgreSQL),

LogLite's rapid decompression (0.46 s) occasionally compensates for the absence of query support, resulting in comparable total time (18.34 s). However, this effect depends on raw JSON's query efficiency and diminishes with larger datasets. In contrast, queryable compressors demonstrate consistent advantages across most datasets. LogTrie achieves particularly strong performance on larger datasets, outperforming μSlope by **22.6%** on MongoDB and by **46.6%** on Windows. μSlope requires full decompression for unsupported queries, whereas LogTrie sustains stable performance across all query categories.

4.4 Ablation Studies

Since the entire system is built around a Trie-based structure, We conduct ablation experiments to examine how individual design components in LogTrie affect compression efficiency. The analysis focuses on three structural aspects of the Trie: field ordering, block partitioning, and type-aware compression. Specifically, *No Type-awareness* applies uniform Zstandard compression to all content, *No Block Partitioning* merges all records into a single block, and *No Field Ordering* randomizes field sequence during Trie construction.

Results in Fig. 3 demonstrate that type-aware encoding provides the most substantial improvements, increasing compression ratios by **16.8%** on PostgreSQL and **6.8%** on HPC. Field ordering contributes moderate but consistent gains, enhancing compression by **9.4%** on Click. The effect of block partitioning varies across datasets, yielding an **83.5%** improvement on PostgreSQL but slight gains on HPC. Interestingly, the configuration without partitioning even surpasses the partitioned setting on Thunderbird, suggesting that the optimal block size is highly data-dependent. This observation points to a trade-off between contextual redundancy and locality, which we further examine through parameter analysis in the next section.

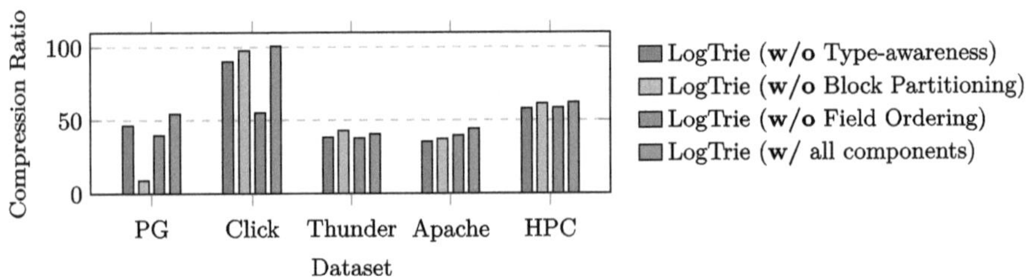

Fig. 3. Ablation study on 100,000-record samples evaluating core components.

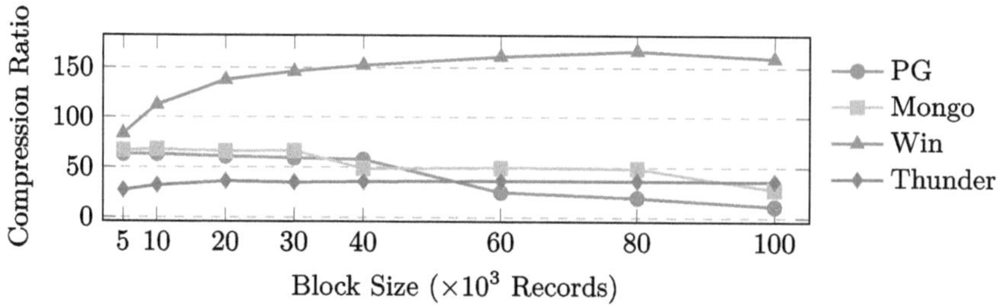

Fig. 4. Compression ratio vs. block size using 240,000-record samples.

4.5 Impact of the Block Size

Figure 4 illustrates how block size influences compression efficiency across multiple datasets. Distinct patterns emerge across workloads. On Windows, compression increases almost linearly with block size, indicating that structural redundancy benefits from broader contextual sharing. Thunderbird, in contrast, reaches a plateau near 20K records, implying that moderate block sizes already capture most schema regularities. Meanwhile, PostgreSQL and MongoDB exhibit a sharp decline beyond 40K records, with compression ratios decreasing by more than 55%. Such degradation reflects weakened field-level locality and increased redundancy across partitions as block size grows excessively large.

4.6 Parallel Performance

We assess the parallel scalability of LogTrie under multi-threaded execution, as individual compressed blocks can be processed independently without inter-thread synchronization.

Figure 5 shows the results on an 8-core system. The speedup results in (a) exhibit near-linear scaling up to four threads, with group queries reaching a 5.50× speedup using eight threads, equivalent to 68.8% of ideal linear scaling. The reduced efficiency beyond four threads is mainly due to memory bandwidth limits and synchronization overhead, effects commonly observed in parallel workloads. As shown in (b), throughput increases from 15 to 82.5 queries per second, consistent with the speedup ratio. Similar scaling trends across query types indicate that the block-level execution model effectively utilizes available cores while maintaining low contention.

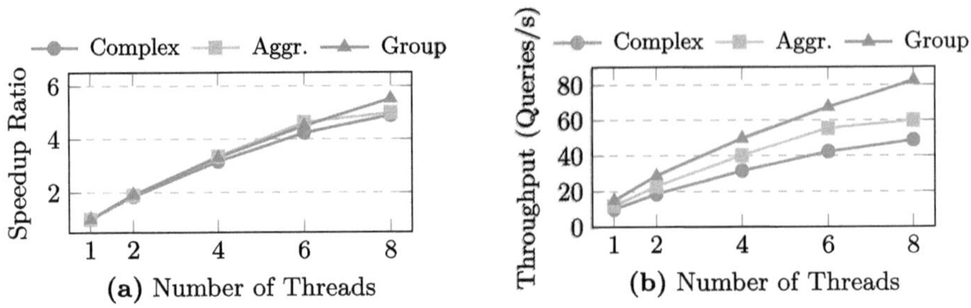

Fig. 5. Parallel performance analysis on an 8-core system. (a) Speedup ratios show near-linear scaling up to 4 threads, with diminishing returns due to memory bandwidth saturation. (b) Throughput measurements demonstrate consistent scaling patterns, with group queries achieving the highest absolute throughput.

5 Conclusions and Future Work

This paper presented **LogTrie**, a compression and query framework for large-scale JSON logs that reconciles the long-standing tension between compact storage and interactive analysis. By combining Trie-based structural representation with type-aware value compression, LogTrie achieves both high compression efficiency and direct queryability over compressed data. Across 18 real-world datasets, LogTrie achieves an average compression of 59.56× and demonstrates up to 46.6% reduction in query latency compared with μSlope.

Future work will explore more principled partitioning strategies informed by data distribution and access locality, extend the query engine to support richer operations under partial decompression, and integrate LogTrie into real-world logging pipelines to assess its end-to-end benefits in production environments. In addition, we will consider applying LogTrie into time-series data compression [13].

Acknowledgments. This paper is supported by the National Science Foundation of China (grant no. 62072419). Peiquan Jin is the corresponding author.

References

1. Agarwal, R., Khandelwal, A., Stoica, I.: Succinct: enabling queries on compressed data. In: NSDI, pp. 265–278 (2015)
2. Amazon web services: Amazon opensearch service developer guide: petabyte-scale domains (2025). https://docs.aws.amazon.com/opensearch-service/latest/developerguide/petabyte-scale.html
3. Cheng, Y.D., Wu, C.H.: An adaptive run-length encoding scheme for data compression. In: ICC, pp. 1462–1466 (1994)
4. Collet, Y.: Zstandard - real-time data compression algorithm. In: DCC (2016). https://facebook.github.io/zstd/

5. Deutsch, P., Gailly, J.L.: ZLIB compression library (1995). https://zlib.net/
6. Dharmaraj, U., Thilagam, P.: Leveraging structural and semantic measures for JSON document clustering. J. Univ. Comput. Sci. **29**, 222–241 (03 2023)
7. Edgedelta: What is logging as a service? (2023). https://edgedelta.com/company/knowledge-center/what-is-logging-as-a-service
8. Fei, X., Xu, K., Zheng, Z., et al.: Seal: Storage-efficient causality analysis on enterprise logs with query-friendly compression. In: USENIX Security, pp. 1749–1766 (2021)
9. Foundation, C.N.C.: The CNCF cloud native survey 2023. CNCF Research (2023). https://www.cncf.io/reports/cncf-annual-survey-2023/
10. Fredkin, E.: Trie memory. Commun. ACM **3**(9), 490–499 (1960)
11. He, P., Zhu, J., Zheng, Z.: LogZip: extracting hidden structures via iterative clustering for log compression. In: ASE, pp. 846–857 (2019)
12. He, S., He, P., Chen, Z., et al.: A survey on automated log analysis for reliability engineering. ACM Comput. Surv. **54**(6) (2021)
13. Jiang, Z., Jin, P.: TopcDB: two-phase compactable lossy compression for timeseries databases. In: ISPA, pp. 739–746 (2025)
14. Kurpicz, F.: Engineering compact data structures for rank and select queries on bit vectors. In: SPIRE, pp. 257–272 (2022)
15. Lemire, D., Boytsov, L.: Efficient data compression in column-oriented database systems. Inf. Syst. **38**(3), 383–397 (2013)
16. Li, X., Zhang, H., Le, V.H., et al.: LogShrink: effective log compression by leveraging commonality and variability of log data (2023)
17. LogPAI: LogHub: a comprehensive log dataset repository (2025). https://github.com/logpai/loghub. Accessed 8 Feb 2026
18. Navarro, G., Providel, E.: Fast, small, simple rank/select on bitmaps. Algorithmica **28**(3), 517–529 (2012)
19. Pavlov, I.: LZMA software development kit (2008). https://www.7-zip.org/sdk.html
20. SentinelOne: Log analytics and management in modern it environments (2024). https://www.sentinelone.com/cybersecurity-101/cybersecurity/log-analytics/
21. Shi, W., Cao, J., Zhang, Q., et al.: Edge computing: vision and challenges. IEEE Internet Things J. **3**, 637–646 (2016)
22. Tang, B., Yang, S., Shen, Z., Zhang, W., Lin, X., Tian, Z.: LogLite: lightweight plug-and-play streaming log compression. Proc. VLDB Endow. **18**(11), 3757–3770 (2025)
23. Vestrum, R., et al.: Delta encoding: an efficient data compression technique. IEEE Data Eng. Bull. **23**(3), 30–38 (2000)
24. Wang, R., Gibson, D., Rodrigues, K., et al.: μSlope: High compression and fast search on semi-structured logs. In: OSDI, pp. 529–544 (2024)
25. Zhang, H., Lim, H., et al.: SuRF: practical range query filtering with fast succinct tries. In: SIGMOD, pp. 323–336 (2018)

Giant: An I/O-Optimized Graph-Based Index for High-dimensional Vector Search via Page Group Expansion

Jiawei Du, Jinlei Hu, Chengxiao Gong, and Jianxi Chen(✉)

Wuhan National Laboratory for Optoelectronics, Huazhong University of Science and Technology, Wuhan, Hubei, China
{jawe1102,hujinlei,gcx,chenjx}@hust.edu.cn

Abstract. Modern vector search systems increasingly depend on high-capacity and cost-efficient NVMe SSDs for storage, as data dimensionality and scale continue to grow. Graph-based indices are widely adopted for their superior search accuracy; however, the graph traversal triggers intensive random I/O on SSDs, significantly degrading search performance. Existing graph-based retrieval methods attempt to reduce I/O overhead through page expansion. However, these approaches introduce two key limitations: 1) *Insufficient page expansion.* As vector dimensions grow, fewer nodes fit within a single SSD page, reducing the number of accessible nodes for expansion and thus limiting traversal efficiency. 2) *Inefficient utilization of I/O time.* Existing candidate pool expansion strategy exhibits diminishing returns in reducing I/O overhead, while its computation increases linearly, resulting in inefficient utilization of I/O time. To address these issues, we propose *Giant*, an I/O-optimized graph index for high-dimensional vector search based on a novel *page group expansion mechanism.* This mechanism is supported by two core techniques: a memory-resident group index enabling deeper node expansion, and a page expansion list offering diverse supplementary candidates. Experiments demonstrate that Giant reduces average search latency by 12%–43% and achieves 1.10×−1.41× higher throughput compared with state-of-the-art methods.

Keywords: Approximate Nearest Neighbor Search · Graph-based vector index · High-dimensional vector search

1 Introduction

High-dimensional vectors are widely used to represent unstructured data such as images, videos, and audio [22,23,26]. Vector search aims to identify the k nearest vectors to a given query vector from a dataset. Finding the exact k nearest neighbors requires scanning the entire dataset, which incurs substantial time overhead. Therefore, practical applications typically adopt Approximate Nearest Neighbor Search (ANNS), which trades off some recall rate for

H. Jung et al. (Eds.): DASFAA 2026, LNCS 16535, pp. 656–672, 2026.
https://doi.org/10.1007/978-981-92-0363-5_40

lower latency [9,12,26]. ANNS technique has been applied in various scenarios, including Retrieval-Augmented Generation (RAG) for Large Language Models (LLMs), recommendation systems, and search engines [22,23,26].

ANNS achieves rapid retrieval by constructing an index over the dataset. Among various indexing techniques, graph-based approaches have garnered significant attention due to their low latency and high accuracy [9,26]. In a graph-based ANNS structure, each node represents a vector, and edges connect vectors that are spatially close [4–6,13,17,18]. During the search phase, the algorithm starts from an entry node and traverses the graph by evaluating neighbors of the current node to progressively locate vectors closer to the query.

As datasets continue to grow in scale, it becomes infeasible to store all raw vectors and index data entirely in memory [3,9,13,26]. For graph-based indexes, a common solution is to leverage Solid State Drive (SSD) for storing index and raw data while maintaining Product-Quantized (PQ) vectors in memory to accelerate distance computation [13,26]. During the search phase, these methods maintain a candidate pool sorted by approximate distances computed using PQ vectors. The top-ranked candidates are selected for expansion, and the corresponding SSD pages are read to fetch their raw vectors and adjacency lists. The distances between the query and the retrieved neighbors are then computed, and their IDs and distances are inserted back into the candidate pool for subsequent iterations. However, this procedure introduces significant query latency because each traversal step requires SSD access to obtain both the raw vectors and adjacency lists. This leads to a large number of inefficient random read operations, which become a major performance bottleneck [3].

Existing techniques aim to reduce query latency in SSD-based graph indexes by minimizing I/O operations [3,13]. Starling [26] represents the state-of-the-art approach, employing two principal strategies. Firstly, it organizes neighboring nodes on a single SSD page, enhancing I/O efficiency to allow multiple nodes on the search path to be retrieved with a single flash read operation. Secondly, it expands the candidate pool during I/O by adding the neighbors of as many nodes as possible from the previously accessed page (an approach referred to as page expansion in this paper). However, our preliminary experiments in Sec.3 show that Starling faces two major limitations that significantly hinder query efficiency: 1) *it demonstrates poor performance on high-dimensional vector data.* As vector dimensionality increases, fewer nodes fit within a single SSD page, reducing the number of available nodes for expansion. 2) *its page expansion strategy does not efficiently utilize I/O time.* Starling expands all nodes from the previously read page by default, yet our analysis shows that expanding only a subset of these nodes can achieve comparable performance. This indicates a substantial amount of redundant computation during the page expansion process.

To address the above problems, we propose Giant, a graph-based ANNS system with two key components: <u>G</u>roup <u>I</u>ndices that reside in memory and p<u>A</u>ge expansio<u>N</u> lis<u>T</u>s stored in the unused space of pages. Based on these components, we propose a novel candidate pool expansion method named Page Group expansion that combines breadth and depth expansion strategies to fully utilize the overlap between I/O operations and computation.

First, we propose a *memory-resident group index* based on a key observation: within an SSD page, nodes ranked higher by distance to the query tend to have neighbors that are more likely to be high-quality candidates. Based on this insight, Giant performs deeper expansions for these high-ranking nodes instead of merely adding their neighbors to the candidate pool. To enable this deeper exploration without incurring additional I/O overhead, Giant constructs a graph index in memory. While prior works build a small in-memory navigation graph [9,13,26], the small navigation is typically used only to select entry points before query processing, as it only samples a very small fraction of nodes. In contrast, Giant aims to perform a deeper expansion within the region surrounding the target node. To achieve this, Giant first divides nodes into groups of spatially related nodes, ensuring that neighboring nodes are allocated to the same group whenever possible. Within each group, Giant prioritizes placing neighboring nodes on the same page and constructs a Vamana graph index [13] for each group separately. Moreover, by reordering the on-disk index and in-memory PQ codes, Giant eliminates the mappings between groups, pages, and nodes, thereby reducing memory consumption.

Second, we propose the *page expansion lists*, which are stored in the unused space at the end of pages. For each node within a page, Giant first selects its neighbors from the node's associated group to form a candidate set. Then, Giant uses the neighbors of these candidates as the initial expansion list. Finally, Giant filters this expansion list to preserve node diversity, ensuring a broader search space while avoiding redundancy. The filtered node IDs are written into the unused space at the end of the page to form the *page expansion list*. These lists supplies additional optional candidates during candidate pool expansion, enhancing the flexibility and effectiveness of the search process.

Based on the two components mentioned above, we propose a hybrid candidate pool expansion strategy named *Page Group Expansion*. During each I/O operation, the search algorithm first computes the distances to the neighbors of a small subset of nodes within the most recently accessed pages. It then evaluates the distances of nodes contained in the expansion lists. Finally, the algorithm performs a deeper search within the group indices corresponding to the top-ranked nodes from the recently read pages, which may contain additional candidates that are closer to the query.

To evaluate the performance of Giant, we conduct experiments on four high-dimensional datasets and compare it with existing SSD-based methods. The results show that Giant achieves lower latency and higher throughput than the baseline methods across all datasets. Specifically, under different target recall levels, Giant achieves an average latency reduction of 12%–43% and a throughput (QPS, queries per second) improvement of 1.10×–1.41× compared to state-of-the-art methods across all datasets.

In summary, we make the following contributions:

- We identify key limitations in existing SSD-based graph indexes, showing that they perform poorly on high-dimensional vector data and fail to effectively exploit the overlap between I/O operations and computation during page expansion.
- We proposed Giant, an I/O-optimized graph index for high-dimensional vector search with group index and page expansion list. Based on these components, Giant proposes a novel Page Group Expansion strategy to reduce I/O overhead.
- We evaluated the performance of Giant on various datasets. Experiment results show that Giant improves the throughput by up to 1.41× and reduces the latency by up to 43% compared to state-of-the-art methods.

2 Related Work

Memory-Based ANNS. Early vector search methods primarily include hash-based, tree-based, and quantization-based algorithms. Hash-based approaches accelerate retrieval by mapping similar data points to the same hash bucket with high probability [11,15,24]. However, their search quality heavily depends on hash function design, and they suffer from issues such as excessive index size, hash collisions, and boundary problems. Tree-based methods [10,14,19] build indexes through spatial partitioning, but their partitioning efficiency degrades significantly as dimensionality increases, limiting their applicability to low-dimensional spaces. Quantization-based methods [7,8,16] compress vectors to reduce storage and computational costs, but quantization errors lead to reduced search recall.

Graph-based algorithms have emerged as the dominant research direction due to their high accuracy and performance. Foundational work such as NN-Descent [4] provides a general graph construction method that progressively refines graph structures through local search principles. Building upon this, NSW [18] introduces small-world properties by constructing short-diameter graphs via local connections and highway mechanisms. HNSW [17] further advances this by adopting skip-list-inspired hierarchical structures, significantly reducing search time complexity. To address scalability and efficiency in large-scale scenarios, researchers have proposed novel graph structures and pruning strategies. For instance, NSG [6] aims to enhance scalability by reducing graph size and optimizing paths, proposing an approximate MRNG structure. SSG (NSSG) [5] further optimizes this with innovative pruning strategies that guarantee theoretical performance across diverse queries. Additionally, some works combine graph indexing with other techniques, such as IVF-HNSW [2], which employs HNSW to accelerate centroid search in inverted file (IVF) indexes.

Flash-Based ANNS. As large-scale graph indexes cannot fit entirely in memory, they are usually stored on SSDs, resulting in substantial random disk access during search operations. DiskANN [13] compresses in-memory vectors using Product Quantization, stores graphs and full-precision vectors on disk, and proposes the Vamana graph algorithm to construct graphs with smaller diameters, thereby minimizing disk reads. However, DiskANN still suffers from low disk bandwidth utilization and high latency. SPANN [3] adopts an inverted index approach, storing centroids in memory and inverted lists on disk, while employing hierarchical clustering and query-aware schemes to reduce disk access. Nevertheless, it faces challenges with high space overhead and bandwidth contention [23].

To further address I/O efficiency and latency issues, subsequent works have introduced more sophisticated I/O optimization strategies. Starling [26] employs a query-sensitive dynamic entry vertex selection strategy to shorten search paths and improves I/O utilization by co-locating nearby nodes on the same SSD page. It also designs the Page expansion strategy to overlap I/O operations with CPU computation. PipeANN [9] addresses the mismatch between best-first search and SSD characteristics by proposing the PipeSearch algorithm, which uses asynchronous I/O and dynamic pipeline width to hide I/O latency, improve pipeline utilization, and resolve sequential bottlenecks in traditional search methods.

Other Efficient Optimization Techniques. Several works leverage advanced hardware or heterogeneous computing capabilities to enhance ANNS performance. SmartSSD [22] offloads indexing and partial computation to intelligent solid-state drives to reduce data movement and host resource contention. FusionANNS [23] utilizes GPU acceleration for distance computation by storing PQ vectors on the GPU and original vectors on the SSD, addressing GPU memory constraints and high CPU-GPU communication overhead. CXL-ANNS [12] leverages the CXL protocol to construct distributed memory pools and designs intelligent caching and prefetching strategies to mitigate performance degradation from CXL remote memory access.

3 Motivation

Starling expands the candidate pool during I/O through page expansion strategy. Our experimental results in Fig. 1 reveal two primary issues that significantly impair query efficiency.

- *Insufficient Page Expansion.* As data dimensionality increases, each page can store fewer nodes due to the larger vector size. This causes the locality of nodes within pages to fail, as neighboring nodes in the graph cannot be placed on the same page. Moreover, during each I/O operation, only a small number of nodes can be used for expansion, leading to few candidates available for selection in the next I/O operation. Figure 1(a) illustrates that the increasing dataset dimension leads to a lower latency reduction ratio for Starling over DiskANN.
- *Inefficient Utilization of I/O Time.* Starling uses a breadth-first-search-style (BFS) candidate pool expansion strategy. The page usage ratio parameter specifies the fraction of nodes from the previously read page that are expanded in each hop's I/O operation, with a default value of 1.0. Our experiments show

that the number of distance computations required for candidate pool expansion increases linearly as this parameter increases. However, increasing this ratio yields diminishing returns on reducing the number of I/Os, as illustrated in Fig. 1(b). This indicates that using all I/O time for BFS candidate pool expansion is highly inefficient and results in substantial unnecessary computation. This occurs because candidates exhibit similarity, and nodes ranked lower in the candidate pool have few effective candidates among their neighbors.

To address these issues, we propose Giant, an I/O-optimized graph index for high-dimensional vector search based on a novel page group expansion mechanism.

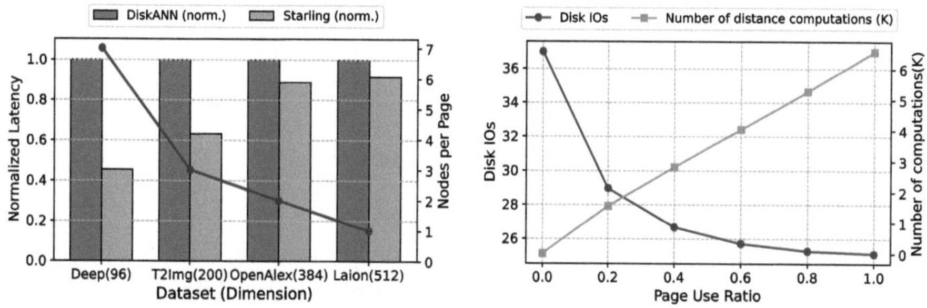

Fig. 1. (a) The performance gain of Starling over DiskANN diminishes as the number of nodes per page decreases with increasing data dimensionality. (b) The number of hops required by Starling to achieve 90% recall on the Deep-100M dataset when using different page usage ratio parameters. Starling uses a default ratio of 1.0, which introduces substantial redundant computation.

4 Methodology

4.1 Overview

Figure 2 presents an overview of Giant's architecture. Giant comprises an offline phase that constructs group indexes and page expansion lists, and an online phase that employs them for page group expansion.

Offline Phase: Firstly, Giant partitions the graph nodes into different groups, allocating adjacent nodes to the same group whenever possible. It then applies the same principle to further place the nodes in the same group into multiple SSD pages and constructs a Vamana graph index for each group (§4.2). Secondly, Giant reorders the index layout on disk and the PQ codes in memory according to the group order, and assigns new IDs to each node (§4.2). Finally, Giant builds a expansion list for each page and writes it to the end of the page (§4.3).

Online Phase: Based on the group index and expansion lists constructed in the offline phase, Giant designs a page group expansion strategy. This strategy

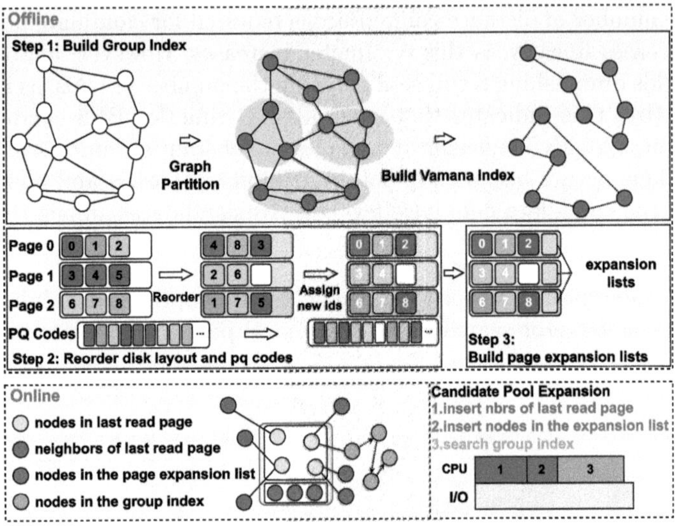

Fig. 2. The Overview of Giant.

performs page expansion and group expansion while waiting for I/O operations to discover more high-quality candidates. By adjusting the ratio between page expansion and group expansion, we ensure that computation can be completed within the I/O time (§4.4).

4.2 Memory-Resident Group Index

Group Index Construction. Giant first calculates the number of pages per group based on the number of nodes each page can accommodate, ensuring that the number of nodes in each group is limited to the range representable by the uint8 type. We extend the BNF graph partitioning method proposed in Starling [26], but elevate the partitioning granularity from SSD pages to groups. After graph partitioning is completed, nodes within the same group still need to be distributed across multiple SSD pages, as the I/O granularity during search remains at the page level rather than the group level. Specifically, for each node in a group, we count how many of its neighbors are contained in each page within the group and place the node in the page that contains the most neighbors. If multiple pages contain the same number of neighbors, we prioritize placement in the page with fewer existing nodes. Finally, we use the Vamana [13] method to construct a group index for the nodes in each group, setting the degree of the group index to a small value to conserve memory. We store the intra-group position of each node's neighbors in the group index using the uint8 data type.

Disk Layout and PQ Code Reconstruction. Two challenges arise if we continue to place nodes into their corresponding pages as done in Starling. First, the original method employs real node IDs for approximate distance computation, while the group index stores nodes' intra-group locations. This requires storing both the real IDs within each group and the corresponding ID-to-group mappings in memory. Moreover, similar to Starling, we must store node-to-page mappings to locate the appropriate page when accessing a node. These requirements introduce additional memory overhead. Second, during group index search, each hop requires table lookups to map intra-group positions to actual IDs before computing PQ distances, incurring substantial lookup time. This is unacceptable for our group expansion, which requires multi-hop searches across multiple indices. The lookup overhead would cause search time to exceed I/O time.

To address these challenges, Giant reorganizes the order of nodes on disk and the order of PQ codes in memory according to group sequence, and assigns new IDs to each node. We then update the neighbor list of each node according to the mapping from old IDs to new IDs. It is important to note that if there are unfilled pages, the corresponding ID slots must be reserved when assigning new IDs, such that the relationships among the new ID, group ID, page ID, and intra-group position can be expressed by Eq. 1 and Eq. 2:

$$id_{new} = gid \times n_g + pos \tag{1}$$

$$pid = \left\lfloor \frac{id_{new}}{n_p} \right\rfloor \tag{2}$$

where *gid* and *pid* denote the group id and page id respectively, n_g and n_p represent the number of nodes per group and per page respectively, and *pos* indicates the position of the node within its group. This allows the search process to directly determine each node's page and group based on the new ID. Moreover, the intra-group positions and the results from group expansion can be directly converted to new IDs for PQ distance computation and candidate pool insertion, eliminating the need for additional mapping tables and lookup operations. The following example illustrates the whole process.

Example 1. In Step 2 of Fig. 2, assume each group contains two pages with three nodes per page. Nodes 1–9 are initially stored sequentially across pages. We first reorder them by group. For example, if the first group consists of nodes 4, 8, 3, 2, and 6, these nodes are placed sequentially into Page 0 and Page 1, with PQ codes arranged in the same order. Giant then assigns new IDs based on this ordering: original node 4 is mapped to new ID 0, original node 8 is mapped to new ID 1, and so forth. Since Page 1 contains only two nodes, ID 5 is left unassigned, and the corresponding PQ code slot is also remain unused.

4.3 Page Expansion List

The effectiveness of page expansion in locating high-quality candidates stems from the presence of nodes close to the query within the already read pages, and the fact that these nodes may have neighbors residing within the same page. Consequently, by expanding from other nodes within the page, we increase

Algorithm 1 BuildPageExpansionList($P, G, N_{nodes}, \delta, K_{check}, S_{node}$)

Input: Page P, Group G, Set of Nodes in P (N_{nodes}), PQ Distance Threshold δ, Check Size K_{check}, space occupied by each node S_{node}.
Output: Expansion List E_P for page P.
$C_{total} \leftarrow \lfloor (4096 \bmod S_{node})/4 \rfloor$ ▷ Total capacity from unused space
$C_{per\_node} \leftarrow \lfloor C_{total}/|N_{nodes}| \rfloor$ ▷ Capacity/Pruning Threshold allocated per node
$E_P \leftarrow \emptyset$ ▷ Initialize final expansion list for page P
for each node $n_t \in N_{nodes}$ **do**
    $N_{intra} \leftarrow \text{Neighbors}(n_t) \cap G$ ▷ Intra-group neighbors of n_t (Candidates)
    $E_{n_t} \leftarrow \emptyset$ ▷ Individual expansion list for n_t
    for each candidate n_c in N_{intra} **do**
        $E_{n_t} \leftarrow E_{n_t} \cup \text{Neighbors}(n_c)$ ▷ Add neighbors of candidate n_c
    end for
    $R \leftarrow \emptyset$ ▷ Set of retained nodes for n_t's expansion list
    for each node n_e in E_{n_t} **do**
        if $|R| \geq C_{per\_node}$ **then** ▷ Check against per-node capacity/threshold
            break ▷ Discard subsequent nodes
        end if
        $is\_diverse \leftarrow$ **True**
        for each retained node n_{ki} in Last(R, K_{check}) **do**
            $d_{PQ} \leftarrow \text{PQ\_Distance}(n_e, n_{ki})$ ▷ Calculate PQ distance
            if $d_{PQ} < \delta$ **then**
                $is\_diverse \leftarrow$ **False**
                break
            end if
        end for
        if $is\_diverse$ **then**
            $R \leftarrow R \cup \{n_e\}$ ▷ Retain node
        end if
    end for
    $E_P \leftarrow E_P \cup R$ ▷ Add retained nodes to final page expansion list
end for
return E_P

the likelihood of discovering nodes even closer to the query. However, in high-dimensional datasets, a single page can accommodate only a very small number of nodes, which severely limits the pool of nodes available for page expansion. Fortunately, the number of nodes within a group does not change significantly as the data dimensionality increases, and numerous neighbor pairs also exist within these groups.

Building on this, we can leverage the intra-group neighbors of the page's nodes as initial candidates for page expansion. We then add the neighbors of these candidates to the page expansion list. Given the limited storage capacity for IDs at the tail of each page, we must subsequently prune this list. The complete procedure for constructing the page expansion list is detailed in Algorithm 1. Specifically, we first calculate how many node IDs each page's unused space can accommodate, and then evenly distribute this capacity among all nodes in the page (Lines 3–4). Subsequently, for each node n_t in the current page, we identify its intra-group neighbors N_{intra} as expansion candidates (Line 7), and then add the neighbors of these candidates to the node's individual expansion list E_{n_t} (Lines 9–11). We then evaluate the node expansion list E_{n_t} to determine which nodes to retain. Once the number of retained nodes $|R|$ exceeds the per-node threshold C_{per_node}, subsequent nodes in the node expansion list are directly discarded (Lines 14–16). To evaluate a node n_e, we calculate its PQ distance d_{PQ} to the most recently retained nodes in R, specifically checking the last K_{check} nodes (Lines 18–24). If all these distances d_{PQ} exceed our predefined threshold δ, it signifies that the current node n_e is significantly diverse from the already retained nodes, and thus the node is retained (Lines 25–27).

4.4 Page Group Expansion Strategy

Page Expansion. Page expansion is the strategy proposed in Starling to expand the candidate pool during I/O operations. Giant enhances the original Page Expansion strategy with the expansion list in each page. Specifically, since increasing the proportion of expanded nodes from previously read pages beyond a certain threshold does not significantly improve recall, as illustrated in Fig. 1(b), we propose expanding only a small subset of nodes from previously read pages. Then Giant selects a subset from the head of the page expansion lists, computes their approximate distances, and inserts them into the candidate pool.

Group Expansion. After page expansion, Giant starts the group expansion process. Nodes from the previously read page are first sorted according to their real distances to the query. Then, following the ascending order of these distances, Giant iteratively adds groups containing each node's neighbors to the set of candidate groups for further search, using each neighbor as the group's entry node. Because each node's neighbor list is already sorted in ascending order of distance, neighbors from the front of the list are given higher priority for group expansion. This process continues until the number of selected groups reaches a predefined threshold, which ensures that the subsequent group-index search can be completed within the I/O time of a single hop. During the group expansion, Giant records the closest nodes to the query discovered in each visited page along the search path and inserts them to the candidate pool.

This hybrid expansion strategy combines the advantages of breadth-first and depth-first exploration: expansion lists provide diverse candidate nodes representing different exploration directions, while group index search enables deep exploration of regions likely to contain closer neighbors. Through this approach, Giant can more quickly discover high-quality candidates, thereby reducing the number of hops required to achieve the target recall.

5 Experiments

5.1 Experimental Setup

Basic Configuration. We conduct experiments on a server with the following configuration:

- **CPU**: 2× 28-core Intel Xeon Gold 5420+ @ 2.0 GHz
- **RAM**: 256 GB (8× 32 GB DDR5 4400 MT/s)
- **SSD**: 1× Samsung PM1743 1.92 TB
- **OS**: Ubuntu 20.04.6 LTS with Linux kernel 5.15.0

Datasets and Parameters. We evaluate Giant using three high-dimensional public datasets, with the specific configurations detailed in Table 1. The Text2Image [1] dataset is a large-scale cross-modal retrieval collection where image embeddings are used as the database, and text embeddings serve as the queries. Laion-400M [21] consists of 400 million image-text pairs. For our experiments, we extract 100 million image vectors from this dataset to construct the graph index and randomly sample 10,000 text vectors to use as query vectors. OpenAlex [20] is a comprehensive multilingual scholarly literature database containing over 250 million works, offering rich information on authors, institutions, and topics. We use the embedding vectors from the OpenAlex dataset provided by the open-source community[1], which were generated with the multilingual-e5-small model [25]. The Wiki dataset[2] contains nearly 250 million paragraph embeddings generated by the Cohere Embed V3 model[3], covering content in over 300 languages. For OpenAlex and Wiki dataset, we extract the first 100 million vectors as the base set and randomly sample 10,000 vectors from the remaining ones as query vectors.

Table 1. Dataset Information and Parameters

Name	#Vectors	Type	Dim	Metric	R	L	PQ size
Text2Image	100M	float32	200	IP	48	128	64
OpenAlex	100M	float32	384	L2	64	128	96
Laion	100M	float32	512	Cosine	128	256	128
Wiki	100M	int8	1024	L2	64	128	128

Baseline Methods. We compare Giant with three disk-resident ANNS methods, including the graph-based methods DiskANN [13] and Starling [26], and the IVF-based method SPANN [3]. We do not evaluate SPANN on the Laion dataset because it incurs enormous storage overhead. PipeANN [9] was also not included in the comparison because it is orthogonal to our work—PipeANN aims to reduce latency by minimizing waiting time among I/Os in the same batch, whereas our approach achieves this by reducing the number of query hops. Hardware-accelerated methods are also excluded from the comparison.

[1] https://huggingface.co/datasets/GlobalCampus/openalex-multilingual-embeddings.

[2] https://huggingface.co/datasets/Cohere/wikipedia-2023-11-embed-multilingual-v3-int8-binary.

[3] https://docs.cohere.com/docs/cohere-embed.

Evaluation Metrics. Consistent with previous research, we adopt Recall@10 as the metric for evaluating the performance of different methods, which represents the proportion of true nearest neighbors included in the top 10 results returned by ANNS. To comprehensively evaluate the performance of various methods, we compare their performance curves of latency, throughput (QPS), and average I/O count under different recall rates. All experiments are conducted using 32 threads, with each query being processed individually by one thread.

5.2 Overall Performance

In this section, we evaluate the latency, throughput, and average number of I/O operations for various methods. We plot the variations of these metrics under different target recall levels to facilitate a comprehensive comparison. Figure 3 presents the experimental results. We did not plot the latency curve for SPANN because its latency under multithreading is significantly higher than the other methods, which would make the comparison among the other methods less distinguishable in the figure.

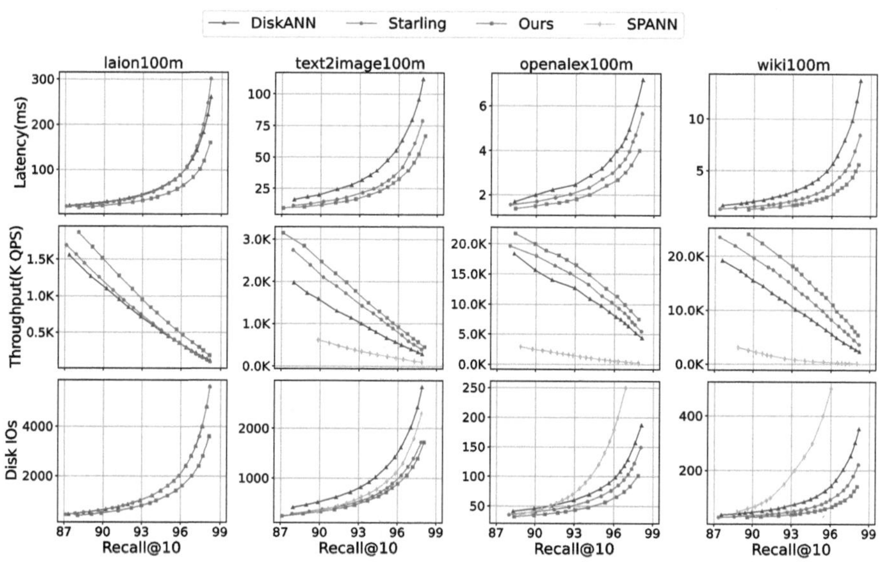

Fig. 3. The Overall Search performance of Different Methods

QPS. Our method consistently demonstrates a significant advantage in QPS across all recall points, with the performance gap widening as the recall requirement increases: At 90% recall, our method achieves a QPS improvement of 1.10×–1.25× over Starling, with an average of 1.17×. This enhancement grows to an average of 1.24 times at 95% recall. Crucially, at the highest recall level of

98%, our method delivers 1.42× the average QPS compared to Starling, particularly demonstrating a 1.71× improvement on the Laion dataset. As the figure shows, SPANN's QPS falls far behind the graph-based methods, due to its simultaneous submission of multiple large-granularity read requests, which causes I/O bandwidth contention across threads.

Latency and Disk I/Os. The efficiency of our method is further validated by the substantial reduction in both query Latency and Mean I/Os. At 90% recall, the average latency is reduced by approximately 17% compared to Starling. This reduction becomes more pronounced at higher recall levels, reaching an average of 21% at 95% recall and a substantial 30% at 98%. The average I/O cost mirrors the latency trend. At 90% recall, the average number of I/Os is reduced by 15% compared to Starling, decreasing to 20% at 95% recall and further to 25% at 98%. This consistently lower I/O cost confirms that our method can achieve the same or better recall quality with substantially fewer disk accesses, leading to lower search latency.

The performance advantage comes from the page group expansion strategy introduced in Giant. It allows more high-quality candidates to be added to the candidate pool while waiting for I/O operations, which in turn provides better options for subsequent I/O and reduces unnecessary accesses. Notably, Giant achieves more significant performance gains on higher-dimensional datasets, where there are typically fewer nodes available for expansion per page. The page expansion lists in Giant help to overcome this limitation. Additionally, as the target recall increases, the advantage of Giant becomes more evident. Higher recall requires longer search paths, which lead to more ineffective I/O operations. The expansion strategy in Giant effectively mitigates these inefficiencies.

5.3 Breakdown Analysis

We conduct breakdown experiments to evaluate the candidate pool expansion strategy proposed in Giant on the Wiki and LAION datasets, as shown in Fig. 4. When the page expansion list is incorporated on top of Starling, the query throughput at 90% recall increases to 1.06× and 1.14×, respectively. The improvement is more pronounced on the LAION dataset, since each page in LAION contains fewer nodes and therefore offers limited nodes for page expansion. The page expansion list constructed by Giant effectively compensates for this limitation. For the Wiki dataset, the page expansion list also provides more diverse candidates for page expansion, leading to a moderate yet consistent performance improvement.

Building upon this, we further adopt Group Expansion and evaluate searches with different numbers of Group Indexes. As illustrated in Fig. 4, when the number of searched group indexes is 5 and 10, the query throughput on the Wiki dataset increases to 1.13×, 1.21×, respectively, compared to the baseline. The corresponding improvements on the LAION dataset reach 1.20× and 1.23×, respectively. These results demonstrate that Group Expansion enables the candidate pool to include more high-quality nodes, thereby enhancing the overall query performance.

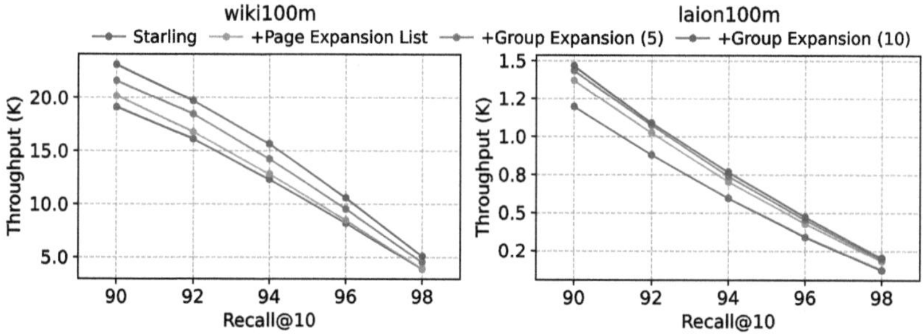

Fig. 4. Breakdown analysis of Giant.

5.4 Memory Cost

Starling stores bidirectional node-page mappings as unsigned integers, a navigation graph for dynamic entry node selection, and vector PQ codes in memory. Giant constructs a graph index for each group stored as uint8 values, which incurs additional memory overhead. However, Giant eliminates the memory overhead of mapping tables through reordering of disk indexes and PQ codes. Both methods also require storing the navigation graph and PQ codes in memory. Their memory overheads can be calculated using Eqs. 3 and 4:

$$M_{\text{Starling}} = N \cdot \text{sizeof(unsigned)} \cdot 2 + M_{\text{nav}} + M_{\text{pq}} + M_{\text{others}} \tag{3}$$

$$M_{\text{Giant}} = N \cdot R_{\text{group}} \cdot \text{sizeof(uint8)} + M_{\text{nav}} + M_{\text{pq}} + M_{\text{others}} \tag{4}$$

where N denotes the total number of nodes, R_{group} indicates the maximum degree of the group index, M_{nav} and M_{pq} represent the memory required for the navigation graph and PQ codes respectively, and M_{others} represents other data structures required during the search. Therefore, by limiting the degree of the group index to 8, we can achieve nearly the same memory overhead as Starling.

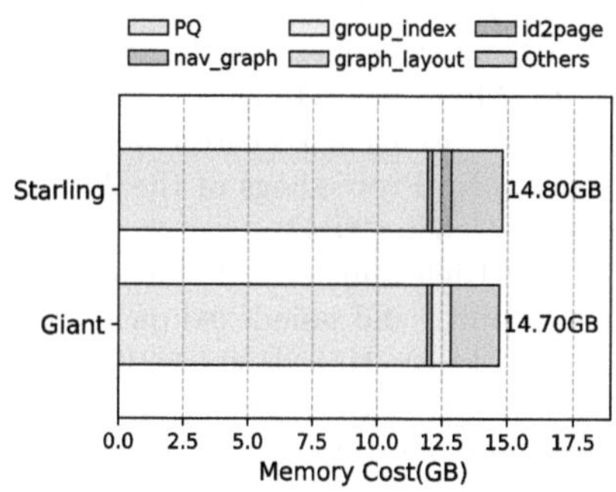

Fig. 5. Memory Cost Breakdown.

Figure 5 details the memory consumption of the different components in the two methods. The results reveal that the memory overhead of the group index in Giant is comparable to that of the bidirectional mapping of nodes and pages in Starling (represented by *id2page* and *graph_layout* in Fig. 5). Since the costs of the remaining components in both methods are also nearly the same, Giant and Starling exhibit nearly identical total memory consumption.

6 Conclusion

We propose Giant, an I/O-optimized graph-based index for high-dimensional vectors search, which successfully mitigates the limitations of existing methods through its two core components: memory-resident group index and page expansion list. These components enable our proposed Page Group Expansion method, which utilizes I/O time to discover more high-quality candidates, thereby reducing the number of I/O hops for a query. Experiments on high-dimensional datasets demonstrate Giant's superiority over state-of-the-art baselines, achieving significant latency reduction and substantial throughput improvement.

Acknowledgments. This work is funded by the National Key Research and Development Program (No. 2024YFB4505104), National Natural Science Foundation of China (NSFC No. U22A2027).

Disclosure of Interests. The authors have no competing interests to declare that are relevant to the content of this article.

References

1. Baranchuk, D., Babenko, A.: Benchmarks for billion-scale similarity search. Webpage (2021). Retrieved March **16**, 2023
2. Baranchuk, D., Babenko, A., Malkov, Y.: Revisiting the inverted indices for billion-scale approximate nearest neighbors. In: Proceedings of the European Conference on Computer Vision (ECCV), pp. 202–216 (2018)
3. Chen, Q., Zhao, B., Wang, H., Li, M., Liu, C., Li, Z., Yang, M., Wang, J.: Spann: highly-efficient billion-scale approximate nearest neighborhood search. Adv. Neural. Inf. Process. Syst. **34**, 5199–5212 (2021)
4. Dong, W., Moses, C., Li, K.: Efficient k-nearest neighbor graph construction for generic similarity measures. In: Proceedings of the 20th International Conference on World Wide Web, pp. 577–586 (2011)
5. Fu, C., Wang, C., Cai, D.: High dimensional similarity search with satellite system graph: efficiency, scalability, and unindexed query compatibility. IEEE Trans. Pattern Anal. Mach. Intell. **44**(8), 4139–4150 (2021)
6. Fu, C., Xiang, C., Wang, C., Cai, D.: Fast approximate nearest neighbor search with the navigating spreading-out graph. arXiv preprint arXiv:1707.00143 (2017)

7. Gao, J., Gou, Y., Xu, Y., Yang, Y., Long, C., Wong, R.C.W.: Practical and asymptotically optimal quantization of high-dimensional vectors in Euclidean space for approximate nearest neighbor search. Proc. ACM Manage. Data **3**(3), 1–26 (2025)
8. Gao, J., Long, C.: RABITQ: quantizing high-dimensional vectors with a theoretical error bound for approximate nearest neighbor search. Proc. ACM Manage. Data **2**(3), 1–27 (2024)
9. Guo, H., Lu, Y.: Achieving {Low-Latency}{Graph-Based} vector search via aligning {Best-First} search algorithm with {SSD}. In: 19th USENIX Symposium on Operating Systems Design and Implementation (OSDI 25), pp. 171–186 (2025)
10. Hu, H., Qiu, J., Wang, H., Liang, B., Zou, S.: DIDS: double indices and double summarizations for fast similarity search. Proc. VLDB Endow. **17**(9), 2198–2211 (2024)
11. Huang, Q., Feng, J., Zhang, Y., Fang, Q., Ng, W.: Query-aware locality-sensitive hashing for approximate nearest neighbor search. Proc. VLDB Endow. **9**(1), 1–12 (2015)
12. Jang, J., Choi, H., Bae, H., Lee, S., Kwon, M., Jung, M.: {CXL-ANNS}:{Software-Hardware} collaborative memory disaggregation and computation for {Billion-Scale} approximate nearest neighbor search. In: 2023 USENIX Annual Technical Conference (USENIX ATC 23), pp. 585–600 (2023)
13. Jayaram Subramanya, S., Devvrit, F., Simhadri, H.V., Krishnawamy, R., Kadekodi, R.: DISKANN: fast accurate billion-point nearest neighbor search on a single node. In: Advances in Neural Information Processing Systems, vol. 32 (2019)
14. Lampropoulos, K., Zardbani, F., Mamoulis, N., Karras, P.: Adaptive indexing in high-dimensional metric spaces. Proc. VLDB Endow. **16**(10), 2525–2537 (2023)
15. Liu, Y., Cui, J., Huang, Z., Li, H., Shen, H.T.: SK-LSH: an efficient index structure for approximate nearest neighbor search. Proc. VLDB Endow. **7**(9), 745–756 (2014)
16. Lu, Z., Chen, J., Lian, D., Zhang, Z., Ge, Y., Chen, E.: Knowledge distillation for high dimensional search index. Adv. Neural. Inf. Process. Syst. **36**, 33403–33419 (2023)
17. Malkov, Y.A., Yashunin, D.A.: Efficient and robust approximate nearest neighbor search using hierarchical navigable small world graphs. IEEE Trans. Pattern Anal. Mach. Intell. **42**(4), 824–836 (2018)
18. Malkov, Y., Ponomarenko, A., Logvinov, A., Krylov, V.: Approximate nearest neighbor algorithm based on navigable small world graphs. Inf. Syst. **45**, 61–68 (2014)
19. Muja, M., Lowe, D.G.: Scalable nearest neighbor algorithms for high dimensional data. IEEE Trans. Pattern Anal. Mach. Intell. **36**(11), 2227–2240 (2014)
20. Priem, J., Piwowar, H., Orr, R.: OpenAlex: a fully-open index of scholarly works, authors, venues, institutions, and concepts. arXiv preprint arXiv:2205.01833 (2022)
21. Schuhmann, C., et al.: LAION-400M: open dataset of clip-filtered 400 million image-text pairs. arXiv preprint arXiv:2111.02114 (2021)
22. Tian, B., Liu, H., Duan, Z., Liao, X., Jin, H., Zhang, Y.: Scalable billion-point approximate nearest neighbor search using SmartSSDs (2024). https://www.usenix.org/conference/atc24/presentation/tian
23. Tian, B., et al.: Towards high-throughput and low-latency billion-scale vector search via CPU/GPU collaborative filtering and re-ranking, pp. 171–185 (2025). https://www.usenix.org/conference/fast25/presentation/tian-bing, gSCC: 0000000
24. Tian, Y., Zhao, X., Zhou, X.: DB-LSH 2.0: locality-sensitive hashing with query-based dynamic bucketing. IEEE Trans. Knowl. Data Eng. **36**(3), 1000–1015 (2023)

25. Wang, L., Yang, N., Huang, X., Yang, L., Majumder, R., Wei, F.: Multilingual E5 text embeddings: a technical report. arXiv preprint arXiv:2402.05672 (2024)
26. Wang, M., et al.: Starling: an I/O-efficient disk-resident graph index framework for high-dimensional vector similarity search on data segment. Proc. ACM Manage. Data **2**(1), 1–27 (2024)

Author Index

H. Jung et al. (Eds.): DASFAA 2026, LNCS 16535, pp. 673–675, 2026.
https://doi.org/10.1007/978-981-92-0363-5

MIX
Papier aus verantwortungsvollen Quellen
Paper from responsible sources
FSC® C105338
FSC
www.fsc.org

If you have any concerns about our products,
you can contact us on
ProductSafety@springernature.com

In case Publisher is established outside the EU,
the EU authorized representative is:
Springer Nature Customer Service Center GmbH
Europaplatz 3, 69115 Heidelberg, Germany

Printed by Libri Plureos GmbH
in Hamburg, Germany